Handbook of Automotive Power Electronics and Motor Drives

ELECTRICAL AND COMPUTER ENGINEERING
A Series of Reference Books and Textbooks

FOUNDING EDITOR

Marlin O. Thurston
Department of Electrical Engineering
The Ohio State University
Columbus, Ohio

1. Rational Fault Analysis, *edited by Richard Saeks and S. R. Liberty*
2. Nonparametric Methods in Communications, *edited by P. Papantoni-Kazakos and Dimitri Kazakos*
3. Interactive Pattern Recognition, *Yi-tzuu Chien*
4. Solid-State Electronics, *Lawrence E. Murr*
5. Electronic, Magnetic, and Thermal Properties of Solid Materials, *Klaus Schröder*
6. Magnetic-Bubble Memory Technology, *Hsu Chang*
7. Transformer and Inductor Design Handbook, *Colonel Wm. T. McLyman*
8. Electromagnetics: Classical and Modern Theory and Applications, *Samuel Seely and Alexander D. Poularikas*
9. One-Dimensional Digital Signal Processing, *Chi-Tsong Chen*
10. Interconnected Dynamical Systems, *Raymond A. DeCarlo and Richard Saeks*
11. Modern Digital Control Systems, *Raymond G. Jacquot*
12. Hybrid Circuit Design and Manufacture, *Roydn D. Jones*
13. Magnetic Core Selection for Transformers and Inductors: A User's Guide to Practice and Specification, *Colonel Wm. T. McLyman*
14. Static and Rotating Electromagnetic Devices, *Richard H. Engelmann*
15. Energy-Efficient Electric Motors: Selection and Application, *John C. Andreas*
16. Electromagnetic Compossibility, *Heinz M. Schlicke*
17. Electronics: Models, Analysis, and Systems, *James G. Gottling*
18. Digital Filter Design Handbook, *Fred J. Taylor*
19. Multivariable Control: An Introduction, *P. K. Sinha*
20. Flexible Circuits: Design and Applications, *Steve Gurley, with contributions by Carl A. Edstrom, Jr., Ray D. Greenway, and William P. Kelly*
21. Circuit Interruption: Theory and Techniques, *Thomas E. Browne, Jr.*
22. Switch Mode Power Conversion: Basic Theory and Design, *K. Kit Sum*
23. Pattern Recognition: Applications to Large Data-Set Problems, *Sing-Tze Bow*
24. Custom-Specific Integrated Circuits: Design and Fabrication, *Stanley L. Hurst*
25. Digital Circuits: Logic and Design, *Ronald C. Emery*
26. Large-Scale Control Systems: Theories and Techniques, *Magdi S. Mahmoud, Mohamed F. Hassan, and Mohamed G. Darwish*
27. Microprocessor Software Project Management, *Eli T. Fathi and Cedric V. W. Armstrong (Sponsored by Ontario Centre for Microelectronics)*
28. Low Frequency Electromagnetic Design, *Michael P. Perry*
29. Multidimensional Systems: Techniques and Applications, *edited by Spyros G. Tzafestas*
30. AC Motors for High-Performance Applications: Analysis and Control, *Sakae Yamamura*

31. Ceramic Motors for Electronics: Processing, Properties, and Applications, *edited by Relva C. Buchanan*
32. Microcomputer Bus Structures and Bus Interface Design, *Arthur L. Dexter*
33. End User's Guide to Innovative Flexible Circuit Packaging, *Jay J. Miniet*
34. Reliability Engineering for Electronic Design, *Norman B. Fuqua*
35. Design Fundamentals for Low-Voltage Distribution and Control, *Frank W. Kussy and Jack L. Warren*
36. Encapsulation of Electronic Devices and Components, *Edward R. Salmon*
37. Protective Relaying: Principles and Applications, *J. Lewis Blackburn*
38. Testing Active and Passive Electronic Components, *Richard F. Powell*
39. Adaptive Control Systems: Techniques and Applications, *V. V. Chalam*
40. Computer-Aided Analysis of Power Electronic Systems, *Venkatachari Rajagopalan*
41. Integrated Circuit Quality and Reliability, *Eugene R. Hnatek*
42. Systolic Signal Processing Systems, *edited by Earl E. Swartzlander, Jr.*
43. Adaptive Digital Filters and Signal Analysis, *Maurice G. Bellanger*
44. Electronic Ceramics: Properties, Configuration, and Applications, *edited by Lionel M. Levinson*
45. Computer Systems Engineering Management, *Robert S. Alford*
46. Systems Modeling and Computer Simulation, *edited by Naim A. Kheir*
47. Rigid-Flex Printed Wiring Design for Production Readiness, *Walter S. Rigling*
48. Analog Methods for Computer-Aided Circuit Analysis and Diagnosis, *edited by Takao Ozawa*
49. Transformer and Inductor Design Handbook: Second Edition, Revised and Expanded, *Colonel Wm. T. McLyman*
50. Power System Grounding and Transients: An Introduction, *A. P. Sakis Meliopoulos*
51. Signal Processing Handbook, *edited by C. H. Chen*
52. Electronic Product Design for Automated Manufacturing, *H. Richard Stillwell*
53. Dynamic Models and Discrete Event Simulation, *William Delaney and Erminia Vaccari*
54. FET Technology and Application: An Introduction, *Edwin S. Oxner*
55. Digital Speech Processing, Synthesis, and Recognition, *Sadaoki Furui*
56. VLSI RISC Architecture and Organization, *Stephen B. Furber*
57. Surface Mount and Related Technologies, *Gerald Ginsberg*
58. Uninterruptible Power Supplies: Power Conditioners for Critical Equipment, *David C. Griffith*
59. Polyphase Induction Motors: Analysis, Design, and Application, *Paul L. Cochran*
60. Battery Technology Handbook, *edited by H. A. Kiehne*
61. Network Modeling, Simulation, and Analysis, *edited by Ricardo F. Garzia and Mario R. Garzia*
62. Linear Circuits, Systems, and Signal Processing: Advanced Theory and Applications, *edited by Nobuo Nagai*
63. High-Voltage Engineering: Theory and Practice, *edited by M. Khalifa*
64. Large-Scale Systems Control and Decision Making, *edited by Hiroyuki Tamura and Tsuneo Yoshikawa*
65. Industrial Power Distribution and Illuminating Systems, *Kao Chen*
66. Distributed Computer Control for Industrial Automation, *Dobrivoje Popovic and Vijay P. Bhatkar*
67. Computer-Aided Analysis of Active Circuits, *Adrian Ioinovici*
68. Designing with Analog Switches, *Steve Moore*
69. Contamination Effects on Electronic Products, *Carl J. Tautscher*
70. Computer-Operated Systems Control, *Magdi S. Mahmoud*

71. Integrated Microwave Circuits, *edited by Yoshihiro Konishi*
72. Ceramic Materials for Electronics: Processing, Properties, and Applications, Second Edition, Revised and Expanded, *edited by Relva C. Buchanan*
73. Electromagnetic Compatibility: Principles and Applications, *David A. Weston*
74. Intelligent Robotic Systems, *edited by Spyros G. Tzafestas*
75. Switching Phenomena in High-Voltage Circuit Breakers, *edited by Kunio Nakanishi*
76. Advances in Speech Signal Processing, *edited by Sadaoki Furui and M. Mohan Sondhi*
77. Pattern Recognition and Image Preprocessing, *Sing-Tze Bow*
78. Energy-Efficient Electric Motors: Selection and Application, Second Edition, *John C. Andreas*
79. Stochastic Large-Scale Engineering Systems, *edited by Spyros G. Tzafestas and Keigo Watanabe*
80. Two-Dimensional Digital Filters, *Wu-Sheng Lu and Andreas Antoniou*
81. Computer-Aided Analysis and Design of Switch-Mode Power Supplies, *Yim-Shu Lee*
82. Placement and Routing of Electronic Modules, *edited by Michael Pecht*
83. Applied Control: Current Trends and Modern Methodologies, *edited by Spyros G. Tzafestas*
84. Algorithms for Computer-Aided Design of Multivariable Control Systems, *Stanoje Bingulac and Hugh F. VanLandingham*
85. Symmetrical Components for Power Systems Engineering, *J. Lewis Blackburn*
86. Advanced Digital Signal Processing: Theory and Applications, *Glenn Zelniker and Fred J. Taylor*
87. Neural Networks and Simulation Methods, *Jian-Kang Wu*
88. Power Distribution Engineering: Fundamentals and Applications, *James J. Burke*
89. Modern Digital Control Systems: Second Edition, *Raymond G. Jacquot*
90. Adaptive IIR Filtering in Signal Processing and Control, *Phillip A. Regalia*
91. Integrated Circuit Quality and Reliability: Second Edition, Revised and Expanded, *Eugene R. Hnatek*
92. Handbook of Electric Motors, *edited by Richard H. Engelmann and William H. Middendorf*
93. Power-Switching Converters, *Simon S. Ang*
94. Systems Modeling and Computer Simulation: Second Edition, *Naim A. Kheir*
95. EMI Filter Design, *Richard Lee Ozenbaugh*
96. Power Hybrid Circuit Design and Manufacture, *Haim Taraseiskey*
97. Robust Control System Design: Advanced State Space Techniques, *Chia-Chi Tsui*
98. Spatial Electric Load Forecasting, *H. Lee Willis*
99. Permanent Magnet Motor Technology: Design and Applications, *Jacek F. Gieras and Mitchell Wing*
100. High Voltage Circuit Breakers: Design and Applications, *Ruben D. Garzon*
101. Integrating Electrical Heating Elements in Appliance Design, *Thor Hegbom*
102. Magnetic Core Selection for Transformers and Inductors: A User's Guide to Practice and Specification, Second Edition, *Colonel Wm. T. McLyman*
103. Statistical Methods in Control and Signal Processing, *edited by Tohru Katayama and Sueo Sugimoto*
104. Radio Receiver Design, *Robert C. Dixon*
105. Electrical Contacts: Principles and Applications, *edited by Paul G. Slade*

106. Handbook of Electrical Engineering Calculations, *edited by Arun G. Phadke*
107. Reliability Control for Electronic Systems, *Donald J. LaCombe*
108. Embedded Systems Design with 8051 Microcontrollers: Hardware and Software, *Zdravko Karakehayov, Knud Smed Christensen, and Ole Winther*
109. Pilot Protective Relaying, *edited by Walter A. Elmore*
110. High-Voltage Engineering: Theory and Practice, Second Edition, Revised and Expanded, *Mazen Abdel-Salam, Hussein Anis, Ahdab El-Morshedy, and Roshdy Radwan*
111. EMI Filter Design: Second Edition, Revised and Expanded, *Richard Lee Ozenbaugh*
112. Electromagnetic Compatibility: Principles and Applications, Second Edition, Revised and Expanded, *David Weston*
113. Permanent Magnet Motor Technology: Design and Applications, Second Edition, Revised and Expanded, *Jacek F. Gieras and Mitchell Wing*
114. High Voltage Circuit Breakers: Design and Applications, Second Edition, Revised and Expanded, *Ruben D. Garzon*
115. High Reliability Magnetic Devices: Design and Fabrication, *Colonel Wm. T. McLyman*
116. Practical Reliability of Electronic Equipment and Products, *Eugene R. Hnatek*
117. Electromagnetic Modeling by Finite Element Methods, *João Pedro A. Bastos and Nelson Sadowski*
118. Battery Technology Handbook, Second Edition, *edited by H. A. Kiehne*
119. Power Converter Circuits, *William Shepherd and Li Zhang*
120. Handbook of Electric Motors: Second Edition, Revised and Expanded, *edited by Hamid A. Toliyat and Gerald B. Kliman*
121. Transformer and Inductor Design Handbook, *Colonel Wm T. McLyman*
122. Energy Efficient Electric Motors: Selection and Application, Third Edition, Revised and Expanded, *Ali Emadi*
123. Power-Switching Converters, Second Edition, *Simon Ang and Alejandro Oliva*
124. Process Imaging For Automatic Control, *David M. Scott and Hugh McCann*
125. Handbook of Automotive Power Electronics and Motor Drives, *edited by Ali Emadi*

Handbook of Automotive Power Electronics and Motor Drives

Edited by

Ali Emadi
Illinois Institute of Technology
Chicago, Illinois, U.S.A.

Taylor & Francis
Taylor & Francis Group

Boca Raton   London   New York   Singapore

A CRC title, part of the Taylor & Francis imprint, a member of the
Taylor & Francis Group, the academic division of T&F Informa plc.

Published in 2005 by
CRC Press
Taylor & Francis Group
6000 Broken Sound Parkway NW
Boca Raton, FL 33487-2742

© 2005 by Taylor & Francis Group
CRC Press is an imprint of Taylor & Francis Group

No claim to original U.S. Government works
Printed in the United States of America on acid-free paper
10 9 8 7 6 5 4 3 2 1

International Standard Book Number-10: 0-8247-2361-9 (Hardcover)
International Standard Book Number-13: 978-0-8247-2361-3 (Hardcover)
Library of Congress Card Number 2004063490

This book contains information obtained from authentic and highly regarded sources. Reprinted material is quoted with permission, and sources are indicated. A wide variety of references are listed. Reasonable efforts have been made to publish reliable data and information, but the author and the publisher cannot assume responsibility for the validity of all materials or for the consequences of their use.

No part of this book may be reprinted, reproduced, transmitted, or utilized in any form by any electronic, mechanical, or other means, now known or hereafter invented, including photocopying, microfilming, and recording, or in any information storage or retrieval system, without written permission from the publishers.

For permission to photocopy or use material electronically from this work, please access www.copyright.com (http://www.copyright.com/) or contact the Copyright Clearance Center, Inc. (CCC) 222 Rosewood Drive, Danvers, MA 01923, 978-750-8400. CCC is a not-for-profit organization that provides licenses and registration for a variety of users. For organizations that have been granted a photocopy license by the CCC, a separate system of payment has been arranged.

Trademark Notice: Product or corporate names may be trademarks or registered trademarks, and are used only for identification and explanation, without intent to infringe.

Library of Congress Cataloging-in-Publication Data

Handbook of automotive power electronics and motor drives / edited by Ali Emadi.
    p. cm. — (Electrical engineering and electronics; 125)
  Includes bibliographical references and index.
  ISBN 0-8247-2361-9 (alk. paper)
    1. Automobiles--Electronic equipment. 2. Power electronics. 3. Electric motors. 4. Electric driving. I. Emadi, Ali. II. Series.

TL272.5.H296 2005
629.2′7--dc22   2004063490

Taylor & Francis Group is the Academic Division of T&F Informa plc.

Visit the Taylor & Francis Web site at
http://www.taylorandfrancis.com

and the CRC Press Web site at
http://www.crcpress.com

Dedication
To my family

Preface

Lighting loads and the starter motor were the initial electrical loads in automobiles. However, the electric power requirement in automobiles has been increasing since the introduction of electrical systems in cars during the past few decades. In fact, demands for reduced fuel consumption and emissions as well as higher performance and reliability push the automotive industry to seek electrification of ancillaries and engine augmentations. As a result, there is an increasing need to replace the conventional mechanical, hydraulic, and pneumatic loads by electrically driven systems. In addition, the need for improvement in comfort, convenience, entertainment, safety, communications, maintainability, supportability, survivability, and operating costs necessitates more electric automotive systems. In advanced automobiles, throttle actuation, power steering, antilock braking, rear-wheel steering, air-conditioning, ride-height adjustment, active suspension, and electrically heated catalyst all benefit from the electrical power system. Therefore, electrical systems with larger capacities and more complex configurations are required to facilitate increasing electrical demands in advanced cars. In these systems, most of the loads as well as generation and distribution systems are in the form of power electronic converters and electric motor drives.

Handbook of Automotive Power Electronics and Motor Drives provides a comprehensive reference in automotive electrical systems for engineers, students, researchers, and managers who work in automotive-related industry, government, and academia.

This handbook consists of five parts. Part I starts with an introduction to automotive power systems. Part II presents semiconductor devices, sensors, and other components used or projected to be used in automobiles. Part III explains different power electronic converters. Electric machines and associated drives are introduced in Part IV. Different advanced electrical loads are described in Part V. In addition, Part V deals with the battery technology for automotive applications.

I would like to acknowledge the efforts and assistance of the staff of Taylor & Francis, especially Nora Konopka, Jessica Vakili, and Susan Fox.

Ali Emadi

Editorial Advisors

Dr. Mehrdad Ehsani
Texas A&M University
College Station, Texas

Dr. Ali Keyhani
Ohio State University
Columbus, Ohio

Dr. John M. Miller
J-N-J Miller Design Services, P.L.C.
Cedar, Michigan

Dr. Balarama V. Murty
Research and Development Center
General Motors
Warren, Michigan

Dr. Paul R. Nicastri
Ford Motor Company
Dearborn, Michigan

Dr. Dean Patterson
University of South Carolina
Columbia, South Carolina

Dr. Pragasen Pillay
Clarkson University
Potsdam, New York

Dr. Kaushik Rajashekara
Delphi Corporation
Kokomo, Indiana

Dr. Muhammad H. Rashid
University of West Florida
Pensacola, Florida

Dr. Mohammad Shahidehpour
Illinois Institute of Technology
Chicago, Illinois

Dr. Timothy L. Skvarenina
Purdue University
West Lafayette, Indiana

Contributors

Mohan Aware
Visvesvaraya National Institute of Technology
Nagpur, India

Ramesh C. Bansal
Birla Institute of Technology and Science
Pilani, India

William Cai
Remy International, Inc.
Anderson, Indiana

Mario Mañana Canteli
University of Cantabria
Cantabria, Spain

X. Chen
University of Windsor
Windsor, Ontario, Canada

Dell A. Crouch
Delphi Corporation
Indianapolis, Indiana

Chris Edrington
University of Missouri-Rolla
Rolla, Missouri

M. Ehsani
Texas A&M University
College Station, Texas

Babak Fahimi
University of Missouri-Rolla
Rolla, Missouri

Daluwathu Mulla Gamage
University of Melbourne
Melbourne, Australia

V.K. Garg
Ford Motor Company
Dearborn, Michigan

Yimin Gao
Texas A&M University
College Station, Texas

Roberto Giral-Castillón
Universitat Rovira i Virgili
Tarragona, Spain

Saman Kumara Halgamuge
University of Melbourne
Melbourne, Australia

Mohammad S. Islam
Delphi Corporation
Saginaw, Michigan

James P. Johnson
Caterpillar Inc.
Washington, Illinois

Mehrdad Kazerani
University of Waterloo
Waterloo, Ontario, Canada

Contributors

Ali Keyhani
Ohio State University
Columbus, Ohio

Byoung-Kuk Lee
Korea Electrotechnology Research
 Institute (KERI)
Changwon, South Korea

Weng Keong Kevin Lim
University of Melbourne
Melbourne, Australia

Wenzhe Lu
Ohio State University
Columbus, Ohio

Luis Martínez-Salamero
Universitat Rovira i Virgili
Tarragona, Spain

Javier Maixé-Altés
Universitat Rovira i Virgili
Tarragona, Spain

A. Masrur
U.S. Army TACOM
Warren, Michigan

John M. Miller
J-N-J Miller Design Services
Cedar, Michigan

Sayeed Mir
Delphi Corporation
Saginaw, Michigan

Khaled Nigim
University of Waterloo
Waterloo, Ontario, Canada

Sung Chul Oh
Korea University of Technology and
 Education
Chungnam, South Korea

Pragasen Pillay
Clarkson University
Potsdam, New York

D.M.G. Preethichandra
Kyushu Institute of Technology
Kitakyushu, Japan

Bogdan Proca
Ohio State University
Columbus, Ohio

Kaushik Rajashekara
Delphi Corporation
Kokomo, Indiana

Hossein Salehfar
University of North Dakota
Grand Forks, North Dakota

Tomy Sebastian
Delphi Corporation
Saginaw, Michigan

Zheng John Shen
University of Central Florida
Orlando, Florida

A. Soltis
Opal-RT Technologies, Inc.
Ann Arbor, Michigan

Ana V. Stankovic
Cleveland State University
Cleveland, Ohio

Albert Z.H. Wang
Illinois Institute of Technology
Chicago, Illinois

Harry Charles Watson
University of Melbourne
Melbourne, Australia

Chung-Yean Won
Sungkyunkwan University
Kyung Ki Do, South Korea

Table of Contents

PART I. Automotive Power Systems   1

1. Conventional Cars   3
   Roberto Giral-Castillón, Luis Martínez-Salamero, and Javier Maixé-Altés
   1.1   Introduction   3
   1.2   Evolution of the Distribution Electrical System   3
         1.2.1   Control Strategy and Wiring Topology   5
         1.2.2   Power Bus Topology   5
         1.2.3   Components   6
   1.3   The Conventional System of Electrical Distribution in Automobiles   6
         1.3.1   Battery and Its Charging System   6
         1.3.2   Motor Starter System   6
         1.3.3   Management System   6
   1.4   Wiring System   6
         1.4.1   Fuses   8
                 1.4.1.1   Polymeric Positive Temperature Coefficient Devices 9
                 1.4.1.2   Smart Power Switches   11
         1.4.2   Behavior Comparison among the Different Protection Devices 12
   1.5   Load Control: Automotive Control Network Protocols   13
         1.5.1   Controller Area Network (CAN)   14
         1.5.2   Local Interconnect Network (LIN)   14
         1.5.3   Byteflight   15
         1.5.4   Time Triggered Protocol (TTP/C)   15
   1.6   New Architectures   15
         1.6.1   Electric Security   15
         1.6.2   Voltage Effect on the Components   16
   1.7   Alternative Architectures   16
         1.7.1   High Frequency AC Bus System   16
         1.7.2   Dual-Voltage DC Bus   17
   References   18

2. Hybrid Electric Vehicles   21
   John M. Miller
   2.1   Parallel Configuration   26
   2.2   Series Configuration   29

	2.3	Combination Architectures	31
	2.4	Grid Connected Hybrids	33
	References		35

3. Hybrid Drivetrains — 37
M. Ehsani and Yimin Gao

- 3.1 Concept of Hybrid Vehicle Drivetrain — 37
- 3.2 Series Hybrid Drivetrain — 38
- 3.3 Parallel Hybrid Drivetrains — 40
  - 3.3.1 Parallel Hybrid Drivetrains with Torque Coupling — 40
    - 3.3.1.1 Torque Coupler — 40
    - 3.3.1.2 Drivetrain Configuration and Operating Characteristics — 41
  - 3.3.2 Parallel Hybrid Drivetrain with Speed Coupling — 46
    - 3.3.2.1 Speed Coupler — 46
    - 3.3.2.2 Drivetrain Configurations and Operating Characteristics — 47
- 3.4 Drivetrains with Selectable Torque Coupling and Speed Coupling — 48
- 3.5 Parallel-Series Hybrid Drivetrain with Torque Coupling and Speed Coupling — 50
- 3.6 Fuel Cell-Powered Hybrid Drivetrain — 50
- References — 52

4. Electric Vehicles — 55
Ramesh C. Bansal

- 4.1 Introduction — 55
- 4.2 Hybrid Electric Vehicles — 56
- 4.3 Main Components of an EV — 57
  - 4.3.1 Motors — 57
- 4.4 Main Safety Components in an EV — 58
- 4.5 Instrumentation — 59
- 4.6 Main Auxiliaries in an EV — 60
- 4.7 Types of Power Storage Used in EVs — 63
  - 4.7.1 Batteries — 63
  - 4.7.2 Types of Batteries Available Today — 64
  - 4.7.3 Flywheels — 67
  - 4.7.4 Ultracapacitors — 67
- 4.8 Emissions Performance — 68
- 4.9 Solar Cars — 69
- 4.10 Fuel Cell Cars — 69
  - 4.10.1 Introduction — 69
  - 4.10.2 Fuel Cell Cars — 71
- Bibliographical Survey on Electric Vehicles — 72
- References — 73

5. Optimal Power Management and Distribution in Automotive Systems — 97
Zheng John Shen, X. Chen, A. Masrur, V.K. Garg, and A. Soltis

- 5.1 Introduction — 97
- 5.2 Automotive Power/Energy Management and Distribution Architecture — 99

Table of Contents

	5.2.1	Power Generation	99
	5.2.2	Energy Storage	100
	5.2.3	Power Bus	100
	5.2.4	Electrical Load	101
	5.2.5	Power Electronics	101
	5.2.6	PMC	101
5.3		Optimization-Based Power Management System Strategy	101
	5.3.1	Dynamic Resource Allocation	103
	5.3.2	Practical Constraints of Vehicle Components	103
	5.3.3	Uninterruptible Power Availability	103
	5.3.4	Power Quality	103
	5.3.5	System Stability	104
	5.3.6	Fault Diagnosis and Prognosis	104
5.4		Case Study: Game-Theoretic Optimal Hybrid Electric Vehicle Management and Control Strategy	104
	5.4.1	System Dynamics	105
	5.4.2	Strategy Design	106
	5.4.3	Game-Theoretic Approach	107
	5.4.4	Simulation Results	110
5.5		Summary	112
		References	112

PART II. Automotive Semiconductor Devices, Components, and Sensors — 115

6. Automotive Power Semiconductor Devices — 117
Zheng John Shen

6.1	Introduction		117
6.2	Diodes: The Rectification, Freewheeling, and Clamping Devices		120
	6.2.1	Rectifier Diodes	121
	6.2.2	Freewheeling Diodes	121
	6.2.3	Zener Diodes	124
	6.2.4	Schottky Diode	125
6.3	Power MOSFETs: The Low-Voltage Load Drivers		125
	6.3.1	MOSFET Basics	127
	6.3.2	MOSFET Characteristics	129
6.4	IGBTs: The High-Voltage Power Switches		139
	6.4.1	IGBT Basics	141
	6.4.2	IGBT Power Modules	146
	6.4.3	Ignition IGBT	147
6.5	Power Integrated Circuits and Smart Power Devices		148
6.6	Emerging Device Technologies: Super-Junction and SiC Devices		150
6.7	Power Losses and Thermal Management		154
6.8	Summary		156
	References		156

7. Ultracapacitors — 159
John M. Miller

7.1	Theory of Electronic Double Layer Capacitance	161
7.2	Model and Cell Balancing	168

7.3	Sizing Criteria	174
7.4	Converter Interface	177
7.5	Ultracapacitors in Combination with Batteries	182
	References	187

8. Flywheels — 189
John M. Miller

8.1	Flywheel Theory	189
8.2	Flywheel Applications in Hybrid Vehicles	192
8.3	Energy Storage System Outlook	193
	References	194

9. ESD Protection for Automotive Electronics — 195
Albert Z.H. Wang

9.1	Introduction	195
9.2	ESD Failures and ESD Test Models	196
9.3	On-Chip ESD Protection	201
	References	211

10. Sensors — 213
Mario Mañana Canteli

10.1	Introduction	213
10.2	Architecture of Electronic Control Units	214
10.3	Voltage and Current Measurement	218
10.4	Temperature	221
10.5	Acceleration	222
10.6	Pressure	223
10.7	Velocity, Position, and Displacement	223
10.8	Other Sensors	224
10.9	Reliability Constraints in Automotive Environment	226
10.10	Conclusions	226
	References	227

PART III. Automotive Power Electronic Converters — 229

11. DC-DC Converters — 231
James P. Johnson

11.1	Why DC-DC Converters?	231
11.2	DC-DC Converter Basics	233
11.3	DC-DC Converter Types	233
11.4	Buck, Boost, and Buck-Boost Converter Commonalities	234
11.5	The Buck Converter	237
11.6	The Boost Converter	239
11.7	The Buck-Boost Converter	240
11.8	Isolated Inverter Driven Converters	241
11.9	Push-Pull Converter	242
11.10	Half-Bridge	243
11.11	Full-Bridge	244
11.12	Other Converter Types	244

				11.13	Control	245

		11.13	Control	245
		11.14	Essential Converter Circuits	247
		11.15	Important Points to Consider	250
		11.16	Simulation vs. Analytical Methods	251
		11.17	Loss Calculations	251
		11.18	Power Device Selections	251
		11.19	EMI	252
		11.20	Other Practical Converter Development Considerations	252
		References		253
12.	AC-DC Rectifiers			255
	Byoung-Kuk Lee and Chung-Yean Won			
	12.1	Diode AC-DC Rectifier		255
		12.1.1	Main Characteristics and Circuit Configuration	255
		12.1.2	Analysis of Three-Phase Full-Bridge Diode Rectifier	255
			12.1.2.1 Circuit without Input Inductors and DC-Link Capacitor	255
			12.1.2.2 Circuit with Input Inductors and DC-Link Capacitor	256
			12.1.2.3 Commutation Analysis Considering Effect of the Input Inductance	257
		12.1.3	Analysis of Input Phase Current and Output Current of Diode Rectifier	259
		12.1.4	Calculation of DC-Link Power	259
		12.1.5	Calculations of DC-Link Capacitor According to Various Load Conditions	260
			12.1.5.1 Case of Continuous Full Load Condition	260
			12.1.5.2 Case of Overload Condition	261
			12.1.5.3 Case of Motor Accelerating Condition	261
		12.1.6	Design of Dynamic Breaking Unit	263
			12.1.6.1 Design Procedure of Dynamic Breaking Resistor	263
	12.2	Thyristor AC-DC Rectifier		264
		12.2.1	Topology and Operation Modes	264
		12.2.2	Fire Angle Control Scheme	264
			12.2.2.1 Linear Fire Angle Control Scheme	265
			12.2.2.2 Cosine Wave Crossing Scheme	267
			12.2.2.3 PLL Scheme	267
		12.2.3	Analysis of Three-Phase Full-Bridge Thyristor Rectifier	268
			12.2.3.1 Equivalent Circuit and Output Voltage	268
			12.2.3.2 Influence of Input Inductance	268
			12.2.3.3 Selection of Input Inductance	271
	References			271
13.	Unbalanced Operation of Three-Phase Boost Type Rectifiers			273
	Ana V. Stankovic			
	13.1	System Description and Principles of Operation		274
	13.2	Analysis of the PWM Boost Type Rectifier under Unbalanced Operating Conditions		275
		13.2.1	Harmonic Reduction in the PWM Boost Type Rectifier under Unbalanced Operating Conditions	278

13.3 Control Methods for Input/Output Harmonic Elimination of the PWM
Boost Type Rectifiers under Unbalanced Operating Conditions   279
    13.3.1 Control Method for Input/Output Harmonic Elimination under
        Unbalanced Input Voltages and Balanced Input Impedances   279
        13.3.1.1 Theoretical Approach   279
        13.3.1.2 Control Method   282
        13.3.1.3 The Physical Meaning of the Proposed Solution
               in d-q Stationary Frame   283
    13.3.2 Control Method for Input/Output Harmonic Elimination of
        the PWM Boost Type Rectifier under Unbalanced Input
        Voltages and Unbalanced Input Impedances   287
        Derivation   287
        13.3.2.1 Control Method   291
13.3 Conclusion   293
References   294

14. DC/AC Inverters   295
Mohan Aware

14.1 DC-to-AC Conversion   295
14.2 Types of Inverters   298
14.3 Voltage Source Inverters   299
    14.3.1 Single-Phase Inverters   300
        14.3.1.1 Half-Bridge Inverters   300
        14.3.1.2 Full-Bridge Inverter   301
    14.3.2 Three-Phase Inverters   304
        14.3.2.1 Six-Step Operation   304
        14.3.2.2 Voltage and Frequency Control   309
        14.3.2.3 Motoring and Regeneration Mode   308
14.4 Current Source Inverters   308
14.5 Control Techniques   310
    14.5.1 Voltage Control Technique   310
        14.5.1.1 Sinusoidal PWM (SPWM) Technique   311
        14.5.1.2 Modulating Function PWM Techniques   312
        14.5.1.3 Voltage Space-Vector PWM Techniques   313
        14.5.1.4 Programmed PWM Techniques   317
    14.5.2 Current Control Technique   318
        14.5.2.1 Hysteresis Current Control   318
        14.5.2.2 Ramp-Comparison Current Control   319
        14.5.2.3 Predictive Current Control   320
        14.5.2.4 Linear Current Control   321
14.6 Multilevel Inverters   321
14.7 Hard Switching Effects   325
    14.7.1 Switching Loss   325
    14.7.2 Device Stress   325
    14.7.3 EMI Problems   325
    14.7.4 Effect on Insulation   325
    14.7.5 Machine Bearing Current   325
    14.7.6 Machine Terminal over Voltage   326
14.8 Resonant Inverters   326

Table of Contents

		14.8.1	Soft-Switching Principle	326
		14.8.2	Resonant Link DC Converter (RLDC)	327
	14.9	Auxillary Automotive Motors Control		338
		14.9.1	Commutator Motors	329
		14.9.2	Switched Field Motors	330
	References			331

15. AC/AC Converters — 333
Mehrdad Kazerani
- 15.1 Introduction — 333
- 15.2 AC/AC Converter Topologies — 334
  - 15.2.1 Indirect AC/AC Converter — 334
  - 15.2.2 Direct AC/AC Converter — 336
    - 15.2.2.1 Naturally Commutated Cycloconverter (NCC) — 336
    - 15.2.2.2 Forced-Commutated Cycloconverter (Matrix Converter) — 339
- 15.3 Summary — 345
- References — 346

16. Power Electronics and Control for Hybrid and Fuel Cell Vehicles — 347
Kaushik Rajashekara
- 16.1 Introduction — 347
- 16.2 Hybrid Electric Vehicles — 347
  - 16.2.1 Series Hybrid Vehicle Propulsion System — 348
  - 16.2.2 Parallel Hybrid Vehicle Propulsion System — 349
    - 16.2.2.1 Toyota Prius — 350
    - 16.2.2.2 Crankshaft-Mounted Integrated Starter-Generator System — 352
    - 16.2.2.3 Side-Mounted Integrated Starter-Generator — 353
- 16.3 Fuel Cell Vehicles — 354
  - 16.3.1 Fuel Cell Vehicle Propulsion System — 355
  - 16.3.2 Fuel Cell Vehicle Propulsion System Considerations — 358
- 16.4 Power Electronics Requirements — 359
- 16.5 Propulsion Motor Control Strategies — 360
  - 16.5.1 Slip Frequency Control — 362
  - 16.5.2 Vector Control of Propulsion Motor — 362
  - 16.5.3 Sensorless Operation — 363
- 16.6 APU Control System in Series Hybrid Vehicles — 364
- 16.7 Fuel Cell for APU Applications — 366
- References — 369

PART IV. Automotive Motor Drives — 371

17. Brushed-DC Electric Machinery for Automotive Applications — 373
Babak Fahimi
- 17.1 Fundamentals of Operation — 374
  - 17.1.1 Introduction — 374
  - 17.1.2 Torque Production in Brushed DC-Motor Drives — 377
  - 17.1.3 Impact of Temperature on Performance of a BLDC Drive — 379
- 17.2 Series Connected DC-Motor Drives — 383

18. Induction Motor Drives — 387
Khaled Nigim
- 18.1 Introduction — 387
- 18.2 Torque and Speed Control of Induction Motor — 388
- 18.3 Basics of Power Electronics Control in Induction Motors — 389
- 18.4 Induction Motor VSD Operating Modes — 390
- 18.5 Fundamentals of Scalar and Vector Control for Induction Motors — 393
  - 18.5.1 Scalar Control — 393
    - 18.5.1.1 Open Loop Scalar Control — 393
    - 18.5.1.2 Closed Loop Scalar Control — 393
  - 18.5.2 Fundamentals of Field-Oriented Control (Vector Control) in Induction Motors — 394
    - 18.5.2.1 Field-Oriented Control — 394
    - 18.5.2.2 Direct Torque Control — 397
- 18.6 Induction Motor Drives for Electric Vehicles — 399
- 18.7 Conclusion — 401
- References — 402
- Appendix: Induction Motor Model in the Stationary Frame — 403

19. DSP-Based Implementation of Vector Control of Induction Motor Drives — 405
Hossein Salehfar
- 19.1 Introduction — 405
- 19.2 Space Vector Control — 405
- 19.3 Experimental Results — 410
- 19.4 Conclusions — 413
- References — 413

20. Switched Reluctance Motor Drives — 415
Babak Fahimi and Chris Edrington
- 20.1 Introduction — 415
- 20.2 Historical Background — 416
- 20.3 Fundamentals of Operation — 417
- 20.4 Fundamentals of Control in SRM Drives — 424
  - 20.4.1 Open Loop Control Strategy for Torque — 425
    - 20.4.1.1 Detection of the Initial Rotor Position — 426
    - 20.4.1.2 Computation of the Commutation Thresholds — 427
    - 20.4.1.3 Monitoring of the Rotor Position and Selection of the Active Phases — 428
    - 20.4.1.4 A Control Strategy for Regulation of the Phase Current at Low Speeds — 429
- 20.5 Closed Loop Torque Control of the SRM Drive — 430
- 20.6 Closed Loop Speed Control of the SRM Drive — 433
- 20.7 Industrial Applications: Vehicular Coolant System — 434
- References — 436

21. Noise and Vibration in SRMs — 437
William Cai and Pragasen Pillay
- 21.1 Introduction — 437
- 21.2 Numerical Models of SRM Stator Modal Analysis — 438

	21.3	Finite Element Results of the Stator Modal Analysis	438
	21.4	Design Selection of Low Vibration SRMs	440
	21.5	The Effects of a Smooth Frame on the Resonant Frequencies	445
	21.6	Conclusions	447
	References		447

22. Modeling and Parameter Identification of Electric Machines — 449
    Ali Keyhani, Wenzhe Lu, and Bogdan Proca
    Nomenclature — 449
    22.1 Introduction — 450
    22.2 Case Study: The Effects of Noise on Frequency-Domain Parameter Estimation of Synchronous Machine — 450
        22.2.1 Problem Description — 450
        22.2.2 Parameters Estimation Technique — 451
            22.3.2.1 Estimation of D-Axis Parameters from the Time Constants — 451
            22.3.2.2 Estimation of Q-Axis Parameters — 453
        22.2.3 Study Process — 453
        22.2.4 Analysis of Results — 454
            22.2.4.1 D-Axis Parameter Estimation — 454
        22.2.5 Conclusions — 459
    22.3 Maximum Likelihood Estimation of Solid-Rotor Synchronous Machine Parameters — 460
        22.3.1 Introduction — 460
        22.3.2 Standstill Synchronous Machine Model for Time-Domain Parameter Estimation — 460
            22.3.2.1 D-Axis Model — 460
            22.3.2.2 Q-Axis Model — 461
        22.3.3 Effect of Noise on the Process and the Measurement — 461
        22.3.4 Maximum Likelihood Parameter Estimation — 462
        22.3.5 Estimation Procedure Using SSFR Test Data — 464
        22.3.6 Results — 465
    22.4 Modeling and Parameter Identification of Induction Machines — 468
        22.4.1 Model Identification — 469
        22.4.2 Parameter Estimation — 472
            22.4.2.1 Estimation of Stator Resistance — 473
            22.4.2.2 Estimation of L_1, L_m, and R_r — 474
        22.4.3 Sensitivity Analysis — 476
            Observation — 478
        22.4.4 Parameter Mapping to Operating Conditions — 478
            22.4.4.1 Magnetizing Inductance, L_m — 479
            22.4.4.2 Leakage Inductance, L_l — 480
            22.4.4.3 Rotor Resistance, R_r — 480
        22.4.5 Core Loss Estimation — 483
            22.4.5.1 Calculation of Rotor Losses at Frequencies of Interest — 483
            22.4.5.2 Calculation of Friction and Windage Losses Using ANN — 483
            22.4.5.3 Calculation of Core Losses — 485
            22.4.5.4 Calculation of Core Resistance — 486

		22.4.6	Model Validation	486
			22.4.6.1 Steady-State Power Input	486
			22.4.6.2 Dynamic	486
		22.4.7	Conclusions	487
	22.5	Modeling and Parameter Identification of Switched Reluctance Machines		490
		22.5.1	Introduction	490
		22.5.2	Inductance Model of SRM at Standstill	491
			22.5.2.1 Three-Term Inductance Model	491
			22.5.2.2 Four-Term Inductance Model	492
			22.5.2.3 Five-Term Inductance Model	493
			22.5.2.4 Voltages and Torque Computation	494
		22.5.3	Parameter Identification from Standstill Test Data	494
			22.5.3.1 Standstill Test Configuration	494
			22.5.3.2 Standstill Test Results	495
		22.5.4	Inductance Model of SRM for On-Line Operation	497
		22.5.5	Two-Layer Recurrent Neural Network for Damper Current Estimation	499
			22.5.5.1 Structure of Two-Layer Recurrent Neural Network	499
			22.5.5.2 Training of Neural Network	501
		22.5.6	Estimation Results and Model Validation	501
		22.5.7	Conclusions	501
	References			503
	Appendix A			508
	Appendix B			510
23.	Brushless DC Drives			515
	James P. Johnson			
	23.1	BLDC Fundamentals		515
	23.2	Control Principles and Strategies		517
	23.3	Torque Production		519
	23.4	Advantages and Disadvantages		521
	23.5	Torque Ripple		523
	23.6	Design Considerations		525
	23.7	Finite Element Analysis and Design Considerations for BLDC		525
	23.8	Permanent Magnets		526
	23.9	BLDC Simulation Model		528
	23.10	Sensorless		535
	References			536
24.	Testing of Electric Motors and Controllers for Electric and Hybrid Electric Vehicles			537
	Sung Chul Oh			
	24.1	Introduction		537
	24.2	Current Status of Standardization of Electric Vehicles		538
		24.2.1	Electric Vehicles and Standardization	538
		24.2.2	Standardization Bodies Active in the Field	539
			24.2.2.1 The International Electrotechnical Commission	539
			24.2.2.2 The International Organization for Standardization	539
			24.2.2.3 Other Regional Organizations	539

	24.2.3	Standardization of Vehicle Components	540
	24.2.4	Standardization Activities in Japan	540
		24.2.4.1 Z108-1994: Measurement of Range and Energy Consumption (at Charger Input)	541
		24.2.4.2 Z109-1995: Acceleration Measurement Test	541
		24.2.4.3 Z110-1995: Test Method for Maximum Cruising Speed	541
		24.2.4.4 Z111-1995: Measurement for Reference Energy Consumption (at Battery Output)	541
		24.2.4.5 Z901-1995: Electric Vehicle: Standard Form of Specifications (Form of Main Specifications)	541
		24.2.4.6 Z112-1996: Electric Vehicle: Standard Measurement of Hill Climbing Ability	541
		24.2.4.7 E701-1994: Combined Power Measurement of Motor and Controller	542
		24.2.4.8 E702-1994: Power Measurement of Motors Equivalent to On-Board Application	542
		24.2.4.9 Japanese Standards Concerning Vehicle Performance and Energy Economy	542
24.3	Test Procedure Using M-G Set		542
	24.3.1	Electric Motor	542
	24.3.2	Controller	543
	24.3.3	Application of Test Procedure	543
	24.3.4	Analysis of Test Items for the Type Test	543
		24.3.4.1 Motor Test	543
		24.3.4.2 Controller Test (Controller Only)	544
24.4	Test Procedure Using Eddy Current-Type Engine Dynamometer		544
	24.4.1	Test Strategy	544
	24.4.2	Test Procedure	545
	24.4.3	Discussion on Test Procedure	545
24.5	Test Procedure Using AC Dynamometer		546
	24.5.1	Test Strategy	546
	24.5.2	Test Items	547
	24.5.3	Test Procedure	547
24.6	Testing of Electric Motor/Controller in Vehicle Environment		548
	24.6.1	Concept of Hardware in the Loop	548
	24.6.2	HIL Application to Motor/Controller	548
	24.6.3	Test Description	550
	24.6.4	Test Results	550
24.7	Conclusion		552
References			553

PART V. Other Automotive Applications — 555

25. Integrated Starter Alternator — 557
    William Cai
    25.1 ISA Subsystem in Vehicle Systems — 558
    25.2 Powertrain Coupling Architecture — 558
        25.2.1 Crankshaft-Mounted ISA Configuration — 559
        25.2.2 Offset-Mounted ISA Configuration — 560

	25.3	Features and Performances of the ISA System		562
		25.3.1 State of the Art		563
		25.3.2 Features of the ISA Subsystem		564
			25.3.2.1 Initial Cranking and Stop/Start	564
			25.3.2.2 High-Efficient Large-Power Generation	566
			25.3.2.3 Launching Torque Assistant	567
			25.3.2.4 Braking Energy Regeneration	568
			25.3.2.5 Low Loss and Cost via High System Voltage	568
			25.3.2.6 Active Damping Oscillation and Absorbing Vibration	569
			25.3.2.7 Cylinder Shutoff	571
			25.3.2.8 Power APU and Other Electric Loads	571
	25.4	Components in the ISA Subsystem		571
		25.4.1 Electric Machine with Dual-Voltage Output		572
		25.4.2 36 V Battery with 12 V Intermediate Terminal		572
		25.4.3 Typical ISA Electrical System		572
		25.4.4 Multifunction Inverter with a Neutral Inductor		573
		25.4.5 Electric Machine		574
			25.4.5.1 Specifications of the ISA Electric Machine	574
			25.4.5.2 Types of ISA Electric Machines	577
			25.4.5.3 Application Comparison of ISA Electric Machines	594
		25.4.6 DC-AC Inverter and AC-DC Rectifier		595
			25.4.6.1 Configuration of Three-Phase Converter	595
			25.4.6.2 Inverter Configuration of the SRM	598
		25.4.7 DC-to-DC Converter		599
			25.4.7.1 Buck Mode of the DC-to-DC Converter	599
			25.4.7.2 Boost Mode of the DC-to-DC Converter	599
			25.4.7.3 Multifunction Inverter	600
	25.5	ISA System Issues		602
		25.5.1 Energy Storage and ISA System		602
		25.5.2 ISA Cooling Styles		605
			25.5.2.1 Air Cooling	605
			25.5.2.2 Liquid Cooling	606
		25.5.3 Other Issues		607
	25.6	Summary		607
	References			608
26.	**Fault Tolerant Adjustable Speed Motor Drives for Automotive Applications**			**611**
	Babak Fahimi			
	26.1	Introduction		611
		26.1.1 Self-Organizing Controllers		612
			26.1.1.1 Hierarchy of Control Methods in Induction Motor Drives	614
			26.1.1.2 Smooth Transition between Various Control Methods	615
			26.1.1.3 Reconstruction of the Phase Currents	619
	26.2	Digital Delta Hysteresis Regulation		620
		26.2.1 Current Reconstruction Algorithm for DDHR		621
	References			623

Table of Contents

27. Automotive Steering Systems — 625
Tomy Sebastian, Mohammad S. Islam, and Sayeed Mir
  27.1 Introduction — 625
  27.2 Steering System — 625
    27.2.1 Manual Steering — 626
    27.2.2 Hydraulically Assisted Steering — 627
    27.2.3 Electrohydraulic Power Steering — 628
    27.2.4 Electric Power Steering — 629
  27.3 Advanced Steering Systems — 630
    27.3.1 Four-Wheel Steering — 631
    27.3.2 Future-Generation Steering Systems — 631
  References — 631

28. Current Intensive Motor Drives: A New Challenge for Modern Vehicular Technology — 633
Babak Fahimi
  28.1 Background — 633
  28.2 Magnetic Design of Current Intensive Motor Drives — 634
  28.3 Stability Considerations in Multiconverter Systems — 637
  28.4 Energy Transfer — 639
  28.5 Impact on Control — 640

29. Power Electronics Applications in Vehicle and Passenger Safety — 641
D.M.G. Preethichandra and Saman Kumara Halgamuge
  29.1 Introduction — 641
  29.2 Power Electronics in Vehicle Safety — 641
    29.2.1 The CAN Bus Used to Network Vehicle Power Electronic Modules — 642
    29.2.2 Engine Safety Systems — 644
    29.2.3 Antitheft Alarm Systems — 648
    29.2.4 Adaptive Cruise Control (ACC) — 649
    29.2.5 Reverse Sensing and Parking System — 650
  29.3 Power Electronics in Passenger Safety — 650
    29.3.1 Seatbelt Control Systems — 651
    29.3.2 Power Window Safety Systems — 652
    29.3.3 Airbags — 653
    29.3.4 Driver Assistance Systems and Stress Monitoring — 653
  29.4 Conclusions — 654
  Acknowledgments — 654
  References — 655

30. Drive and Control System for Hybrid Electric Vehicles — 657
Weng Keong Kevin Lim, Saman Kumara Halgamuge, and Harry Charles Watson
  30.1 Introduction — 657
  30.2 Control Strategy — 659
    30.2.1 Thermostat Series Control Strategy — 660
    30.2.2 Series Power Follower Control Strategy — 660
    30.2.3 Parallel ICE Assist Control Strategy — 661
    30.2.4 Parallel Electrical Assist Control Strategy — 662

	30.2.5	Adaptive Control Strategy	664
	30.2.6	Fuzzy Logic Control Strategy	665
30.3	Power Electronic Control System and Strategy		669
30.4	Current HEVs and Their Control Strategies		672
	30.4.1	Honda Insight	672
	30.4.2	Toyota Prius	673
30.5	Conclusion		674
References			674

31. Battery Technology for Automotive Applications — 677
Dell A. Crouch

- 31.1 Introduction — 677
  - 31.1.1 Battery Technology — 678
    - 31.1.1.1 Valve Regulated Batteries — 680
  - 31.1.2 Present Automotive Battery Requirements — 680
    - 31.1.2.1 Battery Performance Requirements — 681
    - 31.1.2.2 Battery Charging Requirements — 681
    - 31.1.2.3 Battery Termination Standards — 682
- 31.2 Future Automotive Batteries — 682
- 31.3 Combinations of Batteries and Ultracapacitors — 685
- 31.4 Battery Monitoring and Charge Control — 685
- 31.5 Conclusion — 686
- References — 687

Index — 689

Part I

Automotive Power Systems

1

Conventional Cars

Roberto Giral-Castillón, Luis Martínez-Salamero, and Javier Maixé-Altés
Universitat Rovira i Virgili, Tarragona, Spain

1.1 INTRODUCTION

Automobile history begins after the development of important scientific discoveries in the fields of electricity, mechanics, thermodynamics, and materials.

Somehow, it could be considered as automobile antecedents: the discovery of static electricity by Thales of Miletus in 600 B.C., the inventions of Otto Von Guerick (1672), Andreas Gordon (1742), Franklin Youngest child (1747), and Nicholas Cugnot, who constructed in France the first automobile made of wood. However, the first important landmark in the history of the automobile can be placed in 1908 with the starting of the manufacture line of the Ford Model T.

That was the moment when the foundations of the later-named Second Industrial Revolution appeared, which allowed several million people to access a low-cost way of transport. In fact, since then transportation has constituted one of the strategic axes of the industrial development. Proof of the automobile's impact in society is that, at the moment, many of the main world industrial companies center their activity in automobile manufacturing.

1.2 EVOLUTION OF THE DISTRIBUTION ELECTRICAL SYSTEM

The automobile's electrical consumption has grown year by year. From the beginning until the end of the 1950s the growth was smooth. Later on, coinciding with the establishment of the 12 V battery as a supply standard in order to satisfy the increasing exigencies of comfort and security, growth was bigger.

An aspect to consider in the historical evolution of the automobile is the minimal penetration of power electronics in the automobile field until a few years ago. However, in the immediate future a massive presence of power electronics in the automobile is expected, mainly due to the new architectures of the automobile supply systems.

Nowadays the technical committees of the main societies that group the different industries of the automobile sector have already defined the outline of the services and systems that will gradually be included in automobiles during the next 15 years.

One of the axes of change is to increase electrical consumption to be able to elevate the level of comfort and security. Another vector of improvement is the diminution of fuel consumption per kilometer.

It is clear that, to be able to integrate both aspects, it must use an electrical supply system that fulfills these three general requirements:

1. It makes an optimal transformation of the mechanical energy into electrical energy.
2. It distributes the electrical energy with minimum losses.
3. It supplies the required services (loads) with maximum efficiency.

These facts start to force several future situations:

- Only the alternator will have a mechanical connection with the internal combustion engine. All the other motors, from the conditioned air compressor to the refrigeration water pump, will be electrically driven.
- High-efficiency lamps like the HID ones will increasingly substitute the conventional high-beam lights. Moreover, neon lamps or LEDs will be used for signaling and fluorescent lights for interior illumination.
- Wherever it becomes possible, the use of AC motors, with or without regulation, will be attempted.

Table 1.1 shows a prediction of the peak-power and the average-power consumption for automobiles commercialized between 2005 and 2010 [1].

In the 12 VDC conventional supply system (Figure 1.1) the load is connected directly to the battery by means of manual switches, or by means of relays for those loads that need more power or that are placed far from the control panel. All circuits pass through one or more fuse boxes. This system involves complex and heavy wire harnesses. To get an idea, a standard car contains around 2 km of cable, whose weight can be bigger than 30 kg. Furthermore, this system usually implies a huge assembly time, reliability problems, and lack of space to route the wire harnesses.

Another inherent problem of the conventional system is that the battery voltage can vary between 8 V and 16 V, and the loads connected to this battery must accept this operation range. Consequently, these loads prepared to work with such a wide input voltage range increase their price.

Due to the evidence that the conventional systems will not be able to satisfy the future needs of the modern automobiles, the automobile industry is considering different alternatives, which are basically focused on three directions: control strategy and wiring topology, power bus topology, and components.

Conventional Cars

Table 1.1 Prediction of the Power Consumption in the Near Future

Type of Load	W (peak)	W (average)
Electromechanical valves	2400	800
Water pump	300	300
Engine-cooling fan	800	300
Power steering	1000	100
Heated windshield	2500	250
Catalytic converter	3000	60
Active suspension	12,000	360
Communications	100	100
TOTAL		2220

Adapted from J. Kassakian, H.C. Wolf, J.M. Miller, C.J. Hurton. Automotive electrical systems circa 2005. *IEEE Spectrum*: 22–27, 1996.

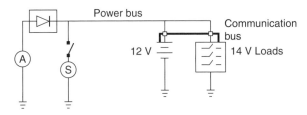

Figure 1.1 12 V conventional electrical system.

1.2.1 Control Strategy and Wiring Topology

An alternative that some car manufacturers are already implementing is the use of a single multiplexed twisted-pair cable to control multiple loads, either with a star topology or with a ring topology. A future improvement will involve the use of the power distribution line for the transmission of the load control orders.

1.2.2 Power Bus Topology

Inside the DC supply options, any increase of the supply voltage involves the reduction of the section of wires and improves the efficiency of certain loads, as it happens with DC motors. Obviously, this voltage has limits due to electrical safety regulations and the increment of the isolation cost of the whole system. Another considerable alternative involves the electrical distribution by means of an AC bus, which at first brings an interesting flexibility to supply different loads with their required voltages. This is possible mainly due to the possibility to connect these loads to the power bus using transformers. The use of this type of bus requires a careful selection not only of the root mean square (rms) voltage, but also of the frequency, which will have an important repercussion both in the technology of the supply system and in the electromagnetic interference.

1.2.3 Components

Another aspect, which can be modified to obtain an improvement in the services and systems of immediate future cars, is to replace many of the components currently used in the automobiles with others of better characteristics. An example could be the substitution of the conventionally used permanent magnet DC motors by brushless DC motors, by induction motors, or by variable reluctance motors. Another example could be the substitution of the incandescent and halogen lamps by High Intensity Discharge (HID) lamps, as well as the solid state relays that soon will start to replace the present electromechanical relays.

1.3 THE CONVENTIONAL SYSTEM OF ELECTRICAL DISTRIBUTION IN AUTOMOBILES

An electrical distribution system groups electric generators, electrical loads, their interconnection elements, and the management and protection systems of this interconnection.

This system has the primary responsibility to generate electrical energy from the mechanical energy coming from the rotational shaft of the explosion motor. Second, it has the responsibility to distribute this energy to those loads that demand it and to store the rest.

As Figure 1.1 shows, three main subsystems can be identified in a conventional distribution system of any car.

1.3.1 Battery and Its Charging System

The battery is an electrical storage device whose function is to store the energy, usually thanks to an electrochemical process, that is generated by the generator device and is not consumed by the loads. Moreover, it allows the supply of the loads even when the motor is stopped and provides the peak of energy demanded when the motor starts. The charging system, usually composed by an alternator, a rectifier, and a voltage regulator, keeps the optimal level of charge of the battery to supply the loads.

1.3.2 Motor Starter System

Nowadays a small DC motor, with less than 3 kW of power, forces the crank of the explosion motor until the appearance of ignition of the air-fuel mixture in the cylinders. The ignition usually appears after a few seconds, and during the first hundreds of milliseconds the consumed current can exceed 500 A.

1.3.3 Management System

Today automobiles have a very complex management system. One part has the responsibility of the interconnection and the protection of the loads: body electrical systems, lighting systems, in-car entertainment, and so on. Another part controls and defines the optimal parameters for a correct behavior of the motor. Finally, the chassis control system looks after the brake management, the suspension, and, in general, the active security systems of the automobile.

1.4 WIRING SYSTEM

The wiring in a vehicle is responsible for making the electrical connection among the different elements of the vehicle electrical architecture. It is made of wires, terminations, clamping elements, protections, anti-moisture and anti-dirt elements, and vibration absorbers.

Conventional Cars

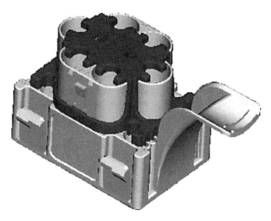

Figure 1.2 A typical power connector (courtesy of ETC-Lear Corp., Tarragona, Spain).

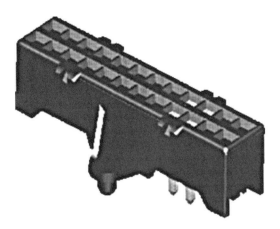

Figure 1.3 A typical signal connector (courtesy of ETC-Lear Corp., Tarragona, Spain).

Wiring can be classified basically by the car zone in which it is installed, by its function, and by its critical level. These are some examples of the classification criteria:

- By the vehicle zone: passenger compartment, engine compartment, boot, mobile parts
- By its function: cockpit, airbag, ABS, engine, injection, doors, lighting
- By its critical level: comfort, safety, communications, drive train

Figures 1.2 and 1.3 show a power connector and a signal connector, respectively.

Today cable harnesses are as short as possible and are divided into small sections connecting junction boxes and integrated modules. One advantage of dividing the wiring in sections is its ease of repairing in case of malfunction. Figure 1.4 shows an example of this kind of wiring system.

Most automotive manufacturers use compact modules called junction boxes to centralize by zones the control of the load connection. Printed circuit boards (PCBs) with cooper thickness among 70 μm and 400 μm are commonly used for the junction boxes.

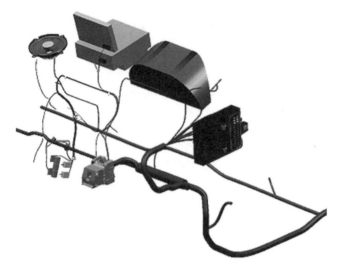

Figure 1.4 An integrated harness (courtesy of ETC-Lear Corp., Tarragona, Spain).

Figure 1.5 A junction box (courtesy of ETC-Lear Corp., Tarragona, Spain).

This technology makes these modules very reliable. A current tendency is to increase the management capabilities of these modules by adding to them smart-FETs and microcontrollers. A modern car has several junction (or service) boxes; usually one is located in the engine compartment, another in the passenger compartment, and another in the rear of the car. Figure 1.5 presents a junction box with a large integration of electronic devices. The communications bus nodes are usually also placed into the junction boxes. They will be discussed later.

1.4.1 Fuses

A fuse is a device usually used as a protection method that prevents the damage of these circuits connected to a power source. That is why the fuse must be always placed between the power source and the protected load. An accurate election of the fuse is a very important decision that affects not only the security, but also the reliability and the efficiency of an electric system. Conventional fuses are activated by the temperature that the fuse reaches

Conventional Cars

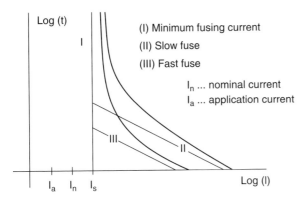

Figure 1.6 I-t curves of two different types of fuses.

in an excessive current situation and can be subdivided in two groups: blow-out fuses and those constructed with materials with a positive temperature coefficient. In both cases, the fusion of the device is caused by the calorific power generated by the power dissipation.

A fuse is designed to correctly dissipate the heat generated by a current under normal conditions, reaching a stationary temperature quite below the melting temperature; when the equilibrium disappears and the temperature exceeds the fuse melting temperature, the fuse melts.

It is well known that the heat generated in the fuse needs time to be transferred to the environment and that the dissipated power is $I^2 R_{fuse}$. It can be deduced from this that the activation energy needed by a concrete fuse at a particular room-temperature is always the product $I^2 t$.

According to this, it is normal that the I-t curves look like the ones depicted in Figure 1.6, where curve (II) represents a slow blow-out fuse and curve (III) represents a faster one.

Conventional fuses are destroyed when any working point of the I-t curve is reached. To come back to a normal operational situation, after the disappearance of the problem that caused the overcurrent, a new fuse must be put in place. This requirement represents a big problem that new electric systems try to overcome.

At the moment, two different circuit-protection devices that offer reusability are being developed: the Polymeric Positive Temperature Coefficient (PPTC) fuse and the Smart Power Switch (SPS).

1.4.1.1 *Polymeric Positive Temperature Coefficient Devices*

The PPTC circuit-protection devices are made of a composite of semi-crystalline polymer and conductive particles. At normal temperatures, the conductive particles form low-resistance networks in the polymer. However, if the temperature rises above the device switching temperature (T_{Sw}), either from high current through the part or from an increase in the ambient temperature, the crystals in the polymer melt and become amorphous. The increase in volume during the amorphous phase causes a separation of the conductive particles and results in a large nonlinear device resistance.

The resistance typically increases by three or more orders of magnitude, as shown in Figure 1.7. This increased resistance protects the equipment in the circuit by reducing the amount of current that can flow under the fault condition to a low steady-state level. The device will remain in its latched (high-resistance) position until the fault is cleared

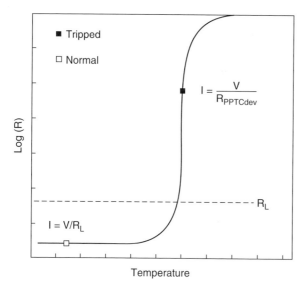

Figure 1.7 Typical resistance curve of a PPTC fuse.

and the power to the circuit is removed; then the conductive composite cools and recrystallizes, restoring the PPTC to a low-resistance state and the circuit and the affected equipment to normal operating conditions.

Some of the critical parameters to consider when designing PPTC devices into a circuit include device hold current and trip current, the effect of ambient conditions on device performance, device reset time, leakage current in the tripped state, and automatic or manual reset conditions.

Figure 1.8 illustrates the hold- and trip-current behavior of PPTC devices as a function of temperature. Region A shows the combinations of current and temperature at which the PPTC device will trip and protect the circuit. Region B shows the combinations of current and temperature at which the device will allow normal operation of the circuit. In region C, it is possible for the device to either trip or to remain in the low-resistance state, depending on the individual device resistance and its environment.

Because PPTC devices can be thermally activated, any change in the temperature around the device could affect the performance of the device. As the temperature around a PPTC device increases, less energy is required to trip the device, and thus its hold current (I_{HOLD}) decreases. Ceramic as well as polymeric PTC manufacturers provide thermal derating curves and I_{HOLD}-vs.-temperature tables to help designers in selecting devices with the appropriate rating.

The heat-transfer environment of the device can significantly affect device performance. In general, by increasing the heat transfer of the device, there is a corresponding increase in power dissipation, time-to-trip, and hold current. The opposite occurs if the heat transfer from the device is decreased. Furthermore, changing the thermal mass around the device changes the time-to-trip of the device.

The time-to-trip of a PPTC device is defined as the time needed, from the onset of a fault current, to trip the device. Time-to-trip depends on the size of the fault current and the ambient temperature.

If the heat generated is greater than the heat transferred to the environment, the device will increase in temperature, resulting in a trip event. The rate of temperature rise

Conventional Cars

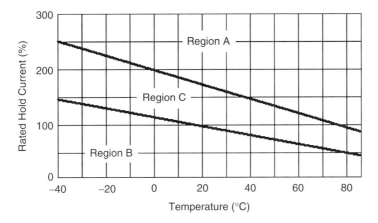

Figure 1.8 Hold and trip behavior of PPTC fuse (adapted from L. Jones, K. Kinsman, A. Cilluffo. PPTC design considerations for automotive circuits. www.ce-mag.com/archive/02/07/Jones.html).

and the total energy required to make a device trip depends on the fault current and heat-transfer environment.

1.4.1.2 Smart Power Switches

The SPSs are semiconductor devices that in addition to certain sensing and protection functions have a power metal oxide semiconductor field effect transistor (MOSFET) that acts as a switch. Figure 1.9 presents the basic structure of a high-side SPS from Infineon (PROFET). The majority of available SPS have switching control and additional features.

The switching control allows the load connection and disconnection like a mechanical switch; this is the main reason why the SPS will soon replace them.

Additional features include short-circuit protection, overcurrent protection of loads, overvoltage protection, current sense, load state diagnosis, and overtemperature protection, among others.

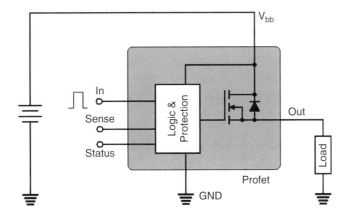

Figure 1.9 Basic structure of a PROFET (Infineon).

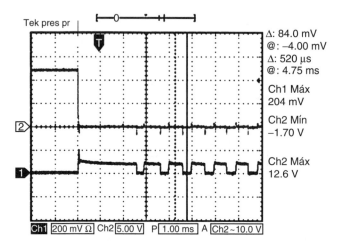

Figure 1.10 14 V–GND short-circuit of Infineon's BTS640. Channel 2 (top): voltage in the 14 V bus (5 V/div.). Channel 1 (bottom): current in the battery (100 A/div.).

The possibility of some SPS to sense the flowing current allows the implementation of a circuit protection system without the placement of conventional fuses. This system includes a comparator that opens the switch when the sensed current would overcomes a reference. However, only the SPS behavior as a direct protector against short-circuits will be discussed next.

Generally, the SPS protection modes rely on the internal temperature and the current limitation. In the temperature limitation mode, when the semiconductor temperature exceeds a limit (usually around 175°C), the SPS shuts down after 100 µs. After this, some SPS reconnect themselves when the temperature falls down below a critic level. This is the case of the BTS640 (Infineon); Figure 1.10 shows the BTS640 reconnection attempts in a short-circuit between the 14 V bus and GND. Other SPS need an external reconnection signal (latching mode).

The current limitation mode becomes active when the voltage V_{DS} overcomes a certain voltage value, from which the MOSFET goes into a current limitation state that is maintained independently of the V_{DS} value [3].

1.4.2 Behavior Comparison among the Different Protection Devices

Figure 1.11 shows the I^2t curves of a blow-out fuse, a PPTC, and a SPS. It can be observed that the blow-out fuse is the fastest device. For high currents, the SPS becomes also a fast device but the PPTC remains as the slowest.

The SPS curve at high currents presents a remarkable edge due to the appearance of a current limitation effect. This is a specific characteristic of SPS that allows the protection of both device and load.

Figure 1.12 presents approximately the time behavior of the three types of fuses at a high current condition. It can be seen that the blow-out fuse is the fastest device, but it is also the one that allows more current to pass before opening the circuit. The second fastest device is the SPS; furthermore, it is the one that presents a major limitation to the flowing current. Finally, the PPTC is the slowest device and, before shutting down, it allows almost as much current as the blow-out fuse. This is the reason why the PPTC, not considering its capability of reuse, is at the moment the least recommended device among them all.

Conventional Cars

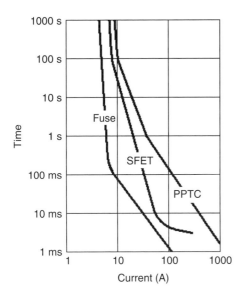

Figure 1.11 I-t curves of a blow-out fuse, a PPTC fuse, and an SPS (SFET).

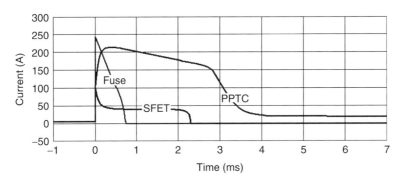

Figure 1.12 Approximate temporal behavior of the three types of fuses (adapted from A. Graf, H. Estl. HL Application Note. Fuse function with PROFET High Side Power Switches, 1999 and Figure 1.10).

1.5 LOAD CONTROL: AUTOMOTIVE CONTROL NETWORK PROTOCOLS

The concept of remote load switching using a multiplexed communications bus has been evolving for several years as a promising approach to significantly reduce the wiring harness complexity. Although multiplexed networks are already beginning to appear in production vehicles, wide implementation of this approach has been hindered by the high cost of the smart semiconductor switches that are crucial to performing the load switching functions.

The power bus is distributed around the vehicle, and the electrical loads are connected directly to this bus at the load points. Smart switches are used to control load switching. Load switches and smart switches are integrated in-load control modules.

Table 1.2 Classification of Automotive Networks

Network Classification	Speed	Application
Class A	< 10 kbit/s	Convenience features
Class B	10–125 kbit/s	General information
Class C	125 kbit/s–1 Mbit/s	Real-time control
Class D	> 1 Mbit/s	Multimedia applications

A code is sent on the control bus to switch a particular load on or off. The interface circuit in the desired load control module responds to this code and controls the switch to turn the load on or off as commanded. A code is sent by the load control module to the central control unit of the vehicle to indicate the action that takes place at the load.

In the U.S. the Society of Automotive Engineers (SAE) has published a series of documents describing recommended practices for vehicle networking. The SAE has also formally classified vehicle networks based on their bit transfer rates [4] [5]. Table 1.2 illustrates the SAE classification categories.

There is a wide range of automotive networks reflecting defined functional and economic niches. High-bandwidth networks are used for vehicle multimedia applications, where cost is not excessively critical. When data must be exchanged at a much faster rate, as is the case with the power-train control, class B or C bus systems must be used. Comfort electronics, systems such as power windows, and some instrumentation modules require only modest response time, which only just surpass human perception time.

Next, some of the most used buses inside an automobile will be described.

1.5.1 Controller Area Network (CAN)

The CAN is the most used bus in Europe. It was developed by Robert Bosch in 1987 as a class C vehicle inner controller. The world's main semiconductor companies have integrated CAN controllers in their microprocessors and microcontrollers. There are two basic versions: CAN 1.0 and CAN 2.0. Both present a two-wire communication system. CAN has a great flexibility, allowing the inclusion of a new node very easily, only affecting the lower priority nodes.

Bus efficiency depends on its own communication charge, so it is recommended not to overcome 70% in order to avoid the bus saturation. Bus speed varies between 10 kbit/s and 100 kbit/s in simple operations of control and connection related with comfort. Between 100 kbit/s and 1 Mbit/s the bus is devoted to critical real-time functions, such as engine management, antilock brakes, and cruise control.

CAN is a robust bus with good features as a general control bus. However, it is not recommended for more specialized functions such as the x-by wire.

1.5.2 Local Interconnect Network (LIN)

LIN is a class A protocol that allows a data transfer speed of 20 kbit/s. It was developed by Audi, BMW, Daimler-Chrysler, Motorola, Volcano, Volvo, and Volkswagen. The bus includes only one transmission line that allows point-to-point communication between a lonely master node and several slave nodes. This avoids collisions.

Conventional Cars

When a new slave node must be connected there is no insertion problem due to the simplicity of the bus. There can be up to 16 slave nodes.

The data frame is character-based, allowing an easy implementation with just a simple microcontroller and a virtual UART (via software). The bus is really efficient for simple applications, such as car seats, door locks, sunroofs, rain sensors, and door mirrors.

1.5.3 Byteflight

Byteflight is a flexible time-division multiple-access (TDMA) protocol for safety-related applications developed by BMW, ELMOS, Infineon, Motorola, and Tyco EC.

It reserves a fixed cycle time for the synchronous transmissions (the first 250 µs), and the rest of the time is for the asynchronous transmissions.

It has a high-speed data transfer rate, but because the protocol is not fault-tolerant it is not recommended for critical functions.

1.5.4 Time Triggered Protocol (TTP/C)

TTP uses the TDMA and is designed for real-time fault-tolerant distributed systems. The bus presents a cyclic behavior, and access is determined with a table that each bus node possesses. A maximum of 64 nodes is allowed and it can achieve transference rates up to 25 Mbit/s.

One of its drawbacks appears in those systems that have not planned their expansion; when a new node has to be included, the rest must be updated. Moreover, the bus speed is fixed by the speed of the slowest microcontroller of the system. These disadvantages are probably the reasons why this bus is not often used.

1.6 NEW ARCHITECTURES

Due to the inability of the conventional supply systems to cover all the electrical needs of the modern vehicles, the automobile industry is developing several solutions.

Some options involve an elevation of the voltage in a DC supply system. As mentioned previously, this would mean a diameter reduction of the wires and also an improvement of the efficiency of some loads. However, there is a voltage limit imposed by the electric security normative and the costs of the isolation.

In 1988, the SAE's Electric System Group recommended to the automobile industry the use of a dual-voltage system or a conventional system with higher voltage. A few years later, the main automobile companies signed the agreement.

All the different alternatives that the automobile manufacturers are conceiving for the new supply systems share the common characteristic of an elevation of the supply voltage.

1.6.1 Electric Security

The main security considerations taken into account in the automobile electric supply systems are the wire fire risks due to overcurrent, sparks, and electric discharges. The most restrictive aspect is determined by the effect of the supply voltage on the human body. The electric discharge on a human body can be indirect (the reaction to the sensation of suffering an electric discharge) or direct (the effect of current passing through human tissues).

It must be remarked that the effects of the contact of a human body with a voltage power supply depend on:

- Current path through the human body
- Humidity level of the surface in contact
- Contact area
- Impedance of the element that comes into contact with the power supply
- Current flow duration through the human body

The most complete reference about the AC and DC voltage effects on a human body appears in the IEC 479-1:1984. In 1991, these references inspired a SAE recommendation [6], where the DC and 50 Hz AC voltage limits were defined as 65 V and 50 V, respectively. Moreover, for the high frequencies, a 75 V limit (25 kHz) and an 87.5 V limit (30 kHz) were also established.

1.6.2 Voltage Effect on the Components

There is an indirect effect of the voltage level on the volume, weight, and price of the wiring system.

It has been estimated that with a 24 V power supply the weight reduction could be between 42% and 58%, and with 42 V (36 V battery) the weight reduction could be between 47% and 67% [7].

The increment of the voltage level would have a pernicious effect on the electromechanical relays due to an increase in the erosion of the terminals. However, with the use of solid-state relays this problem would disappear, especially in the 24 V system.

Brush-type DC motors would take a great benefit from the voltage level increment, increasing its efficiency.

With regard to the illumination systems, the tendency is to substitute the conventional incandescent and halogen lamps by HID lamps, which benefit from the voltage level increment.

Another aspect to consider is that each power supply system generates voltage peaks during the battery charging process (alternator + rectifier). These voltage peaks are proportional to the nominal supply voltage, which implies that a higher supply voltage requires components with higher breakdown voltage.

1.7 ALTERNATIVE ARCHITECTURES

1.7.1 High Frequency AC Bus System

In 1994 the main companies related with the automobile sector founded a work team that, under the coordination of the Massachusetts Institute of Technology (MIT), elaborated the guidelines for a future (before 2015) automobile power supply standard.

The proposal of this committee [1] can be summarized in the following points: 12 V battery, 48 V_{rms} – 25 kHz AC bus, and loads supplied by the bus through a ferrite core transformer.

The main advantage of this system is the facility to obtain the load required voltages just using high frequency transformers. For those cases that would require a DC voltage, a rectifier and a filter should be used.

This system means a radical change with respect to DC buses, so it can be assumed that its introduction to automobiles will take a long time.

1.7.2 Dual-Voltage DC Bus

Some of the most interesting alternatives that the automotive companies are actually developing are the DC dual-voltage systems.

This generic name groups several solutions that have in common the availability of two DC voltage levels in the supply system. Usually one level is the conventional 12 V, while the other is 24 V or 36 V [8]. These systems keep several of the 12 V services included in present cars and allow the connection of higher power loads at a higher supply voltage, taking profit of a reduction of the wire diameter. It should be stated also that some loads, such as brush-type DC motors, improve their efficiency with a higher supply voltage.

The dual-voltage system, because it allows a smooth transition from conventional architectures to higher voltage ones, has become the main power supply alternative system. Among many of the dual-voltage DC bus topologies that have been developed, Figure 1.13 shows an accepted architecture that can be implemented soon.

The system of Figure 1.13 has two batteries. The 36 V battery is connected directly to the alternator and supplies the starting motor and the high power loads. The remaining loads are supplied from a 12 V battery, which is recharged by a DC/DC converter.

Eventually, the 12 V battery can help the 36 V battery in transient situations of high current demand, as in the starting of the explosion motor. To allow this situation, it is obvious that the DC/DC converter must be bidirectional.

A second dual-voltage architecture is shown in Figure 1.14. This concept uses a more complicated alternator with two sets of stator windings to deliver power separately to the 42 V and 14 V buses.

The comparison between the architectures in Figures 1.13 and 1.14 reveals some important cost-performance tradeoffs. The Figure 1.13 architecture, incorporating the DC/DC converter, offers greater energy controllability compared to the architecture of Figure 1.14. However, the cost of the DC/DC converter is expected to be considerably higher than that of the simpler phase-controlled converter depicted in Figure 1.14.

In dual-voltage architectures, automotive industries have proposed to achieve a feasible, reliable, low-cost, and efficient power distribution system. In this way, we can summarize some important challenges opened by automotive industries and power electronics research laboratories.

One of those challenges consists of integrating the starting motor and the alternator in only one device. To reduce fuel consumption, this device has an automatic option consisting of starting-stopping the car engine according to the car motion-standstill cycles.

Dual-voltage architectures also have some drawbacks that must be solved. Among them, the increased possibility of short-circuits is especially important. In present-day

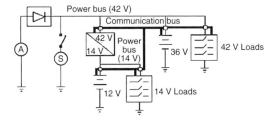

Figure 1.13 Dual-voltage architecture with two batteries and DC/DC converter.

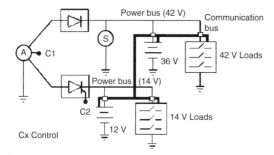

Figure 1.14 Dual-voltage architecture supplied by an alternator with two stators.

14 V-systems, only the short-circuits between the 14 V line and ground are possible. In dual systems, short-circuits can happen between each of the two supply lines (14 V and 42 V) and ground, but also between the two lines directly. The last case is especially dangerous for the 14 V loads because they have to support severe overvoltages until the protection mechanisms start to react.

Adapting the car lighting system to the new dual-voltage standards represents another subject of special interest. The car-lights power consumption can reach 600 W. Assuming a zero-power increment in the lighting system, the increase in the voltage supply means a current reduction that leads to a decrease in the halogen lamps filament section to ensure a good performance. Under such circumstances, the average life span of the lamps is appreciably reduced due to the increment of the filament fatigue by vibration, and consequently using halogen lamps for 42 V is inadvisable.

Different alternatives for the lighting system are now under research: to introduce new kinds of lamps (LEDs arrays, HID lamps), to supply the conventional 14 V halogen lamps by means of switching DC/DC converters (expensive today), and finally, to supply the lamps with a 14 V_{RMS} PWM signal having an amplitude of 42 V and an appropriate duty ratio. This is a less expensive solution but it can increase electromagnetic interference (EMI) levels.

An increase in the voltage of the power distribution system affects the entire vehicle configuration, and also the design of components and devices. As a result, new specifications for the system and components are required.

The situation today shows no clear consensus in automotive industries about which (if any) is the best architecture for the new electric power distribution system. Nevertheless, is quite clear that the power distribution system must be changed to fit all the new specifications and loads. Vehicle manufacturers and subsystem providers might work together to reach an optimum solution satisfying both the new power requirements and their legitimate interests.

REFERENCES

1. J. Kassakian, H.C. Wolf, J.M. Miller, C.J. Hurton. Automotive electrical systems circa 2005. *IEEE Spectrum*: 22–27, 1996.
2. L. Jones, K. Kinsman, A. Cilluffo. PPTC design considerations for automotive circuits. www.ce-mag.com/archive/02/07/Jones.html.
3. A. Graf, H. Estl. HL Application note ANPS039E, Fuse function with PROFET Highside Power Switches. Siemens Power Semiconductors, Munich, 1999.

4. G. Leen, D. Heffernan, A. Dunne. Digital networks in the automotive vehicle. *Computing and Engineering Journal,* Vol. 10, No. 6: 257–266, 1999.
5. G. Leen, D. Heffernan. Expanding automotive electronic systems. *Computer,* Vol. 35, No. 1: 88–93, 2002.
6. Electrical Systems Group, *Recommendations Regarding Voltage Levels and Shock Hazard for Higher Voltage Vehicle Electrical Power Systems in Automotive Applications.* SAE Motor Vehicle Council, Warrendale, PA, 1990.
7. M.F. Matouka. Design considerations for higher voltage automotive electrical systems, Future Transportation Technology Conference, Portland, OR, August 1991.
8. J.V. Hellmann, R.J. Sandel. Dual high-voltage vehicle electrical systems, Future Transportation Technology Conference, Portland, OR, August 1991.
9. J.M. Miller, D. Goel, D. Kaminski, H. Pschöner, T.M. Jahns. Making the case for a next generation automotive electrical system, IEEE-SAE International Congress on Transportation Electronics, Dearborn, MI, Oct. 1998.

2

Hybrid Electric Vehicles

John M. Miller
J-N-J Miller Design Services, Cedar, Michigan

The automotive industry has entered the 21st century with far more power train diversity than when it entered the 20th century, a time when the newly invented internal combustion engine (ICE) competed with steam engines, battery electric drives, and horse-drawn carriages. Now, the internal combustion engine may have new competition in the form of hydrogen fuel cells, but successors to those early power plants persist even to this day. Steam engines, the original external combustion engine, are still around in the form of Stirling engines. Battery electric power trains are still here, but in decline, for much the same reasons as a century ago. A trip to Mackinaw Island, MI, and several other notable historical sites will convince the visitor that horse-drawn carriages may not just be a transportation mode of the past, but a viable option for getting places, particularly where internal combustion engines are banned. And, of course, there is always the bicycle.

With the 21st century still in its infancy we are at a period in personal transportation history when the automobile and light truck consume some 40 to 46% of the total petroleum supply in North America. This is because the transportation sector gets more than 97% of its energy for car and truck power plants from petroleum, and half of the supply is from off-shore. Add to this the fact that air quality concerns are high over ground-level smog and carbon monoxide (CO). In total, highway vehicles are responsible for 27% of all hydrocarbons (HCs), environmentally threatening volatile organic compound emissions. Internal combustion engines also emit some 32% of nitrogen oxides released into the atmosphere annually, not to mention nearly two thirds of all the CO.

Against this backdrop of tightening liquid fossil fuel supply, a high reliance on off-shore oil, and the global concerns over transportation sector emissions, it is clear that new solutions to personal transportation power plants are demanded. The push for a hydrogen economy is one promising alternative that has potential to free the transportation industry from its reliance on petroleum, but the problem persists in how to produce hydrogen from

Table 2.1 Internal Combustion Engine Attributes

Attribute	Units	Hydrogen Fuel Cell*	Hydrogen ICE	Gasoline ICE	Diesel ICE
Power density–volumetric	kW/l	0.5	+	55	45
Power density–gravimetric	kW/kg	0.5	+		0.63
Stack/engine thermal efficiency	%	60	58	< 60++	60
FC/engine brake system efficiency	%	48	38	30	45
Emissions		< Tier 2	SULEV	SULEV	ULEV
Cost	$/kW	$300 mature, $35 target	?	$35	$30+++

* DOE 2004 target
+ 35% lower than conventional ICE due to fuel pre-ignition (knock limited combustion)
++ Ultimate limit with innovations (see Table 2.2)
+++ PNGV stated goal for 2004

natural gas within the confines of the automobile. In 2003, a Mercedes-Benz A-class vehicle using a Ballard fuel cell was delivered to UPS for use as a delivery vehicle in the Ann Arbor, MI, area. In 2004 Daimler-Chrysler put into service the first medium-duty fuel cell commercial vehicle, along with a fueling infrastructure to support them. But fuel cells are not the only alternative to our reliance on petroleum and the need to reduce emissions. Ford Motor Co. has already unveiled its Model U, successor to the Model T of a century ago, which burns hydrogen instead of gasoline or diesel fuel in its modified internal combustion engine. Because hydrogen does not have the energy density of gasoline (3000 times more volume at STP) the Model U relies on a hybrid power train that uses battery storage to augment the vehicle's internal combustion engine when peak performance is demanded. General Motors Corporation has updated its earlier introduction of a new vehicle platform, the Autonomy, that used a fuel cell power plant by adding complete x-by-wire functionality. The resulting Hy-wire platform essentially "re-invents the automobile around fuel cells and drive by-wire technologies," according to GM.

Costs of fuel cell power plants are still very high. Preproduction 50 kW fuel cell power plants may cost more than $100,000 each. Today's gasoline-fueled internal combustion engines average only $35/kW. The U.S. Department of Energy has set a near-term target of $150/kW for fuel cell power plants. Table 2.1 illustrates the present goals or realized targets for internal combustion engine attributes [1,2,3]. In Table 2.1 it is interesting to note that, regardless of fuel cell, hydrogen-fueled ICE, or advanced gasoline/diesel ICE, 60% represents a practical upper bound to thermal efficiency. Higher thermal efficiencies are reported, but at much higher compression ratios than can be achieved in mass-produced engines. Homogeneous charge compression ignition (HCCI), for example, shows promise of very high efficiency, but such combustion technologies that rely on activated radicals for ignition are difficult to control.

Conventional gasoline- and diesel-fueled internal combustion engines will remain practical over the long term, especially with anticipated enhancements and future innovations in combustion processes and engine system ancillaries such as valve trains, combustion

Hybrid Electric Vehicles

Table 2.2 Internal Combustion Engine Innovations

Engine	Engine System
Variable compression ratio	Camless (electronic or hydraulic) valve trains
Variable cylinder displacement	Variable valve timing and lift (VVTL)
Direct Injection (DISI) – stoichiometric or stratified charge	Intake — variable valve timing (VVT)
Homogeneous charge compression ignition (HCCI)	Variable intake phasing
Common rail injection	Exhaust — variable valve timing
Supercharging	Dual equal VVT
Turbocharging — motor boosted or variable geometry	Dual-independent VVT

chamber geometry, fuel injection, and induction air treatments. Table 2.2 summarizes some of the more notable innovations now being worked on or already in limited production.

Until the fuel cell power plant costs come down from their present $3000/kW to even a mature cost of $300/kW as noted in Table 2.1, there will not be a huge demand for such technology. There are also many secondary technical issues to be resolved before fuel cell power plants are commercially viable: (1) hydrogen storage technology, (2) natural gas reformers, (3) water treatment, and (4) electrical energy storage systems. Compressed hydrogen at 3600 to 5000 psi in fiber-reinforced tanks is the most practical system today. Storage in metal hydrides is heavy and inefficient in terms of mass of hydrogen storage to storage tank mass (~ 6%). Liquified hydrogen yields higher fuel density but cryogenic systems, hydrogen evaporation, and stand time are major developmental issues. Natural gas reformers as a means to generate hydrogen on-board are highly sought after and represent the ultimate in fuel storage system efficiency. A reformer is essentially a chemical processing plant on-board. Water treatment is a perplexing issue and one that must be solved for fuel cell vehicles to operate in northern climates year round. A by-product of combustion, water vapor is present in the fuel cell stack membrane electrode assembly (MEA) and must be condensed and stored. Water is needed in the hydrogen humidifier, and this requires special treatment for the same reasons. At the present state of fuel cell technology this power plant has relatively slow dynamics compared to a gasoline ICE. Target ramp-up rates for power (10 to 90%) are now 10 sec with near-term goals of 3 sec and a long-term goal of 1 sec. Not only this, but fuel cell stack start-up in cold climates is presently at 3 min or more. During the fuel cell warm-up period, as well as during transient operation, some means of electrical energy storage is necessary. Batteries are used to supply power during warm-up and during vehicle launch.

Regardless of which type of power plant is used, fuel cell stack or internal combustion engine, the next generation of personal transportation vehicles will rely on hybridization to realize gains in fuel efficiency and reduced emissions. Burning less fuel produces lower emissions. Hybrid-electric power trains are becoming well-accepted near-term solutions to the global concerns over CO_2. This means that a paradigm shift is now in progress within the global automotive industry. It will no longer be possible to survive as a company without shifting vehicle powerplants from conventional gasoline- or diesel-fueled engines to much more thermally efficient ICEs either through innovations or through burning hydrogen in the ICE. The compression ignited direct injected (CIDI) engine (or "diesel" engine, as it is commonly known) offers the highest thermal efficiency of all internal combustion engine technologies. This is not contrary to Table 2.1.

It should be pointed out that a hydrogen ICE is in reality a CIDI engine, because hydrogen fuel has such extreme flammability and an extremely low ignition point (< 0.05 mJ compared to 0.25 mJ for 50% probability of ignition for gasoline). Homogeneous charge compression ignition combustion processes are in reality a precursor to CIDI. Once started, HCCI functions as a diesel engine operating at compression ratios exceeding 20:1 and with equivalence ratios that are typically 0.35.* It is noteworthy that the hydrogen ICE operates with equivalence ratios of 0.12 to 0.4 with idling operation near the upper end of this range.

Even with all of the innovations discussed above it will not be possible to reach emissions regulated targets without further improvements to fuel economy. The hybrid electric power train is now accepted as the lowest cost near-term solution. Hybridization provides the following power train features:

- Idle stop
- Torque augmentation
- Regenerative braking

Idle stop and its extensions, early fuel shut off (EFSO) and decel fuel shut off (DSFO), conserve fuel by not burning it for nonpropulsion events. Fuel conservation is obtained by shutting off the engine fuel supply during decelerations and by simultaneously regulating the power train-mounted electric motor-generator (M/G) so that vehicle deceleration does not exceed some prescribed limit, such as 0.05 g, as the vehicle speed decreases.** Early fuel shut off can be semantically confused with DSFO but it refers more to a prescribed engine speed below which fuel to the engine is inhibited. For example, ESFO could be enabled should the engine speed drop below 1200 rpm (coasting on a downhill grade, for instance).

Torque augmentation is one of the major performance enhancers offered by hybridization. Since one or more electric motor-generators are present in the power train, it is only a matter of scheduling the M/G torque to coincide with certain driveline events such as transmission shifting, vehicle launch, lane changing, and passing maneuvers. M/G torque response, particularly under field-oriented control, is far faster than ICE response and faster still than fuel cell system response. During gear shifting the M/G can add torque to the driveline, thus filling in for lack of engine-supplied torque. During vehicle launch the M/G can augment the engine torque, particularly since downsized engines are virtually a prerequisite of hybridization, and provide the customer with expected acceleration or, better still, improved performance. The Honda Civic hybrid does exactly this. The engine is far too undersized to deliver the performance drivers expected, so the M/G is sized to augment the driveline over the entire engine operating speed range.

Regenerative braking, or energy recuperation, is the principal means through which kinetic energy of the vehicle is returned to electric energy storage rather than burned off as heat in the brake pads. But there are practical limits to how much and how fast regenerative braking can be applied. Smooth and seamless brake feel is the result of a fine

---

* Equivalence ratio, ϕ, is defined as actual fuel/air divided by stoichiometric fuel/air.
** Where g = acceleration due to gravity.

Table 2.3 Electric Machines and Power Electronics Goals (PNGV)

	Units	Target
Electric Machines		
Peak specific power	kW/kg	1.6
Volumetric power density	kW/l	5.0
Specific cost	$/kW	4
Part load efficiency (@ 20% peak torque)	%	96
Power Electronics		
Peak specific power	kVA/kg	5
Volumetric power density	kW/l	12
Specific cost	$/kVA	7
Part load efficiency (@ 20% current)	%	97–98

balance between M/G energy recuperation and the vehicle's foundation brakes. The best brake system for a hybrid is what is known as series regenerative braking system (RBS). With series RBS the M/G extracts braking energy without application of the service brakes, then when higher braking forces are required, or if the brake pedal is depressed faster than a prescribed threshold, the service brakes are engaged so that total braking effort is delivered. A less costly and less efficient approach is parallel RBS. With parallel RBS the vehicle's service brakes and M/G retarding torque are managed simultaneously to deliver the desired total braking effort. Parallel RBS can be envisioned as M/G braking the front axle and foundation brakes on the rear axle. The vehicle's antilock brake system (ABS) controller then manages the distribution of front-rear braking efforts so that vehicle longitudinal stability is not lost.

Hybrid electric power trains require a large investment in M/G and power electronics technology. Package space is extremely restricted so that even with a ground-up design for a hybrid there is precious little space to put 20 to 100 kW electric machines and the power electronics to drive them. Such machines must not only have the highest power density but they must also be robust and efficient. Packaging an M/G into the vehicle driveline means that repair and replacement would entail significant tear-down if a failed M/G was packaged inside the transmission. The power electronics must be of the highest power density both gravimetrically and volumetrically. Design targets for hybrid vehicle electric machines (i.e., M/Gs) and their associated power electronics are summarized in Table 2.3 and derived from the U.S. DOE's Partnership for a New Generation of Vehicles (PNGV)* program.

To put these numbers in some perspective, consider the following examples. A high-volume Lundel alternator rated 120 A continuous at 14.2 VDC has a specific power of 0.24 kW/kg vs. the hybrid vehicle target of 1.6 kW/kg. In the case of power electronics a volumetric target of 12 kW/l is still very aggressive but highly dependent on bus capacitor content, since such components typically consume up to 60% of the power electronics

---

* Partnership for a New Generation of Vehicles, 1993 through 2002, when it was replaced with the Freedom Car fuel cell vehicle initiative.

volume. Looking at this from another vantage point, DC/DC converters for point of load bus regulation typically deliver from 50 to 100 W/in³ (3 kW/l to 6 kW/l). Consider Vicor Corporation's new V-I Chip, a small brick for 48 to 12 V conversion capable of 200 W continuous output at 95% efficiency in a package that measures 1.25" W × 0.85" D × 0.25" H (0.265 in³). This is a new benchmark in power converter volumetric power density of 753 W/in³ or 46 kW/l! The Vicor converter consists of a pair of resonant converters each operating at 1.75 MHz interleaved to yield an output ripple frequency of 3.5 MHz. This converter achieves nearly 4 times the volumetric power density as the hybrid vehicle targets. Unfortunately, such feats are only possible at relatively low powers. When switching hundreds of amperes it is not feasible to operate at megahertz frequencies, so realizing 12 kW/l remains a challenging target that demands considerable innovation in power packaging.

The remainder of this chapter explains various hybrid vehicle power train architectures and their relative merits for passenger sedans and light trucks. Typical passenger sedans are defined as four- to five-occupant vehicles having a curb mass of 1200 to 1500 kg. Light trucks, on the other hand, are vehicles such as minivans, sport utility vehicles, and pick-ups in the category of 1400 to 2400 kg.

2.1 PARALLEL CONFIGURATION

The most common hybrid propulsion system configuration is the parallel architecture shown schematically in Figure 2.1. In this configuration an M/G is added to the existing power train in either a side-mounted or in-line package. Side-mounted M/G configurations are known as belt-ISGs, for integrated starter generators. This is the most economical and least intrusive of the hybridization approaches. With a belt-ISG the functionality of idle stop, torque augmentation, and energy recuperation are available, but to a limited degree owing to its relatively low power levels. The belt and pulley system provides a degree of gearing between the M/G and engine crankshaft. Toyota's Crown mild hybrid is an example of a production belt-ISG system designed for 42 V operation at power levels of < 10 kW.

In-line hybrid configurations in the parallel architecture as shown in Figure 2.1 have the M/G integrated into the vehicle's transmission in a direct drive fashion. Depending on power level, such implementations range from 10 kW ISG to > 30 kW full hybrids operating off of a 300 V battery. Table 2.4 summarizes the benefits and ratings of this architecture for internal combustion engines (gasoline- or diesel-fueled from 60 through 200 kW output) and typical fuel economy gains on the U.S. metro-highway drive cycle.

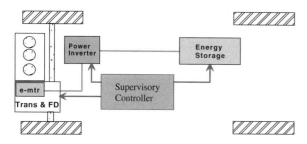

Figure 2.1 Parallel hybrid architecture.

Hybrid Electric Vehicles

Table 2.4 M/G Ratings for Various Hybrid Architectures

Architecture	System Voltage	Typical Power Requirement (kW)	Electric Fraction (%)	Relative Fuel Economy Gain (%)
Parallel	14 V, 42 V, 144 V, 300 V	3–40	5–20	5–40
Series (engine-generator, or fuel cell)	216 V, 274 V, 300 V, 350 V, 550 V, 900 V	> 50	100	> 75
Combination (power split)	216 V, 274 V, 300 V, 400 V, 550 V	15–35 and 35–65	50	> 75
Grid connected	300 V, 550 V	40–80	25–65	> 100

Electric fraction, a metric used to characterize what proportion of the total peak driveline power is supplied via the electric M/G, is defined as the ratio of M/G peak power to the sum of M/G peak power plus ICE peak power. For example, a parallel hybrid having a 60 kW ICE and a 40 kW M/G has an electric fraction of $E_f = 40\%$. Fuel economy gains are incremental over the base vehicle fuel economy for the same drive cycle. Generally speaking, engine downsizing becomes possible when the electric fraction exceeds approximately 20%. Total vehicle performance must be maintained with a downsized engine, and this is only possible if sufficient electrical energy storage is present to sustain peak electrical demands during vehicle acceleration. The amount of electrical power that must be processed in order to meet vehicle acceleration targets is dependent on the M/G peak power rating and the ability of the energy storage system to deliver the needed power. Traction system bus voltage and energy storage system capacity dictate what current magnitudes must be processed by the inverter power semiconductors. Figure 2.2 is a schematic for the hybrid vehicle electrical power processor.

Three-phase, fully controlled bridge configurations are the most prevalent power processing architecture because, for the same total power, it takes fewer cables and connectors. The actual mass of copper used, for example, in a two-phase or four-phase

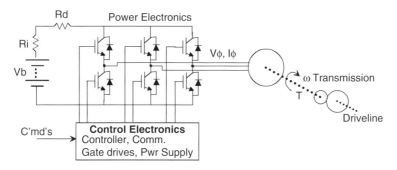

Figure 2.2 Power electronics for hybrid traction drive system.

Table 2.5 m-Phase System Cable Currents Relative to Three-Phase System

# Phases, m=	# Cables	Total System Power, phase	Total System Power, Line-Line	Cable Current, I_m/I_3
1	2	$= 1 \times V_\phi I_\phi$	$= 1.0 \times V_{ll} I_\phi$	3
2	4	$= 2 \times V_\phi I_\phi$		1.5
3	3	$= 3 \times V_\phi I_\phi$	$= 1.732 \times V_{ll} I_\phi$	1.0
4	6	$= 4 \times V_\phi I_\phi$		0.75
5	5	$= 5 \times V_\phi I_\phi$	$= 2.629 \times V_{ll} I_\phi$	0.60
m		$= m \times V_\phi I_\phi$		3/m

system is the same as in three-phase, but in either of those cases it requires four cables and connectors, rather than only three. This is the principal reason that utility power transmission is performed as a three-phase network rather than two-phase or some higher phase order. Table 2.5 is a comparison of the number of cables and the cable currents relative to a three-phase power distribution system. It can be seen that high phase order systems, for the same phase voltage, have correspondingly lower phase currents. Therefore, cable size is smaller, but total cable mass (neglecting insulation) is the same. The real benefit is that three-phase systems outperform lower phase order systems in terms of number of cables and cable size. A two-phase system, for example, will generate a rotating magnetic field in an electric machine, but delivering the power to the machine requires one more cable and 50% higher currents in the cables than if a three-phase system were used.

In Table 2.5 a four-phase system is shown to require six cables. This is because in four phases the individual phase voltages (and currents) of adjacent phases are 90° out of phase, but every other pair of phases are exactly 180° out of phase. This means that split-phase winding can be used for each pair of phases, the center tap being common between a pair. The Scott-T transformer was used in the past to convert a two-phase system into a three-phase system (albeit, not completely balanced).

All of the hybrid propulsion system power processors (i.e., inverters) in this chapter will therefore be assumed to be three-phase systems. As can be seen in Figure 2.5 the power from an on-board electrical energy storage system is delivered to the input of the inverter. Inverter switch modulation then synthesizes balanced three-phase voltages relative to the machine neutral point. Note that inverter voltages are switched relative to the negative power bus, but with a modulation envelope that, although not unique, results in balanced line-to-line voltages. The power switching devices are required to handle 150% of peak phase current amplitude so that short-term overdrive is possible. The traction drive capability curve is therefore limited by the ability to generate a maximum line-to-line voltage, and by the current rating of the power switches. The inverter line-to-line voltage limit is set by the storage system and by the type of modulation used. The inverter currents are dictated by the M/G power rating and by the available line-to-line voltage. Inverter switching device current rating is then specified by the calculation of peak current required by the M/G.

As shown in Figure 2.1 there are several contributors to power processor and hybrid propulsion system inefficiency. Energy storage systems have inherent internal resistance that results from terminations, current collectors in each cell (regardless of electrochemical cell, fuel cell, or ultracapacitor cell), and losses in the electrolyte. This cumulative resistance effect is shown schematically as R_i. Transporting power from the energy storage system to the power processor is a task performed by the electrical distribution network, in this case, a pair of high-current cables. The resistance of these cables is shown schematically as R_d. The combined effect of R_i and R_d is that inverter input voltage, U_i, is decreased from its open circuit value, U_{oc}, by approximately 32%. This means that under full load conditions, inverter input voltage will drop to approximately 68% of its unloaded value. Inverter modulation, particularly in terms of bus voltage utilization, must take this into consideration. Compensation for bus voltage variation is performed by measurement of the bus voltage.

The inverter itself has losses in terms of conduction and switching loss of each semiconductor component, whether active or passive. The active, or controlled, devices such as transistors and thyristors have conduction losses resulting from non-ideal behavior of on-state voltage drop and switching loss due to the time it takes the device to clear any stored charge and to recover forward voltage blocking capability. Uncontrolled switching devices such as diodes have similar conduction and switching loss. Passive components such as the bus ripple capacitors, device snubbers (if used), and line inductors all possess series resistance and hence contribute to overall losses. Hybrid vehicle traction inverters will commonly have efficiencies ranging from 95 to 97%, depending on voltage and power rating.

The M/G has inherent losses in terms of copper loss in the windings, rotor loss in armature bars (if an induction machine), core losses due to both eddy currents and hysteresis in the core materials, and proximity losses in the copper due to an uneven distribution leakage flux across conductor bundles within slots in the machines stator. All of these factors result in machine losses, so that typical M/G efficiency ranges from 88 to 92% for induction machines and from 93 to 95% for permanent magnet synchronous machines.

The mechanical transmission incurs losses due to the meshing of gear teeth, viscous and coulomb losses in the sliding surfaces of bearings, and churning losses in air and oil. A useful approximation is that each gear mesh contributes approximately 2% of loss. Mechanical transmissions also generate audible noise due to gear meshing and structural resonances. Great care has been taken by the automotive industry to minimize gear box noise so that modern transmissions are very quiet, even at peak loading.

Beyond the mechanical transmission there are further losses associated with the propeller shaft to the final drive and then through the drive shafts to the driven wheels. In front-wheel-drive vehicles having a transverse-mounted engine, the final drive is integrated into the transmission along with the hybrid M/G. The final drive contributes additional gear meshing and, hence, loss. Depending on transmission construction and vehicle drive-line architecture, the propeller shaft may or may not be used. Front-wheel-drive vehicles, for example, use either a gear set or a chain drive to connect the transmission to the final drive (also known as a differential).

The parallel hybrid architecture is widely used because the vehicle propulsion architecture is not that different from a conventional vehicle. It is as if the hybrid system were an overlay. Unfortunately, integrating an M/G into the vehicle's transmission requires considerable redesign, cost, and time.

2.2 SERIES CONFIGURATION

The series hybrid architecture has yet to catch on in gasoline or diesel electric power plants. Certainly, the fuel cell hybrid is a series hybrid architecture, but for now we focus

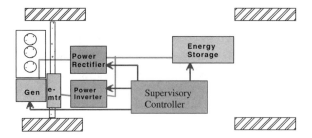

Figure 2.3 Series hybrid architecture.

on the more conventional vehicles. Series architecture means that power from the heat engine is delivered via an electric-only transmission path to the driven wheels. In the parallel architecture there are two paths of power flow to the driven wheels, one mechanical from the heat engine, and one electrical from the electrical energy storage system.

Figure 2.3 depicts the series hybrid architecture. In this configuration the ICE is used to spin a dedicated generator, or starter-alternator (SA), that generates electrical power in much the same fashion as a utility plant. If the transmission path is DC, as it most generally is when electrical energy storage is present, the generator output is rectified and fed to the traction inverter for synthesis into variable voltage and variable frequency (VVVF) power as demanded by the traction M/G.

The transmission path is not required to be DC. There have been and continue to be investigators looking into the prospects of series hybrid propulsion using an AC distribution link from generator to traction inverter [4]. In Reference 4 the authors describe a series hybrid architecture consisting of an engine-driven, doubly fed induction machine (DFIG) with an AC link to the induction machine traction inverter. With this architecture the DFIG functions as a variable voltage and variable frequency power source to directly supply the induction machine traction motor. With inverter control the VVVF source is capable of delivering the necessary V/f control to the load induction machine even with the ICE prime mover operating at constant speed.

Nowadays, the most common series hybrid architecture is the fuel cell vehicle. A fuel cell power plant provides direct conversion of fuel to electricity; hence the power path from engine to driven wheels is electric only. Figure 2.4 shows the essential components of a fuel cell hybrid.

In this architecture the fuel cell is buffered from the DC distribution power path by a DC/DC converter. This converter serves two primary roles: (1) to prevent the DC network from back feeding the fuel cell stack when the vehicle is regenerating into its energy

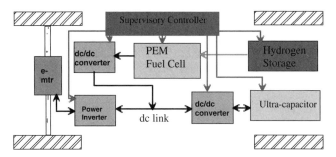

Figure 2.4 Fuel cell hybrid architecture.

storage system, and (2) as a means to regulate the DC bus as the fuel cell is being loaded. Fuel cell stacks are soft voltage sources, with an output of 1.2 V open circuit decreasing to approximately 0.7 V/cell when loaded. This cell potential behavior is very similar to that of nickel-metal hydride (NiMH) cells, only with a fuel cell the available current is a function of input reactant flow rates.

It is this dependency of fuel cell output on reactant supply and flow rate that limits the fuel cell's ability to respond quickly to load transients. Today's fuel cell stack has transient response of roughly 10 to 20 seconds. Near-term targets are to decrease this to 2 seconds under normal ambient conditions of temperature, pressure, and humidity and to further decrease the response time to 1 second eventually. In cold climates the fuel cell warm-up time may be 5 to 10 minutes, so that auxiliary electrical power must be supplied to sustain the fuel cell reactant flows and also to propel the vehicle. This latter point is the reason for the battery and ultracapacitor energy storage systems shown in Figure 2.4.

The high-voltage battery is necessary to not only operate the fuel cell ancillaries such as air compressor, humidifier, and fuel valves, but also to supply the traction motor during launch until the fuel cell stack can ramp up its output power. In the case of rapid unloading of the fuel cell, the excess electrical power is routed to the ultracapacitor transient energy storage system along with energy recuperated during regenerative braking. The combination of battery and ultracapacitor for such applications takes advantage of the strengths of each technology. The battery is necessary for its high energy density necessary to sustain prolonged discharge during fuel cell warm-up and vehicle operations. The ultracapacitor is needed for its extremely high power-to-energy ratio so that fast power pulses can be absorbed without incurring bus overvoltage transients. The DC/DC converter to the ultracapacitor is needed if maximum utilization is to be made of this energy storage medium [5]. In this work the authors note that with the proper sizing of battery and ultracapacitor banks, substantial weight and volume reductions were achieved (40% weight and 21% volume reductions).

The series hybrid propulsion system has yet to find acceptance in gasoline and diesel electric passenger sedans and light trucks. Part of the reason may be the fact that this architecture is heavier and more costly than its parallel and combination architecture alternatives. However, in large vehicles such as locomotives, where weight is less of an issue, the series architecture is used extensively. For the near term, the fuel cell hybrid will be the primary application for the series architecture.

2.3 COMBINATION ARCHITECTURES

The combination of parallel and series hybrid architectures into a single package is quickly becoming the *de jure* standard in passenger vehicle hybridization. As the name implies, the combination hybrid architecture is neither fully parallel nor fully series. Figure 2.5 captures the essentials of the combination architecture.

In Figure 2.5 a pair of M/Gs are integrated into the vehicle power train. M/G1 performs the function of ICE starter-alternator and M/G2 acts as the primary traction motor for vehicle propulsion. To understand the combination architecture it is necessary to envision the action of an epicyclic gear set (i.e., a planetary gear). In this gear arrangement a central sun gear has three or four pinion gears revolving about it attached to a carrier, hence the name carrier gears. Lastly, the pinions attached to the carrier are also designed to mesh with the outside gear, or ring gear. Figure 2.6 illustrates how the planetary gear set is applied in the combination architecture along with a schematic of its working principles.

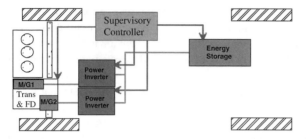

Figure 2.5 Combination, or power split, hybrid architecture.

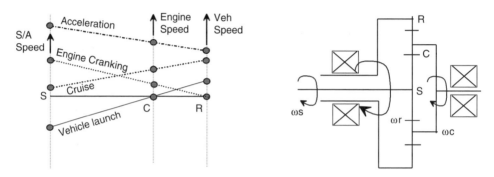

Figure 2.6 Power split propulsion essentials.

In Figure 2.6 on the right-hand side is shown the arrangement of the sun, carrier, and ring gears. Inputs and outputs between the sun and ring are always counter-revolving regardless of what state the carrier is in (i.e., grounded or with speed input). In the combination architecture the ICE crankshaft is typically connected to the carrier and an electric machine, M/G1, at its sun gear. The M/G1 takes on the function of ICE starting and generating; hence, it will be referred to as the starter-alternator, S/A. The ring gear will be tied to the vehicle's driven wheels via a chain drive to the final drive and from there via half-shafts to the wheels.

The Toyota Prius is a prime example of the combination architecture. In the Toyota Hybrid System used in the Prius a single planetary gear serves as the speed summing junction in a hybrid drive train that has no clutches. Because there are no clutches a means must be found to isolate the ICE from the torque-speed status of the driven wheels. Referring to the left-hand side of Figure 2.6, a "stick" diagram illustrates how the speeds of the sun, carrier, and ring gears interact for various driving conditions. These will now be reviewed by taking its operation apart one stick at a time and focusing on the ring gear speed (i.e., M/G2).

- Engine cranking. The driven wheels in contact with the road are stationary. Hence, the driveline is not revolving and the ring gear is at 0 speed. M/G2 cannot be used to crank the engine because it is not free to rotate. Therefore, M/G1, the S/A, is used to crank the engine by reacting against the stationary ring gear. With a power split architecture, the engine does not need to run to launch the vehicle as described next.

- Vehicle launch. Assume the engine is off, that is, the carrier gear is grounded (i.e., at 0 speed). In this mode the M/G2 can act alone to launch the vehicle on level terrain or on a grade. The S/A speed is revolving counter to the ring gear direction of speed. The engine can also be started at any point during launch, typically when the vehicle speed exceeds 18 kph.
- Cruise. During cruise the engine delivers mechanical power directly to the wheels in much the same manner as a parallel hybrid. Engine power is also delivered to the S/A, which now functions as a generator, delivering its electrical output to M/G1 for power summing.
- Acceleration. A design goal of power split is to minimize the operating speed range of the ICE to thereby confine it to a more fuel-efficient set of torque-speed points. Power split accomplishes this by having the S/A spin faster in generating mode so that its output, when delivered to M/G2 along with engine output, is sufficient to match the vehicle's road load.

An advantage of a combination hybrid architecture is the relative simplicity of its transmission; that is, there are no shift points, no clutches to engage and disengage, only a fixed gear ratio between the ICE and wheels. A second advantage is that the engine speed can be confined to a narrow operating range through control of the M/G1 attached to the planetary set sun gear. M/G1 must have low inertia in order to execute the wide speed range and direction shifts necessary to match road load to the engine. A third advantage is that the engine can be further downsized because there are now two electric M/Gs to assist in propulsion, although M/G2 performs the bulk of electric assist.

A disadvantage of a combination architecture is the relatively high complexity of having two AC drive systems integrated into one transmission, along with their dedicated inverters and controllers. This represents twice the power electronics content of a parallel hybrid, and nearly twice the electronics content of a series hybrid. However, the smooth and seamless operation of a power split transmission to deliver the performance and feel that appeal to customers makes this complexity worth it, but not necessarily affordable. Manufacturers of the power split hybrid architecture continue to search for ways to further minimize the electric machine content and the power electronics content. The electric machines need to have high power density (kW/kg) and wide constant power speed range (CPSR). The interior permanent magnet machine (IPM) and the variable reluctance machine (VRM) are two machine types that can satisfy high specific power and wide CPSR of > 5:1.

Downsizing the power electronics in a power split hybrid architecture continues to receive much attention. There is little spare package volume within a hybrid vehicle's engine compartment for any power electronics boxes, let alone a pair of boxes. Innovative packaging and thermal management schemes are sought that will minimize the volume taken by this pair of inverter modules.

In passing, it should be noted that the combination architecture, because of its fixed ratio transmission, has no provision for a reverse gear. Reverse in the power split driveline is accomplished completely by reverse rotation of M/G2 in motoring mode.

2.4 GRID CONNECTED HYBRIDS

The last category of hybrids to discuss is the class of vehicles referred to as grid connected hybrids. All the previous hybrid architectures could be grouped within a classification of charge sustaining. That is, the energy storage system in these vehicles is designed to remain within a fairly confined region of state of charge (SOC). The hybrid propulsion algorithm is designed so that on average, the SOC of the energy storage system will more or less return to its initial condition after a drive cycle. Grid connected hybrids, on the other hand, are designed as charge depleting. This in effect means that part of the "fuel" consumed during a drive is delivered by the utility. Fuel efficiency is then calculated based on actual fuel consumed by the ICE and its gasoline equivalent of the kWh of energy delivered by the utility during recharge.

The fact that grid connected hybrids must be recharged has some connotation of inconvenience, but surprisingly, many potential customers surveyed by the North American Electric Power Research Institute (EPRI) found that this would not be a deterrent, considering the fuel savings potential of such vehicles. Others have pointed out that a large population of such grid connected vehicles would have the potential to act as highly distributed co-generation, in effect, acting as peaking units on the utility during morning and evening peak demand conditions. The company AC Propulsion Inc. has coined the terminology V2G for vehicle-to-grid to describe the connected car. The concept is sound, particularly when the vehicle in question is a hydrogen-fueled hybrid. In this case the vehicle would consume hydrogen during peak utility loading times and deliver from 25 to 50 kW or more of electricity to the grid on demand. Investigators have also developed a novel hydrogen infrastructure so that such vehicles in a parking lot would be supplied with hydrogen as they were being tapped as co-generators. In this era of short supply, rotating brownouts, and potential blackouts, the concept of V2G seems appealing. Unfortunately, it would take tens of thousands of such vehicles delivering full output to match the output of a single utility generator of even moderate rating. But the concept is sound.

Figure 2.7 shows the overall structure of a fuel cell vehicle having provision to be connected to the utility. In this scenario, both a hydrogen supply to the vehicle's fuel stack (injection point downstream of the reformer shown) and electrical connection to the traction inverter output via an EMI common mode and transverse mode filter bank (could be part of the infrastructure) would be necessary.

Figure 2.7 illustrates the high power electronics content of such a vehicle. As already discussed, the fuel cell would be interfaced and buffered from the traction system by a dedicated converter, and the ultracapacitor bank has another dedicated converter to optimize the energy throughput. There is, of course, an inverter needed at the traction motor to synthesize VVVF for the drive motor (or to match the utility). In addition, there are other ancillary power electronics units necessary to run the adjustable speed air compressor for the fuel cell stack and a DC/DC converter rated 3 to 5 kW at high voltage (300 to 550 V) to down-convert to either 14 or 42 V regulated.

Figure 2.7 also highlights fuel cell supporting components for water management, air humidification, and stack thermal management.

Before closing this chapter on hybrids it is instructive to observe from Figure 2.7 that the concept of interfacing the vehicle power plant to the electric traction system via a DC/DC converter has considerable merit. Consider the Toyota Motor Corporation Hybrid Synergy Drive (HSD), first unveiled at the 2003 Detroit International Auto Show for initial release on its RX400H in 2005 as their second generation power split hybrid technology, the THS-II. This is pointed out because the Hybrid Synergy Drive retains the 274 V NiMH battery pack but has a traction inverter that operates off of a 500 V bus. An interface DC/DC buck/boost converter processes the full traction system power. The HSD vehicle is claimed to achieve 52 mpg on the U.S. metro-highway drive cycle, compared to 48 mpg for its predecessor. The 8% increase is due entirely to operating at higher voltage. A similar

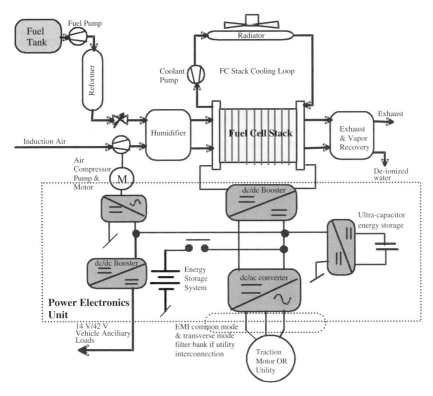

Figure 2.7 Grid connected hybrid architecture.

strategy could be applied to any of the other hybrid propulsion architectures. Operating at higher voltage is not new. Many investigators have proposed operating the traction system at upwards of 900 V to even 5 kV. The issue of exceeding 600 V lies in the manner in which cabling and switch gear must be treated, as well as personal safety in the presence of high voltage (not to mention cost).

Hybrid propulsion architectures that do not require the battery to be designed at voltages above 300 V have benefits in that reliability can be maintained at present levels or increased because additional terminations and cell balancing hardware along with monitoring hardware would not be increased. Instead, the battery would interface to the DC bus through a high-power DC/DC converter (i.e., DC transformer), thereby reducing the size and cost of the semiconductor switching devices and inverter package.

REFERENCES

[1] B.C. Smith. *Positioning the State of Michigan as a Leading Candidate for Fuel Cell and Alternative Powertrain Manufacturing.* Center for Automotive Research report to the Michigan Economic Development Corporation. August 2001.

[2] G. Schmidt. *Future Powertrain Technology for the North American Market: Diesel & Hydrogen.* Global Powertrain Conference keynote address. Ann Arbor, MI. September 2002.

[3] R. Natkin et al. *Ford Hydrogen Internal Combustion Engine Design and Vehicle Development Program.* Global Powertrain Conference and Exposition. Ann Arbor, MI. September 24–26, 2002.

[4] P. Caratozzolo, M. Serra, J. Riera. *Energy Management Strategies for Hybrid Electric Vehicles.* IEEE International Electric Machines and Drives Conference. Monona Terrace Conference Center. Madison, WI. June 1–4, 2003.

[5] R.M. Schupbach, J.C. Balda. *Design Methodology of a Combined Battery-Ultracapacitor Energy Storage Unit for Vehicle Power Management.* IEEE Power Electronics Specialists Conference. 2003.

3

Hybrid Drivetrains

M. Ehsani and Yimin Gao
Texas A&M University, College Station, Texas

Taking advantage of the high energy density of petroleum fuels, the conventional internal combustion engine (ICE) vehicles provide good performance and long operating range. However, they also have the disadvantages of poor fuel economy and environmental pollution. The main reasons for the poor fuel economy of conventional ICE vehicles are due to (1) mismatched engine fuel efficiency characteristics with the vehicle operation requirements, (2) significant amount of energy dissipation in braking, especially while operating in urban areas, and (3) low efficiency of hydraulic transmission in these automobiles in stop-go driving pattern. Battery-powered electric vehicles, on the other hand, possess some advantages over conventional ICE vehicles, like high energy efficiency and zero environmental pollution. But the performance, especially the operating range per battery charge, is far less than the ICE vehicles, due to the lower energy content of batteries vs. the energy content of gasoline. Hybrid electric vehicles (HEVs), which use two power sources, primary power source and secondary power source, have both of the advantages of ICE vehicles and electric vehicles and can overcome their disadvantages [1, 2]. This chapter discusses the basic concept and operation principles of the HEV powertrain

3.1 CONCEPT OF HYBRID VEHICLE DRIVETRAIN

Any vehicle powertrain is basically required to (1) develop sufficient power to meet the demands of vehicle performance, (2) carry sufficient energy on-board to support vehicle driving in the given range, (3) demonstrate high efficiency, and (4) emit less environmental pollutants. Broadly, a vehicle may have more than one energy source and energy converter (power source), such as gasoline (diesel)-IC engine system, hydrogen-fuel cell-electric motor system, and chemical battery-electric motor system. A hybrid vehicle drivetrain

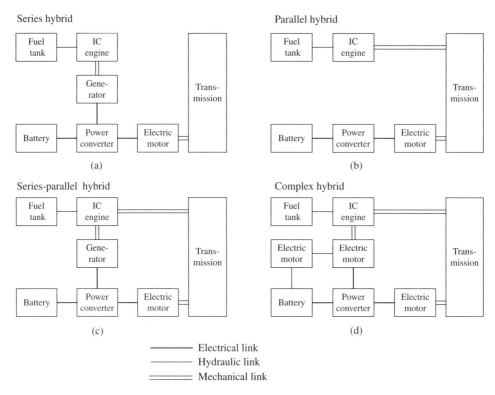

Figure 3.1 Classification of HEV.

consists of two power sources. One is the primary power source and the other is secondary power. For the purpose of recapturing part of the braking energy, hybrid drivetrain has at least one bidirectional energy source, typically chemical battery-electric motor system. At present, IC engines are the first selection for the primary power source, and perhaps fuel cells in the future.

Architecture of a hybrid vehicle is loosely defined as the connection between the components that define the energy flow routes and control ports. Traditionally, HEVs were classified into two basic types: series and parallel. It is interesting to note that, in 2000, some newly introduced HEVs could not be classified into these kinds [5]; hereby, HEVs are newly classified into four kinds: series hybrid, parallel hybrid, series-parallel hybrid, and complex hybrid, which are functionally shown in Figure 3.1 [4]. In Figure 3.1, a fuel tank-IC engine and a battery-electric motor are taken as the examples of the primary power source (steady power source) and secondary power source (dynamic power source), respectively. Of course, the IC engine can be replaced by other types of power sources, such as fuel cells. Similarly, the batteries can be replaced by ultracapacitors, flywheels, or their combinations.

3.2 SERIES HYBRID DRIVETRAIN

A series hybrid drivetrain is a drivetrain in which two power sources feed a single powerplant (electric motor) that propels the vehicle, as shown in Figure 3.2. The typical

Hybrid Drivetrains

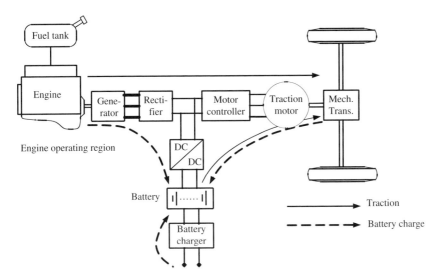

Figure 3.2 Series hybrid electric drivetrain.

primary power source is IC engine coupled to an electric generator. The output of the electric generator is connected to an electric power bus through an electronic converter (rectifier). The secondary power source is a battery pack, which is connected to the bus by means of a power electronics converter (DC/DC converter). The electric power bus is connected to the controller of the electric traction motor. The traction motor can be controlled either as a motor or a generator, and either in forward or reverse motion. This drivetrain may need a battery charger to charge the batteries by wall plug-in from the power network.

Series hybrid electric drivetrain potentially has the following operation modes:

1. Pure electric mode: The engine is turned off and the vehicle is propelled only from the batteries.
2. Pure engine mode: The vehicle traction power only comes from the engine-generator, while the batteries neither supply nor draw any power from the drivetrain. The electric machines serve as an electric transmission from the engine to the driven wheels.
3. Hybrid mode: The traction power is drawn from both the engine-generator and the batteries.
4. Engine traction and battery charging mode: The engine-generator supplies power to charge the batteries and to propel the vehicle.
5. Regenerative braking mode: The engine-generator is turned off and the traction motor is operated as a generator. The power generated is used to charge the batteries.
6. Battery charging mode: The traction motor receives no power and the engine-generator charges the batteries.
7. Hybrid battery charging mode: Both the engine-generator and the traction motor operate as generators to charge the batteries.

Series hybrid drivetrains offer several advantages:

1. The engine is fully mechanical decoupled from the driven wheels. Thus, it can be operated at any point on its speed-torque characteristic map. Therefore, it can be potentially operated solely within its maximum efficiency region as shown in Figure 3.4. The efficiency and emissions of the engine can be further improved by optimal design and control in this narrow region, which is much easier and allows greater improvements than an optimization across the entire range. Furthermore, the mechanical decoupling of the engine from the driven wheels allows the use of a high-speed engine. This makes it difficult to directly power the wheels through a mechanical link, such as gas turbines or powerplants with slow dynamics like the Stirling engine.
2. Because electric motors have near-ideal torque-speed characteristics, they do not need multigear transmissions [7]. Therefore, the construction is greatly simplified and the cost is reduced. Furthermore, instead of using one motor and a differential gear, two motors may be used, each powering a single wheel. This provides the speed decoupling between the two wheels like a differential but also acts as a limited slip differential for traction control purposes. The ultimate refinement would use four motors, thus making the vehicle an all-wheel-drive without the expense and complexity of differentials and drive shafts running through the frame.
3. Simple control strategies may be used as a result from the mechanical decoupling provided by the electrical transmission.

However, series hybrid electric drivetrains suffer some disadvantages:

1. The energy from the engine is converted twice (mechanical to electrical in the generator and electrical to mechanical in the traction motor). The inefficiencies of the generator and traction motor add up and the losses may be significant.
2. The generator adds additional weight and cost.
3. The traction motor must be sized for maximum requirement since it is the only powerplant propelling the vehicle.

3.3 PARALLEL HYBRID DRIVETRAINS

Series hybrid drivetrain couples primary and secondary power sources together electrically. However, a parallel hybrid drivetrain couples them together mechanically, in which the engine supplies its power mechanically to the wheels like in a conventional IC engine-powered vehicle. It is assisted by an electric motor that is mechanically coupled to the transmission. The powers of the engine and electric motor are coupled together by mechanical coupling, as shown in Figure 3.3. The mechanical coupling of the engine and electric motor power leaves room for several different configurations.

3.3.1 Parallel Hybrid Drivetrains with Torque Coupling

3.3.1.1 Torque Coupler

The mechanical coupling in Figure 3.3 may be torque or speed coupling. In the torque-coupling configuration, the torques of the engine and electric motor are added together in a mechanical torque coupler. Figure 3.4 conceptually shows a mechanical torque coupler,

Hybrid Drivetrains

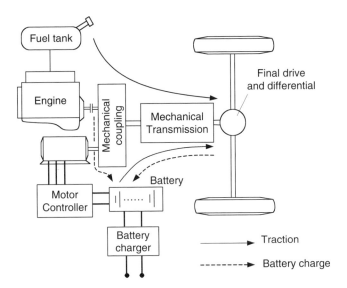

Figure 3.3 Configuration of a parallel hybrid drivetrain.

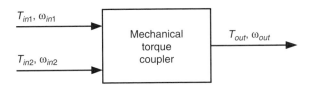

Figure 3.4 Torque-coupling device.

which has two inputs (one is from the engine, one is from the electric motor) and one output to a mechanical transmission.

If loss is ignored, the output torque and speed can be described by:

$$T_{out} = k_1 T_{in1} + k_2 T_{in2}, \tag{3.1}$$

and

$$\omega_{out} = \frac{\omega_{in1}}{k_1} = \frac{\omega_{in2}}{k_2}, \tag{3.2}$$

where k_1 and k_2 are the constants determined by the parameters of torque-coupling device. Figure 3.5 lists some typically used mechanical torque-coupling devices.

3.3.1.2 Drivetrain Configuration and Operating Characteristics

There are several configurations in torque-coupling hybrid drivetrains. These are mainly classified into two-shaft and one-shaft designs. In each category, the transmission can be placed in different positions and designed with different gears, resulting in different tractive characteristics. An optimum design will depend mostly on the tractive requirements, engine size and engine characteristics, motor size and motor characteristics, and so on.

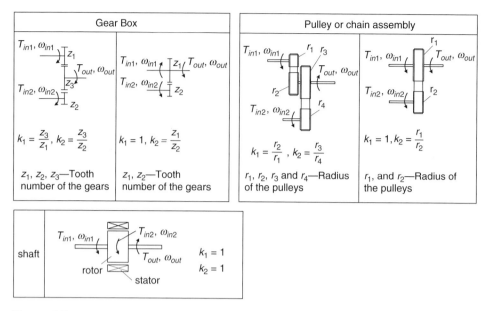

Figure 3.5 Commonly used mechanical torque-coupling devices.

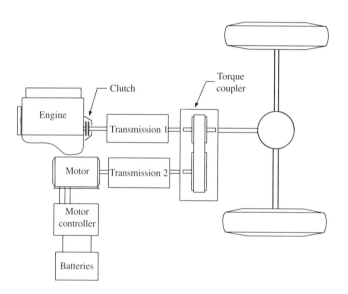

Figure 3.6 Two-axle configuration.

Figure 3.6 shows a two-shaft configuration design, in which two transmissions are used; one is placed between the engine and torque coupling and other between the motor and torque coupling. Both transmissions may be single or multigears. Figure 3.7 shows the speed-tractive effort profiles with different transmission parameters. It is clear that the design with two multigear transmissions results in rich speed-tractive effort profiles. The performance and overall efficiency of the drivetrain is certainly superior to other designs, because two multigear transmissions provide more opportunities for both the engine and electric traction system (electric machine and batteries) to operate in their optimum region. This design also provides great flexibility in the design the of engine and electric motor

Hybrid Drivetrains

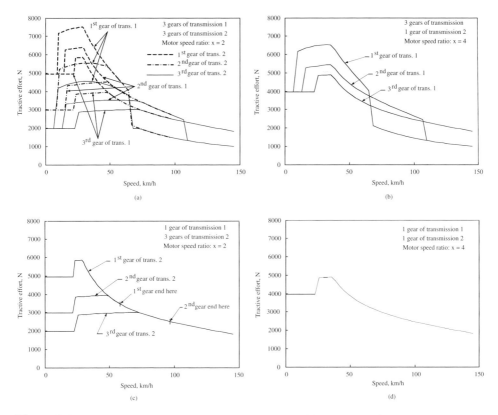

Figure 3.7 Tractive effort along with vehicle speed with different transmission schemes.

characteristics. However, two multigear transmissions will significantly complicate the drivetrain.

In Figure 3.6, a multigear transmission 1 and a single-gear transmission 2 may be employed. The speed-tractive effort profiles are shown in Figure 3.7(b). Employing a single-gear transmission inherently takes the advantage of the high torque characteristic of electric machine at low speeds. The multigear transmission 2 is used to overcome the disadvantages of the IC engine speed-torque characteristics (flat torque output along speed). The multispeed transmission 1 also tends to improve the efficiency of the engine and reduces the speed range of the vehicle, in which an electric machine must alone propel the vehicle, consequently reducing the battery discharging energy.

In contrast to the above design, Figure 3.7(c) shows the speed-tractive effort profiles of the drivetrain, which has a single transmission 1 for the engine and a multispeed transmission 2 for the electric motor. This configuration is considered to be an unfavorable design, because it does not use the advantages of both powerplants.

Figure 3.7(d) shows the speed-tractive effort profile of the drivetrain, which has two single-gear transmissions. This arrangement results in a simple configuration and control. The limitation to the application of this drivetrain is the requirement of the maximum tractive effort. When powers of the engine, electric motor, batteries, and transmission parameters are properly designed, this drivetrain would serve the vehicle with satisfactory performance and efficiency.

Another configuration of the two-shaft parallel hybrid drivetrain is shown in Figure 3.8, in which the transmission is located between the torque coupler and drive shaft. The

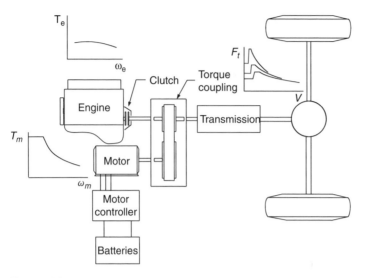

Figure 3.8 Two-shaft configuration.

transmission functions to enhance the torques of both engine and electric motor with the same scale. Design of constant k_1 and k_2 in the torque coupling allows the electric motor to have a different speed range than the engine; therefore, a high speed motor can be used. This configuration would be suitable in the case when a relative small engine and electric motor are used, where a multigear transmission is needed to enhance the tractive effort at low speeds.

The simple and compact architecture of the torque-coupling parallel hybrid drivetrain may be the single-shaft configuration, where the rotor of the electric motor functions as the torque coupler ($k_1 = 1$ and $k_2 = 1$ in Equation (3.1) and Equation (3.2)), as shown in Figures 3.9 and 3.10. A transmission may be placed between either the electric motor and drive shaft, or the engine and the electric motor. The former configuration is referred to as a "pre-transmission" (the motor is in ahead of the transmission, Figure 3.9), and the latter is referred to as "post-transmission" (the motor is in behind the transmission, Figure 3.10).

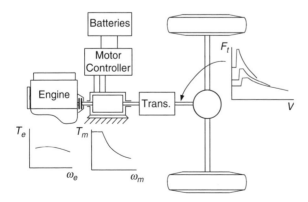

Figure 3.9 Pretransmission single-shaft torque combination parallel hybrid drivetrain.

Hybrid Drivetrains

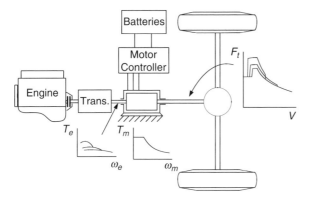

Figure 3.10 Post-transmission single-shaft torque combination parallel hybrid drivetrain.

In the pre-transmission configuration, both the engine torque and motor torque are modified by the transmission. The engine and motor must have the same speed range. This configuration is usually used in the case with a small motor, referred to as a mild hybrid drivetrain, in which the electric motor functions as an engine starter, electrical generator, engine power assistant, and regenerative braking.

However, in the post-transmission configuration as shown Figure 3.10, the transmission only modifies the engine torque while the motor torque is directly delivered to the driven wheels. This configuration may be used in the drivetrain where a large electric motor with a long constant power region is employed. The transmission is only used to change the engine operating points to improve the vehicle performance and engine operating efficiency. It should be noted that the batteries cannot be charged from the engine by running the electric motor as a generator when the vehicle is on standstill because the motor is rigidly connected to the driven wheels.

Another torque-coupling parallel hybrid drivetrain is the separated axle architecture, in which one axle is powered by the engine and another powered by the electric motor (Figure 3.11). The tractive efforts from the two powertrains are added through the vehicle chassis and the road. The operating principle is similar to the two-shaft configuration shown in Figure 3.6. Both transmissions for the engine and electric motor may be single-gear or multigear. This configuration has similar tractive effort characteristics as shown in Figure 3.7.

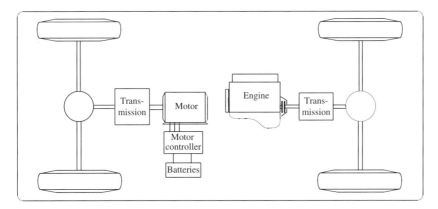

Figure 3.11 Separated axle torque combination parallel hybrid drivetrain.

The separated axle architecture offers some of the advantages of a conventional vehicle. It keeps the original engine and transmission unaltered and adds an electrical traction system on the other axle. It also is four-wheel-drive, which optimizes traction on slippery roads and reduces the tractive effort on a single tire.

However, the electric machines and the eventual differential gear system occupy a lot of space and may reduce the available passenger and luggage space. This problem may be solved if the motor transmission is single-gear and the electric motor is replaced by two small-size electric motors that can be placed within two driven wheels. It should be noted that the batteries cannot be charged from the engine when the vehicle is at standstill.

3.3.2 Parallel Hybrid Drivetrain with Speed Coupling

3.3.2.1 Speed Coupler

The power from two powerplants may be coupled together by coupling their speeds, as shown in Figure 3.12. The characteristics of a speed coupling can be described by:

$$\omega_{out} = k_1 \omega_{in1} + k_2 \omega_{in2}, \tag{3.3}$$

and

$$T_{out} = \frac{T_{in1}}{k_1} = \frac{T_{in2}}{k_2}, \tag{3.4}$$

where k_1 and k_2 are constants associated with the actual design.

Figure 3.13 shows two typical speed couplers: One is a planetary gear unit and the other is an electric motor with a floating stator (called transmotor in this book). A planetary

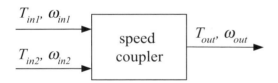

Figure 3.12 Speed coupling.

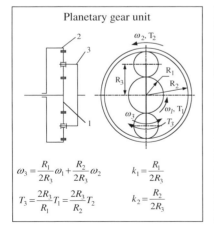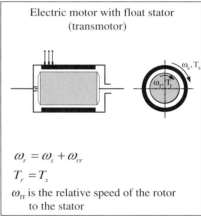

Figure 3.13 Typical speed coupling devices.

Hybrid Drivetrains

gear unit is a three-port unit consisting of the sun gear, the ring gear, and the carrier, labeled 1, 2, and 3, respectively. The speed and torque relationship between the three ports indicates that this unit is a speed-coupling device, in which the speeds of the sun gear and of the ring gear are added together and output through the carrier. The constants k_1 and k_2 depend only on the radius of each gear or the number of teeth of each gear.

Another interesting device in speed coupling is an electric motor (called transmotor in this book), in which the stator, generally fixed to a stationary frame, is released as a power-input port. The other two ports are the rotor and the airgap, through which the electric energy is converted into mechanical energy. The motor speed, in common terms, is the relative speed of the rotor to the stator. Because of action and reaction effects, the torque action on the stator and rotor is always the same and results in the constants $k_1 = 1$ and $k_2 = 1$.

3.3.2.2 Drivetrain Configurations and Operating Characteristics

Similar to t, the speed-coupling units can be used to constitute various hybrid drivetrains. Figure 3.14 and Figure 3.15 show two examples of hybrid drivetrains with speed coupling of the planetary gear unit and an electric transmotor. In Figure 3.14, the engine supplies its power to the sun gear through a clutch and transmission. The transmission is used to modify the speed-torque characteristics of the engine so as to match the traction requirements. The electric motor supplies its power to the ring gear through a pair of gears. Lock 1 and lock 2 are used to lock the sun gear and ring gear to the standstill frame of the vehicle in order to satisfy the different operation mode requirements. The following operation modes can be satisfied:

1. Hybrid traction: When lock 1 and lock 2 are released (the sun gear and ring gear can rotate) and both the engine and electric machine supply positive speed and torque (positive power) to the driven wheels.
2. Engine alone traction: When lock 2 locks the ring gear to the vehicle frame and lock 1 is released. Only the engine supplies power the driven wheels.
3. Motor alone traction: When lock 1 locks the sun gear to the vehicle frame (engine is shut off or clutch is disengaged) and lock 2 is released. Only the electric motor supplies its power to the driven wheels.

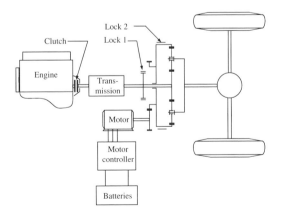

Figure 3.14 Hybrid drivetrain with speed coupling of planetary gear unit.

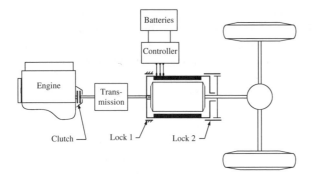

Figure 3.15 Hybrid drivetrain with speed coupling of electric transmotor.

4. Regenerative braking: Lock 1 is set in locking state, the engine is shut off or clutch is disengaged, and the electric machine is controlled in regenerating operation (negative torque). The kinetic or potential energy of the vehicle can be absorbed by the electric system.
5. Battery charging from the engine: When the controller sets a negative speed for the electric machine, the electric machine absorbs energy from the engine.

The drivetrain, consisting of the transmotor as shown in Figure 3.15, has a similar structure to that in Figure 3.14. Lock 1 and lock 2 are used to lock the stator to the vehicle frame and the stator to the rotor, respectively. This drivetrain can fulfill all the operation modes mentioned above. The operating modes analysis is left to the readers.

The main advantage of the hybrid drivetrain with speed coupling is that the speeds of the two powerplants are decoupled; therefore, the speed of both the powerplants can be chosen freely. This advantage is important to the powerplants such as the Stirling engine and the gas turbine engine, in which their efficiencies are sensitive to speed and less sensitive to torque.

3.4 DRIVETRAINS WITH SELECTABLE TORQUE COUPLING AND SPEED COUPLING

By combining torque and speed coupling together, one may constitute a hybrid drivetrain in which torque- and speed-coupling states can be alternatively chosen. Figure 3.16 [9] shows such an example. When the torque-coupling operation mode is chosen as the current mode, lock 2 locks the ring gear of the planetary unit to the vehicle frame while clutch 1 and 3 are engaged and clutch 2 is disengaged. The powers of the engine and the electric motor are added together by adding their torques together [refer to Equation (3.1)], and then delivered to the driven wheels. In this case, the engine torque and the electric motor are decoupled, but their speeds have a fixed relationship as described by Equation (3.2). When the speed-coupling mode is chosen as the current operating mode, clutch 1 is set engaged, whereas clutch 2 and 3 are disengaged, and lock 1 and 2 release the sun gear and the ring gear. The speed of the yoke, connected to the drive wheels, is the combination of engine speed and motor speed [refer to Equation (3.3)]. But the engine torque, the electric motor torque, and the torque on the driven wheels are kept in a fixed relationship as described by Equation (3.4).

Hybrid Drivetrains

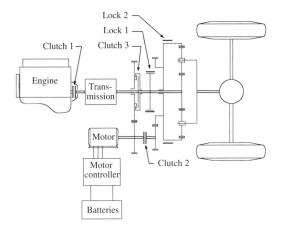

Figure 3.16 Alternative torque and speed hybrid drivetrain with a planetary gear unit.

With the option to choose between the torque coupling and speed coupling, the powerplants have more opportunities to determine their operation manner and operation region so as to optimize their performance. For instance, when vehicle speed is lower than a certain speed, V_b, below which the engine cannot operate steady with torque coupling, speed-coupling mode may be used. In this case, the electric motor operates as a generator with a negative speed and converts part of the engine power into electric power and stores it into batteries. When the vehicle speed is higher than this speed, V_b, and lower than a speed, V_u, torque-coupling operation mode would be suitable for high acceleration or hill climbing. However, when the vehicle speed is higher than the speed V_u, the speed-coupling mode would be used to prevent high engine speeds which cause high fuel consumption. In this case, the electric motor operates with a positive speed and delivers its power to the drivetrain.

The planetary gear unit traction motor in Figure 3.16 can be replaced by a transmotor to constitute a similar drivetrain as shown in Figure 3.17. When clutch 1 is engaged to couple the engine shaft to the rotor shaft of the transmotor, clutch 2 is disengaged to release the engine shaft from the rotor of the transmotor and the lock is activated to set the stator of the transmotor to the vehicle frame. The drivetrain then works in the torque-coupling mode. On the other hand, when clutch 1 is disengaged and clutch 2 is engaged and the lock is released, the drivetrain works in the speed-coupling mode.

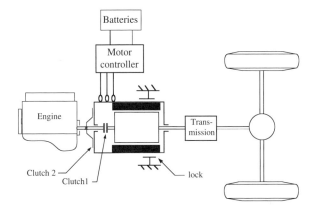

Figure 3.17 Alternative torque and speed coupling hybrid drivetrain with transmotor.

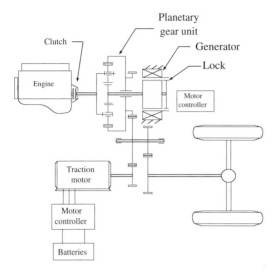

Figure 3.18 Integrated speed and torque coupling hybrid drivetrain (Toyota Prius).

3.5 PARALLEL-SERIES HYBRID DRIVETRAIN WITH TORQUE COUPLING AND SPEED COUPLING

Another good example that uses both torque coupling and speed coupling on a drivetrain is the one developed and implemented in Toyota Prius by Toyota Motor Company [9]. This drivetrain is schematically illustrated in Figure 3.18. A small motor or generator (few kilowatts) is connected through a planetary gear unit (speed coupling). The planetary gear unit splits the engine speed into two speeds [refer to Equation (3.3)]; one is output to the small motor through its sun gear and the other to the driven wheels through its ring gear and an axle-fixed gear unit (torque coupling). A big traction motor (few tens of kilowatts) is also connected to this gear unit to constitute a torque-coupling parallel driveline. At low vehicle speeds, the small motor runs with a positive speed and absorbs part of the engine power. As the vehicle speed increases and the engine speed is fixed to a given value, the motor speed decreases to 0. This is called synchronous speed. At this speed, the lock will be activated to lock the rotor and stator together. Then, the drivetrain is a parallel drivetrain. When the vehicle has a high speed, in order to avoid too high of an engine speed, which usually leads to high fuel consumption, the small motor can be operated with a negative speed and delivers power to the drivetrain. High fuel economy can be achieved when the planetary gear and the small motor are used to adjust the engine speed in order to operate at the optimum speed range.

The small motor and the planetary gear unit in Figure 3.18 can be replaced by an individual transmotor, as shown in Figure 3.19 [11]. This drivetrain has similar characteristics as the drivetrain in Figure 3.18. It will do well for the reader to analyze its operating mode.

3.6 FUEL CELL-POWERED HYBRID DRIVETRAIN

Fuel cells are considered to be the most promising power source for transportation application. Compared with internal combustion engine vehicle, fuel cells have the advantages of high energy efficiency and much lower emissions, due to the direct conversion from free energy in the fuel into electrical energy without undergoing combustions. However,

Hybrid Drivetrains

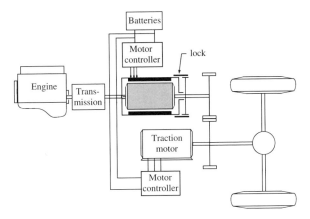

Figure 3.19 Integrated speed and torque coupling hybrid drivetrain with a transmotor.

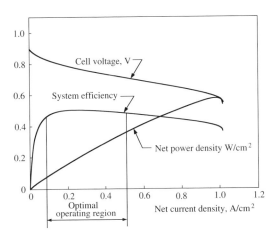

Figure 3.20 Typical operating characteristics of a fuel cell system.

vehicles powered by fuel cells alone bear some disadvantages, such as heavy and bulky power unit caused by the low power density of the fuel cell system, long start-up time, and slow power response. Furthermore, both extremely large power output of the fuel cell system in sharp acceleration and extremely low power output in low-speed driving will lead to a low efficiency, as shown in Figure 3.20.

Hybridization of fuel cell system with a peaking power source (PPS) (battery, ultracapacitor, flywheel, and so on) is the effective technology to overcome the disadvantages of vehicles powered by fuel cells alone. The fuel cell-powered hybrid drivetrain is more like a series hybrid drivetrain by replacing the engine/generator system by a fuel cell system, as shown in Figure 3.21. The battery pack in Figure 3.21 may be replaced by any other type of peaking power source, such as ultracapacitor and flywheel system.

The drivetrain shown in Figure 3.21 can operate with the modes for traction as:

- Fuel cell alone traction: If the load power is less than the minimum power of the optimal operating power range of the fuel cell system as shown in Figure 3.20 and the peaking power source and the peaking power source (battery) is already fully charged, the fuel cell system alone powers the vehicle.
- Hybrid traction: When the load power is larger than the maximum power of the optimal operating power range of the fuel cell system as shown in Figure 3.20,

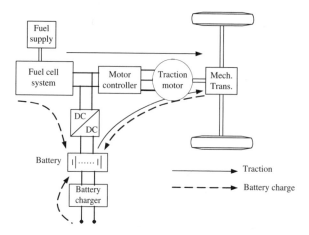

Figure 3.21 Configuration of the fuel cell power hybrid drivetrain.

the fuel cell system is set to operate at this maximum power and peaking power source supply an extra power to meet the load power demand.
- Charging of the peaking power source: When the peaking power source is not fully charged and the load power is smaller than the maximum power of the optimal operating power range, the fuel cell system is set to operate at this maximum power. One part of the fuel cell power propels the vehicle and another part charges the peaking power source.

When the vehicle is experiencing braking, the fuel cell system is set at idle, and the traction motor is operated as a generator to fulfill regenerative braking.

Due to the presence of peaking power source, the fuel cell system is not required to deliver large power to meet the peak load power, thus the power rating of fuel cell system can be reduced. Furthermore, the peaking power source will prevent the fuel cell system from operating with both very high and very low power output. Thus the operating efficiency of fuel cell system can be improved.

REFERENCES

[1] M. Ehsani, K.L. Butler, Y. Gao, K.M. Rahman, and D. Burke, "Toward a Sustainable Transportation without Sacrifice of Range, Performance, or Air Quality: The ELPH Car Concept," FISITA, International Federation of Automotive Engineering Society, Automotive Congress, September 27–October 1, 1998, Paris, France.

[2] M. Ehsani, K.L. Butler, Y. Gao, and K.M. Rahman, "Next Generation Passenger Cars with Better Range, Performance, and Emissions: The ELPH Car Concept," Horizon in Engineering Symposium, Texas A&M University–Engineering Program Office, College Station, TX, September 23–25, 1998.

[3] The Electrically Peaking Hybrid System and Method, M. Ehsani, U.S. Patent No. 5,586,613, December 24, 1996.

[4] C.C. Chan and K.T. Chau, *Modern Electric Vehicle Technology*, Oxford University Press, Inc., New York, 2001.

[5] K. Yamaguchi, S. Moroto, K. Kobayashi, M. Kawamoto, and Y. Miyaishi, "Development of a New Hybrid System-Duel System," SAE paper No. 960231, Society of Automotive Engineers, Warrendale, PA, 1997.

[6] Y. Gao, K.M. Rahman, and M. Ehsani, "The Energy Flow Management and Battery Energy Capacity Determination for the Drive Train of Electrically Peaking Hybrid," SAE paper No. 972647, Society of Automotive Engineers, Warrendale, PA, 1997.

[7] Y. Gao, K.M. Rahman, and M. Ehsani, "Parametric Design of the Drive Train of an Electrically Peaking Hybrid (ELPH) Vehicle," SAE paper No. 970294, Society of Automotive Engineers, Warrendale, PA, 1997.

[8] Y. Gao, L. Chen, and M. Ehsani, "Investigation of the Effectiveness of Regenerative Braking for EV and HEV," SAE paper No. 1999-01-2901, Society of Automotive Engineers, Warrendale, PA, 1999.

[9] Y. Gao and M. Ehsani, "New Type of Transmission for Hybrid Vehicle with Speed and Torque Summation," U.S. patent (pending).

[10] http://www.toyota.com, Toyota Motor Company, September 2003.

[11] Y. Gao and M. Ehsani, "Series–Parallel Hybrid Drive Train with an Electric Motor of Floating Stator and Rotor," U.S. patent (pending).

[12] S. Moore and M. Ehsani, "A Charge-Sustaining Parallel HEV Application of the Transmotor," SAE paper, No. 1999-01-0919, Society of Automotive Engineers, Warrendale, PA, 1997.

[13] Y. Gao and M. Ehsani, "Systematic Design of Fuel Cell Powered Hybrid Vehicle Drive Trains," SAE paper No. 2001-01-2532, Society of Automotive Engineers, Warrendale, PA, 2001.

4

Electric Vehicles

Ramesh C. Bansal
Birla Institute of Technology and Science, Pilani, India

4.1 INTRODUCTION

An electric vehicle (EV) is a vehicle powered by an electric motor, instead of an internal combustion engine (ICE), and the motor is run using the power stored in the batteries. The batteries have to be charged frequently by plugging into any main (120 V or 240 V) supply. EV has a much longer history than most people realize. EVs were seen soon after Joseph Henry introduced the first DC-powered motor in 1830. The first known electric car was a small model built by Professor Stratingh in the Dutch town of Groningen in 1835. The first EV was built by in 1834 by Thomas Davenport in the U.S., followed by Moses Farmer, who built the first two-passenger EV in 1847. There were no rechargeable electric cells (batteries) at that time. An EV did not become a viable option until the Frenchmen Gaston Plante and Camille Faure respectively invented (1865) and improved (1881) the storage battery.

EVs are known as zero emissions vehicles (ZEVs) and are much environment friendly than gasoline- or LPG-powered vehicles. As EVs have fewer moving parts, maintenance is also minimal. With no engine there are no oil changes, tune-ups, or timing and there is no exhaust. EVs are also far more energy efficient than gasoline engines and they are very quiet in operation.

EVs have been in continual use since the 1900s in various applications. Today these quiet vehicles with no tailpipe emissions are no longer limited to golf carts. New advances in battery technology, system integration, aerodynamics, and research and development by major vehicle manufacturers has led to the producing of electric vehicles that will play a practical role on city streets. EVs don't have ICEs in them. Instead, electrical energy is stored in a storage battery or ultracapacitor, converted from chemical energy in a fuel cell, or converted from mechanical energy in a flywheel. This electrical energy is used to power an electric motor, which then turns the wheels and provides propulsion. Since no fuel is burned in an EV, they don't produce the pollution that ICE vehicles do. EVs are not new: The first automobiles to be built were all EVs, using energy stored in rudimentary lead-acid batteries to drive DC electric motors. Only in the 1910s did gasoline-powered vehicles begin to make serious inroads in the automobile market.

Unfortunately, EVs have a serious disadvantage that played a large role in the takeover by ICE vehicles: limited range. By limited range we mean that EVs could only travel on the order of 50 miles or so on a single charge, and that only under good conditions (lead-acid batteries lose energy capacity when they become cold, so in cold weather the range of a vehicle could be reduced by as much as 50% or even more). Also, recharging takes hours. ICE vehicles can go much farther on a tankful of gas and could be quickly refueled.

Now, urban air quality issues, coupled with a rising awareness of the problems associated with the world's (and particularly America's) appetite for oil, have created interest in EVs. Power electronics have revolutionized motor drives, bringing within the realm of possibility electric drivetrains with extremely high performance, and the motors themselves have been improved, offering higher reliability and better performance with reduced cost. Unfortunately, the weak link in the chain remains — electrochemical batteries. There are some new battery chemistries floating around that offer good promise in terms of range, but none has the desired combination of three features: fast charging/discharging (high power density), large storage capacity (high energy density), and low cost. Traditional lead-acid batteries have been improved some over the years, but their energy and power densities remain disappointingly low, particularly when compared with gasoline.

This chapter contains two sections. The first section briefly discusses EVs, solar cars, and fuel cell cars. The chapter briefly outlines hybrid vehicles and presents main components, instrumentation, and main auxiliaries of EV. Various types of power storage used in EVs (e.g., batteries, flywheels, and ultracapacitors) are briefly discussed in the first section. The second section presents a bibliographical survey on electric vehicles.

4.2 HYBRID ELECTRIC VEHICLES

There is another alternative to EVs: hybrid electric vehicles (HEVs). There are many different potential HEV configurations, but in general, an HEV has an electric drivetrain like an EV, plus a fuel-burning engine of some type that can recharge the batteries periodically. The advantage of an HEV is that the fuel-burning engine, in general, is most efficient in only a small range of operating conditions (speed and load). Also, at this most efficient operating point, the fuel-burning engine usually produces its lowest levels of emissions. Unfortunately, while driving, the engine in the car has to run under a wide range of speeds and loads, and thus it is far less efficient and produces much greater emissions than it would if it could run at its most efficient point all the time. Electric drivetrains are also most efficient at only one point, but the reduction in efficiency for other speeds and loads is far less. Therefore, an HEV can run the fuel-burning engine at its most efficient point for battery charging and can use the electric drivetrain to take up all the slack under other conditions. In this way, emissions are much less than for the fuel-burning engine driving the car by itself, and fuel economy can be significantly improved. Hybrid technologies extend the usable range of EVs beyond what an all-electric vehicle can achieve with batteries only. Being a hybrid would allow the vehicle to operate on only batteries within an urban/polluted area, and then switch to its engine outside the urban area. There are two types of hybrids: parallel hybrids and series hybrids.

Electric Vehicles

PARALLEL HYBRIDS

Parallel hybrids, also known as power assist, have a direct mechanical connection between the hybrid power unit (HPU) and the wheels as in a conventional vehicle, but there is an electric motor driving the wheels as well. For example, a parallel vehicle could use the power created from an ICE for highway driving and the power from the electric motor for accelerating. Some benefits of a parallel configuration vs. a series configuration are the vehicle has more power (because both the engine and the motor supply power simultaneously), most parallel vehicles do not need a generator, and the power is directly coupled to the road, thus, it can be more efficient.

SERIES HYBRIDS

Series hybrids, also known as range extenders, use the heat engine with a generator to supply electricity for the battery pack and electric motor. Series HEVs have no mechanical connection between the HPU and the wheels; therefore, all motive power is transferred electrically to an electric motor that drives the wheels. One of the benefits of a series configuration over a parallel configuration is the engine never runs idle, which reduces vehicle emissions. The engine drives a generator to run at optimum performance, which allows a variety of options when mounting the engine and vehicle components. Some series hybrids do not even need a transmission.

4.3 MAIN COMPONENTS OF AN EV

4.3.1 MOTORS

The motor is the main component of an EV. It is very important to select proper type of motor with suitable rating. For example, it is not accurate to simply refer to a 10 hp motor or a 15 hp motor, because horsepower varies with volts and amps, and peak horsepower is much higher than the continuous rating. It is also confusing to compare electric motors to gas engines, since electric motors are given a continuous rating under load, and gas engines are rated at their peak horsepower under unloaded condition. For accurate identification, a motor should be identified by name or model number. Following are the commonly used motors in EVs:

Series Wound Brushed DC Motors

Series wound brushed DC motors (the field winding and armature are connected in series) are the best for the road-going EVs today, as they have a high torque, are cheap compared to other types, have wide availability, and require simple controllers as compared to other types.

AC Motors

AC motors operate at high rpm that have to be stepped down, are expensive, and require a complex speed control mechanism.

Brushless DC Motors

Brushless DC motors require expensive controllers, but need very little maintenance.

Permanent Magnet Motors

Permanent magnet motors are very efficient, but only in a very narrow rpm band, and quickly lose their efficiency in the varying speeds of normal driving.

Shunt Wound DC Motors

Shunt and compound motors are more expensive to build and have poorer acceleration than series motors.

SPEED CONTROLLER

The speed controller is crucial to the efficiency and smooth operation of the electric car. Speed controllers are rated according to the voltage and amperage ranges. Pulse width modulation (PWM) DC motor controllers work by "pulsing" the current delivered to the motor. Just like a piston water pump, the individual pulses are smoothed to produce a continuous flow. They are usually air-cooled or water-cooled.

DC/DC CONVERTER

An electric car normally uses a 12 V auxiliary battery to power all of the original 12 V accessories: lights, horn, and so on. However, unlike a gas car, there is no alternator to keep this battery charged. One option is to use deep-cycle 12 V batteries, as heavy duty as possible. This is not adequate if night driving is intended. As the battery drains in use, the headlights will grow dimmer and the turn signals flash more slowly. It can also affect the running of the car if some of the drive system components do not get the signal they require from the auxiliary battery. Another option is to tap 12 V from two of the main pack batteries, but it causes the pack to discharge unevenly, affecting performance and battery life. The better option may be a DC/DC converter. This taps the full battery pack voltage and cuts it down to a regulated 13.5 V output, similar to an alternator. By tapping the full pack, there is no uneven discharge. Current requirement is so low that there is little effect on range. This also eliminates the need for a separate 12 V charging circuit for the auxiliary battery.

It is not recommended to eliminate the auxiliary battery entirely, for safety reasons. If the DC/DC converter fails at night, or the battery pack falls below the low voltage shut-off of the converter, the auxiliary battery will have enough charge to bring the car home. Since the battery is not being asked to start a gas engine, however, a much smaller starting-type battery than the original can be substituted.

4.4 MAIN SAFETY COMPONENTS IN AN EV

CIRCUIT BREAKERS

A circuit breaker provides a fail-safe manual interruption of the battery power in event of a drive system malfunction. It also provides a convenient way to shut off battery power during routine servicing of the system. It must be installed in a location where it can be operated by the driver. It allows the driver to manually isolate the power train components from the battery pack in an emergency or while working on the car, even if the ignition switch fails. In addition, the circuit breaker will trip automatically in case of a malfunction creating a high current surge, and can be easily reset when the malfunction is corrected. AC and DC breakers have different characteristics, because it is more difficult to break direct current than to break alternating current and it must also be designed to withstand the rigors of an automotive environment.

IGNITION KEY MAIN CONTACTOR

Contactors are used to switch high currents remotely by means of a low-level control voltage. In EVs high voltages, inductive loads, and extremely high current loads are encountered. To switch a current under these conditions requires specifically designed equipments. Contactors have continuous duty coils, silver-cadmium-oxide contacts to prevent welding, and magnetic blowouts, which extinguish electrical arcing. A diode protects other components from current spikes when the field is collapsed. There are many types of contactors such as main contactors, single/double pole contactors, and reversing contactors. The main contactor is an easy way to manually isolate the power train components from the battery pack in an emergency, during charging, or while working on the car.

POTBOX

The potbox is the interface between the throttle pedal and the speed controller. It sends a variable resistance signal to the controller to specify the amount of electricity to be released to the motor. It interfaces directly with any vehicle's existing throttle control cable or linkage. It comes with many safety features, such as deadman switches for emergency disconnect and high pedal lockout to prevent unintentional acceleration.

FUSIBLE LINK

A fusible link should be inserted in the traction battery circuit in each pack in the vehicle. It will break the circuit in case of a short circuit.

4.5 INSTRUMENTATION

Gauges

Gauges allow monitoring of an EV's performance. The primary requirements of gauges are reliability, accuracy, and readability. Many display formats do not live up to these criteria. Panel meters are not designed to be used in vehicles. They cannot function reliably for extended periods in the conditions of vibration, jostling, dust, and heat that are normal in a car. Light emitting diode (LED) gauges are difficult to read in the sunshine. In addition, digital gauges are harder to interpret at a glance than needle gauges. Any gauge that is not backlit will be useless at night. The following are the types of gauges normally used in an electric car.

Ammeter

The ammeter gives a continuous reading of current usage. When an EV is coasting or sitting still, it will read zero. During full acceleration, it will peg at the top of the scale, then gradually fall off as the motor achieves its optimum rpm level. In reality, the ammeter is a mill voltmeter. This makes it possible to wire the gauge with lightweight wire instead of 2/0 cable. The shunt is calibrated to put out a specific voltage in mill volts for a specific amperage input. It is usually mounted near the controller.

State-of-Charge Gauge

This gauge measures the voltage in the battery pack and reads in percentages from full to empty. Under acceleration, it will draw down to a lower voltage, so it should be read when foot is off the throttle. This gauge is more accurate than a segmented bar scale display or a sampling meter, which may tend to read too low.

High-Voltage Meter

This gauge measures battery pack voltage and displays it in volts, to give a running performance display.

Low-Voltage Meter

This gauge allows monitoring the charge level of auxiliary battery. It is normally required for cars with DC/DC converters.

4.6 MAIN AUXILIARIES IN AN EV

Chargers

A good charger is crucial to EV performance. Early crude chargers simply slugged voltage into the batteries until the charger was turned off. This caused the batteries to gas heavily toward the end of the charge, and it shortened battery life. Later chargers used timers to taper the charge down and shut off. These are better, but still did not fully synchronize with the needs of the batteries. Modern chargers can sense the level of charge in the battery pack, and taper the charging current accordingly. The final, low-current part of the charge cycle serves to equalize the charge in all the batteries.

There are differences between 220 V and 110 V input chargers. A 220 V will charge the pack faster, but it is bulkier and heavier, not really suitable for on-board mounting. A 110 V charger will charge more slowly, but it is small and light enough to be mounted on-board so the driver can take advantage of opportunity charging anywhere there is 110 V power.

Relays

When installing high-voltage heaters, air conditioners, power steering pumps, and so on in an EV, the appropriate relay is required to switch these devices on and off with a 12 V DC control voltage. Since these loads are highly inductive, a relay with magnetic blowouts and adequate spacing is required to interrupt the current without creating the contacts weld from the arcs. There are many types of relays such as switching relay and charger interlock relay. A double-pole, double-throw relay is used to disable the vehicle during charging, so that it can't be driven away while still plugged in. When activated by AC input, this relay breaks the circuit between the ignition key and the potbox microswitch. It is normally available with 12 V DC or 120 V AC pull-down coils.

Terminal Blocks

Terminal blocks are often used when there are multiple connections feeding to a common power source or ground. There are two types: Small gauge is suitable for low current 16 ga. wiring; large gauge is a single stud suitable for connection to 2/0 cable by a lug.

Fuse Block

A fuse block serves the same purpose as a terminal block, with the added feature of fusing each connection.

Cables

It is important that the proper cable and lugs are employed when interwiring high-current circuits such as batteries, motors, and controllers. It is also important that lugs are crimped properly onto cable ends to ensure a good mechanical bond. If a lug is soldered on and becomes loose, the heat generated under high currents can cause solder melt. Using an improper cable size can melt the battery terminals and wires or cause fire or even explosion. The high-current connections in an EV need to be made with 2/0 cable. However, not all 2/0 cable is created equal. Some of it will have dozens of strands of copper, each about as thick as a mechanical pencil lead. This is suitable for stationary wiring in a building, but hard to work with in a vehicle, because it is too stiff. The cables used in EVs have several thousand copper strands the thickness of hairs. Both types of cable have the same amount of copper in them, but the fine strand version is very flexible for bending into tight places. It also has a durable insulation protecting it.

Cable Grommets

As cables or wires are passed through a hole in sheet metal, a grommet should be used to protect the wires. These grommets fit closely around the cables, eliminating movement and providing a weatherproof opening.

Connectors

There are some places in an EV where a secure connection is needed and which can also be easily separated. One such example is the wiring to an on-board charger. Hardwiring the charger into the car is awkward and makes it difficult to remove for repairs or modifications. The solution is the connector. This is a heavy-duty plastic case that snaps two connections together securely and separates easily. The case is indexed so that it cannot be snapped together with mismatched polarity. These connectors are an industry standard.

Belleville Washers

Proper battery connections are essential. Loose connections may lead to heat generated damage, melted terminals, or fire. However, since lead is so soft, it is difficult to maintain a snug connection without simply deforming the lead. Split ring washers exert too much pressure for this application, and the pressure is unevenly distributed. Belleville washers are precision constant-pressure devices. They have a slightly concave shape, which flattens when tightened into place. With proper use of these washers, loose connections are avoided, and battery maintenance is reduced.

Heat Shrink Tube

All connections between cables and lugs should be weatherproofed and insulated. Sometimes it is also useful to insulate other, smaller connections and terminals as well. The tubing, slipped over the connection, can be quickly shrunk to a tight fit with a heat gun or hand-held propane torch. It is available in sizes 1/4″, 1/2″, and 3/4″ diameters, which indicate the finished diameter after shrinking.

Flexguard

Wires gathered into proper looms not only look better, they are also better protected from snagging or abrasion. This enhances the vehicle's reliability. Flexguard provides an easy-to-use loom sheath. It is a corrugated flexible tube with a slit down the length. Wires can be slipped into it, and individual wires can be routed out again at any point along the length. It is easy to add or remove wires at any time, or to work on wiring at any point on the loom. It can be tie-wrapped to the chassis at intervals. Flexguard caps snap around the ends to finish the loom.

Noalox

Corrosion is the enemy of an EV. Noalox anticorrosion compound protects connections. If a lug is half-filled with Noalox before being crimped to the cable, then the connection is covered with heat shrink tube, the connection will remain corrosion-free for years. Noalox is also handy to coat on battery terminals before bolting on lugs.

HEAT SINK GREASE

Controllers heat sink through their bases and are intended to be mounted on a flat piece of aluminum for this purpose. A layer of heat sink grease between the controller and the plate is essential for good contact and thermal transfer.

ADAPTORS

The adaptor plate mounts the electric motor to the original manual transmission and clutch. The electro-automotive adaptors are precision machined. The adaptor comes in four parts. The transmission profile plate is machined from aluminum and mimics the original engine-mounting surface. The motor spacer ring is also aluminum. It mounts to the profile plate, and recreates the original spacing between the flywheel and transmission.

TAPER LOCK HUB

The crucial hub is machined from steel. It is a taper lock fit, the industrial standard for high-rpm, high-torque applications. The hub mounts to the flywheel. Its inner surface is tapered and cone-shaped. This slides over a matching tapered bushing around the motor shaft. The bushing has a split in it, and is a slightly larger diameter than the inside of the hub. As the bushing is drawn into the hub by six bolts, it is compressed around the motor shaft, squeezing closed the split. Once in place, it can be removed only with a proper pulling tool.

VACUUM POWER BRAKE SYSTEM

On most small cars, the brakes are designed with a large safety margin and will easily handle the extra weight added by the conversion. However, disc brakes need a power assist, even in conventional cars. The power assist relies on vacuum from the engine manifold. This vacuum source is lost in the conversion, and needs to be replaced. This can be accomplished with a vacuum pump, reservoir, and switch connected to the car's original power brake unit.

Electric Vehicles

4.7 TYPES OF POWER STORAGE USED IN EVS

4.7.1 BATTERIES

The largest hurdle holding back battery-powered electric vehicle commercialization is the battery. Batteries typically account for one third or more of vehicle weight and one fourth or more of the life-cycle cost of an electric vehicle. Major improvements in batteries are expected because, until recently, little effort has been put into designing and building batteries of the size needed for vehicles.

The list of electric vehicle battery candidates includes batteries with solid, liquid, and gaseous electrolytes; high and ambient temperatures; replaceable metals; and replaceable liquids. At least 20 distinct battery types have been suggested as candidates. But what looks promising in a small cell often falls short when scaled up for a vehicle. The reality is that the underlying science of battery technology is highly complex and not well understood, rendering the engineering of large batteries difficult.

Many research efforts are under way to develop and commercialize advanced batteries. The most prominent is the United States Advanced Battery Consortium (U.S. ABC). Launched in 1991, U.S. ABC seeks to increase the energy and power capability, extend the life, and reduce costs of batteries as they are scaled up to sizes suitable to electric vehicles. Improved versions of today's lead-acid batteries dominated through the late 1990s, because this type is relatively inexpensive and relatively reliable. In Europe, the nickel-cadmium battery and the high-temperature (250–320°C) sodium-nickel-chloride battery are also being used initially. Beyond lead-acid batteries, there is no consensus as to which batteries will prove superior for electric vehicles, though it is widely believed that nickel-metal hydride (NiMH) batteries will predominate over much of the next decade, and lithium-based batteries will dominate thereafter. This section discussed a few of the important terms related with battery technology and various types of batteries in use or under development.

Which Battery to Use

This depends on the purpose of use of the EV built. The higher the voltage, the better the acceleration, and a higher top speed can be achieved. Of course, it also depends on the type of motor and the controller used. For example, a normal-sized EV using a voltage system of around 96 to 120 V with 6 V deep-cycle batteries will give more amp hours and weigh more. Therefore, it will have a fairly high range but poor performance. The same car using a voltage system of 96 to 120 V with 12 V batteries will give high performance, but lower range. That is because the battery pack will be lighter and in turn the car will be lighter.

How Many Volts

Deep-cycle batteries are normally available in three voltage sizes: 6, 8, and 12 V. For range, 6 V batteries are used because their specific energy is higher. For performance, 12 V batteries are used. 12 V batteries are very popular with the newer components being used in cars with 144 V systems. The 8 V battery packs offer a good balance between the range of the 6 V and the acceleration capabilities of the 12 V.

How Much Energy

The energy in a battery pack (wired in series) is the amp hour rating times the pack voltage. The amp hour rating is how many amps a battery can supply over a given time. Most batteries are measured over a 20-hour period. This is a standard that is used to compare batteries and can be found in the specifications on most batteries. The 20 A hour rating has to be adjusted for EV use. The faster one draws current from a battery, the lesser will be the capacity.

Effect on Batteries when Temperature Changes

Temperature has a direct effect on the performance of a lead-acid battery. The concentration of sulfuric acid inside the battery increases and decreases with temperature. A battery being used in 32°F weather will only operate at 70% of its capacity. Likewise, a battery being used in 110°F weather will operate at 110% of its capacity. The most efficient temperature that battery manufacturers recommend is 78°F. Because the temperature factor is important in colder climates, insulated battery boxes or thermal management systems are used.

Maintenance of Batteries

Maintaining deep-cycle batteries is very simple and requires little time. Each battery has three flooded cells, or six, which require watering with distilled water once a month. In addition, the batteries need to be cleaned once every month with a solution of distilled water and baking soda to prevent ion tracking. Ion tracking is a condition in which dirt or moisture on top of the battery forms a conductive path from the terminal to another terminal or to metal such as a battery rack. This can cause the ground-fault interrupter to trip on some battery chargers when the car is charging. Ion tracking is more prominent when the batteries are not stored in a protective enclosure or battery box.

Some Technical Terms Related to Batteries

Several terms are used while discussing the characteristics of batteries. These are important in determining which battery should be used for a particular application. Below is a short description of these characteristics.

Specific Energy. Specific energy is an important factor in determining range. Specific energy is the total amount of energy in watt-hours (Wh) that the battery can store per kilogram of its mass for a specified rate of discharge.

Energy Density. Energy density refers to the amount of energy a battery has in relation to its size. Energy density is the total amount of energy (in Wh) a battery can store per liter of its volume for a specified rate of discharge. Batteries that have high energy density are smaller in size.

Specific Power or Power Density. Specific power or power density is an important factor for determining acceleration. Specific power is the maximum number of watts per kilogram (W/kg) a battery delivers at a specified depth of discharge. Specific power is highest when the battery is fully charged. As the battery is discharged, the specific power decreases and acceleration also decreases. Specific power is usually measured at 80% depth of discharge.

Cycle Life. Cycle life is the measure of the total number of times a battery can be discharged and charged during its life. When the battery can no longer hold a charge over 80%, its cycle life is considered to be finished.

Battery Cost. Cost is expressed in currency units, per kilowatt-hour (kWh).

4.7.2 Types of Batteries Available Today

Deep-Cycle Lead-Acid

Most EVs today are fitted with deep-cycle lead-acid batteries The most common makes are Trojan, U.S. Battery, Alco, Deka, Exide, and GNB. Deep-cycle batteries have tall lead plates and are designed for deep discharge cycles. They have a life span of 400 to 800 cycles. Normal batteries, like those used in ICE vehicles, are not suitable for EV use and they will quickly wear out after 30 cycles.

Horizon Lead-Acid

In recent tests the Horizon lead-acid battery from Electrosource in Austin, TX, powered a car 110 miles on a charge. To develop this battery, Electrosource invented a patented process to extrude lead onto fiber-glass filaments that are woven into grids in the battery's electrode plates. The results are greater power capacity, longer life cycle, deep discharge without degeneration, rapid recharge, and high specific energy. These batteries are sealed and maintenance free.

Nickel Cadmium/Nickel Iron

Another type of battery is the nickel cadmium (Ni-Cd) and nickel iron (Ni-Fe). Ni-Cds are already in use in some Japanese and European companies producing EVs. They are more expensive than lead-acid batteries because nickel is costly. Ni-Cd advantages are higher energy density and a cycle life of 1000 charges. Although they can be recharged very quickly, they have a tendency to overheat. Cadmium is highly toxic, so recycling efforts have to be managed very carefully. Although cadmium supplies are not very high it can be produced from sources such as copper, lead, zinc, and cadmium recycling. Ni-Fe batteries have high energy density and are capable of over 1000 deep-discharge cycles before recharging. They need to be 11% overcharged to be charged. The result of the overcharging is water loss and a build-up of hydrogen, which is a safety concern.

Nickel-Metal Hydride

A nickel-metal hydride battery is composed of nontoxic recyclable materials and is environmentally friendly. The NiMH has twice the range and cycle life of today's lead-acid batteries. It is composed of nickel hydroxide and a multicomponent, engineered alloy consisting of vanadium, titanium, nickel, and other metals. It is sealed, is maintenance free, and can be charged as quickly as 15 minutes. It can withstand overcharging and overdischarge abuse. Ovonic Battery of Troy, MI, is currently manufacturing the NiMH battery.

Sodium Sulfur

Sodium sulfur (NaS) batteries are still under development by Ford Motor Company. Ford Ecostars are fitted with these batteries and have a range up to 150 miles per charge. The NaS battery uses a ceramic beta-alumina electrolyte tube with sodium negative electrodes and molten sulfur positive electrodes within a sealed insulated container. To keep the sulfur in a molten state the battery must be kept at a temperature of 300 to 350°C. The batteries have built-in heaters to keep the sodium and sulfur from solidifying. Presently, the battery costs 7 times more than lead-acid batteries, but their price is expected to drop during high-volume production because the materials to build the battery are plentiful and cheap. The main disadvantage of the NaS battery is the high temperature, which has raised safety hazards. Also, the battery must be charged every 24 hours to keep the sodium and sulfur from solidifying.

Sodium Nickel Chloride

Sodium nickel chloride ($NaNiCl_2$) batteries are under development by AEG Anglo Batteries GmbH (Ulm, Germany). The battery operates at a temperature of 300°C and is claimed by its manufacturer to be safe in accidents and will operate even if one of its cells fails. The battery can be cooled down and reheated without damage; however, no current can be drawn from the battery if the temperature is below 270°C. Costs to produce the battery are very high. BMW and Mercedes Benz are testing EVs with $NaNiCl_2$ batteries.

Lithium-Iron

The lithium-ion battery was predicted to be a long-term battery. The battery is being developed for Nissan by Sony Corporation. The promising aspects of the battery are its low memory effect, high specific energy of 100 Wh/kg, high specific power of 300 W/kg, and a battery life of 1000 cycles. The battery is 28.8 V and consists of eight metal cylindrical cells encased in a resin module. Each battery has a built-in cell controller to ensure that each cell is operating within a specified voltage range of 2.5 to 4.2 V during charging and discharging. The cell controller communicates with the vehicle's battery controller to optimize power and energy usage. The disadvantages of the lithium-ion battery are its very high cost and the ventilation system required to keep the batteries cool. The manufacturing costs are high because the battery uses an oxidized cobalt material for the anode, a highly purified organic material for the electrolyte, and a complex cell control system.

Lithium Metal Sulfide

The lithium metal sulfide battery is an elevated-temperature battery based on a lithium alloy/molten salt/metal-sulfide electrochemical system. This system provides high specific power for better acceleration. Other advantages include its small size, low weight, and low cost per kilowatt hour. The battery is composed of iron disulfide and a lithium-aluminum alloy that is completely recyclable. SAFT America, Cockeysville is currently researching this type of battery.

Lithium-Polymer

The lithium-polymer battery is based on thin film technology. The battery is expected to cost 20% more than lead-acid but deliver twice the energy, with a life span of 50,000 miles. It has an operating temperature between 65 and 120°C. It can be fast charged in less than 90 minutes but can be damaged by overcharging. The major challenge confronting this technology is scaling up its size to properly power an EV.

OTHER BATTERY TECHNOLOGIES

Zinc-Air. The zinc-air battery developed by the Israeli firm Electric Fuel, Inc. is being used to power 40 test vans. After the battery pack or cassette is discharged it is taken out of the vehicle and replaced with another cassette. The cassette replacement is done in a matter of minutes with highly automated equipment set up at various geographical locations. The discharged cassette has its electrodes replaced with fresh ones, and the cassette is used for another vehicle. The spent electrodes are recycled and used to make new electrodes. The battery has an energy density 10 times that of lead-acid.

Aluminum-Air. Aluminum-air is a battery that has aluminum plates added every 200 miles to replace the used aluminum. The aluminum plates react with oxygen in a sodium hydroxide electrolyte solution to form sodium aluminate. The sodium aluminate produces a by-product of aluminum, aluminum trihydroxide, which is removed and replaced with fresh aluminum. Aluminum-air batteries will probably be used by larger vehicles because of their size.

Nickel-Hydrogen.   The nickel-hydrogen battery is currently under development by Johnson Controls, Milwaukee, WI. It is very expensive but has a long life and is safe to operate. It is currently being used in spacecraft and deep-water vehicles.

Nickel-Zinc.   The nickel-zinc battery has more power than lead-acid batteries but has a shorter discharge cycle, making it only useful for short commutes.

Zinc-Chloride.   The zinc-chloride battery has high energy but must have a complex system to recapture the chlorine released during recharging.

Zinc-Bromide.   The zinc-bromide battery was developed by Johnson Controls. The battery stores electricity by plating zinc onto a surface and then unplating it. A bromide electrolyte solution, which is 80% water, is pumped through the battery to cause the plating and unplating reactions. Pure bromide is extremely toxic. Safety issues were raised about the battery in a 1992 EV race when it was involved in an accident. A hose that carries the bromide electrolyte became unconnected from the battery and leaked onto the race track, releasing irritating fumes.

4.7.3 Flywheels

The concept of using flywheels to store energy has been around for many years. Although building a flywheel system to power a car has been challenging, the principle behind the idea is fairly straightforward. A typical flywheel system uses a small motor to begin rotating the wheel. After the flywheel gains momentum, it spins freely on its own up to 100,000 rpm, or 1,666 revolutions per second! Magnets on the flywheel's axle speed past wire coils on the flywheel to produce electricity that powers motors attached to the vehicle's wheels. Each flywheel stores about 4 kilowatt-hours of electricity and produces about 25 horsepower. A typical car will require several flywheels to be comparable in range and performance to today's gasoline-powered car.

To keep the flywheel spinning in a frictionless environment, each flywheel spins on an axle with magnetic bearings in a vacuum-sealed aluminum box. The whole unit can weigh up to 90 pounds. The flywheels are about 12 inches in diameter, 3 inches wide, and made with carbon fiber. The challenge facing scientists has been keeping the wheel together at phenomenal speeds, since the gravitational force of the wheel can pull the resin and fibers apart. However, great strides have been made, and U.S. Flywheel Systems of Newbury Park, CA, will unveil its first prototype flywheel-powered vehicle. American Flywheel Systems, Seattle, WA and SatCon Technology, Burlington are also building prototype flywheel systems for EVs and hybrid vehicles.

4.7.4 Ultracapacitors

Ultracapacitors behave like very high-power, low-capacity batteries but store electric energy by accumulating and separating unlike-charges physically, as opposed to batteries, which store energy chemically in reversible chemical reactions. Power densities of 2000 to 4000 W/kg have been demonstrated by ultracapacitors in the laboratory. The fact that ultracapacitors can provide high power for accelerations and can accept high power during regenerative braking makes them ideally suited for the load leveling required in a hybrid electric vehicle.

As with chemical batteries and flywheels, ultracapacitors depend on improvements in materials to achieve performance targets. There are five distinct material combinations mainly to be developed: carbon/metal fiber composites, foamed (aerogel) carbon, carbon particulate with a binder, doped conducting polymer films on carbon cloth, and mixed metal oxide coatings on metal foil. Current trends indicate that higher energy densities will be achievable with a carbon composite electrode using an organic electrolyte, rather than with carbon/metal fiber composite electrode devices with an aqueous electrolyte.

One key aspect of ultracapacitors is that they have excellent cycle life. When fully developed for vehicles, they could be expected to last as long as the car. This is because it is possible to cycle ultracapacitors very quickly and deeply without seeing the large decrease in cycle life that most chemical batteries experience. They also have a high cycle efficiency compared with chemical batteries. The primary obstacle with ultracapacitors at this point is their low specific energy, which is in the range of 5–10 Wh/kg. Ultracapacitors also have the unique feature that their voltage is directly proportional to their state of charge (the measure of how much energy is left in them). Therefore, either their operating range must be limited to high state-of-charge regions, or control electronics must be used to compensate for the widely varying voltage. One of the keys to successfully developing ultracapacitors for vehicle applications is the development of interface electronics to allow the ultracapacitor to optimally load-level the batteries.

4.8 EMISSIONS PERFORMANCE

The amount of pollution created by EVs depends mostly on the source of the electricity used to charge them. This makes it difficult to determine if electric vehicles pollute less than internal combustion engine vehicles without considering where they are to be deployed and by what sources of electricity they are to be powered. An EV that is charged with energy from a clean source, like hydroelectric power, will produce very little pollution, while one charged with energy from an unclean source, like coal or oil, may produce more pollution than an internal combustion engine vehicle. The sources of energy for most regions fall somewhere between these two extremes. While not ready to be used everywhere, electric vehicles have the potential to pollute much less than internal combustion engine vehicles.

A battery electric vehicle (BEV) produces zero vehicular emissions. However, emissions are produced at the generation site when the source fuel is converted into electrical power. The emissions of electric cars therefore depend on the emissions profile of regional generating plants.

Some researchers conclude that, in regions serviced by coal-fired plants, a switch to EVs may actually increase emissions of sulfur oxides (SOx) and particulate matter (PM), and perhaps increase emissions of carbon dioxide (CO_2). Conclusions, however, are usually based on the existing mix of coal-fired plants, and often they do not consider the effect of newer and cleaner plant designs. Studies generally conclude that emissions of SOx, PM, and CO_2 are reduced in regions that rely on natural gas, and virtually eliminated in regions supplied by hydroelectric and nuclear power. According to Electric Power Research Institute (EPRI), substituting EVs for conventional vehicles (CVs) would reduce urban emissions of nonmethane organic gases (NMOG) by 98%, lower nitrogen oxide (NOx) emissions by 92%, and cut carbon monoxide (CO) emissions by 99%. In addition, EPRI estimates that, on a nationwide basis, EVs in the U.S. will produce only half the CO_2 of conventional vehicles. In another study of four U.S. cities, BEVs reduced hydrocarbon (HC) and CO emissions by approximately 97%, regardless of the regional source fuels mix. In comparison to large generating plants, conventional cars produce large amounts of HC and CO emissions, mainly because of cold starts and short trips that do not allow vehicles to become fully warmed up.

The environmental benefits of an HEV depend on the design of the hybrid power system. Designs using a combustion engine for on-board electrical generation and an operating schedule that is heavily biased toward the engine/generator system (genset) produce the greatest amount of harmful emissions. But even in this worst-case scenario, emission levels are lower than those of a typical CV. This is due to the fact that a hybrid vehicle genset is either switched off, and therefore producing zero emissions, or it is operating at a predetermined output where it produces the fewest emissions and achieves the best fuel economy per unit of output. Typically, a hybrid genset is not throttled for variable output, as is the engine in a conventional vehicle. This leads to more effective emission controls because it is technically easier to control combustion-engine emissions when the engine runs continuously and at a constant output. When the hybrid-operating schedule is biased more toward the energy storage system (relies more on the battery, rather than the genset), emission levels become more like those of a BEV, and with fuel-cell hybrids, vehicular emissions are virtually eliminated.

4.9 SOLAR CARS

Solar-powered vehicles (SPVs), such as cars, boats, bicycles, and even airplanes, use solar energy to either power an electric motor directly or charge a battery, which powers the motor. They use an array of solar photovoltaic (PV) cells (or modules made of cells) that convert sunlight into electricity. The electricity either goes directly to an electric motor powering the vehicle or to a special storage battery. The PV array can be built (integrated) onto the vehicle body itself, or fixed on a building or a vehicle shelter to charge an electric vehicle battery when it is parked.

A solar car is basically a type of EV, but the main difference from an EV is that the batteries are charged using solar panels fitted on the surface of the car. We probably won't be able to harness the sun's rays to power our regular vehicles because current photovoltaic cells are too inefficient and it would need a large area of cells to create even a small amount of electricity.

SPVs that have a built-on PV array differ from conventional vehicles (and most EVs) in size, weight, maximum speed, and cost. The practicality of these types of SPVs is limited because solar cells only produce electricity when the sun is shining. Even then, a vehicle completely covered with solar cells receives only a small amount of solar energy each day, and converts an even smaller amount of that to useful energy only about 1500 to 2000 watts of electricity. Even state-of-the-art PV cells only get up to around 24% efficiency. At present, most SPVs with built-on PV arrays are only used as research, development, and educational tools, or to participate in the various SPV races held around the world.

4.10 FUEL CELL CARS

4.10.1 INTRODUCTION

The need for clean, efficient vehicle power systems that operate on nonpetroleum-based fuels has been spurred by the continuing dependence on foreign oil and the deterioration of air quality in our urban areas. Shortly after the OPEC oil embargo in the early 1970s, the U.S. government embarked on a program to introduce electric and alternatively fueled vehicles. Currently, the transportation sector accounts for over 63% of the petroleum used in the nation. Early electric vehicle attempts relied upon the storage battery for the vehicle power and energy. While the battery is excellent at producing power, it suffers in the areas of total amount of energy that can be stored and the length of time to recharge. Thus, present battery-powered electric vehicles can't go far enough and take too long to recharge. The fuel cell has emerged as a strong candidate to circumvent the limitations of storage batteries.

This technology has proven to be an excellent producer of electrical power in a variety of applications and can derive its energy in the same way that a diesel or gasoline engine does, from a refillable liquid fuel tank. The fuel cell powerplant for transportation can operate on nonpetroleum liquid fuels, which could significantly reduce this nation's dependence on oil imports. Additionally, it has emission levels well below any projected clean air standards. The fuel cell is quiet, clean, and more efficient than internal combustion engines, and requires much less maintenance.

A fuel cell produces electricity by reacting hydrogen and oxygen with a catalyst to form water. The chemical energy is converted to electrical energy with high efficiency, negligible pollution, and little noise. Essentially, it is the reverse of water electrolysis. All fuel cells are based upon the hydrogen–oxygen reaction. Hydrogen is fed to the anode of the fuel cell, and oxygen (usually from ambient air) to the cathode. Although the electrochemical reaction varies by fuel cell type, that reaction drives a flow of electrons from the anode to cathode through an external electric circuit. The fuel cell concept was first demonstrated over 150 years ago; however, it was only in the last 30 years that useful applications were developed.

The hydrogen fuel cell is often referred to as the reverse battery. The battery unites hydrogen and oxygen to form water while producing electricity. The main draw back of the fuel cell is its size. However, the battery is clean, can be recharged in a matter of minutes, and has a range comparable to that of a gasoline-powered car. Other fuel cells include phosphoric acid, alkaline (which is used by the space shuttle), solid oxide, and proton exchange membrane. Other types of alternative fuels are ethanol-powered flexible fuel vehicles (FFVs), methanol powered flexible fuel vehicles, natural gas-powered vehicles (NGVs), LPG-powered vehicles (LPG), hydrogen/hythane, reformulated gasoline, and biofuels: biodiesel and soy diesel.

Following are the limitations of hydrogen as a fuel cell to be used in cars:

1. Fuel-cell cars need to have adequate number of new or retrofitted hydrogen filling stations. While hydrogen can be refined from water, no facilities have the capacity to fuel America's 200 million-plus vehicle fleet. Other running prototype fuel cell cars thus far have used gaseous or liquid hydrogen, or an on-board chemical factory (a "reformer") to produce hydrogen. All these technologies have their limitations.
2. Hydrogen gas has to be stored in giant, cylindrical, pressurized tanks that take up the space of at least two seats and most of the cargo area in a typical minivan, which is much larger than other storage forms.
3. Liquid hydrogen contains many more molecules in a smaller tank and so could give a hydrogen vehicle the range of a gas-powered car. But it evaporates into a gas at only a few degrees above absolute zero. So the smaller fuel tank needs rampart-thick insulation: more wasted space.
4. Reformers, which produce hydrogen from hydrogen-rich methane or common gasoline, may solve many distribution problems. But the reformers produce pollution much like today's engines (though less of it), and so negate a key

advantage of hydrogen. They take up almost as much room in the car as a gaseous hydrogen tank, produce infernal heat, and take up to 30 minutes to warm up.

5. Fuel cells can also use hydrogen extracted chemically from natural gas, alcohols, naphtha (coal tar), and other hydrocarbons. This is normally done by a reformer between the fuel tank and the fuel cell that produces hydrogen-rich gas from the fuel, which is then used in the fuel cell. Several car manufacturers seem to be opting to use methanol or gasoline as fuel for the fuel cells because of a lack of a hydrogen supply infrastructure. In the meantime, neither methanol nor gasoline will produce hydrogen without also producing emissions. The reformer will demand a lot of energy, require a long start-up time, and give poorer energy output than the use of pure hydrogen. In addition, methanol is also extremely poisonous.

4.10.2 FUEL CELL CARS

Hydrogen-based fuel cell cars are the only known alternative that can combine zero-emissions with the comfort and driving distance we are accustomed to with today's cars. Fuel cells generate electricity by converting oxygen and hydrogen into water without combustion. Fuel cells are superior to combustion engines because of their dependability, energy efficiency, and environmental friendliness.

The modern fuel cell's beginnings started with the Apollo space program, when NASA needed a reliable and simple electrical supply on-board the space capsules. They continued development of the fuel cell, which at that time was looked upon as a curious invention from 1839. On the Apollo lunar mission, the energy supply on-board the space capsule was based on fuel cells with hydrogen, and the "waste product" (water) was drunk by the crew. Fuel cells are quite common in space travel today. After the successful voyage to the moon, in-depth research into the fuel cell was initiated to develop the fuel cell for use in vehicles. But the cells were quite expensive, especially because of the large amount of platinum needed as a catalyst, and gas prices were also low at the time. But because of increasing environmental awareness and breakthroughs in a technique that reduced the need for platinum, research into development of fuels cells has increased in the last decade.

There are several types of fuel cells with different characteristics and applications. Fuel cells are usually classified by the electrolytes they use. Fuel cell technology is now the latest modern technology. Daimler Chrysler, GM and Toyota, Ford, and Honda have indicated that they will introduce mass-produced fuel-cell cars in 2005. The Canadian company Ballard Power Systems has been a leading company in the development of the proton exchange membrane (PEM) fuel cells and has perhaps come furthest with this type of fuel cell. Other researchers include DeNora, Energy Partners, International Fuel Cells, Toyota, and Panasonic.

There are two main ways to use fuel cells in a car: fuel cells alone, and a hybrid solution. When only fuel cells are used, a powerful fuel-cell system is needed. In addition, it is advantageous to use ultracondensators to store braking power for later use in acceleration. Daimler Chrysler's and Ford's prototypes use this method, which indicates that they foresee the fuel cell becoming economically feasible.

The other method is to use fuel cells in combination with batteries. In this case the fuel cells act as battery chargers, while the batteries take the heavy loads and let the fuel cells operate more smoothly. Toyota's prototype car operates on this principle. Even though this is the most inexpensive solution today, it is by no means the only solution for the future. This solution also has a definite advantage: Most people drive less than 30 km a day, which is a distance batteries alone can cover. In this way, a car can be used as a regular electric car for everyday use and only need be topped off with hydrogen when going on longer trips.

As a natural result of competition with fuel-cell cars, more effective combustion engines will also be developed: Regardless of this, fuel cells will outperform combustion engines because people are moving more and more to urban areas where traffic often goes slowly or at a standstill. When a fuel cell car stands still, just as an electric car, it does not use energy, and even at low speeds it is fuel efficient. A car with a combustion engine is most efficient at 80–90 km/h, but is extremely inefficient at low speeds. On the average, a new medium-sized car with a combustion engine has a fuel efficiency of 12%. A typical PEM fuel cell car with hydrogen may have a system efficiency of over 40%, even at low speeds.

BIBLIOGRAPHICAL SURVEY ON ELECTRIC VEHICLES

Electric vehicles, encompassing electric motors, batteries, and power electronics, have important advantages over today's gasoline-powered internal combustion engines. They are quieter, virtually nonpolluting, and more energy efficient, reliable, and durable. Major advances have been made in various electric drive technology components since the late 1980s. For example, advances in power electronics have resulted in small, lightweight DC/AC inverters that, in turn, make possible AC drives that are cheaper, more compact, more reliable, easier to maintain, more efficient, and more adaptable to regenerative braking than the DC systems used in virtually all electric vehicles through the early 1990s. The electric vehicle motor-controller combination is now smaller and lighter than a comparable internal combustion engine, as well as cheaper to manufacture and maintain. A large number of papers have appeared on EVs during past 15 years. This chapter presents an extensive bibliography on EVs. Papers have been arranged chronologically in descending order. References can be broadly classified in the following categories:

S. No.	Application area of EV	Reference S. No.
1.	Battery	9, 23, 30, 39, 40, 63, 65–67, 80, 84, 104, 115, 116, 118, 119, 124, 126, 134, 140, 149, 150, 161, 181, 189, 190, 193, 198, 200, 206–208, 214, 217, 226, 234, 262, 263, 267, 275, 281, 296, 298, 300, 305, 306, 340, 342, 347, 348, 356, 358, 359, 367, 369, 373, 376, 378, 404, 408, 412, 415, 416, 423, 425, 447, 449, 455, 456
2.	Control of EV	4, 7, 11, 12, 15, 20, 28, 35, 36, 46, 52, 76, 90, 96, 111, 114, 115, 141, 159, 171, 199, 202, 223, 229, 244, 249, 254, 255, 261, 284, 285, 290–292, 328, 336, 338, 350, 352, 363, 370, 377, 380, 383, 384, 393
4.	Charging of batteries/EVs	26, 41, 103, 132, 142, 160, 180, 218, 274, 282, 395, 397, 434, 435, 439, 442, 454
3.	Design	8, 22, 59, 94, 117, 122, 131, 143, 146, 153, 233, 242, 243, 248, 259, 265, 271, 278, 295, 308, 324, 327, 331, 335, 344, 351, 357, 392, 399, 450
4.	Efficiency analysis	71, 73, 102, 145, 257, 269, 409, 440, 445
5.	Energy management	197, 210, 222, 246, 428, 443

6.	Motor's application	24, 47, 51, 56, 62, 74, 78, 83, 85, 86, 88, 91, 92, 99, 107, 108, 112, 127, 135, 155, 166, 167, 173, 178, 182, 191, 195, 219, 227, 235, 238, 247, 258, 264, 270, 272, 273, 280, 287, 288, 302, 307, 312, 313, 318–320, 328, 345, 353, 387, 391, 407, 411, 424, 457
7.	Modeling and simulation	10, 11, 25, 29, 34, 49, 61, 89, 109, 110, 123, 130, 139, 151, 154, 157, 163, 164, 177, 184, 194, 212, 225, 230, 256, 293, 332, 346, 364, 430
8.	Torque analysis	6, 17, 31, 68, 179, 216, 231, 232, 322, 366
9.	Fuel cell	19, 75, 125, 213
10.	Ultracapacitors	3, 38, 341, 453

REFERENCES

2003

1. F. Tahami, R. Kazemi, and S. Farhanghi, "A novel driver assist stability system for all-wheel-drive electric vehicles," *IEEE Trans. Vehicular Technology*, Vol. 52, No. 3, May 2003, pp. 683–692.
2. J.R. Wagner, D.M. Dawson, and Z. Liu, "Nonlinear air-to-fuel ratio and engine speed control for hybrid vehicles," *IEEE Trans. Vehicular Technology*, Vol. 52, No. 1, January 2003, pp. 184–195.

2002

3. A.G Simpson and G.R. Walker, "Life-cycle costs of ultracapacitors in electric vehicle applications," *Proc. IEEE 33rd Annual Power Electronics Specialists Conf. (PESC 02)*, Vol. 2, 23–27 June 2002, pp. 1015–1020.
4. P. Khatun, C.M. Bingham, N. Schofield, and P.H. Mellor, "An experimental laboratory bench setup to study electric vehicle antilock braking/traction systems and their control," *Proc. IEEE 56th Vehicular Technology Conf. (VTC)*, Vol. 3, 24–28 September 2002, pp. 1490–1494.
5. R.M. Schupbach and J.C. Balda, "A versatile laboratory test bench for developing powertrains of electric vehicles," *Proc. IEEE 56th Vehicular Technology Conf. (VTC)*, Vol. 3, 24–28 September 2002, pp. 1666–1670.
6. K.E. Bailey, S.R. Cikanek, and N. Sureshbabu, "Parallel hybrid electric vehicle torque distribution method," *Proc. American Control Conf.*, Vol. 5, 8–10 May 2002, pp. 3708–3712.
7. R. Masaki, S. Kaneko, M. Hombu, T. Sawada, and S. Yoshihara, "Development of a position sensorless control system on an electric vehicle driven by a permanent magnet synchronous motor," *Proc. Power Conversion Conf. (PCC)*, Osaka, Vol. 2, 2–5 April 2002, pp. 571–576.
8. W. Chen, S. Round, and R. Duke, "Design of an auxiliary power distribution network for an electric vehicle," *Proc. 1st IEEE Int. Workshop Electronic Design, Test and Applications*, 29–31 January 2002, pp. 257–261.
9. E. Hansen, L. Wilhelm, N. Karditsas, I. Menjak, D. Corrigan, S. Dhar, and S. Ovshinsky, "Full system nickel-metal hydride battery packs for hybrid electric vehicle applications," *Proc. 7th Annual Battery Conf. Applications and Advances*, 15–18 January 2002, pp. 253–260.
10. H. Xiaoling and J.W. Hodgson, "Modeling and simulation for hybrid electric vehicles. I. Modeling," *IEEE Trans. Intelligent Transportation Systems*, Vol. 3, No. 4, December 2002, pp. 235–243.
11. R. Saeks, C.J. Cox, J. Neidhoefer, P.R. Mays, and J.J. Murray, "Adaptive control of a hybrid electric vehicle," *IEEE Trans. Intelligent Transportation Systems*, Vol. 3, No. 4, December 2002, pp. 213–234.

12. L.A. Tolbert, F.Z. Peng, T. Cunnyngham, and J.N. Chiasson, "Charge balance control schemes for cascade multilevel converter in hybrid electric vehicles," *IEEE Trans. Industrial Electronics*, Vol. 49, No. 5, October 2002, pp. 1058–1064.
13. H. Oman, "Electric car progress," *IEEE Aerospace and Electronic Systems Magazine*, Vol. 17, No. 6, June 2002, pp. 30–35.
14. C.C. Chan, "The state of the art of electric and hybrid vehicles," *Proc. IEEE*, Vol. 90, No. 2, February 2002, pp. 247–275.
15. H. Huang and L. Chang, "Error-driven PI control of EV propulsion systems based on induction motors," *Proc. IEEE 56th VTC*, Vol. 3, 24–28 September 2002, pp. 1686–1690.
16. A.M. Trzynadlowski, W. Zhiqiang, J. Nagashima, and C. Stancu, "Comparative investigation of PWM techniques for General Motors' new drive for electric vehicles," *Proc. 37th IAS Annual Industry Applications Conf.*, Vol. 3, 13–18 October 2002.
17. R.C. Baraszu and S.R. Cikanek, "Torque fill-in for an automated shift manual transmission in a parallel hybrid electric vehicle," *Proc. American Control Conf.*, Vol. 2, 8–10 May 2002, pp. 1431–1436.
18. S.R. Cikanek and K.E. Bailey, "Regenerative braking system for a hybrid electric vehicle," *Proc. American Control Conf.*, Vol. 4, 8–10 May 2002, pp. 3129–3134.
19. M.C. Pera, D. Hissel, and J.M. Kauffmann, "Fuel cell systems for electrical vehicles," *Proc. IEEE 56th VTC*, Vol. 4, 6–9 May 2002, pp. 2097–2102.
20. B.A. Kalan, H.C. Lovatt, and G. Prout, "Voltage control of switched reluctance machines for hybrid electric vehicles," *Proc. 33rd PESC 02*, Vol. 4, 23–27 June 2002, pp. 1656–1660.
21. R. Pusca, Y. Ait-Amirat, A. Berthon, and J.M. Kauffmann, "Modeling and simulation of a traction control algorithm for an electric vehicle with four separate wheel drives," *Proc. 56th VTC*, Vol. 3, 24–28 September 2002, pp. 1671–1675.
22. B.A. Kalan, H.C. Lovatt, M. Brothers, and V. Buriak, "System design and development of hybrid electric vehicles," *Proc. IEEE 33rd PESC*, Vol. 2, 23–27 June 2002, pp. 768–772.
23. B. Tsenter, "Battery management for hybrid electric vehicle and telecommunication applications," *Proc. 7th Annual Battery Conf. on Applications and Advances*, 15–18 January 2002, pp. 233–237.
24. J. Gan, K.T. Chau, C.C. Chan, and J.Z. Jiang, "A new surface-inset, permanent-magnet, brushless DC motor drive for electric vehicles," *IEEE Trans. Magnetics*, Vol. 36, No. 5, September 2000, pp. 3810–3818.
25. F. Briault, M. Helier, D. Lecointe, J.C. Bolomey, and R. Chotard, "Broad-band modeling of a realistic power converter shield for electric vehicle applications," *IEEE Trans. Electromagnetic Compatibility*, Vol. 42, No. 4, November 2000, pp. 477–486.
26. M.M. Morcos, N.G. Dillman, and C.R. Mersman, "Battery chargers for electric vehicles," *IEEE Power Engineering Review*, Vol. 20, No. 11, November 2000, pp. 8–11, 18.
27. R.B. Sepe, Jr., and N. Short, "Web-based virtual engineering laboratory (VE-LAB) for collaborative experimentation on a hybrid electric vehicle starter/alternator," *IEEE Trans. Industry Applications*, Vol. 36, No. 4, July–August 2000, pp. 1143–1150.
28. B.M. Baumann, G. Washington, B.C. Glenn, and G. Rizzoni, "Mechatronic design and control of hybrid electric vehicles," *IEEE/ASME Trans. Mechatronics*, Vol. 5, No. 1, March 2000, pp. 58–72.
29. H. Xiaoling and J.W. Hodgson, "Modeling and simulation for hybrid electric vehicles. II. Simulation," *IEEE Trans. Intelligent Transportation Systems*, Vol. 3, No. 4, December 2002, pp. 244–251.
30. H. Oman, "Artificial hearts, batteries, and electric vehicles," *IEEE Aerospace and Electronic Systems Magazine*, Vol. 17, No. 8, August 2002, pp. 34–39.
31. M.G. Simoes and P. Vieira, Jr., "A high-torque low-speed multiphase brushless machine-a perspective application for electric vehicles," *IEEE Trans. Industrial Electronics*, Vol. 49, No. 5, October 2002, pp. 1154–1164.
32. G. Maggetto and J. Van Mierlo, "Electric and electric hybrid vehicle technology: a survey," *Proc. IEEE Seminar on Electric, Hybrid and Fuel Cell Vehicles*, 11 April 2000, pp. 1/1–111.

33. T.B. Gage, "Lead-acid batteries: key to electric vehicle commercialization. Experience with design, manufacture, and use of EVs," *Proc. 5th Annual Battery Conf. on Applications and Advances*, 11–14 January 2000, pp. 217–222.
34. S. Onoda, S.M. Lukic, A. Nasiri, and A. Emadi, "A PSIM-based modeling tool for conventional, electric, and hybrid electric vehicles studies," *Proc. IEEE 56th VTC*, Vol. 3, 24–28 September 2002, pp. 1676–1680.
35. X. Ma, "Propulsion system control and simulation of electric vehicle in Matlab software environment," *Proc. 4th World Congress on Intelligent Control and Automation*, Vol. 1, 10–14 June 2002, pp. 815–818.
36. Y. Hori, "Future vehicle driven by electricity and control-research on four wheel motored 'UOT Electric March II'," *Proc. 7th Int. Workshop on Advanced Motion Control*, 3–5 July 2002, pp. 1–14.
37. E. Takahara, H. Sato, and J. Yamada, "Series and parallel connections change over system for electric double layer capacitors (EDLCs) to electric vehicle energy saving," *Proc. PCC*, Osaka, Vol. 2, 2–5 April 2002, pp. 577–581.
38. F. Gagliardi and M. Pagano, "Experimental results of on-board battery-ultracapacitor system for electric vehicle applications," *Proc. IEEE Int. Symposium on Industrial Electronics (ISIE)*, Vol. 1, 8–11 July 2002, pp. 93–98.
39. C.A. Nucci, M. Paolone, and A. Venturoli, "Analysis of Ni-Zn batteries performance for hybrid light-vehicle applications," *Proc. ISIE*, Vol. 1, 8–11 July 2002, pp. 123–128.
40. X. Wang and T. Stuart, "Charge measurement circuit for electric vehicle batteries," *IEEE Trans. Aerospace and Electronic Systems*, Vol. 38, No. 4, October 2002, pp. 1201–1209.

2001

41. R. Bass, R. Harley, F. Lambert, V. Rajasekaran, and J. Pierce, "Residential harmonic loads and EV charging," *IEEE Power Engineering Society Winter Meeting*, Vol. 2, 28 January–1 February 2001, pp. 803–808.
42. E. Hall, S.S. Ramamurthy, and J.C. Balda, "Optimum speed ratio of induction motor drives for electrical vehicle propulsion," *Proc. 16th Annual IEEE Applied Power Electronics Conf. and Exposition (APEC)*, Vol. 1, 4–8 March 2001, pp. 371–377.
43. L. Browning and S. Unnasch, "Hybrid electric vehicle commercialization issues," *Proc. 16th Annual Battery Conf. on Applications and Advances*, 9–12 January 2001, p. 45.
44. B.K. Lee and M. Ehsani, "Advanced BLDC motor drive for low-cost and high-performance propulsion system in electric and hybrid vehicles," *Proc. IEEE Int. Electric Machines and Drives Conf. (IEMDC)*, 17–20 June 2001, pp. 246–251.
45. C.T. Liu, J.W. Chen, and C.W. Su, "Optimal efficiency operations of a disc permanent magnet linear machine for electric vehicle application," *Proc. IEMDC*, 17–20 June 2001, pp. 588–590.
46. R. Zhang and Y. Chen, "Control of hybrid dynamical systems for electric vehicles," *Proc. American Control Conf.*, Vol. 4, 25–27 June 2001, pp. 2884–2889.
47. L. Tutelea, E. Ritchie, and I. Boldea, "Permanent magnet in-wheel synchronous motor for electric vehicles," *Proc. 5th Int. Conf. Electrical Machines and Systems (ICEMS)*, Vol. 2, 18–20 August 2001, pp. 831–834.
48. M. Ceraolo and G. Pede, "Techniques for estimating the residual range of an electric vehicle," *IEEE Trans. Vehicular Technology*, Vol. 50, No. 1, January 2001, pp. 109–115.
49. S. Fish and T.B. Savoie, "Simulation-based optimal sizing of hybrid electric vehicle components for specific combat missions," *IEEE Trans. Magnetics*, Vol. 37, No. 1, January 2001, pp. 485–488.
50. F.A. Wyczalek, "Hybrid electric vehicles: year 2000 status," *IEEE Aerospace and Electronic Systems Magazine*, Vol. 16, No. 3, 2001, pp. 15–25.
51. M. Ehsani, K.M. Rahman, M.D. Bellar, and A.J. Severinsky, "Evaluation of soft switching for EV and HEV motor drives," *IEEE Trans. Industrial Electronics*, Vol. 48, No. 1, 2001, pp. 82–90.

52. J.W. Park, D.H. Koo, J.M. Kim, and H.G. Kim, "Improvement of control characteristics of interior permanent-magnet synchronous motor for electric vehicle," *IEEE Trans. Industry Applications*, Vol. 37, No. 6, 2001, pp. 1754–1760.
53. F. Charfi, K. Al Haddad, and F. Sellami, "Study of degradation regimes in the power electronic systems installed in electric vehicles," *Proc. Canadian Conf. on Electrical and Computer Engineering*, Vol. 2, 13–16 May 2001, pp. 759–763.
54. G. Majumdar, K.H. Hussein, K. Takanashi, M. Fukada, J. Yamashita, H. Maekawa, M. Fuku, T. Yamane, and T. Kikunaga, "High-functionality compact intelligent power unit (IPU) for EV/HEV applications," *Proc. 13th Int. Symposium on Power Semiconductor Devices and ICs (ISPSD '01)*, 4–7 June 2001, pp. 315–318.
55. R. Dyerson, A. Pilkington, and O. Tissier, "Technological development and regulation: capability alignment and the electric vehicle," *Proc. Portland Int. Conf. on Management of Engineering and Technology (PICMET '01)*, Vol. 1, 29 July–2 August 2001, p. 49.
56. G. Barba, L. Glielmo, V. Perna, and F. Vasca, "Current sensorless induction motor observer and control for hybrid electric vehicles," *Proc. IEEE 32nd Annual Power Electronics Specialists Conf. (PESC)*, Vol. 2, 17–21 June 2001, pp. 1224–1229.
57. O. Briat, J.M. Vinassa, C. Zardini, and J.L. Aucouturier, "Experimental study of sources hybridization electromechanical storage system integration into an electric vehicle structure," *Proc. 32nd PESC*, Vol. 2, 17–21 June 2001, pp. 1237–1242.
58. W. Dong. J.Y. Choi. F.C. Lee, D. Boroyevich, and J. Lai, "Comprehensive evaluation of auxiliary resonant commutated pole inverter for electric vehicle applications," *Proc. IEEE 32nd PESC*, Vol. 2, 17–21 June 2001, pp. 625–630.
59. J. Cilia, C. Spiteri Staines, V. Buttigieg, C. Caruana, and M. Apap, "Design of an electric vehicle for the Maltese islands," *Proc. 8th IEEE Int. Conf. on Electronics, Circuits and Systems (ICECS)*, Vol. 2, 2–5 September 2001, pp. 647–650.
60. A. Piccolo, L. Ippolito, V.Z. Galdi, and A. Vaccaro, "Optimisation of energy flow management in hybrid electric vehicles via genetic algorithms," *Proc. IEEE/ASME Int. Conf. on Advanced Intelligent Mechatronics*, Vol. 1, 8–12 July 2001, pp. 434–439.
61. J. Soo-Il, K.B. Kim, S.T. Jo, and J.M. Lee. "Driving simulation of a parallel hybrid electric vehicle using receding horizon control," *Proc. ISIE*, Vol. 2, 12–16 June 2001, pp. 1180–1185.
62. S. Shinnaka, S. Takeuchi, A. Kitajima, F. Eguchi, and H. Haruki, "Frequency-hybrid vector control for sensorless induction motor and its application to electric vehicle drive," *Proc. 16th APEC*, Vol. 1, 4–8 March 2001, pp. 32–39.
63. W.R. Warf, M.S. Duvall, B. Moran, and A.A. Frank, "Battery system requirements for battery dominant hybrid electric vehicles," *Proc. 16th Annual Battery Conf. on Applications and Advances*, 9–12 January 2001, pp. 47–53.
64. T. Moore, "Hybrid EVs: making the grid connection," *Proc. 16th Annual Battery Conf. on Applications and Advances*, 9–12 January 2001, pp. 37–43.
65. R.C. Balch, A. Burke, and A.A. Frank, "The effect of battery pack technology and size choices on hybrid electric vehicle performance and fuel economy," *Proc. 16th Annual Battery Conf. on Applications and Advances*, 9–12 January 2001, pp. 31–36.
66. A.A. Pesaran and M. Keyser, "Thermal characteristics of selected EV and HEV batteries," *Proc. 16th Annual Battery Conf. on Applications and Advances*, 9–12 January 2001, pp. 219–225.
67. D.B. Edwards and C. Kinney, "Advanced lead-acid battery designs for hybrid electric vehicles," *Proc. 16th Annual Battery Conf. on Applications and Advances*, 9–12 January 2001, pp. 207–212.
68. C.L. Chu, M.C. Tsai, and H.Y. Chen, "Torque control of brushless DC motors applied to electric vehicles," *Proc. IEMDC*, 17–20 June 2001, pp. 82–87.
69. Q. Wang T. Backstrom, and C. Sadarangani, "A novel drive strategy for hybrid electric vehicles," *Proc. IEMDC*, 17–20 June 2001, pp. 79–81.
70. A. Pilkington, R. Dyerson, O. Tissier, "Patent data as indicators of technological development: investigating the electric vehicle," *Proc. Portland Int. Conf. on Management of Engineering and Technology (PICMET '01)*, Vol. Supplement, 29 July–2 August 2001, pp. 497–502.

71. R.B. Sepe, Jr., C.M. Morrison, J.M. Miller, and A.R. Gale, "High efficiency operation of a hybrid electric vehicle starter/generator over road profiles," *Proc. 36th IAS Annual Meeting IEEE Industry Applications Conf.*, Vol. 2, 30 September–4 October 2001, pp. 921–925.
72. W. Dong, J.Y. Choi, Y. Li, D. Boroyevich, F.C. Lee, J. Lai, and S. Hiti, "Comparative experimental evaluation of soft-switching inverter techniques for electric vehicle drive applications," *Proc. 36th IAS Annual Meeting IEEE Industry Applications Conf.*, Vol. 3, 30 September–4 October 2001, pp. 1469–1476.
73. C.M. Ta, C. Chakraborty, and Y. Hori, "Efficiency maximization of induction motor drives for electric vehicles based on actual measurement of input power," *Proc. 27th Annual Conf. of the IEEE Industrial Electronics Society (IECON '01)*, Vol. 3, 29 November–2 December 2001, pp. 1692–1697.
74. M. Patterson and M. Baxter, "A step-sine motor controller for electric vehicle applications," *Proc. Electrical Insulation, Electrical Manufacturing and Coil Winding Conf.*, 16–18 October 2001, pp. 479–482.
75. F. Koyanagi, Y. Uriu, and R. Yokoyama, "Possibility of fuel cell fast charger and its arrangement problem for the infrastructure of electric vehicles," *Proc. IEEE Power Tech., Porto*, Vol. 4, 10–13 September 2001, p. 6.
76. J. Xu, Y. Xu, and R. Tang, "Development of full digital control system for permanent magnet synchronous motor used in electric vehicle," *Proc. 5th Int. Conf. on Electrical Machines and Systems (ICEMS)*, Vol. 1, 18–20 August 2001, pp. 554–556.
77. X. Ma and Q. Chen, "A novel scheme of propulsion system using soft-switched high-frequency AC-AC converter for electric vehicle," *Proc. 5th ICEMS*, Vol. 1, 18–20 August 2001, pp. 496–499.
78. Y. Xu, J. Xu, W. Wan, and R. Tang, "Development of permanent magnet synchronous motor used in electric vehicle," *Proc. 5th ICEMS 2001*, Vol. 2, 18–20 August 2001, pp. 884–887.
79. K. Lu and E. Ritchie, "Preliminary comparison study of drive motor for electric vehicle application," *Proc. 5th ICEMS*, Vol. 2, 18–20 August 2001, pp. 995–998.
80. K. Xu, X. Guo, and R. Yamamoto, "Development and application of new type plate material NPPS with burrs for electric vehicle battery use," *Proc. 2nd Int. Symposium on Environmentally Conscious Design and Inverse Manufacturing*, 11–15 December 2001, pp. 422–425.
81. R. Dettmer, "Hybrid vigour [hybrid electric vehicles]," *IEEE Review*, Vol. 47, No. 1, January 2001, pp. 25–28.
82. L. Solero, O. Honorati, F. Caricchi, and F. Crescimbini, "Nonconventional three-wheel electric vehicle for urban mobility," *IEEE Trans. Vehicular Technology*, Vol. 50, No. 4, July 2001, pp. 1085–1091.
83. D.H. Cho, H.K. Jung, and C.G. Lee, "Induction motor design for electric vehicle using a niching genetic algorithm," *IEEE Trans. Industry Applications*, Vol. 37, No. 4, July–August 2001, pp. 994–999.
84. L. Solero, "Nonconventional on-board charger for electric vehicle propulsion batteries," *IEEE Trans. Vehicular Technology*, Vol. 50, No. 1, January 2001, pp. 144–149.
85. J. Faiz and M.B.B. Sharifian, "Different techniques for real-time estimation of an induction motor rotor resistance in sensorless direct torque control for electric vehicle," *IEEE Trans. Energy Conversion*, Vol. 16, No. 1, March 2001, pp. 104–109.
86. S.S. Ramamurthy and J.C. Balda, "Sizing a switched reluctance motor for electric vehicles," *IEEE Trans. Industry Applications*, Vol. 37, No. 5, September–October 2001, pp. 1256–1264.
87. T. Harakawa and T. Tujimoto, "Efficient solar power equipment for electric vehicles: improvement of energy conversion efficiency for charging electric vehicles," *Proc. IEEE Int. Vehicle Electronics Conf. (IVEC)*, 25–28 September 2001, pp. 11–16.
88. J. Malan and M.J. Kamper, "Performance of a hybrid electric vehicle using reluctance synchronous machine technology," *IEEE Trans. Industry Applications*, Vol. 37, No. 5, September–October 2001, pp. 1319–1324.
89. T. Markel and K. Wipke, "Modeling grid-connected hybrid electric vehicles using ADVISOR," *Proc. 16th Annual Battery Conf. on Applications and Advances*, 9–12 January 2001, pp. 23–29.

2000

90. J.W. Park, D.H. Koo, J.M. Kim, and H.G. Kim, "Improvement of control characteristics of interior permanent magnet synchronous motor for electric vehicle," *Proc. IEEE Industry Applications Conf.,* Vol. 3, 8–12 October 2000, pp. 1888–1895.
91. J. Malan and M.J. Kamper, "Performance of hybrid electric vehicle using reluctance synchronous machine technology," *Proc. IEEE Industry Applications Conf.,* Vol. 3, 8–12 October 2000, pp. 1881–1887.
92. F.L. Luo and H.G. Yeo, "Advanced PM brushless DC motor control and system for electric vehicles," *Proc. IEEE Industry Applications Conf.,* Vol. 2, 8–12 October 2000, pp. 1336–1343.
93. E. Yamada and Z. Zhao, "Applications of electrical machine for vehicle driving system," *Proc. 3rd Int. Power Electronics and Motion Control Conf. (PIEMC),* Vol. 3, 15–18 August 2000, pp. 1359–1364.
94. N. Tatoh, Y. Hirose, M. Nagai, K. Sasaki, N. Tatsumi, K. Higaki, H. Nakata, T. Tomikawa, A.M. Phillips, M. Jankovic, and K.E. Bailey, "Vehicle system controller design for a hybrid electric vehicle," *Proc. IEEE International Conf. on Control Applications,* 25–27 September 2000, pp. 297–302.
95. D. Kottick, L. Tartakovsky, M. Gutman, and Y. Zvirin, "Results of electric vehicle demonstration program," *Proc. 21st IEEE Convention of the Electrical and Electronic Engineers in Israel,* 11–12 April 2000, pp. 318–321.
96. M. Shino, N. Miyamoto, Y.Q. Wang, and M. Nagai, "Traction control of electric vehicles considering vehicle stability," *Proc. 6th Int. Workshop on Advanced Motion Control,* 30 March–1 April 2000, pp. 311–316.
97. T.E. Lipman, "Manufacturing and life-cycle costs of battery electric vehicles, direct-hydrogen fuel cell vehicles, and direct-methanol fuel cell vehicles," *Proc. 35th Intersociety Energy Conversion Engineering Conf. and Exhibit (IECEC),* Vol. 2, 24–28 July 2000, pp. 1352–1358.
98. S. Rahman and Y. Teklu, "Role of the electric vehicle as a distributed resource," *IEEE Power Engineering Society,* Vol. 1, 23–27 January 2000, pp. 528–533.
99. K.M. Rahman, B. Fahimi, G. Suresh, A.V. Rajarathnam, and M. Ehsani, "Advantages of switched reluctance motor applications to EV and HEV: design and control issues," *IEEE Trans. Industry Applications,* Vol. 36, No. 1, January–February 2000, pp. 111–121.
100. K. Yoshida, Y. Hita, and K. Kesamaru, "Eddy-current loss analysis in PM of surface-mounted-PM SM for electric vehicles," *IEEE Trans. Magnetics,* Vol. 36, No. 4, July 2000, pp. 1941–1944.
101. K.W.E. Cheng, Y.P.B. Yeung, C.Y. Tang, X.D. Xue, and D. Sutanto, "Topology analysis of switched reluctance drives for electric vehicles," *Proc. 8th Int. Conf. on Power Electronics and Variable Speed Drives* (IEEE Conf. Publ. No. 475), 18–19 September 2000, pp. 512–517.
102. W. Dong, J.Y. Choi, Y. Li, H. Yu, J. Lai, D. Boroyevich, and F.C. Lee, "Efficiency considerations of load side soft-switching inverters for electric vehicle applications," *Proc. 15th Annual IEEE Applied Power Electronics Conf. and Exposition (APEC),* Vol. 2, 6–10 February 2000, pp. 1049–1055.
103. C.P. Henze, "Electric vehicle charger with printed circuit board magnetic components," *Proc. 15th Annual IEEE APEC,* Vol. 2, 6–10 February 2000, pp. 796–802.
104. C.C. Hua and M.Y. Lin, "A study of charging control of lead-acid battery for electric vehicles," *Proc. ISIE,* Vol. 1, 4–8 December 2000, pp. 135–140.
105. D. O'Sullivan, M. Willers, M.G. Egan, J.G. Hayes, P.T. Nguyen, and C.P. Henze, "Power-factor-corrected single-stage inductive charger for electric-vehicle batteries," *Proc. IEEE 31st Annual PESC,* Vol. 1, 18–23 June 2000, pp. 509–516.
106. A. Brahma, Y. Guezennec, and G. Rizzoni, "Optimal energy management in series hybrid electric vehicles," *Proc. American Control Conf.,* Vol. 1, No. 6, 28–30 June 2000, pp. 60–64.
107. S.S. Ramamurthy, J.C. Balda, and T. Ericsen, "Sizing a switched reluctance motor for electric vehicles," *Proc. IEEE Industry Applications Conf.,* Vol. 1, 8–12 October 2000, pp. 71–78.
108. T.G. Chang. J.K. Kim, and J.S. Lee, "On-board dynamics failure detection of the two-motor-driven electric vehicle system," *Proc. 52nd VTC,* Vol. 5, 24–28 September 2000, pp. 2047–2053.

109. Y. Yang, M. Parten, J. Berg, and T. Maxwell, "Modeling and control of a hybrid electric vehicle," *Proc. 52nd VTC,* Vol. 5, 24–28 September 2000, pp. 2095–2100.
110. J.Y. Routex, S. Gay-Desharnais, and M. Ehsani, "Modeling of hybrid electric vehicles using gyrator theory: application to design," *Proc. 52nd VTC,* Vol. 5, 24–28 September 2000, pp. 2090–2094.
111. R. Saeks, C. Cox, P. Mays, J. Murray, "Adaptive control of a hybrid electric vehicle," *Proc. IEEE Int. Conf. on Systems, Man, and Cybernetics,* Vol. 4, 8–11 October 2000, pp. 2405–2410.
112. J.M. Hernandez and R.A.P. Moreno, "Design of a variable speed drive with dynamic braking for induction motor for electric vehicles," *Proc. 7th IEEE Int. Power Electronics Congress,* 15–19 October 2000, pp. 211–214.
113. E. Hara, M. Teramoto, and M. Takayama, "The electric vehicle sharing demonstration: the ITS/EV project urban rent-a-car system in the Yokohama Minato Mirai 21 area," *Proc. IEEE Intelligent Vehicles Symposium,* 3–5 October 2000, pp. 116–121.
114. S.E. Lyshevski, A.S.C. Sinha, M. Rizkalla, M. El-Sharkawy, A. Nazarov, P.C. Cho, W. Wylam, J. Mitchell, and M. Friesen, "Analysis and control of hybrid-electric vehicles with individual wheel brushless traction motors," *Proc. of the American Control Conf.,* Vol. 2, 28–30 June 2000, pp. 996–1000.
115. S.I. Sakai, H. Sado, and Y. Hori, "Anti-skid control with motor in electric vehicle," *Proc. 6th Int. Workshop on Advanced Motion Control,* 30 March–1 April 2000, pp. 317–322.
116. H.L. Chan, "A new battery model for use with battery energy storage systems and electric vehicles power systems," *IEEE Power Engineering Society Winter Meeting,* Vol. 1, 23–27 January 2000, pp. 470–475.
117. B. Hredzak and P.S.M. Chin, "Design of a novel multidrive system with reduced torque pulsations for an electric vehicle," *IEEE Power Engineering Society Winter Meeting,* Vol. 1, 23–27 January 2000, pp. 208–212.
118. R.E. MacDougall, J.D. Bertolino, K.L. Rodden, and E.T. Alger, "Lab testing of battery charge management systems for electric and hybrid electric vehicle battery packs to evaluate cycle life improvement," *Proc. the 15th Annual Battery Conference on Applications and Advances,* 11–14 January 2000, pp. 237–242.
119. C.A. Vincent, "Battery systems for electric vehicles," *Proc. IEEE Seminar Electric, Hybrid and Fuel Cell Vehicles* (Ref. No. 2000/050), 11 April 2000, pp. 4/1–4/4.
120. R. Dyerson and A. Pilkington, "Switched on? Patterns of patent activity and the development of the electric vehicle," *Proc. IEEE International Conf. on Management of Innovation and Technology (ICMIT),* Vol. 2, 12–15 November 2000, pp. 726–731.
121. F.A. Wyczalek, "Hybrid electric vehicles — year 2000," *Proc. 35th Intersociety Energy Conversion Engineering Conf. and Exhibit (IECEC),* Vol. 1, 24–28 July 2000, pp. 349–355.
122. C.T. Liu and J.W. Chen, "Design of a disc permanent magnet linear synchronous machine for electric vehicle application," *IEEE Power Engineering Society Winter Meeting,* Vol. 1, 23–27 January 2000, pp. 253–256.
123. J.S. Lee, Y.J. Ryoo, Y.C. Lim, P. Freere, T.G. Kim, S.J. Son, and E.S. Kim, "A neural network model of electric differential system for electric vehicle," *Proc. 26th Annual Conf. of the IEEE Industrial Electronics Society (IECON),* Vol. 1, 22–28 October 2000, pp. 83–88.
124. O. Caumont, P. Le Moigne, C. Rombaut, X. Muneret, and P. Lenain, "Energy gauge for lead-acid batteries in electric vehicles," *IEEE Trans. Energy Conversion,* Vol. 15, No. 3, September 2000, pp. 354–360.
125. M.J. Riezenman, "Fuel cells for the long haul, batteries for the spurts [electric vehicles]," *IEEE Spectrum,* Vol. 38, No. 1, January 2001, pp. 95–97.
126. S. Gair, A. Firth, and J. Hajto, "Design and build of a battery powered electric vehicle incorporating 'Free Light' optical polymers," *Proc. IEEE Seminar Electric, Hybrid and Fuel Cell Vehicles* (Ref. No. 2000/050), 11 April 2000, pp. 9/1–9/5.
127. A. Jack, B. Mecrow, and C. Weimer, "Switched reluctance and permanent magnet motors suitable for vehicle drives — a comparison," *Proc. IEEE Seminar Electric, Hybrid and Fuel Cell Vehicles* (Ref. No. 2000/050), 11 April 2000, pp. 6/1–6/5.

1999

128. A. Pilkington, "Complementary innovation: systems and technologies towards the electric vehicle," *Proc. Portland Int. Conf. on Management of Engineering and Technology (PICMET)*, Vol. 1, 25–29 July 1999, p. 365.
129. S. Sakai, H. Sado, and Y. Hori, "New skid avoidance method for electric vehicle with independently controlled 4 in-wheel motors," *Proc. ISIE*, Vol. 2, 12–16 July 1999, pp. 934–939.
130. W.S. Kim, Y.S. Kim, J.K. Kang, and S.K. Sul, "Electromechanical readhesion control simulator for inverter-driven railway electric vehicle," *Proc. 34th IEEE Industry Applications Conf.*, Vol. 2, 3–7 October 1999, pp. 1026–1032.
131. J.Y. Choi, D. Boroyevich, and F.C. Lee, "A SVM strategy and design of a ZVT three-phase inverter for electric vehicle drive applications," *Proc. 34th IEEE Industry Applications Conf.*, Vol. 1, 3–7 October 1999, pp. 65–71.
132. E.W.C. Lo, D. Sustanto, and C.C. Fok, "Harmonic load flow study for electric vehicle chargers," *Proc. 30th IEEE Int. Conf. on Power Electronics and Drive Systems (PEDS)*, Vol. 1, 27–29 July 1999, pp. 495–500.
133. X. Yan and D. Patterson, "Improvement of drive range, acceleration and deceleration performance in an electric vehicle propulsion system," *Proc. 30th IEEE Int. Conf. PEDS*, Vol. 2, 27 June–1 July 1999, pp. 638–643.
134. M. Maskey, M. Parten, D. Vines, and T. Maxwell, "An intelligent battery management system for electric and hybrid electric vehicles," *Proc. IEEE 49th VTC*, Vol. 2, 16–20 May 1999, pp. 1389–1391.
135. C.T. Liu and H.R. Pan, "Electromagnetic force analyses of a disc permanent magnet linear synchronous machine for electric vehicle application," *Proc. Int. Conf. Electric Machines and Drives (IEMD)*, 9–12 May 1999, pp. 425–427.
136. J.G. Hayes and M.G. Egan, "A comparative study of phase-shift, frequency, and hybrid control of the series resonant converter supplying the electric vehicle inductive charging interface," *Proc. 40th Annual Applied Power Electronics Conf. and Exposition (APEC)*, Vol. 1, 14–18 March 1999, pp. 450–457.
137. J.W. Park, D.H. Koo, J.M. Kim, and H.G. Kim, "High-performance drive unit for 2-motor driven electric vehicle," *Proc. 40th Annual APEC*, Vol. 1, 14–18 March 1999, pp. 443–449.
138. W.R. Cawthorne, P. Famouri, J. Chen. N.N. Clark, T.I. McDaniel, R.J. Atkinson, S. Nandkumar, C.M. Atkinson, and S. Petreanu, "Development of a linear alternator-engine for hybrid electric vehicle applications," *IEEE Trans. Vehicular Technology*, Vol. 48, No. 6, November 1999, pp. 1797–1802.
139. K.L. Butler, M. Ehsani, and P. Kamath, "A Matlab-based modeling and simulation package for electric and hybrid electric vehicle design," *IEEE Trans. Vehicular Technology*, Vol. 48, No. 6, November 1999, pp. 1770–1778.
140. S.R. Ovshinsky, S.K. Dhar, M.A. Fetcenko, D.A. Corrigan, B. Reichman, K. Young, C. Fierro, S. Venkatesan, P. Gifford, and J. Koch, "Advanced materials for next generation NiMH portable, HEV and EV batteries," *IEEE Aerospace and Electronic Systems Magazine*, Vol. 14, No. 5, May 1999, pp. 17–23.
141. R. Saeks, C. Cox, J. Neidhoefer, and D. Escher, "Adaptive critic control of the power train in a hybrid electric vehicle," *Proc. IEEE Midnight-Sun Workshop on Soft Computing Methods in Industrial Applications (SMCia/99)*, 16–18 June 1999, p. 109.
142. Y. Sugii, K. Tsujino, and T. Nagano, "A genetic-algorithm based scheduling method of charging electric vehicles," *Proc. IEEE International Conf. on Systems, Man, and Cybernetics (SMC)*, Vol. 4, 12–15 October 1999, pp. 435–440.
143. B. Carroll, D. Klug, L. Dimitrov, S. Marmara, R. Montemayor, A. Schaffer, P. Shaub, E. Tse, and M. Thompson, "Design of a motor speed controller for a lightweight electric vehicle," *Proc. Electrical Insulation Conf. and Electrical Manufacturing and Coil Winding*, 26–28 October 1999, pp. 559–562.
144. R.B. Sepe, Jr., and N. Short, "Web-based virtual engineering laboratory (VE-LAB) for real-time control of a hybrid electric vehicle starter/alternator," *Proc. 18th Digital Avionics Systems Conf.*, Vol. 2, 24–29 October 1999, pp. 8.B.3-1–8.B.3-7.

145. R.B. Sepe, Jr., J.M. Miller, and A.R. Gale, "Intelligent efficiency mapping of a hybrid electric vehicle starter/alternator using fuzzy logic," *Proc. 18th Digital Avionics Systems Conf.,* Vol. 2, 24–29 October 1999, pp. 8.B.2-1–8.B.2-8.
146. R. Sacks and C. Cox, "Design of an adaptive control system for a hybrid electric vehicle," *Proc. IEEE Int. Conf. on SMC,* Vol. 6, 12–15 October 1999, pp. 1000–1005.
147. S.E. Lyshevski, "Diesel-electric drivetrains for hybrid-electric vehicles: new challenging problems in multivariable analysis and control," *Proc. IEEE Int. Conf. on Control Applications,* Vol. 1, 22–27 August 1999, pp. 840–845.
148. R.B. Sepe, Jr., M. Chamberland, and N. Short, "Web-based virtual engineering laboratory (VE-LAB) for a hybrid electric vehicle starter/alternator," *Proc. 34th IEEE Industry Applications Conf.,* Vol. 4, 3–7 October 1999, pp. 2642–2648.
149. J.T. Brown and M.G. Klein, "Bipolar nickel-metal hydride battery for hybrid electric vehicles," *Proc. 14th Battery Conf. on Applications and Advances,* 12–15 January 1999, pp. 19–24.
150. R. Hobbs, R. Newnham, D. Karner, and F. Fleming, "Development of predictive techniques for determination of remaining life for lead-acid batteries under fast charge," *Proc. 14th Battery Conf. on Applications and Advances,* 12–15 January 1999, pp. 177–188.
151. F. Koyanagi, T. Inuzuka, Y. Uriu, and R. Yokoyama, "Monte Carlo simulation on the demand impact by quick chargers for electric vehicles," *IEEE Power Engineering Society Summer Meeting,* Vol. 2, 18–22 July 1999, pp. 1031–1036.
152. J.M. Miller, B. Wu and E. Strangas, "DSP applications in hybrid electric vehicle powertrain," *Proc. American Control Conf.,* Vol. 3, 2–4 June 1999, pp. 2137–2138.
153. Z. Rahman, K.L. Butler, and M. Ehsani, "Designing parallel hybrid electric vehicles using V-ELPH 2.01," *Proc. American Control Conf.,* Vol. 4, 2–4 June 1999, pp. 2693–2697.
154. W. Turner, M. Parten, D. Vines, J. Jones, and T. Maxwell, "Modeling a PEM fuel cell for use in a hybrid electric vehicle," *Proc. IEEE 49th VTC.,* Vol. 2, 16–20 May 1999, pp. 1385–1388.
155. C.T. Liu and J.W. Chen, "Operating characteristics analysis of a new disc permanent magnet linear synchronous machine for the electric vehicle," *Proc. Int. Conf. Electric Machines and Drives (IEMD),* 9–12 May 1999, pp. 583–585.
156. C.C. Chan, "Engineering philosophy of electric vehicles," *Proc. IEMD,* 9–12 May 1999, pp. 255–257.
157. F. Sun L. Sun J. Zhu, and X. Yu, "Simulation study on synthetical performance of electric vehicles," *Proc. IEEE International Vehicle Electronics Conf. (IVEC '99),* 6–9 September 1999, pp. 338–342.
158. L. Wang and X. Wen, "Dynamic match and optimizing design of electric vehicle powertrain," *Proc. IVEC '99,* 6–9 September 1999, pp. 387–390.
159. S. Sakai, H. Sado, and Y. Hori, "Motion control in an electric vehicle with four independently driven in-wheel motors," *IEEE/ASME Trans. Mechatronics,* Vol. 4, No. 1, March 1999, pp. 9–16.
160. J.G. Hayes, M.G. Egan, J.M.D. Murphy, S.E. Schulz, and J.T. Hall, "Wide-load-range resonant converter supplying the SAE J-1773 electric vehicle inductive charging interface," *IEEE Trans. Industry Applications,* Vol. 35, No. 4, July–August 1999, pp. 884–895.
161. H. Oman, "Making batteries last longer [for electric vehicles]," *IEEE Aerospace and Electronic Systems Magazine,* Vol. 14, No. 9, September 1999, pp. 19–21.
162. J.M. Miller, A.R. Gale, A.V. Sankaran, "Electric drive subsystem for a low-storage requirement hybrid electric vehicle," *IEEE Trans. Vehicular Technology,* Vol. 48, No. 6, November 1999, pp. 1788–1796.
163. C.T. Liu, J.W. Chen, and K.S. Su, "Analytical modeling of a new disc permanent magnet linear synchronous machine for electric vehicles," *IEEE Trans. Magnetics,* Vol. 35, No. 5, September 1999, pp. 4043–4045.
164. G. Rizzoni, L. Guzzella, and B.M. Baumann, "Unified modeling of hybrid electric vehicle drivetrains," *IEEE/ASME Trans. Mechatronics,* Vol. 4, No. 3, September 1999, pp. 246–257.
165. H. Sakamoto, K. Harada, S. Washimiya, K. Takehara, Y. Matsuo, and F. Nakao, "Large airgap coupler for inductive charger [for electric vehicles]," *IEEE Trans. Magnetics,* Vol. 35, No. 5, September 1999, pp. 3526–3528.

166. A. Vamvakari, A. Kandianis, A. Kladas, and S. Manias, "High fidelity equivalent circuit representation of induction motor determined by finite elements for electrical vehicle drive applications," *IEEE Trans. Magnetics*, Vol. 35, No. 3, May 1999, pp. 1857–1860.
167. A. Davis, Z.M. Salameh, and S.S. Eaves, "Comparison of a synergetic battery pack drive system to a pulse width modulated AC induction motor drive for an electric vehicle," *IEEE Trans. Energy Conversion*, Vol. 14, No. 2, June 1999, pp. 245–250.
168. H. Huang and L. Chang, "Electrical two-speed propulsion by motor winding switching and its control strategies for electric vehicles," *IEEE Trans. Vehicular Technology*, Vol. 48, No. 6, March 1999, pp. 607–618.
169. P.S. Kim, and B.Y. Hong, "A method for future cost estimation of hybrid electric vehicle," *Proc. IEEE Int. Conf. on Power Electronics and Drive Systems (PEDS)*, Vol. 1, 27–29 July 1999, pp. 315–320.
170. D.W. Dees, "Overview of electrochemical power sources for electric and hybrid/electric vehicles," *Proc. Int. Conf. Electric Machines and Drives (IEMD)*, 9–12 May 1999, pp. 258–259.
171. H. Sado, S. Sakai, and Y. Hori, "Road condition estimation for traction control in electric vehicle," *Proc. IEEE International Symposium on Industrial Electronics (ISIE)*, Vol. 2, 12–16 July 1999, pp. 973–978.
172. S.Q. Xiao, P.M. Chun, Weng Fei Bing, "Application of BIT design for electric vehicle (EV)," *Proc. IEEE Int. Conf. on PEDS*, Vol. 1, 27–29 July 1999, pp. 516–518.
173. L.U. Gokdere, K. Benlyazid, E. Santi, C.W. Brice, and R.A. Dougal, "Hybrid electric vehicle with permanent magnet traction motor: a simulation model," *Proc. IEMD*, 9–12 May 1999, pp. 502–504.
174. X. Xu, "Automotive power electronics-opportunities and challenges [for electric vehicles]," *Proc. IEMD*, 9–12 May 1999, pp. 260–262.
175. F.A. Wyczalek, "Market mature 1998 hybrid electric vehicles," *IEEE Aerospace and Electronic Systems Magazine*, Vol. 14, No. 3, March 1999, pp. 41–44.
176. L.A. Pecorelli Peres, G. Lambert-Torres, and L.A. Horta Nogueira, "Electric vehicles impacts on daily load curves and environment," *Proc. Int. Conf. on Electric Power Engineering, Power Tech Budapest*, 29 August–2 September 1999, p. 55.
177. L. Guzzella and A. Amstutz, "CAE tools for quasi-static modeling and optimization of hybrid powertrains," *IEEE Trans. Vehicular Technology*, Vol. 48, No. 6, November 1999, pp. 1762–1769.

1998

178. J. Malan, M.J. Kamper, and P.N.T. Williams, "Reluctance synchronous machine drive for hybrid electric vehicle," *Proc. ISIE*, Vol. 2, 7–10 July 1998, pp. 367–372.
179. M.G. Simoes and P. Vieira, Jr., "A high torque low-speed multiphase brushless machine — a perspective application for electric vehicles," *Proc. 26th Annual Conf. of the IEEE Industrial Electronics Society (IECON)*, Vol. 2, 22–28 October 2000, pp. 1395–1400.
180. A. Heider and H.J. Haubrich, "Impact of wide-scale EV charging on the power supply network," *Proc. IEEE Colloquium on Electric Vehicles — A Technology Roadmap for the Future* (Digest No. 1998/262), 5 May 1998, pp. 6/1–6/4.
181. J.G. Hayes, "Battery charging systems for electric," *Proc. IEEE Colloquium on Electric Vehicles — A Technology Roadmap for the Future* (Digest No. 1998/262), 5 May 1998, pp. 4/1–4/8.
182. S. Gair, "Motor drives and propulsion systems," *Proc. IEEE Colloquium on Electric Vehicles — A Technology Roadmap for the Future* (Digest No. 1998/262), 5 May 1998, pp. 3/1–3/6.
183. M.E. Rizkalla, C.F. Yokomoto, A.S.C. Sinha, M. El-Sharkawy, S. Lyshevski, and J. Simson, "A new EE curriculum in electric vehicle applications," *Proc. Midwest Symposium on Circuits and Systems*, 9–12 August 1998, pp. 186–189.
184. M.S.W. Chan, K.T. Chau, and C.C. Chan, "Modeling of electric vehicle chargers," *Proc. 26th Annual Conf. of the IEEE Industrial Electronics Society (IECON)*, 31 August–4 September 1998, pp. 433–438.

Electric Vehicles

185. H. Oman, "Solar energy for electric vehicles: systems analysis," *Proc. 17th AIAA/IEEE/SAE Digital Avionics Systems Conf. (DASC),* Vol. 2, 31 October–7 November 1998, pp. I42/1–I42/6.
186. S.K. Biradar, R.A. Patil, and M. Ullegaddi, "Energy storage system in electric vehicle," *Power Quality '98,* 1998, pp. 247–255.
187. I. Edward, S. Wahsh, and M.A. Badr, "Analysis of PMSM drives for electric vehicles," *Proc. 37th SICE Annual Conf.,* 29–31 July 1998, pp. 979–984.
188. S. Sakai and Y. Hori, "Robustified model matching control for motion control of electric vehicle," *Proc. 5th Int. Workshop on Advanced Motion Control (AMC),* Coimbra, 29 June–1 July 1998, pp. 574–579.
189. B. Dickinson, J. Baer, O.A. Velev, and D. Swan, "Performance, management and testing requirements for hybrid electric vehicle batteries," *Proc. 13th Annual Battery Conf. on Applications and Advances,* 13–16 January 1998, pp. 133–139.
190. W.B. Gu, C.Y. Wang, and B.Y. Liaw, "Integrated simulation and testing of electric vehicle batteries," *Proc. 13th Annual Battery Conference on Applications and Advances,* 13–16 January 1998, pp. 141–146.
191. G. Friedrich and M. Kant, "Choice of drives for electric vehicles: a comparison between two permanent magnet AC machines," *IEEE Proc. — Electric Power Applications,* Vol. 145, No. 3, May 1998, pp. 247–252.
192. J.P. Cornu, "High-energy batteries for EVs — a whole solution to clean up the environment," *Proc. IEEE Colloquium on Electric Vehicles — A Technology Roadmap for the Future* (Digest No. 1998/262), 5 May 1998, pp. 5/1–515.
193. D. Phillips, "Transportation [Technology 1998 analysis and forecast]," *IEEE Spectrum,* Vol. 35, No. 1, January 1998, pp. 84–89.
194. C. Chen and X. Xu, "Modeling the conducted EMI emission of an electric vehicle (EV) traction drive," *Proc. IEEE Int. Symposium on Electromagnetic Compatibility,* Vol. 2, 24–28 August 1998, pp. 796–801.
195. J.M. Miller, A.R. Gale, P.J. McCleer, F. Leonardi, and J.H. Lang, "Starter-alternator for hybrid electric vehicle: comparison of induction and variable reluctance machines and drives," *Proc. 33rd IAS Annual IEEE Industry Applications Conf.,* Vol. 1, 12–15 October 1998, pp. 513–523.
196. J.Y. Choi, M.A. Herwald, D. Boroyevich, and F.C. Lee, "Effect of switching frequency of soft switched inverter on electric vehicle system," *Proc. Power Electronics in Transportation,* 22–23 October 1998, pp. 63–69.
197. W.C. Morchin, "Energy management in hybrid electric vehicles," *Proc. 17th AIAA/IEEE/SAE Digital Avionics Systems Conf. (DASC),* Vol. 2, 31 October–7 November 1998, pp. I41/1–I41/6.
198. M. Bojrup, P. Karlsson, M. Alakula, and B. Simonsson, "A dual-purpose battery charger for electric vehicles," *Proc. 29th Annual IEEE Power Electronics Specialists Conf. (PESC),* Vol. 1, 17–22 May 1998, pp. 565–570.
199. S.E. Lyshevski and C. Yokomoto, "Control of hybrid-electric vehicles," *Proc. American Control Conf.,* Vol. 4, 24–26 June 1998, pp. 2148–2149.
200. K. Sugimori and H. Nishimura, "A novel contact-less battery charger for electric vehicles," *Proc. 29th Annual IEEE Power Electronics Specialists Conf. (PESC),* Vol. 1, 17–22 May 1998, pp. 559–564.
201. B. Hredzak, P.S.M. Chin, and S. Gair, "Latest developments in direct EV wheel drives," *Proc. Int. Conf. on EMPD,* Vol. 2, 3–5 March 1998, pp. 714–717.
202. K. Yoshimoto, A. Kawamura, and N. Hoshi, "Traction control of anti-directional-twin-rotary motor drive based on electric vehicle driving simulator," *Proc. 29th PESC,* Vol. 1, 17–22 May 1998, pp. 578–582.
203. B. Kattentidt, "Intelligent electromechanical drive systems for motor cars," *Proc. 24th IECON,* Vol. 4, 31 August–4 September 1998, pp. 2295–2300.
204. M. Parent, "Distributed motion control of an electric vehicle," *Proc. 5th Int. Workshop on Advanced Motion Control (AMC),* Coimbra, 29 June–1 July 1998, p. 573.
205. X. Jing, I. Celanovic, and D. Borojevic, "Device evaluation and filter design for 20 kW inverter for hybrid electric vehicle applications," *Proc. Power Electronics in Transportation,* 22–23 October 1998, pp. 29–36.

206. G.D. Sugavanam, M.M. Morcos, and N.G. Dillman, "Performance of multiphase microprocessor-based battery charger for electric vehicles," *Proc. IEEE Canadian Conf. on Electrical and Computer Engineering,* Vol. 1, 24–28 May 1998, pp. 69–72.
207. J.O. Osborn, "The use of advanced batteries for meeting California's ZEV requirement," *Proc. 13th Annual Battery Conf. on Applications and Advances,* 13–16 January 1998, pp. 37–40.
208. I. Menjak, P.H. Gow, D.A. Corrigan, S. Venkatesan, S.K. Dhar, R.C. Stempel, and S.R. Ovshinsky, "Advanced Ovonic high-power nickel-metal hydride batteries for hybrid electric vehicle applications," *Proc. 13th Annual Battery Conf. on Applications and Advances,* 13–16 January 1998, pp. 13–18.
209. K. Hansen, R. Bailey, M. Rykiel, and H. Reed, "Battery management: the tortoise and hare approach to racing electric vehicles," *Proc. 13th Annual Battery Conf. on Applications and Advances*, 13–16 January 1998, pp. 357–362.
210. M.G. Simoes, N.N. Franceschetti, and J.C. Adamowski, "Drive system control and energy management of a solar powered electric vehicle," *Proc. 13th Annual Applied Power Electronics Conf. and Exposition (APEC),* Vol. 1, 15–19 February 1998, pp. 49–55.
211. C.S. Namuduri and B.V. Murty, "High power density electric drive for a hybrid electric vehicle," *Proc. 13th APEC,* Vol. 1, 15–19 February 1998, pp. 34–40.
212. B.K. Powell, K.E. Bailey, and S.R. Cikanek, "Dynamic modeling and control of hybrid electric vehicle powertrain systems," *IEEE Control Systems Magazine,* Vol. 18, No. 5, October 1998, pp. 17–33.
213. T. Gilchrist, "Fuel cells to the fore [Electric vehicles]," *IEEE Spectrum,* Vol. 35, No. 11, November 1998, pp. 35–40.
214. G.I. Hunt, "The great battery search [Electric vehicles]," *IEEE Spectrum,* Vol. 35, No. 11, November 1998, pp. 21–28.
215. D. Hermance and S. Sasaki, "Hybrid electric vehicles take to the streets," *IEEE Spectrum,* Vol. 35, No. 11, November 1998, pp. 48–52.
216. H.D. Lee and S.K. Sul, "Fuzzy-logic-based torque control strategy for parallel-type hybrid electric vehicle," *IEEE Trans. Industrial Electronics,* Vol. 45, No. 4, August 1998, pp. 625–632.
217. P.T. Staats, W.M. Grady, A. Arapostathis, and R.S. Thallam, "A statistical analysis of the effect of electric vehicle battery charging on distribution system harmonic voltages," *IEEE Trans. Power Delivery,* Vol. 13, No. 2, April 1998, pp. 640–646.
218. F. Koyanagi and Y. Uriu, "A strategy of load leveling by charging and discharging time control of electric vehicles," *IEEE Trans. Power Systems,* Vol. 13, No. 3, August 1998, pp. 1179–1184.
219. G. Dancygier and J.C. Dolhagaray, "Motor control law and comfort law in the Peugeot and Citroen electric vehicles driven by a DC commutator motor," *Proc. 7th Int. Conf. on Power Electronics and Variable Speed Drives* (IEEE Conf. Publ., No. 456), 21–23 September 1998, pp. 370–374.
220. G. Magetto, "Electric vehicle technology: a worldwide perspective," *Proc. IEEE Colloquium on Electric Vehicles — A Technology Roadmap for the Future* (Digest No. 1998/262), 5 May 1998, pp. 1/1–110.
221. L.M. Tolbert, F.Z. Peng, and T.G. Habetler, "Multilevel inverters for electric vehicle applications," *Proc. Power Electronics in Transportation,* 22–23 October 1998, pp. 79–84.
222. H. Oman, "On-board energy and power management on electric vehicles: effect of battery type," *Proc. 17th AIAA/IEEE/SAE Digital Avionics Systems Conf. (DASC),* Vol. 2, 31 October–7 November 1998, pp. I43/1–I43/6.
223. Y. Hori, Y. Toyoda, and Y. Tsuruoka, "Traction control of electric vehicle: basic experimental results using the test EV UOT electric march," *IEEE Trans. Industry Applications,* Vol. 34, No. 5, September–October 1998, pp. 1131–1138.
224. M.J. Riezenman, "Engineering the EV future," *IEEE Spectrum,* Vol. 35, No. 11, November 1998, pp. 18–20.
225. P.H. Mellor, A.J. Brown, J. Li, and K. Atallah, "Modeling of power electronic converters used in electric vehicles," *Proc. IEEE Colloquium on Power Electronic Systems Simulation* (Ref. No. 1998/486), 23 November 1998, pp. 1/1–1/3.

1997

226. J. Gill, "Advanced EV/HEV battery pack testing using the ABC-150 power system," *Proc. 12th Annual Battery Conf. on Applications and Advances*, 14–17 January 1997, pp. 127–131.

227. M. Ehsani, K.M. Rahman, M.D. Bellar, and A. Severinsky, "Evaluation of soft switching for EV and HEV motor drives," *Proc. 23rd Int. Conf. on Industrial Electronics, Control and Instrumentation (IECON)*, Vol. 2, 9–14 November 1997, pp. 651–657.

228. A. Palumbo, T. Waggoner, J. Major, and B. Piersol, "Electric and hybrid electric vehicles: new product development through university/industry collaboration," *Proc. Electrical Insulation, and Electrical Manufacturing and Coil Winding Conf.*, 22–25 September 1997, pp. 445–448.

229. Y. Hori, Y. Toyoda, and Y. Tsuruoka, "Traction control of electric vehicle based on the estimation of road surface condition-basic experimental results using the test EV UOT Electric March," *Proc. Power Conversion Conf.*, Nagaoka, Vol. 1, 3–6 August 1997, pp. 1–8.

230. J.M. Lee and B.H. Cho, "Modeling and simulation of electric vehicle power system," *Proc. 32nd IECEC*, Vol. 3, 27 July–1 August 1997, pp. 2005–2010.

231. H.D. Lee and S.K. Sul, "Diesel engine ripple torque minimization for parallel type hybrid electric vehicle," *Proc. 32nd IAS Annual IEEE Industry Applications Conf.*, Vol. 2, 5–9 October 1997, pp. 942–946.

232. N. Mutoh, S. Kaneko, T. Miyazaki, R. Masaki, and S. Obara, "A torque controller suitable for electric vehicles," *IEEE Trans. Industrial Electronics*, Vol. 44, No. 1, February 1997, pp. 54–63.

233. M. Ehsani, K.M. Rahman, and H.A. Toliyat, "Propulsion system design of electric and hybrid vehicles," *IEEE Trans. Industrial Electronics*, Vol. 44, No. 1, February 1997, pp. 19–27.

234. P.T. Staats, W.M. Grady, A. Arapostathis, and R.S. Thallam, "A procedure for derating a substation transformer in the presence of widespread electric vehicle battery charging," *IEEE Trans. Power Delivery*, Vol. 12, No. 4, October 1997, pp. 1562–1568.

235. H. Huang and L. Chang, "Tests of electrical-two-speed propulsion by induction motor winding switching for electric vehicles," *Proc. IEEE 47th Vehicular Technology Conf.*, Vol. 3, 4–7 May 1997, pp. 1907–1911.

236. H. Huang and L. Chang, "An error-driven controller for electric vehicle propulsion systems," *Proc. IEEE Canadian Conf. on Electrical and Computer Engineering*, Vol. 2, 25–28 May 1997, pp. 744–747.

237. F.G. Pavuza, G. Beszedics, W. Toriser, M. Wawra, and W. Winkler, "A PLD-controlled multispeed drive circuit for small electric vehicles," *Proc. IEEE Canadian Conf. on Electrical and Computer Engineering*, Vol. 2, 25–28 May 1997, pp. 757–760.

238. D.J. Patterson, "High efficiency permanent magnet drive systems for electric vehicles," *Proc. 23rd Int. Conf. on Industrial Electronics, Control and Instrumentation (IECON)*, Vol. 2, 9–14 November 1997, pp. 391–396.

239. D.C. Katsis, M.A. Herwald, J.Y. Choi, D. Boroyevich, and F.C. Lee, "Drive cycle evaluation of soft-switched electric vehicle inverter," *Proc. 23rd IECON*, Vol. 2, 9–14 November 1997, pp. 658–663.

240. F.G. Pavuza, G. Beszedics, W. Toriser, M. Wawra, and W. Winkler, "3-speed drive for small electric vehicles," *Proc. 32nd Intersociety Energy Conversion Engineering Conf. (IECEC)*, Vol. 3, 27 July–1 August 1997, pp. 2011–2013.

241. L. Lisowski and G. Baille, "Specifications of a small electric vehicle: modular and distributed approach," *Proc. IEEE/RSJ Int. Conf. on Intelligent Robots and Systems (IROS)*, Vol. 2, 7–11 September 1997, pp. 919–925.

242. B.R. Borchers and J.A. Locker, "Electrical system design of a solar electric vehicle," *Proc. Electrical Insulation, Electrical Manufacturing and Coil Winding Conf.*, 22–25 September 1997, pp. 699–704.

243. S. Kawano, H. Murakami, N. Nishiyama, Y. Ikkai, Y. Honda, and T. Higaki, "High-performance design of an interior permanent magnet synchronous reluctance motor for electric vehicles," *Proc. Power Conversion Conf.*, Nagaoka, Vol. 1, 3–6 August 1997, pp. 33–36.

244. M. Mori, T. Mizuno, T. Ashikaga, and I. Matsuda, "A control method of an inverter-fed six-phase pole change induction motor for electric vehicles," *Proc. Power Conversion Conf.,* Nagaoka, Vol. 1, 3–6 August 1997, pp. 25–32.
245. N. Hoshi, K. Yoshimoto, and A. Kawamura, "Driving characteristics of antidirectional twin-rotary motor on electric vehicle simulator," *Proc. Power Conversion Conf.,* Nagaoka, Vol. 1, 3–6 August 1997, pp. 19–24.
246. S. Vaez, V.I. John, and M.A. Rahman, "Energy saving vector control strategies for electric vehicle motor drives," *Proc. Power Conversion Conf.,* Nagaoka, Vol. 1, 3–6 August 1997, pp. 13–18.
247. J. Jung, K. Nam, C. Han, J. Chung, and D. Ahn, "A new vector control scheme considering iron loss for electric vehicle induction motors," *Proc. 32nd IAS Annual IEEE Industry Applications Conf.,* Vol. 1, 5–9 October 1997, pp. 439–444.
248. Y. Honda, T. Nakamura, T. Higaki, and Y. Takeda, "Motor design considerations and test results of an interior permanent magnet synchronous motor for electric vehicles," *Proc. 32nd IAS Annual IEEE Industry Applications Conf.,* Vol. 1, 5–9 October 1997, pp. 75–82.
249. G. Guidi, H. Kubota, and Y. Hori, "Induction motor control for electric vehicle application using low-resolution position sensor and sensorless vector control technique," *Proc. Power Conversion Conf.,* Nagaoka, Vol. 2, 3–6 August 1997, pp. 937–942.
250. R. Mital, R.K. Sievers, and T.K. Hunt, "Ultraclean burner for an AMTEC system suitable for hybrid electric vehicles," *Proc. 32nd Intersociety Energy Conversion Engineering Conf. (IECEC),* 27 July–1 August 1997, pp. 961–966.
251. K. Nasr, M. Tavakoli, M. Thompson, and C. Jordan, "High-speed electric vehicle," *Proc. 32nd Intersociety Energy Conversion Engineering Conf. (IECEC),* Vol. 3, 27 July–1 August 1997, pp. 2024–2029.
252. M. Staackmann, Y.L. Bor, and D.Y.Y. Yun, "Dynamic driving cycle analyses using electric vehicle time-series data," *Proc. 32nd Intersociety Energy Conversion Engineering Conf. (IECEC),* Vol. 3, 27 July–1 August 1997, pp. 2014–2018.
253. G. Metrakos, H. Hong, and T. Krepec, "Strategy for increasing hybrid electric vehicle reversibility," *Proc. American Control Conf.,* Vol. 1, 4–6 June 1997, pp. 694–698.
254. S.R. Cikanek, K.E. Bailey, and B.K. Powell, "Parallel hybrid electric vehicle dynamic model and powertrain control," *Proc. American Control Conf.,* Vol. 1, 4–6 June 1997, pp. 684–688.
255. G.A. Hubbard and K. Youcef-Toumi, "System level control of a hybrid-electric vehicle drivetrain," *Proc. American Control Conf.,* Vol. 1, 4–6 June 1997, pp. 641–645.
256. G.A. Hubbard and K. Youcef-Toumi, "Modeling and simulation of a hybrid-electric vehicle drivetrain," *Proc. American Control Conf.,* Vol. 1, 4–6 June 1997, pp. 636–640.
257. C.C. Chan, R. Zhang, K.T. Chau, and J.Z. Jiang, "Optimal efficiency control of PM hybrid motor drives for electrical vehicles," *Proc. 28th Annual IEEE Power Electronics Specialists Conf. (PESC),* Vol. 1, 22–27 June 1997, pp. 363–368.
258. M. Terashima, T. Ashikaga, T. Mizuno, K. Natori, N. Fujiwara, and M. Yada, "Novel motors and controllers for high-performance electric vehicle with four in-wheel motors," *IEEE Trans. Industrial Electronics,* Vol. 44, No. 1, February 1997, pp. 28–38.
259. H. Shimizu, J. Harada, C. Bland, K. Kawakami, and L. Chan, "Advanced concepts in electric vehicle design," *IEEE Trans. Industrial Electronics,* Vol. 44, No. 1, February 1997, p. 1418.
260. C.C. Chan and K.T. Chau, "An overview of power electronics in electric vehicles," *IEEE Trans. Industrial Electronics,* Vol. 44, No. 1, February 1997, pp. 3–13.
261. D.K. Jackson, A.M. Schultz, S.B. Leeb, A.H. Mitwalli, G.C. Verghese, and S.R. Shaw, "A multirate digital controller for a 1.5 kW electric vehicle battery charger," *IEEE Trans. Power Electronics,* Vol. 12, No. 6, November 1997, pp. 1000–1006.
262. W.A. Lynch and Z.M. Salameh, "Realistic electric vehicle battery evaluation," *IEEE Trans. Energy Conversion,* Vol. 12, No. 4, December 1997, pp. 407–412.
263. P.T. Staats, W.M. Grady, A. Arapostathis, and R.S. Thallam, "A statistical method for predicting the net harmonic currents generated by a concentration of electric vehicle battery chargers," *IEEE Trans. Power Delivery,* Vol. 12, No. 3, July 1997, pp. 1258–1266.

264. S. Henneberger, U. Pahner, K. Hameyer, and R. Belmans, "Computation of a highly saturated permanent magnet synchronous motor for a hybrid electric vehicle," *IEEE Trans. Magnetics,* Vol. 33, No. 5, September 1997, pp. 4086–4088.
265. N.H. Kutkut and K.W. Klontz, "Design considerations for power converters supplying the SAE J-1773 electric vehicle inductive coupler," *Proc. 12th Annual Applied Power Electronics Conf. and Exposition (APEC),* Vol. 2, 23–27 February 1997, pp. 841–847.
266. T. Backstrom, C. Sadarangani, and S. Ostlund, "Integrated energy transducer for hybrid electric vehicles," *Proc. 8th Int. Conf. on Electrical Machines and Drives* (Conf. Publ. No. 444), 1–3 September 1997, pp. 239–243.
267. J.B. Olson and E.D. Sexton, "A high-power spiral-wound lead-acid battery for hybrid electric vehicles," *Proc. 12th Annual Battery Conf. on Applications and Advances,* 14–17 January 1997, pp. 145–149.
268. T. Erekson, "Electric and hybrid electric vehicles: new markets for electrical manufacturers," *Proc. Electrical Insulation, Electrical Manufacturing and Coil Winding Conf.,* 22–25 September 1997, pp. 441–444.
269. S. Imai, N. Takeda, and Y. Horii, "Total efficiency of a hybrid electric vehicle," *Proc. Power Conversion Conf.,* Nagaoka, Vol. 2, 3–6 August 1997, pp. 947–950.
270. A. Kawamura, N. Hoshi, T.W. Kim, T. Yokoyama, and T. Kume, "Analysis of antidirectional-twin-rotary motor drive characteristics for electric vehicles," *IEEE Trans. Industrial Electronics,* Vol. 44, No. 1, February 1997, pp. 64–70.
271. H.C. Lovatt, V.S. Ramsden, and B.C. Mecrow, "Design of an in-wheel motor for a solar-powered electric vehicle," *Proc. 8th Int. Conf. on Electrical Machines and Drives* (Conf. Publ. No. 444), 1–3 September 1997, pp. 234–238.

1996

272. B.J. Chalmers and I. Musaba, "Performance characteristics of permanent-magnet and reluctance machines to meet EV requirements," *Proc. IEEE Colloquium on Machines and Drives for Electric and Hybrid Vehicles* (Digest No. 1996/152), 28 June 1996, pp. 5/1–5/5.
273. N. Schofield and M.K. Jenkins, "High performance brushless permanent magnet traction drives for hybrid electric," *Proc. IEEE Colloquium on Machines and Drives for Electric and Hybrid Vehicles* (Digest No. 1996/152), 28 June 1996, pp. 4/1–4/6.
274. M. Vaidya, E.K. Stefanakos, B. Krakow, H.C. Lamb, T. Arbogast, and T. Smith, "Direct DC/DC electric vehicle charging with a grid-connected photovoltaic system," *Proc. 25th IEEE Photovoltaic Specialists Conf.,* 13–17 May 1996, pp. 1505–1508.
275. B. Schumm, Jr., "Rechargeable zinc/manganese dioxide alkaline cells response to electric vehicle type testing," *Proc. 31st Intersociety Energy Conversion Engineering Conf. (IECEC),* Vol. 2, 11–16 August 1996, pp. 1181–1185.
276. J.R. Goldstein, B. Koretz, and Y. Harats, "Field test of the Electric Fuel™ zinc-air refuelable battery system for electric vehicles," *Proc. 31st Intersociety Energy Conversion Engineering Conf. (IECEC),* Vol. 3, 11–16 August 1996, pp. 1925–1929.
277. F.A. Wyczalek, "Hybrid electric vehicles in Europe and Japan," *Proc. 31st IECEC,* Vol. 3, 11–16 August 1996, pp. 1919–1924.
278. F. Profumo, M. Madlena, and G. Griva, "State variables controller design for vibrations suppression in electric vehicles," *Proc. 27th PESC,* Vol. 2, 23–27 June 1996, pp. 1940–1947.
279. W.R. Young, Jr., "Tracking electric vehicles with GPS," *Proc. Southcon/96 Conf.,* 25–27 June 1996, pp. 285–289.
280. K. Jezernik, "Robust induction motor control for electric vehicles," *Proc. 4th Int. Workshop on Advanced Motion Control (AMC),* Vol. 2, 18–21 March 1996, pp. 436–440.
281. J.L. Arias, D.L. Harbaugh, E.D. Drake, and D.W. Boughn, "A sealed bipolar lead-acid battery for small electric vehicles," *Proc. 11th Annual Battery Conference on Applications and Advances,* 9–12 January 1996, pp. 179–182.

282. P.T. Krein, "Electrostatic discharge issues in electric vehicles," *IEEE Trans. Industry Applications,* Vol. 32, No. 6, November–December 1996, pp. 1278–1284.
283. F.A. Wyczalek, "Ultralight electric vehicle parameters," *IEEE Aerospace and Electronic Systems Magazine,* Vol. 11, No. 1, January 1996, pp. 40–44.
284. J.J. Bardyn, T. Kunemund, A. Papaspyridis, E. Protonotarios, J.F. Sarrau, and L. Schrader, "Energy control system for electro mobile," *Proc.3rd IEEE Int. Conf. on Electronics, Circuits, and Systems (ICECS),* Vol. 1, 13–16 October 1996, pp. 614–617.
285. M.K. Park, I.H. Suh, S.J. Byoun, S.R. Oh, "An intelligent coordinated control system for steering and traction of electric vehicles," Proc. *IEEE 22nd Int. Conf. on Industrial Electronics, Control, and Instrumentation (IECON),* Vol. 3, 5–10 August 1996, pp. 1972–1977.
286. Y.H. Kim, Y.S. Yoo, and H.D. Ha, "Interface circuits for regenerative operation of an electric vehicle," *Proc. IEEE 22nd IECON,* Vol. 1, 5–10 August 1996, pp. 53–58.
287. M.A. Rahman and R. Qin, "A permanent magnet hysteresis hybrid motor drive for electric vehicles," *Proc. IEEE 22nd IECON,* Vol. 1, 5–10 August 1996, pp. 28–33.
288. M. Terashima, T. Ashikagi, T. Mizuno, K. Natori, N. Fujiwara, and M. Yada, "Novel motors and controllers for high-performance electric vehicle with four in-wheel motors," *Proc. 22nd IECON,* Vol. 1, 5–10 August 1996, pp. 20–27.
289. C.C. Chan and K.T. Chau, An overview of electric vehicles — challenges and opportunities," *Proc. IEEE 22nd IECON,* Vol. 1, 5–10 August 1996, pp. 1–6.
290. R.L. Proctor, "Accurate and user-friendly state-of-charge instrumentation for electric vehicles," *Proc. Northcon/96,* 4–6 November 1996, pp. 379–384.
291. J.G. Ingersoll and C.A. Perkins, "The 2.1 kW photovoltaic electric vehicle charging station in the city of Santa Monica, California," *Proc. 25th IEEE Photovoltaic Specialists Conf.,* 13–17 May 1996, pp. 1509–1512.
292. C.P. Quigley, R.J. Ball, A.M. Vinsome, and R.P. Jones, "Predicting journey parameters for the intelligent control of a hybrid electric vehicle," *Proc. IEEE Int. Symposium on Intelligent Control,* 15–18 September 1996, pp. 402–407.
293. Y. Chen, L. Song, and J.W. Evans, "Modeling studies on battery thermal behaviour, thermal runaway, thermal management, and energy efficiency," *Proc. 31st Intersociety Energy Conversion Engineering Conf. (IECEC),* Vol. 2, 11–16 August 1996, pp. 1465–1470.
294. B.K. Bose, M.H. Kim, and M.D. Kankam, "Power and energy storage devices for next-generation hybrid electric vehicle," *Proc. 31st IECEC,* Vol. 3, 11–16 August 1996, pp. 1893–1898.
295. G.H. Chen and K.J. Tseng, "Design of a permanent-magnet direct-driven wheel motor drive for electric vehicle," *Proc. 27th PESC,* Vol. 2, 23–27 June 1996, pp. 1933–1939.
296. A.M. Schultz, S.B. Leeb, A.H. Mitwalli, D.K. Jackson, and G.C. Verghese, "A multirate digital controller for an electric vehicle battery charger," *Proc. 27th PESC,* Vol. 2, 23–27 June 1996, pp. 1919–1925.
297. J.G. Hayes, J.T. Hall, M.C. Egan, and J.M.D. Murphy, "Full-bridge, series-resonant converter supplying the SAE J-1773 electric vehicle inductive charging interface," *Proc. 27th PESC,* Vol. 2, 23–27 June 1996, pp. 1913–1918.
298. S.H. Berisha, G.G. Karady, R. Ahmad, R. Hobbs, and D. Karner, "Current harmonics generated by electric vehicle battery chargers," *Proc. 27th PESC,* Vol. 1, 8–11 January 1996, pp. 584–589.
299. W.R. Young, Jr. and W. Wilson, "Efficient electric vehicle lighting using LEDs," *Proc. Southcon/96 Conf.,* 25–27 June 1996, pp. 276–280.
300. B.J. Masserant and T.A. Stuart, "Online computation of T_j for EV battery chargers," *Proc. IEEE Workshop on Computers in Power Electronics,* 11–14 August 1996, pp. 152–156.
301. T. Furuya, Y. Toyoda, and Y. Hori, "Implementation of advanced adhesion control for electric vehicle," *Proc. 4th Int. Workshop on Advanced Motion Control (AMC),* Vol. 2, 18–21 March 1996.
302. N. Hoshi and A. Kawamura, "Experimental discussion on the permanent magnet type antidirectional-twin-rotary motor drive for electric vehicle," *Proc. 4th Int. Workshop on AMC,* Vol. 2, 18–21 March 1996, pp. 425–429.

Electric Vehicles

303. R. Severns, E. Yeow, G. Woody, J. Hall, and J. Hayes, "An ultracompact transformer for a 100 W to 120 kW inductive coupler for electric vehicle battery charging," *Proc. 11th Annual Applied Power Electronics Conf. and Exposition (APEC),* Vol. 1, 3–7 March 1996, pp. 32–38.
304. R.F. Nelson, "Bolder TMF technology applied to hybrid electric vehicle power-assist operation," *Proc. 11th Annual Battery Conf. on Applications and Advances*, 9–12 January 1996, pp. 173–177.
305. W.B. Craven, "Horizon sealed lead-acid battery in electric vehicle application," *Proc. 11th Annual Battery Conf. on Applications and Advances,* 9–12 January 1996, pp. 159–162.
306. K. Kurani, D. Sperling, and T. Turrentine, "The marketability of electric vehicles: battery performance and consumer demand for driving range," *Proc. 11th Annual Battery Conf. on Applications and Advances,* 9–12 January 1996, pp. 153–158.
307. B.J. Chalmers, L. Musaba, and D.F. Gosden, "Variable-frequency synchronous motor drives for electric vehicles," *IEEE Trans. Industry Applications*, Vol. 32, No. 4, July–August 1996, pp. 896–903.
308. Z. Zhang, F. Profumo, and A. Tenconi, "Improved design for electric vehicle induction motors using an optimisation procedure," *IEEE Proc. Electric Power Applications,* Vol. 143, No. 6, November 1996, pp. 410–416.
309. H.W. Gaul, T. Huettl, and C. Powers, "Radiated emissions testing of an experimental electric vehicle," *Proc. IEEE Int. Symposium on Electromagnetic Compatibility,* 19–23 August 1996, pp. 338–342.
310. K.K. Humphreys, M. Placet, and M. Singh, "Life-cycle assessment of electric vehicles in the United States," *Proc. 31st Intersociety Energy Conversion Engineering Conf. (IECEC)* Vol. 3, 11–16 August 1996, pp. 2124–2127.
311. D. Naunin, "Electric vehicles," *Proc. IEEE Int. Symposium on Industrial Electronics (ISIE),* Vol. 1, 17–20 June 1996, pp. 11–24.
312. G. Henneberger, "Brushless motors for electric and hybrid vehicles," *Proc. IEEE Colloquium on Machines and Drives for Electric and Hybrid Vehicles* (Digest No. 1996/152), 28 June 1996, pp. 2/1–2/4.
313. F.A. Wyczalek, "Hybrid electric vehicles (EVS-13 Osaka)," *Proc. Northcon/96,* 4–6 November 1996, pp. 409–412.
314. A. Kawamura, N. Hoshi, T.W. Kim, T. Yokoyama, and T. Kume, "Survey of antidirectional twin-rotary motor drive characteristics for electric vehicles," *Proc. IEEE 22nd Int. Conf. on Industrial Electronics, Control, and Instrumentation (IECON)*, Vol. 1, 5–10 August 1996, pp. 41–46.
315. M. Ehsani, K.M. Rahmann, and H.A. Toliyat, "Propulsion system design of electric vehicles," *Proc. IEEE 22nd IECON,* Vol. 1, 5–10 August 1996, pp. 7–13.
316. M.E. Murphy, "A Neighborhood Electric Vehicle (NEV) developed in Oregon for the 21st century," *Proc. Northcon/96*, 4–6 November 1996, pp. 369–372.
317. E. Chikuni and M.M. El-Missiry, "Highlights of the electrical vehicle research programme at the University of Zimbabwe," *Proc. 4th IEEE AFRICON Conf.*, Vol. 2, 24–27 September 1996, pp. 957–961.
318. C.C. Chan, K.T. Chau, J.Z. Jiang, W. Xia, M. Zhu, and R. Zhang, "Novel permanent magnet motor drives for electric vehicles," *IEEE Trans. Industrial Electronics,* Vol. 43, No. 2, April 1996, pp. 331–339.
319. R.I. Hodkinson, "Brushless DC motors for electric and hybrid vehicles," *Proc. IEEE Colloquium on Machines and Drives for Electric and Hybrid Vehicles* (Digest No. 1996/152), 28 June 1996, pp. 1/1–1/4.
320. C.C. Chan and K.T. Chau, "An advanced permanent magnet motor drive system for battery-powered electric vehicles," *IEEE Trans. Vehicular Technology*, Vol. 45, No. 1, February 1996, pp. 180–188.
321. M. Ehsani, G. Yimin, and K.L. Butler, "Application of electrically peaking hybrid (ELPH) propulsion system to a full-size passenger car with simulated design verification," *IEEE Trans. Vehicular Technology*, Vol. 48, No. 6, pp. 1779–1787, November 1996.

1995

322. D. Casadei, G. Serra, and A. Tani, "Direct flux and torque control of induction machine for electric vehicle applications," *Proc. 7th Int. Conf. on Electrical Machines and Drives* (Conf. Publ. No. 412), 11–13 September 1995, pp. 349–353.

323. D.A. Sanders, I.J. Stott, and M..J. Goodwin, "Assisting a disabled person in navigating an electric vehicle through a doorway," *IEEE Colloquium on New Developments in Electric Vehicles for Disabled Person*, 17 March 1995, pp. 5/1–5/6.

324. B.C. Keoun, "Designing an electric vehicle conversion," *Proc. Southcon/95 Conf.*, 7–9 March 1995, pp. 303–308.

325. D. Lynch, "The Calstart consortium [Electric vehicle industry]," *IEEE Spectrum*, Vol. 30, No. 7, July 1993, pp. 54–57.

326. V. Wouk, "Hybrids: then and now," *IEEE Spectrum*, Vol. 32, No. 7, July 1995, pp. 16–21.

327. Z. Zhang, F. Profumo, and A. Tenconi, "Improved design for electric vehicles induction motors using optimization procedure," *Proc. 7th Int. Conf. on Electrical Machines and Drives* (Conf. Publ. No. 412), 11–13, September 1995, pp. 21–25.

328. S. Wall, "Vector control: a practical approach to electric vehicles," *Proc. IEEE Colloquium on Vector Control and Direct Torque Control of Induction Motors*, 27 October 1995, pp. 5/1–5/7.

329. B.J. Chalmers, L. Musaba, and D.F. Gosden, "Variable-frequency synchronous motor drives for electric vehicles," *Proc. 30th Annual IEEE Industry Applications Conf.*, Vol. 1, 8–12 October 1995, pp. 717–724.

330. T. Nakajima, S. Yoshida, A. Uenishi, T. Shirasawa, S. Ukita, and Y. Kimura, "New intelligent power module for electric vehicles," *Proc. 30th Annual IEEE Industry Applications Conf.*, Vol. 2, 8–12 October 1995, pp. 954–958.

331. H. Shimizu, J. Harada, and C. Bland, "The role of optimized vehicle design and power semiconductor devices to improve the performance of an electric vehicle," *Proc. 7th Int. Symposium on Power Semiconductor Devices and ICs (ISPSD)*, 23–25 May 1995, pp. 8–12.

332. L.R. Georges, A.G. Kladas, and S.N. Manias, "Finite element based production motor model for adaptive field-oriented drive for electrical vehicle," *Proc. IEEE Int. Symposium on Industrial Electronics (ISIE)*, Vol. 2, 10–14 July 1995, pp. 639–642.

333. G.A. Karvelis, A. Nagel, S.N. Manias, and F. Schope, "Reduction of EMI emission of an on-board charger for electric vehicles," *Proc. ISIE*, Vol. 1, 10–14 July 1995, pp. 115–120.

334. C. Ellers, "Key to saving Midtown USA, mass transit and electric vehicles," *Proc. IEEE Technical Applications Conf. and Workshops Northcon '95*, 10–12 October 1995, pp. 406–410.

335. N.H. Kutkut, D.M. Divan, D.W. Novotny, R. Marion, "Design considerations and topology selection for a 120 kW IGBT converter for EV fast charging," *Proc. 26th Annual IEEE Power Electronics Specialists Conf. (PESC)*, Vol. 1, 18–22 June 1995, pp. 238–244.

336. P.M. Kelecy and R.D. Lorenz, "Control methodology for single stator, dual-rotor induction motor drives for electric vehicles," *Proc. 26th PESC*, Vol. 1, 18–22 June 1995, pp. 572–578.

337. K.W. Klontz, D.M. Divan, and D.W. Novotny, "An actively cooled 120 kW coaxial winding transformer for fast charging electric vehicles," *IEEE Trans. Industry Applications*, Vol. 31, No. 6, November–December 1995, pp. 1257–1263.

338. A. Esser, "Contactless charging and communication for electric vehicles," *IEEE Industry Applications Magazine*, Vol. 1, No. 6, November–December 1995, pp. 4–11.

339. D.R. KeKoster, K.P. Morrow, D.A. Schaub, and N.F. Hubele, "Impact of electric vehicles on select air pollutants: a comprehensive model," *IEEE Trans. Power Systems*, Vol. 10, No. 3, August 1995, pp. 1383–1388.

340. S.K. Sul and S.J. Lee, "An integral battery charger for four-wheel-drive electric vehicle," *IEEE Trans. Industry Applications*, Vol. 31, No. 5, September–October 1995, pp. 1096–1099.

341. E.J. Dowgiallo and J.E. Hardin, "Perspective on ultracapacitors for electric vehicles," *IEEE Aerospace and Electronic Systems Magazine*, Vol. 10, No. 8, August 1995, pp. 26–31.

342. H. Oman and S. Gross, "Electric-vehicle batteries," *IEEE Aerospace and Electronic Systems Magazine*, Vol. 10, No. 2, February 1995, pp. 29–35.

Electric Vehicles

343. W.R. Young, Jr. and W. Wilson, "Monitoring electric vehicle performance," *Proc. Southcon/95 Conf.*, 7–9 March 1995, pp. 298–302.
344. A. Harson, P.H. Mellor, and D. Howe, "Design considerations for induction machines for electric vehicle drives," *Proc. 7th Int. Conf. on Electrical Machines and Drives* (Conf. Publ. No. 412), 11–13 September 1995, pp. 16–20.
345. G. Henneberger, K.B. Yahia, and M. Schmitz, "Calculation and identification of a thermal equivalent circuit of a water-cooled induction motor for electric vehicle applications," *Proc. 7th Int. Conf. on Electrical Machines and Drives* (Conf. Publ. No. 412), 11–13 September 1995, pp. 6–10.
346. K.E. Bailey and B.K. Powell, "A hybrid electric vehicle powertrain dynamic model," *Proc. American Control Conf.*, Vol. 3, 21–23 June 1995, pp. 1677–1682.
347. D.H. Swan, B. Dickinson, M. Arikara, and M. Prabhu, "Construction and performance of a high-voltage zinc bromine battery in an electric vehicle," *Proc. 10th Annual Battery Conf. on Applications and Advances,* 10–13 January 1995, pp. 135–140.
348. D.B. Karner, "Current events in vehicle battery safety," *Proc. 10th Annual Battery Conf. on Applications and Advances,* 10–13 January 1995, pp. 167–169.
349. H.R. Berenji and E.H. Ruspini, "Automated controller elicitation and refinement for power trains of hybrid electric vehicles," *Proc. 4th IEEE Conf. on Control Applications,* 28–29 September 1995, pp. 329–334.
350. H. Huang and L. Chang, "Continuous defuzzification of fuzzy logic controller in electric vehicle induction motor drive systems," *Proc. IEEE Int. Conf. Man and Cybernetics, Intelligent Systems for the 21st Century,* Vol. 3, 22–25 October 1995, pp. 2466–2471.
351. A.H. Wijenayake, J.M. Bailey, P.J. McCleer, "Design optimization of an axial gap permanent magnet brushless DC motor for electric vehicle applications," *Proc. 30th IEEE Industry Applications Conf.,* Vol. 1, 8–12 October 1995, pp. 685–692.
352. T. Uematsu and R.G. Hoft, "Resonant power electronic control of switched reluctance motor for electric vehicle propulsion," *Proc. 26th Annual IEEE Power Electronics Specialists Conf. (PESC),* Vol. 1, 18–22 June 1995, pp. 264–269.
353. F. Huang, X. Jiang, and Y. Wang, "A dedicated permanent magnet synchronous motor drive system for electric vehicle," *Proc. 26th PESC,* Vol. 1, 18–22 June 1995, pp. 252–257.
354. S. Chen and T.A. Lipo, "Soft-switched inverter for electric vehicle drives," *Proc. 10th Annual Applied Power Electronics Conf. and Exposition (APEC),* No. 0, 5–9 March 1995, pp. 586–591.
355. F. Caricchi, F. Crescimbini, and A. Di Napoli, "20 kW water-cooled prototype of a buck-boost bidirectional DC/DC converter topology for electrical vehicle motor drives," *Proc. 10th APEC,* No. 0, 5–9 March 1995, pp. 887–892.
356. N.R. Cox, "A universal power converter for emergency charging of electric vehicle batteries," *Proc. 10th APEC,* No. 0, 5–9 March 1995, pp. 965–969.
357. T. Uematsu and R.S. Wallace, "Design of a 100 kW switched reluctance motor for electric vehicle propulsion," *Proc. 10th APEC,* No. 0, 5–9 March 1995, pp. 411–415.
358. N.L.C. Steele, "Recycling electric vehicle batteries in California," *Proc. 10th Annual Battery Conf. on Applications and Advances*, 10–13 January 1995, pp. 101–106.
359. G. Hopper, T. Turrentine, and E. Gallagher, "Battery mix and aggregate battery demand for electric vehicles in California," *Proc. 10th Annual Battery Conf. on Applications and Advances*, 10–13 January 1995, pp. 107–111.
360. M.G. Jayne, "Requirements of electric vehicles for disabled persons," *IEEE Colloquium on New Developments in Electric Vehicles for Disabled Persons*, 17 March 1995, pp. 1/1–1/4.
361. D.J. McShane, "Combined drive/chargers for electric vehicles: advanced technology to sidestep the batteries," *IEEE Colloquium on Advances in Control Systems for Electric Drives,* 24 May 1995, pp. 1/1–1/7.
362. P.T. Krein, "Electrostatic discharge issues in electric vehicles," *Proc. 30th Annual IEEE Industry Applications Conf.,* Vol. 2, 8–12 October 1995, pp. 1245–1250.
363. D.L. Buntin and J.W. Howze, "A switching logic controller for a hybrid electric/ICE vehicle," *Proc. American Control Conf.*, Vol. 2, 21–23 June 1995, pp. 1169–1175.

364. M. Marchesoni, L. Puglisi, and A. Rebora, "Analysis and modeling of the auxiliary quasi-resonant DC-link inverter applied to AC electric vehicle drives," *Proc. IEEE Int. Symposium on Industrial Electronics (ISIE)*, Vol. 2, 10–14 July 1995, pp. 558–563.
365. C.N. Spentzas and D.B. Koulocheris, "PEV, a software to estimate the power and torque requirements and evaluate the performance of electric vehicles," *Proc. ISIE*, Vol. 2, 10–14 July 1995, pp. 553–557.
366. B. Asaii, D.F. Gosden, and S. Sathiakumar, "A simple high-efficient torque control for the electric vehicle induction machine drives without a shaft encoder," *Proc. 26th Annual IEEE Power Electronics Specialists Conf. (PESC)*, Vol. 2, 18–22 June 1995, pp. 778–784.
367. N.H. Kutkut, H.N.L. Wiegman, D.M. Divan, and D.W. Novotny, "Design considerations for charge equalization of an electric vehicle battery system," *Proc. 10th Annual Applied Power Electronics Conf. and Exposition (APEC)*, No. 0, 5–9 March 1995, pp. 96–103.
368. G.A.J. Elliott, J.T. Boys, and A.W. Green, "Magnetically coupled systems for power transfer to electric vehicles," *Proc. Int. Conf. on Power Electronics and Drive Systems*, 21–24 February 1995, pp. 797–801.
369. P.N. Ross, "A novel zinc-air battery for electric vehicles," *Proc. 10th Annual Battery Conf. on Applications and Advances*, 10–13 January 1995, pp. 131–133.
370. C. Kopf, "Adaptive control of the unique mobility EV drive system to account for time-varying battery parameters," *Proc. 10th Annual Battery Conf. on Applications and Advances*, 10–13 January 1995, pp. 159–165.

1994

371. L. Chang, "Comparison of AC drives for electric vehicles — a report on experts' opinion survey," *IEEE Aerospace and Electronic Systems Magazine*, Vol. 9, No. 8, August 1994, pp. 7–11.
372. K. Rajashekara, "History of electric vehicles in General Motors," *IEEE Trans. Industry Applications*, Vol. 30, No. 4, July–August 1994, pp. 897–904.
373. D.H. Swan, B. Dickinson, M. Arikara, and G.S. Tomazic, "Demonstration of a zinc bromine battery in an electric vehicle," *IEEE Aerospace and Electronic Systems Magazine*, Vol. 9, No. 5, May 1994, pp. 20–23.
374. H. Singh, H.S. Bawa, S. Barada, B. Bryant, and L. Anneberg, "Fuzzy logic approach in determining the range of electric vehicle," *Proc. 37th Midwest Symposium on Circuits and Systems*, Vol. 2, 3–5 August 1994, pp. 1519–1522.
375. M.R. Seal, "Viking 23 zero emissions in the city, range and performance on the freeway," *Proc. Northcon/94 Conf.*, 11–13 October 1994, pp. 264–268.
376. H. Oman, "New electric-vehicle batteries," *Proc. Northcon/94 Conf.*, 11–13 October 1994, pp. 326–330.
377. F. Caricchi, F. Crescimbini, G. Noia, and D. Pirolo, "Experimental study of a bidirectional DC/DC converter for the DC link voltage control and the regenerative braking in PM motor drives devoted to electrical vehicles," *Proc. 9th APEC*, 13–17 February 1994, pp. 381–386.
378. D. Coates and C. Fox, "Multiple advanced battery systems for electric vehicles," *Proc. 9th Annual Battery Conf. on Applications and Advances*, 11–13 January 1994, pp. 110.
379. K.W. Klontz, D.M. Divan, and D.W. Novotny, "An actively cooled 120 kW coaxial winding transformer for fast charging electric vehicles," *Proc. IEEE Industry Applications Society Annual Meeting*, 2–6 October 1994, pp. 1049–1054.
380. P.M. Kelecy and R.D. Lorenz, "Control methodology for single inverter, parallel connected dual induction motor drives for electric vehicles," *Proc. 25th Annual IEEE Power Electronics Specialists Conf. (PESC)*, 20–25 June 1994, pp. 987–991.
381. S. McCrea, "Preparing for the post-petroleum era: what electric vehicle advocates and others need to know about alternative fuel vehicles," *Proc. Southcon/94 Conf.*, 29–31 March 1994, pp. 53–58.
382. C.R. Suggs, "Electric vehicles — driving the way to a cleaner future," *Proc. Southcon/94 Conf.*, 29–31 March 1994, pp. 28–30.

383. D.A. Cobb and T.C. Edwards, "Climate control system for electric vehicles," *Proc. Southcon/94 Conf.*, 29–31 March 1994, pp. 37–41.
384. A.M. Sharaf, A.S. Abd El Ghaafar, and S.S. Shokralla, "A novel efficient rule-based controller for electric vehicles switch mode battery charger," *Proc. Canadian Conf. on Electrical and Computer Engineering,* 25–28 September 1994, pp. 117–120.
385. J.G. Bolger, "Urban electric transportation systems: the role of magnetic power transfer," *Proc. WESCON/94 'Idea/Microelectronics' Conf.,* 27–29 September 1994, pp. 41–45.
386. F. Silver, "Electric vehicle drivetrain components," *Proc. WESCON/94 Idea/Microelectronics Conf.,* 27–29 September 1994, p. 33.
387. C.C. Chan, J.Z. Jiang, G.H. Chen, X.Y. Wang, and K.T. Chau, "A novel polyphase multipole square-wave permanent magnet motor drive for electric vehicles," *IEEE Trans. Industry Applications*, Vol. 30, No. 5, September–October 1994, pp. 1258–1266.
388. A. Lashgari, "Electric vehicle infrastructure market sustaining demand," *Proc. 9th Annual Battery Conf. on Applications and Advances,* 11–13 January 1994, pp. 86–95.
389. M.E. Murphy, "Neighborhood electric vehicles: the simple things that move you," *Proc. Northcon/94 Conf.,* 11–13 October 1994, pp. 260–263.
390. W.R. Young, Jr., "Electric vehicles of yesterday carry us into tomorrow," *Proc. Southcon/94 Conf.,* 29–31 March 1994, pp. 14–16.
391. J.G.W. West, "DC, induction, reluctance and PM motors for electric vehicles," *Power Engineering Journal*, Vol. 8, No. 2, April 1994, pp. 77–88.
392. D.F. Gosden, B.J. Chalmers, and L. Musaba, "Drive system design for an electric vehicle based on alternative motor types," *Proc. 5th Int. Conf. on Power Electronics and Variable-Speed Drives,* 26–28 October 1994, pp. 710–715.
393. S. Matsumura, S. Omatu, and H. Higasa, "Improvement of speed control performance using PID type neurocontroller in an electric vehicle," *Proc. IEEE Int. Conf. on System Neural Networks*, Vol. 4, 27 June–2 July 1994, pp. 2649–2654.
394. Y. Murai, H. Ishikawa, and T.A. Lipo, "New series resonant DC link inverter for electric vehicle drives," *Proc. IEEE Industry Applications Society Annual Meeting*, 2–6 October 1994, pp. 443–447.
395. S.J. Lee and S.K. Sul, "An integral battery charger for four-wheel-drive electric vehicle," *Proc. IEEE Industry Applications Society Annual Meeting,* 2–6 October 1994, pp. 448–452.
396. A. Kawamura, T. Yokoyama, and T. Kume, "Antidirectional twin-rotary motor drive for electric vehicles," *Proc. IEEE Industry Applications Society Annual Meeting*, 2–6 October 1994, pp. 453–459.
397. H.C. Lamb, E.K. Stefanakos, T. Smith, B. Krakow, C. Hernandez, R. Rodriguez, and M. Kovac, "Efficient photovoltaic charging of electric vehicles," *Proc. Southcon/94 Conf.,* 29–31 March 1994, pp. 47–52.
398. B.K. Powell and T.E. Pilutti, "A range extender hybrid electric vehicle dynamic model," *Proc. 33rd IEEE Conf. on Decision and Control*, Vol. 3, 14–16 December 1994, pp. 2736–2741.
399. J. Gallagher and D. Seals, "Design considerations for the power electronics of an electric vehicle propulsion inverter," *Proc. WESCON/94 'Idea/Microelectronics' Conf.,* 27–29 September 1994, pp. 34–40.
400. K. Berringer and G.L. Romero, "High-current power modules for electric vehicles," *Proc. Power Electronics in Transportation,* 20–21 October 1994, pp. 59–65.
401. P.T. Krein, T.G. Roethemeyer, R.A. White, and B.R. Masterson, "Packaging and performance of an IGBT-based hybrid electric vehicle," *Proc. Power Electronics in Transportation*, 20–21 October 1994, pp. 47–52.
402. C.W. Ellers, "Electric transportation: the challenge is yours," *Proc. Northcon/94 Conf.,* 11–13 October 1994, pp. 331–334.
403. J.W. Yerkes, "Electric vehicle developments in Europe and Japan," *Proc. Northcon/94 Conf.*, 11–13 October 1994, pp. 254–259.
404. D.H. Swan, B. Dickinson, M. Arikara, and G.S. Tomazic, "Demonstration of a zinc bromine battery in an electric vehicle," *Proc. 9th Annual Battery Conf. on Applications and Advances,* 11–13 January 1994, pp. 104–109.

405. U. Schaible and B. Szabados, "A torque-controlled high-speed flywheel energy storage system for peak power transfer in electric vehicles," *Proc. IEEE Industry Applications Society Annual Meeting*, 2–6 October 1994, pp. 435–442.
406. J.T. Jacobs, "Range extension for electric vehicles," *Proc. Southcon/94 Conf.*, 29–31 March 1994, pp. 31–36.
407. G.H. Holling, "A novel approach to controlling the phase angle of a variable switched reluctance motor for electric vehicle propulsion using the statistic matrix norm," *Proc. 33rd IEEE Conf. on Decision and Control*, Vol. 3, 14–16 December 1994, pp. 2760–2765.
408. D.L. Harbaugh, "Electric-vehicle commuter car battery requirements," *Proc. WESCON/94 'Idea/Microelectronics' Conf.*, 27–29 September 1994, pp. 28–32.
409. E. Cerruto, A. Consoli, A. Raciti, and A. Testa, "Fuzzy logic based efficiency improvement of an urban electric vehicle," *Proc. 20th Int. Industrial Electronics, Control and Instrumentation (IECON)*, Vol. 2, 5–9 September 1994, pp. 1304–1309.

1993

410. L. Chang, "Recent developments of electric vehicles and their propulsion systems," *IEEE Aerospace and Electronic Systems Magazine*, Vol. 8, No. 12, December 1993, pp. 3–6.
411. E. Holl, G. Neumann, B. Piepenbreier, and H.J. Tolle, "Water-cooled inverter for synchronous and asynchronous electric vehicle drives," *Proc. 5th European Conf. on Power Electronics and Applications*, 13–16 September 1993, pp. 289–293.
412. C. Bourne and R. Ball, "A hybrid battery-powered drivetrain for a high performance electric vehicle," *IEEE Colloquium on Systems for Electric Racing Vehicles*, 26 October 1993, pp. 4/1–4/3.
413. F.A. Wyczalek, "Heating and cooling battery electric vehicles — the final barrier," *IEEE Aerospace and Electronic Systems Magazine*, Vol. 8, No. 11, November 1993, pp. 9–14.
414. C.M. Jefferson, "Power management in electric vehicles," *Proc. IEEE Colloquium on Systems for Electric Racing Vehicles*, 26 October 1993, pp. 5/1–5/5.
415. T. Allen, P.H. Mellor, P.J. Monkhouse, and D. Howe, "Assessment of battery technologies for electric vehicles," *Proc. IEEE Colloquium on Systems for Electric Racing Vehicles*, 26 October 1993, pp. 7/1–7/5.
416. A.M. Sharaf and L.A. Snider, "An intelligent low distortion battery charger controller for fast controlled charging of electric vehicles," *Proc. 2nd Int. Conf. on Advances in Power System Control, Operation and Management (APSCOM)*, 7–10 December 1993, pp. 209–214.
417. A.J. Mitcham, A.G. Jack, and B.C. Mecrow, "High-performance permanent magnet drive for electric vehicles," *Proc. IEEE Colloquium on Motors and Drives for Battery Powered Propulsion*, 15 April 1993, pp. 8/1–8/4.
418. J. Paterson and M. Ramsay, "Electric vehicle braking by fuzzy logic control," *Proc. IEEE Annual Industry Applications Society Annual*, 2–8 October 1993, pp. 2200–2204.
419. X. Xu and V.A. Sankaran, "Power electronics in electric vehicles: challenges and opportunities," *Proc. IEEE Industry Applications Society Annual Meeting*, 2–8 October 1993, pp. 463–469.
420. K. Rajashekara, "History of electric vehicles in General Motors," *Proc. IEEE Industry Applications Society Annual Meeting*, 2–8 October 1993, pp. 447–454.
421. T. Tsunoda, T. Matsuda, Y. Nakadaira, H. Nakayama, and Y. Sasada, "Low-inductance module construction for high speed, high-current IGBT module suitable for electric vehicle application," *Proc. 5th Int. Symposium on Power Semiconductor Devices and ICs (ISPSD)*, 18–20 May 1993, pp. 292–295.
422. C.C. Chan, J.Z. Jiang, G.H. Chen, X.Y. Wang, and K.T. Chau, "A novel polyphase multipole square-wave permanent magnet motor drive for electric vehicles," *Proc. 8th Annual Applied Power Electronics Conference and Exposition (APEC)*, 7–11 March 1993, pp. 315–321.

423. S.J. Visco and L.C. De Jonghe, "Thin-film polymer batteries for electric vehicles," *Proc. WESCON/93 Conf.*, 28–30 September 1993, pp. 512–517.
424. J. Fetz and K. Obayashi, "High-efficiency induction motor drive with good dynamic performance for electric vehicles," *Proc. 24th Annual IEEE Power Electronics Specialists Conf. (PESC)*, 20–24 June 1993, pp. 921–927.
425. H. Schmidt and C. Siedle, "The charge equalizer — a new system to extend battery lifetime in photovoltaic systems, UPS, and electric vehicles," *Proc. 15th Int. Telecommunications Energy Conf. (INTELEC)*, Vol. 2, 27–30 September 1993, pp. 146–151.
426. D. Platt and B.H. Smith, "Twin-rotor drive for an electric vehicle," *IEEE Proc. Electric Power Applications*, Vol. 140, No. 2, March 1993, pp. 131–138.
427. G.W. Davis, G.L. Hodges, and F.C. Madeka, "The development of a series hybrid electric vehicle for near-term applications," *IEEE Aerospace and Electronic Systems Magazine*, Vol. 8, No. 11, November 1993, pp. 15–20.
428. J.W. Pavlat and R.W. Diller, "An energy management system to improve electric vehicle range and performance," *IEEE Aerospace and Electronic Systems Magazine*, Vol. 8, No. 6, June 1993, pp. 3–5.
429. W.A. O'Brien, "Electric vehicles (EVs): A look behind the scenes," *IEEE Aerospace and Electronic Systems Magazine*, Vol. 8, No. 5, May 1993, pp. 38–41.
430. A. El-Antably, X. Luo, and R. Martin, "System simulation of fault conditions in the components of the electric drive system of an electric vehicle or an industrial drive," *Proc. IECON*, 15–19 November 1993, pp. 1146–1150.
431. C.C. Chan and K.T. Chau, "Power electronics challenges in electric vehicles," *Proc. IECON*, 15–19 November 1993, pp. 701–706.
432. C.C. Chan, "Present status and future trends of electric vehicles," *Proc. 2nd Int. Conf. on Advances in Power System Control, Operation and Management (APSCOM)*, 7–10 December 1993, pp. 456–469.
433. H. Fan, G.E. Dawson, and T.R. Eastham, "Model of electric vehicle induction motor drive system," *Proc. Canadian Conf. on Electrical and Computer Engineering*, 14–17 September 1993, pp. 1045–1048.
434. K.W. Klontz, A. Esser, R.R. Bacon, D.M. Divan, D.W. Novotny, and R.D. Lorenz, "An electric vehicle charging system with 'universal' inductive interface," *Proc. Power Conversion Conf.*, Yokohama, 19–21 April 1993, pp. 227–232.
435. K.W. Klontz, A. Esser, P.J. Wolfs, and D.M. Divan, "Converter selection for electric vehicle charger systems with a high-frequency high-power link," *Proc. 24th PESC*, 20–24 June 1993, pp. 855–861.
436. R.A. Weinstock, P.T. Krein, and R.A. White, "Optimal sizing and selection of hybrid electric vehicle components," *Proc. 24th PESC*, 20–24 June 1993, pp. 251–256.
437. C.C. Chan, "An overview of electric vehicle technology," *Proc. IEEE*, Vol. 81, No. 9, September 1993, pp. 1202–1213.
438. S. Rahman and G.B. Shrestha, "An investigation into the impact of electric vehicle load on the electric utility distribution system," *IEEE Trans. Power Delivery*, Vol. 8, No. 2, April 1993, pp. 591–597.
439. A. Esser, "Contactless charging and communication system for electric vehicles," *Proc. IEEE Industry Applications Society Annual Meeting*, 2–8 October 1993, pp. 1021–1028.
440. A.K. Adnanes, R. Nilsen, R. Loken, and L. Norum, "Efficiency analysis of electric vehicles, with emphasis on efficiency optimized excitation," *Proc. IEEE Industry Applications Society Annual Meeting*, 2–8 October 1993, pp. 455–462.
441. P.H. Brasch, "A precision DC energy monitor for electric vehicles," *Proc. WESCON/93 Conf.*, 28–30 September 1993, pp. 526–530.
442. J.K. Nor, "Art of charging electric vehicle batteries," *Proc. WESCON/93 Conf.*, 28–30 September 1993, pp. 521–525.

1992

443. E.J. Cairns, "A new mandate for energy conversion: zero emission (electric) vehicles," *Proc. IEEE 35th Int. Power Sources Symposium*, 22–25 June 1992, pp. 310–313.
444. M.J. Riezenman, "Electric vehicles — architecting the system," *IEEE Spectrum*, Vol. 29, No. 11, November 1992, pp. 94–96.
445. M.J. Riezenman, "Electric vehicles — pursuing efficiency," *IEEE Spectrum*, Vol. 29, No. 11, November 1992, pp. 22–24, 93.
446. M.J. Riezenman, "Electric vehicles," *IEEE Spectrum*, Vol. 29, No. 11, November 1992, pp. 18–21.
447. A.J. Appleby and R.C. Kainthla, "Rechargeable nontoxic manganese-zinc batteries," *Proc. IEEE 35th Int. Power Sources Symposium*, 22–25 June 1992, pp. 14–17.

1991

448. K. Desmond, "Electric vehicles — quietly making progress," *IEEE Review*, Vol. 37, No. 4, 18 April 1991, pp. 145–147.
449. S. Venkatesan, M.A. Fetcenko, S.K. Dhar, and S.R. Ovshinsky, "Advances in the development of ovonic nickel metal hydride batteries for industrial and electric vehicles," *IEEE Aerospace and Electronic Systems Magazine*, Vol. 6, No. 11, November 1991, pp. 26–30.
450. P. Singh, P. von Glahn, and W. Koffke, "The design and construction of a solar electric commuter car," *Proc. 22nd IEEE Photovoltaic Specialists Conf.*, Vol. 1, 7–11 October 1991, pp. 712–716.

1990

451. P.G. Patil, "Prospects for electric vehicles," *IEEE Aerospace and Electronic Systems Magazine*, Vol. 5, No. 12, December 1990, pp. 15–19.
452. M. Eghtesadi, "Inductive power transfer to an electric vehicle-analytical model," *Proc. 40th IEEE Vehicular Technology Conf.*, 6–9 May 1990, pp. 100–104.
453. A.F. Burke, J.E. Hardin, and E.J. Dowgiallo, "Application of ultracapacitors in electric vehicle propulsion systems," *Proc. 34th Int. Power Sources Symposium*, 25–28 June 1990, pp. 328–333.

1989

454. M. Fujinaka, "The practically usable electric vehicle charged by photovoltaic cells," *Proc. 24th Intersociety Energy Conversion Engineering Conf. (IECEC)*, 6–11 August 1989, pp. 2473–2478.
455. J.E. Quinn, P.C. Symons, D.R. Brown, D.J. Riley, G. Atherton, and S.C. Rampton, "Projected costs for sodium-sulfur electric vehicle batteries," *Proc. 24th IECEC*, 6–11 August 1989, pp. 1335–1340.
456. A.S. Homa and E.J. Rudd, "The development of aluminum-air batteries for electric vehicles," *Proc. 24th IECEC*, 6–11 August 1989, pp. 1331–1334.

1988

457. B.K. Bose and P.M. Szczesny, "A microcomputer-based control and simulation of an advanced IPM synchronous machine drive system for electric vehicle propulsion," *IEEE Trans. Industrial Electronics*, Vol. 35, No. 4, November 1988, pp. 547–559.

5

Optimal Power Management and Distribution in Automotive Systems

Zheng John Shen
Univeristy of Central Florida, Orlando, Florida

X. Chen
University of Windsor, Windsor, Ontario, Canada

A. Masrur
U.S. Army TACOM, Warren, Michigan

V. K. Garg
Ford Motor Company, Dearborn, Michigan

A. Soltis
Opal-RT Technologies, Inc., Ann Arbor, Michigan

5.1 INTRODUCTION

With the increase in electrical/electronic content in vehicles, the on-board electric power requirement is likely to increase from 1 kW to 5 kW for non-propulsion loads and to 100 kW or more for propulsion loads in the near future. This is based on the trend observed during the past decade, and this dramatic increase in power demand is mainly due to the emergence of various electric and hybrid drivetrains. However, other emerging automotive technologies for internal combustion engine (ICE) vehicles, such as variable engine valve, active suspension, x-by-wire (e.g., steering-by-wire and brake-by-wire), and heated catalytic converter, have also made considerable contributions to this trend. Table 5.1 lists the peak and average power requirements of various electrical loads that current or future automotive power systems are required to meet.

As indicated, such a large amount of power makes the issue of appropriate power management and distribution a much more complicated and difficult matter compared to the management of the existing low-electrical-power vehicles. In a conventional ICE vehicle, electric power is simply managed through the control of alternator and voltage

Table 5.1 An Example of Automotive Electric Power Requirements

Electric Loads	Peak Power (kW)	Average Power (kW)
Electric vehicle hybrid electric vehicle propulsion	30–100	10–30
Fuel-cell electric vehicle air compressor	12	8
Active suspension	12	0.36
Integrated starter generator	4–8	2–4
Electric AC compressor	4	1
Variable engine valve	3.2	1
Heated catalytic converter	3	0.1
Heated windshield	2.5	0.25
Electric power steering	1.5	0.1
Engine cooling fan (ICE)	0.8	0.4
Engine coolant pump (ICE)	0.5	0.4

regulator. Early electric and hybrid vehicles included the control of battery state of charge as an important component of the on-board power management strategy. Nevertheless, the overall power management scheme for these vehicles remains fairly simple and straightforward. However, with the increasing power level and system complexity, this can no longer be the case. Hence, optimal power and energy management strategy is becoming increasingly necessary.

The primary function of automotive electric power management is to prioritize real-time power requests from the loads and allocate power resources available from the generation and storage devices in an optimized manner for maximum vehicle efficiency and performance. The use of energy storage devices in vehicles, such as batteries, ultracapacitors, and flywheels, has essentially mandated the extension of the power management concept into one that covers both power and energy (i.e., time integral of power) perspectives. Good power and energy management strategies can help reduce the weight, size, and cost; improve the performance of the vehicle and improve system reliability during limited resource availability.

This chapter introduces a generic power/energy management and distribution architecture and an optimization-based integrated power management/distribution system approach, which are applicable to a wide variety of vehicles including conventional ICE-powered, electric, hybrid, and fuel cell vehicles. As an example of this approach, a game-theoretic optimal hybrid vehicle management strategy is also presented. The function of all major components or subsystems such as power generators, storage devices, power buses, power control units (PCUs), central power management controllers (PMCs), and electrical loads are discussed and integrated system approaches are presented, aiming to achieve maximum energy efficiency, high performance, and low emissions. The proposed power management strategy features optimal allocation and prioritization of power resources to meet the demand of various loads, uninterruptible power availability, high power quality and system stability, and fault diagnosis and prognosis. In addition, the power management strategy must take into consideration practical component constraints such as limited battery cycle life, variable power output, and efficiency of primary power sources and generators.

Optimal Power Management and Distribution in Automotive Systems

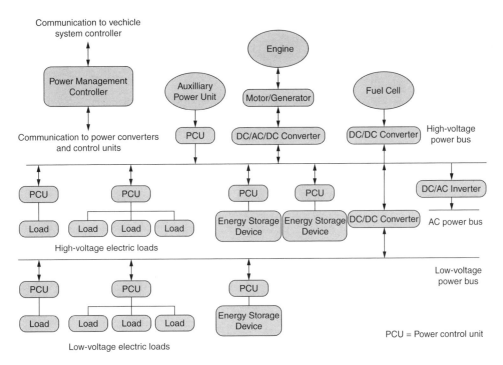

Figure 5.1 Generic automotive power/energy management and distribution system.

5.2 AUTOMOTIVE POWER/ENERGY MANAGEMENT AND DISTRIBUTION ARCHITECTURE

An automotive electrical power/energy management and distribution system architecture is conceptually depicted in Figure 5.1.

This generic architecture is applicable to ICE, electric, fuel-cell, or any hybrid-type vehicles. For a particular vehicle configuration, not all components or subsystems presented in Figure 5.1 are necessary, and some minor changes in the system topology may be required. The power management system can be divided into the following subsystems:

- Power generation
- Energy storage
- Power bus
- Electrical load
- Power electronics
- PMC

5.2.1 Power Generation

The main function of the electric power generation subsystem is to convert chemical, electrochemical, mechanical, and other forms of energy into electrical energy. Examples of primary power sources include, but are not limited to, ICEs (petrol, diesel, hydrogen, other alternative fuels), other types of engines (Stirling and turbine), and fuel cells. Figure 5.1

shows an auxiliary power generation unit (APU), which can provide temporary power when the primary power generator is not in operation or can supplement the primary power generator during normal operation. In military vehicular applications, the APU is also used to provide power during silent watch, when the vehicle may be using its computers and other peripherals with the engine shut down.

Fuel-cell devices convert electrochemical energy directly into electrical energy, while all engine types of power sources have to rely on various electric machines to generate electrical energy. It should be pointed out that all power sources and generators have their own optimal operating ranges in terms of power output and efficiency. Operation outside the desirable ranges may result in low system performance and efficiency. Therefore, it is a major task for the power management system to control and coordinate the operation of the power generation subsystem to achieve maximum system efficiency.

5.2.2 ENERGY STORAGE

The main function of the energy storage subsystem is to provide readily available electric power/energy when the primary power sources have no or insufficient power output, and store surplus electric energy generated by the primary power sources or recovered by regenerative braking. The most common energy storage devices are the various types of batteries, ranging from lead-acid, metal-hydride to lithium-ion types. The key consideration for battery selection is specific power (power/weight), specific energy (energy/weight), cost, and cycle lifetime. Other energy storage devices, such as ultracapacitors, flywheels, and hydraulic/pneumatic accumulators, are normally used along with batteries to compensate for the limited battery power capability. While these devices typically have a limited energy capability, they can greatly improve the system power capability to respond to a burst-mode (high power but short duration) power request. Evidently, the proper control of the energy storage subsystem presents both a challenge and opportunity for the power and energy management system.

5.2.3 POWER BUS

Two direct current (DC) power buses are illustrated in Figure 5.1. One is designated as the high-voltage power bus, the other as the low-voltage power bus. Both loads and power supplies are connected to these power buses through various PCUs and power converters. This is significantly different from the point-to-point or direct connection between the source and load in traditional ICE vehicles. Usually, the high-voltage power bus provides electrical energy to propulsion motors and other high-power loads, while the low-voltage bus supplies energy to low power accessory type of loads such as lamps, small motors, and microcontroller units.

The use of two or more power buses at different voltages in a vehicle can simultaneously address power and safety requirements. It should be pointed out that the "high" and "low" voltages are in relative terms and largely depend on the vehicle configuration, For example, an electric vehicle may have a high-voltage bus at 300 V and a low-voltage bus at 42 V. Yet another ICE vehicle may have a high-voltage bus at 42 V and a low-voltage bus at 14 V. There are one or more DC/DC converters linking the two power buses to transfer energy back and forth. In addition, an alternating current (AC) power bus (110 or 220 V) is sometimes included to provide power access for external plug-in appliances or machine tools.

5.2.4 Electrical Load

Automotive electrical loads can be divided into two classes: propulsion and nonpropulsion loads. The propulsion loads usually include one or more electric machines such as induction or synchronous machines to serve as the traction motor or generator. The propulsion load demands the highest power level in a vehicle, ranging from < 10 kW (mild hybrid) to 100 kW (full electric) or more. Nonpropulsion loads include all other electric loads in a vehicle, such as lamps, heaters, solenoids, and small motors for fans and pumps. Some nonpropulsion loads, such as power steering and special heaters indicated in Table 5.1, actually demand a large amount of electric power. In general, automotive loads, such as starter motors and cold lamps (for example, gas discharge lamps), tend to have a peak power several times higher than its average power consumption. It is important for the power management system to ensure that the power request from all loads are met in a timely manner. In the meantime, it is impractical and cost prohibitive to offer a continuous power capacity many times higher than the average power demand, just to meet momentary peaks in power needs. Thus, how to prioritize load-power requests and allocate limited power resources becomes a major task for the power and energy management system.

5.2.5 Power Electronics

Power electronic components include all power converters and PCUs. These power converters and PCUs control the power flows between the power sources, loads, and power buses, following the commands of the central PMC. For example, a bidirectional DC/AC converter (inverter) is used to control the motor/generator machine, which is mechanically linked to the engine. Depending on the control signal received, the DC/AC converter can run the electric machine as a generator and serves as either a "rectifier" to output DC current or an "inverter" to convert the DC bus voltage into three-phase voltage to drive the machine as a motor or starter. The PCUs are usually power semiconductor load switches, which can communicate with the central PMC via a data network such as controller area network (CAN). One PCU can control one or more loads as shown in Figure 5.1.

5.2.6 PMC

The PMC serves as the central control unit for the power/energy management and distribution system to command, control, and coordinate various components. The PMC communicates with all power converters and PCUs by sending control commands to, and receiving sensor signals and status reports from, these units. It also communicates with the vehicle system controller to interact with other systems of the vehicle. The overall power and energy management algorithm is implemented in the PMC. Even though only one PMC is shown in Figure 5.1, a secondary backup PMC might be necessary for some systems.

5.3 OPTIMIZATION-BASED POWER MANAGEMENT SYSTEM STRATEGY

An optimization-based, integrated system approach for automotive electrical power and energy management and distribution systems is presented in this section. This approach

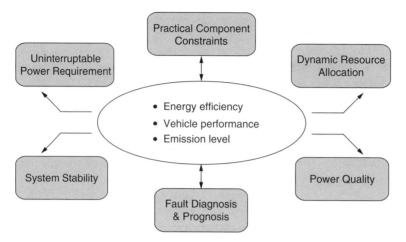

Figure 5.2 Automotive electrical power and energy management: an integrated systems approach.

intends to achieve optimal performance of the overall vehicle system rather than that of each individual subsystem. It is imperative to carefully study all power subsystems including power sources, energy storage devices, and various electric loads, and develop a well-defined problem statement. Important factors to be considered include determining appropriate objectives, constraints, and inter-relationships between these subsystems. Figure 5.2 depicts such a system approach.

Three fundamental objectives have been identified for automotive electrical power and energy management and distribution systems:

- Achieving maximum energy efficiency
- Providing a high level of vehicle performance
- Maintaining a low emission level

The ultimate goal of system optimization is to achieve maximum energy efficiency (for a predefined or chosen performance and emission level), that is, the efficiency of converting chemical (petrol, hydrogen, and other fuels) or electrochemical (battery) energy into mechanical motion of a vehicle. In conventional ICE or hybrid vehicles, this is defined as fuel economy (miles per gallon). Maximum energy efficiency needs to be achieved without compromising vehicle performance aspects, such as driving range, acceleration and comfort/convenience features, and emission levels.

The optimization of a power and energy management system can be formulated into a mathematical model, and the operation of each power subsystem or component can be controlled by a set of decision variables. The appropriate measure of system performance can be expressed as a mathematical function of these decision variables, which is called the "objective function." The objective function for a particular vehicle power system can be developed by combining all three previously mentioned objectives, with each carrying a certain weight. Any restrictions on the values that can be assigned to the decision variables are called "constraints." The essence of the optimization problem is then to choose the values of the decision variables so as to maximize (or minimize, as the case may be) the objective function, subject to the specific constraints.

Optimal Power Management and Distribution in Automotive Systems

The constraints in automotive electric power management systems are imposed by several key factors, including the following (see Figure 5.2):

- Dynamic resource allocation requirement
- Practical component constraints
- Uninterruptible power availability requirement
- Power quality requirement
- System stability requirement
- Fault diagnosis and prognosis requirement

5.3.1 Dynamic Resource Allocation

The peak and average power demands differ considerably for various types of electric loads in a vehicle, as previously discussed. It is impractical to design a power system that can simultaneously meet the peak power demands of all electric loads. Automotive power and energy management systems should not only optimize the available sources on-board, but also match them properly with the loads. It prioritizes and schedules the load request and allocates power/energy dynamically based on need and criticality.

5.3.2 Practical Constraints of Vehicle Components

Components in vehicle power systems usually have practical limitations and impose constraints to the optimization of power and energy management algorithms. Battery — the most common energy storage device in vehicles — has a limited power and energy capacity and, more importantly, cycle life and sensitivity to temperature. Electric machines are used in vehicle drivetrains both as motors and generators. The power output and efficiency of these electric machines are not fixed but are functions of machine speed, torque, and temperature. These limitations and constraints have a direct impact on the optimization of the power management system.

5.3.3 Uninterruptible Power Availability

Some on-board electric loads in a vehicle are mission- and safety-critical. They mandate uninterruptible power availability at all times. Examples include steer-by-wire, brake-by-wire, "Identification Friend or Foe" systems (particularly in military systems), and safety restraint subsystems. It is advantageous to have one or more backup batteries in the automotive power system to meet the power demand for these mission-critical loads for at least a short time period in case of a primary power source interruption. The backup or secondary battery usually has a limited energy and power capacity to minimize its weight and cost. The charging and discharging cycles of the backup battery should be minimized to maintain a long operational life. It should also be disconnected via a PCU when not needed to maintain its state of charge. The on-board power and energy management system needs to meet these requirements.

5.3.4 Power Quality

It is important to maintain the power quality of automotive power buses to guarantee the safety and proper operation of all electric loads and PCUs. Automotive environments are extremely noisy and subject to various types of transients caused by switching large currents through inductive loads. A worst-case scenario is the so-called "load-dump event," where the battery is inadvertently disconnected while still undergoing a charging process by an alternator or generator. Load dump can also occur when a large load is suddenly disconnected either purposely or inadvertently, leading to voltage spikes. If not suppressed, a high-voltage spike would appear on the power bus for a certain time period and potentially endanger all electronic modules connected on the power bus. The bus voltage can also decrease below a desirable range possibly due to the low-voltage battery level at low temperatures (or due to sudden overload). The on-board power and energy management system needs to address these issues.

5.3.5 System Stability

DC power systems that employ multiple switch-mode power converters are known to be prone to instability because of the high degree of sensitivity to parameter and load variation. In particular, automotive power systems may be susceptible to large signal stability concerns. Large signal stability refers to the ability of the system to transit from one steady-state operating point to another following a disturbance, such as change in power demand, loss of power sources, short circuits, and open circuits. In addition, the power system dynamic is affected by interconnection between its components.

The stability of automotive power systems depends on several factors, such as switching off power electronic converters, nonlinearities of magnetic components, self-protection operation of power electronic circuits, and temperature variations. A well-designed automotive power management system can maintain system stability by managing the loads properly according to the operating conditions of power sources and distribution systems, and ensure that the system always operates around its nominal power.

5.3.6 Fault Diagnosis and Prognosis

With the increasing complexity of electrical/electronic architectures, it is desirable to provide fault diagnosis and prognosis to the automotive electric power management and distribution system. These features allow detection of fault conditions, such as short circuits and open circuits, and graceful degradation of the system performance to avoid hard, fail-stop behaviors. Overall fault detection or tolerance can be accomplished by the automotive power management and distribution system, which monitors and isolates a fault condition, and possibly reconfiguring the system to minimize the impact of the fault condition. In a more sophisticated scheme, historical data on the behaviors of various components can be recorded and analyzed. This information can be used to predict the state or condition of these components or subsystems and can provide early warnings for the components or modules that are degrading but have not yet failed catastrophically.

5.4 CASE STUDY: GAME-THEORETIC OPTIMAL HYBRID ELECTRIC VEHICLE MANAGEMENT AND CONTROL STRATEGY

Interests in hybrid electric vehicles originate from the general concerns about emission pollution and fuel efficiency that are associated with the conventional gasoline- or diesel-powered vehicles. Hybrid electric vehicles are powered by at least two energy sources. Typically, these sources are an electric motor (EM) and an internal combustion engine. In this section, a control and management strategy is proposed for parallel HEVs, based on the ideas from game-theoretic optimization approach [3, 9, 10]. This strategy has a

Optimal Power Management and Distribution in Automotive Systems

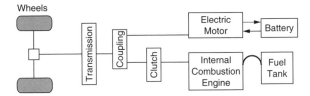

Figure 5.3 A single-shaft parallel arrangement.

pronounced effect on the performance of the vehicle powertrain, in terms of fuel consumption and torque delivery. To further narrow the scope, a parallel, single-shaft powertrain shall be assumed. It is argued that the strategies introduced could generally be applied to other powertrain arrangements, making appropriate amendments. In this control strategy, the game "players" would be the individual power sources, such as EM and ICE, and the "strategies" of players would be their alternating states. The objective of the players is to maximize their payoff, where the payoff is a function of the powertrain efficiency.

5.4.1 System Dynamics

We focus on a parallel single-shaft arrangement, as shown in Figure 5.3.

The components in this powertrain operate concurrently and are to be controlled by a core control system. Before the control strategy can be explored, it is necessary to define the governing mechanical relationships that are present in this system. To completely define the operating point of the powertrain in question, we require the value of three variables [7]:

T_c: Torque from electric motor
T_{ice}: Torque from IC engine
k: gear ratio

In fact, using Equations 5.1 and 5.2, it can be seen that only two of the three variables are independent when defining system states, as the third one can be calculated using the other two.

$$\omega_\omega(t) = \frac{\omega_{ice}(t)}{R(k)} = \frac{\omega_e(t)}{R(k)\rho} \tag{5.1}$$

$$T_\omega(t) = \eta_{gb} R(k)\left(T_{ice}(t) + \rho\eta_e T_e(t)\right). \tag{5.2}$$

where
  t = sample time (s)
  ω_ω = speed of the wheel (revolution/s)
  ω_{ice} = speed of ICE
  ω_e = speed of EM
  $R(k)$ = reduction ratio in terms of k (gear ratio)

T_ω = Torque at the wheels (N·m)
T_{ice} = Torque of ICE
T_e = Torque of EM
η_{gb} = efficiency of gearbox
η_e = efficiency of EM reductor (coupling)
ρ = constant reduction ratio at EM reductor

It must be noted that is a constant ratio calculated such that both engine and motor achieve their maximum speed simultaneously. Additionally, the gear number, at any given time, is taking values from an admissible set, typically (1, 2, 3, 4, 5, …). During simulation, $\omega_\omega(t)$ can be given by a "driving cycle," and then $T_\omega(t)$ can be calculated using the vehicle's software model. Matlab/Simulink models can mimic the mechanical behavior we have described for varying system specifications. In addition to the mechanical system, the battery is another dynamic system that requires accurate modeling. The battery's charging/discharging rates and other parameters will have an important effect on the capabilities and the control of the powertrain as a whole. The state of charge (SOC) $x(t)$ of a battery can be expressed in the following equation [7]:

$$x(t+1) = x(t) + P_e\left(\omega_e(t), T_e(t)\right) A_{cc}\left(x(t), T_e(t)\right) \Delta \tag{5.3}$$

where

t = sample time (s)
$x(t)$ = SOC
P_e = power at battery level (W)
A_{cc} = battery charge acceptance rate
Δ = sampling period (s)

5.4.2 Strategy Design

The interaction between discrete and continuous components can cause difficulties in controller design. A discrete controller issuing commands that are incompatible with the state of the continuous system could cause catastrophic results [9]. Our solution is to generate continuous controllers and consistent discrete abstractions for the closed loop system. We are proposing to use a hierarchical structure that utilizes a centralized information sharing system (Figure 5.4). This structure avoids myopic decisions [3]. The top

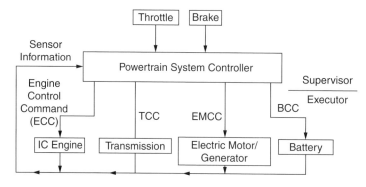

Figure 5.4 Hierarchical structure of powertrain control system.

level of this hierarchy is management/supervision, while assuming conventional all lower level controllers. Therefore, we divide the controller structure into only two levels: supervision and execution.

Dealing with the supervising controller level, the dynamics of the system are best described by discrete-event equations. For example, we could define the control algorithm as a simple thermostat on/off strategy: If the battery SOC is greater than a certain level, use battery (motor) only, otherwise use engine only. Obviously such a strategy is very simplistic and makes no effort to maximize the global system efficiency. However, treating the system states as a series of discrete events, or "flags," is a useful tool that we can exploit.

As the first step, the following sub-optimal yet practical strategy is set:

1. We shall divide the system trajectory (set of state variables as functions of time t) into small sections in time such that each section has a corresponding sampling period.
2. Optimization shall occur over each sampling period to maximize the payoff function (or minimize a cost function). It is important to note that we are now optimizing over a time period rather than over the entire "trip." The control solution may not, as a result, be absolutely optimal over the entire process. However, such a strategy is practical, as in reality it is rare to have a completely predefined drive cycle.

5.4.3 Game-Theoretic Approach

Game theory is a distinct approach to study human behaviors. Although it was originally used for economic and behavior applications, it has recently been a subject of research in engineering applications ranging from network traffic optimization to shop floor control. In terms of applying the game-theoretic approach, the following questions, associated with the HEV control problem, have to be addressed [10]:

1. What does it mean to choose strategies "rationally" when outcomes depend on the strategies chosen by others and when information is incomplete?
2. In "games" that allow mutual gain (or mutual loss), is it "rational" to cooperate to realize the mutual gain (or avoid the mutual loss) or is it "rational" to act aggressively in seeking individual gain regardless of mutual gain or loss?
3. If the answer to 2 is "sometimes," in what circumstances is "aggression" rational and in what circumstances is "cooperation" rational?
4. Do ongoing relationships differ from on/off encounters in this connection?

A cooperative game can be characterized by a tuple $(X, u_i, J_i: i \in X)$, where X is the set of players, u_i and $J_i(u_1, u_2, ..., u_n)$ are the corresponding strategy and cost functional of the ith player. The cooperative game optimization problem is then defined as find a set of optimal strategies $(u_i^*, i = 1, 2, ..., n)$ such that:

$$J_i\left(u_1, \cdots, u_{i-1}, u_i^*, u_{i+1}, \cdots, u_n\right) \geq J_i\left(u_1^*, \cdots, u_i^*, \cdots, u_n^*\right),$$

$$\forall u_j \neq u_j^*, j \neq i, i = 1, 2, \ldots, n,$$

In proposing a game theoretic solution to the powertrain management problem, it is necessary to apply the outlined issues directly to defining the "game." Since we know a great deal about the dynamic system that we are controlling, it is possible to dismiss many game classes as being not applicable.

The strategy is privy to all relevant sensor and executor level information, so the powertrain control problem can be treated as a cooperative game. To define the game, a payoff matrix is created for "players" (the powertrain elements). As a simplification, we treat the powertrain control system as a finite state machine. The possible strategies of the control system are as follows:

1. Provide tractive power with the ICE only
2. Provide tractive power with the EM only
3. Use some ICE power to drive the EM as a generator to charge the battery, and use the remaining ICE power to provide tractive power
4. Slow the vehicle (providing no tractive power), and let the wheels drive the EM as a generator to charge the battery
5. Provide tractive power with both the ICE and EM

We consider strategies 1–4 to be single value states while strategy 5 takes a finite set of values. For example, it could take only three values: 75/25, 50/50, and 25/75 splits between ICE and EM powers, stated as a percentage of the total torque demand. Each value serves as a strategy for both "players" and will have various "payoffs" for different inputs. The control strategy will determine the equilibrium for the relevant time period and dictate the system response.

The powertrain control "game" is then represented as a tuple of the form $(X, S_i, P_i, i \in X)$, where X is the set of players, S_i is the set of strategies, and P_i is a function that describes the payoff for each player in playing the strategies. Efficiency maps have been used to determine optimal power distribution through the powertrain system as shown in Figure 5.5 [14].

In the powertrain control game, the two players (ICE and EM) share common goals while still having their own payoff criteria:

Shared Payoff: brake and throttle pedals must act accurately and seamlessly to maximize efficiency of the entire powertrain
Electric Motor Agenda: keep SOC as steady as possible within a given range
Internal Combustion Engine Agenda: minimize consumption of fuel and reduce emission pollutants

The payoff function for powertrain control can be defined as follows:

$$P_{ice} = E_{ice}(k, T_{ice}) + X(T_{ice}, \omega_{ice}) + \lambda_{ice}(v) \tag{5.4}$$

$$P_e = E_e(k, T_e) + Ax + \lambda_e(v) \tag{5.5}$$

where
P_{ice} = payoff function for ICE "player"
P_e = payoff function for EM "player"

Optimal Power Management and Distribution in Automotive Systems

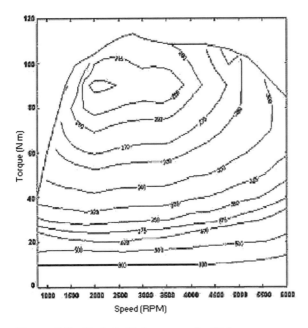

Figure 5.5 Typical efficiency map for EM.

E_{ice} = efficiency function (based on efficiency map) for ICE
E_e = efficiency function (based on efficiency map) for EM
X = predicted emissions cost for next ICE state + operating point
A = multiplicative constant to account for importance of SOC to EM payoff function
x = state of charge of battery
λ_{ice} = state-to-state transition function for ICE
λ_e = state-to-state transition function for EM
ν = current state of the system

Therefore, the payoff is a function of the current system state, the SOC, and the current system operating point. It is important to distinguish between the current state and the current operating point. The current state is the previous decision of the control. The current operating point consists of the current speeds of the different axles and the associated torques as well as the gear number. The accuracy of the operating point (discretization error) is an issue; however, it is postulated to have little effect as long as reasonable precision is used. A sample payoff matrix is shown in Table 5.2.

Table 5.2 Payoff Matrix

ICE Only	EM Only	Change w/ICE	Regen. break	Hybrid 25/75	Hybrid 50/50	Hybrid 75/25
77,82	86,78	86,40	10,8	91,84	85,83	88,84

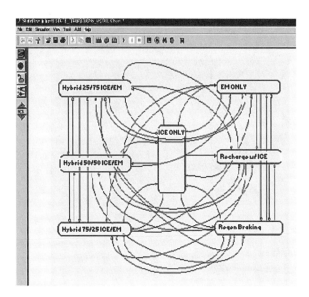

Figure 5.6 Stateflow machine used for HEV parallel powertrain control.

It is important to note that the entries in this matrix are of the form P_{ice}, P_e and that this matrix is only valid for a given SOC, current state, and operating point, and that the values within the matrix are actually dependent on these. The far right entry in the table would suggest that the ICE supplies 75% of the total requested torque while the EM supplies 25%.

Inspecting Table 5.2 qualitatively, it is important to note that we choose a state where the battery is at full SOC, and traveling at a speed such that the efficiency of the ICE and EM have a comparable waveform.

Using the proposed design, a Simulink model of the control system for HEV is generated using Stateflow for a generic model of the entire powertrain, as illustrated in Figure 5.6.

The designed control module is inserted into a parallel HEV powertrain laid out in Simulink for off-line simulation. The payoff functions described in Equations 5.4 and 5.5 are used as criteria for the event-based state transitions. The design includes several constants that are not defined with numerical values. Using this implementation strategy, a parametric analysis was performed to determine legitimate values for these multiplicative constants *A* and *X*.

5.4.4 SIMULATION RESULTS

Off-line simulations are performed using the design proposed in Section 5.3. The Simulink models return relevant numerical data for comparison of vehicle powertrain specifications such as NOx emissions and fuel consumption information. Table 5.3 highlights some of these findings. Results for a conventional ICE vehicle of similar dimensions are also shown as a reference.

The drive cycles used for the sample data are displayed in Figure 5.7. Figure 5.8 displays several waveforms obtained after running a generic, comparable HEV model using our game-theoretic control algorithm. The default control strategy included in ADVISOR [5] is used as our benchmark for comparison.

Optimal Power Management and Distribution in Automotive Systems

Table 5.3 Performance Results from Simulation

Control	Fuel Economy (Drive Cycle)	HC	CO	NOx
Game-Theoretic	44.6 (Urban)	0.425	2.050	0.444
	27.6 (Bus)	1.815	3.742	0.372
	77.1 (45 mph)	1.73	6.15	0.85
ADVISOR HEV (Parallel Powertrain)	40.4	0.52	2.468	0.407
	21.9	1.779	8.63	0.845
	70.2	2.884	11.07	1.09
Conventional ICE	35.5	0.588	2.517	0.415
	19.4	1.931	8.641	0.994
	55.6	3.641	13.474	2.137

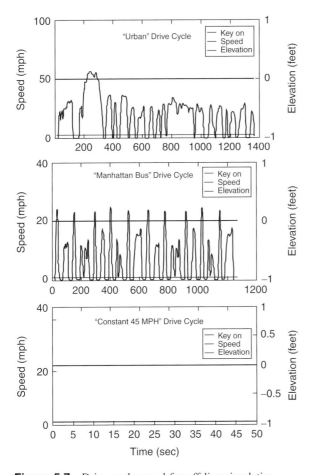

Figure 5.7 Drive cycles used for off-line simulation.

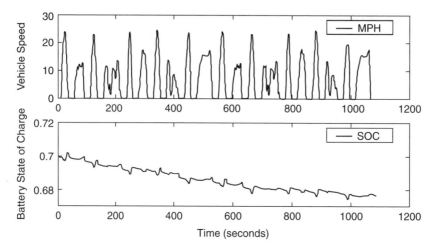

Figure 5.8 Matlab plots for given drive cycle and battery state of charge.

In comparing performance results for each control case, vehicle parameters such as weight and engine size were kept constant between cases wherever possible. This is to promote a fair comparison of the powertrain control methods.

5.5 SUMMARY

In this chapter, we have discussed power management and distribution strategies in automotive systems. It is argued that, due to the complexity of various power devices in vehicles, high power or low power, optimal management and control is needed for optimizing power usage. Such a strategy is very important for both vehicle efficiency and emission reduction. To echo our arguments, a game-theoretic optimal HEV power management and control strategy is presented and the simulation results show that, with the optimal cooperation between the internal combustion engine and the electric motor, both fuel efficiency and emission reduction can be satisfactorily achieved.

REFERENCES

1. K.E. Bailey, S.R. Cikanek, and N. Sureshbabu, "Parallel Hybrid Electric Vehicle Torque Distribution Method," *Proc. American Control Conference*, pp. 3708–3712, 2002.
2. K.E. Bailey and B.K. Powell, "Dynamic Modeling and Control of Hybrid Electric Vehicle Powertrain Systems," *IEEE Control Systems Magazine*, pp. 17–33, 1998.
3. D. Ben-Arieh and M. Chopra, "Evolutionary Game-Theoretic Approach for Shop Floor Control, " *Proc. IEEE International Conference on Systems, Man, and Cybernetics*, Vol. 1, pp. 463–468, 1998.
4. A. Brahma, Y. Guezennec, and G. Rizzoni, "Optimal Energy Management in Series Hybrid Electric Vehicles," *Proc. American Control Conference*, pp. 60–64, 2000.
5. S. Burch, M. Cuddy et al., "ADVISOR 2.1 Documentation," *National Renewable Energy Laboratory*, 1999.
6. S.R. Cikanek and K.E. Bailey, "Regenerative Braking System for a Hybrid Electric Vehicle," *Proc. American Control Conference*, pp. 3129–3133, 2002.

7. S. Delprat et al., "Control strategy optimization for an hybrid parallel powertrain," *Proc. American Control Conference*, pp. 1315–1320, 2001.
8. A. Emadi and M. Ehsani, "Electrical System Architectures for Future Aircraft," *Society of Automotive Engineering (SAE)*, 1999-01-2645, 1999.
9. J. Lygeros, "Hierarchical, Hybrid Control of Large Scale Systems," Ph.D. dissertation, University of California, Berkeley, 1996.
10. R.A. McCain, "Strategy and Conflict: An introductory sketch of game theory," Drexel University, 1999.
11. P. Nicastri, "42 V Powernet Implementation Issues," *Proc. IEEE Workshop on Power Electronics in Transportation (WPET)*, pp. 3–9, 2000.
12. E. Panizza et al., "Control Systems for Efficiency Improvements of Electric Vehicles," *SAE 94C049*, 1994.
13. C.A. Rabbath, H. Desira, and K. Butts, "Effective Modeling and Simulation of Internal Combustion Engine Control Systems," *Proc. American Control Conference*, pp. 1321–1326, 2001.
14. M. Salman et al., "Control Strategies for Parallel Hybrid Vehicles," *Proc. American Control Conference*, pp. 524–528, 2000.
15. J. Shen, A. Abul, V. Garg, and J. Monroe, "Automotive electric power and energy management — A system approach," *Global Automotive Manufacturing and Technology*, pp. 53–59, April 2003.
16. G.A. Williams, "The Challenges and Opportunities for Electrical Power Systems," *SAE 944072*, 1994.
17. R. Zhang and Y. Chen, "Control of Hybrid Dynamical Systems for Electric Vehicles," *Proc. American Control Conference*, pp. 2884–2889, 2001.

PART II
Automotive Semiconductor Devices, Components, and Sensors

6

Automotive Power Semiconductor Devices

Zheng John Shen
University of Central Florida, Orlando, Florida

6.1 INTRODUCTION

Power electronics plays an increasingly important role in improving vehicle performance, fuel economy, emission, safety, and comfort. Power semiconductor devices are widely used in automotive power electronic systems, and often dictate the efficiency, cost, and size of these systems. Active power semiconductor switches such as MOSFETs and IGBTs serve as load drivers for motors (ranging from 75 kW AC traction motors to 1W DC motors), solenoids, ignition coils, relays, heaters, lamps, and other automotive loads. Diodes are used in automotive systems to rectify AC current generated by the alternator, provide freewheeling current path for IGBTs or MOSFETs in DC/AC inverters and DC/DC converters, and suppress voltage transients. An average vehicle nowadays has over 50 actuators, which are often controlled by power MOSFETs or other power semiconductor devices. It was estimated that the power semiconductor content was around $100 to $200 in a mid-range conventional passenger vehicle in 2000. The power semiconductor contents may increase by three- to fivefold in an electric or hybrid vehicle.

Despite the wide variety of applications, the use of power semiconductor devices in automotive systems can be classified into three basic configurations: the low-side switching, the high-side switching, and the half-bridge switching, as shown in Figure 6.1. In the low-side switching configuration, the power switch is located between the load and the negative terminal of the battery and controls energy flow by opening or closing the power return path. Note that the negative terminal of the battery, usually connected to the vehicle chassis, forms the ground for automotive electrical systems. The control of the power

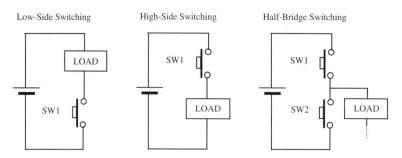

Figure 6.1 Basic automotive load switching configurations: low-side, high-side, and half-bridge switching.

switch is simple and straightforward, since the control terminal is in reference to ground. In the high-side switching configuration, the power switch is located between the load and the positive terminal of the battery and controls energy flow by opening or closing the power supply path. The high-side switching configuration offers a unique advantage of protecting against inadvertent load short events in which the load may be short-circuited to ground (i.e., vehicle chassis). Because the power switch is located upstream, a load short does not cause a catastrophic battery short as in the case of the low-side switching configuration. However, the control of the power switch in the high-side configuration is more complicated and costly than the low-side configuration, since now the control terminal of the power switch is in reference to the load instead of the ground. A control voltage higher than the battery voltage is generally needed to turn on and off the power switch. As a result, some type of charge pump or level shifting circuitry is needed. An alternative solution is to use a complementary type of power switch such as a P-channel MOSFET. In either case, there will be an increase in component cost. Half-bridge configuration is used to form H-bridge or three-phase bridge circuits. H-bridge circuits are typically used in bidirectional DC motor control and single-phase DC/AC inverter applications. Three-phase bridge inverters are used in electric drivetrain and electric power steering motor control applications.

An ideal power semiconductor switch should have the following characteristics:

- Block large forward and reverse voltages in off mode
- Conduct large currents in on mode
- Switch between on and off instantaneously without incurring switching losses
- Ease of control
- Rugged and reliable
- Low EMI during switching
- Low cost

However, no single semiconductor device possesses all these ideal properties in reality. Depending on actual applications, there may be one or more types of devices available that offer performance sufficiently close to these ideal device properties within the specified voltage, current, and switching frequency ranges. However, it is the deviations of these practical devices from the ideal power switch, or the nonideal properties, that essentially dictate the performance and cost-effectiveness of power electronic systems. The operation of all practical power semiconductors is limited by a series of ratings that

Automotive Power Semiconductor Devices

Table 6.1 Automotive Power Systems and Voltage Ratings of Power Semiconductor Devices

Automotive Power Systems	Battery Voltage	Nominal Op. Voltage	Max. Op. Voltage	Max. Dynamic Over-voltage	Power Device Voltage Ratings
14 V Car/light truck	12 V	14 V	24 V (Jump start)	—	30–60 V
28 V Heavy truck	24 V	28 V	34 V (Regulator failure)	—	75 V
42 V Powernet	36 V	42 V	50 V	58V (Load Dump)	75–100 V
Hybrid vehicle	150–200 V	150–200 V	—	—	600 V
Electric or fuel cell vehicle	300–400 V	300–400 V	—	—	600 V

define the operating boundaries of the device. These ratings include limits on the maximum forward and reverse voltages, maximum peak and continuous currents, maximum power dissipation, and maximum device junction temperature.

The voltage rating of a power device is primarily related to the maximum forward or reverse voltage that the device can sustain in the automotive power electronic circuit. Internal combustion engine (ICE) based conventional passenger vehicles typically use a 12 V lead-acid battery to supply electric power to all on-board electrical/electronic components. The DC bus voltage is actually close to 14 V, and commonly referred to as the 14 V system. The maximum operating voltage for 14 V systems is specified at 24 V, representing a double battery jump-start condition. For some heavy trucks, a 28 V power bus is used in which the battery voltage is 24 V. The maximum operating voltage for 28 V systems is 34 V, simulating the condition when the voltage regulator inside the alternator fails. Recently, the 42 V electrical system (with a 36 V battery), or the 42 V PowerNet, has been developed to meet the ever-increasing on-board electrical power demand. The maximum operating voltage for 42 V systems to be supplied by the generator is specified as 50 V (including ripples). Hybrid electric vehicles typically operate on a DC bus of 150 to 200 V, while pure electric or fuel cell vehicles use a 300–400 V DC bus. The typical voltage ratings of power devices in various automotive power systems are summarized in Table 6.1.

Note that the voltage ratings of power semiconductor devices are considerably higher than the specified maximum operating voltage or battery voltage of the power systems. This is because the voltage rating of automotive power electronics is mainly determined by the survivability of these devices to the commonly encountered overvoltage transients in the automotive environment, instead of just the maximum operating voltages. The transients on the automobile power supply range from the severe, high energy transients generated by the alternator/regulator subsystem to the low-level noise generated by the switching of inductive loads such as ignition coils, relays, solenoids, and DC motors. A typical automotive electrical system has all of these elements necessary to generate undesirable transients. It is critical that automotive power semiconductor devices have sufficient voltage ratings to sustain these electrical transients. (Later on we will discuss another device attribute, the avalanche capability, to sustain relatively low energy transients.)

The current ratings of a power semiconductor are mainly related to the energy dissipation and the junction temperature in the device. The maximum continuous current is usually defined as the current that the device is capable of conducting continuously without exceeding the maximum junction temperature. A maximum pulsed current is often specified for the allowable peak current the device can safely handle under a 10 μs pulsed condition, which is significantly higher than the continuous current rating. With the advancement of silicon processing and device technologies, the continuous and pulsed current ratings of power devices are sometimes limited not by the junction temperature, but rather by the device package (mainly the current-carrying capability of wire bonds).

The maximum power dissipation specifies the power dissipation limit that takes the junction temperature to its maximum rating while the ambient temperature is being held at 25°C. The maximum junction temperature represents the maximum allowable junction temperature of the device under normal operation. It is based on long-term reliability data. Exceeding this value will shorten the device's long-term operating life. Currently 150°C is specified as the maximum junction temperature for most power semiconductor devices. However, there are a limited number of automotive power devices on the market from several semiconductor manufacturers that are rated at 175°C or even 200°C.

Commercially available power semiconductor devices can be categorized into several basic types such as diodes, thyristors, bipolar junction transistors (BJT), power metal oxide semiconductor field effect transistors (MOSFET), insulated gate bipolar transistors (IGBT), and gate turn-off thyristors (GTO). In addition, there are power integrated circuits (ICs) and smart power devices that monolithically integrate power switching devices with logic/analog control, diagnostic, and protective functions. In this chapter, we will focus our discussion on the three most commonly used devices in automotive power electronics: diodes, low-voltage power MOSFETs, and high-voltage IGBTs. In addition, we will briefly discuss power ICs and smart power devices as well as two emerging device technologies: the silicon superjunction power devices and SiC power devices.

6.2 DIODES: THE RECTIFICATION, FREEWHEELING, AND CLAMPING DEVICES

Diodes are the simplest semiconductor device, comprising of a PN or Schottky junction with two external terminals. Diodes are used in applications that require current to flow in one direction only. In automotive applications, diodes are mainly used to perform the following functions:

- Rectify AC current from the alternator to DC current that charges the battery.
- Allow load current freewheeling as the anti-parallel diodes for IGBTs or MOSFETs in inverter or converter applications. A freewheeling diode provides an alternative path for the inductive load current when the main IGBT or MOSFET turns off, and prevents high Ldi/dt voltage spikes from damaging the main switching devices.
- Suppress voltage transients when being reverse biased (Zener diodes).

In attempts to improve its static and dynamic performance, numerous diode types have been developed. Appropriate selection of diode types is required for different automotive applications.

Automotive Power Semiconductor Devices

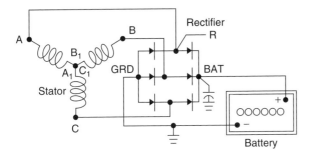

Figure 6.2 Simplified rectifier bridge in alternator charging circuit.

6.2.1 RECTIFIER DIODES

The first type of automotive application of diodes is rectification. Figure 6.2 shows a simplified automotive rectifier circuit that converts the AC current generated by the automotive alternator (a three-phase AC generator) into DC current to charge the battery and provide electric power to on-board electronic modules. For avalanche alternator rectifier applications, the basic parameters characterizing the diodes are the maximum average forward rectified current, peak forward surge current, maximum blocking voltage, forward voltage drop at the rated current, and reverse avalanche energy capability for load dump protection. The high temperature derating of the maximum output rectifier current also deserves special consideration. Avalanche alternator rectifiers are a very mature semiconductor product with good reliability and durability.

6.2.2 FREEWHEELING DIODES

The second type of automotive application of diodes is freewheeling. For freewheeling applications, fast-recovery rectifiers are used in conjunction with active power switches such as IGBTs or MOSFETs in various types of power converters. In addition to the voltage rating, current rating, and forward voltage drop, the recovery characteristics of freewheeling diodes dictate the selection of rectifiers for fast switching power circuits. A power diode requires a finite time to switch from off-state (reverse bias) to on-state (forward bias) and vice versa. Both the recovery times and the shapes of the waveforms are affected by the intrinsic properties of the diode and by the external circuit.

Figure 6.3 shows a power diode switching from a blocking state to on-state and then subsequently switching back to the blocking state. The switching properties are often provided in the manufacturer's data sheet for diode current with a specified time rate of change di/dt, which is determined by the testing setup. Figure 6.4 illustrates a typical testing circuit for characterizing diode recovery properties. Since the time constant L/R is much longer than the transient recovery times of the diode under test, the load current can be considered as constant during the testing time window. An active switch (usually a power MOSFET) is used to pre-charge the load current and force the diode into forward and reverse recovery processes. A source inductor L_S along with its own freewheeling diode is used to define the rate of change of the diode current di_F/dt. This testing circuit reproduces the operating condition for a freewheeling diode in DC/DC buck converters or full-bridge and three-phase inverters.

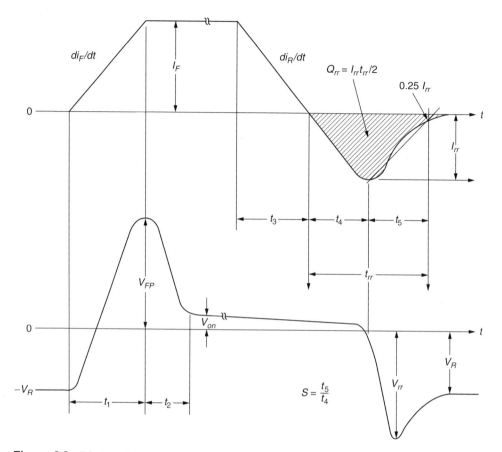

Figure 6.3 Diode switching waveforms [1].

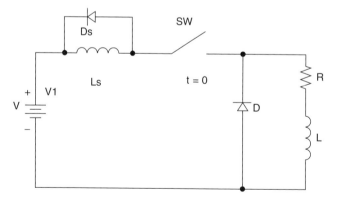

Figure 6.4 Diode forward and reverse recovery test circuit.

The recovery process from off- to on-state is termed "forward recovery," during which a visible voltage overshoot V_{FP} is observed across the diode, as shown in Figure 6.3. This voltage overshoot is typically several tens of volts, large enough to adversely impose higher electric stress and turn-off switching losses on the main power switching device. During forward recovery testing, the switch is closed for the time $t < 0$. The load current steadily ramps up and the diode D is reverse biased at $-V_R$ (off-state). At $t = 0$, the switch is opened. The diode becomes forward biased, providing a path for the load current in R and L. The diode current i_D rises to the peak load current I_F after a short time t_1 (rise time), and the diode voltage drop falls to its steady state value V_{on} after another time t_2 (fall time). The diode forward recovery time $t_{fr} = t_1 + t_2$ is the time needed for the charge in the diode to change from one equilibrium state (off) to the other (on), during which the depletion layer of the diode is first discharged and then the diode is forward biased. Conductivity modulation then takes place due to the growth of excessive carriers in the diode, accompanied by a reduction of series resistance. The voltage overshoot V_{FP}, which is mainly due to the large series resistance of the diode prior to full conductivity modulation, will gradually drop to the steady state diode forward voltage drop V_{on}.

The recovery process from on to off is termed "reverse recovery." During reverse recovery testing, the switch is closed for some time to charge the load. The load current steadily ramps up and the diode D is reverse biased. Next the switch is opened, and the diode D provides a freewheeling path for the load current. After the diode reaches a steady state with a low voltage drop, the switch is closed again. This is the onset of the reverse recovery process of the diode. The diode current i_D will gradually decrease while the current from the voltage source i_S gradually increases. The rate of change of the diode current di_R/dt with respect to time (opposite to that of the source current di_S/dt) is mainly determined by the source inductance L_S. During time interval t_3, excess carriers stored in the drift region are reduced in number by recombination with the decreasing forward current. However, even when the diode current i_D becomes zero, there are still excess carriers remaining in the drift region and the diode still shows a low forward voltage drop. During time interval t_4, the excess carriers are completely removed and the diode becomes unbiased. Note that a significant reverse recovery current i_{rr} is developed during this process. This reverse recovery current continues to sweep charges out of the PN junction and form a depletion region during time interval t_5. The depletion layer is needed to support the reverse voltage across the diode when it is in off-state. During this time period, the reverse diode current demanded by the stray inductance of the external circuit cannot be supported by the excess carrier distribution because too few carriers are left. The diode current simply ceases its growth in the negative direction and quickly falls to zero. The reverse current reaches its maximum value I_{rr} at the end of the interval t_4.

The time interval $t_{rr} = t_4 + t_5$ shown in Figure 6.3 is often termed the reverse recovery time of the diode. The total time integral of the reverse current i_{rr} during t_{rr} is termed the total reverse charge Q_{rr}, and is often estimated by $Q_{rr} = I_{rr}t_{rr}/2$. Diode data sheets often give detailed information on t_{rr}, Q_{rr}, and I_{rr}. These parameters are important in almost all power electronic circuits where freewheeling diodes are used. For example, the reverse recovery current of freewheeling diodes in a three-phase inverter is added to the load current that the active IGBTs or MOSFETs in the same leg have to conduct during turn-on process. In many cases, it is the reverse recovery characteristic of the freewheeling diodes that determines the turn-on loss of the active switch and, to a large extent, the total power loss of an inverter power module.

In addition, a "soft factor" parameter S ($S = t_5/t_4$) is often defined as the ratio between t_5 and t_4, as indicated in Figure 6.3. S is a good indicator of how soft or snappy the diode's switching behavior is. A soft diode is always desired to minimize electromagnetic interference (EMI) generated by the power electronic circuit. Note that all these reverse recovery parameters and other device parameters such as breakdown voltage and on-state voltage drop are interrelated to each other and often present conflicting requirements for

Table 6.2 Major Device Parameters of a Fast-Recovery Diode

Parameters	Values	Units	Conditions
$I_{F(AV)}$ Max. average forward current	20	A	@T_C = 97°C, half sine waves
V_{RRM} Max. reverse voltage	600	V	
V_F Forward voltage drop	1.2	V	@20 A, T_J = 25°C
t_{rr} Reverse recovery time	160	ns	I_F @20 A
I_{rr} Reverse recovery current	10	A	@100 A/µs
Q_{rr} Reverse recovery charge	1.25	µC	@25°C
S Snap Factor	0.6	—	

device design and fabrication. Various types of fast-recovery diodes with different tradeoffs among these device parameters are available on the market. Table 6.2 summarizes the major parameters of a commercial fast-recovery diode as an example. Circuit designers need to carefully select the most suitable diodes for their application.

6.2.3 ZENER DIODES

The third type of automotive application of diodes is to suppress voltage transients and clamp bus voltages in electronic modules. Diodes for this type of application operate in a reverse avalanche mode and are often called Zener diodes. As previously mentioned, it is important to protect automotive electronic equipment from the severe, high-energy transients generated by the alternator/regulator subsystem and the low-level noise generated by the switching of inductive loads such as ignition coils, relays, solenoids, and DC motors. The most severe transient encountered in the automotive environment is the so-called load dump transient. Load dump typically generates an exponentially decaying voltage that occurs in the event of a battery disconnect while the alternator is still generating charging current. The amplitude of load dump voltage depends on the alternator speed and the level of the alternator field excitation at the moment of battery disconnection. A voltage spike of 25 to 125 V can be easily generated in a 14 V system. The time duration of load dump transient ranges from several to several hundred milliseconds, depending on the configuration of the power generation and regulation system. A Zener diode central suppressor can serve as the principal transient suppression device for the entire vehicle. In this case, the Zener diode is connected directly across the main power supply and ground. It must absorb the entire available load dump energy and withstand the maximum operating voltage. Another approach is the so-called distributed transient suppression scheme, in which one or more Zener diodes is used in each of the electronic modules connected to the power bus. Locally generated transients can also be suppressed with this approach.

For clamping and transient suppression applications, the basic diode parameters include the breakdown voltages at different current levels (e.g., 100 mA and 90 A), repetitive peak reverse surge current, and reverse avalanche energy capability. Figure 6.5 shows a simplified load dump test setup and a typical load dump pulse current. In addition, the temperature dependence of the device parameters should be taken into consideration since automotive applications, especially the underhood applications, mandate an operating ambient temperature range of –40 to 125°C.

Automotive Power Semiconductor Devices

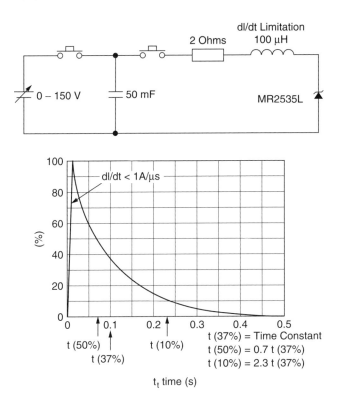

Figure 6.5 Load dump test circuit and typical load dump current waveform (courtesy of ON Semiconductor, Phoenix, AZ).

6.2.4 SCHOTTKY DIODE

As an alternative to PN diodes, Schottky diodes may present a viable solution for some automotive applications. Schottky diodes have several distinctive advantages over conventional PN diodes: low forward voltage drop (0.3–0.6 V vs. 0.7–1 V for PN diodes) and the absence of reverse recovery charge. However, silicon Schottky diodes suffer from low breakdown voltage (typically less than 200 V) and high leakage current, especially at high temperatures. Nevertheless, Schottky diodes may find applications in on-board switching mode power supplies, reverse battery protection circuits, and other low-voltage automotive applications. High-voltage Schottky diodes, possibly made from nonsilicon materials, are very desirable for freewheeling applications in high power inverter modules to reduce switching losses.

6.3 POWER MOSFETS: THE LOW-VOLTAGE LOAD DRIVERS

For power electronic applications with a voltage rating below 200 V, power MOSFETs are the device of choice. Power MOSFETs have replaced traditional power bipolar junction transistors (BJTs) as the load switch in many applications including automotive systems, because of their low on-state resistance, high switching speed, ease of control, and superior safe operating area (SOA) and device ruggedness. In addition, MOSFETs are preferred even for 200–600 V applications, which may require a high switching speed but only a

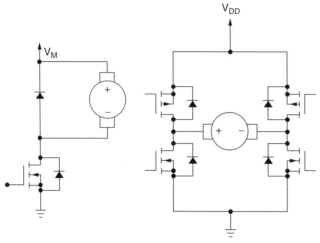

Low Side DC Motor Drive   H-Bridge DC Motor Drive

Figure 6.6 Power MOSFETs in a low-side and H-bridge DC motor drive circuit.

moderate power level. Figure 6.6 shows a low-side and an H-bridge DC motor drive circuit using power MOSFETs as the load drivers.

MOSFETs can be found in a wide variety of automotive subsystems including the following:

- Powertrain subsystems
  - Fuel injection solenoid driver
  - Electronic throttle control
  - Transmission gearshift solenoid driver
  - Cruise control
  - Electric radiator fan motor control
  - Integrated starter generator (ISG) motor control
  - Variable timing engine valve control
  - Glow plug current switch for diesel engines
- Chassis and safety subsystems
  - Electric power steering (EPS) and steer-by-wire motor control
  - Anti-lock braking system (ABS) and traction control system (TCS) solenoid drivers
  - Brake-by-wire motor control
  - Airbag activation switches
  - Active suspension motor control
- Body/comfort/convenience subsystems
  - Climate control motor and solenoid drivers
  - Light control switches
  - High intensity discharge (HID) lamp control circuits
  - Power door/window motor control
  - Power seat motor control
  - Windshield wiper control
  - Windshield and mirror heater drivers

Automotive Power Semiconductor Devices

- Electric power generation and distribution subsystems
  - Switching mode voltage regulator
  - Multiplex wiring and smart junction boxes (SJBs)
  - DC/DC converters between different bus voltages
  - Power suppliers for on-board logic and analog electronics

Automotive MOSFETs cover a wide range of voltage and current ratings. For conventional 14 V automotive systems, 30 and 40 V MOSFETs are usually used in half- and full-bridge circuits, while 55 and 60 V MOSFETs are typically used in single-ended, high- or low-side load control circuits. For 24 V battery systems (heavy-duty trucks) or the new 42 V PowerNet, 75 V devices are the top choice. For applications requiring higher boost voltages or shorter inductive load recovery times, 100 to 150 V MOSFETs can be used. MOSFETs with a breakdown voltage greater than 400V can be found in engine driver trains, HID headlight control circuitry, and some DC/DC converters. The current rating of automotive MOSFETs ranges from a few hundred mA to well over 100 A. Switching frequencies are typically in the 10–100 kHz range.

6.3.1 MOSFET Basics

A power MOSFET is a three-terminal device where the gate (i.e., the control terminal) controls the main current flow between the two output terminals: the drain and source. The source terminal is usually common to the gate and drain terminals. Power MOSFET output characteristics, that is, the drain current i_D as a function of drain-to-source voltage v_{DS} with gate-to-source voltage V_{GS} as a parameter, are shown in Figure 6.7. In power electronic applications, the MOSFET is used to switch back and forth between the cutoff region (off-state) and the Ohmic region (on-state). The MOSFET is in off-state when the gate-source voltage V_{GS} is less than the threshold voltage $V_{GS(th)}$, which is typically a few volts. The device is an open circuit and must have sufficiently high breakdown voltage BV_{DSS} to sustain the bus voltage applied to the circuit as well as voltage transients

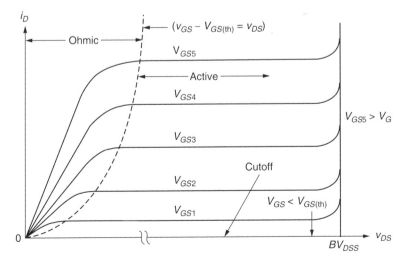

Figure 6.7 Output characteristics of Power MOSFET.

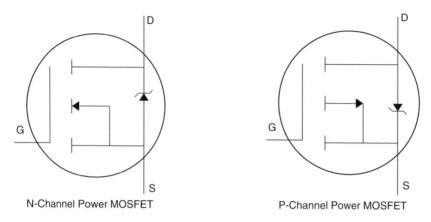

Figure 6.8 Circuit symbols of N-channel and P-channel power MOSFETs.

experienced in the harsh automotive environment. When the MOSFET is driven by a large V_{GS}, it is driven into the Ohmic region and behaves like a resistor. The on-state resistance of the MOSFET, usually referred to as the $R_{DS(on)}$, is generally considered the most important device parameter for the MOSFET. $R_{DS(on)}$ basically determines the conduction power loss and the maximum current and power ratings of the MOSFET.

There are two major types of MOSFETs: the N-channel and P-channel MOSFETs. N-channel MOSFETs, due to their smaller chip size and lower cost, dominate nearly all automotive applications. P-channel MOSFETs, which are simple to control for high-side switching circuits, can be found in some applications such as dashboard instrument and transmission control modules. Their circuit symbols are shown in Figure 6.8. The arrows indicate the direction of current flow if the body-source PN junction were forward biased. An N-channel MOSFET has the arrow pointing into the MOS channel, while the arrow of a P-channel MOSFET points outwardly. A diode is also included in a MOSFET symbol, which represents an integral part of a power MOSFET and is commonly referred to as the body diode.

Low-voltage power MOSFETs can be fabricated with either planar or trench technology. Figure 6.9 illustrates the cross-sectional views of a planar and trench MOSFET.

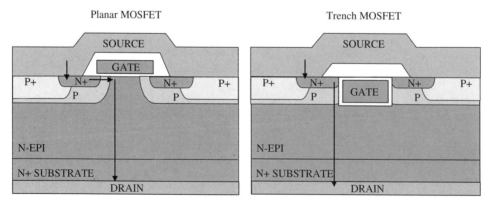

Figure 6.9 Cross sections of planar and trench power MOSFETs.

Automotive Power Semiconductor Devices

In the planar MOSFET, the drain electrode is formed on the back side (bottom side) of the silicon die, while the source and gate electrodes are formed on the top surface. The MOS channels are formed between the P-body and N+ source diffusions under the planar poly-silicon gate. Planar MOSFET technology is a mature technology with an inherently wide SOA, excellent device ruggedness, and high avalanche energy rating. However, the on-resistance per unit silicon area, or the specific on-resistance, of planar MOSFET is generally higher than that of the trench MOSFET. In a trench MOSFET, the poly-silicon gate is located inside a vertical trench, and the MOS channels are formed along the sidewalls of the trench. This constitutes the main structural difference between the two technologies. The trench MOSFET technology offers higher cell density, low specific on-resistance, and consequently smaller chip size and lower cost than the planar MOSFET technology. Trench MOSFETs used to suffer from reduced SOA and device ruggedness. However, tremendous improvement has been made in these areas, and now trench MOSFETs are gaining considerable popularity in automotive applications.

6.3.2 MOSFET Characteristics

Next, we will examine the data sheet of a commercial power MOSFET, International Rectifier's IRF2804, to gain familiarity with characteristics of power MOSFETs. IRF2804 is an automotive MOSFET rated at 40 V and 75 A. It can be packaged into an industry-standard leaded TO-220, surface mount D²PAK, or TO-262 package. Other commonly used packages for automotive applications include TO-247, DPAK, Micro-8, and SOT-223 to accommodate various silicon die sizes.

Table 6.3 lists the absolute maximum ratings of IRF2804. Among these parameters, the maximum continuous drain current I_D is rated at 75 A at a case temperature of 25°C, which is limited by the wire bonds of the package instead of the silicon junction temperature. All power transistors have a specified maximum peak current rating. This is conservatively set at a level that guarantees reliable operation and it should not be exceeded. It is often overlooked that, in a practical circuit, peak transient currents can be well in excess of the expected normal operating current. For example, high in-rush currents can be generated during the start-up of a motor or the turn-on of a cold incandescent lamp. High transient current can also be experienced by the MOSFET during the turn-on process, as a result of the reverse recovery of the companion freewheeling diode.

Maximum gate-to-source voltage is rated at ±20 V. Voltages above this limit may damage the gate oxide and permanently destroy the MOSFET. Caution should be exercised when handling MOSFET products during testing and assembly processes to avoid electrostatic discharge (ESD) damage. ESD transients can be as high as several thousands volts. Power MOSFET parts should be left in their anti-static shipping bags or conductive foam, or they should be placed in metal containers or conductive tote bins, until required for testing or connection into a circuit. The person handling the device should ideally be grounded through a suitable wrist strap. Devices should be handled by the package, not by the leads. Work stations and testing equipment should be placed on electrically grounded table and floor mats.

The single pulse avalanche energy E_{AS} and repetitive avalanche energy E_{AR} indicate the survivability of the MOSEFT under transient overvoltage stress commonly encountered in the automotive environment. The avalanche energy capability, also referred to as the unclamped inductive switching (UIS) capability, is another important device parameter for automotive applications (arguably the second most important device parameter, right after the on-state resistance $R_{DS(on)}$).

Table 6.3 Absolute Maximum Ratings of IRF2804

	Parameter	Max.	Units
I_D @ T_C = 25°C	Continuous drain current, V_{GS} @ 10 V (silicon limited)	280	A
I_D @ T_C = 100°C	Continuous drain current, V_{GS} @ 10 V (see Figure 6.9)	200	
I_D @ T_C = 25°C	Continuous drain current, V_{GS} @ 10 V (package limited)	75	
I_{DM}	Pulsed drain current ①	1080	
P_D @ T_C = 25°C	Maximum power dissipation	330	W
	Linear derating factor	2.2	W/°C
V_{GS}	Gate-to-source voltage	±20	V
E_{AS}	Single pulse avalanche energy (thermallly limited) ②	670	mJ
E_{AS} (tested)	Single pulse avalanche energy tested value ②	1160	
I_{AR}	Avalanche current ①	See Figures 6.12a,b, 6.15, 6.16	A
E_{AR}	Repetitive avalanche energy ②		mJ
T_J T_{STG}	Operating junction and storage temprature range	−55 to +175	°C
	Soldering temperature, for 10 sec	300 (1.6 mm from case)	
	Mounting torque, 6-32 or M3 screw	10 lbf · in (1.1 N · m)	

Source: International Rectifier data sheets and application notes: http://www.irf.com/.

Figure 6.10 shows the UIS testing circuit to specify E_{AS} and E_{AR}, which also resembles a practical low-side load switching circuit with an inductive load such as a solenoid or DC motor. Also shown in Figure 6.11 are the typical UIS waveforms of drain-to-source current and voltage of the device under test (DUT) MOSFET. A 0.24 mH load inductor L and a 25 gate resistor R_G are used in this UIS testing circuit.

A pulsed voltage signal is sent to the gate of the DUT MOSFET and subsequently turns it on. A power supply V_{DD} then starts to charge the inductor L through a driver MOSFET, which is already in on-state. Once the load current I_{AS} reaches a certain level, the driver MOSFET first turns off to isolate the power supply from the DUT. Note that the load current I_{AS} does not change instantaneously because the freewheeling diode provides an alternative current path. Immediately after the turn-off of the driver MOSFET, the DUT MOSFET is turned off. The electromagnetic energy stored in the inductor tends to maintain the current flow and forces the drain voltage V_{DS} to rise rapidly to exceed the breakdown voltage $V_{(BR)DSS}$ of the DUT MOSFET. The inductor will be fully discharged through the avalanched MOSFET, or more accurately, the avalanched body diode of the MOSFET. Increasing the gate pulse width t_p will raise the peak load current I_{AS} and the total amount of electromagnetic energy E_{AS} stored in the inductor, which is given by:

$$E_{AS} = \frac{1}{2} L \times I_{AS}^2$$

Automotive Power Semiconductor Devices

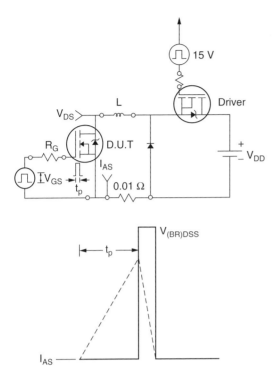

Figure 6.10 Circuit setup and typical waveforms of MOSFET UIS test (courtesy of International Rectifier, El Segundo, CA).

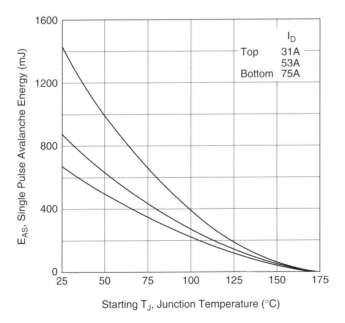

Figure 6.11 Avalanche energy E_{AS} as a function of the starting junction temperature for three different load currents (courtesy of International Rectifier, El Segundo, CA).

Note that this energy is completely dissipated in the avalanched MOSFET, resulting in an increase in its junction temperature. The avalanche energy rating E_{AS} of the MOSFET is the amount of energy allowed to increase the device junction temperature to the maximum junction temperature rating T_{JMAX}, which is 175°C for this device. Repetitively exceeding T_{JMAX} may cause concerns on the long-term reliability of the semiconductor device, but not necessarily instantaneous device failures. This is why the tested value of single pulse avalanche energy rating $E_{AS(tested)}$ is much higher than the thermally limited E_{AS}. However, once the junction temperature of the MOSFET increases to a range of 330 to 380°C, several internal device failure mechanisms may be triggered to cause the collapse of the device breakdown voltage and eventually the catastrophic failure of the device. One of the failure mechanisms is the activation of a parasitic BJT as an integral part of the MOSFET structure, as shown in Figure 6.9. The NPN parasitic bipolar transistor is formed among the N+ source ("emitter"), the P-body ("base"), and the N-type drain ("collector"). During normal operation of the MOSFET, this BJT is inactive since its base and emitter are shorted by the source metal of the MOSFET. However, when the body diode between the drain and P-body of the MOSFET is in avalanche mode, a large avalanche current will flow through the P-body and induce a voltage drop across the P-body resistance ("base resistance"). If this voltage drop exceeds 0.7 V, the emitter-base PN junction will be forward biased at certain locations and initiate the BJT activation process. Once activated, the BJT will demonstrate a snapback negative resistance characteristic, resulting in the collapse of the MOSFET breakdown voltage. It may subsequently lead to current crowding in localized areas and molten silicon or metal interconnects in these "hot spots."

Another failure mechanism is that the body PN diode simply loses its voltage blocking capability if the intrinsic carrier concentrations approach these of the P and N doping regions when the junction temperature rises to the 360 to 380°C range. The thermally generated carriers will generate an extremely high leakage current and lead to voltage snapback, localized current crowding, and catastrophic device failure, similar to the case of parasitic BJT activation.

Heat is generated during the avalanche process, but at the same time removed from the semiconductor junction by a thermal conduction mechanism in the silicon and package. These two competing processes eventually determine the actual junction temperature rise. When the avalanche energy pulse is short (high power pulse), there may not be enough time for the heat removal process to make a difference since it typically has a much larger time constant. Depending on the load inductance, load current, breakdown voltage, and starting junction temperature (the junction temperature at the beginning of UIS test), the MOSFET may sustain significantly different amounts of avalanche energy. Therefore, it is not enough to simply quote UIS energy without specifying the test conditions. Figure 6.11 shows IRF2804's single pulse avalanche energy E_{AS} as a function of the starting junction temperature for three different load currents.

In practical power electronic circuits, self-inflicted overvoltage transients can be produced when a power MOSFET is switched off, similar to the aforementioned UIS testing condition. Figure 6.12 shows how a voltage spike is produced when switching the device off, as a result of inductance in the circuit. The faster the device is switched, the higher the overvoltage will be.

Usually, the main inductive component of the load is clamped by a freewheeling diode, as shown in Figure 6.13. However, the power MOSFET is still subject to overvoltage transients produced by stray circuit inductance. Furthermore, the freewheeling diode may not provide an instantaneous clamping action, due to its forward recovery characteristic.

Automotive power MOSFETs have an avalanche energy rating that allows them to withstand these inductive spikes, assuming that the data sheet limits for energy and

Automotive Power Semiconductor Devices

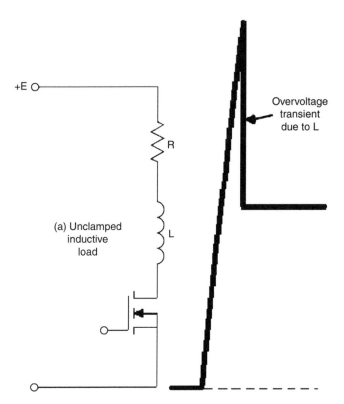

Figure 6.12 Drain-source overvoltage transient when turning off a MOSFET with unclamped inductive load (courtesy of International Rectifier, El Segundo, CA).

temperature are not exceeded. However, it is still desirable to reduce the voltage transients by minimizing stray circuit inductance with careful printed circuit board layout design.

Table 6.4 lists the electrical specifications of IRF2804. Among these parameters, the drain-to-source breakdown voltage $V_{(BR)DSS}$ is rated at a minimum of 40 V at a leakage current of 250 µA, a current level used to specify the breakdown voltage for most power devices regardless of their actual device size. $V_{(BR)DSS}$ has a positive temperature coefficient. The typical drain-to-source on-resistance $R_{DS(on)}$ is at 1.8 mΩ, measured at a gate voltage V_{GS} of 10 V and a drain current I_D of 75 A.

Power MOSFETs, like any other majority-carrier devices, have a tradeoff relationship between the breakdown voltage and on-resistance. This is because the low doping concentration of the N-epi region needed to provide a high breakdown voltage for the body PN diode also adds a high series resistance to the total on-resistance of the MOSFET, as shown in Figure 6.9. On-resistance increases dramatically with increasing voltage rating, as shown in Figure 6.14. A simple equation is often used to describe this tradeoff: $R_{DS(on)} \propto V_{(BR)DSS}^{2.7}$. For this reason, power MOSFETs are limited to low-voltage or high-voltage but low-power applications. This is also why the IGBT, a minority-carrier power device, dominates the high-voltage, high-power applications, as we will discuss in the next section. $V_{GS(th)}$ is the gate-to-source voltage at which the magnitude of drain current has been increased to a typical 250 µA. This parameter has a negative temperature coefficient.

A power MOSFET has three internal parasitic capacitances from terminal to terminal, as shown in Figure 6.15. C_{gs} is the dielectric capacitance between the polysilicon gate electrode and the source metal electrode. C_{gd} is the feedback capacitance between the polysilicon gate electrode and the drain electrode, which is often referred to as the Miller

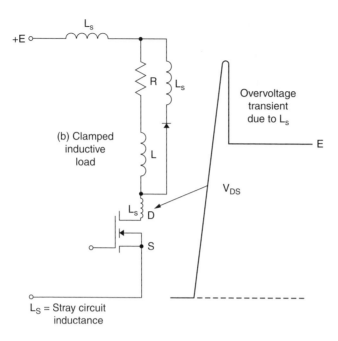

Figure 6.13 Drain-source overvoltage transient when turning off a MOSFET with an inductive load clamped by a freewheeling diode (courtesy of International Rectifier, El Segundo, CA).

capacitance. C_{ds} is the capacitance of the body diode between the drain and the P-body. However, MOSFET data sheets usually specify the input capacitance C_{iss}, output capacitance C_{oss}, and reverse transfer capacitance C_{rss} instead of C_{gs}, C_{gd}, and C_{ds}. C_{iss} is the capacitance between the gate and source terminals, with the drain terminal short-circuited to the source terminal for AC current. C_{oss} is the capacitance between the drain and source terminals, with the gate terminal short-circuited to the source terminal for AC current. C_{rss} is the capacitance between the drain and gate terminals, with the source terminal connected to the guard terminal of a three-terminal capacitor bridge. C_{iss}, C_{oss}, and C_{rss} are related to C_{gs}, C_{gd}, and C_{ds} by the following equations:

$$C_{iss} = C_{gs} + C_{gd}$$

$$C_{rss} = C_{gd}$$

$$C_{oss} = C_{ds} + C_{gd}$$

Note that all MOSFET capacitances are voltage dependent rather than of constant values. Figure 6.16 shows C_{iss}, C_{oss}, and C_{rss} as a function of drain-to-source voltage V_{DS}. The capacitances are measured with a 1 MHz capacitance bridge.

The turn-on or turn-off switching of a power MOSFET is essentially a charging or discharging process of the internal capacitors. MOSFET data sheets usually specify a turn-on delay time $t_{d(on)}$, rise time t_r, turn-off delay time $t_{d(off)}$, and fall time t_f, for a resistive load switching circuit, as shown in Figure 6.17. The definitions of these time parameters are given by the waveforms shown in Figure 6.18.

Table 6.4 Electrical Specifications of IRF2804

	Parameter	Min.	Typ.	Max.	Units	Conditions
$V_{(BR)DSS}$	Drain-to-source breakdown voltage	40	—	—	V	$V_{GS} = 0$ V, $I_D = 250$ μA
$\Delta BV_{DSS}/\Delta T_J$	Breakdown voltage temp. coefficient	—	0.031	—	V/°C	Reference to 25°C, $I_D = 1$ mA
$R_{DS(on)}$ SMD	Static drain-to-source on-resistance	—	1.5	2.0	mΩ	$V_{GS} = 10$ V, $I_D = 75$ A ④
$R_{DS(on)}$ TO-220	Static drain-to-source on-resistance	—	1.8	2.3		$V_{GS} = 10$ V, $I_D = 75$ A ④
$V_{GS(th)}$	Gate threshold voltage	2.0	—	4.0	V	$V_{DS} = V_{GS}$, $I_D = 250$ μA
gfs	Forward transconductance	130	—	—	S	$V_{DS} = 10$ V, $I_D = 75$ A
I_{DSS}	Drain-to-source leakage current	—	—	20	μA	$V_{DS} = 40$ V, $V_{GS} = 0$ V
		—	—	250		$V_{DS} = 40$ V, $V_{GS} = 0$ V, $T_J = 125$°C
I_{GSS}	Gate-to-source forward leakage	—	—	200	nA	$V_{GS} = 20$ V
	Gate-to-source reverse leakage	—	—	−200		$V_{GS} = -20$ V
Q_g	Total gate charge	—	160	240	nc	$I_D = 75$ A
Q_{gs}	Gate-to-source charge	—	41	62		$V_{DS} = 32$ V
Q_{gd}	Gate-to-drain ("Miller") charge	—	66	99		$V_{GS} = 10$ V ④
$t_{d(on)}$	Turn-on delay time	—	13	—	ns	$V_{DD} = 20$ V
t_r	Rise time	—	120	—		$I_D = 75$ A

Table 6.4 (continued) Electrical Specifications of IRF2804

Parameter	Min.	Typ.	Max.	Units	Conditions
$t_{d(off)}$ Turn-off delay time	—	130	—		$R_G = 2.5\ \Omega$
t_f Fall time	—	130	—		$V_{GS} = 10\ V$ ④
L_D Internal drain inductance	—	4.5	—	nH	Between lead,
L_S Internal source inductance	—	7.5	—		6 mm (0.25 in.) from package and center of die contact
C_{iss} Input capacitance	—	6450	—	pF	$V_{GS} = 0\ V$
C_{oss} Output capacitance	—	1690	—		$V_{DS} = 25\ V$
C_{rss} Reverse transfer capacitance	—	840	—		$f = 1.0$ MHz, see Figure 6.5
C_{oss} Output capacitance	—	5350	—		$V_{GS} = 0\ V, V_{DS} = 1.0\ V, f = 1.0$ MHz
C_{oss} Output capacitance	—	1520	—		$V_{GS} = 0\ V, V_{DS} = 32\ V, f = 1.0$ MHz
C_{oss} eff. Effective output capacitance	—	2210	—		$V_{GS} = 0\ V, V_{DS} = 0\ V$ to $32\ V$

Note: Statis @ $T_J = 25°C$ (unless otherwise specified).

Source: International Rectifier data sheets and application notes: http://www.irf.com/.

Automotive Power Semiconductor Devices

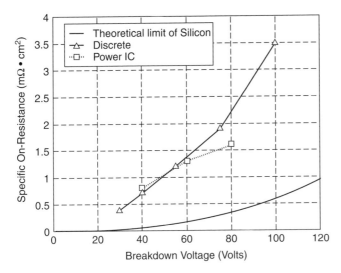

Figure 6.14 Tradeoff relationship between on-resistance and breakdown voltage of MOSFET.

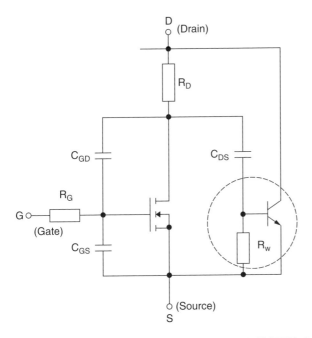

Figure 6.15 An equivalent circuit of power MOSFET showing parasitic capacitances and a parasitic NPN transistor.

However, these switching time parameters are difficult to use in designing actual drive circuitry for the MOSFET. Gate charge provides a better indication of the switching capability of power MOSFETs. This is why the data sheet shown in Table 6.4 also specifies a total gate charge Q_g, gate-source charge Q_{gs}, and gate-drain charge Q_{gd}. Q_g is defined as the total gate charge required to charge the MOSFET's input capacitance C_{iss} to the applied gate voltage. Q_{gs} is defined as the gate charge required to charge the MOSFET's input

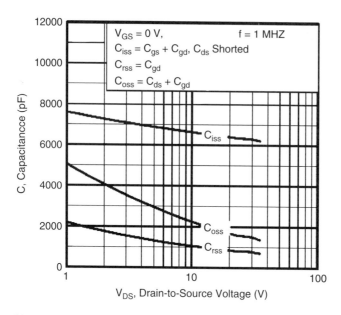

Figure 6.16 MOSFET capacitance as a function of drain-source voltage (courtesy of International Rectifier, El Segundo, CA).

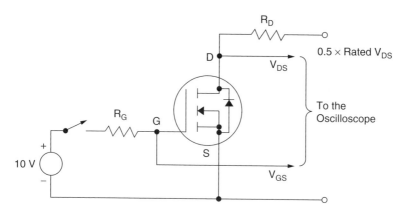

Figure 6.17 Resistive load switching circuit to characterize MOSFET switching times (courtesy of International Rectifier, El Segundo, CA).

capacitance C_{iss} to a gate voltage sufficiently large to conduct a specified drain current. Q_{gd} is defined as the gate charge required to charge C_{rss} to the same voltage of C_{iss}. Figure 6.19 and Figure 6.20 show the gate charge test circuit and the gate charge waveform, respectively.

In addition, MOSFET data sheets usually include characteristics of the body diode, similar to what we have discussed in Section 6.2. This is because the body diode sometimes serves as the companion freewheeling diode for the MOSFET in certain applications. However, for many applications, an external freewheeling diode is used instead, since the reverse recovery characteristics of the "free" body diode are usually not satisfactory. There have been efforts in the past to improve the performance of the body diode to make it

Automotive Power Semiconductor Devices

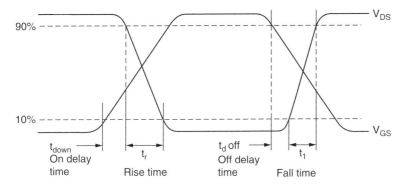

Figure 6.18 Definition of MOSFET switching times.

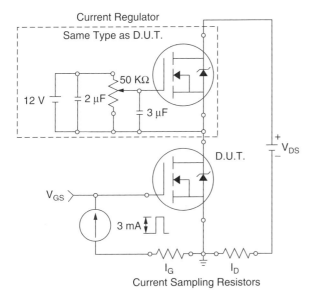

Figure 6.19 Power MOSFET gate charge test circuit (courtesy of International Rectifier, El Segundo, CA).

more "usable." Furthermore, thermal resistance and transient thermal impedance information is also provided by the data sheet, which we will discuss in Section 6.7 shortly.

6.4 IGBTs: THE HIGH-VOLTAGE POWER SWITCHES

IGBTs are the device of choice for high-voltage (400–1200 V) and medium- to high-current (10–1000 A) automotive power switching applications because of their superior current conduction capability over high-voltage power MOSFETs. 600–900 V IGBT inverter power modules are exclusively used in electric and hybrid propulsion systems with a power rating greater than 20 kW. Figure 6.21 shows a three-phase IGBT inverter driving an AC motor. 400–600 V discrete IGBTs are also widely used as the ignition coil

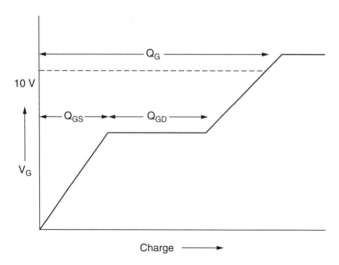

Figure 6.20 Power MOSFET gate charge test waveform.

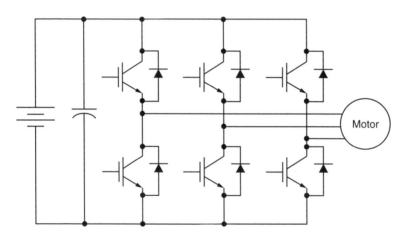

Figure 6.21 Three-phase IGBT inverter driving an AC motor.

driver in conventional vehicles powered by internal combustion engines. In the future, IGBTs may also find applications in inter-bus DC/DC converters and nonpropulsion motor drives in electric, hybrid, and fuel cell vehicles.

The IGBT is a switching transistor controlled by voltage applied to the gate terminal. Device operation and structure are similar to that of a power MOSFET. The principal difference is that the IGBT relies on conductivity modulation to reduce on-state conduction losses. The IGBT has high input impedance and fast turn-on speed like a MOSFET, but exhibits an on-state voltage drop and current-carrying capability comparable to that of a bipolar transistor while switching much faster. IGBTs have a clear advantage over MOSFETs in high-voltage applications where conduction losses must be kept low. Although turn-on speeds are very fast, turn-off of the IGBT is slower than a MOSFET. The IGBT exhibits a current fall time or "tailing." The tailing restricts the IGBT to operating at moderate frequencies (less than 50 kHz) in conventional PWM switching applications. Since most automotive motor drive applications operate within this frequency range, IGBTs easily outperform high-voltage MOSFETs and offer a 2–3 times reduction in silicon area and cost.

6.4.1 IGBT Basics

The structure of an IGBT is similar to that of a power MOSFET. One difference between a MOSFET and an IGBT is the substrate of the wafer material. An N-channel IGBT is fabricated on a P-type substrate, while an N-channel MOSFET is made on an N-type substrate, as shown in Figure 6.22. In both devices, an N-epi layer of high resistivity is needed to support the required high breakdown voltage. It is this highly resistive epi-region that is responsible for the high on-state resistance of the MOSFET. However, the N-epi region is placed on the P+ substrate in the IGBT to form a PN junction. When forward biased, this PN junction injects a large number of holes into the N-epi region, which is flooded with excess electrons and holes. The conductivity of the N-epi region is therefore increased by orders of magnitude. This is referred to as conductivity modulation, the reason why IGBTs offer much higher current conducting capability than their MOSFET counterparts.

Figure 6.23 shows the circuit symbol and equivalent circuit of an IGBT. Notice that the IGBT has a gate like a MOSFET but has an emitter and a collector like a BJT. The operation of the IGBT is best understood by referring to the cross section of the device and its equivalent circuit. Current flowing from collector to emitter must pass through the PN junction between the P+ substrate and N-epi region. The voltage drop across this PN junction is similar to that of a typical diode and results in an offset voltage in the output characteristic. It is this offset voltage that makes IGBTs not as cost effective as power MOSFETs for low-voltage applications (i.e., less than 100 V). When the gate voltage goes above a threshold voltage, a MOS channel is formed. The internal PNP transistor, formed by the P+ substrate (emitter), N-epi region (base), and P-body (collector), will turn on. The PNP transistor is in a Darlington configuration with the internal MOSEFT, with the PNP base current being supplied by the MOS channel current. The base region is conductivity modulated, and the forward voltage drop $V_{CE(on)}$ of the IGBT is close to that of the collector-to-emitter voltage drop of the PNP transistor. For power switching applications, $V_{CE(on)}$ is a critical parameter, as it determines the conduction loss. For a voltage rating of 600 V, the $V_{CE(on)}$ of an IGBT is roughly one third the $V_{DS(on)}$ of a MOSFET,

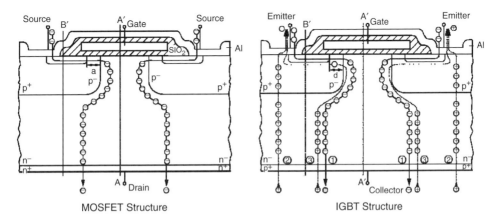

Figure 6.22 Comparison of MOSFET and IGBT device structures (courtesy of Semikron International, Germany).

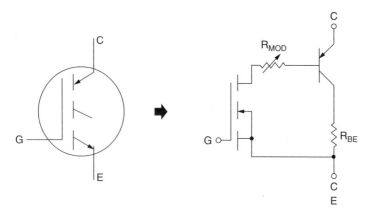

Figure 6.23 Circuit symbol and simplified equivalent circuit of IGBT.

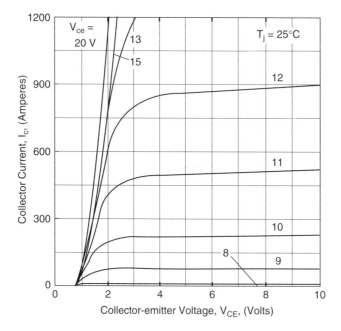

Figure 6.24 IGBT output characteristics (courtesy of Powerex Inc., Youngwood, PA).

assuming both devices conduct a current density of 100 A/cm². $V_{CE(on)}$ of an IGBT is a function of collector current, gate-to-emitter voltage, and junction temperature. Figure 6.24 shows typical output characteristics of an IGBT at various gate voltages. Figure 6.25 shows $V_{CE(on)}$ of the IGBT as a function of collector current at a junction temperature of 25°C and 125°C, respectively. Note that the temperature coefficient of $V_{CE(on)}$ is negative at low collector currents but becomes positive at high collector currents.

When the gate voltage goes below the threshold voltage, the MOS channel disappears and the base current supply of the PNP transistor is cut off. With the PNP transistor being turned off, the excess electrons and holes in the N-epi region are either swept out of the region or recombined. The IGBT is subsequently in off-state, and the reverse voltage is blocked by the PN diode formed between the P-body and the N-epi region. The turn-off

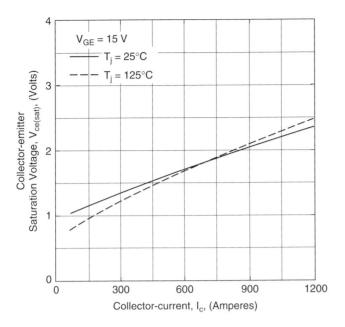

Figure 6.25 $V_{CE(on)}$ of the IGBT as a function of collector current at a junction temperature of 25°C and 125°C (courtesy of Powerex Inc., Youngwood, PA).

time of an IGBT is slow because many minority carriers are stored in the N-epi region. When the gate is initially brought below the threshold voltage, the N-epi contains a very large concentration of electrons, and there is a significant injection of electrons and holes across the junction between the P+ substrate and N-epi region. As the electron concentration in the N-epi region decreases, the injection current decreases, leaving the rest of the electrons to recombine with the holes. Therefore, the turn-off of an IGBT has two phases: an injection phase where the collector current falls very quickly, and a recombination phase in which the collector current decreases more slowly (or "tailing"). This tailing time can be reduced by carrier lifetime control techniques such as electron irradiation. Switching fall times of 100–300 ns can be achieved. Figure 6.26 depicts a half-bridge inductive switching test circuit used to characterize the switching characteristics of the lower IGBT working along with the upper freewheeling diode. Figure 6.27 shows typical turn-on and turn-off waveforms of collector current $i_C(t)$ and collector voltage $v_{CE}(t)$ of the IGBT.

The IGBT energy loss during turn-on E_{on} per switching cycle can be determined by integration of the power dissipation $p(t) = i_C(t)v_{CE}(t)$ during turn-on transition time window. The turn-on energy comprises the effects of the reverse recovery current of the freewheeling diode. The IGBT energy loss during turn-off E_{off} per switching cycle can be determined by integration of the instantaneous power dissipation $p(t) = i_C(t)v_{CE}(t)$ during turn-off transition time window. The turn-off energy E_{off} is determined to a large extent by the tailing current of the IGBT. The typical values of E_{on} and E_{off} of an IGBT are often provided by the data sheet as a function of the collector current, collector voltage, gate series resistance, and junction temperature, as shown in Figure 6.28. Average switching power dissipation can be calculated by multiplication of the switching frequency with E_{on} and E_{off}.

For hard switching applications such as motor control inverters, the SOA (safe operating area) of IGBT is a very important factor in protecting IGBTs against overvoltage or overcurrent transients. The SOA is defined as the loci of points defining the maximum

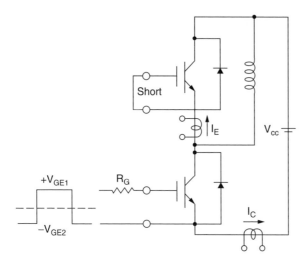

Figure 6.26 IGBT half-bridge inductive switching test circuit.

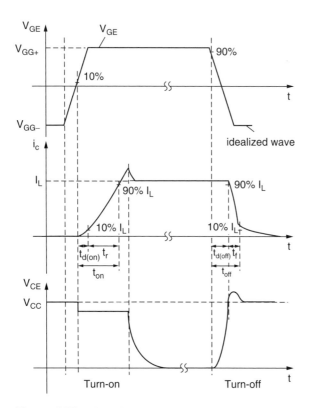

Figure 6.27 IGBT inductive switching waveforms.

allowable simultaneous occurrence of collector current and collector-to-emitter voltage during switching operation. The switching operation for a typical inverter bridge circuit will generate the current and voltage loci shown in Figure 6.29. The turn-on current overshoot is due to the reverse recovery of the freewheeling diode, while the turn-off

Automotive Power Semiconductor Devices

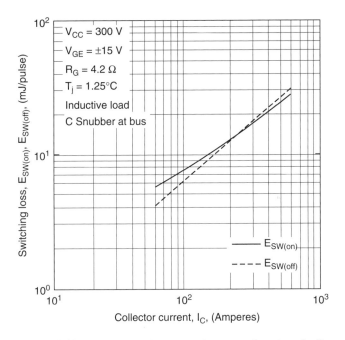

Figure 6.28 IGBT switching energy losses as a function of collector current (courtesy of Powerex Inc., Youngwood, PA).

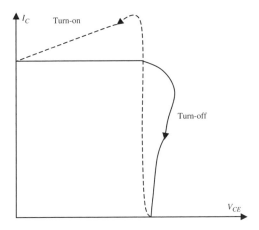

Figure 6.29 IGBT current and voltage loci during an inductive switching.

voltage overshoot is mainly caused by the parasitic lead inductance of the IGBT. IGBTs typically offer a square SOA of the rated voltage and 1–2 times rated current. However, due to the parasitic PNPN thyristor structure, the IGBT demonstrates less avalanche energy capability than the MOSFET and needs to be clamped well below the rated breakdown voltage. Short-circuit SOA (SCSOA) is yet another important device ruggedness parameter. During fault conditions such as a shorted load, the IGBT must survive a limited amount of time until the protection circuitry detects and shuts down the system. The standard specification for SCSOA is surviving for 10 µs at a starting junction temperature of 125°C.

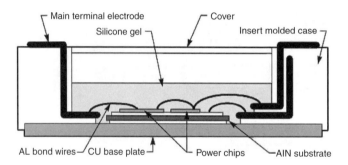

Figure 6.30 Cross-section and photo of an IGBT module (courtesy of Powerex Inc., Youngwood, PA).

6.4.2 IGBT Power Modules

For electric propulsion applications, IGBT power modules with a voltage rating of 600–1200 V and a current rating of 50–1000 A are commonly used. IGBT modules consist of multiple IGBT and freewheeling diode chips and are typically in half-bridge or three-phase bridge configuration. Figure 6.30 shows the cross section and photo of a half-bridge IGBT module. Both IGBT and diode chips are soldered to the metal surface of an isolated substrate, which is subsequently soldered on a heat-sinking copper base plate. The insulated substrate electrically isolates the silicon chips from the module base, and at the same time provides excellent thermal conductivity. The commonly used isolated substrates for power modules include ceramics such as Al_2O_3, AlN, BeO, and SiC, with copper films bonded on both sides by direct copper bonding (DCB) or active metal brazing (AMB). These substrate materials provide good thermal conductivity, high isolation voltage, low coefficient of thermal expansion, and improved partial discharge capability. The top sides of the IGBT and diode chips are connected to the external terminal electrodes by thin aluminum bond wires. The module is housed in an epoxy molded case and potted with silicon gel to provide mechanical support and a contamination barrier.

In addition, passive components such as gate resistors, current sensors, or temperature sensors may be integrated into the module. Moreover, gate drive, protection, and diagnostic circuitry can also be integrated into the power module, transforming the module into an intelligent power module.

The difference in thermal expansion coefficients of silicon chip, aluminum bond wires, copper metal films, ceramic substrates, solder joints, and the copper base plate causes thermal stress during production and operation of IGBT power modules. This may

Automotive Power Semiconductor Devices

lead to thermal-mechanical fatigue, and eventually failure of the IGBT module, as the junction temperature of the IGBT module fluctuates in motor control applications. Recently, various types of pressure contact power modules have been developed to overcome the reliability problem.

The selection of IGBT modules for automotive drivetrain applications is essentially based on voltage, current, SOA, switching speed, and reliability considerations. Under no circumstances should the maximum ratings of voltage, current, and junction temperature of the IGBT module be exceeded. Most electric, hybrid, or fuel cell vehicles operate on a DC power bus of 150 to 300 V. However, overvoltage transients caused by parasitic inductance and *di/dt* need to be taken into consideration. 600 V IGBT modules are commonly used for these applications to provide sufficient design margins. The current rating is selected to ensure the total power dissipation of the IGBTs and freewheeling diodes of a power module does not raise the junction temperature above the maximum rating. Additionally, the sum of the load current and the diode reverse recovery current should not exceed the maximum current rating of the IGBT for SOA considerations. For automotive drivetrain applications, module selection should be based on the peak load current, which can be several times higher than the average current to handle infrequent but real operating conditions such as engine cranking and regenerative braking.

6.4.3 IGNITION IGBT

Discrete IGBTs with integrated collector-gate clamp diodes (or self-clamped IGBTs) are widely used as the ignition coil driver in internal combustion engine vehicles. An IGBT has many advantages over a traditional Darlington bipolar power transistor as an ignition switching device, such as simpler driver circuit design, built-in reverse battery protection, and better SOA. Figure 6.31 shows a high-voltage ignition coil driver with an internally clamped IGBT that has been specially designed for automotive applications. The IGBT

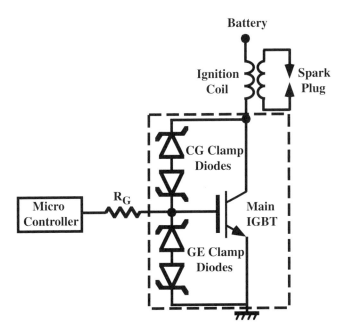

Figure 6.31 Clamped IGBT in automotive ignition coil drive application.

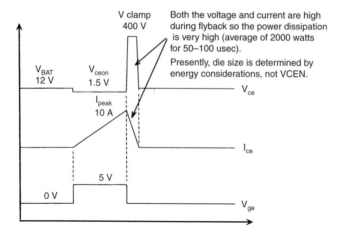

Figure 6.32 Ignition IGBT switching waveforms.

utilizes polysilicon diodes to clamp the voltage between its collector and gate and offers a cost-effective solution for ignition coil drive.

Figure 6.32 shows the collector current and voltage waveforms of the IGBT during switching operation. The IGBT is first turned on to ramp the current in the primary of the coil to a preset value (e.g., 7–12 A). Then the IGBT is turned off, and the energy stored in the primary generates a voltage spike of a few hundred volts. The voltage on the secondary rises toward 20–40K V until the spark plug arcs over and ignites the fuel mixture. Because of the integrated collector-gate clamp diodes, the IGBT is able to sustain a considerable amount of energy in a faulty condition of spark plug disconnection ("open secondary"). The collector-gate clamp voltage is typically in a range between 350 and 400 V for most ignition systems.

6.5 POWER INTEGRATED CIRCUITS AND SMART POWER DEVICES

It is highly desirable to monolithically integrate control, diagnostic, and protective functions with power switches such as MOSFETs or IGBTs to reduce cost and improve reliability of the power electronic system. Power integrated circuit technologies, such as Motorola's SMARTMOS, integrate power MOSFETs (lateral or up-drain vertical DMOS) with CMOS logic, bipolar/MOS analog circuitry on the same chip, providing great flexibility and convenience for circuit designers. For example, Motorola's MC33291L is an eight-output, low-side power switch with 8-bit serial input control, which interfaces directly with a microcontroller to control various inductive or incandescent automotive loads. MC33291L features diagnostic and protection functions such as overvoltage, overcurrent, and overtemperature shut-down as well as fault status report via its serial output pin. Each output has an internal 53 V clamp and can be independently shut down. Figure 6.33 and Figure 6.34 show the simplified application schematic and block diagram, respectively.

The output current rating of automotive power ICs is usually limited to a few amperes due to cost-effectiveness consideration and package thermal limitation. For applications requiring higher current capability, another class of intelligent products, the smart power devices, can be used. Smart power devices integrate a limited number of intelligent

Automotive Power Semiconductor Devices

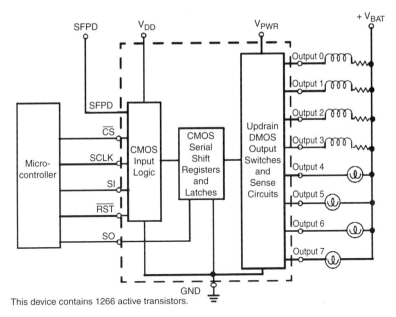

Figure 6.33 Simplified application schematic of MC33291L (courtesy of Motorola Inc.).

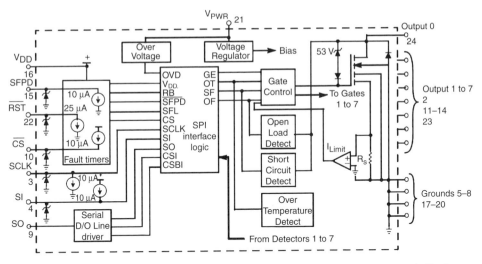

Figure 6.34 Simplified function block diagram of MC33291L (courtesy of Motorola Inc.).

functions with a single power switching device to offer a very cost-effective solution for many automotive applications, such as the replacement of electromechanical relays and fuses in the on-board electrical distribution system. Several semiconductor manufacturers offer smart power products under various trade names, such as PROFET from Infineon Technologies and TOPFET from Philips Semiconductors. For example, Infineon's BTS441R is a 20 mΩ, single-channel, high-side smart power switch, which integrates a vertical power MOSFET with charge pump and ground referenced CMOS circuitry. BTS441R features short-circuit protection, overcurrent limiting, overload protection, thermal

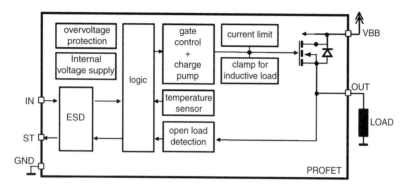

Figure 6.35 Simplified function block diagram of BTS441R (courtesy of Infineon Technologies, Germany).

shutdown, overvoltage protection (including load dump), reverse battery protection, loss of ground or battery protection, and ESD protection. It also provides diagnostic feedback on open load detection and thermal shutdown. Note that BTS441R has a nominal current rating of 21 A. Figure 6.35 shows the function block diagram of BTS441R.

6.6 EMERGING DEVICE TECHNOLOGIES: SUPER-JUNCTION AND SIC DEVICES

In the past two decades, the development of novel device concepts and the improvement in fabrication processes have spurred tremendous progress in power semiconductor technology. However, the performance of state-of-the-art power semiconductor devices is quickly approaching the theoretical limits of silicon. Power MOSFETs demonstrate excellent switching performance, but are limited by their current conducting capability due to the concurrent blocking voltage requirement. Recently a revolutionary device concept termed the super-junction (SJ) transistor has been reported to break through the theoretical limit set by the tradeoff between the on-state resistance and the breakdown voltage of the conventional power MOSFET. The concept is also known in commercial applications as COOLMOS or MDmesh.

The super-junction transistor is based on the concept of charge compensation across a PN junction made of alternatively stacked, heavily doped N and P regions (pillars or columns) and relaxation of the peak electric field by diverging into three dimensions. Figure 6.36 compares the device structures and electric field distributions of a one-dimensional PN junction and a three-dimensional super-junction. For the same doping profiles, the electric field across the super-junction is much lower than that in the conventional PN junction. Therefore, a much higher breakdown voltage is achieved in a super-junction transistor. In other words, a super-junction transistor can offer a high breakdown voltage and a low on-resistance (due to the heavy doping in the N and P pillars) all at the same time. Both theoretical analysis and experiment results have indicated that super-junction MOSFETs have 5–100 times lower specific on-resistance than the conventional power MOSFETs. Figure 6.37 demonstrates the specific on-resistance as a function of the breakdown voltage of both conventional and super-junction MOSFETs. It is clearly shown that super-junction MOSFETs have a specific on-resistance several times to several orders of magnitude lower than their conventional counterparts. IGBT data is also plotted in

Automotive Power Semiconductor Devices

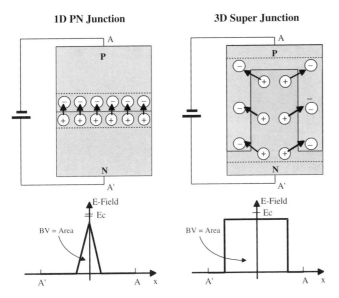

Figure 6.36 Comparison of 1D PN junction and 3D super junction: device structures and electric field distributions.

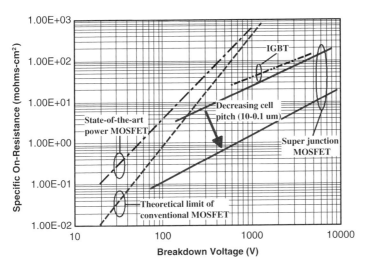

Figure 6.37 Specific on-resistance as a function of breakdown voltage for conventional MOSFET, SJ-MOSFET, and IGBT.

Figure 6.37. It is amazing to observe that the SJ MOSFET, a majority-carrier device, can actually handle more power than the conductivity-modulated IGBT, and does so at a much lower switching loss because of the absence of storage charges.

The super-junction concept, while elegant and amazingly simple in principle, is extremely difficult and challenging to realize in practice. This is due to the requirement of forming three-dimensional device structures with a very high aspect ratio (for example, N or P pillars of several tens of micrometers in depth and a few micrometers in width).

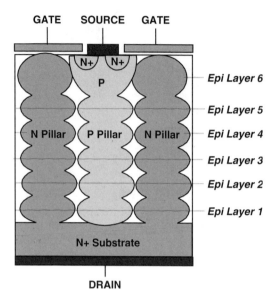

Figure 6.38 Super-junction MOSFET fabricated with the multi-epitaxial process.

The only commercially successful fabrication technology reported so far was based on a buried layer, multi-epitaxial process to form the required P and N pillars. Figure 6.38 provides the cross-sectional view of a power MOSFET fabricated with such a process.

The multi-epitaxial process requires many extra photo masks and results in very high fabrication cost per unit silicon area. The use of super-junction MOSFETs are currently limited to very high switching frequency power electronic applications. Super-junction MOSFETs can be potentially used in electric drivetrain inverters, inter-bus DC/DC converters, nonpropulsion motor control, and other automotive applications, if their costs become more competitive in the future. Super-junction MOSFETs will offer lower switching losses in these applications and allow use of simpler, lighter, and cheaper thermal management components.

Another emerging device technology is SiC-based power devices. SiC is generally considered the most promising semiconductor material to replace silicon in future power electronic systems. SiC power devices offer the following benefits over their silicon counterparts:

- The higher breakdown electric field strength of SiC allows a much thinner drift region and thus a much smaller specific on-resistance of SiC devices than their silicon counterparts. For example, a 600 V SiC Schottky diode offers a specific on-resistance of 1.4 mΩ-cm^2, which is much lower than 73 mΩ-cm^2 for a Si Schottky diode. This means that SiC devices will have a much smaller chip size.
- The low on-resistance of SiC devices for a voltage rating of 600–2000 V allows the use of majority-carrier devices like MOSFET and Schottky diodes rather than minority-carrier devices such as IGBT and PiN diodes. This results in a much reduced switching losses absence of charge storage effect. Lower switching losses will further allow higher switching frequency and subsequently

Automotive Power Semiconductor Devices

smaller and less expensive passive components such as filter inductors and capacitors.
- The larger bandgap results in higher intrinsic carrier concentration and higher operating junction temperature. In principle, SiC devices could operate at a junction temperature as high as 300°C, as compared to 150°C maximum junction temperature of silicon devices. The increased operating temperature will reduce the weight, volume, cost, and complexity of thermal management systems.
- The very high thermal conductivity of SiC reduces the thermal resistance of the device die.
- The higher bandgap also results in much higher Schottky metal-semiconductor barrier height as compared to Si. This leads to extremely low leakage currents even at elevated junction temperatures due to reduced thermionic electron emission over the barrier.

Significant progress has been made in research and development of SiC material and device technologies during the past decade. Various types of SiC switching devices and diodes have been developed and reported. As a major milestone, SiC Schottky diodes became commercially available in 2002. Figure 6.39 shows a comparison of the reverse recovery characteristics of a 600 V SiC diode and Si diode at 25°C and 150°C. Such improvement in freewheeling diode characteristics can lead to significant reduction in switching losses of power electronics.

On the other hand, there are still many technical and nontechnical barriers that prevent SiC technology from large-scale commercialization in the near future. Many technical barriers, including the high defect density and cost of SiC wafers, exist and

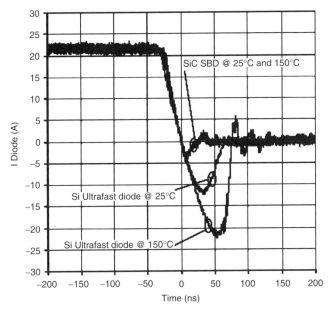

Figure 6.39 Comparison of the reverse recovery characteristics of a 600 V SiC diode and Si diode at 25°C and 150°C (courtesy of Cree Inc., Durham, NC).

prevent SiC from wide-spreading commercialization. Currently the cost of SiC devices ranges from 5 to 10 times that of silicon devices with the same voltage and current ratings. It is projected that the cost ratio between SiC and silicon will decrease to 3 in 3–5 years and eventually to 1 in 8–12 years. Unlike military or space applications, automotive applications are extremely sensitive to component costs. One critical question is how to fully exploit these benefits to achieve system-level improvement in the weight, size, efficiency, performance, and even possibly the overall cost of the power electronic subsystem. In another words, the dramatic increase in component costs due to the introduction of SiC devices must be fully justified. We must rethink the ways power electronics are designed and constructed in electric vehicles, including but not limited to converter topology, switching frequency, maximum operating temperature, thermal management, system partitioning, and package/assembly techniques.

6.7 POWER LOSSES AND THERMAL MANAGEMENT

Power devices are thermally limited. They must be mounted on a heat sink that is adequate to keep the junction temperature within the rated limit under the "worst case" condition of maximum power dissipation and maximum ambient temperature.

It must be remembered that in a switching application, the total power is due to the conduction losses and the switching losses. Switching time and switching losses of power MOSFETs are essentially independent of temperature, but the conduction losses increase with increasing temperature, because $R_{DS(on)}$ increases with temperature. IGBTs, on the contrary, have switching losses that are highly dependent on temperature, while conduction losses are relatively insensitive to temperature variation. This must be taken into account when sizing the heat sink or other thermal management design.

The transistor conduction power P_C is given approximately by:

$$P_C = I_{on} V_{on}$$

The switching energy loss depends upon the voltage and current being switched and the type of load. The total switching loss P_S is the total switching energy loss per cycle E_S multiplied by the switching frequency f:

$$P_S = E_S f$$

where E_S is the sum of the turn-on energy E_{on} and turn-off energy E_{off} of the power transistor for each switching cycle.

The total power dissipation P_T is the sum of the conduction power P_C and the switching power P_S:

$$P_T = P_C + P_S$$

The junction temperature T_J can be calculated as:

$$T_J = P_T R_{\theta JA} + T_A$$

where $R_{\theta JA}$ is the junction-to-ambient thermal resistance of the power device, which is usually provided by the device data sheet. For example, the IRF2804 MOSFET in D²PAK

Automotive Power Semiconductor Devices

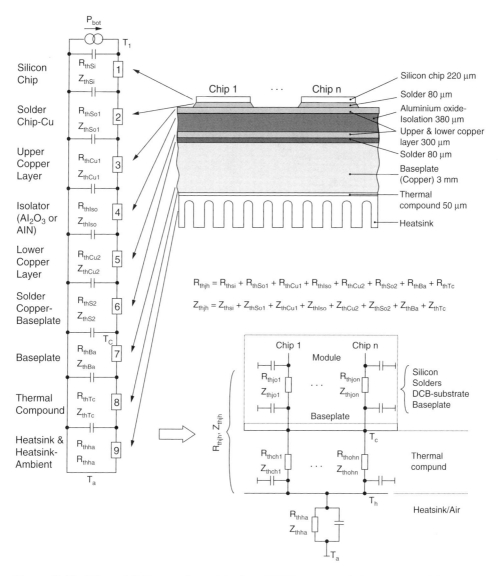

Figure 6.40 Thermal RC network model of an IGBT power module (courtesy of Semikron International, Germany).

package has a R_{JA} of 62°C/W without using any heat sink. The internal junction-to-case thermal resistance R_{JC} is often provided by the data sheet as well. Note that the calculation is based on a simple steady state assumption and may only serve as the first-order estimation. Practical operation of device and circuit is more complex, and device transient thermal impedance often needs to be taken into consideration. For discrete power devices, a simple air-cooled heat sink is usually sufficient to keep the device junction temperature below the rated value.

However, for IGBT or MOSFET power modules, thermal management is a far more challenging task due to the large power dissipation involved and the complexity of the module package. Liquid cooling techniques are often used. Figure 6.40 shows an equivalent thermal RC network for a power module, where the thermal resistances and capacitances characterize both static and transient heat dissipation and transfer. The factors that determine the thermal resistances and impedances include:

- Silicon chip (surface area, thickness, geometry, and location)
- DCB substrate (material, thickness, top-side structure)
- Chip-to-substrate attachment (solder or adhesive)
- Base plate (material, thickness)
- Module assembly (surface quality, thermal contact to heat sink, thickness and quality of thermal grease)

The largest contribution to the total thermal resistance is from the insulated substrate (56%), followed by the copper base plate (20%) and solder (10%). Due to the large amount of power dissipation in the IGBT module, liquid cooling techniques (water or glycol-based coolant) are usually used to keep the junction temperature below the rated value. Liquid cooling systems typically consist of a cold plate, circulation pump, heat exchanger, and connecting hoses. The coolant temperature is mostly kept below 70°C.

6.8 SUMMARY

Power electronics continue to change modern automobile design. As low-cost electric or hybrid drivetrains replace internal combustion engines, reliable motors replace belts and pulleys, starters and alternators combine into a single unit, and electronic systems replace hydraulic steering and braking systems, there will be new demands for high-performance, low-cost, and extremely rugged power semiconductor devices. In the past two decades, tremendous progress has been made in power semiconductor device technology to better serve the automotive industry. Low-voltage trench MOSFETs with improved avalanche capability are replacing the conventional planar MOSFETs, resulting in major cost reduction. New generations of IGBT modules offer lower power loss and a longer lifetime. More intelligence functions are integrated into power switches in the forms of power ICs or smart power devices. Emerging device technologies such as super-junction devices and SiC devices may penetrate into the automotive market in the near future. On the other hand, the importance of technology reuse and application-specific product development based on existing technology should not be underestimated because of the volume and cost sensitivity of the automotive markets. In any event, one can be certain that power semiconductor devices will be essentially at the heart of every cost-effective electronic solution for future vehicle systems.

REFERENCES

1. N. Mohan, T. Undeland, and W. Robbins, *Power Electronics: Converters, Applications and Designs*, Wiley, New York, 1989.
2. R. Jurgen (Ed.), *Automotive Electronics Handbook*, 2nd Edition, McGraw-Hill, New York, 1999.
3. International Rectifier data sheets and application notes: http://www.irf.com/.
4. ON Semiconductor data sheets and application notes: http://www.onsemi.com.
5. Motorola data sheets and application notes: http://e-www.motorola.com/.
6. Powerex data sheets and application notes: http://www.pwrx.com/.

Automotive Power Semiconductor Devices

7. Infineon Technologies data sheets and application notes: http://www.infineon.com/.
8. STMicroelectronics data sheets and application notes: http://www.stm.com/.
9. Philips Semiconductors data sheets and application notes: http://www.semiconductors.philips.com/.
10. Semikron data sheets and application notes: http://www.semikron.com/.
11. Cree data sheets and application notes: http://www.cree.com/.

7

Ultracapacitors

John M. Miller
J-N-J Miller Design Services, Cedar, Michigan

The ability to transiently store and redeliver high power levels is essential to the proper functioning of advanced hybrid propulsion systems. The hybridized vehicle owes its existence to the ability of its energy storage subsystem to cycle energy on demand and at the level demanded by operating conditions. The cumulative throughput energy, or cycle life energy, of the energy storage subsystem to a large extent determines how cost-effective the hybrid function will be. Advanced battery systems are an essential component of today's hybrid, because sufficient energy must be stored to sustain the vehicle ancillaries and customer amenities when the internal combustion engine is off. However, the energy cycle life of batteries is quite limited. A nickel metal hydride battery may be able to cycle its full capacity perhaps 5000 to 6000 times before becoming worn out. Over its lifetime, perhaps 20 MWh of energy throughout is possible by restricting the cycling to 2 to 3% of capacity. The term "wear out" may appear inappropriate to use on a nonmechanical component, but battery electrochemistry does indeed wear out. Wear out is defined as the point at which the battery is now longer capable of delivering 80% or more of its capacity. A good example of wear out is apparent in the automotive lead-acid battery system after some time in service. In hot climates the combination of temperature and energy cycling will cause the lead and lead dioxide plates to become sulfated, thereby diminishing the capacity of the battery. At some point the battery will no longer retain access to the full plate area, consequently its energy storage and power deliver capability will be restricted. The battery will have been worn out. Similar behavior is seen in nickel metal- and lithium-based battery systems (e.g., "rocking-chair" electrochemistries). In these advanced batteries the electrodes will become contaminated or the electrolyte poisoned, thus limiting its storage capacity.

In Table 7.1 a comparison is made of battery systems and ultracapacitors (UCs) in terms of gravimetric and volumetric energy densities, and energy throughput. The inter-

Table 7.1 Characteristics of Advanced Energy Storage System Technologies

Type	Battery-EV					Hybrid Vehicle					Temp Range °C
	Energy Wh/kg	Power W/kg	Cycles # @80% DOD	P/E #	Energy-Life #Wh/kg	Energy Wh/kg	Power W/kg	Cycles # @80% DOD	P/E #	Energy-Life #Wh/kg	
VRLA	35	250	400	7	11,200	25	80	300	3.2	6,000	−30, +70
TMF						30	800	?	27	?	0, +60
NiMH	70	180	1200	2.6	67,200	40	1000	5500	25	176,000	0, +40
Li-Ion	90	220	600	2.4	43,200	65	1500	2500	23	130,000	0, +35
Li-Pol	140	300	800	2.1	89,600						0, +40
EDLC						4	9000	500 k	2250	1,600,000	−35, +65

Ultracapacitors

esting point in this table is the column labeled energy cycle life. Energy cycle life is the product of that fraction of the storage system capacity in kWh that is being absorbed or delivered to the vehicle's propulsion system times the number of times this cycling occurs. Some distinction is made as to whether the battery system is being applied to a battery-electric vehicle or a hybrid vehicle. If a battery-EV, then cycling to 80% or more of capacity is a natural consequence of this application. If a hybrid vehicle, then cycling to 4% is more characteristic. This has led to the development of energy batteries and power batteries. These two major categories are shown side by side in Table 7.1.

Energy batteries are optimized for deep cycling and high specific energy. Power batteries are optimized for high pulse power and shallow cycling. In terms of battery electrochemistry, the distinction between energy and power is characterized by thick electrodes (high energy) and thin electrodes (high pulse power), respectively. Thick, porous electrodes have more available surface area for reactants, but the high porosity means that the dynamics of accessing the energy are slower than for a thin electrode. This also means that high pulse power systems are more fragile than energy optimized systems.

Referring again to Table 7.1 we see that the replacement interval of a given energy storage system (ESS) technology can be inferred by taking its energy cycle life number and dividing by the vehicle's average consumption in kWh/mile. For the average hybrid sedan the average energy consumption is approximately 0.4 kWh/mile over a metro highway drive cycle. For a sport utility vehicle (SUV) this average moves upwards to approximately 0.6 kWh/mile. For a battery-EV having 500 kg of energy storage this calculation gives a wear-out interval of roughly 84,000 miles for a nickel metal hydride (NiMH) battery and 112,000 for Li-Polymer. If an ultracapacitor of the same mass is used, its comparable "range" would be an astonishing 2,000,000 miles before the onset of wear-out. Of course, 500 kg of ultracapacitors such as the Maxwell Technologies TC2700 rated 2700 F at 2.5 V/cell has a specific energy of only 2.55 Wh/kg, yielding a total energy of a respectable 1.7 kWh. This energy calculation is obtained by dividing the 500 kg mass constraint by 0.7 kg/cell for the ultracapacitors. In most mild hybrid vehicle propulsion systems today the traction system voltage is maintained below 60 V. The calculation above yields a total of 714 cells, which, if three strings of cells are paralleled, gives a total voltage of 59 V and a capacity of 34 F. More will be said of ultracapacitor energy storage systems in subsequent sections of this chapter.

This chapter discusses the theory and application of energy storage technologies that can be used in combination with battery systems to divert the "cycling" portion of energy exchange from the average. Energy cycling (along with or in combination with temperature) is what wears out battery systems. If a technology that has inherent high cycling capability is mated with a battery system, then the combination can be made lighter and more compact than either technology alone. This chapter assumes the application of either ultracapacitors or flywheels in combination with battery systems.

7.1 THEORY OF ELECTRONIC DOUBLE LAYER CAPACITANCE

Electronic double layer capacitors (EDLCs) represent a particular subclass of electrolytics. A brief overview of capacitor types is in order before proceeding with a discussion of EDLC. Capacitors can be characterized as electrostatic or electrolytic by virtue of the medium that sustains the electric field between a pair of electrodes. Figure 7.1 illustrates the two broad groups of capacitors along with the EDLC as a particular class of electrolytic.

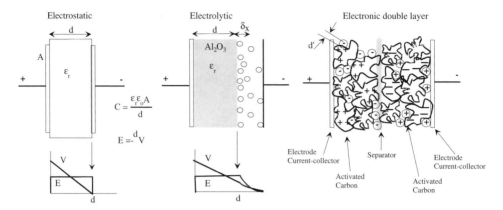

Figure 7.1 Classification of capacitor types.

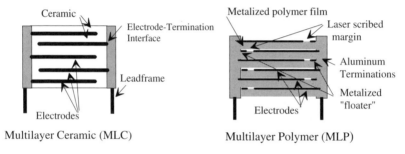

Figure 7.2 Comparison of stacked multilayer capacitor types: (a) ceramic, MLC; (b) polymer, MLP.

The electrostatic capacitor shown in Figure 7.1 is generally constructed of metal film electrodes having a dielectric material sandwiched in between. Dielectrics such as glass, mylar (Dupont's trade name for polyester), mica, polypropylene, polycarbonate, and many other materials are in common usage. The class of polyester (polyethylene teraphthlate dielectric) metal film capacitors are popular in high-frequency applications owing to their low dissipation factor, capacitance stability with age, and ability to withstand 220°C during solder reflow [2–5]. Figure 7.2 illustrates the two main types of electrostatic capacitors, the multilayer ceramic (MLC) and multilayer polymer (MLP). The electrostatic ceramic capacitor, MLC, generally consists of 20 to several hundred layers of metalized ceramic. A palladium layer is deposited on a ferro-electric ceramic and serves as the capacitor plate. The ceramic itself is relatively thick, but somewhat fragmented, and without use of rare earth metalization generally leads to field failures due to fracturing. Multilayer polymer capacitors, on the other hand, are far more durable and can withstand higher pulse currents and also survive reflow soldering at 220°C.

The metal film in an MLP capacitor is some 100–200 Angstroms thick, resulting in a specific plate resistance of 1.0 to 5.0 Ohm/square. The dielectric is ultra-thin polyethylene teraphthlate (PET), a polyester film that is 1.2 to 0.9 μm thick. This extremely thin film makes it possible to stack several thousand layers to form a capacitor suited for surface mount and reflow solder applications. For illustration, a 4 μF, 10 V MLP capacitor, available from ITW Paktron as its Capstick product, is constructed using 4000 layers of PET dielectric and rated for 11.5 Arms ripple current at 500 kHz with 7 VDC bias.

To further illustrate the physical construction and electrical characteristics of an MLP capacitor, consider a 1 cm^2 plate area device consisting of 4000 stacked films. PET polyester is a thermoplastic suited to injection molding that it is highly resistant to virtually all known chemicals and solvents. PET has a dielectric constant of 3.5 at 1 MHz, a withstand capability of 50 V/mil, and a dissipation factor of 0.016 at 1 MHz. These characteristics reveal that for a 10 V device the PET film must be at least 0.5 µm thick. The big three companies involved in capacitor grade PET film manufacturing, Hoechst-Diafoil (Europe), Dupont (U.S.), and Toray (Japan), have had 0.6 µm film in development since the mid-1990s. Equation 7.1 is the classical equation for an electrostatic capacitor, here in the form of a stacked plate device.

$$C = n\varepsilon_r\varepsilon_0 \frac{A}{d}$$

$$\varepsilon_r = 3.5 \tag{7.1}$$

$$\varepsilon_0 = 8.854 * 10^{-12}$$

For $n = 4000$ plates in the stack and for the calculated film thickness $d = 0.5$ µm the capacity of this MLP component comes out to 0.25 µF.

Before considering the electrolytic capacitor it is worthwhile to examine the tantalum capacitor. Long used as a high-capacitance, compact, decoupling capacitor in electronic circuitry, the tantalum capacitor construction is a precursor to EDLCs. Figure 7.3 illustrates the basic construction of a tantalum capacitor. In this component the anode is constructed of tantalum powder with a tantalum wire inserted that is then pressed into a pellet. When the pellet is sintered all the individual tantalum particles become fused (i.e., interconnected) together and to the tantalum wire in a highly porous structure. The cathode electrolyte, manganese dioxide (MnO_2), penetrates far into the porous channels in the tantalum particle pellet, forming a contact within this highly porous structure. Upon polarization a very thin film of tantalum oxide forms on the tantalum particle surfaces, thereby insulating them from the cathode electrolyte. This dielectric (Ta_2O_5) is extremely thin, on the order of nanometers, and results in a very high A/d factor in Equation 7.1. The issue with tantalum capacitors has been their high equivalent series resistance (ESR) owing to the porous web of interconnected tantalum particles.

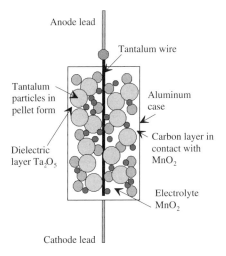

Figure 7.3 Tantalum capacitor construction.

The electrolytic capacitor shown diagrammatically in Figure 7.1 is similar in construction to the electrostatic capacitor except for the presence of a conductive electrolyte salt. This electrolyte is in direct contact with the metal current collector, or cathode, that is also the containment vessel, generally a can. The anode structure is akin to the tantalum capacitor system, but in this case instead of a pellet of pressed tantalum particles, a thin etched metal foil and Kraft paper separator to insulate adjacent layers is coiled up around the anode lead. Before insertion into the electrolytic capacitor can, the anode foil assembly is first formed by immersion into a forming salt with electric potential applied. During the forming process a thin alumina dielectric layer is grown on the aluminum foil. After forming the anode assembly is inserted into the cathode can and filled with electrolyte.

During operation the electrolytic capacitor is designed to withstand a maximum surge voltage that depends on two parameters: 1) the thickness of the tantalum oxide dielectric layer that was formed on the pellet of interconnected tantalum particles, and 2) the dissociation potential of the electrolyte. Normal applied DC bias, or working voltage, is generally 60 to 80% of the design voltage. The presence of an etched foil dramatically increases anode surface area (cathode surface is the conductive electrolyte). Unlike the crisply defined electric field within an electrostatic capacitor, in the electrolytic capacitor the electric field extends more or less uniformly across the alumina (Al_2O_3) dielectric with a monotonically decreasing tail as it extends into the electrolyte. As a consequence, the electrolytic capacitor's working voltage is further restricted by the nature of electrolyte dissociation.

As was the case for the tantalum capacitor, the electrolytic capacitor exhibits high ESR, due in part to the required voltage withstand capability of its electrolyte salt composition. In the aluminum electrolytic, ESR is a consequence of cathode conductivity, whereas in the tantalum the ESR is a consequence of the sintered tantalum particle conductivity.

Because the electric field extends across the isotropic dielectric and on into the electrolyte the accumulated charge is no longer a linear function of applied potential. The effect is small, but not negligible. Equation 7.2 describes the impact of this on capacitor current.

$$i = C \frac{d}{dt}\left\{\int E dx\right\} + \int E dx \left\{\frac{dC}{dt}\right\} \tag{7.2}$$

Solution of Equation 7.2 requires knowledge of how the electric field decays within the electrolyte. In general usage, the effect of the electric field fringing on voltage is only somewhat pronounced so that the integral term in Equation 7.2 may be replaced by the voltage stress on the capacitor, V.

Temperature has corresponding impact on electrolytic capacitors. At elevated temperatures the electrolyte is less capable of sustaining high working voltages so that some measure of voltage derating must be applied. Equation 7.3 summarizes the voltage withstand capability of an aluminum electrolytic subjected to elevated temperatures. In this example an electrolytic rated 55 V at 65°C would only be capable of withstanding 41 V at 125°C.

$$V(T) = 550\bigl(1 - \gamma(T - 65)\bigr)$$
$$\gamma = 0.00416 \tag{7.3}$$

Ultracapacitors

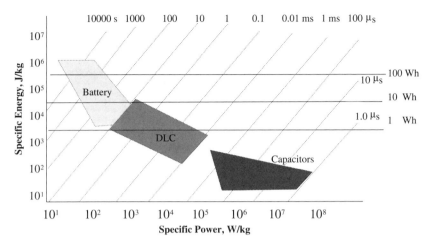

Figure 7.4 Comparison of energy storage system specific energy and specific power.

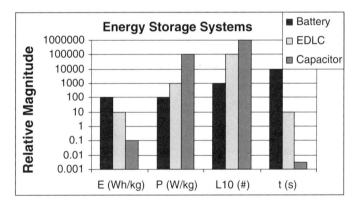

Figure 7.5 Energy storage system relative ranking in powers of ten.

In summary, electrostatic and electrolytic capacitors have theoretical energy densities of 4.34 kJ/kg and 12 kJ/kg, respectively. A polyester (Mylar) capacitor under an impressed field of 6.5×10^8 V/m yields the specific energy stated for an electrostatic unit. Aluminum electrolytics have typical specific energies of 1.5 (kJ/kg). Specific power density, on the other hand, for such capacitors is extremely high, on the order of 0.5 MW/kg and higher, but for pulse discharge times of only microseconds as noted in Figure 7.4.

To put the relative comparisons of these energy storage system technologies in a somewhat different perspective, consider Figure 7.5. In Figure 7.5 the three principal electrical energy storage technologies are charted over 9 orders of magnitude on a logarithmic scale. Capacitor technologies clearly excel in pulse power and cycle life, whereas electrochemical energy storage excels in energy density and charge/discharge duration time.

The chart shown in Figure 7.5 is a visual representation of the relative ranking of energy storage system technologies summarized in Table 7.2. In this table the specific energy, E in Wh/kg, and specific power, P in W/kg, are compared in a powers-of-ten binning approximation along with L_{10} cycle life (e.g., a metric used to signify a point at

Table 7.2 Energy Storage System Relative Ranking in Powers of Ten

	Units	Electrochemical Cell	Electronic Double Layer Capacitor	Electrolytic/Electrostatic
E	(Wh/kg)	10^2	10^1	10^{-1}
P	(W/kg)	10^2	10^3	10^4
L_{10}	(# cycles)	10^3	10^5	10^6
t	(s)	10^4	10^1	10^{-3}

which 10% of the starting units have begun to fail in terms of number of cycles). A perspective on energy storage system response time is given by a relative ranking of charge and discharge times under the label of time constant, τ in seconds.

Electronic double layer capacitors, or ultracapacitors as they are commonly referred to, were first commercially developed and patented in 1961 by Standard Oil of Ohio (SOHIO). In the SOHIO work, the EDLC effect was exhibited within a set of carbon electrodes in a nonaqueous solvent [6]. The capacitance value in an EDLC system is dependent on the surface properties of the activated carbon used for the electrodes in terms of its specific area, m^2/g, and its pore size distribution. Adding to the capacitance value is the electrolyte's ionic accessibility to all pores available in the activated carbon and to the electrolyte properties themselves. In studies of EDLC, the electrode pore size to a very large extent determines achievable capacitance. Figure 7.6 illustrates the concept of pore size and electrolyte ion access. In this figure activated carbon is shown with pore sizes ranging from nanometer through micrometer average diameters. These pore size distributions can be grouped as micro, meso, and macro pores with typical surface areas of the composite on the order of 1000 to 2000 m^2/g.

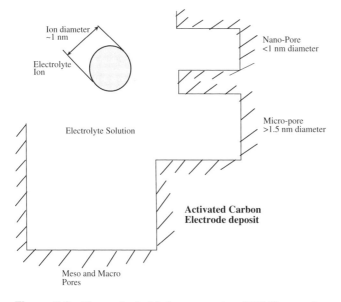

Figure 7.6 Electronic double layer capacitor (EDLC) pore size.

Ultracapacitors

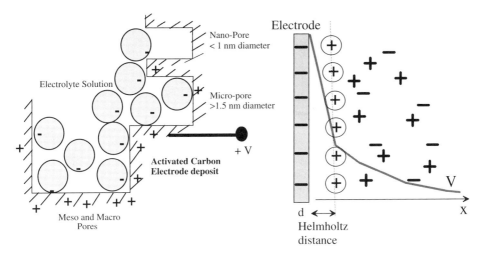

Figure 7.7 EDLC with potential applied.

The capacity of an EDLC is higher than a simple calculation of the available electrode surface area, because the ions that fill the various pores do so in a tiered, or layered, manner rather than forming a single double layer. Early investigators noted that an EDLC was formed as a compact layer of ionized solvent molecules against the pore walls without any further layering out into the solution. This early model led to the designation of a Helmholtz distance as the means to define the charge separation in a EDLC. The Helmholtz distance was defined as the distance from the ionic centers to the surface of the pore wall and this distance would be on the order of tens of Angstroms, i.e., atomic distances, hence the ability to develop enormous capacitance. Later, investigators modified the Helmholtz model to account for a variable thickness layer of ions and one wherein the ions were free to move about. This model, attributed to Gouy and Chapman [6], became known as the diffuse double layer model. In later work, Stern [6] conjectured that the EDLC is due to an inner Helmholtz compact layer and an outer diffuse layer by combining the two previous models. Figure 7.7 illustrates the movement of ionic charges into pores under the presence of an applied potential. It can be seen in Figure 7.7 that the potential function is highly nonlinear beyond the inner Helmholtz plane as the charges become diffuse and free to move about.

Solvents used in ultracapacitor manufacture may be either organic or aqueous. Organic electrolytes are known for their high working voltage, compactness, and high energy density with wide working temperature, and they are amenable to packaging in metal cans. Aqueous electrolytes, on the other hand, have a low decomposition voltage of only 1 V and have a limited temperature range. However, on the positive side, aqueous solvents promote ultracapacitors that have low ESR because of high ionic mobility, are nonflammable, and are low cost. Referring again to Figure 7.5 we can say that for activated carbon pore sizes of 3 nm (30 Angstrom) both aqueous and organic electrolytes will exhibit the EDLC effect. As pore size decreases to 2 nm there is EDLC effect for only the aqueous electrolyte, and at 1 nm the EDLC effect is lost to both types of electrolytes. This dependence on activated carbon morphology in terms of pore size distribution supports the statement made at the beginning of this section that pore size to a large extent determines the capacitance of an EDLC component.

In the following section the EDLC model is developed based on the foregoing electrode description and capacitance of materials. An EDLC is a highly distributed network of series resistance through the carbon fiber mat and double layer capacitance of pores.

7.2 MODEL AND CELL BALANCING

Electronic double layer capacitors have construction characteristics that resemble aluminum electrolytics in that high surface area (etched metal foil) electrodes are spiral wound with an insulating separator and with tantalums, in that the electrodes are highly porous and accessible to the electrolyte. In the EDLC the carbon-carbon electrodes are composed of a highly porous activated carbon mat in which the pore size distribution is sufficiently larger than the electrolyte ion diameters. EDLC electrodes must be exposed to a primary vacuum to remove any residual gas that may be trapped in the pores prior to immersion in the solvent. Symmetrical design EDLCs (i.e., carbon-carbon electrodes) use either aqueous or organic electrolytes as discussed earlier, but with an ionic conducting separator such as glass paper.

The current collector side of tubular EDLC electrodes are aluminum foil having an activated carbon powder film deposited (akin to a silk screening operation). Some EDLC manufacturers use activated carbon powder composites for the electrodes. Other manufacturers are investigating conducting polymers as electrode materials. The polymer in this case resembles the construction of an absorbent glass mat (AGM) type of lead-acid battery (i.e., starved electrolyte type) in that charge is stored in the bulk of the electrode material.

In all EDLCs the terminal capacitance consists of the series combination of an anode double layer capacitor and a cathode double layer capacitor. Figure 7.8 illustrates the basic EDLC cell model consisting of individual electrode assembly capacitances, series resistance that arises from the highly porous activated carbon mat, and a shunt resistance to model cell leakage.

The EDLC cell voltage is dependent on the type of electrolyte used, whether aqueous or organic. Organic electrolyte EDLCs have higher decomposition voltages and higher specific energy than aqueous electrolytes, but have higher resistance as well, in part due to larger molecule sizes. This low conductivity of organic electrolyte EDLCs results in higher ESR than aqueous electrolyte systems enjoy, but the impact can be mitigated to

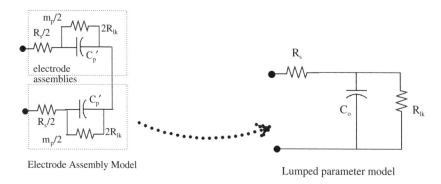

Figure 7.8 EDLC electrode assembly cell model.

Ultracapacitors

Table 7.3 EDLC Electrode Properties

Electrode System	Specific Area, Sp (m²/gm)	Specific Capacitance, Cp (F/gm)
Carbon-metal composite	2000	200
Aerogel	800	160
Anhydrous Ruthenium Oxide	150	150

some extent by employing vapor-grown carbon fiber and through the use of additives such as the solvent acetronitrile, AN. With AN, the ESR is seen to be reduced noticeably when operating at cold temperatures.

In Figure 7.8 the specific electrode capacitance, or capacitance per plate, C_p in (F/gm), can be thought of as the effective capacitance due to a specific surface area, S_p in (m²/gm), over a separation of one Helmholtz distance, d, to the electrolyte. The cell assembly is further constrained to a mass of $m_p/2$ in (gm), or one half the active cell mass. Equation 7.4 describes the specific capacitance of each EDLC electrode in a carbon-carbon system (i.e., symmetrical ultracapacitor).

$$C'_p = \frac{\varepsilon_r \varepsilon_o S_p}{d} \text{ (F/gm)} \quad (7.4)$$

The specific surface area and specific capacitance for various electrode structures is summarized in Table 7.4 condensed from Reference 7.

For an EDLC cell with total active mass, m_p, and recognizing the series combination of two electrode assembly capacitances as shown in Figure 7.8, the lumped parameter model equivalent capacitance is:

$$C_o = \left(\frac{C'_p}{2}\right)\left(\frac{m_p}{2}\right) \text{ (F)} \quad (7.5)$$

Therefore, the lumped parameter equivalent model illustrated in Figure 7.8 consists of a series resistance representing the combined effect of a highly distributed structure and a net capacitance for the cell, C_o, along with a shunt branch representing cell leakage.

To continue the development of the ultracapacitor cell model it is instructive to first review the fundamentals of network synthesis [9]. The EDLC, or ultracapacitor, exhibits multiple time constant responses during charging and discharging that the simple lumped parameter model shown in Figure 7.8 cannot capture. Consider an expanded lumped parameter model in the form of a network synthesis equivalent circuit shown in Figure 7.9. The impedance function, $Z(s)$, in the complex frequency domain (i.e., $s = \sigma + j\omega$) represents the overall terminal response of the three energy storage model. It is convenient to use a third order model to approximate an ultracapacitor because this level of resolution appears sufficient to discern fast response (seconds) from that of slow response (days).

The basic synthesis function corresponding to the network in Figure 7.9 is given as Equation 7.6 where the α's represent reciprocal time constants (i.e., product of Rs and Cs). The impedance function for this network has two 0s and three poles.

Table 7.4 Summary of EDLC Cell Equivalent Distributed Models

	Fast Branch			Medium Branch			Slow Branch		
	Cauer I	Foster II	MIT	Cauer I	Foster II	MIT	Cauer I	Foster II	MIT
ESR_0	0.68 mΩ	ESR_0 0.68 mΩ	R_f 0.68 mΩ	R_{d1} 0.63 mΩ	R_m 0.8 Ω	R_m 0.8 Ω	R_{d2} 3.73 Ω	R_s 2.9 Ω	R_s 2.9 Ω
C_f	2600 F	C_f 2600 F	C_f 2600 F	C_{d1} 247 F	C_m 250 F	C_m 250 F	C_{d2} 563 F	C_s 560 F	C_s 560 F

Ultracapacitors

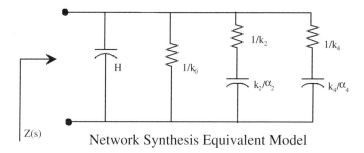

Figure 7.9 EDLC network synthesis representative cell model.

$$\frac{Y(s)}{s} = \frac{1}{sZ(s)} = \frac{(s+\alpha_1)(s+\alpha_3)}{(s+\alpha_2)(s+\alpha_4)} \tag{7.6}$$

The parameter values for the Foster II network equivalent shown in Figure 7.9 are obtained from a partial fraction expansion of Equation 7.6 as follows:

$$Y(s) = Hs + k_0 + \frac{k_2 s}{s+\alpha_2} + \frac{k_4 s}{s+\alpha_4} \quad (\Omega^{-1}) \tag{7.7}$$

The parameter values computed from Equation 7.7 are now reassigned according to the network values on the right-hand side of Figure 7.10. In this figure the network branches are assigned to fast, medium, and slow dynamics according to their respective element time constants. There is a one-to-one correspondence between the branch elements in Figure 7.9 and those on the right-hand side of Figure 7.10.

The evolution of EDLC circuit models can be thought of as passing from the electrode assembly equivalent in Figure 7.8 and its expansion to three time constants, as was done in Figure 7.9, on to the distributed, or π-equivalent, shown on the left-hand side of Figure 7.10. This Cauer I network equivalent is more intuitive in the sense that a ladder network more closely approximates the highly distributed structure that is an EDLC. Working backward in the network synthesis realm, it can be seen that the Foster II three-branch equivalent circuit has an equivalent in the distributed, π-equivalent.

In order to determine the parameter values of the Cauer I equivalent, it is necessary to take the expanded fraction decomposition of Equation 7.6 after reassigning parameter

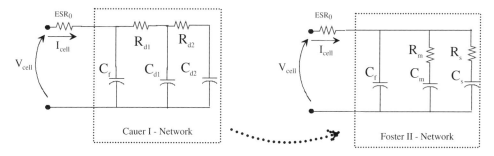

Figure 7.10 EDLC network synthesis distributed parameter equivalent model.

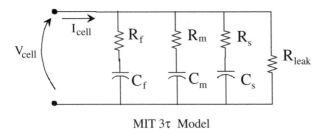

Figure 7.11 MIT three-time constant EDLC cell model.

values to correspond to those of fast, medium, and slow dynamics. This has been done in Equation 7.8 to illustrate the various parameter positions in the overall network admittance function, $Y(s)$. Notice that the capacitance value of the fast branch is the first parameter extracted, and it becomes the input branch of the Cauer I network, followed by a combination of medium and slow branch elements constituting the remaining Cauer I network.

$$Y(s) = \frac{C_f s^3 + \left[(\alpha_m + \alpha_s)C_f + \frac{1}{R_m} + \frac{1}{R_s}\right]s^2 + \left[\alpha_m \alpha_s C_f + \frac{\alpha_s}{R_m} + \frac{\alpha_m}{R_s}\right]s}{s^2 + (\alpha_m + \alpha_s)s + \alpha_m \alpha_s} \quad (7.8)$$

Taking the partial fraction expansion of Equation 7.8 according to Reference 9 results in the parameter values listed in Table 7.4. The network equivalent parameter values used in the above derivations are calculated from the three-time constant MIT model by New et al. of MIT [10] and reported later in Reference 11. The three-time constant MIT model is shown schematically in Figure 7.11.

The MIT three-time constant EDLC cell model parameters are listed in Table 7.4 for comparison to the various network equivalents to this topology. Notice the striking similarity of element values among each of the representative networks according to branch dynamics in Table 7.4. There appears to be very little difference between the MIT and Foster II equivalent, and only a small difference in parameter values with the Cauer I equivalent. This surprising result should instill confidence that the simple nodal branch model according to MIT behaves exactly as the more intuitive π-equivalent or Cauer I model for an ultracapacitor cell.

Ultracapacitor modules used in various energy storage system architectures, particularly for fuel cell and hybrid vehicles, are typically comprised of several hundred individual cells in a string. Ultracapacitor tolerances are standardized in the industry to C_o + 30%/–10%. In a string of 230 such cells it would be very likely that cell capacitances will be distributed about the nominal capacitance, C_o, with some cells 30% higher and some 10% lower, and with values in between. A system having N = 230 cells in a string will have a maximum working voltage of 2.5N = 57 V. However, the voltage will not be evenly distributed across cells 575/N, but rather some cells will have voltages 30% of 2 V, or 0.7 V, lower (the higher capacitance cells) and others will have voltages 10% of 2 V, or 0.2 V, higher. This direct calculation immediately shows that the low-capacitance cells will exceed their maximum working voltage of 2 V by 50 mV (i.e., the 10% low-capacity cells will be stressed by 2.7 V). To assist in maintaining voltage equalization among the cells, it becomes necessary to provide some means of charge transfer from low-capacity cells to high-capacity cells, or else the low-capacity cells will encounter over potential and eventual

Ultracapacitors

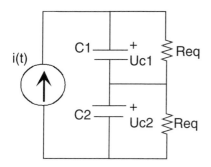

Figure 7.12 Basic cell balancing technique for ultracapacitor modules.

wear out due to electrolyte decomposition. Figure 7.12 illustrates the most basic cell balancing technique, that of a resistive divider across the string of cells.

With a resistive divider network, the resistance values are selected to hold individual ultracapacitor cell potentials to V/N where N is the total number of cells in the string and V is the applied potential. The equalizing resistances should be at least 1 order of magnitude lower than the cell leakage resistance. Referring again to Figure 7.12, we define the capacitor tolerance on the kth cell in an N-cell string as $\delta_k = (+30\%, -10\%)$ that leads to a cell voltage deviation according to Equation 7.9:

$$\Delta V_k = \frac{C_0 \Delta V_0}{C_0 + \delta_k C_0} = \frac{\Delta V_0}{1 + \delta_k} \quad (V) \tag{7.9}$$

For a series string the charge deposited in each cell (series combination of two electrode capacitances) is equal, leading to the reciprocal behavior of voltage deviation noted in Equation 7.9 and made clearer by referring to Equation 7.10:

$$\Delta V = \left\{ \begin{array}{c} \frac{\Delta q}{C_0(1-\delta)} \\ \frac{\Delta q}{C_0} \\ \frac{\Delta q}{C_0(1+\delta)} \end{array} \right\}; \delta : \left\{ \begin{array}{c} < 0 \\ 0 \\ > 0 \end{array} \right\} \quad (V) \tag{7.10}$$

Ultracapacitor tolerancing is nonsymmetrical according to the rationale noted in Equation 7.10. This is because the highest cell voltage over stress condition occurs for capacitance under tolerance conditions.

Enhancements to the resistive balancing techniques include the placement of sharp knee Zener diodes across individual cells in the same manner as the resistive network in Figure 7.12. A variant on the Zener diode dissipative equalization technique is to place 5 V sharp knee Zener diodes across each pair of cells, but with the Zener diode string straddling all the cells in an interlaced fashion.

When nondissipative equalization techniques are employed in ultracapacitor strings it is common to use active element buck-boost stages that are interlaced in a fashion reminiscent of the interlaced Zener diode approach. The concern with nondissipative balancing is the additional cost incurred in placing such a large number of active elements in the module. On the plus side, the transistor switches are low voltage (i.e., trench

MOSFETs suffice) and the current ratings are small since only an amount of charge necessary to balance an individual cell is necessary. With proper cell balancing the full energy storage capacity of the ultracapacitor module becomes accessible.

To appreciate the implications of the need for proper cell balancing consider the module energy storage capacity difference between a string of cells limited by the highest voltage cell ($\delta > 0$) case, and the remaining cells at a corresponding under voltage. Define the resulting module voltage ratio as $\sigma = V_{tol}/V_0$ then if $V_{tol} = (1/(1+ \langle\delta\rangle))V_0$ where $\langle\delta\rangle$ is the average capacitor tolerance of the complete N-cell string, then $\sigma = (1/(1 + \delta))$ and the resulting available pack energy relative to its nominal capacity is:

$$\frac{W_{avail}}{W_0} = (1-\sigma^2)$$

$$\frac{W_{avail}}{W_0} = \left(1 - \frac{1}{(1+\delta)^2}\right) \quad (\#) \tag{7.11}$$

Equation 7.11 means that if the ultracapacitor module has just a 10% average undertolerance in capacity then its available energy relative to its nominal pack energy is reduced by 17%. Proper cell equalization is therefore worth the effort in order to maximize available energy content.

7.3 SIZING CRITERIA

Ultracapacitors are sized according to the demands of the supported application as noted in References 1 and 8. Each vehicle application supported has a characteristic power, P_0, demand over a time interval, T. In Reference 12 Mitsui, Nakamura, and Okamura demonstrate that the charge and discharge efficiency of an ultracapacitor bank is uniquely determined by the ratio of its internal time constant, $\tau = R_i C_0$, where R_i = ESR of the ultracapacitor module (or cell) and the time, T_d, over which the ultracapacitor is fully charged or depleted at a rate determined by the application power, P_0. The power available from the ultracapacitor module is then:

$$P_0 = \frac{\eta_d(1-\sigma^2)W_0}{T} \quad (W) \tag{7.12}$$

where the discharge efficiency in Equation 7.12 is the function proposed in Reference 12 and obtained by solving the following expressions for charge and discharge cases for just the ultracapacitor alone (no power electronic interface).

For the charging case:

$$\eta_c = \frac{W_{stored}}{W_{stored} + W_{loss}}$$

$$\eta_c = \frac{\left(\frac{1}{2}CV_0^2\right)}{\left(\frac{1}{2}CV_0^2 + R_i I^2 T_c\right)} \tag{7.13}$$

Ultracapacitors

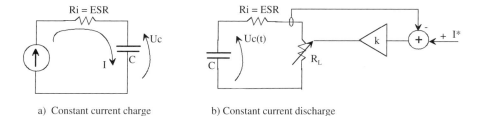

a) Constant current charge   b) Constant current discharge

Figure 7.13 Experimental charge and discharge testing of ultracapacitor modules.

where T_c is the time needed to fully charge the ultracapacitor module at the rated current I. Manipulating Equation 7.13 and making use of the fact that charging is done at constant current leads to:

$$\eta_c = \frac{1}{\left(1 + \frac{2R_i T_c}{V_0^2}\right)\left(\frac{I^2}{C}\right)}$$

$$\frac{I^2}{C} = \frac{V_0^2 C}{T_c^2} \quad (\#) \tag{7.14}$$

$$\eta_c = \frac{1}{\left(1 + 2\frac{\tau}{T_c}\right)}$$

For the discharging case:

$$\eta_d = \frac{W_{Out}}{W_{avail}} = \frac{W_0 - W_{loss}}{W_0} \tag{7.15}$$

Making substitutions into Equation 7.15 similar to what was done in Equation 7.14 leads to:

$$\eta_d = \left(1 - 2\frac{\tau}{T_d}\right) \quad (\#) \tag{7.16}$$

where T_d is the time needed to fully deplete the ultracapacitor module when the current is at its rated value, I. The derivations in Equations 7.12 through 7.16 are derived from the experimental setup illustrated in Figure 7.13. Ultracapacitor charging is accomplished under constant current, and the voltage across the ultracapacitor and charge time is monitored until rated potential is reached. Discharge is performed in a similar manner, only in this case the dissipative element (e.g., an active load) is controlled such that discharge current is maintained constant over the discharge period, T_d.

If the energy exchange is 100%, then according to the work in Reference 1 the resulting capacitor current, I_c, and necessary capacitance, C, to support such a current draw for the stated efficiency are:

Table 7.5 Automotive Ancillary and Accessory Load Supported by 42 V Ultracapacitor Modules

Load	P (kW)	T (s)	W_o (kJ)	I_r (A)	C (F)	I_c (A)	τ (s)	R_i (mΩ)	C_{cell} (F)
Steering	1.2	3	3.6	28.6	6.4	45	0.23	35	128
Brakes	2	3	6	47.6	10.7	75	0.23	21	214
Throttle	0.1	0.2	0.02	2.4	0.04	3.8	0.015	417	0.72
Turbo	2.5	2	5	59.5	8.9	94	0.15	17	178
E-Cat	3	8	24	71.4	42.7	112	0.6	14	854
ISG-launch	8	10	80	190	142	299	0.75	5.3	2850
ISG-regen.	10	5	50	238	89	374	0.38	4.2	1780
HVAC	1.5	180	270	35.7	480	56	13.5	28	9600

$$I_c = \frac{8W_0}{6\eta_d V_{c0} T}$$
$$C = \frac{8W_0}{3\eta_d V_{c0}^2} \quad (A_{dc}, F) \tag{7.17}$$

Illustrative examples are listed in Table 7.5 for selected ancillary and accessory loads in hybrid automobiles. In Table 7.5 the first five columns characterize the specified load according to its power draw over a specified time interval and the cumulative energy. The column for rated current, I_r, is a calculation based on a 4 V PowerNet electrical distribution system. Columns 6 through 10 describe the application of Equation 7.17 to each specified load, the resulting ultracapacitor module time constant, and internal resistance. The internal resistance is extracted from the calculated time constant, τ, which in turn was extracted from the relationship given in Equation 7.16 by equating it to 85%. Hence, the overall constraints were a system voltage of 4 V and a system efficiency of 85%.

Notice also in Table 7.5, in the last column for calculated cell capacity in an N = 20 cell string, that the values of required capacitance are consistent with available cell capacity available in the market today. It may be argued that distributed modules are more costly than a single dedicated module, or that distributed modules are not necessary. The need to distribute modules within the vehicle's electrical distribution system and to locate them close to the load in question is no different in this macro sense than is the practice to distribute a combination of ceramic and tantalum capacitors in close proximity to integrated circuits (ICs) on a printed wiring board. The further removed the local energy storage component is from the point of load, the more diminished is its ability to provide transient energy supply during load changes. The same rationale applies to the case of high current loads in the vehicle. The battery is simply too remote to provide transient energy when demanded without there being undervoltage intervals at the point of load. The argument that local energy storage modules are not necessary must first contest the need for redundant energy storage supply in the case of safety critical loads (i.e., the first three rows in Table 7.5).

Ultracapacitors

This section has addressed how to size an ultracapacitor module to support a selected function. But the sizing was accomplished by assuming that fully 100% of the ultracapacitor's energy was exchangeable in the process. This is not the case in practice due to the wide voltage swing required. In general a DC/DC converter interface is necessary to isolate the ultracapacitor from the vehicle's electrical distribution system (ancillary and accessory loads) or from the traction battery high-voltage bus (hybrid and fuel cell vehicles).

7.4 CONVERTER INTERFACE

From Equation 7.11 the available energy from an ultracapacitor, using the function for final to initial voltage as a parameter, can be rewritten as Equation 7.18:

$$W_{avail} = (1 - \sigma^2)W_0 \quad (J) \tag{7.18}$$

The available energy in Equation 7.18 is the total content of the ultracapacitor when its working voltage swing is taken as σ. Power electronic drives, particularly traction drives, benefit by operating from a high-voltage bus that is regulated (i.e., as stiff as possible). With a converter interface the ultracapacitor pack may experience a wide working voltage swing, but still deliver constant current into a regulated bus (e.g., a constant power discharge). Figure 7.14 illustrates the interface concept.

Schupbach and Balda in Reference 13 show that the half-bridge buck-boost DC/DC converter outperforms both the Cuk and Sepic topologies when interfacing an ultracapacitor to a vehicle traction drive. In the discussions to follow, the buck-boost will be assumed to be the converter power electronic topology of choice for interfacing an ultracapacitor module to either the vehicle's traction drive or as a distributed module for ancillary and accessory functions. In charging mode (buck) the half-bridge configuration has a duty cycle, d, that matches the working voltage ratio, σ. In other words, $d = \sigma$ during charge. During discharge, the half-bridge converter operates in boost mode with its duty cycle ranging from 0 to 0.5. In other words, during discharge (boost mode) the converter duty cycle $d = (1 - \sigma)$. Figure 7.15 illustrates the change in duty cycle in response to a sequence of a discharge pulse followed by a charge pulse at constant power.

Table 7.6 summarizes the relevant DC/DC converter attributes of inductor ripple current, switch current, and diode current occurring during boost and buck modes of operation. For the required level of converter output current, I_d, in Figure 7.15, the converter currents are shown to have magnitudes that are in proportion to the switch duty cycle, d, mapped to ultracapacitor voltage ratio, σ, and to the specified (design constraints) values for inductor ripple current and output capacitor ripple voltage. In buck mode the ultracapacitor is the output capacitor, and in boost mode either a local electrolytic or the DC link capacitor is the output capacitor.

The correlation of converter active switch duty cycle variation with ultracapacitor voltage ratio, σ, is shown in Figure 7.16. Ultracapacitor voltage is shown decreasing from fully charged to half voltage during boost mode of operation (i.e., discharge), followed by increasing voltage from half the bus voltage to rated bus voltage during buck mode of operation (i.e., charging).

The maximum bus current, I_{dm}, is a design constraint on the interface converter and a key parameter in the relationships listed in Table 7.6. In the process of extracting or delivering energy to the ultracapacitor, the peak inductor current will reach twice the maximum bus current. A similar situation exists for the peak capacitor current, I_{cpp}.

Table 7.6 DC/DC Converter Attributes during Boost/Buck Modes

Attribute	Boost Mode (Discharge)	Buck Mode (Charge)
Definition, UC voltage ratio, σ	$V_c = \sigma_d V_d$ Discharge mode: $\sigma_d = 1.0 \to 0.5$	$V_c = \sigma_c V_d$ Charge mode: $\sigma_c = 0.5 \to 1.0$
Definition, duty cycle, $d =$	$\left(1 - \dfrac{1}{\dfrac{V_{out}}{V_{in}}}\right)$	$\left(\dfrac{V_{out}}{V_{in}}\right)$
Duty cycle, $d =$	$(1-\sigma_d)$	σ_c
Ripple current ratio, $r_L =$	$\left(\dfrac{\Delta I_L}{\langle I_L \rangle}\right)$	$\left(\dfrac{\Delta I_L}{\langle I_L \rangle}\right)$
Inductor ripple current, $\Delta I_L =$	$\dfrac{V_d}{f_s L}\sigma_d(1-\sigma_d)$	$\dfrac{V_d}{f_s L}\sigma_c(1-\sigma_c)$
Inductor average current, $\langle I_L \rangle =$	$\left(\dfrac{I_{dm}}{\sigma_d}\right)$	$\left(\dfrac{I_{dm}}{\sigma_c}\right)$
Inductor rms current, $I_{Lrms} =$	$\dfrac{I_{dm}}{\sigma_d}\sqrt{1+\dfrac{r_L^2}{12}}$	$\dfrac{I_{dm}}{1-\sigma_c}\sqrt{1+\dfrac{r_L^2}{12}}$
Output capacitor pk-pk current, $I_{cpp} =$	$\dfrac{I_{dm}}{\sigma_d}\left(1+\dfrac{r_L}{12}\right)$	$I_{dm} r_L$
Switch rms current, $I_{sw} =$	$\dfrac{I_{dm}}{\sigma_d}\sqrt{1-\sigma_d}\sqrt{1+\dfrac{r_L^2}{12}}$	$\dfrac{I_{dm}}{\sigma_c}\sqrt{\sigma_c}\sqrt{1+\dfrac{r_L^2}{12}}$
Switch pk- current, $I_{swpk} =$	$\dfrac{I_{dm}}{\sigma_d}\left(1+\dfrac{r_L^2}{12}\right)$	$\dfrac{I_{dm}}{\sigma_c}\left(1+\dfrac{r_L^2}{12}\right)$
Diode rms current, $I_{dio} =$	$\dfrac{I_{dm}}{\sigma_d}\sqrt{\sigma_d}\sqrt{1+\dfrac{r_L^2}{12}}$	$\dfrac{I_{dm}}{\sigma_c}\sqrt{1-\sigma_c}\sqrt{1+\dfrac{r_L^2}{12}}$

Adapted from R.M. Schupbach, J.C. Balda, "Comparing DC/DC Converters for Power Management in Hybrid Electric Vehicles," IEEE International Electric Machines and Drives Conference, IEMDC'03, Monona Terrace Convention Center, Madison, WI, June 1–4, 2003.

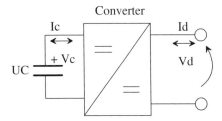

Figure 7.14 Power electronic converter interface.

Ultracapacitors

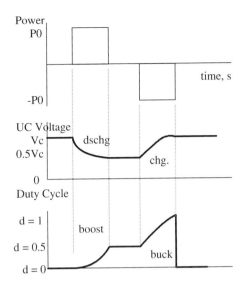

Figure 7.15 Converter duty cycle response due to toggling the power demand.

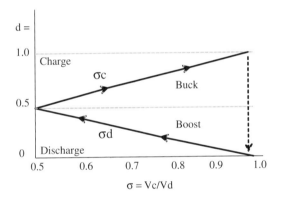

Figure 7.16 Ultracapacitor interface converter duty cycle variation with voltage ratio.

Relationships necessary for calculation of the ultracapacitor with converter interface charge and discharge efficiency are the switch and diode currents listed in Table 7.6. With converter interface we define the two cases of ultracapacitor discharge, η_d, and charge efficiency, η_c, as:

$$\eta_{di} = \frac{W_{sto} - W_{loss}}{W_{sto}}$$

$$\eta_{ci} = \frac{W_{sto}}{W_{sto} + W_{loss}} \quad (7.19)$$

The expressions for energy loss, W_{loss}, are now more involved than in Equations 7.13 and 7.14. With a converter interface the loss expressions must account for converter switch conduction and switching losses and diode loss. Consider the circuit configuration for a

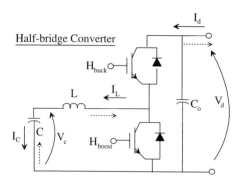

Figure 7.17 Half-bridge DC/DC converter.

half-bridge buck-boost converter shown in Figure 7.17. In this figure the buck mode currents are shown as solid traces and the boost mode currents as dotted traces. In buck mode (charging) the upper switch having gate control function H_{buck} is active along with the lower auto-commutating diode. Conversely, in discharge mode the lower switch having gate control function H_{boost} is active along with the top freewheeling diode.

The following relationships summarize the energy loss in the converter over the duration of a constant power discharge or charge event having duration, T, seconds:

$$W_{Qcond} = (I_{swrms} V_{swON} dT_s)(f_s T) \quad (J) \tag{7.20}$$

where the quantities within the first set of parentheses yields the conduction energy per switching event and the second set of parentheses defines the number of such events over the complete interval, T. Also, in Equation 7.20 the switching time is the reciprocal of switching frequency, f_s, and assuming IGBTs as the active element, we will use on-state voltage, $V_{ON} = 1$ V. Making these substitutions, and using the relationship for switch rms current from Table 7.6, yields the converter switch conduction loss in boost mode of:

$$W_{Qcond\_boost} = V_{ce(ON)} \left(\frac{I_{dm}}{\sigma_d} \sqrt{1-\sigma_d} \sqrt{1+\frac{r_L^2}{12}} \right)(1-\sigma_d)T \quad (J) \tag{7.21}$$

$$W_{Qcond\_boost} = V_{ce(ON)} \left(\frac{I_{dm}}{\sigma_d} (1-\sigma_d)^{\frac{3}{2}} \sqrt{1+\frac{r_L^2}{12}} \right)T$$

During charging (buck mode) the converter switch rms current is not the same, resulting in an expression for conduction loss of:

$$W_{Qcond\_buck} = V_{ce(ON)} \left(I_{dm} \sqrt{\sigma_c} \sqrt{1+\frac{r_L^2}{12}} \right)T \quad (J) \tag{7.22}$$

In both Equations 7.22 and 7.23 the variation of switch conduction loss is approximately proportional to square root of the ultracapacitor voltage ratio, σ. IGBT switching loss can be approximated by assuming a linear transition of switch current and voltage during the IGBT current fall time, t_f. In this analysis, we approximate the IGBT fall time as $t_f = 0.5$ μs, the on-state voltage, $V_{ce(ON)} = 1$ V, and auto-commutating diode on-state

Ultracapacitors

voltage, $V_{DF} = 1$ V. The expressions for IGBT switching loss, using the transition model and substituting the appropriate expression from Table 7.6, results in:

$$W_{Qsw\_boost} = \frac{t_f f_s I_{dm} V_d}{6\sigma_d}\left(\frac{3}{2} + \frac{r_L}{24}\right) T \quad (J) \tag{7.23}$$

$$W_{Qsw\_buck} = \frac{t_f f_s I_{dm} V_d}{6\sigma_c}\left(1 + \frac{\sigma_c r_L}{2}\right) T \quad (J) \tag{7.24}$$

Following a similar procedure for the diode conduction loss, and noting that its loss is the same regardless of boost or buck mode, results in:

$$W_{dio} = V_{DF} I_{dm}\left(\sqrt{1 + \frac{r_L^2}{12}}\right) T \quad (J) \tag{7.25}$$

The ultracapacitor ESR must be taken into account as a loss component by noting that the inductor ripple current also passes through the ultracapacitor in Figure 7.17 as well as through its ESR (not shown). Using Table 7.6 for the inductor ripple current results in the following expressions for energy loss due to capacitor ESR:

$$W_{ESR} = R_i I_{Lrms}^2 T$$

$$W_{ESR\_boost} = R_i \left(\frac{I_{dm}^2}{\sigma_d^2}\right)\left(1 + \frac{r_L^2}{12}\right) T \tag{7.26}$$

$$W_{ESR\_buck} = R_i \left(\frac{I_{dm}^2}{(1-\sigma_c)^2}\right)\left(1 + \frac{r_L^2}{12}\right) T$$

During boost mode (i.e., discharging), the ultracapacitor with interface converter efficiency equates to an expression resembling Equation 7.16, but with a modified time constant. That is:

$$\eta_{di} = 1 - 2\left(\frac{\tau_{eqiv}}{T}\right) \tag{7.27}$$

$$\tau_{equiv} = C\left\{\frac{R_i}{\sigma_d}\left(1 + \frac{r_L^2}{12}\right) + \left(\frac{V_{ce(ON)}(1-\sigma_d)^{\frac{3}{2}}}{I_{dm}} + \frac{\sigma_d V_{DF}}{I_{dm}}\right)\sqrt{1 + \frac{r_L^2}{12}} + \frac{t_f f_s V_d}{I_{dm}}\left(\frac{3}{2} + \frac{r_L}{24}\right)\right\}$$

According to Equation 7.27, the presence of an interface converter with its attendant operating point-dependent losses acts to modify the ultracapacitor equivalent series resistance, ESR = R_i, to a higher value that is dependent on the design properties of the converter (IGBT characteristics and diode on-state voltage).

Following the same methodology the buck mode efficiency (charging) can be calculated and found to be an expression resembling Equation 7.14 but with a modified time constant. The equivalent time constant will differ somewhat from that shown for the boost mode in Equation 7.27 due to differences in the loss components (i.e., expressions in Equations 7.22, 7.24, 7.25, and 7.26). But the end result remains the same: The time constant is extended by the presence of additional loss terms. Regardless of mode, an effectively longer ultracapacitor time constant means that its system efficiency (η_{di} and η_{ci}) are correspondingly reduced.

Present-day ultracapacitors have ESR values that are twice the value necessary if > 85% system efficiency is required at the discharge rates specified in Table 7.5. The ultracapacitor industry is making solid progress in lowering specific resistance by use of electrolyte additives such as acetonitrile, through optimized macropore structure of the electrodes so that ionic mobility is improved, through higher conductivity activated carbon materials for electrodes such as vapor-grown carbon fiber, and through improved current collector foil to activated carbon interfaces.

Ultracapacitor manufacturers and investigators today are looking at carbon nanotube electrodes in which single wall carbon nanotubes grow like wires from the electrode surface [16] claimed to have potential for 5 to 10 times the energy density of conventional activated carbon DLCs.

7.5 ULTRACAPACITORS IN COMBINATION WITH BATTERIES

The combination of a high specific power density ultracapacitor with a high specific energy density advanced battery would appear on the surface to offer the best of both worlds: high power for transient demands and high energy for prolonged sourcing and sinking of power. In Reference 14, Burke notes that in all cases studied the combination of a battery and ultracapacitor increased the usable energy and power capability of the system over that of a battery acting alone. In the study by Burke both 4 V mild hybrid and 20 V full hybrid powertrains are considered. In the full hybrid system studied the battery-alone option required 400 Wh and 25 kW of energy and power, respectively. When combined with a 100 Wh and 50 kW at 90% discharge efficiency ultracapacitor, the battery plus ultracapacitor system was found to deliver usable energy of 800 to 1000 Wh, or more than double its usable energy when acting alone. A further finding of Burke [14] was that lead-acid batteries do not exhibit the expected cycle life when subjected to SOC variations near 50% that would occur in a hybrid vehicle. However, combined with an ultracapacitor, the lead-acid energy storage system seems to benefit the most from the combination. It was also noted that advanced electrochemistries such as alkaline electrolyte-based NiMH and the "rocking-chair" chemistry of lithium-ion systems are already suited to shallow cycling for very high cycle life. Because of their suitability to high throughput energy cycle life, the advanced batteries do not benefit as much from being paired with ultracapacitors. However, the combination does mean that the energy storage system can be operated over a much wider SOC range than when used alone; hence, the energy storage system can be optimized to particular performance requirements.

Schupback and Balda in Reference 15 note that usable energy in a battery-ultracapacitor energy storage system can be estimated by computer simulation of a real-world performance objective such as 0 to 85 (t_{Z85}) mph time that start with the battery at 60% SOC and the ultracapacitor at 90% SOC. The t_{Z85} performance objective effectively sets a constant power discharge target that is defined by the acceleration time target and by the vehicle's mass and road load values. In their work, the authors in Reference 15 found that there were substantial weight and volume reductions possible through a combination of battery with ultracapacitors for the energy storage system — even when considering the efficiency of the interface power electronics. In the cited example, the battery-ultracapacitor

Ultracapacitors

Table 7.7 Vehicle Characteristic Parameters for SUV

Parameter	Units	Value
Curb mass, m_v	kg	2250
Static rolling resistance, R_o	Kg/kg	0.012
Aerodynamic drag coefficient, C_d	#	0.44
Vehicle frontal area, A_f	m^2	2.67
Tire dynamic rolling radius, r_w	m	0.31
Overall gear ratio in 1st, G_{r1}	#	13.67
Driveline efficiency, η_{dl}	#	0.815
Engine power, P_{eng}	kW	151

energy storage unit realized a 40% reduction in weight over the battery-only case and a 21% savings in volume.

Thermal management is another aspect of energy storage system design that benefits from combination of a battery with an ultracapacitor. A battery-only system may operate at 72% turnaround efficiency, but an ultracapacitor can contribute efficiencies near 90% to the mix. Thermal loading is a serious matter for any battery system, particularly for advanced batteries, since considerable heat can be generated and dissipation is limited. Ultracapacitors can be air-cooled far more readily than batteries that are likely to require either liquid cooling or full climate control systems. Energy Conversion Devices, for example, has experimented with liquid cooling of NiMH battery modules by forcing coolant through the electrode assemblies themselves, so that thermal paths from the sources of heat generation become intimately connected to the heat-removal medium.

To illustrate how the battery and ultracapacitor may be sized for a combined unit consider the sport utility vehicle characteristic parameters listed in Table 7.7. The vehicle is subjected to a relatively stringent performance test of 0 to 85 mph with a target elapsed time of 20 sec or under as noted in Reference 15. In this example the vehicle has similar attributes to that used in Reference 15 but some of the parameters are different.

When the SUV having the attributes listed in Table 7.7 is subjected to a simulated acceleration run it is found that the required powerplant must develop 285 Nm of torque at the crankshaft and, furthermore, that its peak power at the first shift point must be 151 kW in order to meet the elapsed time target of 20 sec. Vehicle acceleration performance is found by applying a fourth order Runge-Kutta integration routine to the SUV road load equation, consisting of rolling resistance and aerodynamic drag, plus a contribution in load due to ancillary power demand of a constant 750 W. Driveline inefficiency in the simulation is accounted for by torque converter losses, gear mesh losses, and final drive losses. It is found that for a 151 kW powerplant the SUV will have an elapsed time of 20 sec during a full-throttle 0 to 85 mph acceleration performance as shown in Figure 7.18.

The next step in the energy storage system sizing methodology is to characterize the components of the ESS. It is well known from battery modeling that each electro-chemistry exhibits its own unique energy and power relationship. That is, the available energy from a battery is diminished when the energy is removed at higher rates. The Ragone relationship is traditionally used to characterize such behavior and it can be expressed mathematically as follows.

$$E_{bat} = E_{bat0} - k_{bat}P_{bat} \quad \text{(Wh/kg)} \quad (7.28)$$

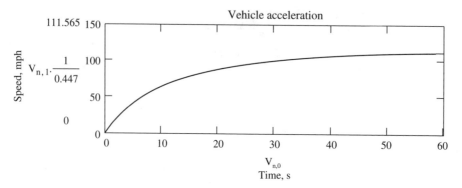

Figure 7.18 Vehicle acceleration for z85 test.

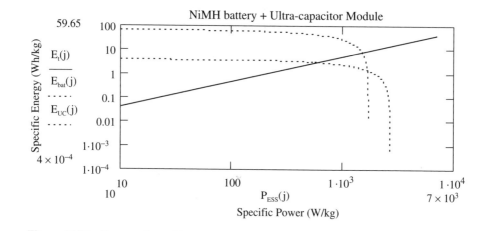

Figure 7.19 Ragone plots of battery plus ultracapacitor ESS.

For a NiMH battery the specific energy, E_{bat}, consists of a static component, E_{bat0} = 65 Wh/kg, and a dynamic coefficient, k_{bat} = 0.035 (h). Similar expressions are written for the ultracapacitor.

$$E_{NiMH} = 65 - 0.035 P_{NiMH} \quad \text{(Wh/kg)} \tag{7.29}$$

$$E_{UC} = 3.5 - 0.0013 P_{UC} \quad \text{(Wh/kg)} \tag{7.30}$$

A Ragone plot of Equations 7.29 and 7.30 along with an expression for the t_{z85} elapsed time constraint is shown as Figure 7.19. The diagonal line in Figure 7.19 is the specific energy corresponding to specific power multiplied by the z85 time of 20 sec.

The intersection of the ultracapacitor trace in Figure 7.19 with the t_{z85} diagonal line represents the optimal specific energy and power of the ultracapacitor for the SUV hybrid application. In this chart it is found that the optimal values are:

Ultracapacitors

$$E_{UCopt} = 2.815$$

$$P_{UCopt} = 527 \quad \text{(Wh/kg, W/kg, Wh/kg)} \tag{7.31}$$

$$E_{batopt} = 41.6$$

Next, it is further assumed that the vehicle powerplant is downsized as a result of hybridization to 90 kW, and that the ESS contributes the difference of P_{ESS} = 61 kW for t_{z85} sec. This represents a need for available energy from the ESS of:

$$W_{ESS} = (P_{driveline} - P_{eng}) \frac{t_{z85}}{3600} \quad \text{(Wh)} \tag{7.32}$$

According to Equation 7.32 the ESS must have a capacity of 339 Wh for the duration of the acceleration event. From the ESS capacity stated in Equation 7.32 and using the optimum values for ultracapacitor specific energy and power given in Equation 7.31, we calculate the optimum ultracapacitor module mass as follows:

$$m_{UC} = \frac{W_{ESS}}{E_{UCopt}} \quad \text{(kg)}$$

$$m_{UC} = \frac{339}{2.815} = 120.4 \tag{7.33}$$

An expression similar to Equation 7.33 can be written for ultracapacitor pack mass based on ESS peak power and the chart-derived optimum specific power noted in Equation 7.31 but this would be redundant to Equation 7.33, since the same optimum mass would be calculated. Due to plotting resolution in Figure 7.19 the optimum values for specific energy and power are actually straddling the optimum point. Performing a calculation of ultracapacitor mass using the specific power value of 527 W/kg results in a mass of 115.7 kg for the pack. Therefore, the ultracapacitor pack in reality would be intermediate between 115.7 kg and 120.4 kg. At this point an average value of 118 kg is a better choice for an ultracapacitor-only case. The Ragone plot of NiMH battery pack specific energy vs. specific power intersects the performance time target well beyond the ultracapacitor intersection and well beyond the practical limitations of NiMH power density. Real-world NiMH is limited to approximately P_{batnom} = 180 W/kg nominal and four times this, or P_{batmax} = 720 W/kg, maximum (corresponds to 15C rate). The battery pack must not exceed terminal current and voltage limitations during discharge or charge mode. For NiMH the limitations at the terminals are:

$$\sigma_{bat} V_r < V_{bat} < V_{max}$$

$$I_{bat} < I_{max} = 15C \quad (V_{dc}, A_{dc}) \tag{7.34}$$

$$\sigma_{bat} = 0.69$$

The battery-only case mass is more complex than calculating the ultracapacitor mass to suit a single performance metric (in the previous case, the energy and power for a 0 to 85 mph hard acceleration). The battery pack initial SOC = 60% for a hybrid, and for

NiMH technology the terminal voltage drops to 69% of its rated value, as noted in Equation 7.34. Furthermore, the NiMH maximum current is limited to less than a 15C rate where C = Ah capacity of the NiMH cells in the pack. If we set the system rated voltage, V_r = 30 V, then the battery-only case discharge current must equal:

$$I_{max} = \frac{P_{ESS}}{\sigma_r V_r} \quad (A)$$

$$I_{max} = \frac{61000}{0.69 * 300} = 295$$

(7.35)

From the assumption on maximum discharge current stated (15C), the capacity of the battery pack at room temperature would therefore be:

$$C_{bat} = \frac{I_{max}}{15} = 19.7 \quad (Ah) \tag{7.36}$$

According to Equation 7.36 a 20 Ah NiMH cell would be necessary. Therefore, a pack that is rated 30 V would require N = 360 cells (at 1 V/cell) and it would have a storage capacity of:

$$W_{bat} = \sigma_{bat} V_r C_{bat} \quad (kWh)$$

$$W_{bat} = 0.69(300)20 = 4.14$$

(7.37)

At a nominal SOC 60% the battery pack would have an available capacity of 2.48 kWh, of which 0.1 kWh would be nominally be exchanged per hybrid event (at 4% SOC deviation per event). A NiMH battery pack capable of delivering a pulse power of P_{ESS}, yet maintain its SOC within the bounds cited, would have a mass of:

$$m_{bat} = \frac{P_{ESS}}{P_{bat\,max}} \quad (kg)$$

$$m_{bat} = \frac{61000}{720} = 84.7$$

(7.38)

The NiMH battery pack mass is some 33 kg lighter than the ultracapacitor pack necessary to perform the z85 acceleration event. However, the battery pack has energy far in excess of that needed for just a single acceleration due to its terminal limitations on voltage and current.

In combination, a NiMH battery pack plus an ultracapacitor module would be iteratively sized based on the vehicle characteristic parameters driving a standard drive schedule. Given this caveat, the sizing of battery and ultracapacitor combinations is performance metric driven as well as cost driven. With NiMH costs at $650/kWh and ultracapacitor costs now approaching $0.01/F, it is practical to combine both and realize weight, volume, and cost savings. One procedure for this has been developed by the authors in Reference 15.

REFERENCES

[1] J.M. Miller, R. Smith, "Ultracapacitor Assisted Electric Drives for Transportation," IEEE International Electric Machines and Drives Conference, IEMDC'03, Monona Terrace Convention Center, Madison, WI, June 1–4, 2003.

[2] I.W. Clelland, R.A Price, "Multilayer Polymer (MLP) Capacitors Provide Low ESR and Are Stable over Wide Temperature and Voltage Ranges," *Proceedings of the 8th Annual European Capacitor and Resistor Technology Symposium*, October 1994.

[3] I.W. Clelland, R.A. Price, "Requirement for Robust Capacitors in High Density Power Converters," IEEE 16th Annual Applied Power Electronics Conference, APEC, Adams Mark Hotel, Dallas, TX, March 2001.

[4] I.W. Clelland, R.A. Price, "Evaluation of SMT Polymer Film Capacitors using Newly Developed, Low Shrinkage PET," *Proceedings of the 18th Annual Capacitor and Resistor Technology Symposium*, March 1998.

[5] B. Carsten, "Optimizing Output Filters Using Multilayer Polymer Capacitors in High Power Density Low Voltage Converters," Power Conversion Intelligent Motion Conference, San Jose, CA, May 8–11, 1995.

[6] M. Endo, T. Takeda, Y.J. Kim, D. Koshiba, K. Ishii, "High Power Electric Double Layer Capacitor (EDLCs); from Operating Principle to Pore Size Control in Advanced Activated Carbons," *Journal of Carbon Science*, Vol. 1, No. 3–4, January 2001, pp.117–128.

[7] A.F. Burke, T.C. Murphy, *Material Research Society Symposium Proceedings*, Vol. 393, 1995, p. 375.

[8] J.M. Miller, P.J. McCleer, M. Cohen, "Ultracapacitors as Energy Buffers in a Multiple Zone Electrical Distribution System," 2003 Global Powertrain Congress, *Advanced Propulsion Systems Proceedings GPC03*, Crowne Plaza Hotel, Ann Arbor, MI, September 23–25, 2003.

[9] A. Budak, *Passive and Active Network Analysis and Synthesis*, Houghton Mifflin Company, Boston, 1974.

[10] D. New, J.G. Kassakian, J. Schindall, "Automotive Applications of Ultracapacitors," MIT/Industry Consortium on Advanced Automotive Electrical/Electronic Components and Systems, *Consortium Project Review*, Winter 2003. See also www.mitconsortium.org.

[11] J. Schindall, J.G. Kassakian, D. Perreault, D. New, "Automotive Applications of Ultracapacitors: Characteristics, Modeling and Utilization," MIT/Industry Consortium on Advanced Automotive Electrical/Electronic Components and Systems, Spring Meeting, Ritz-Carlton Hotel, Dearborn, MI, March 5–6, 2003.

[12] K. Mitsui, H. Nakamura, M. Okamura, "Capacitor-Electronic Systems (ECS) for ISG/Idle Stop Applications," *Proceedings of the Electric Vehicle Symposium*, EVS19, 2002.

[13] R.M. Schupbach, J.C. Balda, "Comparing DC/DC Converters for Power Management in Hybrid Electric Vehicles," IEEE International Electric Machines and Drives Conference, IEMDC'03, Monona Terrace Convention Center, Madison, WI, June 1–4, 2003.

[14] A. Burke, "Cost-Effective Combinations of Ultracapacitors and Batteries for Vehicle Applications," 2nd International Advanced Automotive Battery Conference, AABC, Las Vegas, NV, Session 6, February 2002.

[15] R.M. Schupback, J.C. Balda, "Design Methodology of a Combined Battery-Ultracapacitor Energy Storage Unit for Vehicle Power Management," IEEE Power Electronic Specialists Conference, PESC'03, Hyatt-Regency Hotel, Acapulco, Mexico, June 15–19, 2003.

[16] G. Zorpette, "Super Charged," *IEEE Spectrum*, January 2005, pp. 32–37.

8

Flywheels

John M. Miller
J-N-J Miller Design Services, Cedar, Michigan

8.1 FLYWHEEL THEORY

Flywheel systems have been promoted as "mechanical" batteries and offered as energy storage systems capable of high power and high cycle life. Operation of a flywheel energy storage system (ESS) is basically the same as that for an ultracapacitor, only the energy is stored as kinetic rather than potential energy. There is a direct analogy between angular speed of a flywheel and voltage in an ultracapacitor. The higher the angular speed (e.g., voltage) the greater the energy storage capacity in a flywheel. Issues that plague flywheel systems parallel those of the ultracapacitor: A containment vessel to hold a vacuum for a long term and non-contacting bearings correspond to minimization of ESR in an ultracapacitor. High tensile strength materials that can withstand large hoop stress without rupturing correspond to the dissociation potential of the ultracapacitor electrolyte. Even the governing equations are similar. However, whereas the ultracapacitor stores energy in the same form as it is being used, the flywheel system depends on electromechanical energy conversion in both directions; hence, its efficiency will be lower than that of an ultracapacitor.

The flywheel, on the other hand, does offer unique benefits that make it suitable for spacecraft, specifically, its rotational moment [1]. This enables the flywheel ESS to play a dual role: energy storage for communications and guidance backup during periods of solar eclipse plus platform stabilization (i.e., attitude control). Flywheels for space-borne applications face critical thermal management issues because no convection is present for cooling and only radiant means are available for heat rejection. This means that the flywheel, its bearing structure, and the interface motor-generator (M/G) needed for energy conversion have the highest possible efficiency. Most flywheel energy storage systems operate in the angular speed regime of 50 to 90 krpm. Figure 8.1 is a sketch of fundamental components of a flywheel ESS.

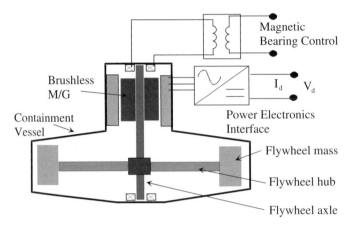

Figure 8.1 Elementary flywheel energy storage system.

With proper design and materials the flywheel is a feasible energy storage unit because it is inherently non-toxic, fully recyclable, and completely rechargeable. The energy throughput life of a flywheel is limited only by the bearing system, the M/G, and containment vessel vacuum integrity.

Modern materials capable of withstanding the high hoop stress resulting from centrifugal forces are generally the lightest weight materials that possess very high tensile stress. This is because the hoop stress varies in direct proportion to the material density, whereas the energy stored grows with angular speed squared. A lightweight material flywheel can store the same energy as a steel flywheel but weigh less. In terms of material properties, the energy storage capacity of a flywheel is given as Equation 8.1:

$$W_{FW} = \frac{K_{max}}{2\rho} \quad (Wh/kg) \tag{8.1}$$

where K_{max} is the limiting tensile stress above which the rim will delaminate and burst. Fibers that offer the highest energy storage capacity range from E-glass (which can store as much energy as high-strength steel) to Kevlar, a type of nylon that can store 7 times more energy than high-strength steel. A common fiber used in today's flywheel components is fused silica glass. A flywheel ESS can also be lighter than an advanced battery for the same energy content.

The flywheel shown diagrammatically in Figure 8.1 has a polar moment of inertia calculated in Equation 8.2, where r_{ro} = outer radius of the rim, r_{ri} = inner radius of the rim, m_r = mass of the rim (glass fiber), m_h = mass of the hub, and define $\upsilon = r_{ri}/r_{ro}$.

$$J_{FW} = \frac{1}{2}m_r(1-\upsilon^2)r_{ro}^2 + \frac{1}{2}m_h\upsilon^2 r_{ro}^2 \quad (kgm^2) \tag{8.2}$$

Flywheel polar moment of inertia is proportional to its rim radius squared. However, a higher rim radius exposes the material to higher tensile stress, leading to failure if speeds are too high. Reasonable numbers for rim radius are 50 to 100 mm, and $\upsilon = 0.65$ is typical.

Energy storage capacity of the flywheel is dependent on the polar moment of inertia and on the speed ratio over which it can be operated. Equation 8.3 summarizes the energy

Flywheels

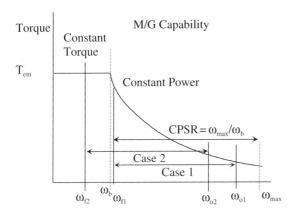

Figure 8.2 M/G torque-speed capability curve for flywheel interface.

capacity of a flywheel system in terms of its available energy and taking into account some typical component efficiencies [2].

$$W_{FWavail} = \frac{\eta}{2} J_{eq}(1-\sigma_f^2)\omega_o^2$$

$$\eta = \eta_I \eta_m \eta_{FW}$$

$$J_{eq} = J_{FW} + J_{MG} \quad (J) \quad (8.3)$$

$$\sigma_f = \frac{\omega_f}{\omega_o}$$

where the component efficiencies used are $\eta_I = 0.96$ for the inverter, $\eta_m = 0.90$ for the motor-generator, $\eta_{FW} = 0.85$ for the flywheel system, and ω_o is the nominal or rated angular velocity. Representative inertias for the system are $J_{FW} = 0.05$ kgm² for the flywheel itself and $F_{MG} = 0.005$ kgm² for the motor-generator. The speed ratio $\sigma_f = \omega_f/\omega_o$ can be viewed as a direct analogy to voltage swing in the ultracapacitor ESS.

For a given value of M/G electromagnetic torque, T_{em}, the terminal power of the flywheel ESS during discharge can be represented as Equation 8.4.

$$P_{ESS} = -\eta\sigma_f\omega_o T_{em} \quad (W) \quad (8.4)$$

It is readily apparent from inspection of Equation 8.4 that the angular speed swing range, σ_f, and the torque rating of the M/G determine the peak power capability of the flywheel ESS. This is due to the fact that the flywheel angular speed can slew no faster than $T_{em}/(J_{FW} + J_{MG})$. In fact, the particular choice of M/G technology and where its corner point speed is set determines the time to charge the flywheel ESS. Figure 8.2 depicts a typical M/G torque-speed capability curve for two cases of flywheel minimum angular speed, ω_f, one when $\omega_f = \omega_b$, the M/G base or corner point speed that describes the entry into constant power regime, and the second case when $\omega_f < \omega_b$, in which case the M/G must first accelerate the flywheel in constant torque up to an angular speed of ω_b and then transition into constant power mode. It should also be pointed out that M/Gs having a constant power speed range (CPSR) that matches or exceeds the angular speed swing range of the flywheel is a necessary condition.

For a given M/G power, P_{MGb}, the time to charge the flywheel is strongly dependent on whether the M/G operates entirely in constant power or if it starts out in constant torque and completes charging under constant power.

$$P_{MGb} = T_{em}\omega_b \quad (W) \tag{8.5}$$

Case 1: $\omega_f = \omega_b$, and P_{MGb} = constant for $\omega_f < \omega < \omega_b$. For this case the time to charge the flywheel from its initial angular speed, ω_f, to its final speed, ω_b, requires t_{f1} seconds where:

$$t_{f1} = \int_{\omega_f}^{\omega_o} \frac{J_{eq}}{T_{em}} d\omega$$

$$t_{f1} = \int_{\omega_f}^{\omega_o} \frac{J_{eq}\omega_b}{P_{MGb}} d\omega \quad (s) \tag{8.6}$$

$$t_{f1} = \frac{J_{eq}}{T_{em}}(\omega_o - \omega_f)$$

Case 2: $\omega_f < \omega_b$, and T_{em} = constant for $\omega_f < \omega < \omega_b$, and P_{MGb} = constant for $\omega_b < \omega < \omega_o$. In this case the flywheel first charges under constant torque at a rate determined by the M/G torque rating and then under constant power as the flywheel angular speed passes through the M/G corner point speed. In this case the time to charge the flywheel, t_{f2}, becomes:

$$t_{f2} = \int_{\omega_f}^{\omega_b} \frac{J_{eq}}{T_{em}} d\omega + \int_{\omega_b}^{\omega_o} \frac{J_{eq}\omega_b}{P_{MGb}} d\omega$$

$$t_{f2} = \frac{J_{eq}}{T_{em}}((\omega_b + \omega_o) - 2\omega_f) \quad (s) \tag{8.7}$$

$$t_{f2} = \frac{J_{eq}}{T_{em}}\omega_b + t_{f1}$$

The flywheel operating initially in constant torque, as shown in Figure 8.2 for case 2, results in a longer charging time, all else being equal. For best response the flywheel should be charged and discharged under constant power operating mode of the interface M/G. This means that M/G technologies having CPSR > 3:1 are recommended for flywheel applications. This speed ratio admits induction, switched reluctance, and interior permanent magnet machines in either radial (drum) or axial (pancake) geometries.

8.2 FLYWHEEL APPLICATIONS IN HYBRID VEHICLES

To date there have not been any applications of flywheel technologies in hybrid vehicles. There are many proposals for such systems, but the fact remains that flywheel energy storage at present is still more costly than electrochemical batteries or ultracapacitors.

Issues with containment and bearings have been solved so that this technology is possible for introduction into mass market hybrid vehicles.

To briefly review the types of flywheel applications in hybrid vehicles consider the following scenarios:

- A flywheel directly coupled to the vehicle drivetrain via a continuously variable transmission (CVT). In this architecture the CVT acts as a speed matching device with infinitely variable ratio so that flywheel angular momentum can be transferred into the vehicle propulsion system.
- A system wherein the traction M/G has rotating rotor and stator with the stator assembly powered via slip rings and where its mass is the flywheel energy storage unit. Such systems have been proposed and shown to offer considerable merit in terms of energy storage and fuel economy improvements. However, the mechanics of charging and discharging the flywheel impose some odd operating modes on the M/G and vehicle engine such as having the M/G in generating mode during a vehicle launch and vice versa.
- A more conventional system in which the flywheel assembly is standalone and packaged in the vehicle in much the same manner as a traction battery. In this architecture the flywheel would by necessity require counter-rotating elements so that it did not impose a torque couple on the vehicle chassis during grade changes or turns. Of course, a gimbaled single flywheel assembly would not contribute any couple into the chassis, but it would require considerably more packaging attention.

8.3 ENERGY STORAGE SYSTEM OUTLOOK

Energy storage systems that retain the working form of energy offer the most promising choices for hybrid vehicle systems. Electrochemical storage batteries do not store energy as accumulated charge (although there exists a double layer capacitive element within their structure), but rather as energy of covalent bonds. Chemical reactions are necessary for such storage to be effective, and in many instances, the reactions may proceed faster in one direction than another, adding a dimension of non-symmetrical behavior to their charge/discharge characteristics. Temperature effects are far more pronounced in electrochemical reactions than in electrostatic storage or in mechanical forms of energy storage.

The spectrum of energy storage mediums can be categorized according to the manner in which the energy is stored, whether as a change in internal energy of the storage medium, or as some form of stress on the materials comprising the storage medium. For instance, capacitors rely on the electric stress across a dielectric (i.e., separation of charge) with the limiting value of stress occurring when the dielectric breaks down. A flywheel is similar and encounters its limiting value of charge when the materials comprising the rim and hub begin to fail in tension. Both of these energy storage forms rely on molecular bond strength to define their limit of storage density (i.e., atomic binding energy). Electrochemical forms of energy storage rely on the strength of ionic bonds to define their limiting energy density. Electrochemical storage in terms of energy/unit mass is typically an order of magnitude or higher than molecular bond strength limited forms of energy storage for conventional materials. However, electrochemical storage in terms of being limited by atomic binding energies is on par with mechanical and electrostatic energy storage systems. Storing energy in covalent bonds (heats of reaction, combination, and separation of atoms), such as in fuels, pushes the energy density some 2 to 3 orders of magnitude higher than electrochemical energy densities. Finally, energy storage in nuclear bonds (i.e., combination and separation of nucleons) pushes the energy storage density some 6 orders of magnitude higher than that of covalent bonds.

Practical energy storage systems for automotive applications are those that can be adapted to the automotive environment and operate efficiently over wide temperature extremes. Today, advanced batteries and fuel cells do not operate efficiently over wide temperature extremes and must be augmented either internally with additives to do so or externally with other types of storage mediums in order to deliver nominal performance.

For reasons that parallel the choice of series electric hybrid for large, massive vehicles, the flywheel is viable in locomotive applications [3]. Operating in a manner reminiscent of the battery, or ultracapacitor, in a hybrid automobile, the flywheel energy storage system in a locomotive supplies (and absorbs) its intermittent power demand. This feature permits the locomotive power plant to be downsized. Moreover, the onboard energy storage, if sufficiently high, would give the locomotive the flexibility to traverse long tunnels without emissions. Electric mode in tunnels is becoming a sought after feature in hybrid city buses.

REFERENCES

[1] B.H. Kenny, P.E. Kascak, R. Jansen, T. Dever, "A Flywheel Energy Storage System Demonstration for Space Applications," IEEE International Electric Machines and Drives Conference, IEMDC2003, Monona Terrace Convention Center, Madison, WI, June 1–4, 2003, pp. 1314–1320.

[2] M. Ehsani, "Power Electronics & Motor Drives for Military Vehicles," U.S. Army Vetronics Institute Seminar Series, Warren, MI, June 2–9, 2003.

[3] R.F. Thelen, J.D. Herbst, M.T. Caprio, "A 2MW Flywheel for Hybrid Locomotive Power," IEEE 58th Vehicular Technology Conference, VTC 2003-Fall, Orlando, FL, October 6–9, 2003.

9

ESD Protection for Automotive Electronics

Albert Z.H. Wang
Illinois Institute of Technology, Chicago, Illinois

9.1 INTRODUCTION

Electronics (other than radios) were first introduced into automobiles in the late 1950s. However, the rebirth and rapid implementation of automotive electronics in modern vehicles really came during the 1970s, mainly due to the introduction of tough governmental regulations for fuel economy and emission control as well as the emergence of low-cost solid-state electronics based on integrated circuit (IC) technologies. Modern automotive electronics can be found everywhere in automobiles, from engine control, driveline control, motion control, to instrumentation for vehicle performance monitoring and on-board diagnosis, to safety and comfort, to various in-vehicle entertainment, communication, and navigation applications. These automotive electronics applications can be roughly characterized into three categories: control, measurements, and communications. New automotive electronics applications and features emerge at a very fast speed in modern automobiles; for example, global positioning systems (GPSs) for navigation, in-motion detection for anticollision, automatic cruise control, in-vehicle theater-quality entertainment, wireless communications, and local interconnect network (LIN)-based in-vehicle networking. Today, the cost of automotive electronics might account for up to 25% of the total vehicle costs.

While automotive electronics have many different applications, generally speaking, they operate under rather hash environments and meet very high reliability standards. Particularly, automotive electronics are prone to electrostatic discharge (ESD) damages; hence, they require very high ESD protection specifications. This chapter is devoted to ESD protection issues for automotive electronics that cover ESD fundamentals, ESD test models, ESD protection structure, and ESD protection circuitry design.

9.2 ESD FAILURES AND ESD TEST MODELS

ESD occurs when two objects of different electrical potential come into close proximity, where electrostatic charges start to transfer between the two objects. Such electrostatic discharge produces large current or voltage surges that may cause damage to electronic parts. Typical ESD pulse has a short duration of around 100 ns and a very short rise time of nanosecond scale. Typical ESD surges have a current height of several amps or a voltage peak up to a few tens of kilovolts. Such strong current or voltage surges can easily damage IC parts and cause immediate circuit malfunction.

Generally, ESD failures can be classified into two categories: hard failure and soft failure. Hard ESD failures result in immediate circuit malfunction; while soft failures typically lead to performance degradation and lifetime problems. There are typically two types of hard ESD failures. The first type is thermal failure in semiconductors, e.g., silicon material or metal interconnects. Such kind of thermal filament is usually caused by overheating in silicon and metal interconnect due to the large ESD current pulse, which generates a large amount of heat locally that cannot be dissipated immediately and raises up the material temperature to beyond its melting temperature threshold. The second type of failure is dielectric rupture due to extremely strong local electric field strength caused by a large ESD voltage surge. The typical ESD failure signature of this kind is seen as damage in CMOS gate oxide films.

Currently, there exist many different kinds of ESD test models proposed by various organizations to characterize ESD failure behaviors. Some of those have already been established as industrial standards. These various ESD test models generally differ in terms of the origins of ESD pulse generation. They are human body model (HBM) [1, 2], machine model (MM) [3–5], charged device model (CDM) [6], a model from the International Electrotechnical Commission (IEC) [7], as well as a transmission line pulsing (TLP) model [8]. These ESD test models are critical to understanding the nature of different ESD phenomena, characterizing ESD failures, developing various ESD testing instrumentations, and evaluating ESD protection circuitry.

A well-known ESD mechanism is associated with human body-caused discharge that is described by the HBM model, which was initially proposed as a military standard [1]. HBM model describes an ESD phenomenon where a charged human body touches an electronic part and the electrostatic charges stored inside the human body discharge into the electronic part, as illustrated in Figure 9.1, where R_o is the charging resistance, C_{ESD} is the human body capacitance, R_{ESD} is the discharging resistance, and the device under test (DUT) is the electronic part. A commonly used HBM discharging equivalent circuit model is given in Figure 9.2, where a discharging inductance, L_{ESD}, is included for real-world consideration. Table 9.1 lists typical values for the HBM circuit model elements.

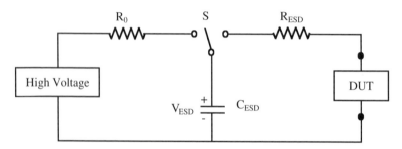

Figure 9.1 An HBM model circuit.

ESD Protection for Automotive Electronics

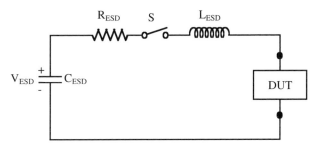

Figure 9.2 An HBM ESD discharge model circuit.

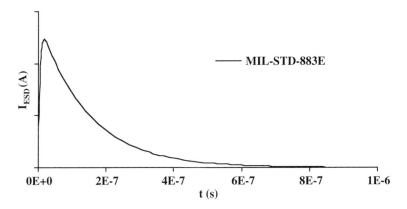

Figure 9.3 A typical HBM ESD discharge waveform following the military standard MIL-STD-883E [1].

Table 9.1 HBM Model Circuit Parameters

Devices	Values
R_o	1–10 MΩ
R_{ESD}	1.5 kΩ
C_{ESD}	100 pF

Figure 9.3 illustrates a typical short-circuit HBM model ESD discharge waveform, with its time constants listed in Table 9.2. Commonly, HBM ESD protection levels are classified as in Table 9.3.

Another common ESD discharge phenomenon is associated with the manufacturing environments where metallic machinery, such as in-line inspection and automatic test equipment (ATE), are heavily used to evaluate IC parts. The machinery can be charged and then discharged into the IC part when touching it. This ESD discharging mechanism is simulated by the MM model [3]. The equivalent circuit model for the MM model can

Table 9.2 HBM Model Short-Circuit Discharge Waveform Specifications

Times	Values (ns)
Rise time, t_r	< 10
Decay time, t_d	150 ± 20

Table 9.3 MIL-STD-883E HBM ESD Classifications

Classes	ESD Protection Levels
Class 1	< 2 kV
Class 2	2–4 kV
Class 3	> 4 kV

be represented by the same circuit model shown in Figure 9.2. However, unlike its HBM model counterpart, MM model features relatively large capacitance of C_{ESD} = 200pF, and negligible discharging R_{ESD} and L_{ESD}. Figure 9.4 illustrates a typical MM model ESD discharging waveform featuring oscillation. The peak currents in MM model are usually much higher than that in HBM model, as given in Table 9.4 as an example.

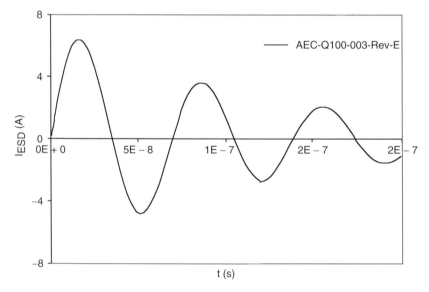

Figure 9.4 A typical oscillatory MM ESD discharge waveform following the AEC-Q100-003-Rev-E Standard [5] with a 400 V ESD source.

ESD Protection for Automotive Electronics

Table 9.4 MM ESD Model Short-Circuit Discharge Waveform Features for a 400 V ESD Source

V_{ESD} (V)	1st Peak I_p (A)	2nd Peak I_{osc} (A)	Main Period (ns)
400	6.0–8.1	67–90% of I_p	66–90

Another quite different ESD event is described by the CDM ESD model [6]. Unlike HBM ESD events, CDM ESD discharge features a self-discharge property, where an IC part is precharged during manufacture or field applications, which then discharges when getting connected to ground. CDM model circuit is the same as the HBM model circuit shown in Figure 9.2. However, CDM model circuit has very small values for its capacitance, inductance, and resistance due to its small size. For example, a typical C_{ESD} is 6.8 pF and L_{ESD} is 50 nH, while R_{ESD} is only a few tens of Ohms. Nevertheless, a CDM ESD discharge waveform is very rapid, with a sub-ns rise time constant, and features a very high current pulse that is also oscillatory. Figure 9.5 illustrates a sample CDM ESD waveform associated with a 500 V ESD source, with its characteristics listed in Table 9.5. The uniqueness for a CDM ESD event is that is extremely fast and involves a large energy in a very short period, which makes ESD protection design very challenging. CDM ESD pulses typically cause gate oxide damages in CMOS ICs.

A different ESD model, initially developed in Europe for equipment-level ESD testing, is called IEC ESD model [7], with its equivalent model circuit similar to that of HBM model in Figure 9.2. However, it has different values for each circuit elements; e.g., C_{ESD} = 150pF, R_{ESD} = 330 Ω, and L_{ESD} = 0H. IEC ESD discharge also features very fast rise time of less than 1 ns. Figure 9.6 illustrates a typical IEC ESD discharge waveform, with its specifications listed in Table 9.6 for a case for 1000 V ESD source.

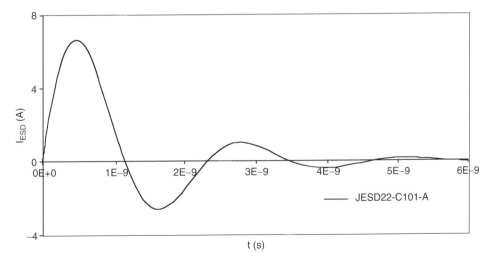

Figure 9.5 A typical CDM ESD discharge waveform following the JESD22-C101-A Standard [9] with a 500 V ESD source.

Table 9.5 CDM ESD Model Short-Circuit Discharge Waveform Features for a 500 V ESD Source

V_{ESD} (V)	1st Peak I_p (A)	2nd Peak I_{osc} (A)	Full Width at Half Height	t_r
500	5.75	< 50% of I_p	1.0 ±0.5 ns	< 0.4 ns

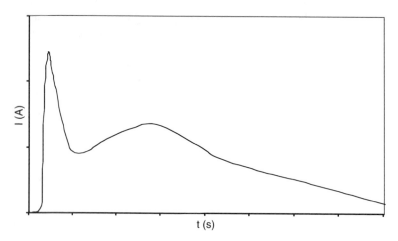

Figure 9.6 A typical IEC ESD discharge waveform.

Table 9.6 IEC ESD Model Short-Circuit Discharge Waveform Specifications for a 1 kV ESD Source

V_{ESD} (V)	1st Peak I_p (A)	Current at 30ns (A)	Duration (ns)	t_r (ns)
1000	7.5	4	~ 80	0.7–1.0

In general, HBM, MM, CDM, and IEC models are standardized ESD models that have similar equivalent model circuit, but with different values for circuit elements as summarized in Table 9.7. Various ESD zapping test equipment is available, developed based upon these ESD test models. Another very useful, yet not standardized, ESD model is the TLP model [8]. Figure 9.7 illustrates a typical TLP model circuit where a piece of transmission line cable is charged, which then discharges into the DUT device through a specially designed network, so that a well-defined square waveform is produced that is used to stress the DUT to simulate a real-world ESD stressing situation [8]. TLP test is very important and informative to practical ESD protection circuit design.

ESD Protection for Automotive Electronics

Table 9.7 Circuit Parameters for Different ESD Model Circuits

ESD Models	C_{ESD} (pF)	R_{ESD} (Ω)	L_{ESD} (μH)
HBM	100	1500	7.5
MM	200	0*	0+
CDM	6.8	0*	0+
IEC	150	330	—

* At Ohms level.
+ Very small.

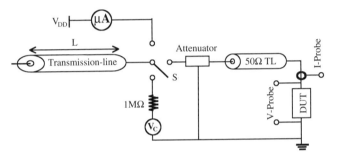

Figure 9.7 A typical TLP testing diagram.

9.3 ON-CHIP ESD PROTECTION

Since ESD transients are very fast and extremely short pulses with enough energy to damage electronic parts, some ESD protection meanings are needed to protect them. In terms of protecting IC parts, there are many different ways for ESD protection, such as ESD-insensitive package and ESD-sensitive materials. The most efficient way is to use on-chip ESD protection structures to protect an IC part. Industrial standards suggest that each IC part should be protected by some ESD protection structures to avoid possible ESD damage. As discussed previously, there are two types of hard ESD failure mechanisms: thermal failure caused by overheating due to large ESD current transients and dielectric failure associated with large electric field density due to high ESD voltage pulses. Accordingly, there should be two basic ESD protection aspects. The first one is to provide a low-impedance active discharge path to shunt the large ESD current transients safely without generating too much heat. The second way is to clamp the IC pad voltage to a sufficiently low level to avoid high electric field density. Generally, there exist two on-chip ESD protection mechanisms as shown in Figure 9.8. Figure 9.8(a) shows a simple diode turn-on type of ESD protection mechanism, where the ESD protection structure is turned on at a specific voltage level to form a low-impedance discharge channel to drain the large ESD current surges. This can be easily implemented by using semiconductor diodes. Figure 9.8(b) illustrate a more efficient ESD discharging method featuring a snapback I-V characteristic, where the ESD protection device is triggered off and driven

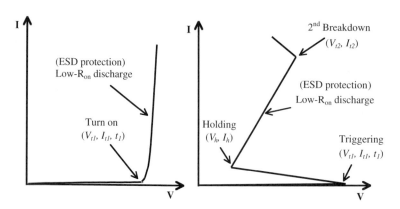

Figure 9.8 Two common ESD protection mechanisms: (a) simple turn-on and (b) I-V snapback.

into a deep snapback, and then forms a low-impedance ESD shunting path to discharge the ESD transients safely. The critical design parameters for an ESD protection structure are their triggering threshold (V_{t1}, I_{t1}, t_1), holding point (V_h, I_h), discharging resistance (R_{ON}), and thermal breakdown threshold (V_{t2}, I_{t2}; i.e., ESD failure point). In practical ESD protection design, one ought to design these parameters carefully, if not resorting to a trial-and-error method. On-chip ESD protection networks can be basic semiconductor devices or ESD protection sub-circuits. This chapter discusses a few commonly used ESD protection structures.

Diodes are the simplest devices to realize the simple turn-on ESD protection mechanism. Diode ESD protection structures can be in various formats, such as forward-biased, reverse-biased (preferably Zener diode), and combined diode strings. Figure 9.9 shows typical I-V characteristics of a diode in forward and reverse biasing mode, both featuring a simple turn-on I-V property. Figure 9.10 illustrates a diode-based ESD protection scheme for an I/O pad on an IC chip. Under ESD protection, a diode operates in a high-current mode, where thermal behavior is important. Triggering voltage, V_{t1}, is mainly determined by the diode forward turn-on voltage or reverse breakdown voltage, depending upon the diode connection mode.

Bipolar junction transistor (BJT), a very efficient ESD protection device, is the operational basis for many ESD protection structures. Figure 9.11 illustrates a common BJT-based ESD protection scheme in its complementary form for an IC pad. Figure 9.12 shows its typical cross-section. The operation follows. Basically, if one can control the base of the BJT transistor and turn it on upon an ESD pulse, it would form an ESD discharging path. Specifically for the given scheme, as an ESD transient appears at the I/O pad with respect to V_{SS}, it reverse-biases the collector junction until it breaks down. Then, the avalanche current is collected by the ground via the external resistor, R, which builds up a potential that eventually turns on the BJT transistor. A low-impedance active discharging path is hence created to shunt the large ESD transient safely, therefore providing ESD protection. ESD protection operation of a BJT structure typically follows the snapback I-V curve as shown in Figure 9.13. When a negative ESD pulse comes, the parasitic collector junction diode will be turned on to shunt the ESD transient. Hence, BJT features an asymmetric ESD discharge I-V characteristic.

MOSFET ESD protection structure is the most popular ESD protection structure in CMOS technologies. Figure 9.14 illustrates a common MOSFET-based ESD protection scheme for an IC pad in its complementary format, with its cross-section shown in

ESD Protection for Automotive Electronics

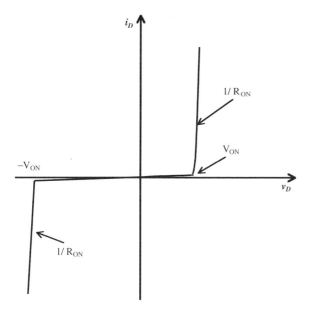

Figure 9.9 Simple turn-on ESD I-V curve of a diode.

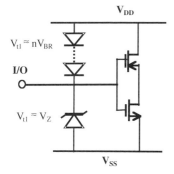

Figure 9.10 A common diode-based ESD protection scheme.

Figure 9.15. The most commonly used connection is the so-called grounded-gate MOSFET (ggMOS), where the gate and source are shorted to the ground. Its ESD protection operation mechanism follows. As a positive ESD pulse appears at the I/O pad with respect to the V_{SS}, it reverse-biases the drain junction of the ggNMOS until it breaks down. Since the gate, source, and body terminals are grounded; the avalanche current will be collected by the ground. Because of the existence of the parasitic lateral well resistance, a voltage potential is built up across the source junction, which eventually turns it on. Hence, the parasitic lateral NPN BJT is triggered to form a low-impedance conduction channel to discharge the large ESD transient safely. Therefore, the core device is the parasitic BJT inside the ggNMOS. On the other hand, if a negative ESD pulse comes to the pad, the parasitic drain junction diode will be turned on to provide the ESD protection. Again, a ggMOS ESD protection structure is an asymmetric device in terms of ESD protection operation. A MOSFET ESD protection structure features a snapback I-V characteristic as

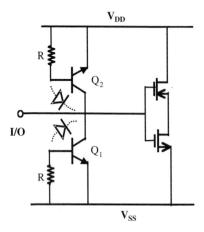

Figure 9.11 A typical BJT ESD protection scheme in complementary format.

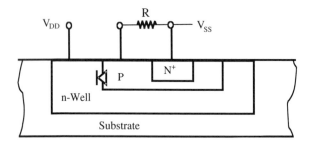

Figure 9.12 Cross-section for a common BJT ESD protection structure.

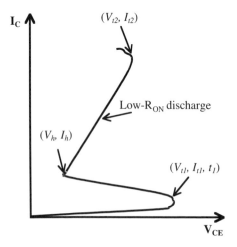

Figure 9.13 Typical I-V curve for a BJT ESD protection structure.

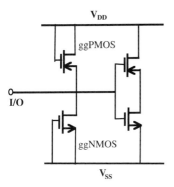

Figure 9.14 A common MOSFET-based ESD protection scheme in its complementary form.

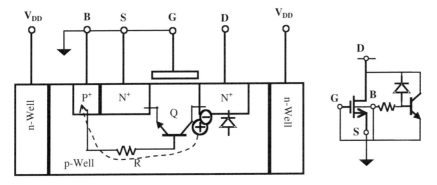

Figure 9.15 Cross-section of a typical MOSFET ESD protection structure.

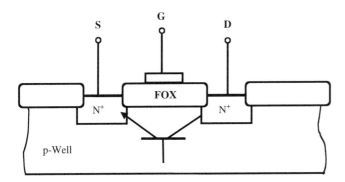

Figure 9.16 Cross-section of a thick-gate MOSFET ESD protection structure.

well. Figure 9.16 shows a different formation of a MOSFET ESD protection structure, whose controlling gate is on top of a thick field oxide layer, as opposed to the thin gate oxide as in normal MOSFET. This is called a thick-gate MOSFET ESD protection structure.

Silicon-controlled rectifier (SCR) can be designed as a very efficient ESD protection structure because of its nature snapback I-V behavior. Figure 9.17 shows the cross-section

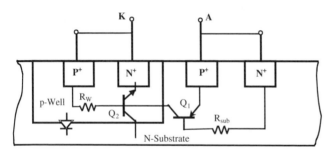

Figure 9.17 Cross-section of a typical SCR ESD protection structure.

of a typical SCR ESD protection structure consisting of two terminals: an anode and a cathode. Figure 9.18 shows its equivalent circuit. In one ESD protection scheme, the anode is connected to the I/O pad and the cathode is connected to the ground. The SCR ESD protection device electrically consists of a pair of parasitic BJT transistors, i.e., a lateral PNP BJT, Q_1, and a vertical NPN device, Q_2, as well as parasitic resistances, as shown in the figures. Its ESD protection operation follows. As an ESD pulse appears at the IC pad, the p-well/n-substrate junction is reverse-biased until it breaks down. The avalanche current is then collected by the cathode and builds up voltage potential across the p-well parasitic resistor, R_w, which eventually turns on Q_2 which, in turn, turns on the Q_1. Hence, the SCR device is triggered to create an active low-impedance shunting path to discharge the ESD transient safely. Its I-V behavior follows a deep snapback characteristic that can clamp the pad voltage to a very low level to avoid dielectric breakdown. In case a negative ESD pulse comes to the pad, the parasitic p-well/n-substrate junction will take the role for ESD discharge. Apparently, the SCR ESD protection structure features an asymmetric I-V characteristic as shown in Figure 9.19. The equivalent triggering voltage for the SCR structure is roughly given by:

$$V_{t1} = BV_{A'K'} \approx BV_{DJ_2}\left(1 - \alpha_1 \frac{I_A}{I} - \alpha_2 \frac{I_K}{I}\right)^{1/n}$$

where BV_{DJ_2} is breakdown voltage of the n-substrate/p-well junction diode, α_1 and α_2 are current gain of the Q_1 and Q_2, n is the ideality factor, and I_A is anode current.

The structures discussed so far are typical single-device ESD protection structures. In practice, a large variety of different ESD protection structures and sub-circuits may be used that can offer better ESD protection performance, depending upon the IC chips under protection. A few examples are given here in this chapter. For more ESD protection circuits and more involving discussions, readers are referred to Reference 10 for details.

For high ESD protection, a large ESD protection structure size is needed in general. To ensure ESD protection performance, one usually uses a multifinger layout structure for large ESD protection devices. For example, Figure 9.20 illustrates a multiple-finger ggNMOS ESD protection structure. Unfortunately, in real-world design, multiple-finger structure does not always perform well due to non-uniform triggering across all the fingers. Hence, early ESD failure often occurs for multiple-finger structure because one finger may be damaged first before any other fingers can be turned on during ESD stress. To ensure uniform finger triggering, a reduced triggering voltage of a ggMOS structure is required, which can be realized by use of a gate-coupled MOSFET (gcMOS) ESD protection, as shown in Figure 9.21 where, instead of grounding the gate, an RC coupling network is

ESD Protection for Automotive Electronics

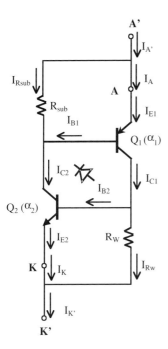

Figure 9.18 Equivalent circuit for the common SCR ESD protection structure.

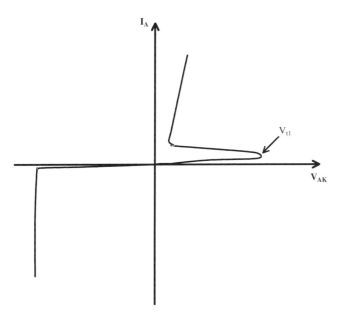

Figure 9.19 A typical asymmetric I-V characteristic for the SCR ESD protection structure.

included in the ESD protection structure. During ESD stress, the gate potential of the MOSFET is raised up, therefore reducing the triggering voltage of the ESD protection structure, leading to a uniform turn-on across all the fingers and, hence, better ESD protection performance.

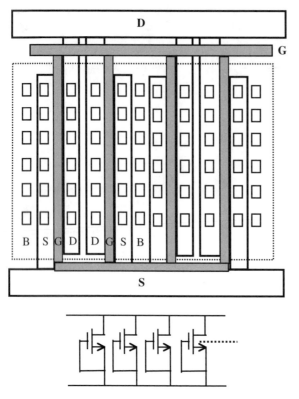

Figure 9.20 A multiple-finger ggNMOS ESD protection structure.

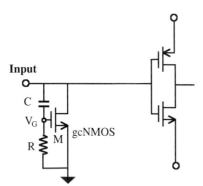

Figure 9.21 A typical gcNMOS ESD protection structure.

Generally, even though SCR ESD protection by itself is very efficient in discharging ESD transients due to its deep snapback I-V characteristic, it is not widely used in practical design due to two reasons; the first is the latch-up problem and the second is that its triggering voltage is usually too high for many ICs because its triggering starts from the breakdown of the p-well/n-well junction. One way to reduce the triggering voltage of an SCR is shown in Figure 9.22, where N$^+$ diffusion is added so that the relatively lower

ESD Protection for Automotive Electronics

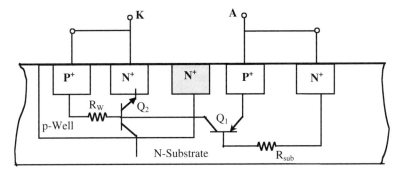

Figure 9.22 An MVSCR ESD protection structure.

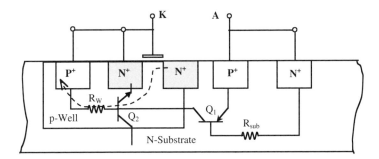

Figure 9.23 A ggNMOS-triggered LVSCR ESD protection structure.

breakdown voltage of the N$^+$/p-well junction determines its triggering voltage, which is lower. Further, a ggNMOS device can be added to a traditional SCR, as shown in Figure 9.23, to even further reduce its triggering voltage, because a ggNMOS normally features a fairly low triggering voltage. Hence, the former is called medium-voltage SCR (MVSCR) and the latter is known as low-voltage SCR (LVSCR). A variety of other modifications can be made to increase SCR ESD protection performance [10]. In addition, many novel ideas can be implemented in designing superior ESD protection circuits so long as they meet the basic criteria; that is, first, to safely discharge the ESD current surge, and second, to clamp the pad voltage to a sufficiently low level. Besides, ESD protection design layout is of critical concern in practical design because ESD protection performance is very geometry-dependent. Also a layout-friendly design is desirable to practical IC chip design in general. As one example, Figure 9.24 shows a novel all-mode ESD protection structure, which has three terminals each connected to I/O pad, V_{SS}, and V_{DD} [11]. As can be seen in the cross-section, this all-mode ESD protection structure consists of six parasitic BJTs that form two bidirectional SCRs, each connected to I/O-to-V_{SS} and I/O-to-V_{DD}, respectively. Therefore, an all-mode SCR (aSCR) ESD protection structure provides a low-impedance active shunting channel to discharge ESD pulses of any polarities, and only one such aSCR per pad is enough to provide complete ESD protection, as illustrated in the scheme of Figure 9.25. The aSCR features a symmetric I-V characteristic, as shown in Figure 9.26.

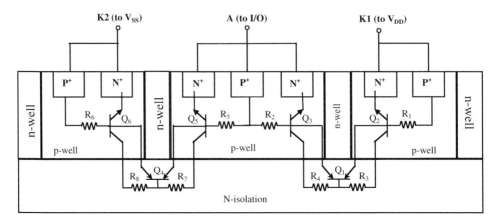

Figure 9.24 Cross-section of an all-mode SCR (aSCR) ESD protection structure.

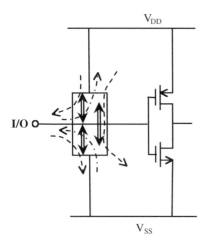

Figure 9.25 An ESD protection scheme using only one aSCR per I/O pad.

In summary, a sea of different ESD protection structures and circuits were developed to provide adequate ESD protection to meet various needs of IC applications. In principle, so long as one can create a low-impedance path to discharge ESD currents safely and clamp the pad voltage to a sufficiently low level, various ESD protection structures can be designed and used to provide the desired on-chip ESD protection. For further information on this topic, readers are referred to Reference 10 for details.

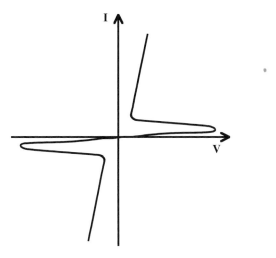

Figure 9.26 A symmetric I-V characteristic for the aSCR.

REFERENCES

1. *MIL-STD-883E, Method 3015.7*, Electrostatic Discharge Sensitivity Classification, Dept. of Defense, Test Method Standard, Microcircuits, 1989.
2. *ESD STM5.1-1998 Revised*, Electrostatic Discharge Sensitivity Testing — Human Body Model Component Level, the ESD Association, 1998.
3. *EIAJ-Standard, IC-121-1988*, Testing Method 20: Electrostatic Destructive Test, Electrostatic Discharge, 1988.
4. *ESD STM5.2-1999 Revised*, Electrostatic Discharge Sensitivity Testing — Machine Model Component Level, the ESD Association, 1999.
5. *AEC-Q100-003 Rev-E*, Machine Model Electrostatic Discharge Test, the Automotive Electronics Council, 2001.
6. *ESD STM5.3.1-1999 Revised*, Electrostatic Discharge Sensitivity Testing — Charged Device Model Component Level, the ESD Association, 1999.
7. *IEC 1000-4-2*, "Electromagnetic Compatibility, Part 4: Testing and Measurement Techniques, Section 2: Electrostatic Discharge Immunity Test," the International Electrotechnical Commission, 1995.
8. Maloney, T. and Khurana, N., "Transmission Line Pulsing Technique for Circuit Modeling of ESD Phenomena," *Proc. EOS/ESD Symp.*, 1985, pp. 49–54.
9. *JEDEC JESD22-C101-A*, Field-Induced Charged-Device Model Test Method for Electrostatic Discharge-Withstand Thresholds of Microelectronic Components, the Electronics Industries Alliance, 2000.
10. Albert Wang, *On-Chip ESD Protection for Integrated Circuits — An IC Design Perspective*, Kluwer Academic Publishers, Boston, 2002.
11. Feng, H.G., Gong, K., and Wang, A.Z., "A Novel On-Chip Electrostatic Discharge Protection Design for RFICs," *J. Microelectronics*, Vol. 32/3, Elsevier Science, March 2001, pp. 189–195.

10

Sensors

Mario Mañana Canteli
University of Cantabria, Cantabria, Spain

10.1 INTRODUCTION

The development of the automobile over the last few years has been closely linked with the development of the electronic systems that they contain. In fact, it could be said today that the electronic component is equally as important as the mechanical component. This electronic development has led to the improvement of all aspects linked, both directly and indirectly, with the automobile: design, manufacture, control of the engine, stability, braking system, and many additional aspects such as navigation systems and external communications.

Within this increasing sophistication of the automobile, sensors play a vital role. They act as refeeding elements of the system and are the keys for supplying the control systems with the appropriate information at all times.

The current systems require dozens of sensors which provide a wide range of information: chemical variables (oxygen, CO, etc.) for the optimal performance of the engine; electrical variables (voltage, current) for correct functioning of the control systems, communications, start-up, and so on; and mechanical variables (pressure, volume, angular speed, position, strain) to control functions, which are critical to the behavior of the vehicle, such as tire pressure, flow of oxygen in the fuel mixture, or wheel slide in antilock braking systems (ABS) [1].

Barron and Powers [2] have estimated that the number of sensors used in engine control applications will increase from approximately 10 in 1995 to more than 30 in 2010.

One of the reasons for the increase in sensor development is the social pressure demanding better and safer vehicles. This requirement translates into more complex vehicles with more electronic components and systems. From an economic point of view, electrical components have a constant impact in car production costs. However, the expansion

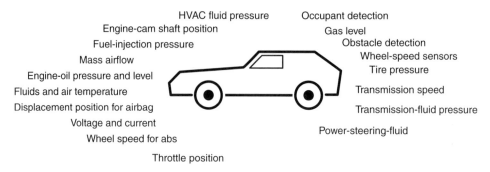

Figure 10.1 Sensors in a late-model car.

of the electronic element inside the car leads to an increase in the production costs of a typical automobile. Since the 1960s, the percentage of electronic systems in cars has gradually increased until reaching 20–30% of the production cost of a complete vehicle [3]. Figure 10.1 summarizes some of the most widely employed sensors in late-model luxury cars.

The competition of today's automobile industry means that sensors, like any other component used in the manufacture of automobiles, must have a low price. In fact, the incorporation of a new technological element in the automobile is reflected in its market price, even in the case of the highest range vehicles. These specific requirements have led to the rapid development of the technology related to the design and manufacture of sensors. Historically, the latter have been rather expensive elements, whose costs have gradually been lowered as measurement technologies have been developed based on integrated circuits that allowed costs to be reduced due to large-scale production.

The sensor market has grown over the last few years to reach a value of around $13 billion in 2000. This market is divided approximately into around 30% for the U.S., 30% for Europe, and 30% for Japan. The remaining 10% is shared by the rest of the world [4]. For the purposes of classification, the analysis of the various types of sensors used in the automobile industry can be made following several different criteria. Application criteria, pressure, force, displacement, velocity, and so on may be used, while sensor technology criteria will include capacitive, piezoresitive, and inductive sensors. Some authors define three major areas for automotive sensor application: powertrain, chassis, and body [5].

In any case, there are five sensor types that make up around 80–90% of the range of application: speed, temperature, acceleration, pressure, and position [4].

Table 10.1 shows the relationships between sensor technology and measured parameters in automotive applications.

10.2 ARCHITECTURE OF ELECTRONIC CONTROL UNITS

Before analyzing the basic concepts of the most widely used sensors in the automobile industry, a brief analysis of measurement system architecture is required. Figure 10.2 shows the block diagram of a generic electronic control unit. It should be noted that, in general, in the electronic development of the automobile industry, there has been a tendency to use a distributed schema, in the sense that individual functions have been implemented that are connected to each other but which function autonomously. A brief introduction to the concept of electronic vehicle control systems and their integration with the rest of

Table 10.1 Relationships between Sensor Technology and Measured Parameters in Automotive Applications

Sensor Type	Force	Pressure	Position	Displacement	Velocity	Acceleration	Shock Vibration	Proximity	Temperature	Flow	Voltage	Current	CO/O₂ NOx
Capacitance	X	X	X	X		X	X	X					
Strain gauge	X	X	X	X	X	X	X						
Piezoelectric	X	X		X	X	X	X						
Potentiometer	X	X	X	X									
Differential transformer	X	X	X	X	X	X							
Eddy current			X	X	X			X					
Inductive			X	X	X		X	X				X	
Hall effect			X	X	X			X					
Magnetoresistive			X	X				X					
Thermocouple									X				
Thermistor									X				
Semiconductor junction									X				
Resistive										X	X	X	
Vortex										X			
Optical			X	X									
Chemical												X	X

Adapted from Strassberg [1] and Jurgen [6].

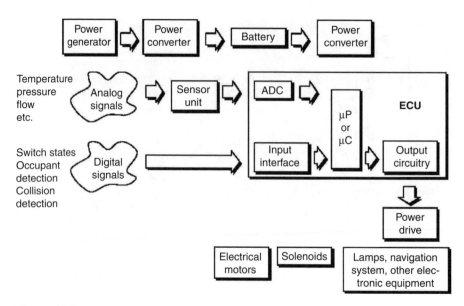

Figure 10.2 Block diagram of an electronic control unit (ECU).

the elements of a vehicle can be found in Reference 7. The main objective of this kind of schema is to maximize the reliability of the system [8].

As can be observed in Figure 10.2, the sensors form an essential element of any control system. They constitute independent elements. In this sense, sensors have evolved from being mere transduction elements, requiring the use of adaptation and compensation circuits, to "intelligent" elements integrating, in a single silicon substrate, the transducer element and all of the electronics necessary for its functioning. This current tendency has made them easier to use, to such a point that they have come to be known as "intelligent" sensors. Figure 10.3 shows the evolution of the sensor elements.

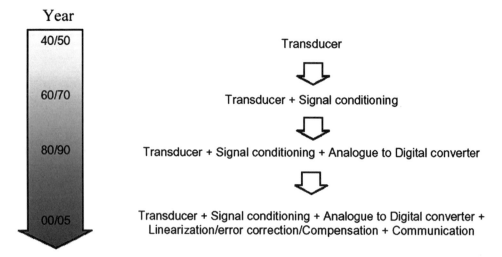

Figure 10.3 Evolution of sensor devices.

Table 10.2 Automotive Sensor Unit Requirements (Dell'Acqua and Timossi [9])

	Sensor Unit
Transduction Element	Accuracy
	Stability in the operating temperature range
	Stability through the lifetime
	Reliability through the lifetime
Electronics	Signal elaboration
	Possibility of calibration
	Compensation for environmental change
	Analog to digital conversion
	Serial communication interface
	Electromagnetic compatibility
	Short circuit polarity reversal and overvoltage protection
Packaging	Robustness
	Resistance to thermal cycling
	Resistance to shocks and vibrations
	Resistance to corrosion (dusts, salt, fuel, water, etc.)
	Easy handling
	Small size
	Low cost

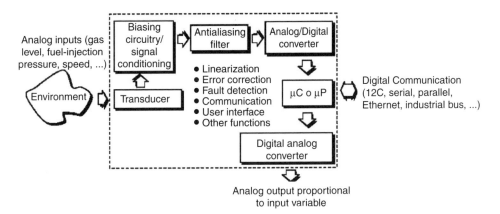

Figure 10.4 Block diagram of a last-generation smart sensor device.

From the point of view of the requirements that a sensor used in the automobile industry must fulfill, some experts [9] establish three groups of requirements, as shown in Table 10.2. In addition, Figure 10.4 shows a block diagram of an intelligent sensor.

Below, a review is made of the basic operating principles of some types of sensors widely used in the automobile industry, which are also destined to play an important role in electric vehicles [10]. Among all of the sensors presented, special attention has been paid to current measurement using the Hall effect due to its widespread application in power converters and electric drives.

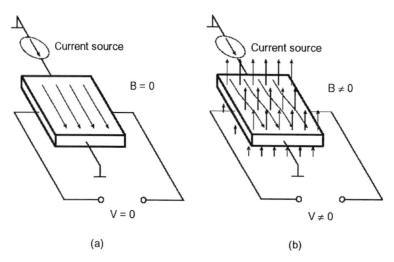

Figure 10.5 Basic structure of Hall sensor: (a) Without magnetic field; (b) With magnetic field.

10.3 VOLTAGE AND CURRENT MEASUREMENT

Current measurement in the electrical systems used in vehicles differs in certain respects from current measurement in other types of electrical installations. In general, electrical systems operate within the range of 12 to 24 V of DC, with consumptions that vary from a few mA to some tens of A. Current measurement can be performed using several different types of sensors, though the current tendency is to use Hall-effect sensors [11].

Voltage measurement, in contrast, is quite simple because it is done using high-precision resistor dividers. Thus, the focus here is on current measurement. Despite the difficulties, there are some technologies available for current measurement, of which shunt resistor and Hall sensors are the most widely used. Hall-effect sensor technology is reviewed in more depth than other technologies, since it forms the base for current and flux measurement in power electronic and motor drives.

Hall-effect devices can be considered good current measuring sensors. From the point of view of their electrical bandwidth, their response ranges from DC to some kHz. Another important advantage is that conductors carrying high currents do not need to be interrupted [12].

Modern Hall-effect sensors usually have four terminals. Two of them are supply terminals and the other two are output terminals. The structure of a basic Hall sensor is shown in Figure 10.5.

The use of this kind of sensor is not a complex task if some points are taken into account. If we consider a stable voltage supply and also a constant temperature, the voltage output is basically proportional to the magnetic flux density perpendicular to the face of the sensor package. Figure 10.6 shows the typical characteristics.

Some of the most important points about the above characteristics are:

- Output voltage range. The maximum and minimum values of the output are set by the supply voltage.
- Output saturation. The output saturates when it is close the maximum or minimum values.

Sensors

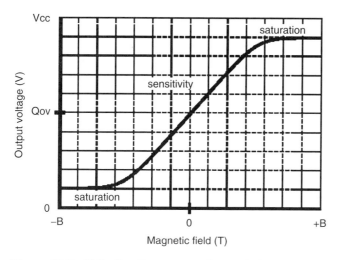

Figure 10.6 Hall-effect linear sensor characteristic.

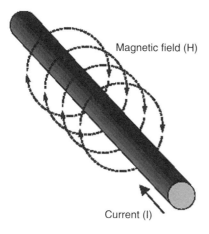

Figure 10.7 Current-carrying conductor and related magnetic field.

- Quiescent output voltage. The output value when the magnetic field applied to the device is zero. Most linear Hall sensors have a quiescent output voltage equal to half the supply voltage.
- Sensitivity. The voltage sensitivity is the proportional ratio between the magnetic flux and the output voltage.

From a commercial point of view, most linear Hall-effect sensors are "radiometric." This means that quiescent output voltage and sensitivity are proportional to the supply voltage. There are commercial solutions covering applications from the low milliampere range into the thousands of amperes. Figure 10.7 shows a current-carrying conductor and the related magnetic field at a certain distance.

If we consider an infinite wire of radius R carrying a current I, the current density J will be:

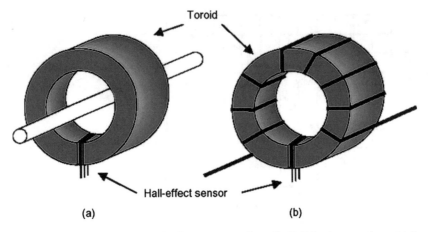

Figure 10.8 (a) Toroid used for flux concentration. (b) Coil for increased sensitivity.

$$J = \frac{I}{\pi R^2}$$

The current density is uniformly distributed over the cross-section of the wire. As the wire is infinite, we can consider from symmetry that magnetic field H has only a tangential component H_ϕ, whose magnitude is a function of distance d only. Using Ampere's circuit law [13]

$$\oint_C \vec{H} d\vec{l} = \int_S J dS$$

if we integrate around a circular contour of radius $r \geq R$ then:

$$\oint_C \vec{H} d\vec{l} = \int_0^{2\pi} H_\phi r d\phi = I$$

so:

$$H_\phi = \frac{I}{2\pi r} \quad r \geq R$$

The Hall sensor position must be carefully analyzed because flux lines are tangential to the conductor and Hall sensors are sensitive to magnetic fields passing perpendicularly through it.

One method for obtaining accurate current measurements is by using a gapped toroid with the current-carrying conductor passing through the toroid. At the same time, the sensor is positioned in the toroid gap. The main objective of the toroid is to concentrate the magnetic field through the sensing element. Figure 10.8(a) shows the setup position of the conductor, toroid, and sensor device.

If the current through the wire is low or if the sensitivity needs to be increased, a toroid coil can be used. Figure 10.8(b) shows a multiple-turns schema for measuring low currents. If the number of turns of wire in the coil is N then

$$H_\phi = \frac{NI}{2\pi r} \quad r \geq R$$

10.4 TEMPERATURE

The measurement of temperature in vehicles imposes requirements in terms of application (air, water, oil, and so on), in the range of temperatures to be measured and in the environment. Thus, inside an internal combustion engine, at the point where the spark plugs are located, the temperatures in the combustion gases can range from 60 to 3000°C. As regards the rest of the elements that make up the engine, the temperatures may vary from 60 to 300°C. It is also necessary to measure the temperature inside and outside the vehicle, which increases the range from −30 to 50°C. In the case of electric vehicles, the measurement of temperature is essential in elements such as the seating area, electric engines, batteries, power electronics, and so on.

There are a large number of technologies dedicated to the measurement of temperature, from thermocouples, thermistors, or resistance temperature detectors (RTDs) to solid-state technology.

Thermocouples are among the most widely used elements in temperature measurement. Basically, they consist of two metal wires of varying composition in contact with each other. The operating principle of a thermocouple is based on the Seebeck effect [14], which consists in the generation of EMF at the union between the two different metals when this union is subjected to a specific temperature. The typical operating range is established between −300 and 1500°C, with a sensitivity that ranges from 5 to 70 µV/°C, which means that a low-offset amplifier is needed in order to produce a usable output voltage. They have nonlinearities in the temperature-to-voltage transfer function, so they often have to be compensated with external circuits or look-up tables.

Thermistors are special resistors whose resistance varies with temperature. The resistance-temperature function is highly nonlinear so that if they are used to measure a wide range of temperatures, it is necessary to perform linearization. The temperature range goes from around −100 to 600°C.

Resistance temperature detectors are also special resistors designed for a practically linear resistance-temperature curve. They exhibit nonlinearities at very low and very high temperatures, but their behavior is predictable and repeatable. The temperature range goes from about −200 to 800°C.

Thermocouples, thermistors, and RTDs require external circuits in order to provide an adequate output voltage with a linear response.

Currently, the tendency is to use monolithic sensors with on-chip signal conditioning. These have a good operation temperature range, typically from −50° to +150°C with a sensitivity of 10 mV/°C. From the point of view of their utilization, the sensor is an integrated circuit and can therefore include signal processing circuitry. The signal conditioning eliminates the need for additional trimming, buffering, or linearization circuits. In addition, this sensor design approach reduces the overall system cost.

Most of the monolithic temperature sensors are ratiometric (the output voltage is proportional to the temperature times the supply voltage). Figure 10.9 shows a simplified block diagram of a ratiometric temperature sensor.

The resistance R_T in Figure 10.9 is a temperature-dependent resistor similar to the classical resistance temperature detectors. It exhibits a resistance whose value is proportional to temperature. The output voltage response follows the law:

$$V_{out} = \frac{V_{cc}}{A}(B + C.T)$$

where V_{cc} is the supply voltage; A, B, and C are constants; and T is the temperature.

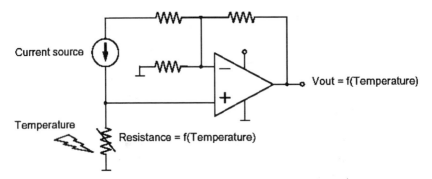

Figure 10.9 Simplified schematic circuit of a ratiometric temperature sensor.

From the point of view of commercial devices, there are a broad range of sensors. It is possible to find temperature sensors with either voltage or current output. Others also include voltage comparators to provide thermostat functions. Some of them have digital I/O with control registers, making them the best interface solution for microprocessor- or microcontroller-based systems.

10.5 ACCELERATION

Acceleration measurement is needed for airbag and active suspension systems. It is also used in ride control systems and antilock brakes in luxury cars. Nowadays the most important application is airbag activation control. Airbag systems require various accelerometers (crash sensors) to provide information to an electronic control unit. The sensors are positioned at different points in the front part of the vehicle. If a crash occurs, the sensors provide a trigger signal to the system that ignites the propellant that inflates the bag [15].

From the point of view of their utilization, it is important to differentiate between crash sensors and accelerometers. The former are digital accelerometers in the sense that their output is active when the acceleration is above a threshold level. Otherwise the output is nonactive. Accelerometers provide an output voltage/current proportional to the acceleration.

There are three basic types of acceleration sensors [16]:

- Piezoelectric. This kind of sensor uses a piezoelectric material. When the crystal is stressed in tension or compression then it develops an electrical charge. The piezoelectric material is attached to a seismic mass so it produces a charge proportional to the acceleration or vibration level.
- Piezoresistive. This type of sensor uses a resistive material whose resistance varies with the acceleration. The sensor device is arranged in a Wheatstone bridge configuration in order to increase the output sensitivity. The frequency range of these sensors is lower than the piezoelectric one, having the advantage of being able to monitor static or DC acceleration levels. The degree of change depends on some physical properties of material: type of piezoresistive material, grain size, doping level, temperature, and others. This kind of devices have been and still are the base of some accelerometers, pressure sensors, and flow sensors.

Sensors

Table 10.3 Basic Parameters of Some Commercial Accelerometers

Principle of Operation	Parameter	Typical Range
Piezoelectric	Frequency response	[2 Hz–10 kHz]
	Measuring range	1000 g
	Sensitivity	[0.001–1.0] V/g
	Output	Digital, voltage, current (4–20 mA)
Piezoresistive	Frequency response	[DC 1 kHz]
	Measuring range	1000 g
	Sensitivity	[0.001–1.0] V/g
	Output	Digital, voltage, current (4–20 mA)
Capacitive	Frequency response	[DC 10 kHz]
	Measuring range	100 g
	Sensitivity	[0.01–1.0] V/g
	Output	Digital, voltage, current (4–20 mA)

- Capacitive. This sensor measures acceleration based on a change in capacitance by means of a capacitor sensing element with a moving plate. It has some advantages with respect to a piezoresistive sensor because it generally offers more sensitivity and more resolution. In general, gaseous dielectric capacitors are relatively insensitive to temperature. They also have a frequency response from DC to some kHz [17].

Table 10.3 summarizes some basic parameters of commercial accelerometers.

10.6 PRESSURE

Pressure sensor applications in the automotive industry have several applications and use different technologies. If tire pressure is not taken into account [18], pressure measurement is mainly dedicated to the measurement of barometric or manifold absolute pressure for engine control systems.

From the point of view of their utilization, these devices have to measure pressures ranging from 0 to 1800 bars in diesel common-rail fuel pressure systems. Consequently, and in a similar way to acceleration measurement, there are several different technologies for pressure measurement [16,19]: piezoelectric, piezoresistive, and capacitive.

10.7 VELOCITY, POSITION, AND DISPLACEMENT

The measurement of velocity, position and displacement is essential to many of the processes taking place during the running of a vehicle: speed control, number of revolutions per minute of the engine, or displacement of the acceleration and brake pedals, to name some of the most typical applications.

Several different technologies are used in the measurement of velocity, position, and displacement:

- Eddy current. This coil-based transducer uses the effect of eddy currents to sense the proximity of conductive materials. The transducer and the target establish a magnetic circuit with a nonlinear relationship between the distance and the impedance of the transducer coil.
- Hall effect. A Hall sensor is a device that, when subjected to a magnetic filed, provides an output voltage proportional to the magnetic field strength. These devices can be used as switches or position transducers.
- Variable reluctance. These sensors are electromagnetic devices consisting of a permanent magnet surrounded by a winding of wire. The sensor is used in conjunction with a ferrous target wheel that has either teeth or notches. When the target wheel rotates near the sensor it changes the magnetic flux, creating a voltage signal in the sensor coil.
- Reed switch. This kind of switch consists of two parts: the reed switch and the actuator magnet. The reed switch will change state when the actuator magnet comes into close proximity to it.
- Wiegand effect. This technology employs specific magnetic properties of specially processed, small-diameter ferromagnetic wire. A uniform voltage pulse is generated when the magnetic field of this wire is suddenly reversed.
- Magnetoresistive. This is based on the change of resistivity of a material due to a magnetic field. The amplitude of that magnetic field depends on the relative position between a magnet and the transducer.
- Potentiometric. These devices use linear or angular potentiometers whose resistance varies linearly or angularly with length or rotation angle. They are not used in velocity measurement.
- Opto. Optical sensors use light emitters and detectors together with a mask wheel in order to measure angular rotation.

Table 10.4 summarizes the different technologies widely used in velocity measurement.

10.8 OTHER SENSORS

There are other sensor technologies used or available for use in automotive applications. Some of them are restricted to classical cars based on combustion engines, but others can also be used in electric or hybrid cars.

- Chemical detectors. These sensors are basically used in emissions control of engines (exhaust gas oxygen and NOx for air/fuel ratio control) [5].
- Fuel level. Ultrasonic, optical, or potentiometer float-arm sensor used for fuel-level measurement [21].
- Rain detector. This sensor is based on infrared technology. It is placed facing the windshield of the car in order to detect water or moisture [22].
- Obstacle detection. Far-distance obstacle detection is used for avoiding collision and for cruise control systems that control both vehicle-to-vehicle spacing and

Table 10.4 Comparative Properties of Different Velocity Sensor Devices

Sensor Type	Zero Speed	Maximum Frequency	Magnets	Nonferrous Wheel	Square Wave Output	Signal/ Noise	Immunity to EMC	Environmental Robustness (Oil, Dirt, Dust, Moisture, Vibration, Temperature, ...)
Eddy current	Yes	500 kHz	No	Yes[a]	Yes	High	Poor	High
Hall effect	Yes/No[b]	1 MHz	Yes	No[c]	Yes	Moderate	Poor	Moderate
Variable reluctance	No	10–100 MHz	Yes	No	No	Low	Moderate	Moderate
Reed switch	Yes	600 Hz	Yes	No	Yes	High	High	Moderate
Wiegand effect	Yes	20 kHz	Yes	No	No	High	—	Moderate
Magnetoresistive	Yes	1 MHz	Yes	No	No	High	High	High
Opto	Yes	10 MHz	No	Any	Yes	Very High	High	Low-Moderate

[a] The target element must be a metal plate.
[b] Sensors with capacitor in feedback circuit cannot achieve zero speed.
[c] Magnetic rotors in a rubber or plastic matrix material are suitable.

Modified from Frank [20].

speed. This kind of sensor is implemented using different technologies. They can be classified in five main groups [23–25]:

- Millimeter-wave radar detectors are strong against rain and dirt but expensive and have to comply with national legislation.
- Laser or IR detectors are comparatively inexpensive but vulnerable to rain and dirt.
- Passive IR detectors are also comparatively inexpensive but vulnerable to dirt and bad weather conditions.
- Ultrasonic detectors have low cost and a simple structure but do not provide good results in the detection of medium to long distances.
- Machine vision detectors have compact size and the ability to detect and classify specific objects. The main problem is their vulnerability to rain, dirt, and night-time.

Flemming's overview of automotive sensors [5] also includes a review of some emerging sensor technologies covering engine combustion sensors, oil quality deterioration sensing, engine/transmission/steering torque sensors, and multiaxis micromachined inertial sensors.

10.9 RELIABILITY CONSTRAINTS IN AUTOMOTIVE ENVIRONMENT

It is well known that automotive components have to be designed with adequate reliability constraints in order to overcome the specific conditions to which a vehicle is exposed.

There are standards covering the testing of automotive electrical/electronic components. In the U.S., the Society of Automotive Engineers (SAE) has defined, among others, the documents SAE J1221 and J575G [26]. From a military point of view, there are also standards defining different aspects related to land vehicles [27]. On an international level, the International Standards Organization (ISO) [28] and the International Electrotechnical Commission (IEC) [29] have defined some documents that govern various aspects of IC and IC-package performance in automotive applications. The Institute of Electrical and Electronic Engineers (IEEE) is working on standards related to electronic components and communications used in vehicles [30]. In addition, some automotive companies have their own technical references.

Table 10.5 summarizes the basic conditions to be taken into account when an automotive environment has to be defined according to SAE.

10.10 CONCLUSIONS

This chapter has presented a basic overview of sensor technology in automotive applications. The automotive environment is extremely harsh because sensors are exposed to wide temperature ranges, mechanical vibrations, high humidity conditions, chemical agents, electromagnetic interferences, and other disturbances, so sensor devices for automotive applications must be highly reliable. At the same time, customers want low costs. This means that a great effort has to be made in the development of microelectromechanical systems (MEMS) technology. In addition, new integrated sensor devices must include some interface requirements such as amplification, calibration, buffering, linearization, temperature compensation, analog to digital conversion, malfunction detection, and network communication.

Hall-effect sensor technology is reviewed in more depth than other technologies because it forms the base for current and flux measurement in power electronics and motor drives.

REFERENCES

[1] D. Strassberg. Automotive-sensor technology drives nonautomotive embedded designs. *EDN,* September 2002, pp. 77–86.
[2] M. Barron, W. Powers. The role of electronic controls for future automotive mechatronic systems. *IEEE/ASME Trans. Mechatronics,* Vol. 1: 80–88, March 1996.
[3] M. Baumler, L.J. Olsson. Prometheus: Sensors for the Automobile of the Future. The Sensor Industry in Europe. G. Tschulena (Ed.). *Proceedings of a Battelle Europe Conf.,* 1989, pp. 77–88.
[4] L. Ristic, R. Roop. Sensing the Real World. In L. Ristic (Ed.), *Sensor Technology and Devices.* Artech House, Norwood, MA, 1994, pp. 1–10.
[5] W.J. Flemming. Overview of Automotive Sensors. *IEEE Sensors Journal.* Vol. 1, No. 4: 296–308, December 2001.
[6] R.K. Jurgen (Ed.). *Automotive Electronics Handbook. 2nd ed.* McGraw-Hill-SAE, New York, 1999.
[7] H. Morris. Control Systems in Automobiles. In Happian-Smith (Ed.), *An Introduction to Modern Vehicle Design.* SAE International, Warrendale, PA, 2002.
[8] M.A. Burchett. An overview of in-vehicle control systems development. 8th International Conference on Automotive Electronics, 1991, pp. 95–97.
[9] R. Dell'Acqua, G.M. Timossi. Sensors technologies for automotive applications. 6th IEEE/CHMT International Electronic Manufacturing Technology Symposium, Japan, 1989, pp. 196–200.
[10] W. Dunn. Automotive sensor applications. IEEE Workshop on Electronic Applications in Transportation, 1990, pp. 25–31.
[11] J. Gilbert, R. Dewey. Linear Hall-effect sensors. Application note 27702A. Allegro MicroSystems, Inc. http://www.allegro.com.
[12] Honeywell. Hall-effect sensing and application. http:/www.honeywell.com/sensing.
[13] R. Plonsey, R. Collin. *Principles and Applications of Electromagnetic Fields.* McGraw-Hill, New York, 1961.
[14] National Semiconductor. *Temperature Sensor Handbook.* http://www.national.com.
[15] N. Barbour, G. Schmidt. Inertial Sensor Technology Trends. *IEEE Sensors Journal*, Vol. 1, No. 4: 332–339, December 2001.
[16] D.S. Eddy, D.R. Sparks. Application of MEMS Technology in Automotive Sensors and Actuators. *Proceedings of the IEEE*, Vol. 86, No. 8: 1747–1755, August 1998.
[17] A. Béliveau, G.T. Spencer, K.A. Thomas, S.L. Roberson. Evaluation of MEMS Capacitive Accelerometers. *IEEE Design & Test of Computers.* October–December: 48–56, 1999.
[18] M. Hill, J.D. Turner. Automotive Tyre Pressure Sensing. *IEE Colloquium on Automotive Sensor.* 1992, pp. 5.1–5.6.
[19] Walter, P.L. *Dynamic Force, Pressure and Acceleration Measurement.* Endevco Corporation. http://www.endevco.com.
[20] R. Frank. Sensors for the Automotive Industry. In L. Ristic (Ed.). *Sensor Technology and Devices.* Artech House, Norwood, MA. 1994. pp. 377–419.
[21] E.N. Goodyer. Novel sensors for measuring fuel flow and level. Sixth International Conference on Automotive Electronics. 1987, pp. 275–279.
[22] K.C. Cheok, K. Kobayashi, S. Scaccia, G. Scaccia. A fuzzy logic-based smart automatic windshield wiper. *IEEE Control Systems Magazine.* December: 28–34, 1996.

[23] M. Kawai. Collision Avoidance Technologies. *Proceedings of the International Congress on Transportation Electronics*, SAE P-283, 1994, pp. 305–316.

[24] R. Stobart, M. Upton. Techniques for distance measurement between vehicles. 28th International Symposium on Automotive Technology and Automation, 1995.

[25] S. Tokoro. Automotive Application Systems of a Millimeter-wave Radar. *Proceedings of the 1996 IEEE Intelligent Vehicles Symposium,* 1996, pp. 260–265.

[26] SAE Standards. The USA Engineering Society for Advancing Mobility in Land Sea Air and Space. http://www.sae.org.

[27] U.S. Military Standards. The U.S. Department of Defense Single Stock Point for Military Specifications, Standards and Related Publications. http://www.dodssp.daps.mil.

[28] ISO Standards. The International Organization for Standardization. http://www.iso.org.

[29] IEC Standards. The International Electrotechnical Commission. http://www.iec.ch.

[30] IEEE Standards. The Institute of Electrical and Electronic Engineers. http://standards.ieee.org.

Part III

Automotive Power Electronic Converters

11

DC-DC Converters

James P. Johnson
Caterpillar Inc., Washington, Illinois

This chapter consists of insight into the design and development of the most commonly used DC-DC converters. The following sections contain DC-DC converter basics, a summary of converter control information, common topologies with necessary design equations including isolated converters, supplemental circuits such as gate drivers and protection circuits, practical considerations in converter design, and other essential converter development details.

11.1 WHY DC-DC CONVERTERS?

Linear voltage regulators convert one level of DC to another, typically using a series pass transistor circuit. The main problem with linear regulators is that the series pass transistor usually operates in the linear region of its characteristic curves, and thus exhibits a much higher loss when compared to a switching converter. As shown in Figure 11.1, the efficiency of this system is highly dependent on the difference in potential across the series pass device. With switching converters, the power semiconductor device operates in either a fully "on" or fully "off" mode, and thus in the on state operates in the nonlinear "low voltage drop" region, exhibiting a lower V × I conduction loss across the device. As pulse width modulation (PWM) of the power switch is used to control the power through the DC-DC converter, its efficiency is less dependent on the voltage difference from input to output, allowing operation with a wider input voltage range and higher efficiencies.

DC-DC converters may include galvanic isolation from input to output in the form of a transformer, or they may be nonisolated. High-frequency switching of a power semiconductor device utilizing PWM is the core of DC-DC converters. The highest frequency switching is most common in the lowest power DC-DC converters, or when

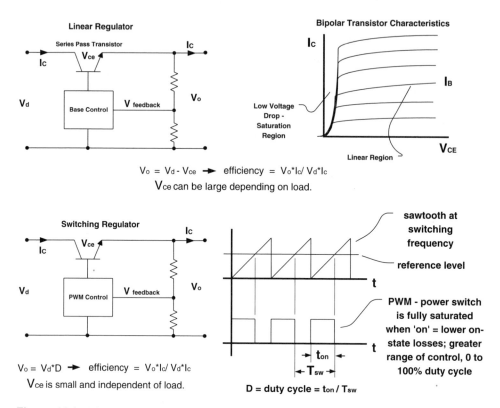

Figure 11.1 Linear regulator compared to PWM switching regulator.

resonant techniques are used. The use of high-frequency switching allows expensive and bulky transformers and filter components (inductors and capacitors) to be reduced in size, reducing cost, weight, and packaging requirements. The limitations of high-frequency switching are switching losses in the power semiconductor devices and the practical limits on the size of high-frequency transformer core materials such as ceramics. Another factor is the higher initial cost of development of larger high-frequency magnetic components and associated packaging. The graph of Figure 11.2 shows which types of converters are used at various power levels. More details of why these limitations on power are made are included with the descriptions of the various types of converters that follow.

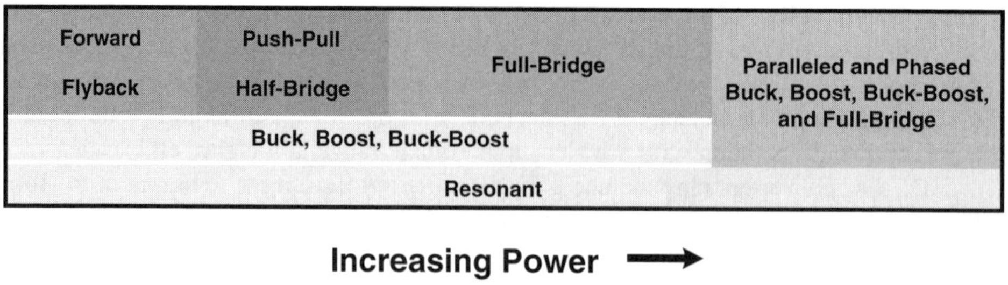

Increasing Power ⟶

Figure 11.2 General power spectrum of DC-DC converter topologies.

DC-DC Converters

11.2 DC-DC CONVERTER BASICS

A DC-DC converter in this chapter is defined as any power electronics system with a primary function of taking, as an input, DC power from a source with a given volts-amps characteristic, and producing as an output DC power with a specified volts-amps characteristic. The volts-amps characteristics may be as simple as maximum and minimum voltages and currents, or more complex characteristic curves that may be dependent on other outside parameters, such as current control requirements, or photovoltaic or fuel cell power output characteristics. Other converter requirements may include voltage regulation requirements; input and output impedance specifications in order to better match filters, loads, and sources; ambient operating temperature range; and requirements concerning vibration, efficiency, durability, reliability, protection, weight, volume, manufacturability, cost, and applicable standards. Typically, a first step in the development or procurement of a DC-DC converter is to create a specification covering each of these areas in detail utilizing technical standards of the IEEE, UL, SAE, NEC, ANSI, and CENELEC as sources of reference. On commercial vehicle applications the most important aspects of the design are durability, reliability, and cost. Vehicle applications require special attention be given to thermal and vibration considerations in the design and packaging, as vehicles can be exposed to harsh environments, and engine compartments and other parts of a vehicle tend to be extremely hot or cold in certain climates.

DC-DC converters are typically used for power supplies for other circuits: battery charging, welders, heaters, upconverters that transfer power from a lower DC voltage bus to a higher DC voltage bus, and downconverters that transfer power from a higher DC voltage bus to a lower DC voltage bus. DC-DC converters are also used in DC motor drives, AC machine field control circuits, and power factor correction circuits. DC-DC converters are often combined with other types of power converters such as inverters and rectifiers to form more complex power converters such as DC-to-AC and AC-to-DC converters. The inverter-driven DC-DC converters, i.e. the push-pull, half-bridge, and full-bridge, consist of a DC-AC stage followed by an AC-DC stage: an inverter-rectifier two-stage system. All of the other types discussed in this chapter are purely DC-DC converters.

11.3 DC-DC CONVERTER TYPES

There are many types of DC-DC converters including buck or step-down, boost or step-up, buck-boost (step-up or -down), flyback, Cúk, Sepic, resonant types, and inverter-driven types: push-pull, half-bridge, and full-bridge. The highest power levels that need DC-DC conversion use paralleled DC-DC converter units that operate with "phasing" that requires the switching pulses of each of the paralleled units be staggered or "phased" through 360 degrees, such that a higher frequency current ripple is present in the combined converter, reducing the filtering requirements, increasing the volume of similar parts used, and thus reducing costs. Alternatively, for higher power systems, an inverter-rectifier system or DC-AC/AC-DC two-stage system is used with a step-up or step-down transformer to provide the DC-DC conversion. In the presentations of the converter types that follow, only the continuous current mode of operation is covered for converters without parasitics. Continuous current operation means that the inductor current in the converter is continuous, never staying at zero current. At the boundary of continuous and discontinuous conduction the current in the inductor will just reach zero at the lowest peak of the current waveform once each cycle. The discontinuous mode of operation is typically avoided, as the disadvantages include higher peak currents, lower power throughput, and greater computational difficulty in the analysis of discontinuous mode and in developing control for the discontinuous mode of operation. Filtering requirements are also greater for the discontinuous mode due to the higher peak requirements for the same throughput as in a like system with continuous current mode operation. Parasitics are the additional resistances, inductances, and capacitances associated with electronic components such as the equivalent series resistance (ESR) of a capacitor or the series resistance of an inductor. As some insight into the converter performance with parasitics will be included in the converter explanations to follow, much of the effect of parasitics can be reduced or eliminated with adequate and modern control methods. For details on discontinuous mode operation refer to References 2 and 8, and see References 2, 8, 9, and 12 for analysis of DC-DC converters with parasitics.

11.4 BUCK, BOOST, AND BUCK-BOOST CONVERTER COMMONALITIES

The buck, boost, and buck-boost converters each consist of a power switch, a diode, and an inductor, and are often accompanied by an output filter capacitor and input filter. The arrangement of the components varies slightly from one topology to the next, as will be discussed in the sections to follow; however, some similarities will first be presented.

In the explanations that follow, and as shown in Figure 11.3, Figure 11.4, and Figure 11.5, V_d is the converter input voltage, V_o is the converter output voltage, ΔI is the peak to peak variation in the inductor current, L is inductance, and t_{on} is the time duration the power switch is turned on in a switching cycle.

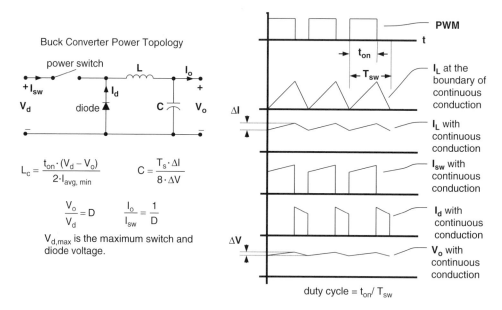

Figure 11.3 Buck converter topology and waveforms.

DC-DC Converters

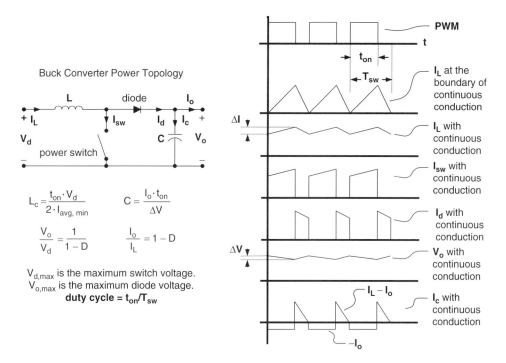

Figure 11.4 Boost converter topology and waveforms.

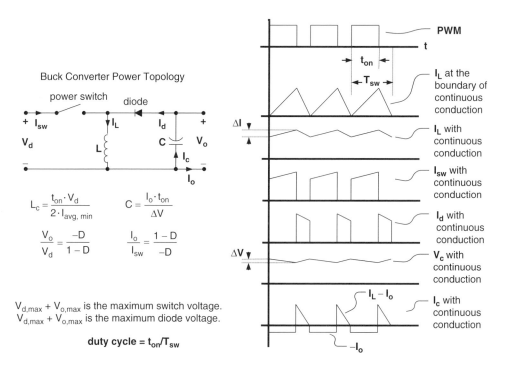

Figure 11.5 Buck-boost converter topology and waveforms.

In the buck, boost, and buck-boost the rate of rise of the inductor current, and the magnitude of ΔI in the inductor, depends on the voltage across the inductor or $V_d - V_o$ in the buck, and V_d in the boost and buck-boost. The ripple current variation in the inductor, ΔI, is such that

$$\Delta I = \frac{V_d - V_o}{L} t_{on}$$

in the buck, and

$$\Delta I = \frac{V_d}{L} t_{on}$$

in the boost and buck-boost. ΔI varies depending on the circuit inductance, the voltage across the inductor, and the duty cycle of switching.

In all three converters, during continuous current mode operation at a steady operating point, the switch turns on each cycle and immediately assumes operation at the same level of current it started at in the previous cycle, I_1, as shown in Figure 11.3, Figure 11.4, and Figure 11.5. The current then rises as the inductor charges to a current, I_2, at which time the switch is turned off. Once the switch is turned off, the diode assumes the I_2 current level and decreases with the inductor field until the I_1 level is reached, at which time a new cycle begins.

To ensure continuous current operation, a minimum average current level requirement for the converter output is chosen. This corresponds to the inductor minimum average current, $I_{avg,min}$. As shown in Figure 11.3, Figure 11.4, and Figure 11.5, at the boundary of continuous conduction $I_{avg,min} = 0.5 \, \Delta I$. During the "on" interval or t_{on}, it is seen that

$$L \frac{\Delta I}{t_{on}} = V_d - V_o$$

in the buck, and

$$L \frac{\Delta I}{t_{on}} = V_d$$

in the boost and buck-boost. Substituting for ΔI, this is manipulated to

$$L_c = \frac{t_{on} \cdot (V_d - V_o)}{2 \cdot I_{avg,min}}$$

in the buck, and

$$L_c = \frac{t_{on} \cdot V_d}{2 \cdot I_{avg,min}}$$

in the boost and buck-boost, which is the critical inductance. The critical inductance is the minimum inductance required to cause continuous conduction with a load such that the current in the inductor is greater than or equal to $I_{avg,min}$. The V_o/V_d ratio of the buck is D, the boost $1/(1 - D)$, and the buck-boost $-D/(1 - D)$, where D is defined as t_{on}/T_{sw}, and T_{sw} is the switching period.

DC-DC Converters

In determining the critical inductance, we must choose the operating point within the range of input and output voltages at which the minimum inductor current will be drawn. Usually some minimum load is required before continuous conduction will occur. As $I_{avg,min}$ is the minimum average inductor current, and the input voltage and D can be known for this condition, L_c can be determined. A larger value of inductance is typically used, e.g., 2 to 10 times L_c, which reduces the ripple current in the inductor and input current. Another consideration in this decision is the slope of the ramping part of the current waveform in the inductor as shown in Figure 11.3, Figure 11.4, and Figure 11.5. If the slope is not great enough, the current mode control current feedback is more prone to noise. If the slope is too great, the peaks of the current will necessarily be higher to produce the desired output power. The higher the peaks, the greater the losses in the power components in the circuit, i.e., switch, diode, inductor, and capacitor. On the other hand, the higher the inductance value for the same average current, the more bulky and expensive the inductor.

The output voltage ripple is considered in determining the output filter capacitance. Knowing that $Q = VC$ in the capacitor and

$$I = C \frac{dV}{dt},$$

and $I \, \Delta t = \Delta Q$, for the buck

$$\Delta Q = \frac{1}{2} \cdot \frac{T_{sw}}{2} \cdot \frac{\Delta I}{2} = C \cdot \Delta V \Rightarrow C = \frac{T_{sw} \cdot \Delta I}{8 \cdot \Delta V}$$

(buck), and for the boost and buck-boost,

$$\Delta Q = I_o \cdot t_{on} = C \cdot \Delta V \Rightarrow C = \frac{I_o \cdot t_{on}}{\Delta V}$$

(boost and buck-boost). The charge considered is defined as the area between the average inductor current and the inductor current waveform for the buck, and the area between the average output current and the diode current in the boost and buck-boost. The areas above and below the average must equal in these analyses and are the charge that must be stored and returned to and from the output capacitor. These values of capacitance are required for a voltage ripple requirement of ΔV with a peak-to-peak current ripple of ΔI in the case of the buck, and for an average output current of I_o in the boost and buck-boost. This is the minimum capacitance required. The actual value of capacitance should be chosen to handle the ripple current in the buck, or output current in the boost and buck-boost, as well as meet the other specifications of the converter. The equations for the average and rms currents in the switch and diode are shown in Figure 11.6 through Figure 11.10.

The maximum currents in the switch, diode, and inductor occur at full load conditions and are trapezoidal waveforms in the switch and diode, and a triangular ripple with a DC component in the case of the inductor.

11.5 THE BUCK CONVERTER

The buck converter and associated waveforms and equations are shown in Figure 11.3. As seen in the waveforms of Figure 11.3, during the "on" interval, or t_{on}, the power switch

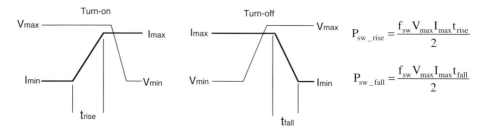

Figure 11.6 Approximate switching losses in MOSFETs and IGBTs.

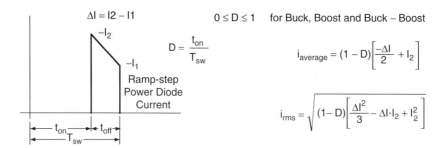

Figure 11.7 Diode current equations for buck, boost, and buck-boost converters.

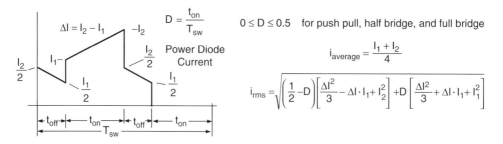

Figure 11.8 Diode current equations for push-pull, half-bridge, and full-bridge converters.

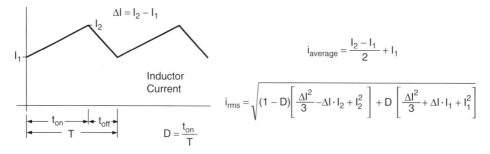

Figure 11.9 Inductor current equations for DC-DC converters in continuous conduction mode.

DC-DC Converters

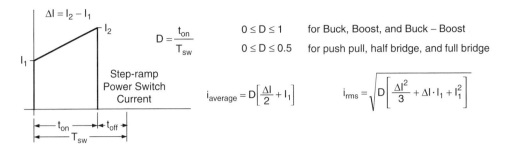

Figure 11.10 Switch current equations for DC-DC converters in continuous conduction mode.

is fully on. This charges the inductor field, with current flowing through the inductor to the output filter capacitor and load. During the "off" interval, or t_{off}, the power switch is fully off. The diode freewheels, allowing the current to continue to flow through the inductor into the output filter capacitor and load as the inductor field collapses. The power switch voltage requirement in the buck is equal to the maximum potential across it or $V_{d,max}$. The diode maximum reverse voltage withstand rating is also $V_{d,max}$. The average inductor current is approximately the average load current, less the ripple current in the capacitor.

Neglecting parasitics and losses, the current ratio in the buck is

$$\frac{I_o}{I_{sw}} = \frac{1}{D}.$$

The average input current is always less than or equal to the average output current.

Adding isolation to the buck converter with a high-frequency transformer, as shown in Figure 11.11, results in a "forward" converter, which is a "buck-derived" converter [10]. As the flux transition in the core of the transformer is unidirectional, a tertiary winding and diode is generally added to reset the core flux to avoid saturation, as shown in Figure 11.11(a). Alternatively, a freewheeling diode and zener may be placed across the primary, as shown in Figure 11.11(b). Without one of these circuit additions, the core may saturate, causing high peaking currents in the switch. Besides galvanic isolation, another advantage the transformer provides is the ability to further magnify or demagnify the voltage with the turns ratio of the transformer. As the core is underutilized, and therefore larger for a given power rating, the forward converter is normally used in only low power converters. The input current that is the switch current in a buck converter is discontinuous, and thus an input filter is normally required to reduce ripple current and harmonic content in the source current.

11.6 THE BOOST CONVERTER

The boost converter and associated waveforms and equations are shown in Figure 11.4. The boost is capable of providing a voltage increase from input to output without a transformer [12]. As seen in Figure 11.4, during the "on" interval, or t_{on}, the power switch is fully on. This charges the inductor field, with current flowing through the inductor and the switch. During the "on" interval, the output filter capacitor supplies current to the load.

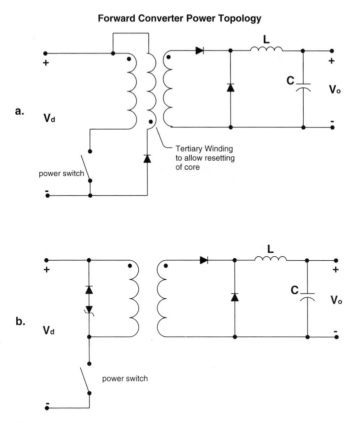

Figure 11.11 Forward converter topologies.

During the "off" interval, or t_{off}, the power switch is fully off and the inductor field collapses. The diode provides a path for the inductor current to flow into the output filter capacitor and load. The power switch voltage requirement in the boost is equal to $V_{d,max}$. The diode maximum reverse voltage withstand rating is $V_{o,max}$. Other considerations in using the boost are that the input current is always greater than the load current, as the current ratio in the boost is $I_o/I_L = 1 - D$, where $0 < D < 1$. Also, the output capacitor must be capable of supplying the total load current and not just the ripple current as in the buck.

The transformer-isolated form of the boost converter is a push-pull converter with an input inductor, referred to as a "current-fed converter" [8]. This circuit has primarily seen use without the secondary rectifier, acting as a current-source inverter.

A nonideal boost that contains parasitic resistances, inductances, and capacitances shows instability above a maximum duty cycle [2].

11.7 THE BUCK-BOOST CONVERTER

The buck-boost converter and associated waveforms and equations are shown in Figure 11.5. The buck-boost is capable of providing a voltage increase or decrease from input to output depending on the duty cycle. The buck-boost creates a negative polarity output voltage. As seen in Figure 11.5, during the "on" interval, or t_{on}, the power switch

DC-DC Converters

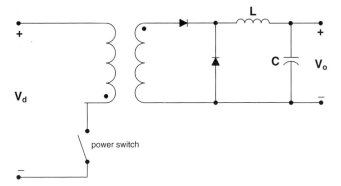

Figure 11.12 Flyback converter topology.

is fully on. This charges the inductor field, with current flowing through the switch and the inductor. During the "on" interval the output capacitor supplies the load with current. During the "off" interval, or t_{off}, the power switch is fully off and the inductor field collapses. The diode provides a path for the inductor current to flow into the output filter capacitor and load. Note that with $0 < D < 0.5$, $V_o < V_d$ or the converter acts as a buck, and for $0.5 < D < 1$, $V_o \geq V_d$ as in a boost operation. The current ratio in the buck-boost is

$$\frac{I_o}{I_{SW}} = \frac{1-D}{D}$$

and thus the input current may be larger or smaller than the output current. The power switch voltage requirement in the buck-boost is equal to the maximum potential across it when it is off, or $V_{d,max} + V_{o,max}$. The diode maximum reverse voltage withstand rating is $V_{d,max} + V_{o,max}$, which is required when the switch is on.

The transformer-isolated buck-boost converter is known as a flyback converter, as shown in Figure 11.12. The magnetizing inductance of the transformer is used as the inductor in the flyback. When the switch is on, the magnetizing inductance charges and current flow is blocked by the diode in the secondary. When the switch is off, the voltage across the inductance reverses, the field collapses, and current flows in the secondary through the diode to the output filter and load. The flyback is usually used at only low power, as the transformer core material is only used in the first quadrant of the hysteresis curve. Also there is a trade-off in the size of the transformer and magnetizing inductance, and the peak currents that tend to be greater in the flyback for the same power rating as other isolated converters that utilize bipolar switching.

11.8 ISOLATED INVERTER DRIVEN CONVERTERS

The push-pull, half-bridge, and full-bridge converters all consist of an inverter, or DC-to-AC converter, followed by a transformer, diode rectifier or DC-to-AC converter, and output LC filter. The inverter part of these converters utilizes bipolar switching, which creates a square-wave AC with PWM. The AC is then passed through the transformer, which steps the voltage up or down, depending on the turns ratio. The secondary output AC is rectified and passed through the output filter in order to attain the output voltage.

The output filter inductor acts as the inductance in a buck converter, and thus the equation for critical inductance for the buck is valid for these "buck-derived" converters as long as the appropriate transformation of the input voltage due to the transformer, modulation, and topology is applied. As with the buck, a minimum average inductor current must be chosen to obtain a minimum inductance for continuous current mode of operation.

11.9 PUSH-PULL CONVERTER

The push-pull converter, waveforms, and equations are shown in Figure 11.13. The two switches alternate, turning on and off with a PWM. Deadtime is included to prevent both switches from conducting simultaneously. When one switch is on, the other switch has twice the input voltage imposed upon it due to transformer action, and thus, ideally, the switch voltage requirement is $2 \times V_d$. When both switches are off, each switch has the input voltage across it. It should be noted that the primary current, and not the secondary current, must be used in feeding the current-mode controller to prevent magnetizing current from saturating the core due to circuit imbalances. The critical inductance is determined as in a buck converter with the added complexity of a transformer. Knowing the transformer ratio, $N = N_2/N_1$, the voltage imposed on the inductor is $V_d \times N \times t_{on} - V_o$.

The output filter capacitor must only handle the ripple current as in the buck. The push-pull transformer requires a center-tapped primary and usually a center-tapped secondary. Thus $2 \times N_1$ turns are required on the primary and with a center-tapped secondary,

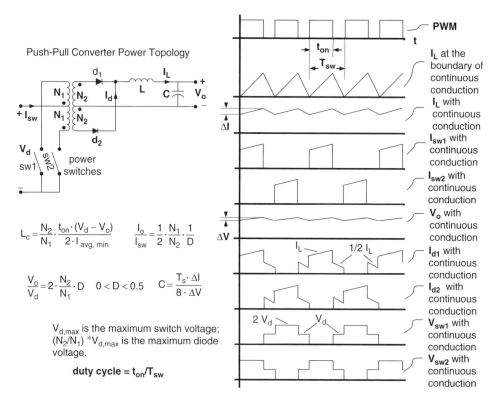

Figure 11.13 Push-pull converter and waveforms.

DC-DC Converters

$2 \times N_2$ turns are required on the secondary. With a single secondary, only N_2 turns are required on the secondary. With a center-tapped secondary, only one diode drop is incurred due to the rectifier, as opposed to a single secondary that, with a full-bridge rectifier, would have two diode drops and twice the diode losses. Conversely, with the center-tapped secondary and full-wave center-tapped rectifier, the diode voltage ratings are twice that required for the full-bridge rectifier.

11.10 HALF-BRIDGE

The half-bridge converter, waveforms, and equations are shown in Figure 11.14. The half-bridge utilizes the centerpoint of a split-bus input capacitor bank as the neutral point.

The switches alternate with a PWM. Special consideration must be made to ensure that sufficient deadtime is included in the PWM to avoid a short-circuit or "shoot-through" condition, which occurs when both switches are on simultaneously. A short-circuit condition causes extremely high-current levels that stress the power semiconductors. Stressing the power devices is something to avoid, as this shortens their life expectancy. The deadtime should be greater than the slower of the slowest turn-on or turn-off times of the power devices with gate signal propagation delays considered, and is usually from 2 to 6 μs at low to moderate power levels. With a transformer ratio of $N = N_2/N_1$, the voltage across the inductor reaches $V_d/2 \cdot N \cdot t_{on} - V_o$. Thus twice the turns ratio is needed for the half-bridge to obtain the same output voltage as the push-pull (and full-bridge). This requirement

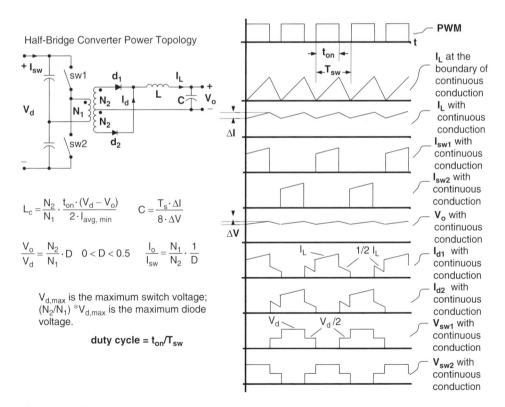

Figure 11.14 Half-bridge converter and waveforms.

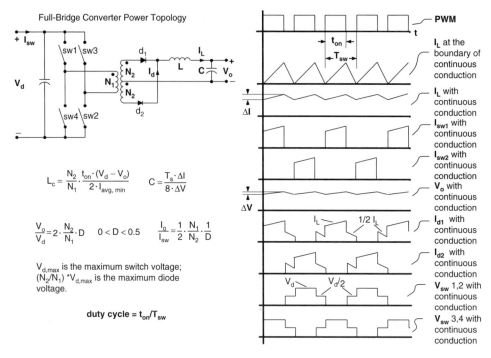

Figure 11.15 Full-bridge converter and waveforms.

means that the switches must handle twice the current for the same output power, as compared to the push-pull and full-bridge. The maximum voltage across each switch is the input bus voltage. Another disadvantage of the half-bridge is the added requirement of a split-bus and thus two capacitor banks that must be rated at $V_d/2$ and high current ratings.

11.11 FULL-BRIDGE

The full-bridge converter, waveforms, and equations are shown in Figure 11.15. The full-bridge also requires deadtime to prevent shoot-through conditions. The inductor voltage reaches $V_d \times N \times t_{on} - V_o$. The maximum voltage across the switches in the full-bridge is V_d. There are two power switches on at any given time in the full-bridge compared to only one device at any given time in the push-pull, and thus a greater switch loss. Other disadvantages are the need for additional circuitry to control the power switches and the higher device count.

11.12 OTHER CONVERTER TYPES

Other less common types of DC-DC converters are resonant types and the Cúk converter.

Resonant converters utilize an LC tank circuit within the converter and frequency modulation (FM) to create an oscillating circuit that provides the benefits of 0 current or 0 voltage switching, eliminating or reducing the switching losses in the converter. This allows higher switching frequencies to be used, allowing a reduction in size of magnetic

DC-DC Converters

and filter components. The drawbacks include increased difficulty with EMI when operating with higher frequencies that require shorter antennae in order to propagate radio frequency (RF) emissions, higher peak currents for the same throughput power as in classical DC-DC converters, greater parts count, and greater control complexity and thus initial development costs. As with conventional DC-DC converters, standard control integrated circuits are available that ease the difficulty of implementation for resonant converters.

The Cúk converter provides a method of obtaining a negative output voltage as in the buck-boost. The primary advantage of the Cúk is that the input and output currents are nearly ripple-free, reducing or eliminating the input and output filtering requirements [2]. The disadvantages include a series capacitor in the topology that must handle large currents, additional parts count in the converter, control complexity, and patent issues.

11.13 CONTROL

In order that a converter act in a manner that meets required control specifications of dynamic and steady-state response, various methods of control theory may be implemented, including classical PID, adaptive, state-feedback, and many others. The control is usually designed to regulate the converter output voltage; however, any output or input parameter could be chosen as the controlled variable. Many integrated circuit devices now exist that reduce the difficulties associated with controlling DC-DC converters. The most predominant type today uses current mode control. An example of this is the Unitrode UC2846 current mode PWM controller IC. Past converter control technology was more oriented toward voltage mode controllers that utilized an output voltage feedback to generate a PWM signal for gating the power semiconductor switch(es), in accordance with a control law designed to regulate the output voltage. The PWM was generated by comparing the output of the error amplifier of the feedback controller with a constant frequency sawtooth waveform that was generated in the controller IC. In current mode controllers the sawtooth waveform is replaced by a sampling of the power switch "step-ramp" current waveform, as illustrated in Figure 11.16.

The outer loop in the current mode control acts on the output voltage creating an error, according to a sampling of the output voltage and a voltage reference such as in the voltage mode control. Using the switch current in an inner control loop allows termination of the PWM pulse on-time when the current reaches a level, allowed according to the control of the output voltage or an overriding maximum current limit configured as part of the IC's supporting circuitry, e.g., the CUR_LIM_ADJ input on the UC2846. The effect in the current mode control is to eliminate the pole associated with the converter output filter inductor (part of a second order filter), as the current through that inductor is controlled by the feed-forward of a sampling of the inductor current, reducing the complexity in the control design. Other advantages include [2]: (1) peak switch current limiting, (2) the possibility of parallel operation of DC-DC converters, as currents in the individual units are individually controlled, (3) the prevention of transformer core saturation in push-pull converters due to imbalance, and (4) feed-forward control, providing an inherent compensation for variations in supply voltage by using the switch current as feedback which, for the same load, naturally increases or decreases with source voltage.

A first step in developing the control circuitry for a DC-DC converter is to develop a model of the converter. Typically either state-space averaging [2] and linear matrix algebra [3], or an equivalent circuit model and circuit solving techniques [4], are used. State-space averaging requires writing the state-space equations for the converter, averaging these according to switch-on and -off times (or duty cycle), and then using linear

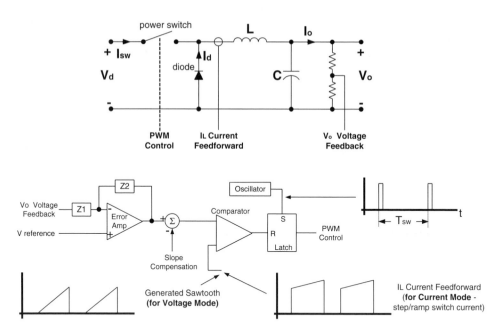

Figure 11.16 Voltage mode and current mode controller block circuit diagram.

algebra, control methods, and matrix methods to manipulate the linearized equations to form the desired converter transfer function for control, which is the output voltage to the duty cycle or control voltage. Also obtainable with these methods are the input-to-output voltage transfer function and the input and output impedances.

The equivalent circuit model technique [4] uses circuit solving techniques to obtain required transfer functions. This technique allows easier inclusion of parasitics compared to state-space methods. Circuit-solving knowledge is required in using this technique.

In current mode control, for buck-derived converters, the small-signal control transfer function from output voltage to control voltage, $v_o/v_c = K \times R_o \times H(s)$, where $K = I_{o,max}/v_{c,max}$, where $I_{o,max}$ is the maximum sensed output current via a transducer in volts (read at the controller IC), and $v_{c,max}$ is the maximum control voltage in volts, where the voltage output of the error amplifier is the control voltage, and the inputs to the error amplifier are the output reference voltage and the actual sensed output voltage. R_o is the load resistance. $H(s)$ is the small-signal transfer function of the power circuit for current mode control, and is

$$H(s) = \frac{1 + sR_{ESR}C}{1 + sR_oC},$$

where R_{ESR} is the output filter capacitance Equivalent Series Resistance, and C is the total output filter capacitance. This transfer function has a 0 due to the output filter capacitance and its ESR, and a pole due to the load resistance and output filter capacitance. The pole due to the output filter inductor is not present in the small-signal analysis due to the current mode control.

Next, a desired control technique must be chosen. Typical approaches involve using Bode plots and frequency domain methods [5]. Essentially, with the plant transfer function known from output voltage to the control voltage or duty cycle ratio, and with a desired

DC-DC Converters

frequency domain response characteristic developed, the control required is the difference between the desired and the actual without control.

The error amplifier of the current mode controller IC is typically used as the controller with the appropriate supporting circuitry. More commonly used types of op-amp-based control circuits and their frequency responses are readily available [2][6][7]. A variety of control references are available for determination of the necessary circuit parameters in the feedback circuit of the error amplifier, such as References 2, 5, 6, and 7. Alternatively, digital control utilizing a microcontroller or microprocessor avoids the nuisances of drift and noise common to analog control circuits. Analog circuits can be made nearly equal to digital with sufficient filtering and thermal compensation circuitry, and possibly at lower cost than the processor-based controller and chip set.

In frequency domain control, the gain margin of the closed-loop system is typically required to be 2 (6 dB) in order to ensure stability at all conditions [8]. The gain at low frequencies should also be as high as possible to reduce steady-state errors [2]. The phase margin should be between 45 and 60° [2]. Smaller phase margins result in an oscillatory system, while larger phase margins result in a system with a more sluggish dynamic response.

11.14 ESSENTIAL CONVERTER CIRCUITS

There are several important parts of a DC-DC converter other than the power circuitry that has been described. These essential circuits include gate drivers with or without isolated supplies, protection circuits, control circuits, and power supplies. A general DC-DC converter system block diagram with essential parts is shown in Figure 11.17.

The gate driver takes the PWM gate signal from the control circuitry and conditions it in such a manner as to provide sufficient drive to a power switch. Gate drivers typically include protection circuits for the power switch, including short-circuit protection and protection from the loss of control voltage. The gate driver must provide sufficient current into the gate or base of the power switch to cause the power device to saturate in a short time, typically in the hundreds of nanoseconds to a few microseconds. Often the gate driver utilizes transformer isolated power supplies to provide the driving power to the switch. This is required in "high-side" drivers, such as the switches in the upper part of the phase legs in the half- and full-bridge converters, and the switch in the buck converter. This inherently gives the push-pull, flyback, forward, boost, and buck-boost advantages,

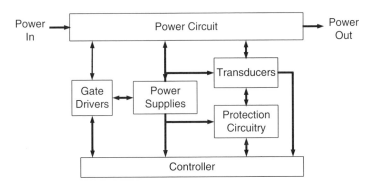

Figure 11.17 Block diagram of complete DC-DC converter.

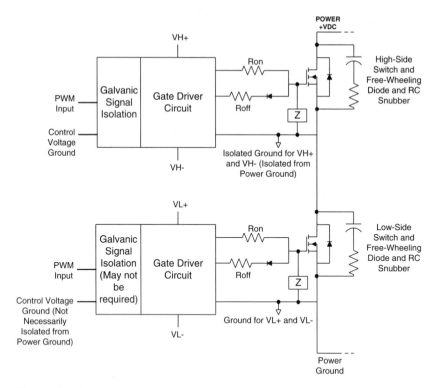

Figure 11.18 Gate driver drawing.

as the gate driver for these converters does not require isolation from the input power source ground.

The maximum current required in the gate driver is

$$\frac{V^+ - V^-}{R_{on}},$$

which is the peak current at turn-on, and where V^+ and V^- are the gate driver positive and negative supply voltages, and R_{on} is the gate driver turn-on resistance as shown in Figure 11.18. The charge is typically considered when a MOSFET or IGBT is turned on. In the following explanation the "source" will refer to either a source connection on a MOSFET or the emitter of an IGBT, and the power switch referred to will be either a MOSFET or IGBT. Knowing $Q = I \times T$ [13], and by knowing the charge required to turn on the device, the average current, I, can be determined. When the turn-on first begins, the gate-to-source capacitance is being charged. When the gate-source voltage reaches the turn-on threshold of the power switch, the gate to drain, or Miller capacitance, begins to require charging as the drain current begins to increase. The Miller capacitance is usually much larger than the gate-to-source capacitance, and as long as a there is a designed safety margin of several amps of additional gate drive current, the gate-to-source capacitance can be neglected. Usually the turn-on time required is chosen. This is T.

During the drain current rise time the gate voltage is constant at the threshold voltage. Once the amps requirement is determined from $Q = IT$, the turn-on gate resistance required is

DC-DC Converters

$$\frac{V^+ - V_{threshold}}{I},$$

where $V_{threshold}$ is the gate-to-source threshold voltage of the power switch. Once the Miller capacitance has been charged, the device is "on" and the gate-source capacitance continues to charge up to the gate driver positive supply voltage. Considering turn-off, an acceptable turn-off time is chosen. Initially the gate-source capacitance is discharged until the device threshold is met. Once the threshold is reached the turn-off time begins according to $Q = IT$, where Q is the Miller capacitance charge. With I determined from Q and T, the turn-off resistance is

$$\frac{V_{threshold} - V^-}{I}.$$

Once the Miller capacitance has discharged, the device is "off" and the gate-source capacitance continues to discharge down to the gate driver negative supply voltage. The gate-source and gate-drain (Miller) capacitances are normally given on the power device data sheet.

The short-circuit protection for a MOSFET typically utilizes a comparator and timing circuit to trip the system while it is gated on. In a MOSFET the current level is directly proportional to the IR drop across the device, which will reach a value indicating an excessive current. In the calculation of the resistance in this formula, the device temperature must be considered, as the normal operating resistance of the MOSFET is typically twice that of its cold value. In the IGBT short-circuit protection, the detection of the bipolar part of the IGBT coming out of saturation, or becoming "desaturated," determines the trip point. If the IGBT collector-emitter voltage when gated on, and after a short delay for the rise time, is greater than a few volts and stays that way for a several microseconds, then the device is determined to be desaturated and a "desat" trip occurs. Short-circuit protection typically acts within 10 µs of the shorted or "desat" condition to prevent device destruction due to an excessive junction temperature due to an extremely high current. Care should be taken in the supporting circuit design to prevent consecutive short-circuit conditions from occurring without sufficient device cooling time in between shorts to prevent device overheating and destruction. Typically a few seconds between short-circuit conditions that exist for under 10 µs is sufficient.

Gate driver inputs are usually opto- or transformer-isolated from the control circuits to reduce EMI transfer and to provide galvanic isolation. Gate driver output leads to the device should be as short as possible to reduce oscillations due to lead inductance–gate capacitance interaction; they are typically less than 6 inches and, if flying leads are used, twisted pair. For high-power and higher voltage systems, fiber optic cables are often used to transfer gate signals to the drivers, allowing greater flexibility in locating the control circuitry with respect to the gate drivers and power devices, while providing galvanic isolation. Pulse transformer isolation is sometimes used to isolate the gate driver from the power device. This requires careful V-s calculations to ensure the pulse transformer is of sufficient size, as well as ensuring the transformer doesn't saturate by providing a bipolar drive signal with 0 average or with a resetting circuit such as a zener–diode pair across the primary.

Other gate driver circuitry may include a Schottky diode on the control signal input to control ground to prevent negative spikes in the system due to reverse recombination in the power diodes from coupling back on the control input and causing a CMOS level gate driver input circuit to erroneously gate-on. Also, a pull-down resistance between gate and source and a small capacitor between gate and source are sometimes used. The pull-down resistor keeps the device off when no control signals are present. The small capacitance, typically around 15 pF, helps to filter high-frequency noise. A larger capacitance could also be used to slow down the turn-on or turn-off switching, to help reduce transients due to reverse recombination spikes during device turn-on (and corresponding diode turn-off) or power switch turn-off overshoot voltages due to leakage or stray inductance. This, however, comes at the expense of additional switching losses due to the longer rise or fall time during switching.

Protection circuits in the DC-DC converter include protection for input power polarity reversal, input overvoltage, input undervoltage, input overcurrent, output overvoltage, output overcurrent, overheating of power devices, short-circuit, loss of control power supplies, and fire protection.

Input power polarity reversal protection may be as simple as a diode in series with the input or a more complex scheme utilizing a comparator and pass-switch or crowbar circuit. Overvoltage, undervoltage, and overcurrent protections typically utilize comparator circuitry with sufficient filtering and hysteresis to allow clean trip points and circuits, which are not prone to false or repeated trips due to noise. Thermal protection may come in the form of thermal switches that are mounted next to the power devices on the heat sink. The thermal switch is normally open or closed and has hysteresis. The temperature rating of the thermal switch is selected based on the power device loss calculations, noting the maximum possible case temperature corresponding to the worst-case junction temperature, at the worst-case operating condition, with sufficient safety margin. Typical thermal switch temperature ratings are around 85 to 95°C. Loss of control power supply is protected against in a variety of ways from dedicated ICs to pull-down resistors on gate driver inputs. Fire protection is protection against catastrophic failure and resulting collateral damage and typically consists of appropriately sized fuses on the input and output of the converter.

Power supplies for the converter are used to provide low-voltage power for the control, protection, and gate driver circuits. Power supplies may themselves be smaller DC-DC converters or may be in the form of linear supplies such as the common 78xx and 79xx types of three terminal regulators.

11.15 IMPORTANT POINTS TO CONSIDER

When starting a converter design, a common method is to develop a spreadsheet with the equations used in the development of the converter, including all component calculations, loss calculations, and control calculations. Each column in the spreadsheet can be used for a variation of the circuit and thus different designs can be compared for performance, cost, component count, heat sink requirements, magnetic component requirements, and other metrics. Reuse of the spreadsheet in future designs reduces design time.

Consider the nonideal elements when designing the converter. Electrolytic capacitors turn into inductors somewhere between 20 and 100 kHz. Above 20 kHz switching frequency, it is best to consider metalized polypropylene film capacitors rather than electrolytics. All elements in the circuit including leads between components consist of resistance, inductance, and capacitances. Power circuit layout is as important as power component selection in a successful converter design. Stray inductances between the power source and the switching device can cause high overvoltage spikes on power device turn-off, which can be a designer's worst nightmare and can be the cause of a design failure.

DC-DC Converters

Inductance is the result of a loop in a circuit. Eliminating distance between power component connections is essential to reducing the size of these loops, which create stray inductance. Laminated busbars greatly reduce stray inductance and are highly recommended as a way to alleviate voltage overshoot problems. An important point in using laminated busbars is to ensure that the bus capacitance and the power semiconductor devices are both directly attached to the laminated bus, as the primary purpose of the laminated bus is to eliminate the inductance loop between the capacitance and the power switch. Another big problem in designs is in poorly designed magnetic components that have larger leakage inductances, also a cause of overshoot voltages.

11.16 SIMULATION VS. ANALYTICAL METHODS

Simulation affords the user a fast and simple way of determining rms, average, and peak currents in the power devices: inductors, transformers, capacitors, and semiconductors. It is a simple matter to learn to simulate DC-DC converters in Pspice, Saber, Psim, or one of many simulation packages that are readily available. Simulations can also provide valuable insight into circuit operations. Calculations of the rms currents are somewhat more difficult in many cases, requiring integration and algebra. Accurate calculations of composite waveforms such as capacitor currents are even more difficult, yet simulations allow not only the knowledge of levels of the overall waveform, but also harmonic content, allowing more detailed and accurate loss and lifetime calculations.

11.17 LOSS CALCULATIONS

Losses in the power circuit consist of switching and conduction losses in the power switch, conduction losses in the diode, capacitor losses, inductor losses, and input filter losses.

11.18 POWER DEVICE SELECTIONS

A practical selection of power semiconductor devices must take several points into consideration.

1. The layout will have stray inductance. This translates into overvoltage spikes on turn-off of the power switches. To be conservative, device voltage ratings should be greater than the theoretically required value. Two times the highest voltage to be seen in the circuit (theoretically) is conservative if the layout is tight.
2. Diodes have a reverse recovery spike associated with their turn-off. For high-frequency switching applications high-frequency (fast) diodes should be used. For an even greater reduction in noisy spikes due to reverse recombination, use fast and soft recovery diodes with a voltage rating at least twice the highest theoretical voltage imposed on the diodes. Snubbers are often required on diodes due to the reverse recovery spike and stray and leakage inductance effects. Increasing the turn-on time of the power switches, and thus slowing down the reverse recovery of the corresponding diode, can help to reduce this spike.

Snubbers are often required, however, to control this spike and thus reduce potentially harmful voltage levels on the diodes and EMI problems elsewhere in the system and surrounding systems due to this spike.
3. Power semiconductor devices exhibit higher losses with higher voltage ratings, and lower losses with higher current ratings. Use the highest current rated device in the package size you plan to use, and the lowest voltage rating you can.

11.19 EMI

Electromagnetic interference (EMI) is often a problem in power converter designs due to high-speed switching of power-level currents, stray and leakage inductances, parasitics, and stray coupling capacitances. EMI filter requirements and noise minimization are important considerations in a converter design. Shielded leads and components reduce noise in circuits that can cause false trips or unintended power switch gating. Printed circuit boards (PCBs) in the power converter should make use of power and ground planes. Layout should include considerations of magnetic fields around power cables and devices, i.e., power switches, diodes, and especially magnetic components and power lines. Signal lines, whether shielded or unshielded, and PCBs should be kept as far from power lines and magnetic components as possible to avoid cross-talk of power switching into the small signal circuits.

11.20 OTHER PRACTICAL CONVERTER DEVELOPMENT CONSIDERATIONS

An accurate bill of material (BOM) includes part nomenclature, part number, manufacturing and ordering information, cost, quantity ordered, quantity on-hand, quantity used, lead-time of components, and overall sub-circuit and per converter costs. In the initial system design, an accurate design BOM can greatly aid in the determination of commercialization feasibility.

To avoid delays in testing, spare parts and spare units should be kept on hand throughout the initial development of first designs and prototypes, especially when the parts are considered long lead-time items.

Testing of a first design may include several stages. Sub-circuits that have not been part of previous systems may need to be tested and characterized prior to their use in the converter. When the system or sub-system is assembled for testing, a first check should include a careful visual examination of all terminations and cleanliness of the components and layout, checks for component damage and proper assembly techniques, and continuity checks. The second step is to provide low-voltage control power to the control, protection, transducer, and other circuits. This voltage should be brought up slowly initially, observing current levels to ensure there are no shorts in the system and nothing emits smoke. This testing ends with a check of low-voltage power throughout the system to ensure it's in the right places. Control and other supplementary circuit checks can then be performed, carefully injecting signals or providing power gradually to transducers.

When all of the small signals and circuits are ready, the power can gradually be brought up, carefully watching voltages and currents across and through power semiconductor devices and filter components. It should be noted that it is crucial that the small-signal control and protection circuits and power supplies are properly and reliably working

DC-DC Converters

prior to subjecting the system to higher power. If the small-signal circuits don't work properly, loss of the larger power devices may result. A well-designed converter protects its own circuitry from destruction, as well as any source or load connected to it that is designed to supply or consume power at or above the ratings of the converter. This protection circuitry should be in hardware as opposed to software for greater reliability. For extremely critical circuitry, a complete parallel converter and appropriate switching circuitry may be used to provide greater backup protection. A UPS may also be utilized in certain applications to maintain control power levels during operation, to allow a safe shutdown if loss of the primary control power occurs.

REFERENCES

[1] Unitrode Application Note, Unitrode Corporation, Lexington, MA.
[2] Mohan, N., Undeland, T.M., and Robbins W.P., *Power Electronics — Converters, Applications, and Design, 2nd Edition*, John Wiley and Sons, Inc., New York, 1995.
[3] Bronson, R., *Matrix Methods*, Academic Press, Inc., San Diego, CA, 1991.
[4] Vorperian, V., "Simplify Your PWM Converter Analysis Using the Model of the PWM Switch," *Current*, Fall 1988, pp. 8–13.
[5] Golten, J., Verwer, A., *Control System Design and Simulation*, McGraw-Hill Book Company Europe, Berkshire, England, 1991.
[6] Nise, N., *Control Systems Engineering*, the Benjamin/Cummings Publishing Company, Inc., Redwood City, CA, 1992.
[7] Venable Industries, *New Techniques for Measuring Feedback Loop Transfer Functions in Current Mode Converters*, Venable Industries.
[8] Mitchell, D.M., *DC-DC Switching Regulator Analysis*, McGraw-Hill Book Company, New York, 1988.
[9] Fisher, M.J., *Power Electronics*, PWS-Kent Publishing Company, 1991.
[10] Severns, R.P., Bloom, G.E., *Modern DC-to-DC Switchmode Power Converter Circuits*, Van Nostrand, New York, 1985, Chapters 5 and 6.
[11] SGS Thomson Microelectronics, *Discrete Power Semiconductor Handbook, 1st Edition*, April 1995.
[12] Rashid, M.H., *Power Electronics — Circuits, Devices, and Applications — 2nd Edition*, Prentice-Hall, Englewood Cliffs, NJ, 1993.
[13] Wild, A., *Engineering Notes*, Caterpillar Inc., 2000.

12

AC-DC Rectifiers

Byoung-Kuk Lee
Korea Electrotechnology Research Institute (KERI), Changwon, South Korea

Chung-Yean Won
Sungkyunkwan University, Kyung Ki Do, South Korea

12.1 DIODE AC-DC RECTIFIER

12.1.1 Main Characteristics and Circuit Configuration

The overall characteristics of the diode rectifier are summarized in Table 12.1 and the circuit configuration is depicted in Figure 12.1.

12.1.2 Analysis of Three-Phase Full-Bridge Diode Rectifier

12.1.2.1 Circuit without Input Inductors and DC-Link Capacitor

With the assumption of no input inductors and DC-link capacitor, the output voltage of the diode rectifier contains 360 Hz ripple component. Also, the input currents can be represented by the input currents of diodes according to the magnitude of line-to-line voltages. The detailed voltage and current waveforms are depicted in Figure 12.2.

As shown in Figure 12.2, the output voltage V_{PN} contains 360 Hz ripple component and the average value of output voltage can be calculated as

$$V_{PN} = \frac{3}{\pi} \int_{-\frac{\pi}{6}}^{\frac{\pi}{6}} \sqrt{2} V_{AB} \cos(\omega t) = 1.35 V_{AB} \qquad (12.1)$$

where V_{AB} is rms value of line-to-line voltage.

Table 12.1 Characteristics of Diode Rectifier

Items	Contents
Output voltage	Fixed according to the input source
Input power factor and current control	Power factor control is impossible Current control is impossible
Major applications	Servo drives, high power drives, universal induction motor drives
Cost and structure	Low cost and simple structure

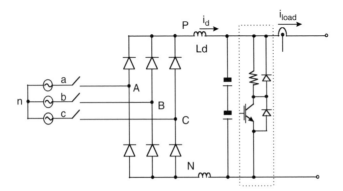

Figure 12.1 Circuit diagram of diode rectifier.

12.1.2.2 Circuit with Input Inductors and DC-Link Capacitor

The equivalent circuit as shown in Figure 12.3 can be obtained by replacement of the output voltage of Figure 12.2 with constant voltage source and considering the input stray inductance and resistance.

Based on the equivalent circuit in Figure 12.3, the voltage V_{PN} can be expressed as

$$V_{PN}(t) = \sqrt{2} V_{AB} \cos(wt) = 2i_D(t) R_S + 2L_S \frac{di_D(t)}{dt} + v_{DC}(t) \tag{12.2}$$

where

$$-\frac{\pi}{6} \leq \omega t \leq \frac{\pi}{6}$$

and the current $i_D(t)$ can be calculated as

$$i_D(t) = i_C(t) + i_{DC}(t) = C_{DC} \frac{dv_{DC}(t)}{dt} + i_{DC}(t) = C_{DC} \frac{dv_{DC}(t)}{dt} + \frac{v_{DC}(t)}{R_{Load}} \tag{12.3}$$

where

AC-DC Rectifiers

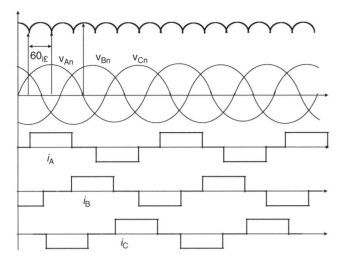

Figure 12.2 Phase voltages and currents waveforms without input inductors and DC-link capacitor.

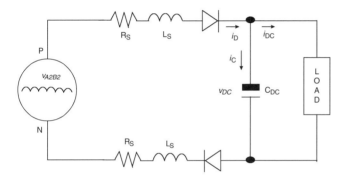

Figure 12.3 Equivalent circuit of three-phase full-bridge diode rectifier.

$$-\frac{\pi}{6} \leq \omega t \leq \frac{\pi}{6}$$

Equation 12.2 and Equation 12.3 can be transferred into a state equation as

$$\begin{bmatrix} \dfrac{di_D}{dt} \\ \dfrac{dv_{DC}}{dt} \end{bmatrix} = \begin{bmatrix} -\dfrac{R_S}{L_S} & -\dfrac{1}{L_S} \\ \dfrac{1}{C_{DC}} & -\dfrac{1}{C_{DC}R_{Load}} \end{bmatrix} \begin{bmatrix} i_D \\ v_{DC} \end{bmatrix} + \begin{bmatrix} \dfrac{1}{2L_S} \\ 0 \end{bmatrix} v_{PN}(t) \qquad (12.4)$$

12.1.2.3 Commutation Analysis Considering Effect of the Input Inductance

Due to the input inductance, the currents, which are flowing through diodes, cannot be changed immediately as a step function, but they are commutated with a constant slop. Therefore, during this commutation period, the voltage difference between two phases

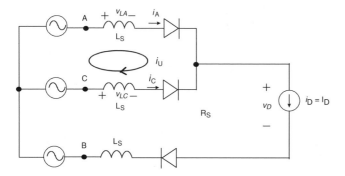

Figure 12.4 Equivalent circuit during the commutation considering input inductance.

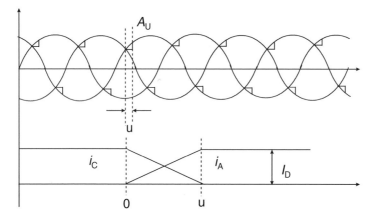

Figure 12.5 Input voltage and current waveforms during the commutation.

appears on the stray inductance. The equivalent circuit during the commutation can be expressed as Figure 12.4 in case of commutation from phase C to phase A, and Figure 12.5 shows the input voltage distortion according to the effect of input inductance.

Based on the assumption that the diode output current is constant, the currents of phase A and C and the voltage drop of the stray inductance can be obtained as

$$\text{If } i_A = i_u, \text{ then } i_C = I_D - i_u$$

$$v_{LA} = L_S \frac{di_A}{dt} = L_S \frac{di_u}{dt}$$

$$v_{LC} = L_S \frac{di_C}{dt} = -L_S \frac{di_u}{dt} \qquad (12.5)$$

$$v_{An} - v_{Cn} = v_{LA} - v_{LC} = 2L_S \frac{di_u}{dt}$$

$$\therefore L_S \frac{di_u}{dt} = \frac{v_{An} - v_{Cn}}{2}$$

AC-DC Rectifiers

As it is noted from Equation 12.5, the line-to-line voltage between phases A and C during the current commutation is 2 times the voltage drop of the tray inductance. Therefore, the magnitude of distortion due to the stray inductance is directly related to the value of A_u and it can be summarized in Equation 12.6.

$$\omega L_S \int_0^{I_D} di_u = \omega L_S I_D = \frac{\sqrt{2} V_{AC}(1-\cos u)}{2} = A_u$$

$$\therefore \text{ Average Value of Voltage Drop} = \frac{A_u}{\frac{\pi}{3}} = \frac{\pi}{3}\omega L_S I_D \qquad (12.6)$$

$$\text{Also, } \cos u = 1 - \frac{2\omega L_S I_D}{\sqrt{2} V_{AC}}$$

where ω = angular frequency.

The voltage distortion of the input source increases the lower frequency harmonic components, resulting in harmful effect on peripheral equipment.

12.1.3 ANALYSIS OF INPUT PHASE CURRENT AND OUTPUT CURRENT OF DIODE RECTIFIER

If one assumes that the output current of the diode rectifier is continuous and the input source is balanced with wye connection, the input current does not have order of 3 or even harmonic components. Using the means of Fourier series, the relation between input current and the output current of diode rectifier can be expressed by

$$i_A = \sum_{n=1,3,5,\ldots} (a_n \cos n\omega t + b_n \sin n\omega t)$$

$$a_n = \frac{2}{\pi} \int_0^{\frac{2\pi}{3}} I_D \cos n\omega t \, d\omega t$$

$$b_n = \frac{2}{\pi} \int_0^{\frac{2\pi}{3}} I_D \sin n\omega t \, d\omega t \qquad (12.7)$$

$$i_A = \frac{2\sqrt{3}}{\pi} I_D \left[\sin \omega t - \frac{1}{5}\sin 5\omega t - \frac{1}{7}\sin 7\omega t \cdots \right]$$

12.1.4 CALCULATION OF DC-LINK POWER

The DC-link power can be calculated by the product of DC-link voltage and current. As shown in the equivalent circuit, the input phase current consists of two components, such as the load current (DC-link output current) and charging/discharging current of DC-link capacitor. Therefore, considering the charging/discharging current, the DC-link power can be calculated as in Equation 12.8.

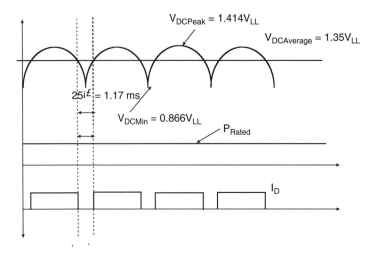

Figure 12.6 Case of continuous full load condition.

$$p_{dc\text{-}link} = v_{DC}(t)i_D(t) = v_{DC}(t)\left(i_{DC}(t) + C_{DC}\frac{dv_{DC}(t)}{dt}\right)$$

$$= v_{DC}(t)i_D(t) + v_{DC}(t)C_{DC}\frac{dv_{DC}(t)}{dt} = p_{dc\text{-}link\ output} + p_{charging/discharging}$$

(12.8)

12.1.5 Calculations of DC-Link Capacitor According to Various Load Conditions

12.1.5.1 Case of Continuous Full Load Condition

As shown in Figure 12.6, the input line-to-line voltage is less than the output of diode rectifier during 1.7 ms, so that the DC-link voltage is fluctuated with largest value. Using this voltage fluctuation, the DC-link capacitance can be calculated as

$$p_{dc\text{-}link\ capacitor} = C_{DC}\frac{dv_{DC}}{dt} \cdot v_{DC}$$

$$p_{load} = p_{rated}$$

$$E_{dc\text{-}link\ capacitor} = \int_0^{1.17 \cdot 10^{-3}} C_{DC}\frac{dv_{DC}}{dt} \cdot v_{DC} dt$$

$$E_{load} = \int_0^{1.17 \cdot 10^{-3}} p_{rated} dt$$

(12.9)

Setting voltage variation ΔV, input line-to-line voltage V_{LL}, then

$$E_{dc\text{-}link\ capacitor} = \frac{C_{DC}}{2}\left(1.35V_{LL}^2 - (1.35V_{LL} - \Delta V)^2\right)$$

$$E_{load} = 1.17 \cdot 10^{-3} \cdot P_{rated}$$

(12.10)

AC-DC Rectifiers

$$E_{dc\text{-link capacitor}} \geq E_{load}$$

$$\therefore C_{DC} \geq \frac{2 \cdot 1.17 \cdot 10^{-3} \cdot P_{rated}}{\left(2 \cdot 1.35 V_{LL} \cdot \Delta V - \Delta V^2\right)}$$

Usually in case of full load condition, the DC-link voltage variation is set as 10% of the DC-link average voltage.

12.1.5.2 Case of Overload Condition

As similar to Equation 12.9 and Equation 12.10, the DC-link capacitance for overload condition can be calculated as

$$p_{dc\text{-link capacitor}} = C_{DC} \frac{dv_{DC}}{dt} \cdot v_{DC}$$

$$p_{load} = p_{peak}$$

$$E_{dc\text{-link capacitor}} = \int_0^{1.17 \cdot 10^{-3}} C_{DC} \frac{dv_{DC}}{dt} \cdot v_{DC} dt$$

$$E_{load} = \int_0^{1.17 \cdot 10^{-3}} p_{peak} dt \tag{12.11}$$

$$E_{dc\text{-link capacitor}} = \frac{C_{DC}}{2} \left(1.35 V_{LL}^2 - \left(1.35 V_{LL} - \Delta V\right)^2\right)$$

$$E_{load} = 1.17 \cdot 10^{-3} \cdot P_{peak}$$

$$E_{dc\text{-link capacitor}} \geq E_{load}$$

$$\therefore C_{DC} \geq \frac{2 \cdot 1.17 \cdot 10^{-3} \cdot P_{peak}}{\left(2 \cdot 1.35 V_{LL} \cdot \Delta V - \Delta V^2\right)}$$

12.1.5.3 Case of Motor Accelerating Condition

In general, the diode rectifier is combined with a PWM converter and inverter for driving electric motors, such as DC motors, induction motors, and Brushless DC (BLDC) motors, so that in this section the much practical consideration of the diode rectifier in variable speed control systems is investigated.

For the simple analysis, the following assumptions are made:

- The load is applied to the diode rectifier as a ramp function.
- The accelerating profile of the motor is a ramp function, having a constant slope.
- The time delay due to the inverter is ignored.
- The time delay due to the input inductor of diode rectifier and motor is considered.
- The accelerating power of motor during the field-weakening region is proportional to the speed, and beyond this region the power is constant. Also, the motor speed at initiating field-weakening is assumed as a quarter of maximum motor speed.

The mechanical output power of a motor can be expressed using generating torque, angular frequency, damping factor, and load torque. Equation 12.12 shows the generating torque of the motor.

$$T_e = k\Phi I_q = (J_M + J_L)\frac{d\omega_m}{dt} + B\omega_m + T_L \quad (12.12)$$

where T_e is generating torque, k is constant, Φ is rotating flux, I_q is torque current, $(J_M + J_L)\,d\omega_m/dt$ is motor accelerating and decelerating torque, B is damping factor, ω_m is mechanical angular frequency of the motor, and T_L is load torque.

Multiplying the angular frequency ω_m on the both sides, then the output power of inverter and motor can be calculated as

$$T_e\omega_m = k\Phi I_q \omega_m = (J_M + J_L)\frac{d\omega_m}{dt}\omega_m + B\omega_m^2 + T_L\omega_m \quad (12.13)$$

The accelerating profile is assumed as a ramp function, and the converter should provide mechanical output of the motor and electrical and mechanical loss, so that the power provided from the converter can be expressed as

$$P_{ACC\,Converter} = T_e \frac{\omega_{max}}{T_{acc}} t \quad (12.14)$$

where $P_{ACC\,Converter}$ is the power providing from a converter during the motor acceleration, T_{acc} is the accelerating time up to the maximum speed, ω_{max} is motor maximum angular frequency, and T_e is generating torque.

The equivalent resistance at load can be expressed by combination of load and DC-link voltage as

$$R_{out} = \frac{V_{DC}^2}{P_{Load}} \quad (12.15)$$

Equation 12.15 is available in case of constant load and DC-link voltage. Under the condition, such that motor is accelerating and DC-link voltage is fluctuated as shown in Figure 12.7, the equivalent resistance at load can be obtained as

$$R_{out} = \frac{v_{DC}^2(t)}{p_{Load}(t)} = \frac{\left(-\dfrac{\Delta V}{t_d}t + 1.35V_{LL}\right)^2}{T_e \dfrac{\omega_{max}}{T_{acc}}(t - t_{mech})} \quad (12.16)$$

The time constant τ and time delay t_d due to the input inductor and equivalent resistance at load can be expressed as

$$\tau = \frac{L_{eff}}{R_{out}(t)} = \frac{2L_S \cdot p_{Load}(t)}{v_{DC}^2(t)} = \frac{2L_S \cdot T_e \dfrac{\omega_{max}}{T_{acc}}(t - t_{mech})}{\left(-\dfrac{\Delta V}{t_d}t + 1.35V_{LL}^2\right)} \quad (12.17)$$

$$t_d = 4\cdot\tau$$

AC-DC Rectifiers

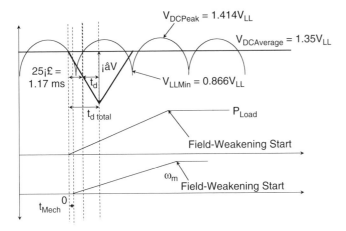

Figure 12.7 Relation between motor accelerating power and speed.

Although Equation 12.17 shows the time delay due to input inductor and load, it becomes very complicated to calculate the DC-link capacitance because that t_d is expressed as a function of time. Therefore, it is simplified using the equivalent resistance at the point of maximum load as

$$\tau = \frac{L_{eff}}{R_{out}(t)} = \frac{2L_S \cdot p_{Load}(t)}{v_{DC}^2(t)} = \frac{2L_S \cdot T_e \frac{\omega_{max}}{T_{acc}} \cdot \frac{T_{acc}}{4}}{(1.35V_{LL} - \Delta V)^2} \quad (12.18)$$

$$t_d = 4 \cdot \tau = \frac{2L_S \cdot T_e \cdot \omega_{max}}{(1.35V_{LL} - \Delta V)^2}$$

As explained above, the two time delay components, which are due to input inductor and DC-link voltage fluctuation, are the important factor to decide the DC-link power.

12.1.6 Design of Dynamic Breaking Unit

12.1.6.1 Design Procedure of Dynamic Breaking Resistor

For the motor acceleration, the torque equation is as follows:

$$T_e = k_T I_{qs,max} = (J_M + J_L)\frac{d\omega_m}{dt} + T_L \quad (12.19)$$

And in case of deceleration, it can be

$$-T_e = -k_T I_{qs,max} = (J_M + J_L)\frac{d\omega_m}{dt} + T_L \quad (12.20)$$

The dynamic breaking resistor is operating when the motor is in breaking mode and DC-link voltage is above the discharging voltage. The power delivering from the motor to the DC-link becomes the decelerating power of the motor. As shown in Figure 12.8, the equivalent circuit during this mode becomes of parallel circuit, which consists of DC-link capacitor, resistor, and current source.

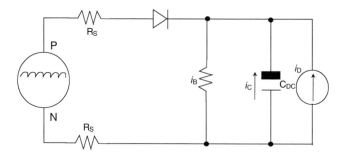

Figure 12.8 Equivalent circuit during the operation of dynamic breaking resistor.

According to Figure 12.8, the maximum value of the dynamic breaking resistor can be obtained based on a starting voltage v_{stat}

$$C_{DC} \frac{dv_{DC}}{dt} \cdot v_{stat} = T_B \omega_m - 0.2 \cdot T_M \omega_m - \frac{v_{stat}^2}{R_{DB}} \leq 0$$

$$\therefore R_{DB} \leq \frac{v_{stat}^2}{T_B \omega_m - 0.2 \cdot T_M \omega_m} \quad (12.21)$$

The power at the moment of breaking should be dissipated through the resistor. Therefore, the average power of motor powers at the moments of initiation ω_1 and completion ω_0 of breaking action should be dissipated. Based on this, the power rating of the breaking resistor can be calculated as

$$P_{DC\_avg} = \left(T_B - 0.2 \cdot T_M\right) \frac{\omega_1 - \omega_0}{2} \quad (12.22)$$

12.2 THYRISTOR AC-DC RECTIFIER

12.2.1 Topology and Operation Modes

Figure 12.9 shows the thyristor AC-DC rectifier configuration for practical applications. The thyristor rectifier can control the magnitude of DC-link voltage by changing fire angle, which means the turn-on instant time of the thyristor. Also, if the fire angle is set beyond 90°, the polarity of output voltage can be changed, which is called inverter mode and is commonly used for breaking mode of the DC motor operation. Figure 12.10 shows the two different operating modes according to the fire angle.

The average value of output voltage according to the fire angle can be calculated as

$$V_{out\_avg} = \frac{3}{\pi} \int_{-\frac{\pi}{6}+\alpha}^{\frac{\pi}{6}+\alpha} V_{LL} \cos\theta \cdot d\theta = \frac{3\sqrt{2}}{\pi} \cos\alpha \cdot V_{LL} \approx 1.35 \cos\alpha \cdot V_{LL} \quad (12.23)$$

12.2.2 Fire Angle Control Scheme

The fire angle should be formed based on the input line-to-line voltage, so that the input voltage needs to be sensed for the synchronization with the fire angle. The synchronization

AC-DC Rectifiers

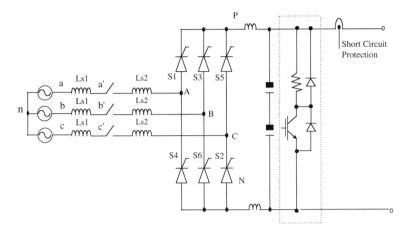

Figure 12.9 Three-phase full-bridge thyristor rectifier.

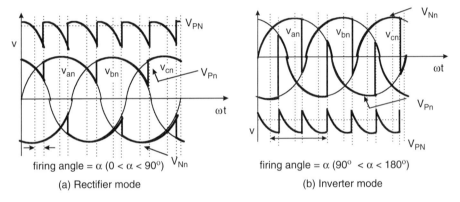

Figure 12.10 Operation modes of thyristor rectifier.

schemes can be mainly divided the following three ways; Table 12.2 compares and summarizes the characteristics of these schemes.

The conduction sequence of thyristors is consecutively conducted with 60° intervals and furthermore thyristors is conducted for 120° from 0 crossing points of line-to-line voltages, as shown in Figure 12.11.

12.2.2.1 Linear Fire Angle Control Scheme

Figure 12.12 shows the linear fire angle control scheme. At first, a saw-tooth waveform, which is synchronized with the input line-to-line voltage and has 2 times the frequency of input voltage frequency, is generated and compared with a voltage signal, which is proportional to the real value of input voltage. Using a comparator and logic circuit, a gate signal is generated at the point of cross point of two signals; the fire angle can be controllable by Equation 12.2.

$$\alpha° = 180° \frac{v_{control}}{v_{st}} \qquad (12.24)$$

Table 12.2 Characteristics of Synchronization Schemes

Schemes	Merits	Demerits
Linear fire angle control	Circuit is simple	Line distortion problem
Cosine wave Crossing control	Controllable during the inverter mode	Line distortion problem
PLL control	Line distortion effect can be minimized	Circuit is complicated and cost is high

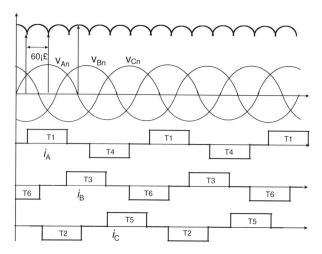

Figure 12.11 Input voltages and conduction sequences of thyristors (fire angle = 0).

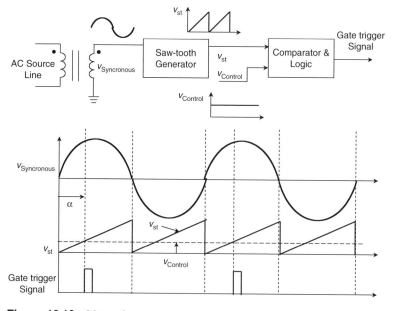

Figure 12.12 Linear fire angle control scheme.

AC-DC Rectifiers

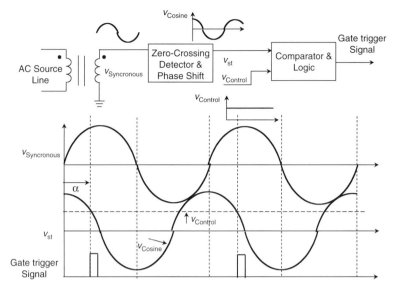

Figure 12.13 Cosine wave crossing scheme.

12.2.2.2 Cosine Wave Crossing Scheme

Compared with the linear fire angle control, the synchronized signal with the input voltage is shifted to 90° and the transformed cosine waveform is compared to the control voltage in order to generate a fire angle. Using this scheme, the fire angle can be more easily controlled during the inverter mode operation of the rectifier. Figure 12.13 shows the cosine wave crossing scheme.

12.2.2.3 PLL Scheme

The above-mentioned two schemes generate the fire angle by shifting the line-to-line voltage. Using this phase shifting can cause the notch (distortion) of the input source voltage to be included in the carrier wave, so that the control reliability can be deteriorated. In order to minimize the influence of the input source's distortion, an input filter is generally used, but it can also be a source of phase delay. The input source's distortion can be effectively minimized by converting the synchronized control signal into a digital signal. This PLL method may be widely used one among of the other schemes. Figure 12.14 shows the overall block diagram of the PLL scheme.

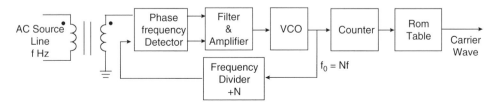

Figure 12.14 PLL scheme.

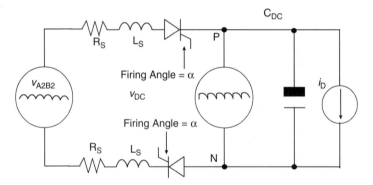

Figure 12.15 Equivalent circuit of three-phase full-bridge thyristor rectifier.

12.2.3 ANALYSIS OF THREE-PHASE FULL-BRIDGE THYRISTOR RECTIFIER

12.2.3.1 Equivalent Circuit and Output Voltage

Figure 12.15 shows an equivalent circuit of three-phase full-bridge thyristor rectifier. It consists of input source voltage, input resistance and inductance, a voltage source controlled by the fire angle, and a constant current source.

The output voltage level can be controlled by changing the fire angle, and the value of voltage variation according to the fire angle can be expressed as

$$V_\alpha = \frac{3}{\pi} \int_0^\alpha \sqrt{2} \cdot V_{LL} \sin(\omega t) d\omega t = 1.35 V_{LL} \left(1 - \cos \alpha \right) \quad (12.25)$$

where V_{LL} is the rms value of line-to-line voltage.

The average value of the output voltage of the thyristor rectifier is

$$V_{d\alpha} = 1.35 V_{LL} - 1.35 V_{LL} \left(1 - \cos \alpha \right) = 1.35 V_{LL} \cos \alpha \quad (12.26)$$

12.2.3.2 Influence of Input Inductance

Due to the input inductance components, the current flowing through thyristors is commutated with a constant slope, and it causes a voltage drop. This voltage drop gives an influence on the input source voltage as well as the output voltage. Figure 12.16 shows an equivalent circuit while the current is commutated from phase C to phase A. This time, it is assumed that the output DC current is constant and the voltage drop of the thyristor can be ignored.

Figure 12.17 shows the phase voltage and current waveforms considering the input inductance. If the output current of the thyristor rectifier is constant and the exchanging ratio (slope) of the conducting current is also constant, the voltage drop due to the input inductance can be expressed as

$$v_{Pn} = v_{An} - v_{LA}$$
$$v_{LA} = L_S \frac{di_A}{dt} \quad (12.27)$$

AC-DC Rectifiers

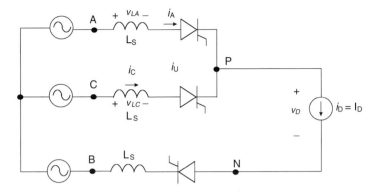

Figure 12.16 Equivalent circuit during current commutation considering input inductance.

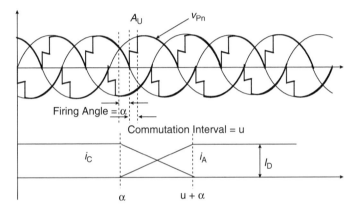

Figure 12.17 Phase voltage and current waveforms considering the input inductance.

The voltage variation A_u and DC-link voltage V_d can be

$$A_u = \int_{\alpha}^{\alpha+u} v_{LA} dt = \omega L_S \int_0^{I_D} di_u = \omega L_S I_D$$

$$\text{Average Value of Voltage Drop} = \frac{A_u}{\frac{\pi}{3}} = \frac{\pi}{3}\omega L_S I_D \qquad (12.28)$$

$$\therefore V_d = 1.35 V_{LL} \cos\alpha - \frac{\pi}{3}\omega L_S I_D$$

And the commutation interval can be calculated as

$$v_{Pn} = v_{An} - L_S \frac{di_A}{dt}$$

$$v_{Pn} = v_{Cn} - L_S \frac{di_C}{dt}$$

$$\therefore v_{Pn\_commutation} = \frac{v_{An} + v_{Cn}}{2} - \frac{L_S}{2}\left(\frac{di_A}{dt} + \frac{di_C}{dt}\right)$$

$$\frac{di_A}{dt} = -\frac{di_C}{dt} \text{ so that}$$

$$v_{Pn} = \frac{v_{An} + v_{Cn}}{2}$$

$$\frac{di_A}{dt} = \frac{v_{AC}}{2}$$

$$v_{AC} = \sqrt{2} V_{LL} \sin \omega t \qquad (12.29)$$

$$\int_0^{I_D} di_A = \sqrt{2} \frac{V_{LL}}{2\omega L_S} \int_\alpha^{\alpha+u} \sin \omega t \, d(\omega t)$$

$$\therefore \cos(\alpha + u) = \cos\alpha - \sqrt{2} \frac{V_{LL}}{2\omega L_S} I_D$$

During the commutation, voltages of two phases might be applied to the input inductance during a certain period. In this case, the magnitude of notch of the input line-to-line voltage becomes double of one of Equation 12.7 as:

$$\text{Magnitude of Notch} = A_n = 2\omega L_S I_D$$

$$\text{Depth of Notch} = \sqrt{2} V_{LL} \sin \alpha \qquad (12.30)$$

$$\text{Width of Notch} = \frac{2\omega L_S I_D}{\sqrt{2} V_{LL} \sin \alpha}$$

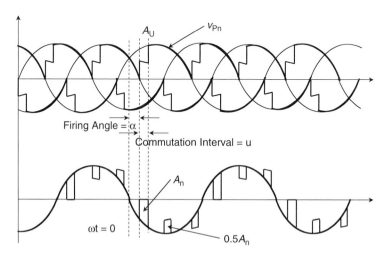

Figure 12.18 Notch of input line-to-line voltage due to commutation and input inductance.

AC-DC Rectifiers

12.2.3.3 Selection of Input Inductance

The input inductances consist of two components, such as line stray inductance and filter inductance. The filter reactor is additionally inserted into the system, and if the notch, which comes from the input filter reactor, is larger than that from the stray inductance, the notch due to the input stray inductance can be ignored. The ratio stray inductance L_{S1} to filter inductance L_{S2} is:

$$\varphi = \frac{L_{S1}}{L_{S1} + L_{S2}} \tag{12.31}$$

As shown in Equation 12.9, it is noted that small value of φ makes the magnitude of notch of line side. In addition, according to *VDE* regulation, the filter inductor is recommended to design as the following value.

$$\omega L_{S2} I_{A1} \geq 0.05 \frac{V_{AB}}{\sqrt{3}} \tag{12.32}$$

where L_{S2} is filter inductance, I_{A1} is rms value of fundamental component of input phase current, and V_{AB} is rms value of input line-to-line voltage.

REFERENCES

[1] J. Kassakian et al., *Principle of Power Electronics*, John Wiley & Sons, New York, 1991.
[2] N. Mohan et al., *Power Electronics,* 2nd Ed., John Wiley & Sons, New York, 1995.
[3] K. Ogata, *Modern Control Engineering,* 2nd Ed., Prentice Hall, Englewood Cliffs, NJ, 1990.
[4] *User's Guide*, Siemens.

13

Unbalanced Operation of Three-Phase Boost Type Rectifiers

Ana V. Stankovic
Cleveland State University, Cleveland, Ohio

The Boost Type Rectifier has been extensively developed and analyzed in recent years. It offers advantages over traditionally used phase controlled thyristor rectifiers because of its capability for nearly instantaneous reversal of power flow, power factor management, and reduction of input harmonic distortion. Unfortunately, the features that PWM Boost Type Rectifier offers are fully realized only when the supply three-phase input voltage source is balanced. Under an unbalanced input voltage supply, there is a deterioration of the rectifier input and output characteristics. The imbalance in input supply may occur frequently, especially in weak systems. Nonuniformly distributed single phase loads, faults, or unsymmetrical transformer windings could cause imbalance in the three-phase supply, both in magnitude and in phase. Regardless of the cause, unbalanced voltages have a severe impact on the performance of the PWM Boost Type Rectifier. Actually, the huge harmonics of lower frequencies, not present in the PWM switching functions, appear at both the input and output ports of the rectifier. The problems include a significant distortion in input current waveforms and increase in the DC capacitor ripple current and voltage. These additional low-frequency components cause additional losses and should be considered in filter design of these converters.

In this chapter the analysis of the PWM Boost Type Rectifier under unbalanced operating conditions is presented. Special emphasis is given to the evaluation of control methods for input/output harmonic elimination of the PWM Boost Type Rectifier under unbalanced operation conditions. The proposed technique maintains a high quality sinusoidal current input and DC current output even though the input voltages remain unbalanced. In addition, the severe unbalanced operating conditions have been considered. Under severe fault conditions in the distribution system, not only input voltages, but also input impedances must be considered as unbalanced. The control methods that maintain high-quality input currents and output DC voltage under severe fault conditions are presented.

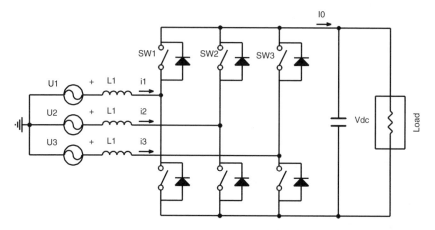

Figure 13.1 PWM Boost Type Rectifier.

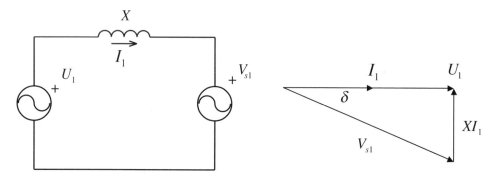

Figure 13.2 Per-phase equivalent circuit and phasor diagram for balanced operating conditions.

13.1 SYSTEM DESCRIPTION AND PRINCIPLES OF OPERATION

Figure 13.1 shows the structure of the PWM Boost Type Rectifier. Per-phase equivalent circuit and phasor diagram of the PWM Boost Type Rectifier under balanced operating conditions is shown in Figure 13.2. Power flow in the PWM converter is controlled by adjusting the phase shift angle δ between the source voltage U_1 and the respective converter reflected input voltage V_{s1} [1,2]. When U_1 leads V_{s1} the real power flows from the AC source into the converter. Conversely, if U_1 lags V_{s1}, power flows from the converter's DC side into the AC source. The real power transferred is given by Equation 13.1.

$$P = \frac{U_1 V_{s1}}{X_1} \sin(\delta) \tag{13.1}$$

The AC power factor is adjusted by controlling the amplitude of V_{s1}. The phasor diagram in Figure 13.2 shows that to achieve a unity power factor, V_{s1} has to be

$$V_{s1} = \sqrt{U_1^2 + (X_1 I_1)^2} \tag{13.2}$$

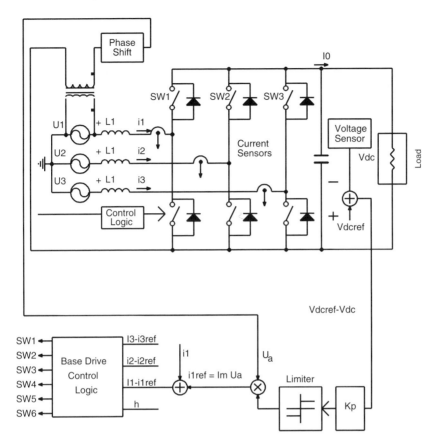

Figure 13.3 Indirect current control of a unity power factor sinusoidal current PWM Boost Type Rectifier.

It has been shown in Reference 3 that in order to control the DC output voltage of the PWM Boost Type Rectifier, the input line currents must be regulated. In typical rectifier controllers presented to date, the DC bus voltage error is used to synthesize a line current reference. Specifically, the line current reference is derived through the multiplication of a term proportional to the bus voltage error by a template sinusoidal waveform. The sinusoidal template is directly proportional to the input voltage, resulting in a unity power factor. The line current is then controlled to track this reference. Current regulation is accomplished through the use of hysteresis controllers [4]. Proposed control method [3] is shown in Figure 13.3.

13.2 ANALYSIS OF THE PWM BOOST TYPE RECTIFIER UNDER UNBALANCED OPERATING CONDITIONS

Unfortunately, the features the PWM Boost Type Rectifier offers are fully realized only when the supply three phase input voltage source is balanced. Under an unbalanced input voltage supply, there is a deterioration of the rectifier input and output characteristics [5]. Actually, the huge harmonics of lower frequencies, not present in the PWM switching functions, appear at both the input and output ports of the rectifier. It was proved that unbalanced input voltages cause an abnormal second harmonic at the DC bus that reflects back to the input, causing the third-order harmonic current to flow. Next, the third-order harmonic current causes the fourth-order harmonic voltage at the output, and so on. This results in the appearance of even harmonics at the DC output and odd harmonics in the input currents. These additional components should be considered in the filtering design of these converters.

The output current I_0 of the matrix converters is a function of the converter transfer function vector T and the input current vector i and is given by

$$I_0 = Ti, \qquad (13.3)$$

The converter transfer function vector T is composed of three independent line to neutral switching functions.

$$T = \begin{bmatrix} SW_1 & SW_2 & SW_3 \end{bmatrix} \qquad (13.4)$$

The input current vector is given by

$$i = \begin{bmatrix} i_1 \\ i_2 \\ i_3 \end{bmatrix} \qquad (13.5)$$

The line to neutral switching functions are balanced and can be represented only by their fundamental components.

$$SW_1(t) = S_1 \sin(wt - \Theta)$$
$$SW_2(t) = S_1 \sin(wt - \Theta - 120^0) \qquad (13.6)$$
$$SW_3(t) = S_1 \sin(wt + 120^0 - \Theta)$$

Therefore, converter synthesized line to neutral voltages can be expressed as

$$V_{s1} = \frac{1}{2} V_{dc} S_1 \sin(wt - \Theta)$$
$$V_{s2} = \frac{1}{2} V_{dc} S_1 \sin(wt - 120^0 - \Theta) \qquad (13.7)$$
$$V_{s3} = \frac{1}{2} V_{dc} S_1 \sin(wt + 120^0 - \Theta)$$

Equation 13.7 shows that the rectifier synthesized voltages are always balanced. For this reason, there will be no negative sequence voltage component present at its terminals.

Figure 13.4 shows the per-phase positive and negative sequence equivalent circuit.

It is obvious that the negative sequence line currents are limited only by the AC reactor and the AC source impedance. It follows that the input currents are unbalanced and given by

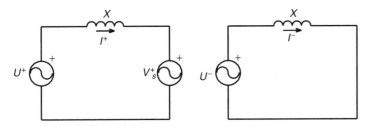

Figure 13.4 Per-phase positive and negative sequence equivalent circuit for unbalanced input voltages.

$$\begin{bmatrix} I_0 \\ I^+ \\ I^- \end{bmatrix} = \begin{bmatrix} 1 & 1 & 1 \\ 1 & a^2 & a \\ 1 & a & a^2 \end{bmatrix} \begin{bmatrix} I_0 \\ I^+ \\ I^- \end{bmatrix} \quad (13.8)$$

where I^0, I^+, I^- are zero, positive, and negative sequence input currents. Since $I_1 + I_2 + I_3 =$ zero, the zero sequence current never flows in this circuit. $I^0 = 0$. Input unbalanced voltages are given by

$$\begin{bmatrix} U_1 \\ U_2 \\ U_3 \end{bmatrix} = \begin{bmatrix} 1 & 1 & 1 \\ 1 & a^2 & a \\ 1 & a & a^2 \end{bmatrix} \begin{bmatrix} U_0 \\ U^+ \\ U^- \end{bmatrix} \quad (13.9)$$

where U^0, U^+ and U^- are zero, positive, and negative sequence input voltages. In the time domain, the fundamental components of the three-phase input currents are given by

$$\begin{aligned} i_1(t) &= I_1 \sin(wt - \varphi_1) \\ i_2(t) &= I_2 \sin(wt - 120^0 - \varphi_2) \\ i_3(t) &= I_3 \sin(wt + 120^0 - \varphi_3) \end{aligned} \quad (13.10)$$

According to Equation 13.3, the output current $I_0(t)$ is given by

$$\begin{aligned} I_0(t) &= I_1 \sin(wt - \varphi_1) S_1 \sin(wt - \Theta) \\ &+ I_2 \sin(wt - 120^0 - \varphi_2) S_1 \sin(wt - 120 - \Theta) \\ &+ I_3 \sin(wt + 120^0 - \varphi_3) S_1 \sin(wt + 120^0 - \Theta) \end{aligned} \quad (13.11)$$

By using a trigonometric identity, $I_0(t)$ becomes

$$\begin{aligned} I_0(t) &= \frac{1}{2} I_1 S_1 \left[\cos(\Theta - \varphi_1) - \cos(2wt - \Theta - \varphi_1) \right] \\ &+ \frac{1}{2} I_2 S_1 \left[\cos(\Theta - \varphi_2) - \cos(2wt - 240^0 - \Theta - \varphi_2) \right] \\ &+ \frac{1}{2} I_3 S_1 \left[\cos(\Theta - \varphi_3) - \cos(2wt + 240^0 - \Theta - \varphi 3) \right] \end{aligned} \quad (13.12)$$

The output current consists of DC and the harmonic current.

$$I_0(t) = I_{dc} + I_{sh}(2wt) \tag{13.13}$$

where $I_{sh}(2wt)$ is the second-order harmonic current and is given by

$$I_{sh}(2wt) = -\frac{S_1 I_1}{2}\cos(2wt - \Theta - \varphi_1) - \frac{I_2 S_1}{2}\cos(2wt - 240^0 - \Theta - \varphi_2) \\ -\frac{I_3 S_1}{2}\left[\cos(2wt + 240^0 - \Theta - \varphi_3)\right] \tag{13.14}$$

Therefore, the output DC voltage will also contain the second-order harmonic, which will reflect back to the input and cause the third-order harmonic current to flow. The third harmonic current will reflect back to the output, causing the fourth-order harmonic to flow. Even harmonics will appear at the output and odd at the input of the converter under unbalanced input voltages. The second- and third-order harmonics are of the primary concern.

13.2.1 HARMONIC REDUCTION IN THE PWM BOOST TYPE RECTIFIER UNDER UNBALANCED OPERATING CONDITIONS

Attempts were made to reduce low-order harmonics at the input and the output of the PWM Boost Type Rectifier under unbalanced input voltages [6]. This study is devoted to the modeling and control of the PWM Boost Type Rectifier in the case of network variations. The positive and negative sequence d and q reference currents are calculated from the following assumptions:

1. Input real power is constant.
2. Reactive power equals 0.

From the above assumptions, four equations are obtained and can be represented in matrix form.

$$P_4 = \begin{bmatrix} \frac{2}{3}P \\ 0 \\ 0 \\ 0 \end{bmatrix} = M_{4\times 4} I_4 = \begin{bmatrix} U_d^+ & U_q^+ & U_d^- & U_q^- \\ -U_q^+ & U_d^+ & -U_q^- & U_d^- \\ U_d^- & U_q^- & U_d^+ & U_q^+ \\ -U_q^- & U_d^- & U_q^+ & -U_d^+ \end{bmatrix} \begin{bmatrix} I_d^+ \\ I_q^+ \\ I_d^- \\ I_q^- \end{bmatrix} \tag{13.15}$$

The solution to this system is thus:

$$I_4 = M_{4\times 4}^{-1}(P_4) \tag{13.16}$$

where U_q^+, U_d^+, U_q^-, U_d^-, I_q^+, I_d^+, I_q^-, I_d^-, are the d and q components of positive and negative sequence input voltages and currents, and P is input constant real power. If the matrix $M_{4\times 4}$ is not singular, which is the case when the network is disturbed, only one solution is possible. Therefore, the input reference d and q components of the input current can be obtained. However, pulsating power will remain on three input impedances, causing the second-order harmonic at the output of the converter. This method only reduces the

Unbalanced Operation of Three-Phase Boost Type Rectifiers

harmonic content at the input and output of the PWM Boost Type Rectifier under unbalanced operating conditions.

13.3 CONTROL METHODS FOR INPUT/OUTPUT HARMONIC ELIMINATION OF THE PWM BOOST TYPE RECTIFIERS UNDER UNBALANCED OPERATING CONDITIONS

Recently, Stankovic and Lipo [7] proposed new methods for input-output harmonic elimination of the three-phase PWM Boost Type Rectifiers under unbalanced operating conditions. The first method is related to harmonic elimination of three-phase Boost Type Rectifiers under unbalanced input voltages and balanced input impedances. The second method is related to harmonic elimination of the PWM Boost Type Rectifiers under severe fault conditions. Under severe fault conditions both input voltages and input impedances might be unbalanced.

13.3.1 CONTROL METHOD FOR INPUT/OUTPUT HARMONIC ELIMINATION UNDER UNBALANCED INPUT VOLTAGES AND BALANCED INPUT IMPEDANCES

Based on the analysis in the open loop configuration, a feed-forward control method has been proposed. In order to control the output DC voltage and eliminate harmonics at the input and the output of the PWM Boost Type Rectifier under unbalance input voltages and balanced input impedances, not only current magnitudes but also their phase angles have to be controlled.

13.3.1.1 Theoretical Approach

The circuit shown in Figure 13.1 is analyzed under unbalanced input voltages and balanced input impedances. The assumptions used in the derivation are:

1. The system is lossless.
2. The switching functions are unbalanced and contain no zero sequence.
3. Only fundamental components of switching functions and input currents are taken into account.

By using symmetrical component theory, line to neutral switching functions SW_1, SW_2, and SW_3 are given by

$$\begin{bmatrix} SW_1 \\ SW_2 \\ SW_3 \end{bmatrix} = \begin{bmatrix} 1 & 1 & 1 \\ 1 & a^2 & a \\ 1 & a & a^2 \end{bmatrix} \cdot \begin{bmatrix} 0 \\ S^+ \\ S^- \end{bmatrix} \quad (13.17)$$

Since the zero sequence current is never present in this circuit, currents I_1, I_2, and I_3 are given by

$$\begin{bmatrix} I_1 \\ I_2 \\ I_3 \end{bmatrix} = \begin{bmatrix} 1 & 1 & 1 \\ 1 & a^2 & a \\ 1 & a & a^2 \end{bmatrix} \cdot \begin{bmatrix} 0 \\ I^+ \\ I^- \end{bmatrix} \quad (13.18)$$

where S^+ and S^- are positive and negative sequence switching functions, I^+ and I^- are positive and negative sequence currents, and $a = 1\angle 120°$. As it was shown earlier, output current I_0 is given by

$$I_0 = SW_1 I_1 + SW_2 I_2 + SW_3 I_3$$
$$= (S^+ + S^-)+(I^+ + I^-)+(a^2 S^+ + aS^-)(a^2 I^+ + aI^-)+(aS^+ + a^2 S^-)(aI^+ + a^2 I^-)$$
$$= 3S^+ I^- + 3S^- I^+$$

In the time domain, the pulsating component of the output current is expressed as:

$$\begin{aligned} I_0(t) &= 3\,\text{Real}\left(\left|S^+\right|e^{j*w*t}e^{j*\theta_s^+}\right)\text{Real}\left(\left|I^-\right|e^{j*w*t}e^{j*\theta_i^-}\right) \\ &+ 3\,\text{Real}\left(\left|S^-\right|e^{j*w*t}e^{j*\theta_s^-}\right)\text{Real}\left(\left|I^+\right|e^{j*w*t}e^{j*\theta_i^+}\right) \\ &= 3\left|S^+\right|\left|I^-\right|\cos\left(wt+\Theta_s^+\right)\cos\left(wt+\Theta_i^-\right)\ldots \\ &+ 3\left|S^-\right|\left|I^+\right|\cos\left(wt+\Theta_s^-\right)\cos\left(wt+\Theta_i^+\right) \end{aligned} \quad (13.19)$$

By using trigonometric identity $\cos(\alpha)\cos(\beta) = 1/2[\cos(\alpha + \beta) + \cos(\alpha - \beta)]$ one arrives at this result:

$$I_0(2wt) = 3/2\left|S^+\right|\left|I^-\right|\cos\left(2wt+\Theta_s^+ + \Theta_i^-\right) + 3/2\left|S^-\right|\left|I^+\right|\cos\left(2wt+\Theta_s^- + \Theta_i^+\right) \quad (13.20)$$

Equation 13.20 shows the presence of the second-order harmonic current at the output of the rectifier. The above equation indicates that the second-order harmonic could be eliminated under the following conditions.

$$\begin{aligned} &1.\ \left|S^+\right|\left|I^-\right| = \left|S^-\right|\left|I^+\right| \\ &2.\ \Theta_s^- + \Theta_i^+ = \pi + \Theta_s^+ + \Theta_i^- \end{aligned} \quad (13.21)$$

where $|S^+|$, $|S^-|$, Θ_s^+ and Θ_s^- are the magnitudes and phase angles of the positive and negative sequence switching functions, respectively. $|I^+|$, $|I^-|$, Θ_i^+ and Θ_i^- are the magnitudes and phase angles of the positive and negative sequence input currents.

The per-phase equivalent circuit under unbalanced input voltages and unbalanced synthesized voltages at the input of the rectifier are shown in Figure 13.5(a). From Figure 13.5(a), two additional equations are obtained and given by

$$ZI^+ = U^+ - V_s^+ \quad (13.22)$$

$$ZI^- = U^- - V_s^- \quad (13.23)$$

where V_s^+ and V_s^- are the positive and negative sequence voltages at the rectifier input and U^+ and U^- are the positive and negative sequence input voltages.

$$V_s^+ = V_{dc} S^+ \big/ \left(2\sqrt{2}\right),\ V_s^- = V_{dc} S^- \big/ \left(2\sqrt{2}\right) \quad (13.24)$$

Unbalanced Operation of Three-Phase Boost Type Rectifiers

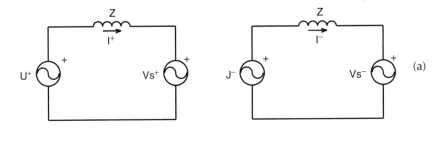

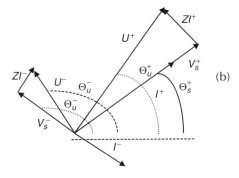

Figure 13.5 (a) Per-phase positive and negative sequence equivalent circuit. (b) Graph representation of the proposed solution for harmonic elimination.

where V_{dc} is output DC voltage. From the power equation, the following relationship can be obtained,

$$P_{dc} = \text{Real}\left(3U^+I^+ + 3U^-I^-\right) \qquad (13.25)$$

where P_{dc} is average output power.

The solution for harmonic elimination is obtained by combining Equation 13.21, through Equation 13.25. The solution is given by the following set of equations:

1. $\dfrac{|S^+|}{|S^-|} = \dfrac{|U^+|}{|U^-|}$ \hfill (13.26)

2. $\Theta_u^- - \Theta_s^- = \Theta_s^+ - \Theta_u^+$ \hfill (13.27)

3. $|S^+| = \dfrac{2\sqrt{2}|U^+|}{V_{dc}} \cos\left(\Theta_u^+ - \Theta_s^+\right)$ \hfill (13.28)

4. $\sin 2\left(\Theta_u^+ - \Theta_s^+\right) = \dfrac{2P_{dc}Z}{3\left(|U^{+2}| - |U^{-2}|\right)}$ \hfill (13.29)

The set of four equations shown above represents the open loop steady-state solution for input/output harmonic elimination of the PWM Boost Type Rectifier under unbalanced input voltages. The steady-state solution is also presented in graph form in Figure 13.5(b).

13.3.1.2 Control Method

Based on the analysis of the open loop configuration presented above, a feed-forward control method is proposed. In order to control the output DC voltage and eliminate harmonics at the input and output of the PWM Boost Type Rectifier under unbalance, not only current magnitudes but also their phase angles have to be controlled. The DC bus error is used to synthesize the magnitudes and the phase angles of the positive and negative sequence reference currents. The sequence components are then transformed into three-phase (abc) quantities that become reference signals for the hysteresis controller [3]. The relationship between input currents and input voltages in the open loop configuration for harmonic elimination is obtained by combining Equations 13.22, 13.23, 13.25, and 13.26 and given by

$$\frac{|U^+|}{|U^-|} = \frac{|I^+|}{|I^-|} \tag{13.30}$$

$$\theta_u^+ - \theta_i^+ = \theta_i^- - \theta_u^- - \pi \tag{13.31}$$

However, it is always true that average power input is equal to average power output and given by the following equation

$$P_{dc} = V_{dc}I_{dc} = 3|U^+||I^+|\cos(\theta_u^+ - \theta_i^+) + 3|U^-||I^-|\cos(\theta_u^- - \theta_i^-) \tag{13.32}$$

The output DC voltage is proportional to the positive and negative sequence currents (magnitudes), and so is the error signal ($V_{ref} - V_{dc}$). In order to satisfy Equation 13.30, the condition for second harmonic elimination, the positive and negative sequence commands for current magnitudes should satisfy Equation 13.33 and Equation 13.34.

$$|I^+| = K_p|U^+|(V_{ref} - V_{dc}) \tag{13.33}$$

$$|I^-| = K_p|U^-|(V_{ref} - V_{dc}) \tag{13.34}$$

Equation 13.33 and Equation 13.34 represent positive and negative sequence magnitude commands. The positive and negative sequence commands for phase angles are derived from the graph in Figure 13.5(b) and Equations 13.33 and Equation 13.34. The following equation can be obtained from the graph shown in Figure 13.5(b):

$$\sin(\theta_u^+ - \theta_i^+) = \frac{Z|I^+|}{|U^+|} \tag{13.35}$$

By substituting Equation 13.33 into Equation 13.35 the following equation is obtained

$$\sin(\theta_u^+ - \theta_i^+) = \frac{ZK_p|U^+|(V_{ref} - V_{dc})}{|U^+|} = ZK_p(V_{ref} - V_{dc}) = K_{p1}(V_{ref} - V_{dc}) \tag{13.36}$$

Unbalanced Operation of Three-Phase Boost Type Rectifiers

Equation 13.36 can be approximated by the following equation

$$\theta_i^+ = \theta_u^+ - K_{p1}(V_{ref} - V_{dc}) \qquad (13.37)$$

Similarly, the negative sequence magnitude commands for phase angles are derived and given by

$$\theta_i^- = \theta_u^- + \pi + K_{p1}(V_{ref} - V_{dc}) \qquad (13.38)$$

Positive and negative sequence current commands are transformed to abc quantities and used as references for hysteresis controller. The transformation is given by

$$i_{1ref}(t) = |I^+|\sin(wt + \theta_i^+) + |I^-|\sin(wt + \theta_i^-) \qquad (13.39)$$

$$i_{2ref}(t) = |I^+|\sin(wt + \theta_i^+ - \frac{2\pi}{3}) + |I^-|\sin(wt + \theta_i^- + \frac{2\pi}{3}) \qquad (13.40)$$

$$i_{3ref}(t) = |I^+|\sin(wt + \theta_i^+ + \frac{2\pi}{3}) + |I^-|\sin(wt + \theta_i^- - \frac{2\pi}{3}) \qquad (13.41)$$

The proposed control method is shown in more detail in Figure 13.6.

13.3.1.3 The Physical Meaning of the Proposed Solution in d-q Stationary Frame

The proposed method for complete harmonic elimination is based on the cancellation of input pulsating power. Under unbalanced operating conditions, the input power consists of a constant and a pulsating part. In the proposed solution, the pulsating portion of input power gets cancelled on three input inductors as shown in Figure 13.7.

The proposed solution for harmonic elimination can be explained in the d-q stationary frame, where the cancellation of input pulsating power becomes more obvious. The positive and negative sequence components of the input voltages and currents are instantaneous quantities and are shown in Figure 13.8. Instantaneous power input is given by

$$p(t) = p_0(t) + p_1(t) + p_2(t) + p_3(t) \qquad (13.42)$$

where

$$p_0(t) = 3/2 \, (u_{pq}i_{pq} + u_{pd}i_{pd} + u_{nq}i_{nq} + u_{nd}i_{nd}) \qquad (13.43)$$

$$p_1(t) = 3/2 \, (u_{pq}i_{nq} + u_{nq}i_{pq} + u_{pd}i_{nd} + u_{nd}i_{pd}) \qquad (13.44)$$

$$p_2(t) = 3/2 \, (u_{pd}i_{pq} - u_{pq}i_{pd}) \qquad (13.45)$$

$$p_3(t) = 3/2 \, (u_{nd}i_{nq} - u_{nq}i_{nd}) \qquad (13.46)$$

The first term, $p_0(t)$, represents the constant portion of input pulsating power, since all its components are products of in-phase q and d quantities of the positive and negative

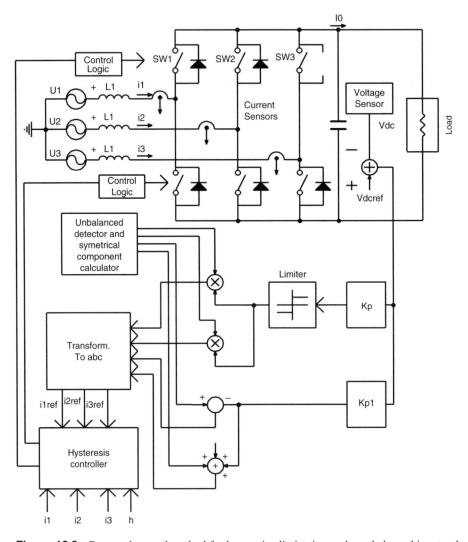

Figure 13.6 Proposed control method for harmonic elimination under unbalanced input voltages and balanced input impedances.

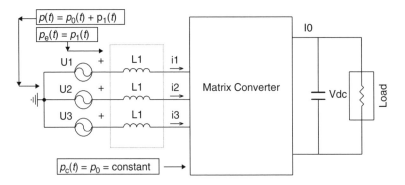

Figure 13.7 Cancellation of the instantaneous pulsating power.

Unbalanced Operation of Three-Phase Boost Type Rectifiers

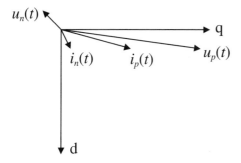

Figure 13.8 The positive and negative sequence input voltages and currents in d-q stationary frame.

sequence input voltages and currents. This term also exists in a balanced three-phase system.

The second term, $p_1(t)$, represents the pulsating portion, since all its components are products of the positive and negative in-phase components of input currents and voltages. It basically represents a mutual interaction between positive and negative sequence quantities, which normally does not exist in a balanced three-phase system.

The third term, $p_2(t)$, represents the interaction between d and q positive sequence quantities. This term exists in a balanced three-phase system.

The fourth term, $p_3(t)$, represents the interaction between d and q negative sequence quantities. This term does not exist in a balanced three-phase system.

Similarly, instantaneous power on the inductor, $p_l(t)$ consists of three parts and can be expressed as:

$$p_{1l}(t) = 3/2\, wL(2i_{pd}i_{nq} - 2i_{pq}i_{nd}) \tag{13.47}$$

$$p_{2l}(t) = 3/2\, wL(i_{pq}i_{pq} + i_{pd}i_{pd}) \tag{13.48}$$

$$p_{3l}(t) = 3/2\, wL(i_{nq}i_{nq} + i_{nd}i_{nd}) \tag{13.49}$$

The solution for second harmonic elimination assumes the following form:

1. $p_1(t) = p_{1l}(t)$ \hfill (13.50)

2. $p_2(t) = p_{2l}(t)$ \hfill (13.51)

3. $p_3(t) = p_{3l}(t)$ \hfill (13.52)

The power going into the converter is given by

$$p_c(t) = p(t) - p_l(t) \tag{13.53}$$

The power going into the converter, as shown in Figure 13.7, is constant since all the other components are shown to cancel out.

Example

In order to demonstrate feasibility, the system shown in Figure 13.6 has been simulated in SABER. The parameters used for the simulation are shown in Table 13.1.

The plot in Figure 13.9 shows the controlled output DC voltage under unbalanced input voltages. It also shows the current commands for the hysteresis controller and actual input line currents. Good control of the DC voltage is evident in spite of the voltage unbalance.

Table 13.1 Simulation Parameters

Parameter	Value	Parameter	Value
Input supply peak voltages	$U_1 = 100$ V $U_2 = 105$ V $U_3 = 104$ V	DC link capacitor, C	100 µF
DC link voltage, V_{dc}	270 V	Output resistive load, R	100 Ω
Per-phase line inductance, L	1 mH	Fundamental frequency, f	50 Hz

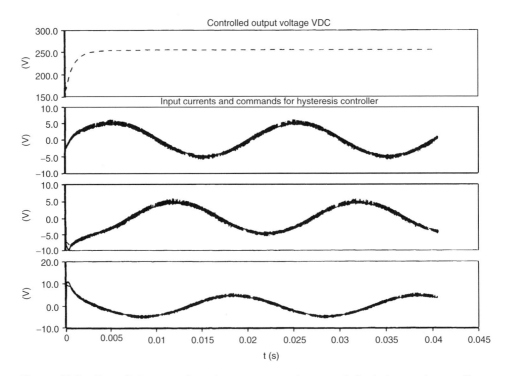

Figure 13.9 Controlled output voltage, input currents, and commands for the hysteresis controllers.

Unbalanced Operation of Three-Phase Boost Type Rectifiers

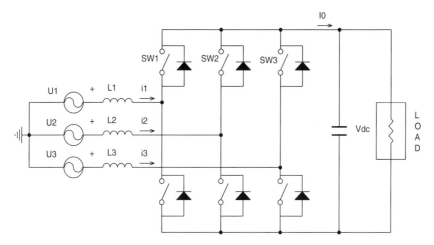

Figure 13.10 The PWM Boost Type Rectifier under unbalanced input voltages and unbalanced input impedances.

13.3.2 CONTROL METHOD FOR INPUT/OUTPUT HARMONIC ELIMINATION OF THE PWM BOOST TYPE RECTIFIER UNDER UNBALANCED INPUT VOLTAGES AND UNBALANCED INPUT IMPEDANCES

Under severe fault conditions in the distribution system, not only input voltages but also input impedances have to be considered unbalanced [8]. The circuit shown in Figure 13.10 is analyzed under the following assumptions:

1. The input voltages are unbalanced.
2. The input impedances are unbalanced.
3. The converter is lossless.

Harmonic elimination can be achieved by generating unbalanced reference commands for three input currents under unbalanced input voltages and unbalanced input impedances.

Derivation

From the circuit shown in Figure 13.8 it follows that

$$U_1 = z_1 I_1 + V_{s1} \tag{13.54}$$

$$U_2 = z_2 I_2 + V_{s2} \tag{13.55}$$

$$U_3 = z_3 I_3 + V_{s3} \tag{13.56}$$

$$I_1 = -I_2 - I_3 \tag{13.57}$$

$$S^* = U_1^* I_1 + U_2^* I_2 + U_3^* I_3 \tag{13.58}$$

$$SW_1 I_1 + SW_2 I_2 + SW_3 I_3 = 0 \tag{13.59}$$

where U_1, U_2, U_3, I_1, I_2, I_3, z_1, z_2, z_3, V_{s1}, V_{s2}, V_{s3}, S, SW_1, SW_2 and SW_3 are input voltages, input currents, input impedances, synthesized voltages at the input of the rectifier, apparent power, and switching functions, respectively, represented as phasors.

Equation 13.59 represents the condition for the second harmonic elimination. Synthesized voltages V_{s1}, V_{s2} and V_{s3} can be expressed as

$$V_{s1} = SW_1 \frac{V_{dc}}{2\sqrt{2}} \tag{13.60}$$

$$V_{s2} = SW_2 \frac{V_{dc}}{2\sqrt{2}} \tag{13.61}$$

$$V_{s3} = SW_3 \frac{V_{dc}}{2\sqrt{2}} \tag{13.62}$$

where V_{dc} is output DC voltage.

By substituting Equations 13.60, 13.61, and 13.62 into 13.54, 13.55, and 13.56 the following set of equations is obtained:

$$U_1 = z_1 I_1 + SW_1 \frac{V_{dc}}{2\sqrt{2}} \tag{13.63}$$

$$U_2 = z_2 I_2 + SW_2 \frac{V_{dc}}{2\sqrt{2}} \tag{13.64}$$

$$U_3 = z_3 I_3 + SW_3 \frac{V_{dc}}{2\sqrt{2}} \tag{13.65}$$

$$I_1 = -I_2 - I_3 \tag{13.66}$$

$$S^* = U_1^* I_1 + U_2^* I_2 + U_3^* I_3 \tag{13.67}$$

$$SW_1 I_1 + SW_2 I_2 + SW_3 I_3 = 0 \tag{13.68}$$

For given input power, S, input voltages, U_1, U_2, U_3 and input impedances z_1, z_2 and z_3, input currents, I_1, I_2 and I_3, can be obtained from the above set of equations. By multiplying Equations 13.63, 13.64, and 13.65 by I_1, I_2 and I_3, respectively, and adding them up the following equation is obtained,

$$U_1 I_1 + U_2 I_2 + U_3 I_3 = z_1 I_1^2 + z_2 I_2^2 + z_3 I_3^2 + \frac{V_{dc}}{2\sqrt{2}} (SW_1 I_1 + SW_2 I_2 + SW_3 I_3) \tag{13.69}$$

Unbalanced Operation of Three-Phase Boost Type Rectifiers

By substituting Equation 13.68 into Equation 13.69 the following equation is obtained:

$$U_1 I_1 + U_2 I_2 + U_3 I_3 = z_1 I_1^2 + z_2 I_2^2 + z_3 I_3^2 \tag{13.70}$$

The set of six equations with six unknowns, 13.63 to 13.68, reduces to three equations with three unknowns and are given by

$$I_1 = -I_2 - I_3 \tag{13.71}$$

$$S^* = U_1^* I_1 + U_2^* I_2 + U_3^* I_3 \tag{13.72}$$

Equations 13.70, 13.71, and 13.72 represent a set of three equations with three unknowns.

By substituting Equation 13.71 into Equations 13.70 and 13.72, the following set of equations is obtained and given by

$$U_1(-I_2 - I_3) + U_2 I_2 + U_3 I_3 = z_1(-I_2 - I_3)^2 + z_2 I_2^2 + z_3 I_3^2 \tag{13.73}$$

$$S^* = -U_1^* I_2 - U_1^* I_3 + U_2^* I_2 + U_3^* I_3 \tag{13.74}$$

Equation 13.73 can be simplified as

$$I_2(U_2 - U_1) + I_3(U_3 - U_1) = (z_1 + z_2)I_2^2 + (z_1 + z_3)I_3^2 + 2z_1 I_2 I_3 \tag{13.75}$$

From Equation 13.74 current, I_2, can be expressed as

$$I_2 = \frac{S^* - I_3(U_3^* - U_1^*)}{U_2^* - U_1^*} \tag{13.76}$$

Finally by substituting Equation 13.76 into Equation 13.75,

$$\frac{S^* - I_3(U_3^* - U_1^*)}{U_2^* - U_1^*}(U_2 - U_1) + I_3(U_3 - U_1) =$$

$$(z_1 + z_2) \frac{S^{*2} - 2S^* I_3(U_3^* - U_1^*) + I_3^2(U_3^* - U_1^*)^2}{(U_2^* - U_1^*)^2} \tag{13.77}$$

$$+ (z_1 + z_2) I_3^2 + 2z_1 \frac{S^* - I_3(U_3^* - U_1^*)}{U_2^* - U_1^*} I_3$$

$$\left[\frac{2z_1(U_3^* - U_1^*)}{U_2^* - U_1^*} - \frac{(z_1 + z_2)(U_3^* - U_1^*)^2}{(U_2^* - U_1^*)^2} - (z_1 + z_3) \right] I_3^2$$

$$+ \left[(U_3 - U_1) - \frac{(U_3^* - U_1^*)(U_2 - U_1)}{U_2^* - U_1^*} - \frac{2z_1 S^*}{U_2^* - U_1^*} + \frac{2S^*(z_1 + z_2)(U_3^* - U_1^*)}{(U_2^* - U_1^*)^2} \right] I_3 \tag{13.78}$$

$$+ \frac{S^*(U_2 - U_1)}{U_2^* - U_1^*} - \frac{(z_1 + z_2)S^{*2}}{(U_2^* - U_1^*)^2} = 0$$

Currents I_2 and I_1 can be obtained from Equations 13.76 and Equation 13.71.

Equation 13.71, Equation 13.76, and Equation 13.78 represent the steady state solution for input currents under both unbalanced input voltages and unbalanced input impedances. An analytical solution represented by Equation 13.78 always exists unless all the coefficients of the quadratic equations are equal to 0.

A Critical Evaluation. The analytical solution that has been obtained is general. In particular, the PWM Boost Type Rectifier can operate with unity power factor and still maintain DC voltage at the output. The only constraint that exists, as far as the level of unbalance is concerned, is governed by constraints of the operation of the PWM bridge itself.

The proposed generalized method for input/output harmonic elimination is valid if and only if $U_i, z_i \neq 0$, where $i = 1, 2, 3$. In other words, the solution exists for all levels of unbalance in input voltages and impedances, except for cases where both voltage and impedance in the same phase are equal to 0. Therefore, the maximum level of input voltage imbalance with balanced input impedances, for which the proposed solution is still valid, is given as

$$U_1 \neq 0$$

$$U_2 = U_3 = 0$$

$$z_1 = z_2 = z_3 \neq 0$$

Under unbalanced input voltages and unbalanced input impedances, the maximum level of imbalance for which the proposed solution is still valid is given as

$$U_1 \neq 0$$

$$U_2 = U_3 = 0$$

$$z_1 = 0$$

$$z_2 \neq z_3 \neq 0$$

From the above discussion, it follows that the three-phase PWM Boost Type Rectifier can operate from the single-phase supply as well (the special case of imbalance of the three-phase system). This is an extremely important result, since it means that the three-phase PWM Boost Type Rectifier can operate from the center-tapped transformer as well and still maintain high-quality DC output voltage and input currents.

In this case, input voltages and impedances are given as

$$|U_1| = |U_3|$$

$$\text{phase}(U_1) = 0°$$

$$\text{phase}(U_3) = -180°$$

$$U_2 = 0$$

$$z_2 \neq 0, z_1 = z_3 = 0$$

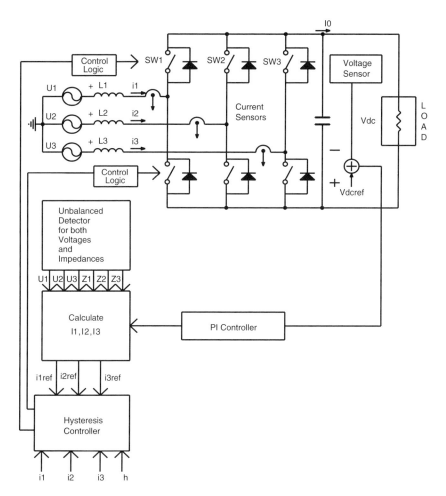

Figure 13.11 The proposed closed loop solution for input/output harmonic elimination under unbalanced input voltages and unbalanced input impedances.

13.3.2.1 Control Method

Based on the analysis of the open loop configuration presented above, a feed forward control method is proposed. Input voltages as well as input impedances have to be measured. Based on this information and a DC bus error, reference currents are calculated according to Equation 13.71, Equation 13.76, and Equation 13.78, which become reference signals for the hysteresis controller [8]. Only one PI controller is utilized, which has been shown to be sufficient for good regulation. The proposed control method is shown in more detail in Figure 13.11.

Based on the same principle of nullifying the oscillation component of the instantaneous power, Yongsug et al. recently proposed a new control method for input/output harmonic elimination under unbalanced operating conditions [9]. The proposed control technique has been done in d-q synchronous frame.

Example. The operation of the three-phase PWM Boost Type Rectifier under severe unbalanced operating conditions is shown in Figure 13.12. The PWM Boost Type Rectifier supplied from the single-phase supply with unbalanced impedances was built in

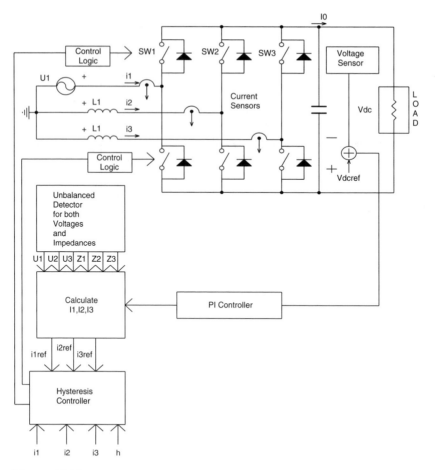

Figure 13.12 PWM Boost Type Rectifier supplied from the single-phase source.

Table 13.2 Simulation Parameters

Parameter	Value	Parameter	Value
Input supply voltages	$U_1 = 100\angle 0$ $U_2 = 0$ $U_3 = 0$	DC link capacitor, C	100 µF
DC link voltage, V_{dc}	220 V	Output resistive load, R	100 Ω
Per-phase line inductance, L	L1 = 0 L2 = L3 = 1 mH	Fundamental frequency, f	60 Hz

SABER. The parameters used in simulation are shown in Table 13.2. The plot on Figure 13.13 shows the controlled output DC voltage. Actual line currents are shown in Figure 13.14. Input currents are unbalanced, as expected, to cancel the pulsating power coming from the input. The converter operates at unity power factor, and it shows stable behavior in spite of the extreme level of unbalance.

Unbalanced Operation of Three-Phase Boost Type Rectifiers

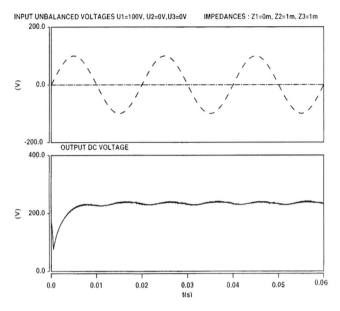

Figure 13.13 Input voltages and output DC voltage when PWM Boost Type Rectifier is supplied from the single-phase supply.

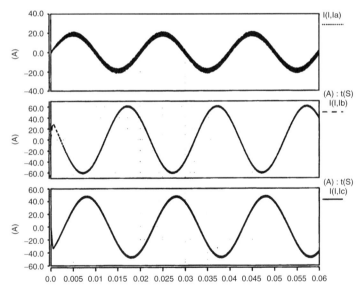

Figure 13.14 Input currents when PWM Boost Type Rectifier is supplied from the single-phase source.

13.3 CONCLUSION

In this chapter, the operation of the PWM Boost Type Rectifier under unbalanced operating conditions has been analyzed. It has been shown that unbalanced operating conditions cause huge low-order harmonics in input currents and output DC voltage. Low-order harmonics cause additional losses. In addition, ripple on the DC link is a known cause of interaction between current regulators, which can even cause system instability [10]. In order to eliminate the harmonics and maintain the high-quality input currents and output voltage without using an additional filter, several control methods have been described.

An analytical solution for input/output harmonic elimination under unbalanced input voltages and balanced input impedance in the open loop configuration has been presented. Based on analysis of the open loop configuration, a closed loop method has been presented. With the proposed technique, high-quality input and output waveforms are maintained under unbalanced input voltages and balanced input impedances. Simulation results show excellent response and stable operation of the PWM Boost Type Rectifier under unbalanced input voltages and balanced input impedances. In addition, severe unbalanced operating conditions have been analyzed. In this case, both input voltages and input impedances have been considered unbalanced. An analytical solution for harmonic elimination under severe unbalanced operating conditions has been shown. Based on the solution in the open loop configuration, the closed loop solution has been developed and described. The operation of the PWM Boost Type Rectifier under severe unbalanced operating condition has been simulated. The results presented in the example show excellent behavior of the PWM Boost Rectifier even though the bridge operated from a single-phase supply (special case of unbalanced three-phase voltages).

REFERENCES

[1] J.W. Wilson. The Forced-Commutated Inverter as a Regenerative Rectifier. *IEEE Transactions on Industry Applications.* Vol. IA-14, No. 4: 335–340, July/August 1978.

[2] A.V. Stankovic. Three-Phase Pulse-Width-Modulated Boost-Type Rectifiers, in *The Power Electronics Handbook.* CRC Press LLC, Boca Raton, FL, Chapter 4.3, 2001.

[3] J.W. Dixon and B.T. Ooi. Indirect Current Control of a Unity Power Factor Sinusoidal Current Boost Type Three-phase Rectifier. *IEEE Trans. Ind. Electron.* Vol. 35, No. 4: 508–515, November/December 1988.

[4] D.M. Brod and D.W. Novotny. Current Control of VSI-PWM Inverters. *IEEE Transactions on Industry Applications.* Vol. IA-21, No. 4: 769–775, November/December 1984.

[5] L. Moran, P.D. Ziogas, and G. Joos. Design Aspects of Synchronous PWM Rectifier-Inverter System Under Unbalanced Input Voltages Conditions. *IEEE Transactions on Industry Applications.* Vol. 28, No. 6: 1286–1293, November/December 1992.

[6] P. Rioual, H. Pouliquen, and J.P. Louis. Regulation of a PWM Rectifier in the Unbalanced Network State. *IEEE-PESC, Conf. Rec:* 641–647, 1993.

[7] A.V. Stankovic and T.A. Lipo. A Novel Control Method for Input-Output Harmonic Elimination of the PWM Boost Type Rectifier Under Unbalanced Operating Conditions. *IEEE Transactions on Power Electronics.* Vol. 16. No 5: 603–611, September 2001.

[8] A.V. Stankovic and T.A. Lipo. A Generalized Control Method for Input-Output Harmonic Elimination for the PWM Boost Type Rectifier Under Simultaneous Unbalanced Input Voltages and Input Impedances. *Proceedings of IEEE PESC,* Vancouver, pp. 1309–1314, 2001.

[9] S. Yongsug, V. Tieras, and T.A. Lipo. A Nonlinear Control of the Instantaneous Power in d-q Synchronous Frame for PWM AC/DC Converter Under Generalized Unbalanced Operating Conditions. *Conference Record of the 2002 IEEE Industry Applications Conference,* Chicago, pp. 1189–1196, 2002.

[10] R.P. Stratford and D.E. Steeper. *Reactive compensation and harmonic suppression for industrial power systems using thyristor converters. Proceedings of IEEE IAS Annual Meeting,* p. 1:21, 1974.

14

DC/AC Inverters

Mohan Aware
Visvesvaraya National Institute of Technology, Nagpur, India

The increasing electrical power demands in advanced cars are for improving the engine performance, passenger comforts, and safety. This growing high-power requirement in automobiles needs efficient management. The major such power management and modulation functions are feasible by power electronics devices.

In automobiles, the electrical and electronic accessories are supported by the battery. This power is modulated by using the converters. The basic principles of DC-to-AC power conversion using power electronics inverter are explained. The different types of converter topologies are described with their operating modes. The current and voltage control techniques used in this conversion process are explained. The new DC-AC conversion techniques using multilevel and resonant DC link inverters are briefly introduced. The auxiliary automotive motor control applications are illustrated.

14.1 DC-TO-AC CONVERSION

The Lundell alternator is a source of electrical power to be utilized in most automobiles. The power available is to be utilized during running and in stationary conditions. The role of the battery is very important in this situation. This stored power is utilized as per the needs and requirements in the vehicles.

Power conversion and control are performed by power electronic converters that are built by network of semiconductor power switches. For automotive applications, the converters are supplied from DC voltage sources. The basic state of a voltage source converter can be distinguished by the operating states. The input and output terminals are directly connected, cross connected, or separated. The appropriate sequence of states results in conversion of the given input source voltage to the desired output load voltage.

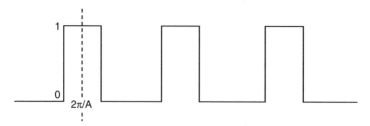

Figure 14.1 Existence function.

Figure 14.2 Isolated switch of converter matrix.

The DC to AC power conversion is realized by inverters. In the converter for any application, two related sets of parameters are of paramount interest. The external terminal performance, defined by the dependent quantities, will determine how well the converter meets application needs and what will be needed to interface it successfully with actual sources and loads. The internal currents and voltages of the converter's switching loops will determine the selection of the active switching devices to be used and also any auxiliary passive components needed to enable the device to function properly. To obtain a precise, quantitative definition of the dependent quantities, some means of formally and quantitatively describing switching pattern is to be identified as an *existence function*. This mathematical expression allows dependent quantity and internal converter waveform to be constructed graphically. The existence function for a single switch assumes unit value whenever the switch is closed and is zero whenever the switch is open. In a converter, each switch is closed and opened according to some repetitive pattern. Hence, its existence function shall be represented as shown in the Figure 14.1. The train of pulses has unit amplitude but neither the pulses nor the intervening zero value periods will necessarily all have the same time duration. The requirement that a repetitive switching pattern exist means that the function must at least consist of repeated groups of pulses. The simplest, or unmodulated, existence functions have pulses all of the same time duration and zero intervals with the same property. The more complex variety, which has differing duration of pulses and various interspersed zero times, is called a modulated existence function. The use of existence function to derive the relation between dependent quantities with the switch is indicated in Figure 14.2.

The switch is connected to V_i, the ith of a set of M-defined voltages, and to I_j, the jth set of N-defined currents in the matrix converter.

The complete equation for the voltage impressed on I_j can be written as:

$$V_j = \sum_{i=1}^{i=M} H_{ij} V_i \tag{14.1}$$

DC/AC Inverters

The equation for the total current flowing in V_i can be written as:

$$I_i = \sum_{j=1}^{j=N} H_{ij} I_j \tag{14.2}$$

where H_{ij} is the existence function having unit value when switch is closed and zero value when it is open. The voltage impressed on a switch, V_s, is the difference between its voltage source, V_i and the voltage impressed on its current source while it is open. Thus the equation for V_s becomes

$$V_s = V_i - V_j \tag{14.3}$$

Thus, the dependent quantities and the switch stress can be completely defined using the existence functions. By defining the existence function in appropriate mathematical expression, the dependent quantity and switch stress waveforms can be drawn. This basic analytic technique to switching converter is used to formulate the dependent quantities.

The existence functions are trains of unit value pulses interspersed by periods of zero value. Since the switching patterns are invariably repetitive, existence functions are periodic. They can be mathematically represented in various ways. The Fourier expression is suitable for expressing the existence functions.

Fourier series expansion of a simple existence function such as that depicted in Figure 14.1 is, for present purposes, best accomplished by setting the zero reference at the midpoint of one of the unit value periods. The repetition frequency of the pulses is f with the time period $T = 1/f$ having angular frequency $\omega = 2\pi f$. If the angular duration of the unit-value period is $2\pi/A$ rad, where $A \geq 1$, then the boundaries of the unit value period with respect to the time zero reference are $-\pi/A$ and π/A rad. The expression for unmodulated existence function is represented as

$$H(\omega t) = 1/A + (2/\pi) \sum_{n=1}^{n=\infty} \left[\sin(n\pi/A)/n \right] \cos(n\omega t) \tag{14.4}$$

If n is an integer multiple of A, then $\sin(n\pi/A) = 0$, and so the all frequency components that satisfy this condition will disappear from the expression.

In DC-to-AC and AC-to-DC converters, two basic techniques for controlling power flow are available. First, consider that if one defined quantity is DC and other is AC, then, when these are multiplied by the existence functions, the wanted component of the dependent quantity at the AC terminals must have the same frequency as the defined quantity if power transfer is to occur. Such a wanted component will exist if one of the oscillatory terms in Equation 14.4 has the same frequency as the defined AC quantity. Since the term for $n = 1$ has the largest amplitude, it is logical to choose it. When the defined AC quantity is multiplied by the existence functions, perfectly smooth DC wanted components must result in the resultant dependent quantity. The term $1/A$ will produce oscillatory components, and the sum of some number of such components can only be an oscillatory component or zero. All other terms will result from products of the type $\cos(n\omega t) \cos(\omega_s t)$, where ω is from the existence functions and ω_s is the angular frequency of the defined AC quantity.

The term ω cannot be varied, or else the wanted components will no longer be generated. Also, A is an invariant integer being existence function. It must be a complete set to avoid the violation of KCL and KVL. Therefore, the control is possible by introducing some sort of modulating function into the arguments of the oscillatory terms of the existence function. It is clear that the required modulating function is simply phase shifted, with respect to existence function taking an external time reference. Usually, the AC-defined quantity is taken as the reference, and the advance or delay angle employed for control is designated. It is also clear that the wanted component of the dependent AC quantity does not suffer amplitude modification when this control technique is employed.

The second technique for implementing control in AC-to-DC and DC-to-AC converters is perhaps more obvious but leads to rather complex expressions for the existence functions and dependent quantities. It consists of making A time dependent so that the term $1/A$ in Equation 14.4 assumes the form $(1/A_0) [1 \pm m \cos(\omega_s t + \varphi)$, where m is a *modulation index* and φ an arbitrary phase angle. Since it involves cyclic variation of the unit value period, this method is termed as pulse width modulation (PWM). The modulation index, m, which controls the amplitude of the wanted components is varied by varying the degree to which the unit value period is varied.

The inverters are built by number of switches depending on the performance and its application. The operation of these circuit configurations is explained by writing the suitable existence functions.

14.2 TYPES OF INVERTERS

In the conversion process, the defined quantities may be either current or voltage. There are two types of AC-to-DC/DC-to-AC converters, referring to the defined DC terminal quantities: current sourced and voltage sourced. In the voltage source inverters (VSI), DC input appears as a DC voltage source (ideally with no internal impedance) to the inverter. In current source inverter (CSI), the DC input appears as a DC current source (ideally with the internal impedance approaching infinite) to the inverter. Since power flow is reversible, they are both inherently capable of two-quadrant operation at their DC terminals; the current source is a unidirectional-current bidirectional-voltage converter, and the voltage-source is a unidirectional-voltage bidirectional-current converter.

Inverters can be built with any number of output phases. In practice, single-phase and three-phase inverters are almost exclusively used. The basic building components of these inverters are power semiconductor switches. In the past, SCRs were used in high- and medium-power inverters. SCR-based inverters require commutating circuits, to turn the SCR off. The commutating circuits increase the inverter size and cost, and reduce reliability and switching frequency of the inverters. Presently, fully controlled semiconductor power switches, mostly MOSFETs/IGBTs (in low- and medium-power inverters) and GTO (in high-power inverters), are almost exclusively used.

A DC-to-AC converter is supplied from a DC source, while the AC output voltage and current have strong fundamental components with adjustable frequency and amplitude. Considering the AC motor requirements, they must satisfy the following basic requirements:

1. Ability to adjust the frequency according to desired speed
2. Ability to adjust the output voltage to maintain air gap flux in the constant torque region
3. Ability to supply a rated current on a continuous basis at any frequency

DC/AC Inverters

14.3 VOLTAGE SOURCE INVERTERS

The voltage source inverters are the most common power electronics converters. The DC input voltage for a voltage source inverter can be obtained from a rectifier, usually of the uncontrolled, diode type, or from another DC source, such as a battery, fuel cell, or solar photovoltaic array.

Most automobiles use 14 V DC electrical systems supported by 12 V batteries, and average electrical power demand in the car is about 1.2 kW. The present 14 V power system has reached its limits of capability. The increasing power demand in modern automobiles is expected to be around 4 kW and supported by enhancing the DC bus voltage to 42 V. The operating voltage and the current rating of the switch in inverter will be increased. The MOSFETs/IGBTs switches will be preferred to build the inverter circuits.

The voltage source inverter supplied from the uncontrolled rectifier is shown in Figure 14.3. The DC link capacitor constitutes the actual voltage source, since voltage across it cannot change instantly.

The inverter output can be single phase or polyphase and can have square wave, sine wave, PWM wave, stepped wave, or quasi-square wave at the output. Voltage-fed converters are used extensively, and some of their applications may be as follows:

- AC motor drives
- AC uninterruptible power supplies (UPS)
- Induction heating
- AC power supply from battery, photovoltaic array, or fuel cell
- Static VAR generator (SVG) or compensator (SVC)
- Active harmonic filters

In voltage-fed converters, the power semiconductor device always remains forward-biased due to DC supply voltage; therefore, self-controlled forward or asymmetric blocking devices, such as GTOs, BJTs, IGBTs, power MOSFETs, and IGCTs, are suitable. Force-commutated thyristor converters were used before, but now they have become obsolete. A feedback diode is always connected across the device to have free reverse current flow. The switch-mode inverters are commonly used for automotive applications and are explained in subsequent sections.

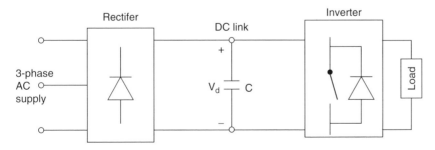

Figure 14.3 Voltage-source inverter supplied from a diode rectifier.

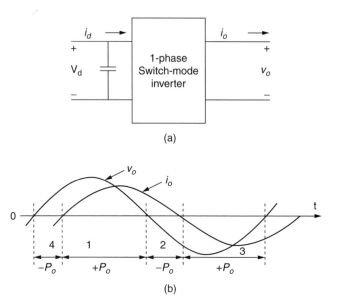

Figure 14.4 Single-phase switch mode inverter.

14.3.1 SINGLE-PHASE INVERTERS

The power flow during the conversion of DC-to-AC by the inverter is presented with inductive load. A single-phase voltage source inverter with its functional block is shown in Figure 14.4(a). The output voltage of the inverter is filtered so that v_o can be assumed to be sinusoidal. Considering the inductive load such as AC motor, load current i_o will lag behind v_o, as shown in the Figure 14.4(b). The output waveform of Figure 14.4(b) shows that during interval 1, v_o and i_o are both positive, whereas during interval 3, v_o and i_o are both negative. Therefore, during intervals 1 and 3, the instantaneous power flow P_o is from the DC side to AC side, corresponding to an inverter mode of operation. In contrast, v_o and i_o are of opposite signs during intervals 2 and 4, and therefore P_o flows from the AC side to the DC side of the inverter, corresponding to the rectifier mode of operation.

14.3.1.1 Half-Bridge Inverters

One of the simplest possible inverter configurations is the single-phase, half-bridge inverter shown in Figure 14.5(a). Circuit consists of a pair of devices S_1 and S_2 connected in series across the DC supply, and the load is connected between point *a* and the center point *o* of a split-capacitor power supply. The devices S_1 and S_2 are closed alternatively for 180° to generate the square-wave output voltage as shown in Figure 14.5(b).

While operating, switches in any leg of the inverter may not be simultaneously on, since they would short the supply source. This condition of shot-through should be avoided. The semiconductor switches, even the fast ones, require finite transition times from one conduction state to the other. Therefore, in practice, to avoid shot-through, a switch is turned off shortly before turn-on of the other switch in the same leg. The interval between the turn-off and turn-on signals is called blanking time or dead time, t_d.

The load is usually inductive, and assuming perfect filtering, the sinusoidal load current will lag behind the fundamental voltage by angle φ, as shown. When supply voltage and load current are of the same polarity, the mode is active, meaning the power is absorbed by the load. On the other hand, when the voltage and current are of opposite polarities

DC/AC Inverters

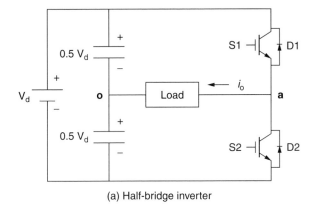

(a) Half-bridge inverter

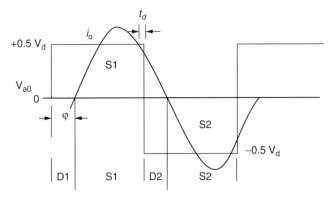

(b) Output voltage and current waves in square-wave mode

Figure 14.5 Single-phase half-bridge inverter.

indicating diode conduction, the power is fed back to the source. However, average power will flow from the source to the load.

14.3.1.2 Full-Bridge Inverter

A single-phase full H-bridge configuration is shown in Figure 14.6(a). The operation of the circuit can be explained with consideration of zero blanking time. While operating, each leg of the inverter can assume two states only: Either the upper (common-anode) switch is on and the lower (common-cathode) switch is off or the other way around. Thus, switching variables a and b can be assigned to individual legs of the inverter and defined as

$$a = \begin{cases} 1 \text{ if } S_1 \text{ is ON and } S_4 \text{ is OFF} \\ 0 \text{ if } S_1 \text{ is OFF and } S_4 \text{ is ON} \end{cases}$$
$$b = \begin{cases} 1 \text{ if } S_2 \text{ is ON and } S_3 \text{ is OFF} \\ 0 \text{ if } S_2 \text{ is OFF and } S_3 \text{ is ON} \end{cases} \quad (14.5)$$

When a given switching variable assumes a value of 1, the positive terminal of the supply source is connected to the corresponding output terminal of the inverter. A value of 0 indicates connection of the negative terminal of the source to the terminal of the inverter. Consequently, the output voltage, v_o, of the inverter can be expressed as

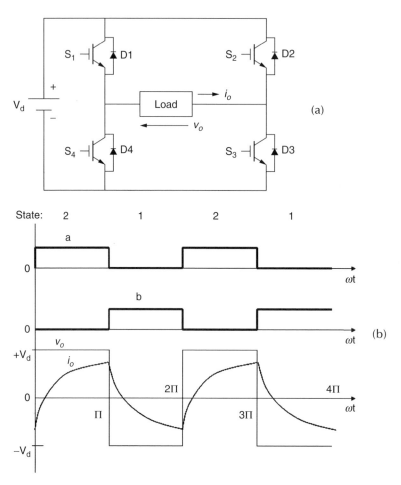

Figure 14.6 (a) Single-phase H-bridge voltage source inverter; (b) switching variables and output voltage and current waveforms in a single-phase voltage-source inverter in the simple square-wave mode.

$$v_0 = V_d (a - b) \qquad (14.6)$$

In this inverter configuration, there are $2^2 (= 4)$ states. These four states are from state 0 through 3, and the voltage can assume three values only: 0, $+V_d$, and $-V_d$, corresponding to states 0 or 3, state 2, and state 1, respectively (see Table 14.1).

Table 14.1 Operation of Single Phase Inverter

Switches Closed	Output Voltage, v_a	States
S_1 and S_3	$+V_d$	2
S_2 and S_4	$-V_d$	1
S_1 and S_2	0	0
S_3 and S_4	0	3

DC/AC Inverters

When the output current, i_o, is of such polarity that it cannot flow through a switch that is turned on, the freewheeling diode parallel to the switch provides a closed path for the current. If, for instance, the output current is positive, as shown in Figure 14.6(b), and switch S_2 is on, the current is forced to close through diode D_2. In a similar way, it is easy to determine conditions under which each of the other three diodes become necessary for continuous conduction of the output current. The freewheeling diodes would not be needed if the load is purely resistive. The simplest version of the so-called square-wave operation mode of the inverter is described by the following control laws:

$$a = \begin{cases} 1 & \text{for } 0 < \omega t \leq \pi \\ 0 & \text{otherwise} \end{cases}$$
$$b = \begin{cases} 1 & \text{for } \pi < \omega t \leq 2\pi \\ 0 & \text{otherwise} \end{cases} \quad (14.7)$$

where ω denotes the fundamental output frequency of the inverter. Only states 1 and 2 are utilized. Waveforms of the output voltage and current in RL load of an inverter in the simple square wave mode are shown in Figure 14.6(b).

The objective of the inverter is to use a DC voltage source to supply a load that requires AC voltage and current. It is useful to describe the quality of the AC voltage or current. The quality of nonsinusoidal wave can be expressed in terms of total harmonic distortion (THD). THD is the ratio of the rms values of all of the nonfundamental frequency terms to the rms value of the fundamental frequency term. Assuming no DC component in the output,

$$\text{THD} = \frac{\sqrt{\left(V_{rms}^2 - V_{1,rms}^2\right)}}{V_{1,rms}} \quad (14.8)$$

where V_{rms} = total rms value of voltage and $V_{1,rms}$ = rms value of fundamental voltage.

The THD of the load current is often more interesting than that of output voltage. This is obtained by substituting the current for voltage in Equation 14.8. Another measure of distortion, such as the ratio of the rms value of the fundamental frequency to the total rms value, is the distortion factor, which can also be applied to describe the output current waveform for inverter.

$$\text{DF} = \frac{I_{1,rms}}{I_{rms}} = \sqrt{\frac{1}{1+(THD)^2}} \quad (14.9)$$

The rms value ($V_{1,rms}$) of the output voltage waveform shown in Figure 14.6(b) is

$$V_{1,rms} = \frac{V_{1,rms,p}}{\sqrt{2}} = \frac{2\sqrt{2}}{\pi} V_d = 0.9 V_d; \text{ and } V_{rms} = V_d \quad (14.10)$$

By substituting the values in Equation 14.8, the value of total harmonic distortion of the output voltage (THD) becomes 0.483. These terms are equally important for output current. In practical inverters, the output current is considered to be high quality if THD does not exceed 0.05 (5%).

The performance of the inverter can be analyzed by basic components of voltage and current waveforms. The performance terms used for inverters are presented as:

- Power efficiency, η, of the converter, defined as

$$\eta = \frac{P_o}{P_i} \quad (14.11)$$

where P_i and P_o denotes the input power and output power of the converter, respectively.

- Conversion efficiency, η_c, of the converter, defined as

$$\eta_c = \frac{P_{o,1}}{P_i} \quad (14.12)$$

for AC output converters.

Symbol $P_{o,1}$ is the AC output power carried by fundamental components of the output voltage and current.

- Input power factor, F_p, of the converter, defined as

$$F_p \equiv \frac{P_i}{S_i} \quad (14.13)$$

where S_i is the apparent input power. The power factor can also be expressed as

$$F_p \equiv k_d \, k_\varphi \quad (14.14)$$

Here, K_d denotes the so-called distortion factor, as defined in Equation 14.9. K is the displacement factor, that is, cosine of the phase shift, θ, between the fundamentals of input voltage and current.

In an ideal power converter, all the three figures of merit defined above would nearly equal unity.

The total harmonic distortion of the output voltage can be minimized by interspersing states 1 and 2 with states 0 and 3 lasting, in the t domain, as shown in Figure 14.7.

The control laws yielding the optimal square-wave operating are

$$a = \begin{cases} 1 & \text{for } \alpha_d < \omega t \leq \pi + \alpha_d \\ 0 & \text{otherwise} \end{cases}$$
$$b = \begin{cases} 1 & \text{for } \pi + \alpha_d < \omega t \leq 2\pi - \alpha_d \\ 0 & \text{otherwise} \end{cases} \quad (14.15)$$

where α_d is 23.2° for optimal square-wave operating mode. This gives 8% reduction in fundamental output voltage, but total harmonic is reduced by as much as 40%, to 0.29. There are various methods to improve the THD of the inverter output waveforms to achieve the desired limits as per the power quality norms.

The quality of the operation of the inverter can further be improved by pulse width modulation. These PWM techniques are dealt in the control scheme.

14.3.2 Three-Phase Inverters

14.3.2.1 Six-Step Operation

Three-phase inverters generally are used to supply three-phase loads. It is possible to supply a three-phase load by means of three separate single-phase inverters, where each

DC/AC Inverters

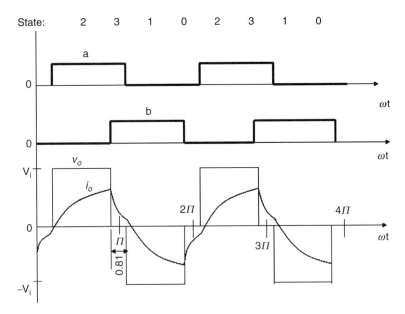

Figure 14.7 Switching variables and output voltage and current waveforms in a single-phase voltage-source inverter in the optimal square-wave mode.

inverter produces an output displaced by 120° with respect to each other. Though this arrangement may be preferable under certain conditions, it requires either a three-phase output transformer or separate access to each of the three phases of the load. In practice, such access is generally not available and it requires 12 switches.

The power circuit of a three-phase voltage source inverter is obtained by adding a third leg to the single-phase inverter, as shown in the Figure 14.8. Assuming, as before, that of the two power switches in each leg (phase) of the inverter one and only one is always on; that is, neglecting the time intervals when both switches are off (blanking time), three switching variables a, b, and c can be assigned to the inverter. The total states of an inverter are 2^3 (= 8), from state 0, when all output terminals are clamped to the negative DC bus, though state 7, when they are clamped to the positive bus. The operating states are shown in Figure 14.9(a).

It is easy to show that the instantaneous line-to-line voltages, V_{AB}, V_{BC}, V_{CA}, are given by

$$\begin{bmatrix} V_{AB} \\ V_{BC} \\ V_{CA} \end{bmatrix} = V_d \begin{bmatrix} 1 & -1 & 0 \\ 0 & 1 & -1 \\ -1 & 0 & 1 \end{bmatrix} \begin{bmatrix} a \\ b \\ c \end{bmatrix} \tag{14.16}$$

In balance three-phase systems, the instantaneous line to neutral output voltages, V_{AN}, V_{BN}, and V_{CN} can be expressed as

$$\begin{bmatrix} V_{AN} \\ V_{BN} \\ V_{CN} \end{bmatrix} = \frac{1}{3} \begin{bmatrix} 1 & 0 & -1 \\ -1 & 1 & 0 \\ 0 & -1 & 1 \end{bmatrix} \begin{bmatrix} V_{AB} \\ V_{BC} \\ V_{CA} \end{bmatrix} \tag{14.17}$$

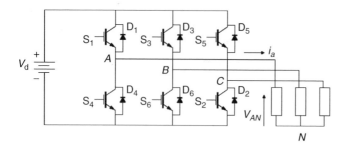

Figure 14.8 Three-phase voltage source inverter.

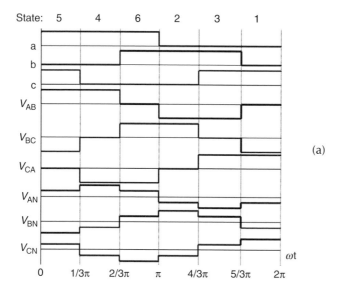

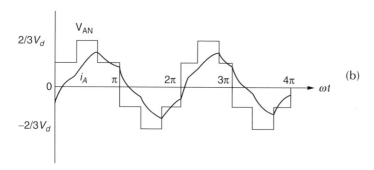

Figure 14.9 (a) Switching variables and output voltage waveforms in a three-phase voltage-source inverter in square wave mode; (b) output voltage and current waveform in three phase voltage source inverter in the square-wave mode.

DC/AC Inverters

Table 14.2 States and Output Voltages of the Voltage Source Inverter

States	Switching Variables	V_{AB}/V_d	V_{BC}/V_d	V_{CA}/V_d
0	000	0	0	0
1	001	0	−1	1
2	010	−1	1	0
3	011	−1	0	1
4	100	1	0	−1
5	101	1	−1	0
6	110	0	1	−1
7	111	0	0	0

From Equations 14.16 and 14.17, it is obtained

$$\begin{bmatrix} V_{AN} \\ V_{BN} \\ V_{CN} \end{bmatrix} = \frac{V_d}{3} \begin{bmatrix} 2 & -1 & -1 \\ -1 & 2 & -1 \\ -1 & -1 & 2 \end{bmatrix} \begin{bmatrix} a \\ b \\ c \end{bmatrix} \quad (14.18)$$

If the 5-4-6-2-3-1-… sequence of states is imposed on an inverter, each state lasting one sixth of the desired period of the fundamental output voltage, the individual line-to-line and line-to-neutral voltages have waveforms shown in Figure 14.9(a). In this square mode of operation, each switch of the inverter is turned on and off once within the cycle of output voltage. The states and output voltages of the three-phase voltage source inverter are presented in Table 14.2.

The output voltage and current waveform for one phase of the inverter are shown in Figure 14.9(b). They are similar to those in a single-phase inverter in the same mode and rich in low-order harmonics, except for the triple ones. A typical input current is shown in Figure 14.10; it can be observed that its fundamental frequency is 6 times higher than that of the output currents. The input current, i_i, is related to the output currents i_A, i_B, and i_C as

$$i_i = a\,i_A + b\,i_B + c\,i_C \quad (14.19)$$

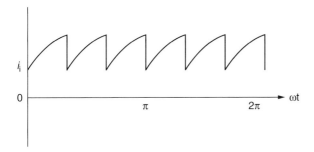

Figure 14.10 Input current waveform in three phase voltage source in the square-wave mode.

14.3.2.2 Voltage and Frequency Control

Normally, the inverter output voltage and frequency are controlled continuously. For machine drive applications, the range of voltage and frequency control is wide. For a regulated AC power supply, the frequency is constant, but the voltage requires control due to supply and load variations. Inverter frequency control is very straightforward. The frequency command may be fixed or variable, and it can be generated from a microprocessor with the help of look-up table, hardware and software counters, and D/A converters. Alternatively, a stable analog DC voltage may represent the frequency command, which can be converted to a proportional frequency through a voltage-controlled oscillator (VCO) and processed through counters and logic circuits. The stability of inverter frequency is determined by the stability of the reference signal frequency and is not affected by load and source variations. The inverter output voltage can, in general, be controlled by the following two methods:

- Inverter input voltage control (pulse amplitude modulation, or PAM)
- Voltage control within the inverter by PWM

These methods have characteristic advantages and disadvantages. However, the PWM method of inverter voltage control is the most common.

These inverters are further divided into the following three general categories:

1. Square-wave inverters
2. Single-phase inverters with voltage cancellation
3. Pulse width modulated inverters

14.3.2.3 Motoring and Regeneration Mode

An inverter can supply average power to the load in the usual inverting or motoring mode, as shown in Figure 14.11(a). The phase current wave i_a is assumed to have perfect filtering by the load, and it is indicated with lagging phase angle $\varphi = \pi/3$. In the first segment, the phase voltage is positive, but the phase current is negative and flowing through the diode D_1, indicating that power is fed back to the source. In the next segment, the IGBT S_1 is carrying the active load current. The next half cycle is symmetrical, and the respective conduction intervals of D_4 and S_4 are shown in Figure 14.8. It may be inferred that if the load is purely resistive or has unity Displacement Power Factor ($\varphi = 0$), each IGBT conducts for an angle of π.

It can be shown that the inverter can also operate in rectification or regeneration mode, pumping average power from the AC to the DC side. Figure 14.11(b) shows waveform for regeneration mode at angle $\varphi = 2\pi/3$, indicating that the feedback interval is considerably larger than the active interval. In the extreme condition, if $\varphi = \pi$, the inverter operates as a diode rectifier with only the diodes conducting.

14.4 CURRENT SOURCE INVERTERS

A current-fed or current-sourced inverter, as the name indicates, likes to see a stiff DC current source at the input. This is in contrast to a stiff voltage source, which is desirable in a voltage-fed inverter. A variable voltage source can be converted to a variable current source by connecting a large inductance in series and controlling the voltage within a feedback current loop. The variable DC voltage can be obtained from a utility supply

DC/AC Inverters

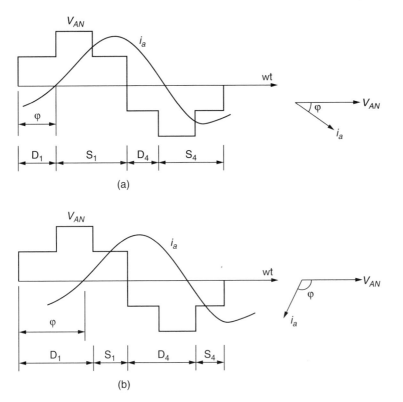

Figure 14.11 Waveforms showing (a) motoring mode and (b) regenerative mode.

through a phase-controlled rectifier, or from a rotating excitation-controlled AC generator through a diode rectifier, or from a battery-type power supply through a DC-DC converter. With a stiff DC current source, the output AC current waves are not affected by the load condition. The power semiconductor devices in a current-fed inverter must withstand reverse voltage; therefore, standard asymmetric voltage blocking devices such as power MOSFETs, BJTs, IGBTs, MCTs, and GTOs cannot be used. Symmetric voltage blocking GTOs and thyristor devices should be used. The forward blocking devices can be used with series diodes. The block diagram of a three-phase current-source inverter is shown in Figure 14.12. The current-source inverter does not require any freewheeling diode, since

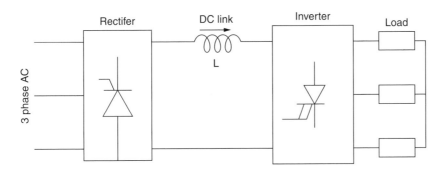

Figure 14.12 Current-source inverter supplied from a controlled rectifier.

current in any half-leg of the inverter cannot change its polarity; it can only flow through the semiconductor power switches. The important feature of the current-source inverter is that it prevents the overcurrents under the short-circuit in the inverter or load. The current-sourced inverters are preferred for medium- and high-power rating drives. In many respects, current-fed inverters are somewhat dual to voltage-fed inverters.

Typical applications of the current-fed inverters are:

- Speed control of large power induction and synchronous motors
- Variable frequency starting of 60 Hz wound-field synchronous motors
- High-frequency induction heating
- Superconducting magnetic energy storage (SMES)
- DC motor drives
- Static VAR compensators
- Active harmonic filters

14.5 CONTROL TECHNIQUES

The DC-AC conversion with inverters in single-phase and three-phase configuration are utilized to realize the output sinusoidal voltage. The voltage source inverter is widely used for low and medium rating AC drives. In voltage source inverters, the controlled variables are the frequency and amplitude of the fundamental output voltage. The pulse width modulation techniques for voltage-source inverters are used for either the output voltage or output current control. The PWM techniques are broadly classified on the basis of voltage-controlled or current-controlled techniques.

14.5.1 Voltage Control Technique

It is possible to control the output voltage as well as optimize the harmonics content by performing multiple switching within the inverter. In voltage controlled voltage source inverters various PWM control techniques are used. There are many possible PWM techniques available for single-phase and three-phase inverters. The relative performance of these and the basic principle to control the output voltage are explained.

The desired characteristics of a PWM technique include:

1. Good utilization of the DC supply voltage; that is, a possibly high value of the voltage gain, K_V, defined here as

$$K_V = \frac{V_{LL,1,P(\max)}}{V_d} \tag{14.20}$$

where $V_{LL,1,p(\max)}$ denotes the maximum peak value of the fundamental line-to-line output voltage available using the technique under consideration and V_d is DC bus voltage

2. Linearity of the voltage control, that is

$$V_{LL,1,P}(M) = M \; V_{LL,1,P(\max)} \tag{14.21}$$

DC/AC Inverters

where *M* denotes the magnitude control ratio. As usual, the magnitude control ratio is defined as the ratio of the actual output voltage (line-to-line or line-to-neutral, peak or rms value) to the maximum available value of this voltage.

3. Low amplitudes of low-order harmonics of the output voltage, to minimize the harmonic content of the output current.
4. Low switching losses in the inverter switches.
5. Sufficient time allowance for proper operation of the inverter switches and control system.

14.5.1.1 Sinusoidal PWM (SPWM) Technique

The sinusoidal PWM technique is very popular for industrial converters. The basic principle of the PWM technique involves the comparison of triangular carrier wave frequency with the fundamental frequency sinusoidal modulating wave. The three-phase reference waveforms, r_A, r_B, r_C, are presented as

$$r_A(\omega t) = F(m, \omega t)$$

$$r_B(\omega t) = F\left(m, \omega t - \frac{2}{3}\pi\right) \quad (14.22)$$

$$r_C(\omega t) = F\left(m, \omega t + \frac{2}{3}\pi\right)$$

where $F(m, \omega t)$ denotes the modulating function employed, as compared with unity-amplitude triangular carrier waveform, $y(\omega t)$. The switching variables *a*, *b*, and *c* are changed from 0 and 1 and 1 to 0 at every sequential intersection of the carrier and respective reference waveform. The sinusoidal modulating function $F(m, \omega t) = m \sin(m, \omega t)$ used in Figure 14.13 can be replaced with any other modulating function, such as the third

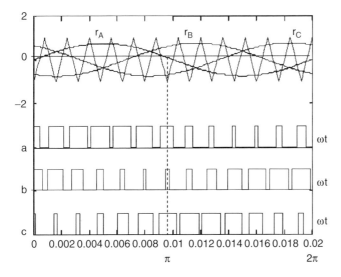

Figure 14.13 Sinusoidal PWM technique.

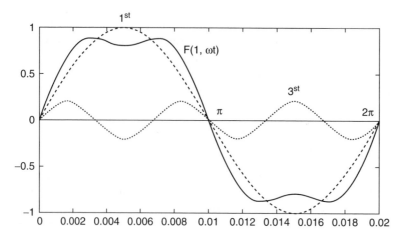

Figure 14.14 Third-harmonic modulating function and its components.

harmonic. The magnitude control ratio, M, of the inverter equals the modulating index m. This is the ratio of peak value of modulating wave to peak value of carrier wave. Ideally, it can be varied between 0 to 1 to give linear relation between the modulating index and output wave.

The switching variables a, b, and c are used to control the switch in the inverter leg so that the output follows the controlled output.

14.5.1.2 Modulating Function PWM Techniques

A PWM technique will be considered that is based on following control laws with various modulating functions. These modulating function PWM strategies yield switching patterns similar to those generated by the SPWM technique. The voltage gain (K_V) as given equals ($\sqrt{3}/2$) 0.866 when the sinusoidal modulating function is used, but increases by employing a different, nonsinusoidal modulating function. Since a modulating function determines the duty ratios of converter switches, its amplitude may not exceed unity. It is possible to build modulating functions consisting of a fundamental and triple harmonics only, such that the fundamental has amplitude greater than unity.

A simple third harmonic modulating function is given by

$$F(m,\omega t) = \frac{2}{\sqrt{3}} m \left[\sin(\omega t) + \frac{1}{6}\sin(3\omega t) \right] \qquad (14.23)$$

and shown in Figure 14.14 with its components, for m = 1. The fundamental is higher than that of sinusoidal modulating function by 15.5%. The third-harmonic modulating function is linear with respect to the modulation index, that is, expressible as $F(m,\omega t) = mF_1(\omega t)$. Use of the third-harmonic modulating function results in a 7% increase of the maximum available DC output voltage in comparison with that of the sinusoidal modulation.

The discontinuous modulating functions are also used. They are nonlinear with respect to modulation index, m. The best known discontinuous modulating functions are given by

DC/AC Inverters

$$F(m,\omega t) = \begin{cases} 2m\cos\left(\omega t - \dfrac{\pi}{3}\right) - 1 & \text{for } 0 \leq \omega t < \dfrac{\pi}{3} \\ 1 & \text{for } \dfrac{\pi}{3} \leq \omega t < \dfrac{2}{3}\pi \\ 2m\sin\left(\omega t - \dfrac{\pi}{6}\right) - 1 & \text{for } \dfrac{2}{3}\pi \leq \omega t < \pi \\ 2m\cos\left(\omega t - \dfrac{\pi}{3}\right) + 1 & \text{for } \pi \leq \omega t \dfrac{4}{3}\pi \\ -1 & \text{for } \dfrac{4}{3}\pi \leq \omega t < \dfrac{5}{3}\pi \\ 2m\sin\left(\omega t - \dfrac{\pi}{6}\right) + 1 & \text{for } \dfrac{5}{3}\pi \leq \omega t < 2\pi \end{cases} \quad (14.24)$$

This modulating function is illustrated in Figure 14.15 for m = 0, 0.25, 0.5, 0.75, and 1. It can be seen that independently of the value of the modulation index, the total time in which the modulating function is 1 or −1 equals one third of the cycle. According to the laws, no switching takes place in each leg of the inverter the total of one third of the cycle, significantly reducing the switching losses in comparison with the inverters employing continuous modulating functions.

14.5.1.3 Voltage Space-Vector PWM Techniques

The space-vector PWM (SVM) method is an advanced, computation-intensive PWM method. The concept of voltage space vector, originally devised for analysis of electrical AC machines, is very well suited for control of modern three-phase power electronic converters.

In three-wire three-phase systems, three-phase to two-phase conversion using parks transformation does not lose any information. For the description of voltage space-vector

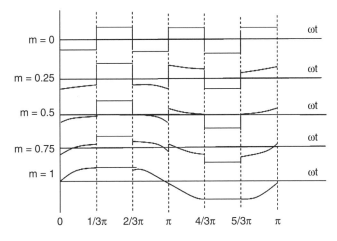

Figure 14.15 Discontinuous modulating function.

PWM it is convenient to convert voltage space vector in d-q format. The voltage space-vector V with the components V_d and V_q are represented as

$$V = V_d + j V_q \qquad (14.25)$$

Then the three-phase sinusoidal and balanced voltages given by the equations

$$V_a = V_m \cos \omega t$$
$$V_b = V_m \cos (\omega t - 2\pi/3) \qquad (14.26)$$
$$V_c = V_m \cos (\omega t + 2\pi/3)$$

are applied to a three-phase induction motor. It can be shown that space-vector V with magnitude V_m rotates in a circular orbit at angular velocity ω, where direction of rotation depends on the phase sequence of the voltages. With the sinusoidal three-phase command voltages, the composite PWM fabrication at the inverter output should be such that the average voltage follows these command voltages with minimum amount of harmonic distortion.

In three-phase bridge inverter, there are 2^3 (= 8) permissible switching states. This gives a summary of the switching states and corresponding phase-to-neutral voltages of an isolated neutral machine. Consider, for example, state 1, when switches S_1, S_6, and S_2 are closed in the inverter (see Figure 14.8). In this state, phase A is connected to the positive bus and phases B and C are connected to the negative bus. The simple circuit solution indicates that V_{AN} = 2/3 V_d, V_{BN} = −1/3 V_d, and V_{CN} = −1/3 V_d. The inverter has six active states (1–6) when voltage is impressed across the load, and two zero states (0 and 7) when the machine terminals are shorted through the lower devices or upper devices, respectively. The set of phase voltages for each switching state can be combined with help of equations to derive the corresponding space vectors. The graphical derivation of V_1 (100) in Figure 14.16 indicates that the vector has a magnitude of 2/3 V_d and is aligned in the horizontal direction as shown. In the same way, all six active vectors and two zero vectors are derived and plotted in Figure 14.17(a).

$$\overline{V} = \frac{2}{3}\left[V_{AN} + aV_{BN} + a^2 V_{CN}\right] \qquad (14.27)$$

$$a = e^{j2\pi/3} \quad \text{and} \quad a^2 = e^{-j2\pi/3}$$

The active vectors V_0 (000) and V_7 (111) are at the origin. For three-phase, square-wave operation of the inverter, it can be easily verified that the vector sequence is V_1, V_2 ... V_6 with each dwelling for an angle of π/3, and there are no zero vectors. Considering the undermodulation region, the inverter transfer characteristics are normally linear. The

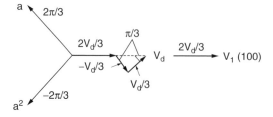

Figure 14.16 Construction of inverter space vector.

DC/AC Inverters

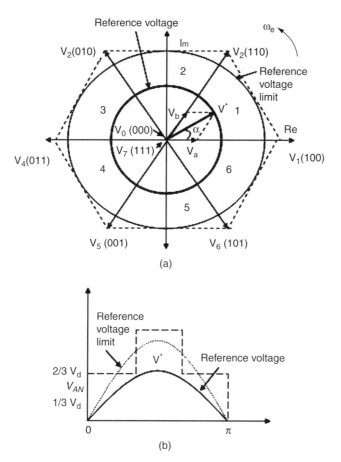

Figure 14.17 (a) Space vectors of a three-phase bridge inverter; (b) corresponding reference phase voltage wave.

modulating command voltages of three-phase inverters are always sinusoidal; therefore, they constitute a rotating space vector V^*, as shown in Figure 14.17(b). The figure shows the phase a component of the reference wave on the six-step phase voltage profile. For the location of the V^* vector shown in Figure 14.17 as an example, a convenient way to generate the PWM output is to use the adjacent vectors V_1 and V_2 of sector 1 on a part-time basis to satisfy the average output demand. The V^* can be resolved as, where V_a, and V_b are the components of V^* aligned in the direction of V_1 and V_2, respectively. Considering the period T_c during which the average output should match the command, we can write a vector addition

$$V^* = V_a + V_b = V_1 \frac{t_a}{T_c} + V_2 \frac{t_b}{T_c} + V_o \text{ or } V_7 \frac{t_c}{T_c} \quad (14.28)$$

where

$$V_a = \frac{2}{\sqrt{3}} V^* \sin\left(\frac{\pi}{3} - \alpha\right) \quad \text{and} \quad V_b = \frac{2}{\sqrt{3}} V^* \sin \alpha$$

$$V^* T_c = V_1 t_a + V_2 t_b + (V_o \text{ or } V_7) t_o \quad (14.29)$$

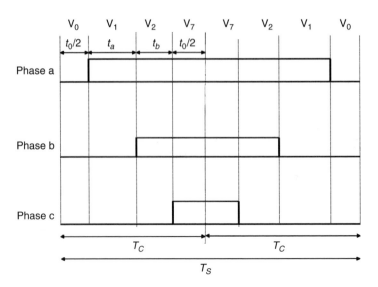

Figure 14.18 Construction of symmetrical pulse pattern for three phases.

where

$$t_a = \frac{V_a}{V_1}T_c; \quad t_b = \frac{V_b}{V_2}T_c \quad \text{and} \quad t_o = T_c - (t_a + t_b)$$

Note that the time intervals t_a and t_b satisfy the command voltage, but time t_o fills up the remaining gap for T_c with the zero or null vector. Figure 14.18 shows the construction of the symmetrical pulse pattern for two consecutive T_c intervals that satisfy Equation 14.29. Here, $T_s = 2T_c = 1/f_s$ (f_s = switching frequency) is the sampling time. Note that the null time has been conveniently distributed between the V_0 and V_7 vectors to describe the symmetrical pulse widths.

In the undermodulation region shown in Figure 14.17(b), the vector V^* always remains within the hexagon. The mode ends in the upper limit when V^* described the inscribed circle of the hexagon. Let us define a modified modulation factor m' given by

$$m' = \frac{\hat{V}^*}{\hat{V}_{1,sw}} \qquad (14.30)$$

where $\hat{V}^*$ = vector magnitude, or phase peak value, and $\hat{V}_{1,sw}$ = fundamental peak value ($2V_d/\pi$) of the square-phase voltage wave. The m' varies from 0 to 1 at the square-wave output. From the geometry of the figure, the maximum possible value of m' at the end of the under-modulation region can be derived. The radius of the inscribed circle can be given as

$$V_m^* = \frac{2}{3}V_d \cos\left(\frac{\pi}{6}\right) = 0.577\,V_d \qquad (14.31)$$

DC/AC Inverters

Therefore, m' at this condition can be derived as

$$m' = \frac{\hat{V}_m^*}{\hat{V}_{1,sm}} = \frac{0.577 V_d}{\left(\dfrac{2}{\pi}\right) V_d} = 0.907 \tag{14.32}$$

This means that 90.7% of the fundamental at the square wave is available in the linear region, compared to 78.55% in the sinusoidal PWM.

14.5.1.4 Programmed PWM Techniques

The best compromise between efficiency and quality of inverter operation is achieved in programmed or optimal switching pattern PWM techniques. The undesired lower order harmonics can be eliminated from the square wave, and the fundamental voltage can be controlled by using selected harmonic elimination method. To understand their principle, consider the switching pattern for phase A, as shown in Figure 14.19. The $a(\omega t)$ waveform has both the half-wave symmetry and quarter-wave symmetry. Consequently, the full-cycle switching pattern is uniquely determined by switching angles α_1 through α_4 in the first quarter-cycle. These angles, whose number, K, here 4 is arbitrary, are the primary switching angles.

The half-wave symmetry of waveforms of switching variables results in the absence of even harmonics as well, while the quarter-wave symmetry allows expressing amplitude of the Kth harmonic, $A_{k,p}$, of $a(\omega t)$ as

$$A_{k,p} = \frac{4}{k\pi}\left[\sum_{i=1}^{k}(-1)^{i-1}\cos(k\alpha_i) - \frac{1}{2}\right] \tag{14.33}$$

It can be seen that the amplitude of each harmonic depends on all the primary switching angles. The most common approach to the definition of optimal values of primary switching angles is called a harmonic-elimination PWM technique. It consists of setting $A_{1,p}$ to $MA_{1,p(\max)}$, where M is the magnitude control ratio and $A_{1,p(\max)}$ denotes the

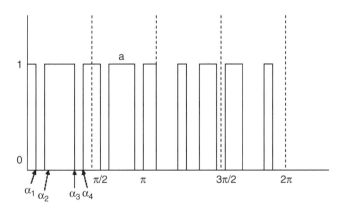

Figure 14.19 Switching pattern with the half-wave and quarter-wave symmetry.

maximum amplitude of the fundamental available with the given number of switching angles. Amplitudes, $A_{5,p}, A_{7,p}, A_{11,p}, \ldots$ of the K-1 lowest order odd and non-triple harmonics are set to zero. For instance, if $K = 5$, the maximum available amplitude of the fundamental, $A_{1,p(max)}$, of $a(\omega t)$ is 0.583 and the 5th, 7th, 11th, and 13th harmonics can be eliminated, leaving the 17th as the lowest-order one.

In practice, computation of optimal switching angles is so time consuming that it cannot be done in real time. Therefore, sets of values of switching angles for each required value of the magnitude control ratio must be stored in the memory of a digital control system of an inverter. Besides the large memory, the disadvantage of programmed PWM techniques manifests itself in disruption of the optimal switching patterns, with rapid changes of the reference frequency or magnitude of the output voltage. Therefore, practical applications of these techniques are limited mostly to an uninterrupted AC power supply, which operates with a constant output frequency and narrow range of voltage control.

14.5.2 Current Control Technique

The methods of inverter switching discussed earlier are of feed forward voltage control PWM techniques. In a machine drive system, control of machine current is important because it influences the flux and developed torque directly. High-performance drives require current control. For a voltage-fed inverter with voltage control PWM, a feedback current loop can be applied to control the machine current.

High-performance current control is a challenging task because in most practical cases the inverter load is unknown and varying. A successful current control technique should ensure:

1. Good utilization of the DC supply voltage, meant here as the feasibility of producing a possibly high current in a given load
2. Low static and dynamic current-control error, thought of as the difference between the reference and actual output currents in the steady state and under transient conditions
3. Low switching losses in the inverter, a requirement that can be translated into a need for possibly infrequent switching of inverter switches
4. Sufficient time allowance for proper operation of the inverter switches and control system

Following four classical approaches to the current control in three-phase control in three-phase inverters are discussed.

14.5.2.1 Hysteresis Current Control

Hysteresis band PWM is basically an instantaneous feedback current control method of PWM, where the actual current continually tracks the command current within a hysteresis band. A block diagram of a voltage-source inverter with the simplest version of the so-called hysteresis control of output currents is shown in Figure 14.20(a). The output currents i_A, i_B, and i_C of an inverter are sensed and compared with the respective reference current waveforms i_A, i_B, and i_C. Current errors Δi_A, Δi_B, and Δi_C are applied to current controllers that produce switching variables a, b, and c for the inverter. The $a = f(\Delta i_A)$ characteristic of a current controller for phase A of the inverter is shown in the Figure 14.20(b). The characteristics constitute a hysteresis loop described as

DC/AC Inverters

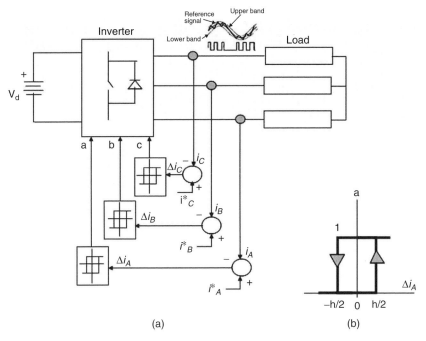

Figure 14.20 Hysteresis current control of inverter.

$$a = \begin{cases} 0 & \text{if } \Delta i_A < \dfrac{-h}{2} \\ 1 & \text{if } \Delta i_A > \dfrac{h}{2} \end{cases} \qquad (14.34)$$

where h denotes the width of the loop. If $-h/2 \leq \Delta i_A \leq h/2$, the value of variable a remains unchanged. The loop width, h, can be thought of as the width of a tolerance band for the controlled current, i_A, since as long as the current error, Δi_A, remains within this band, no action is taken by the controller. If the error is too high (i.e., the actual current is lower from its reference waveform by more than h/2), variable a assumes a value of 1. This makes the voltage V_{AN} equal to or greater than zero, which is a necessary condition for current i_A to increase. Analogously, switching a to zero when the output current is too high causes V_{AN} to be equal to or less than zero, which is conductive for i_A to decrease.

The hysteresis control is excellent for a fast response of an inverter to rapid changes of the reference currents, since the current controllers have negligible inertia and delays. When the amplitude, frequency, and phase of reference currents are simultaneously changed, it is only the time constant of the load that limits the speed of the transition of the current to the new waveform.

14.5.2.2 Ramp-Comparison Current Control

The hysteresis control systems are characterized by unnecessarily high switching frequencies, especially at low values of the magnitude control ratio, since the three current controllers act independently from each other. Also, the somewhat chaotic operation of the inverter is often perceived as a disadvantage. To stabilize the switching frequency, the

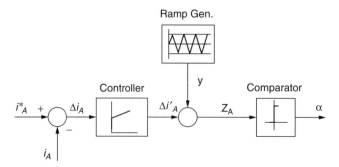

Figure 14.21 Ramp-comparison control scheme for current controlled voltage source inverter.

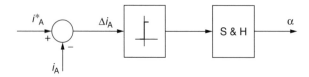

Figure 14.22 Current regulated delta modulation scheme for current controlled voltage source inverter.

so-called ramp-comparison control can be employed. A block diagram of this control scheme is shown in Figure 14.21 for phase A of an inverter.

The current error, $\Delta i_A = i_A^* - i_A$, is applied to the input of a linear controller, usually of the proportional-integral (PI) type. The output signal of the controller is, in turn, compared with a triangular ramp signal, y, similar to that used in the carrier-comparison PWM technique for voltage-controlled inverters. The difference, Z_A, of those two signals activates a comparator, which generates the switching variable a according to the equation

$$a = \begin{cases} 0 & \text{if } Z_A \leq 0 \\ 1 & \text{if } Z_A > 0 \end{cases} \qquad (14.35)$$

Identical control loops are used for the other two phases. The ramp-comparison control can be thought of as the carrier-comparison PWM technique with the processed current error, $\Delta i'_A$, as the modulating function. If, for instance, $i_A < i_A^*$, that is, voltage V_{AN} should be increased to boost current i'_A, switching variable a is modulated by signal i_A^* in such a manner that wide pulses are interspersed with narrow notches.

The ramp-comparison technique, which is usually implemented in an analog system, is so-called current-regulated delta modulator, a common discrete PWM scheme. In the inherently digital delta modulator, the ramp signal is replaced with a sample-and-hold circuit, which allows fixing the switching frequency of the inverter, which is shown in Figure 14.22.

14.5.2.3 Predictive Current Control

An optimal switching pattern for a given set of values of the reference currents could be determined if parameters of the load were known. This assumption underlines the principle

DC/AC Inverters

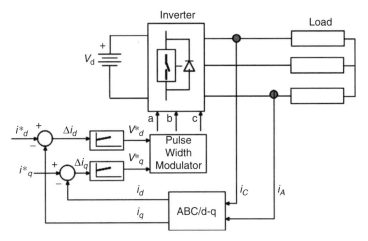

Figure 14.23 Linear control scheme for current controlled voltage source inverter.

of the so-called predictive current controllers. Two basic types of these controllers minimize either the average switching frequency, f_{sw}, or the total harmonic distortion of the controlled currents at a fixed value of f_{sw}. In each step of operation, based on the response of the load to known voltage changes, a predictive controller performs estimation of the load parameters to optimize selection of the next state of the inverter.

14.5.2.4 Linear Current Control

In the three control schemes described above, the current feedback directly enforces switching patterns in the inverter. A different, indirect approach, illustrated in Figure 14.23, involves traditional linear controllers, typically of the PI type. The space-vector version of hysteresis control, the reference current is expressed as a space-vector i^*, whose components, i_d and i_q, serve as reference signals for the respective components, i_d and i_q, of the actual current vector, i. Control errors, Δi_d and Δi_q, are converted by the linear controllers into components, V_d and V_q, of the voltage reference vector V, to be realized by the inverter using the voltage space-vector PWM technique.

14.6 MULTILEVEL INVERTERS

The three-phase voltage source inverters can be classified as two-level inverters. The two-level term has arisen from the fact that the voltage at any output terminal can only assume two values. While operating, the switches of an inverter connect each terminal to either the positive or negative DC bus. Consequently, the line-to-line voltage can assume three values only, and the line-to-neutral voltage five values. This, particularly in the square-wave mode, limits the available control options for minimizing the distortion of the voltage waveform. If more than two voltage levels were obtainable at the output terminals, the output voltage waveforms could be shaped to better resemble the sine wave.

If the fundamental output voltage and corresponding power level of PWM inverter are to be increased to a high value, the DC link voltage must be increased and the devices must be connected in series. By using matched devices in series, static voltage sharing may be somewhat easy, but dynamic voltage sharing during switching is always difficult. This problem may be solved by using multilevel, or neutral-point clamped (NPC), inverters.

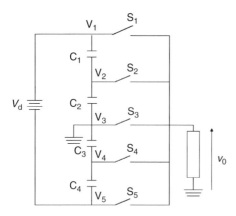

Figure 14.24 Generic multilevel inverter (l = 5).

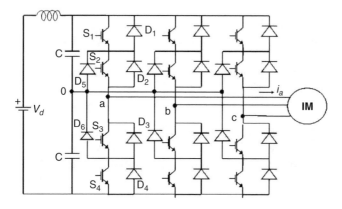

Figure 14.25 Three-level, three-phase inverter.

The basic concept of a multilevel inverter is illustrated by considering a five-level generic inverter, as shown in Figure 14.24. The four capacitors, C_1 through C_4, make up a voltage divider, which constitutes a DC link for the inverter. The center node of the divider and one terminal of the load are grounded. Five switches, S_1 through S_5, of which one and only one is assumed to be on at any time, allow applying any of the five fractional voltages, V_1 through V_5, to the nongrounded load terminal. Thus the output voltage, V_0, could assume any of these five values, including $V_0 = V_3 = 0$. Generally, for a generic n-level inverter output is given as,

$$V_0 = V_j$$

where $j = 1, 2, \ldots, n$.

The unidirectional semiconductor devices in practical multilevel inverters impose somewhat more complicated topologies. The power circuit of the so-called three-level neutral clamped inverter is shown in Figure 14.25 using IGBTs. The DC link capacitor C has been split to create neutral point o.

Each of the three legs of the inverter is composed of four semiconductor power switches, S_1 through S_4; four freewheeling diodes, D_1 through D_4; and two clamping diodes,

DC/AC Inverters

D_5 and D_6. The necessity of clamping diode is easy to demonstrate by considering what would happen if, for instance, diode D_5 was missing (that is, replaced with a short), and switch S_1 was turned on. Clearly, capacitor C_1, serving as a voltage source, would then be shorted. Similarly, diode D_6 prevents shorting C_2 by switch S_4.

Theoretically, the four switches in each inverter leg imply the possibility of 2^4 states of a leg, and 2^{12} states of the whole inverter. In practice only 3 states of a leg are used, which makes for a total of 27 states of the inverter. A ternary switching variables can thus be assigned to each inverter phase and, for phase A, defined as

$$a = \begin{cases} 0 \text{ if } S_1, S_2 \text{ are OFF and } S_3, S_4 \text{ are ON} \\ 1 \text{ if } S_1, S_4 \text{ are OFF and } S_2, S_3 \text{ are ON} \\ 2 \text{ if } S_1, S_2 \text{ are ON and } S_3, S_4 \text{ are OFF} \end{cases} \quad (14.36)$$

It is possible to express the voltage of a given output terminal of the inverter, with respect to neutral o, expressed in terms of the associated switching variable and input voltage. For instance, the voltage, V_A terminal A is

$$V_A = \frac{a-1}{2} V_d \quad (14.37)$$

Consequently, the output line-to-line voltages are given by

$$\begin{bmatrix} V_{AB} \\ V_{BC} \\ V_{CA} \end{bmatrix} = \frac{V_d}{2} \begin{bmatrix} 1 & -1 & 0 \\ 0 & 1 & -1 \\ -1 & 0 & 1 \end{bmatrix} \begin{bmatrix} a \\ b \\ c \end{bmatrix} \quad (14.38)$$

And the line-to-neutral voltage by

$$\begin{bmatrix} V_{AN} \\ V_{BN} \\ V_{CN} \end{bmatrix} = \frac{V_d}{6} \begin{bmatrix} 2 & -1 & -1 \\ -1 & 2 & -1 \\ -1 & -1 & 2 \end{bmatrix} \begin{bmatrix} a \\ b \\ c \end{bmatrix} \quad (14.39)$$

Practical control methods allow only change only from 0 to 1, 1 to 2, and vice versa for each switching variable, while transition from 0 to 2 and 2 to 0 are forbidden. This ensures smooth commutation and reduces the chances for a shoot-through, since out of the four switches in a leg only two change their conduction states simultaneously. The DC voltage is shared by at least two switches, so that their voltage rating can be lower than those of switches in a regular, two-level inverter.

Space vectors of the line-to-neutral voltages across a wye-connected load, corresponding to individual states of the inverter, are shown in Figure 14.26. Comparing the vector diagram with that in Figure 14.17(a) for the two-level inverter, it is easy to guess that the 18-21-24-15-6-7-8-5-2-11-20-19-… state sequence (each state maintained for one 12th of the desired cycle of output voltage) represents the square-wave operation mode.

The corresponding waveforms of switching variables and output voltages of the inverter are shown in Figure 14.27. Although the voltage gain is 1.065, that is, slightly

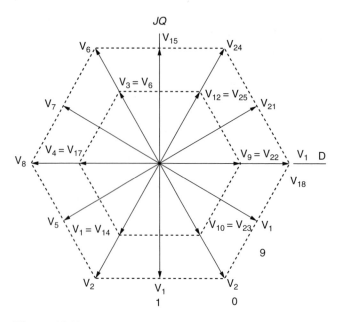

Figure 14.26 Voltage space vectors in a three-level neutral clamped inverter.

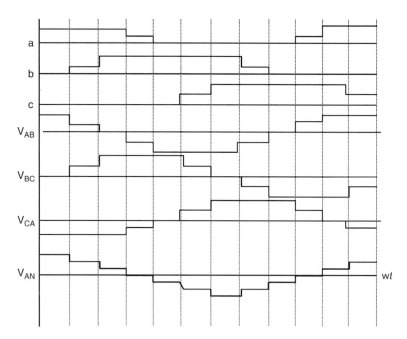

Figure 14.27 Switching variables and output voltage waveforms in three-level neutral clamped inverter in square-wave mode.

lower than 1.1 for a two-level inverter, higher quality of the output waveforms in the three-level inverter is obvious. Even higher quality of operation can be achieved by pulse width modulation, with lower switching frequencies than those typical for two-level inverters.

DC/AC Inverters

14.7 HARD SWITCHING EFFECTS

The conversion of DC-to-AC through the semiconductor switches is obtained by hard switching. The overall performance is subjected to the speed of switching. Due to high-speed power switching, devices generate high-voltage slew rates (dv/dt) and high common mode voltages, causing some serious problems such as premature winding failures, ground leakage currents, shaft voltages and bearing currents, and conducted or radiated electromagnetic interference. They have a number of detrimental effects, which are briefly reviewed here.

14.7.1 Switching Loss

All semiconductor devices are operated through the switching circuitry. The overlapping of voltage and current waves during each turn-on and turn-off switching cause a large pulse of energy loss. With an RC snubber, turn-off loss can be decreased, but the stored energy in the capacitor is lost at turn-on switching. Therefore, with a snubber, the total switching loss may increase. With higher switching frequency, inverter loss increases; that is, its efficiency decreases. The PWM switching frequency of an inverter is limited because of switching loss.

14.7.2 Device Stress

In hard switching, the switching locus moves through the active region of the volt-ampere area, which stresses the device. The reliability of the device may be impaired due to prolonged hard switching operation. A snubber circuit reduces power loss in the device during switching and protects the device from switching stress of high voltage and currents.

14.7.3 EMI Problems

High dv/dt, di/dt, and parasitic ringing effect at the switching of a fast device can create severe EMI problems, which may affect the control circuit and nearby apparatus. The parasitic inductance can be a source of EMI due to large induced voltage and induced common mode coupling current, respectively.

14.7.4 Effect on Insulation

The high dv/dt impressed across the stator winding insulation can create large displacement current $\left(C\dfrac{dv}{dt}\right)$, which can deteriorate machine insulation.

14.7.5 Machine Bearing Current

It is observed that PWM inverter drives with fast switching semiconductor devices are known to cause a machine bearing problem. Fast switching devices create high dv/dt, and inverters can be represented by common mode equivalent circuits with high dv/dt sources. These dv/dt sources create circulating currents to ground through the machine bearing stray capacitances, as indicated in Figure 14.28. The figure shows the stray capacitances

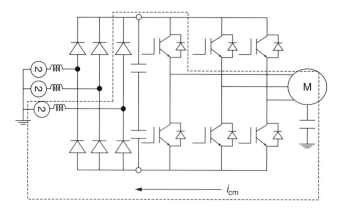

Figure 14.28 Common mode dv/dt induced current flow-through motor bearing.

linking a machine's stator winding with the rotor and the stator. The common mode dv/dt impressed on the stator winding couples to the rotor and creates a circulating current to the ground through the machine shaft and stray capacitances of the insulated bearing. The circulating current through the bearing will increase with higher dv/dt and higher PWM switching frequency. This causes the bearing to wear out and in turn cause damages to the machine.

14.7.6 Machine Terminal over Voltage

The PWM inverters are often required to link a machine with a long cable, in an industrial environment. The high dv/dt at the inverter output boosts the machine terminal voltage by the reflection of the high-frequency traveling wave. High-frequency ringing occurs at the machine terminal with stray circuit parameters. The resulting excessive overvoltage threatens the motor insulation.

The high dv/dt effect is essentially shunted to ground through the low pass filter. The filter also solves bearing current and machine insulation deterioration problems. These hard switches are essential to design the inverter and selection of the device ratings.

14.8 RESONANT INVERTERS

The disadvantage of the hard switching effects as discussed can be practically eliminated in a soft-switched inverter. In fact, resonant inverters featuring soft switching can be used for AC drive applications.

14.8.1 Soft-Switching Principle

The switching devices absorb power when they turn on or off if they go through a transition when both voltage and current are nonzero. As the switching frequency increases, the transition occurs more often and average power loss in the device increases. High switching frequencies are desirable because of the reduced size of filter components and transforms, which reduces the size and weight of the converter. In resonant switching circuits, switching takes place when voltage or current is zero, thus avoiding simultaneous transition of voltage and current and thereby eliminating switching loss. This type of switching is called "soft" switching. The principle of soft switching using zero current switching (ZCS) and zero voltage switching (ZVS) is shown in Figure 14.29.

DC/AC Inverters

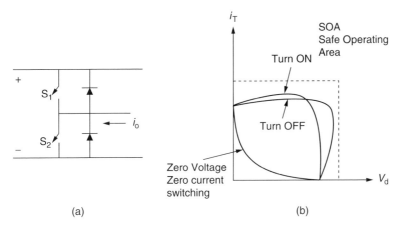

Figure 14.29 Zero voltage zero current and hard switching v-i locus.

It was shown before that in hard switching, the voltage and current overlap to create large switching loss. The main idea in soft switching is to prevent or minimize this overlap so that the switching loss is minimal. In ZCS, when the device turns on, the current build-up can be delayed with series inductance. Similarly, at turn-off, the current may be zero when device voltage builds up.

Resonant converters includes resonant switch converters, load resonant converters, and resonant DC link converters. In general, they may be DC link or AC link types. The DC link types can be classified as resonant link DC (voltage-fed or current-fed) and resonant pole DC inverters. Voltage-fed resonant link inverters can be further classified as free resonance and quasi-resonance types. The AC link types can be classified as resonant link and nonresonant link type. The most common types of topology used for AC motor drives are explained in next section.

14.8.2 Resonant Link DC Converter (RLDC)

A voltage-fed inverter that operates on free running resonance (tens of kilohertz) in the DC link is shown in Figure 14.30. The DC voltage source V_d, obtained from a battery or through a rectifier, is converted to sinusoidal voltage pulses (V_d) with zero voltage gap on the inverter bus through an L_r C_r resonance circuit. The gap permits zero voltage soft switching of the inverter devices. The variable voltage, variable frequency waves at the motor terminal can be fabricated by delta modulation principle using the integral voltage pulses. To establish the resonant bus voltage pulses, an initial current in the resonant inductor L_r is needed for the compensation of the resonant circuit loss and reflected inverter input current. This current can be established to the appropriate values during the zero voltage gap by shorting the inverter devices.

The bus voltage (higher than V_d) can be controlled by either a passive or active clamping technique. In the passive clamping method, a transformer with a series diode in the secondary is connected across the inductor L_r and, as the voltage tends to exceed a threshold value, the diode conducts, pumping the inductor's trapped energy to the source. In this method, the peak DC link voltage V_d can be limited typically to 2.5 V_d. Using the active clamping method, as shown in Figure 14.30, the peak value can typically be limited to 1.5 V_d. At the end of the zero voltage gap, when the desired initial current is reached, the selected devices of the inverter are opened to established the output phase voltages, as dictated by the PWM algorithm.

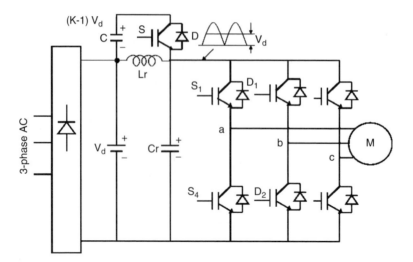

Figure 14.30 Resonant DC link inverter with active voltage clamping.

At the end of resonant cycle, the inverter diode provides a path for negative current in the gap interval. The active clamping circuit consists of a precharged capacitor in series with an IGBT-diode pair, as shown, and its operation can be explained as follows: On releasing DC bus short, the link voltage swings towards its natural peak. However, on reaching the voltage level kV_d, diode D conducts and clamps bus voltage at this level. With D conducting, S is turned on in lossless manner. The trapped inductor current linearly decays to zero and then becomes negative through S to balance the capacitor charge. At current zero, S is turned off to initiate the resonance again until the bus voltage falls to zero. The circuit has the disadvantage of a voltage penalty on the devices, besides the need of extra components and additional loss in the resonant circuit.

14.9 AUXILLARY AUTOMOTIVE MOTORS CONTROL

The trend for comfort and convenience features in today's cars means that more electric motors are required than ever. It is estimated that up to 30 motors may be used in top-range models of modern cars. All these motors are activated or deactivated from the dashboard and supplied from the battery. However, the alternator is used for charging the battery.

Motor design for automobile applications represents an attempt at achieving the optimum compromise between conflicting requirements. The torque/speed characteristics demanded by the application must be satisfied while taking account of the constraints of the materials, of space, and of cost. There are four families of DC motors that are, or have the potential to be, used in automobiles:

- Wound Field DC Commutator Motors
- Permanent Magnet (PM) DC Commutator Motors
- PM Brushless Motors
- Switched Reluctance Motors

DC/AC Inverters

Basic motor drive configuration depends on the type of the motor. These are broadly categorized under commutator and switched field motors.

14.9.1 Commutator Motors

Both permanent magnet and wound field commutator motors can be controlled by a switch in series with the DC, supply as shown in Figure 14.31. When motor is switched off, it may be running. If so, the motor acts as a voltage source and the rotating mechanical energy must be dissipated either by friction or by being transformed into electrical energy and returned to the supply via the inherent anti-parallel diode of the MOSFET.

Reversing the polarity of the supply to a commutator motor reverses the direction of the rotation. This usually requires an H bridge configuration, as shown in Figure 14.32. Its operation is similar to that of the H-bridge inverter circuit shown in Figure 14.6.

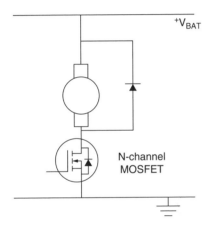

Figure 14.31 Commutating motor switch.

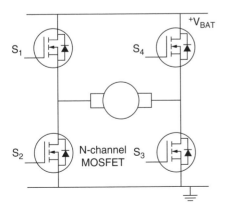

Figure 14.32 H-bridge using MOSFETs.

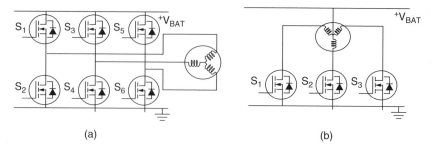

Figure 14.33 (a) Brushless motor drive with six switches; (b) brushless motor drive with three switches.

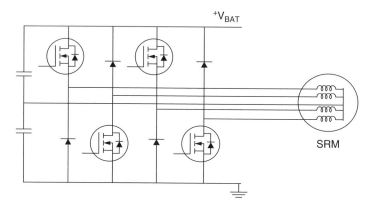

Figure 14.34 Four MOSFET switched reluctance motor drive.

14.9.2 SWITCHED FIELD MOTORS

The PM brushless motors typically require six switches to generate the rotating field, as shown in Figure 14.33(a). Motors that operate at lower power density can be driven from three switches. The circuit in Figure 14.33(b) shows a low-side switch version of such a drive. A similar arrangement with high-side switches would be possible.

Switch reluctance motors may as few as 4 or as many as 12 switches to generate the rotating field. The 4-switch configuration is shown in Figure 14.34. The speed and direction of all switched field motors is controlled by the timing of the pulses. In the case of brushless DC machines, these timing pulses can be derived from a dedicated IC. Rotor position sensing required may be obtained by using magnetoresistive sensors to determine which windings should be energized. Compared to a DC commutator motor, the power switches for a brushless motor are required to perform a tougher duty, because they must switch at every commutation. PWM speed control pushes up the required switching speed.

These strategies of motor controls are implemented by way of analog and digital circuits. The analog control circuit possesses the advantage of fast response, but suffers the disadvantages of complex circuitry, limited function, and difficulty in circuit modification. With the progress in motor control and microelectronics technologies, the development of universal AC drives had become a major trend. High-performance, low-cost digital signal processors (DSPs) are used in digital PWM control and digital current control for AC drives. These control schemes have the advantages of simple circuitry, software control, and flexibility in adoption to various automotive applications.

GLOSSARY

BJT: Bipolar junction transistor
CSI: Current source inverter
DF: Distortion factor
DPF: Displacement power factor
DSP: Digital signal processors
GTO: Gate turn-off thyristor
IGBT: Insulated-gate bipolar junction transistor
M: Modulation index
MCT: Metal oxide silicon-controlled rectifier
MOSFET: Metal oxide silicon field effect transistor
PAM: Pulse amplitude modulation
PWM: Pulse width modulation
SCR: Silicon-controlled rectifier
THD: Total harmonic distortion
VCO: Voltage controlled oscillator
VSI: Voltage source inverter
ZCS: Zero current switching
ZVS: Zero voltage switching

REFERENCES

[1] P. Wood, *Switching Power Converters*, Van Nostrand Reinhold Company, New York, 1981.
[2] D.W. Hart, *Introduction to Power Electronics*, Prentice-Hall, Englewood Cliffs, NJ, 1997.
[3] A.M. Trzynadlowski, *Introduction to Modern Power Electronics*, John Wiley & Sons, New York, 1998.
[4] B.K. Bose, *Modern Power Electronics and AC Drives*, Pearson Education (Singapore), 2002.
[5] N. Mohan, T.M. Undeland, and W.P. Robbins, *Power Electronics: Converters, Application, and Design*, John Wiley & Sons, New York, 1995.
[6] M.H. Rashid, *Power Electronics: Circuits, Devices, and Applications*, Prentice-Hall of India, 1995.
[7] Z. Yu, A. Mohammed, and I. Panahi, "Review of PWM techniques," *Proc. American Control Conf.*, Albuquerque, NM, June 1997, pp. 257–261.
[8] E.R. Cabral de Silva, M.C. Cavalcanti, and C.S. Jacobina, "Comparative study of pulsed DC link voltage converters," *IEEE Trans. on Power Electronics*, Vol. 18, No. 4, July 2003.
[9] J.G. Kassakian, "Automotive electrical systems: The power electronics market of the future," Fifteenth Annual IEEE Applied Power Electronics Conference and Exposition, APEC 2000, Vol. 1, 6–10 February 2000, pp. 3–9.
[10] J.G. Kassakian, H.-C. Wolf, J.M. Miller, and C.J. Huston, "Automotive electrical systems circa 2005," *IEEE Spectrum*, August 1996, pp. 22–27.
[11] "Phillips Semiconductor Application," Phillips Semiconductor Application Laboratory Manual, Hazel Grove, Stockport, Chashire, 1994.

15

AC/AC Converters

Mehrdad Kazerani
University of Waterloo, Waterloo, Ontario, Canada

15.1 INTRODUCTION

In this chapter, AC/AC converter refers to a static frequency changer mainly made up of semiconductor switches, which is used to convert the AC power at a given frequency to the AC power at a desired frequency.

AC-to-AC conversion can be indirect or direct depending on whether or not the process includes an intermediate DC stage (DC link). Frequency conversion can be restricted or unrestricted, depending on whether or not the achievable output frequency range is limited by the input frequency. Depending on the application, an AC/AC converter can accommodate conversion from a variable frequency to a fixed frequency or vice versa.

Since the introduction of the first mercury-arc static frequency changer prototype in the early 1930s [1], the advances in semiconductor technology, power electronics, and control theory have resulted in a huge progress in the performance and applications of AC/AC converters. Static frequency changers find applications in AC motor drives and, thus, in the automotive industry, which is moving towards more electric vehicles (MEVs). With an internal combustion engine (ICE) driving a synchronous or induction generator, the AC/AC converter can convert the AC power produced to variable frequency AC power, which will be used to drive the motor engaged with the wheels at the desired speed.

The following sections introduce different established AC/AC converter topologies together with their advantages, disadvantages, and restrictions. The objective of this chapter is to familiarize the reader with the established AC/AC converter topologies and their potentials in the automotive industry, especially with regard to hybrid electric vehicles (HEVs). The references cited in this chapter provide sources of detailed information for further study.

15.2 AC/AC CONVERTER TOPOLOGIES

AC/AC converter topologies fall under two major categories, indirect and direct. In the following sections, these topologies will be described.

15.2.1 Indirect AC/AC Converter

Indirect or DC-link AC/AC converters are composed of two back-to-back voltage- or current-source converters, connected via a DC-link capacitor or reactor. Figure 15.1 and Figure 15.2 show the schematic diagrams of the voltage- and current-source converter-based indirect AC/AC converters. The switches S_{a+}, S_{b+}, ..., S_{C-} in Figure 15.1 and Figure 15.2 are unidirectional switches, respectively.

The indirect AC/AC converter topology has also been called rectifier-inverter pair. Depending on the mode of operation being "motoring" or "regenerative breaking," one of the two back-to-back converters assumes the role of a "rectifier," and the other converter will act as an "inverter." The two converters in each topology can be controlled using phase-control, pulse-width modulation (PWM), space vector modulation (SVM), or selective harmonic elimination (SHE) technique. The DC terms, V_d and I_d, in voltage- and current-source converter-based topologies, respectively, are regulated by closed-loop control systems to ensure input/output real power balance.

The advantages of the indirect AC/AC converter topology lie in its low number of components, its simplicity of control, the possibility of simultaneous input power factor control and output vector control, the ride-through capability due to the existence of a DC-link energy storage element, and the immunity to the adverse effects of harmonic distortion and imbalance in the input voltage.

The disadvantages of the indirect AC/AC converter topology lie in the presence of the DC-link reactive elements that add to the weight, size, and cost of the scheme, and

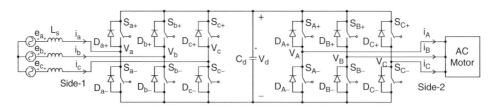

Figure 15.1 Voltage-source converter-based indirect AC/AC converter.

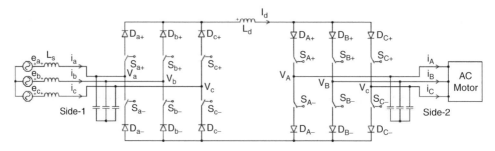

Figure 15.2 Current-source converter-based indirect AC/AC converter.

AC/AC Converters

the fact that a high DC-link voltage ($V_d > 1.634\ V_{LL}$ for two-level PWM voltage-source converter-based topology) or current ($I_d > 1.616\ I$ for three-level PWM current-source converter-based topology) has to be maintained in order to be able to produce the required output AC voltage and current over the entire operating range, thus adding to the losses and device ratings. If the instantaneous input/output power balance is maintained through implementation of a proper control strategy, the need for energy storage in the DC-link can be dramatically reduced, leading to a quasi-direct AC/AC converter [2]. Also, if the DC-link current in the current-source converter-based topology is adjusted based on the load requirements, the losses due to high DC-link currents can be minimized.

Current-source converter-based AC/AC converters are used for medium- to high-power motor drives, especially if a long cable is needed between the converter and the motor in applications such as submerged pumps. In this situation, a voltage-source converter-based AC/AC converter is not appropriate, as the output pulsed-voltage traveling wave can result in destructive overvoltages at the motor terminals. This can damage the insulation of the motor windings in the first few turns unless specially designed motors, which are more expensive than the ordinary ones, are used. Voltage-source converter-based AC/AC converters are more commonly used as AC motor drives and are more likely to be used in hybrid electric vehicles due to their higher efficiency and possibility of using a battery directly across the DC link for energy storage. The lower efficiency of current-source converter-based topology is attributed to the higher conduction losses in this topology due to the presence of diodes, which are placed in series with the switches to enhance the reverse voltage withstand capability, as well as the fact that the DC-link inductors are more lossy than the DC-link capacitors. Figure 15.3 and Figure 15.4 show the block diagrams of the HEV series and parallel designs that can be realized by the voltage-source converter-based AC/AC converter topology.

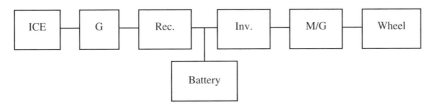

Figure 15.3 HEV serial design based on voltage-source converter-based AC/AC converter.

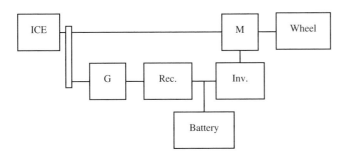

Figure 15.4 HEV parallel design based on voltage-source converter-based AC/AC converter.

Indirect AC/AC converter is an unrestricted frequency changer, as the output frequency is independent of the input frequency. When operating at a moderately high switching frequency, the qualities of the input and output waveforms are very high, with no low-order harmonics and, thus, minimal filtering requirements. Also, the dynamic response will be faster due to higher control bandwidth. In order to enjoy the benefits of high switching frequency while avoiding high switching losses and electromagnetic interference (EMI), soft-switched high-frequency converters have been proposed based on parallel resonant DC-link (RDCL) and series RDCL [3]. Even though the resonant topologies feature better performance and higher power density, they are not as established as the original hard-switched voltage- and current-source converter-based topologies and are known to have lower reliability and higher cost [3].

15.2.2 Direct AC/AC Converter

In direct AC/AC converter topology, the DC-link reactive element is eliminated, allowing for a more compact design. However, the number of components will be higher compared with that of the rectifier-inverter pair. Furthermore, the system lacks ride-through capability.

Direct AC/AC converters are divided into two major categories, naturally and forced-commutated cycloconverters.

15.2.2.1 Naturally Commutated Cycloconverter (NCC)

Figure 15.5 shows the schematic diagram of a three-phase-to-three-phase NCC. NCCs are based on the mature technology of SCRs (thyristors) and are operated under phase-control technique. NCCs are considered to be more efficient, less costly, and lighter than the rectifier-inverter pair AC/AC converters [4]. The high-power handling capability of SCRs makes the NCC the natural choice for high-power frequency changers. The operational limitation of NCC is in the output frequency. In order for the variation of the firing angles of the SCRs in the NCC to follow the typical sequence of commutation between the successive phases, the output frequency has to be lower than or equal to one third of the input frequency, making NCC a restricted frequency changer. This limitation will not be an issue when the input-to-output frequency ratio is high, as is the case when AC power is generated by high-speed generators. This high frequency ratio results in very high-quality output waveforms, as well as very fast dynamic response. On the other hand, operating at high input frequency asks for high-speed SCRs, which are more expensive. The other advantage of NCC is in the inherent zero-current switching (ZCS) capability of SCRs, resulting in low switching losses. In contrast to NCC, a hard-switched rectifier-inverter pair suffers high switching losses.

NCC is based on the dual converter topology and may be operated under circulating-current or noncirculating-current control. Figure 15.6(a) and Figure 15.6(b) show the schematic diagrams of one phase of the circulating current and noncirculating current NCC topologies, respectively. In the circulating current NCC, both P and N converters are operated simultaneously, where the firing angles of the thyristors in the two converters are related through $\alpha_N = 180° - \alpha_P$. As in this case, the instantaneous DC-side voltages of the two converters are different, even though the average values are the same, inductors are used on the DC side between the two converters to make the connection of two unequal voltages possible and limit the circulating current. In the noncirculating current NCC, the P converter is operated at the firing angle α when the load current is positive, whereas the N converter is operated at the firing angle $\alpha_N = 180° - \alpha_P$ when the load current is negative. The noncirculating-current NCC, which has been referred to as the practical

AC/AC Converters

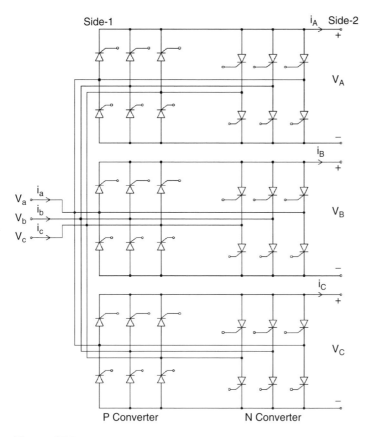

Figure 15.5 Schematic diagram of a three-phase NCC.

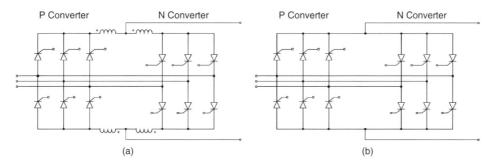

Figure 15.6 (a) Circulating current NCC; (b) noncirculating current NCC.

cycloconverter [5], due to its higher efficiency and more compact design, is potentially more feasible in hybrid electric vehicles. The control of the firing angles of the SCRs in the dual converters is based on the cosine-wave intersection method [6,7]. In this method, a sinusoidal modulating signal of amplitude between 0 and 1 at the desired output frequency is intersected with a timing cosine waveform of amplitude 1 at the input frequency. The firing angle α is then calculated as the angle corresponding to the points of intersection of the modulating signal with the falling edges of the timing signal in different cycles

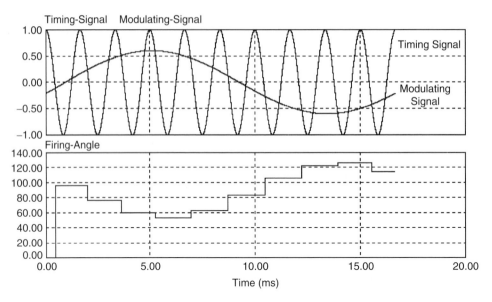

Figure 15.7 Cosine-wave intersection method.

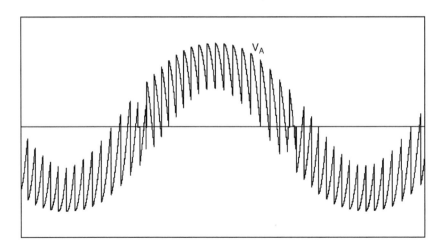

Figure 15.8 Typical output voltage waveform of NCC.

(Figure 15.7). The firing angle is updated once in every cycle of the input signal. This means that in NCC, the control bandwidth is limited by the input frequency. The local average of the output voltage of the noncirculating current NCC is given by 1.35 V_{LL} cos α_P [6], where V_{LL} is the rms value of the line-to-line input voltage and α_P is the firing angle of the P converter valves. In order to realize the desired output voltage, a modulating signal of the form $v_m = M \cos(\omega t + \theta)$ is produced by a closed-loop control circuit, where $0 \leq M \leq 1$, ω is the angular frequency of the output voltage, and θ is the desired phase angle of the output voltage. This signal is input to a $\cos^{-1}$ block whose output is used as the firing angle for the converter valves. As a result, the local average of the output voltage of the dual converter becomes 1.35 $V_{LL} M\cos(\omega t + \theta)$, implying that in local average sense, the dual converter behaves as a linear amplifier that duplicates the modulating signal

AC/AC Converters

at its output terminals with a gain of 1.35 V_{LL}. Figure 15.8 shows a typical output voltage waveform for one output phase of NCC (v_A in Figure 15.5). The fundamental component of the output voltage waveform is clearly seen as the center-line, with high-frequency components making a band around it.

When operating under noncirculating-current control, either the three-phase voltages feeding the dual converters or the output phases have to be isolated from one another to avoid interaction between the dual converters [6]. To avoid isolation transformers and reduce the size and weight of the system, the high-speed generator or the traction motor has to have isolated three-phase windings.

A major disadvantage for NCC is its low and uncontrollable input displacement power factor due to the fact that phase-controlled thyristor converters draw reactive power from the source irrespective of the mode of operation (rectifier or inverter). The need for additional devices for input reactive power compensation and power factor correction makes NCC a poor choice for the hybrid electric vehicle industry. Another major disadvantage for NCC is the low-order harmonic distortion of the input current. Addition of harmonic filters to eliminate harmonics makes the scheme more costly and less competitive.

15.2.2.2 Forced-Commutated Cycloconverter (Matrix Converter)

The conventional three-phase forced-commutated cycloconverter (FCC) or matrix converter is composed of an array of nine bidirectional switches, each connected between one phase of the input and one phase of the output. A matrix converter is an unrestricted frequency changer, which constructs the output voltage waveforms by piecing together selected segments of the input voltage waveforms [8]. Figure 15.9 shows the schematic diagram of a conventional matrix converter.

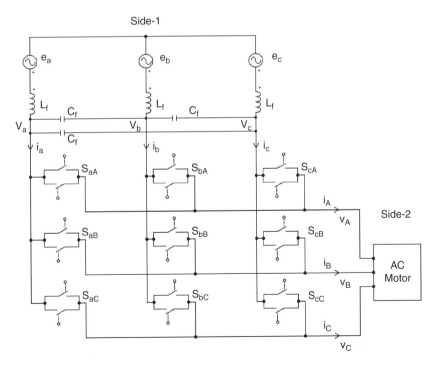

Figure 15.9 Schematic diagram of a conventional matrix converter.

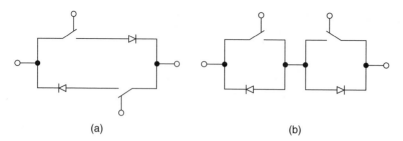

Figure 15.10 Two possible structures for bidirectional switch.

A bidirectional switch is composed of two two-quadrant switch elements connected in one of the two ways shown in Figure 15.10, which is capable of conducting current in both directions in the on-state and blocking voltages of both polarities in the off-state. For this reason, a bidirectional switch is commonly referred to as a four-quadrant switch. To date, no monolithic four-quadrant switch is available off the shelf.

The 9-bidirectional-switch matrix converter topology, when realized by any of the two bidirectional switch configurations of Figure 15.10, uses 18 unidirectional switches and 18 fast-recovery diodes. New advancements in the semiconductor technology have led to the introduction of forward and reverse blocking non-punch-through (NPT) IGBTs, eliminating the need for the diode that has to be placed in series with each unidirectional switch in Figure 15.10(a) when constructing a bidirectional switch [9]. Using the new bidirectional switch configuration based on the NPT IGBT switches reduces the number of semiconductor elements in the 9-bidirectional-switch matrix converter topology to 18 switches and 0 diodes. This brings the matrix converter in a real competitive position with respect to the well-known rectifier-inverter pair topology.

Note that voltage-source converter-based indirect AC/AC converter topology has 12 unidirectional switches, 12 fast-recovery diodes, and one DC-link capacitor. The current-source converter-based rectifier-inverter pair AC/AC converter uses 12 unidirectional switches, 12 fast-recovery diodes, and one DC-link inductor. In current-source converter-based topology, the number of semiconductor devices is reduced to 12 unidirectional switches and 0 diodes if NPT IGBTs are used. However, the need for the DC-link inductor will be a serious disadvantage for the current-source converter-based topology until super-conductive materials for magnetic energy storage become affordable for public applications. Furthermore, a current-source converter-based AC/AC converter is not appropriate for hybrid electric cars, as direct integration of battery for energy storage on the DC-link is not possible. It is said that the success of the matrix converter topology depends on the availability of monolithic bidirectional switches to reduce the semiconductor component count [3].

Due to the special arrangement of the switches, at any moment of time, one and only one of the three switches connecting the three phases of the input to each phase of the output (e.g., S_{aA}, S_{bA}, and S_{cA} in Figure 15.9) has to be on, to avoid short-circuiting of the side-1 voltage sources or interrupting the side-2 inductive load currents. This constraint reduces the number of possible switch-state combinations from $2^9 = 512$ to 27 valid states [10]. From the practical point of view, translating the calculated switch duty ratios into actual gating signals requires distribution of the on-periods of the switches over each switching period according to a special pattern. This distribution pattern is not unique and its choice strongly affects the total harmonic distortions (THDs) of the side-1 and side-2 waveforms.

AC/AC Converters

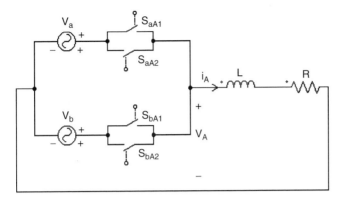

Figure 15.11 Two bidirectional switches S_{aA} and S_{bA} commutating the phase A current [11].

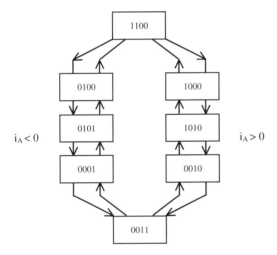

Figure 15.12 Switching sequence diagram for safe operation of S_{aA} and S_{bA} [11].

Furthermore, in order to obtain high-quality waveforms, a high switching frequency has to be adopted. In order to avoid a dead-time or an overlap between the on-states of the bidirectional switches connecting two input phases to one output phase while commutating an inductive load current, a multistep switching strategy is used. Figure 15.11 shows the two bidirectional switches S_{aA} and S_{bA} connecting the input phases a and b to the output phase A. Figure 15.12 shows a special four-step switching sequence proposed in Reference 11 for the control of the two-quadrant switch elements of the bidirectional switches shown in Figure 15.11, to protect the matrix converter switches against hazardous conditions. The 4-bit switch status code shown in Figure 15.12 represents the positions of the switches S_{aA1}, S_{aA2}, S_{bA1}, and S_{bA2}, respectively, where 1 stands for on position and 0 stands for off position. As shown in Figure 15.12, the two-quadrant switch elements of each bidirectional switch are controlled individually to avoid any dead-time or overlap between the on-states of the bidirectional switches. As the switching frequencies are high, the implementation of the four-step switching strategy of Figure 15.12 requires very fast controller and switches.

The main advantages of matrix converters lie in their compact design (due to the elimination of the DC-link reactive element), the possibility of simultaneous power factor correction at the input and vector control at the output, and the high-quality waveforms on both sides. The main disadvantages of matrix converters are: high switching losses, complexity of control, lack of energy storage leading to practically zero ride-through capacity, and direct transfer of harmonic distortion and imbalance in the side-1 voltage and side-2 current to the side-2 voltage and side-1 current.

Based on Figure 15.9, the voltage and current transformations in an FCC are given by

$$\begin{bmatrix} v_{2A} \\ v_{2B} \\ v_{2C} \end{bmatrix} = [S] \begin{bmatrix} v_{1a} \\ v_{1b} \\ v_{1c} \end{bmatrix} \tag{15.1}$$

and

$$\begin{bmatrix} i_{1a} \\ i_{1b} \\ i_{1c} \end{bmatrix} = [S]^T \begin{bmatrix} i_{2A} \\ i_{2B} \\ i_{2C} \end{bmatrix}, \tag{15.2}$$

where $[S]$ is the existence matrix composed of time-varying entries that assume binary values of 1 or 0 for on or off state of the nine bidirectional switches, respectively. As far as the fundamental components of the voltages and currents on both sides are concerned, the matrix $[S]$ can be replaced by the transformation matrix $[H]$ whose elements are the local averages of the corresponding elements of the $[S]$ matrix. The structure of a typical $[H]$ matrix is as follows [12]:

$$H = \begin{bmatrix} h_{11} & h_{12} & h_{13} \\ h_{21} & h_{22} & h_{23} \\ h_{31} & h_{32} & h_{33} \end{bmatrix} = 1/3 \begin{bmatrix} 1 & 1 & 1 \\ 1 & 1 & 1 \\ 1 & 1 & 1 \end{bmatrix} \tag{15.3}$$

$$+ M_1 \begin{bmatrix} \cos[(\omega_1+\omega_2)t+\gamma] & \cos[(\omega_1+\omega_2)t-120°+\gamma] & \cos[(\omega_1+\omega_2)t+120°+\gamma] \\ \cos[(\omega_1+\omega_2)t-120°+\gamma] & \cos[(\omega_1+\omega_2)t+120°+\gamma] & \cos[(\omega_1+\omega_2)t+\gamma] \\ \cos[(\omega_1+\omega_2)t+120°+\gamma] & \cos[(\omega_1+\omega_2)t+\gamma] & \cos[(\omega_1+\omega_2)t-120°+\gamma] \end{bmatrix}$$

$$+ M_2 \begin{bmatrix} \cos[(\omega_1-\omega_2)t-\gamma] & \cos[(\omega_1-\omega_2)t+120°-\gamma] & \cos[(\omega_1-\omega_2)t-120°-\gamma] \\ \cos[(\omega_1-\omega_2)t-120°-\gamma] & \cos[(\omega_1-\omega_2)t-\gamma] & \cos[(\omega_1-\omega_2)t+120°-\gamma] \\ \cos[(\omega_1-\omega_2)t+120°-\gamma] & \cos[(\omega_1-\omega_2)t-120°-\gamma] & \cos[(\omega_1-\omega_2)t-\gamma] \end{bmatrix}.$$

Note that for the elements of $[H]$ matrix to represent the local averages of the elements of $[S]$ matrix, the following existence condition must be satisfied [13]:

$$0 \le h_{ij} \le 1 \quad \text{for} \quad i = 1, 2, 3 \quad \text{and} \quad j = 1, 2, 3 \tag{15.4}$$

Also, for one and only one of the switches connecting the three phases on side-1 to each phase on side-2 to be on at any instant, the following condition must be met [13]:

AC/AC Converters

$$\sum_{j=1}^{3} h_{ij} = 1 \quad \text{for} \quad i = 1, 2, 3 \tag{15.5}$$

Assuming that

$$\begin{bmatrix} v_{1a} \\ v_{1b} \\ v_{1c} \end{bmatrix} = \begin{bmatrix} V_{1m} \cos \omega_1 t \\ V_{1m} \cos(\omega_1 t - 120°) \\ V_{1m} \cos(\omega_1 t + 120°) \end{bmatrix}, \tag{15.6}$$

one can find

$$\begin{bmatrix} v_{2A} \\ v_{2B} \\ v_{2C} \end{bmatrix} = [H] \begin{bmatrix} v_{1a} \\ v_{1b} \\ v_{1c} \end{bmatrix} = 3/2 \, (M_1 + M_2) V_{1m} \begin{bmatrix} \cos(\omega_2 t + \gamma) \\ \cos(\omega_1 t - 120° + \gamma) \\ \cos(\omega_1 t + 120° + \gamma) \end{bmatrix} \tag{15.7}$$

The voltage gain of the matrix converter can therefore be expressed as

$$G_v = \frac{V_{2m}}{V_{1m}} = 3/2 \, (M_1 + M_2) \tag{15.8}$$

Also, assuming that

$$\begin{bmatrix} i_{2A} \\ i_{2B} \\ i_{2C} \end{bmatrix} = I_{2m} \begin{bmatrix} \cos(\omega_2 t + \gamma + \phi_2) \\ \cos(\omega_2 t - 120° + \gamma + \phi_2) \\ \cos(\omega_2 t + 120° + \gamma + \phi_2) \end{bmatrix}, \tag{15.9}$$

one can find

$$\begin{bmatrix} i_{1a} \\ i_{1b} \\ i_{1c} \end{bmatrix} = [H]^T \begin{bmatrix} i_{2A} \\ i_{2B} \\ i_{2C} \end{bmatrix} = I_{1m} \begin{bmatrix} \cos(\omega_1 t + \phi_1) \\ \cos(\omega_1 t + \phi_1 - 120°) \\ \cos(\omega_1 t + \phi_1 + 120°) \end{bmatrix}, \tag{15.10}$$

where

$$I_{1m} = 3/2 \, I_{2m} \sqrt{M_1^2 + M_2^2 + 2 M_1 M_2 \cos 2\phi_2} \tag{15.11}$$

and

$$\phi_1 = -\tan^{-1} \left(\frac{M_1 - M_2}{M_1 + M_2} \tan \phi_2 \right). \tag{15.12}$$

From Equation 15.8 and Equation 15.12, one can find the values for M_1 and M_2 that satisfy the specifications on the desired voltage gain and side-1 displacement angle as follows:

$$M_1 = 1/3\, G_v \left(1 - \frac{\tan \phi_1}{\tan \phi_2}\right) \tag{15.13}$$

and

$$M_2 = 1/3\, G_v \left(1 + \frac{\tan \phi_1}{\tan \phi_2}\right). \tag{15.14}$$

As seen, the conventional 9-bidirectional-switch matrix converter has 3 levers of control; i.e., M_1, M_2, and γ. M_1 and M_2 are used to control the magnitude of the side-2 voltage and the side-1 displacement power factor, whereas γ is used to control the phase angle of the side-2 voltage. All together, the three levers of control are able to control the active power flow as well as the side-1 and side-2 reactive powers independent of each other.

Common control techniques applied to conventional matrix converter topology include scalar control methods that are based on the instantaneous values of the input voltages and transfer function analysis [12,14,15], and space vector modulation [16]. Even though based on Equation 15.8, the maximum attainable voltage gain of the matrix converter, within the limits imposed by Equation 15.4, is 1, due to physical constraints and regardless of the switching strategy used, the highest gain that can be achieved in practice is 0.866 [10]. This gain can be improved to 1.05 using space vector modulation involving overmodulation at the price of higher harmonic contents and higher filter requirements [10]. The principle of matrix conversion has been also used to realize frequency changers made up of unidirectional switches only [17–19]. In these frequency changers, the inherent switching problems of conventional matrix converter is avoided.

As far as the specific application of hybrid electric vehicle is concerned, matrix converters provide input power factor correction, output vector control, bidirectional power flow, compact design (i.e., high-power density), high-quality waveforms, and unrestricted frequency conversion. The fact that matrix converters, as well as NCCs, lack ride-through capacity is not a serious issue in the specific application of parallel HEV designs, as the ICE would be able to supply the power directly to the wheels in case of temporary interruption in the electric power. The fact that there is no place for the integration of an energy storage battery in the matrix converter and NCC system, without an additional converter, remains a disadvantage. However, if a flywheel is used as the energy storage device, this problem can be solved [20]. Figure 15.13 and Figure 15.14 show two possible serial and parallel HEV designs based on matrix converter and NCC.

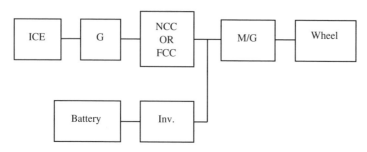

Figure 15.13 Potential serial HEV design using NCC or FCC (matrix converter).

AC/AC Converters

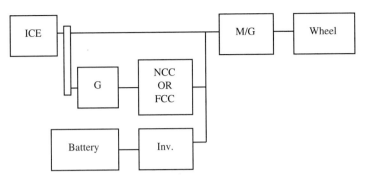

Figure 15.14 Potential parallel HEV design using NCC or FCC (matrix converter).

Note that in all the HEV configurations based on AC/AC converters, the internal combustion engine can be replaced or assisted by a flywheel that stores mechanical energy and can be recharged through an external circuit, ICE, battery, or wheels (during regenerative breaking) [20]. Also, a hydrogen-fueled fuel cell, as a source of electrical energy, can be incorporated in the system.

15.3 SUMMARY

With the evolution of more electric vehicles, the role of power electronics as a power conditioning tool becomes more vital. Hybrid electric vehicles, due to their high efficiency, flexibility, and reliability, are receiving more attention compared with totally electric vehicles. This chapter introduces the structures, principles of operation, advantages, and disadvantages of the three main types of AC/AC converters, i.e., rectifier-inverter pair, naturally commutated cycloconverter, and forced-commutated cycloconverter (or matrix converter), and their potential roles in the automotive industry, especially in hybrid electric vehicles.

The factors affecting the choice of a certain AC/AC converter topology are the cost, weight, size, waveform quality, efficiency, robustness, reliability, filter requirements, ride-through capability, immunity to harmonics distortion and imbalance, reactive power requirements, control complexity, restriction of the input/output frequency ratio, and input power factor correction capability. Even though the rectifier-inverter pair and NCC topologies are well-established and accepted by the industry, there is a high potential for other types of AC/AC converters.

Matrix converters with unrestricted frequency conversion ability, compact design, and input power factor correction capability seem to be a good candidate for AC/AC conversion in hybrid electric vehicles. The lack of ride-through capability will not be a serious issue in the hybrid electric vehicle of parallel design, where a temporary interruption in the electric power can be compensated by the internal combustion engine under a proper control regime. However, the success of matrix converters will depend on the availability of monolithic bidirectional switches. Soft-switched high-frequency converters are another eligible candidate for AC/AC converters in the automotive industry due to their high efficiency and high-power density. However, they have to prove to be cost-effective and reliable before winning against more established competitors.

REFERENCES

[1] P. Bowler. The Application of a Cycloconverter to the Control of Induction Motors. *Proceedings of IEE Conference on Power Applications of Controllable Semiconductor Devices*, No. 17, pp. 137–145, 1965.

[2] L. Malesani, L. Rossetto, P. Tenti, P. Tomasin. AC/DC/AC PWM Converter with Reduced Energy Storage in the DC Link. *IEEE Transactions in Industry Applications*, Vol. 31, No. 2, pp. 287–292, 1995.

[3] S. Bhowmik, R. Spee. A Guide to the Application-Oriented Selection of AC/AC Converter Topologies. *IEEE Transactions on Power Electronics*, Vol. 8, No. 2, pp. 156–163, 1993.

[4] L.J. Jacovides, M.F. Matouka, D.W. Shimer. A Cycloconverter-Synchronous Motor Drive for Traction Applications. *IEEE Transactions on Industry Applications*, pp. 549–561, 1980.

[5] L.J. Lawson. The Practical Cycloconverter. *IEEE Transactions on Industry and General Applications*, Vol. IGA-4, No. 2, pp. 141–144, 1968.

[6] J. Vithayathil. *Power Electronics: Principles and Applications.* New York: McGraw-Hill, pp. 519–527, 1995.

[7] G.K. Dubey. *Power Semiconductor Controlled Drives,* 1st Ed. Englewood Cliffs, NJ: Prentice-Hall, pp. 345–349, 1989.

[8] L. Gyugyi, B.R. Pelly. *Static Power Frequency Changers: Theory, Performance, and Application.* New York: John Wiley & Sons, 1976.

[9] S. Bernet, T. Matsuo, T.A. Lipo. A Matrix Converter Using Reverse Blocking NPT-IGBTs and Optimized Pulse Patterns. *Proceedings of IEEE Power Electronics Specialists Conference (PESC),* Baveno, Italy, pp. 107–112, 1996.

[10] M.H. Rashid. *Power Electronics: Circuits, Devices, and Applications,* 3rd Ed. Englewood Cliffs, NJ: Pearson Prentice-Hall, pp. 536–537, 2003.

[11] N. Burany. Safe Control of 4-Quadrant Switches. *Proceedings of IEEE Industry Applications Society Annual Meeting,* San Diego, CA, pp. 1190–1194, 1989.

[12] M. Venturini, A. Alesina. The Generalized Transformer: A New Bidirectional Sinusoidal Waveform Frequency Changer with Continuously Adjustable Input Power Factor. *Proceedings of IEEE Power Electronics Specialists Conference*, pp. 242–252, 1980.

[13] D.G. Holmes. A Unified Modulation Algorithm for Voltage and Current Source Inverters Based on AC-AC Matrix Converter Theory. *IEEE Transactions on Industry Applications*, Vol. 28, No. 1, pp. 31–40, 1992.

[14] G. Roy, G.E. April. Direct Frequency Changer Operation Under a New Scalar Control Algorithm. *IEEE Transactions on Power Electronics*, Vol. 6, No. 1, pp. 100–107, 1991.

[15] A. Ishiguro, T. Furuhashi, S. Okuma. A Novel Control Method for Forced-Commutated Cycloconverters using Instantaneous Values of Input Line-to-Line Voltages. *IEEE Transactions on Industrial Electronics*, Vol. 38, No. 3, pp. 166–172, 1991.

[16] L. Huber, D. Borojevic. Space Vector Modulated Three-Phase to Three-Phase Matrix Converter with Input Power Factor Correction. *IEEE Transactions on Industry Applications*, Vol. 31, pp. 1234–1246, 1995.

[17] M. Kazerani, B.T. Ooi. Feasibility of both Vector Control and Displacement Factor Correction by Voltage Source Type AC-AC Matrix Converter. *IEEE Transactions on Industrial Electronics*, Vol. 42, No. 5, pp. 524–530, 1995.

[18] M. Kazerani. A Direct AC/AC Converter Based on Current-Source Converter Modules. *IEEE Transactions on Power Electronics*, Vol. 18, No. 5, pp. 1168–1175, 2003.

[19] S. Kim, S.K. Sul, T.A. Lipo. AC to AC Power Conversion Based on Matrix Converter Topology with Unidirectional Switches. *Proceedings of IEEE Applied Power Electronics Conference,* Anaheim, CA, pp. 301–307, 1998.

[20] G.J. Hoolboom, B. Szabados. Nonpolluting Automobiles. *IEEE Transactions on Vehicular Technology*, Vol. 43, No. 4, pp. 1136–1144, November 1994.

16

Power Electronics and Control for Hybrid and Fuel Cell Vehicles

Kaushik Rajashekara
Delphi Corporation, Kokomo, Indiana

16.1 INTRODUCTION

Power electronics is an enabling technology for the development of propulsion systems for hybrid and fuel cell vehicles. In electric, hybrid, and fuel cell vehicles, the challenges are to have a high-efficiency, rugged, smaller, and lower cost inverter and the associated electronics for controlling a three-phase electric machine. Many of the requirements related to power converter and the propulsion motor control strategies of electric and hybrid vehicles apply for fuel cell vehicles also. In addition, in fuel cell vehicles, the power converters match the fuel cell voltage to the propulsion system voltage so that the propulsion system can be designed independent of the fuel cell voltage. This would lead to operation of each system at its most efficient design levels. The power electronics also has a major role in the 42 V architecture-based vehicles. In this chapter, the operating strategies of hybrid and fuel cell vehicles are described with the associated power electronics required for these systems.

16.2 HYBRID ELECTRIC VEHICLES

Hybrid electric vehicles (HEVs) have two or more sources of energy or two or more sources of power on-board the vehicle. The sources of energy can be battery, flywheel, and so on. The sources of power can be engine, fuel cell, battery, ultracapacitor, and so on. Depending on the configuration of the vehicle, two or more of these power or energy sources are used. These hybrid vehicles can reduce the air pollution significantly through increased fuel economy, use of alternative fuels, improved power unit, and after-treatment technology [1]. Hybrid vehicles offer the following benefits:

- Greatest potential for improved fuel efficiency and performance in the mid-term, hence, potential improvements to fuel economy
- Eliminates idle fuel consumption during deceleration and stops
- Operates engine at more efficient points
- On-board power generation using auxiliary power units; hence, improved accessory power generating efficiency
- Addresses carbon dioxide emission reduction through improved fuel efficiency and reduced hydrocarbons and nitrogen oxides, using smaller internal combustion engines
- Uses regenerative braking to recover the kinetic energy of the vehicle
- Enables engine size reduction with the same vehicle performance due to additional power available from the electric motor
- Incorporates technologies that are well known and can move into production quickly
  - Electric motors
  - Internal combustion engines (ICE)
  - Batteries
- Added customer features and benefits including customer differentiation and adaptation

Hybrid vehicles are generally classified as series hybrids and parallel hybrids. In a series hybrid vehicle, generally, an engine drives the generator to produce the power to charge the batteries and to provide power to the propulsion motor. In a parallel hybrid vehicle, the engine and the electric motor are used to drive the vehicle.

The series hybrid vehicle provides more possibilities for the development of low fuel consumption and low emission vehicles but it needs higher power and more efficient motors, an additional generator, a smaller IC engine, and a battery with a high power rating. Adding the cost of all these items could result in an expensive vehicle. The parallel hybrid can offer the lowest cost and the option of using existing manufacturing capability for engines, batteries, and motors. But the parallel vehicle needs complex control systems. Series hybrid vehicles offer lower fuel consumption in the city driving cycle, and parallel hybrid vehicles have lower fuel consumption in the highway driving cycle. There are various configurations of the parallel hybrid vehicles depending on the role of the electric motor/generator and the engine. Hence, hybrid vehicles are also classified as mild hybrids, power hybrids, and energy hybrids depending on the role played by the engine and the electric motor, and the mission that the system is designed to achieve.

16.2.1 SERIES HYBRID VEHICLE PROPULSION SYSTEM

A typical series hybrid propulsion system configuration is shown in Figure 16.1. A series hybrid vehicle is essentially an electric vehicle with an on-board source of power, also called auxiliary power unit (APU), for charging the batteries. Generally, an engine is coupled to a generator to produce the power to charge the batteries. It is also possible to design the system in such a way that the generator could act as a load-leveling device providing propulsion power. In this case, the size of the batteries could be reduced but the size of the generator and the engine need to be increased. The power electronic components required for a typical series hybrid vehicle system are a converter for converting

Power Electronics and Control for Hybrid and Fuel Cell Vehicles

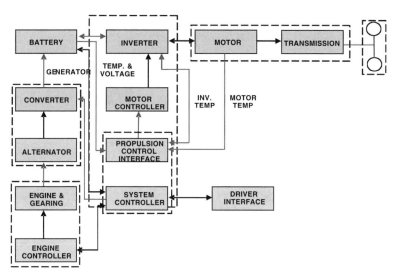

Figure 16.1 Series hybrid vehicle propulsion system.

the generator output to DC for charging the batteries and an inverter for converting the DC to AC to power the propulsion motor. A DC-DC converter is required to charge the 12 V battery in the vehicle. In addition, an electric air-conditioning unit needs an inverter and associated control systems.

The engine driving the generator could be an IC engine, Stirling engine, microturbine, diesel engine, or a natural gas engine. The generator is generally a three-phase permanent magnet or an induction-type machine. The output of the generator is converted to DC using a three-phase bridge-controlled rectifier or using an insulated gate bipolar transistor (IGBT) bridge. The control system for the engine and the generator has to be designed in such a way that the engine operates at the optimum speed to obtain the highest possible efficiency. Depending on the state of charge of the batteries, the output of the converter has to be controlled. Generally, the state of charge of the batteries is kept below about 80% to allow room for the batteries to be charged using regeneration. Also, it is desirable to turn off the engine in the urban areas to operate as a zero emission vehicle. While selecting the APU, consideration should be given to the following items.

- The power required for a prime mover (or engine) to perform the APU function is much lower than that required to perform the propulsion function.
- The power required for an APU unit is about 25 kW to 40 kW depending on the size of the vehicle.
- The prime mover should emit lowest possible emissions.
- The rating of the alternator (or generator) depends on whether the system is range extender type series vehicle or the load leveling type vehicle, where the propulsion power is obtained both from the battery and the APU.

16.2.2 Parallel Hybrid Vehicle Propulsion System

In the parallel hybrid vehicle, the engine and the electric motor can be used separately or together to propel a vehicle. A typical parallel hybrid vehicle propulsion system configuration is shown in Figure 16.2. In this system, the engine is disconnected during starting

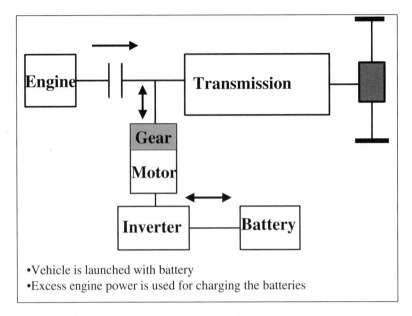

Figure 16.2 A typical parallel hybrid propulsion system.

and deceleration, and vehicle is run using the electric motor. The electric motor, powered by the battery, also assists the engine when the transmission is in high gear. When the engine has excess power, the motor acts as a generator and charges the batteries. The Toyota Prius and the Honda Civic are two examples of parallel hybrid systems that are commercially available [2]. The parallel hybrid can offer the lowest cost and the option of using existing manufacturing capability for engines, batteries, and motors.

16.2.2.1 Toyota Prius

Toyota Prius [3] is an advanced technology parallel hybrid vehicle. It has a high-efficient 1.5 liter engine, a power split device, continuously variable drive train, permanent magnet generator and the PM propulsion motor, two IGBT inverters, and high-power NiMH battery pack. The main attributes of the Toyota Prius are 52 mpg, City; 45 mpg, Highway; SULEV operation; 0–60 mph in approximately 12.5 seconds; and 100 mph top speed. Toyota Prius system has a power split device in the transmission that sends engine power either directly to the wheels or the electric generator. The power split device uses a planetary gear to constantly vary the amount of power supplied from the engine to either the wheels or generator. The transmission is controlled electronically to control engine speed, generator output, and the speed of the electric motor to handle the operation in different driving modes. The system is designed to keep the engine running within its most efficient speed range. Its five main operating modes, which are shown in Figure 16.3, are [3–4]:

1. When pulling away from a stop or under a light load, only the electric motor powers the vehicle.
2. Under normal driving operation, a combination of gasoline and electric power is used. The engine drives the generator and provides power to run the electric motor. The excess power of the generator is used for charging the battery.

Power Electronics and Control for Hybrid and Fuel Cell Vehicles 351

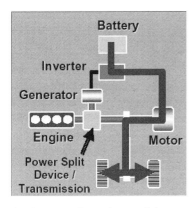
Starting or moving under very light load conditions

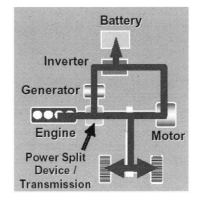
Normal driving operation

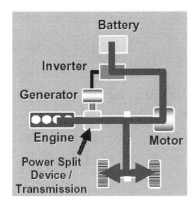
Full throttle acceleration

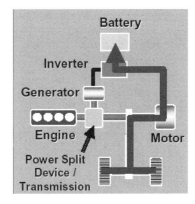
Deceleration and braking conditions

Figure 16.3 Operating modes of the Prius vehicle [3].

3. Under full-throttle acceleration, the full power of the engine and the electric motor, powered by the battery, are used together to provide the maximum torque.
4. During deceleration or braking, the electric motor functions as a generator to recharge the batteries.
5. When charging of the battery is needed, power from the engine is used to drive the generator. This eliminates the need for an external charger or power connection.

The 2004 Toyota Prius is the first Toyota equipped with the new high-voltage/high-power full Hybrid Synergy Drive powertrain [4]. The new drive has a 50% more powerful 50 kW drive-motor, operating at up to 500 V. The generator has a higher peak operating speed that increases electric-mode operation in city and freeway slow-and-go operation. With 50% more electric power available and improved low-end torque from the drive motor, a significant boost in acceleration performance is possible. The Hybrid Synergy Drive enables Prius to be nearly 30% lower in emissions than the first-generation Prius and has higher fuel economy.

16.2.2.2 Crankshaft-Mounted Integrated Starter-Generator System

Many automotive companies are working on the development of crankshaft-mounted integrated starter-generator (ISG) system-based hybrid vehicles. The ISG concept offers the ability to reduce fuel consumption through the use of engine-off during coast-down and idle, early torque converter lockup with torque smoothing, regenerative braking, and electric launch assist. The feature stop-start, which means IC engine off at idle, is integrating the quiet starting and the high power generation into one single machine. This specific feature offers high potential for reducing fuel consumption, exhaust, and noise. In addition, the ISG provides the capability for generating higher power than today's conventional automotive alternators. This higher power would enable features such as electric power steering, electric HVAC, electric valve trains, mobile AC power, and many entertainment features. Delphi Corporation has built and tested an SUV equipped with an Energen-10 ISG system [5,7], shown in Figure 16.4. The vehicle has a parallel hybrid architecture in which the electric machine and IC engine can each provide torque to the drive wheels separately or simultaneously. The electric machine assists the IC engine by providing additional torque in the operating regions where the engine is less efficient. The Energen-10 system replaces the conventional vehicle's flywheel, alternator, and starter motor with an electric machine that fits between the engine and transmission. The system has a power generation capability in the 5 to 10 kW range (hence the 10 in its name). The electric power take-off (PTO) function can provide on-board electric power for powering the appliances on the fly and when the vehicle is parked. The PTO consists of a single phase inverter for converting 42 VDC to 120 V/240 AC power. The typical rating of the inverter is about 2.4 kVA.

The requirements in respect to vehicle starting mode can be very different from the generation mode. The diagrams in Figure 16.5 describe the current level requirements for a 5 kW induction machine in both modes. In order to generate the specified power level at a temperature of 125°C and at a 42 V system voltage, the maximum AC current level is almost reaching 150 A [6]. For starting the engine, a starting torque of about 200 Nm is required at the worst-case condition, i.e., at a temperature of −30°C. To provide that torque with this machine design, stack length, and rotor diameter, a current of more than 450 A has to be supplied. The result is that, between generator and motor functionality, the current level has to be raised by a factor of 3. Although the current requirements for

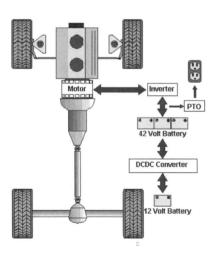

Figure 16.4 ISG based on Energen-10® system architecture (courtesy Delphi Corporation).

Power Electronics and Control for Hybrid and Fuel Cell Vehicles

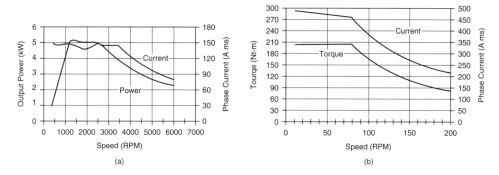

Figure 16.5 (a) Generator requirements at 125°C; (b) starting requirements at −30°C (courtesy Delphi Corporation).

the silicon power devices is low during generation mode, they need to be designed to meet the requirements of starting current. In addition, the battery has to be able to provide that amount of electrical power at the respective ambient temperatures.

16.2.2.3 Side-Mounted Integrated Starter-Generator

Recently, there has been an increasing interest in the side-mounted ISG, that is, the belt-driven integrated starter-generator system [9]. A typical side-mounted starter/generator system based on 42 V architecture is shown in Figure 16.6 [10]. The side-mounted ISG can be realized using the conventional generator of today's vehicle. With the addition of position sensors and a three-phase inverter, the generator can be operated as a motor and can provide enough torque through the belt to the combustion engine to perform a fast and a quiet restart for a warmed-up engine. On smaller engines, it is possible to cold crank the engine, eliminating the conventional starter. Further improvements in the generator and power electronics technology will increase the system efficiency, the power generation, and the cranking torque to fulfill future requirements and allow also a cold-cranking of larger engines. The benefits of this system are low cost, simple implementation, minimal changes in the electrical system, and use of the present belt-driven machine.

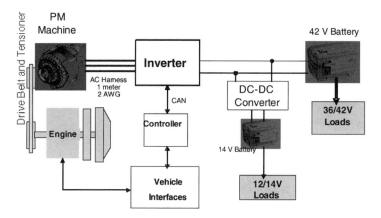

Figure 16.6 Side-mounted starter/generator system.

16.3 FUEL CELL VEHICLES

With the advancement in the technology of fuel cells, there is an increasing interest in automotive industry for using fuel cells for propulsion and for on-board power generation. The advantages of fuel cell vehicles compared to internal combustion engine vehicles are [11]:

- Direct energy conversion (no combustion).
- No moving parts in the energy converter, quiet, and fuel flexibility.
- The fuel cell vehicles can dramatically lower energy use, lower air pollution, and increase the use of alternative fuels.
- The fuel cell efficiency does not decrease sharply as the size of the system is decreased.
- The fuel cell efficiency does not appreciably change if the fuel cell operates at part load.
- Under comparable road load conditions, the fuel cell efficiency is significantly greater than the efficiency of internal combustion engines, especially at part load. At a nominal driving speed of 30 mph, the efficiency of fuel cell electric drive using hydrogen from natural gas is about 2 times higher than that of a conventional engine.

Replacement of ICE with a fuel cell system could save 60% of the primary energy consumption; the CO_2 emission can be reduced by about 75%; and release of toxic substances could be largely reduced. Various types of fuel cells are in the development stage. The proton exchange membrane (PEM, also called polymer electrolyte membrane) and solid oxide fuel cells (SOFC) are mainly considered for automotive applications. The PEM fuel cells are gaining importance as the fuel cell for propulsion applications because of their low operating temperature, higher power density, specific power, longevity, efficiency, relatively high durability, and the ability to rapidly adjust to changes in power demand. The PEM fuel cell operates at about 100°C and has a faster response time for load changes; also, the system can be started in less than a minute. Several companies are developing PEM fuel cells for propulsion applications. The PEM is more suitable for automotive applications for the following reasons:

- PEM can be started easily at ordinary temperatures and can operate at relatively low temperatures, below 100°C.
- Since they have relatively high power density, the size could be smaller. Hence, they could be easily packaged in the vehicles.
- Because of the simple structure compared to other types of fuel cells, their maintenance could be simpler.
- They can withstand the shock and vibrations of the automotive environment because of their composite structure.

But the PEM system has the following disadvantages:

- PEM fuel cell requires pure hydrogen as the fuel, thus complicating the design of the reformer system.
- Any small amount of carbon monoxide in the fuel will poison the electrodes, resulting in severe degradation of performance.

- As there is a continuous generation of water at the cathode and also the requirement of certain level of humidification, a sophisticated water management system is required.
- Platinum metal is required to coat the electrodes to enhance the reactions. Because of the higher cost of platinum, the PEM system is relatively expensive

Traditionally, fuel cells have been mainly considered for the propulsion applications. But recently, they are also being considered for on-board power generation as auxiliary power units to provide the power to the accessory loads during both engine on and off conditions. High-temperature solid oxide fuel cell is particularly suitable for automotive APU applications and also as a range extender in series hybrid vehicles, instead of an engine-driven generator. The advantages of the SOFC system are [12,13]:

- The fuel processor requires a simple partial oxidation reforming process that eliminates the need for an external reformer.
- SOFC has less stringent requirements for reformate quality and uses carbon monoxide directly as a fuel. Hence, a sophisticated reformer is not required.
- SOFC can operate at extremely high temperatures in the order of 700 to 1000°C. As a result, it can tolerate relatively impure fuels, such as those obtained from the gasification of coal.
- Waste heat is high-grade, allowing for smaller heat exchangers and the possibility of co-generation to produce additional power.
- Water management is not a concern because the electrolyte is solid-state and does not require hydration. The by-product is steam rather than liquid water; hence, no need for water management.
- SOFC does not need precious metal catalysts.

The disadvantages of a SOFC system are:

- Because of the high-temperature operation, the starting time of the system is of the order of several minutes. For a 5 kW system, it is of the order of 20 to 30 minutes. Hence, SOFC is not suitable for propulsion applications.
- Packaging of the low-temperature electronics and the high-temperature stack within the same enclosure is a major challenge.

16.3.1 Fuel Cell Vehicle Propulsion System

A fuel cell system designed for vehicular applications must have weight, volume, power density, start-up, and transient response similar to the present-day internal combustion engine-based vehicles. Other requirements are very high performance for short time, rapid acceleration, good fuel economy, easy access, and safety considerations with respect to fuel handling. Cost and expected lifetime are also very important. In order to obtain high-efficiency and high-performance characteristics from a fuel cell-based propulsion system, it is very important to have the best possible system architecture and the control strategy. A typical fuel cell vehicle system is shown in Figure 16.7. The fuel (gasoline or diesel) is processed inside the fuel processor, also called reformer, to obtain the required hydrogen as input to the fuel cell stack. The oxygen required for the fuel cell is generally drawn from the external air. Inside the fuel cell stack, the hydrogen and oxygen are combined to produce direct current electricity and heat. The output voltage of the stack is conditioned using a power conditioner to obtain the required voltage to the inverter. An inverter is used

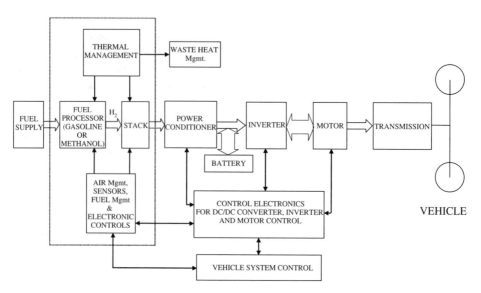

Figure 16.7 A typical fuel cell vehicle propulsion system.

to convert the DC to variable voltage and variable frequency to power the propulsion motor. A battery or an ultracapacitor is generally connected across the fuel cell system to provide supplemental power and for starting the system. A complete fuel cell system consists of several of the following components:

- Reformer to convert the fuel to hydrogen-rich gas, or if it is a direct hydrogen system, a compressed hydrogen storage tank is needed.
- Fuel cell power section, which consists of stacks of fuel cells where the hydrogen gas and oxidants are mixed to produce direct current electricity and heat.
- Air compressor to provide pressurized oxygen to the fuel cell.
- Cooling system to maintain the proper operating temperature.
- Water management system to manage the humidity and the moisture in the system (to keep the fuel cell membrane saturated and at the same time prevent the water being accumulated at the cathode).
- Power conditioner to condition the output voltage of the fuel cell stack.
- Inverter to convert the DC to variable voltage and variable frequency to power the propulsion motor.
- Propulsion motor and transmission.
- Battery or ultracapacitors to provide supplemental power and for starting the system.

A fuel cell propulsion system with a battery pack and a power conditioner is shown Figure 16.8. The battery unit and the fuel cell stack supply the power required for propulsion. If the propulsion unit is designed for a higher voltage than the fuel cell voltage, the power conditioner has to boost the fuel cell stack voltage to the required battery voltage of about 300 V. The power conditioner also charges the propulsion battery. The power conditioner has to be sized based on the maximum power capability of the fuel cell stack. The diode at the output of the fuel cell stack is necessary to prevent the negative current going into the stack. If the negative current is allowed, it is possible that cell reversal could

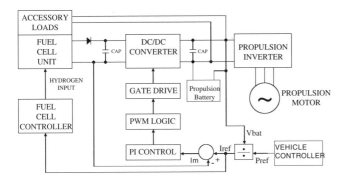

Figure 16.8 Fuel cell converter control system.

occur and damage the fuel cell stack. The ripple current seen by the fuel cell stack due to the switching of the power devices inside the power conditioner has to be low.

The power conditioner controls the output power provided by the stack to the load. The power command is proportional to the required power and is divided by the battery voltage to derive the current reference. The current reference is compared with the measured current and the error is amplified and integrated to derive the duty cycle for controlling the output power of the power conditioner. Controlling the output current of the fuel cell stack controls the power drawn from the fuel cell. This is because the amount of hydrogen generated if reformer is used (or the amount of hydrogen input to the stack in the case of direct hydrogen system) could be better controlled if the fuel cell stack output current is directly controlled. In this control scheme, for a constant current at the stack output, the stack voltage is also constant, and thus the power at the stack output remains constant, for a given operating pressure and temperature. Hence, the power conditioner output power will also be constant. This control scheme avoids the wide variation in the fuel input to the stack. In addition, it enables constant current load that is ideal for fuel cell operation, and constant power at the output of the power conditioner that is optimum for fuel cell hybrid vehicle operation.

The fuel cell can be designed for load sharing operation or for range extender operation. A range extender-type fuel cell could be designed for lower power to only charge the batteries. The battery needs to be designed to provide the full power. Because of the favorable efficiency curve of the fuel cell unit in the partial load range, a system with a smaller battery and a full-power fuel cell stack appears to be more attractive. However, if cost is the major concern, smaller fuel cell and a larger battery could be the better choice. In this type of application, it is possible to use the solid oxide fuel cell instead of the PEM fuel cell. The starting time and the response time of the fuel cell is not a major factor.

Figure 16.9 shows a configuration in which the battery pack voltage is lower than the DC bus voltage of the inverter. The battery is connected to the inverter DC bus through a DC-DC converter. When the vehicle is started, the power to the propulsion motor is provided from the battery by boosting the battery voltage. During rapid acceleration, power is provided by the fuel cell and the battery. Once the vehicle reaches the steady speed, only the fuel cell will be providing the propulsion power (it also charges the battery). In this case, the DC-DC converter will be operating in the buck mode. During regeneration, the battery is charged and the fuel cell will not be providing any power. A similar configuration has been used in the Toyota fuel cell vehicle [25].

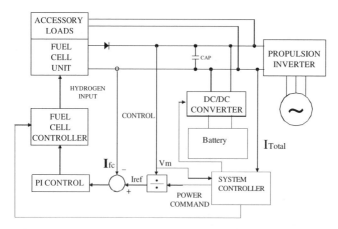

Figure 16.9 Fuel cell system with lower voltage battery and DC-DC converter.

16.3.2 FUEL CELL VEHICLE PROPULSION SYSTEM CONSIDERATIONS

Some of the important issues to be considered in the design of the fuel cell propulsion system are [11,14]:

- System DC voltage. The optimum DC voltage for the stack and for the propulsion drive determines the number of cells to be connected in series for the stack.
- Rate of increase of the output power of the fuel cell stack. Due to the sudden application of the load, the fuel cell may not respond instantly because of the requirement of additional fuel flow and also the change in the rate of fuel flow. If the amount of hydrogen flow to the stack is higher than that required by the electrical load, then energy is wasted in the exhaust. If the fuel flow is less than that required by the electrical load, then the impedance of the stack increases, thus overheating the stack. Hence, it is necessary to match the amount of hydrogen flow to the stack to meet the desired electrical load at the output.
- The sequence of shutting off the entire system.
- Effect on the fuel input if the load is suddenly disconnected.
- Coordination between subsystems for optimum operation.
- Isolation of the fuel cell stack from the drive system.
- Connecting the battery and the fuel cell stack together as one system.
- Charging the capacitors of the inverter and limitation of capacitor inrush current.
- Coordination of battery charging simultaneously using regenerative energy and from the fuel cell.
- Charging of battery from the fuel cell stack alone.
- Coordination of the power delivered from the battery and from the fuel cell stack, particularly if the power conditioner is not used.
- Limiting of battery current during regeneration and charging at the same time.
- Supplying the power to the accessory loads of the vehicle and to the accessory loads of the fuel cell stack.
- Matching the fuel cell output characteristics with the characteristics of the battery and the drive system.

16.4 POWER ELECTRONICS REQUIREMENTS [6,11,15]

The power switching devices, electric motors, and the associated control systems and components play a major role in bringing hybrid and fuel cell vehicles to market with reliability and affordability. The power electronic system should be efficient to improve the range of the electric vehicles and fuel economy in hybrid vehicles. The selection of power semiconductor devices, converters/inverters, control and switching strategies, packaging of the individual units, and the system integration are very important for the development of efficient and high-performance vehicles. Hence, to meet the challenges of the automotive environment, several technical challenges need to be overcome and new developments are needed from device level to system level. Some of the requirements for the major power devices in propulsion application are listed below.

Voltage Ratings: The voltage rating of the devices is based on the battery nominal voltage, the maximum voltage to which the battery is charged and the battery voltage during regenerative mode. If the nominal battery voltage is about 300 V, then the maximum voltage shall be about 370 V. During regeneration, the battery voltage may go up to some set limit, as high as 400 V. In this situation, power devices of 600 V continuous rating should be used. The factors to be considered are the end of charge voltage of the battery and the maximum allowed voltage during regeneration.

Current Ratings: The device power requirement is reflected to its current rating that is determined by the required output power and the number of devices connected in parallel. In electric propulsion applications, the peak power rating of the motor is about 2 to 4 times the continuous power rating. Due to the thermal limitations of the power semiconductor devices, the current rating of the power devices has to be based on the peak power rating of the propulsion motor. If a single device to carry all the current is not available, lower current rated devices could be connected in parallel. When paralleling the devices, on-state and switching characteristics have to be closely matched.

Switching Frequency Requirements: The switching frequency depends on the power of the motor being fed from the inverter. If the inverter is to be operated in **Pulse Width Modulation** (PWM) mode, the device should be able to be switched at a minimum frequency of 20 kHz, so that there would not be any acoustic noise from the inverter. However, it has been observed that switching frequency of about 10 kHz does not pose significant noise problem. Switching at higher frequencies would bring down the size of filters, if any are used. In addition, it will help to meet the EMI limitation requirements, particularly if the same inverter is used for charging the batteries also.

Power Loss Requirements: In electric and hybrid propulsion systems, achieving highest efficiency is a very important factor. The conduction losses in the device should be minimum. The device forward voltage drop, even at higher currents (> 400 A), must be less than 2 V and at the same time be able to be operated at switching frequencies higher than 10 kHz. Similarly, the switching losses should be as low as possible. Having low turn-on and turn-off times of the device could reduce the switching losses. Higher switching frequencies increase the losses in the power converter. On the other hand, lower switching frequencies, due to the higher amplitude of lower frequency components, increase the motor losses. Thus, switching frequencies of about 10 kHz would be an optimum for efficiency, noise, and EMI considerations. In order to minimize the off-state losses, the leakage current of the devices have to be less than 1 mA.

Dynamic Characteristic Requirement: The device should require very little energy to turn on and turn off, allowing a simple circuit to drive the device. The drive input capacitance of the device should be low and drive input resistance has to be very high (of the order of several M ohms). The device should have a large dv/dt (of the order of 20,000 V/µs) and high di/dt (of the order of 10,000 A/µs) capability. The devices should be able to be easily paralleled without the aid of external circuits or prematching of the devices. The antiparallel diode across the main device should have good dynamic characteristics, with low reverse recovery time and the same power handling capability as the main device.

Protection Requirement: The device should be rugged and be rated to withstand a specific amount of unclamped avalanche energy when operated above its maximum rated voltage. It should have some I^2t withstand capability and be able to be protected by fast semiconductor fuses with limiting impedance in series. The Safe Operating Areas (SOA) of the device should be such as to operate with no or minimal use of snubber circuits. The SOA area should be rectangular with boundaries at the maximum voltage and current. It would be preferable to have some self protecting capability in the device. At any instant, if the current through the device exceeds its maximum rated value, it should turn off by itself or its conducting time should become very narrow. This could be done by either internal current sensing or voltage drop sensing of the device.

Packaging Requirement: Devices are available in different packages. It is preferable to have an isolation package so that the devices could be mounted on a heat sink without concern for electrical isolation. Inductance of the package should be as low as possible. A packaging inductance of less than 10 nH would be desirable. The technologies related to device packaging are very important for developing a modular power switch. Wire bonding, device interconnections, etc. are the barriers to development of high current density power units. The technologies such as power connection without wire bonds or minimizing wire bonds, heat-sinking both sides of the die, and interconnect solutions for large scale manufacturing need further work. In addition to the power devices and controllers, several other components such as capacitors, inductors, bus bars, thermal system form a major portion of the power electronics unit. The packaging of all these units as one system has significant challenges.

Cycling and Reliability Requirements: In EV and HEV applications, as the vehicle is being frequently accelerated and decelerated, the devices are subjected to thermal cycling at frequent intervals. The power devices should reliably work under these conditions of stress.

The power converter topologies for propulsion applications are extensively covered in other chapters of this book. The trend is to have topologies with two or more integrated functions such as inverter, charger, and DC/DC converter, and with minimum use of capacitors. Integrated EMI filters for control of EMI generated due to switching of the devices needs to be part of the inverter/converter topology. Fault-tolerant topologies and control techniques need further investigation. The FreedomCAR goals for propulsion motor and power inverter are quite challenging and are given in Table 16.1, which includes the motor, inverter, gearbox, and controller [16].

16.5 PROPULSION MOTOR CONTROL STRATEGIES

The propulsion motor controller must be designed to meet the following requirements:

- **Four-Quadrant Operation:** Forward motoring, reverse motoring, forward regenerative braking, and reverse regenerative braking.
- **Speed:** Wide speed range. Standard range of 0 to about 13,000 rpm with a high-speed range of 0 to 16,000 rpm. (This depends on the type of the vehicle.)

Power Electronics and Control for Hybrid and Fuel Cell Vehicles

Table 16.1 FreedomCAR goals and Technical Targets for Power Electronics and Propulsion Motor (Integrated Inverter/Motor)

Characteristics	2010	2015
Peak power	55 kW for 18 sec	55 kW for 18 sec
Continuous power	30 kW	30 kW
Lifetime	> 15 years	> 15 years
Cost	$12/kW	$10/kW
Specific power at peak load	1.2 kW/kg	1.3 kW/kg
Volumetric power density	3.4 kW/L	3.5 kW/L
Efficiency (10–100% speed, 20% rated torque)	90%	95%

- **Zero Speed Operation:** Full torque must be available at zero speed for sufficient launch acceleration.
- **Efficiency:** The controller should be designed to operate the propulsion system at maximum efficiency in both motoring and regeneration modes. There are several technical challenges in developing an optimum controller that would operate the motor at highest efficiency under all operating conditions. Some of the challenges are:
  - Tracking the efficiency of the system on-line and operating at optimum efficiency point
  - Efficiency optimization within the limited time available in a control cycle
  - On-line compensation for the motor parameter variations
- **Torque:** The output torque must meet the system and motor design parameters throughout the entire speed range. The torque must be constant up to the base speed of the machine.
- **Power:** The power from base speed to maximum speed must be constant (constant power region).
- **Torque Command Hysteresis:** There shall be a hysteresis band available for the torque command when going from zero torque command to some initial value.
- **Fault Detection:** The motor controller must detect the faults and the information must be sent to the system controller and appropriate protective action has to be initiated. The common types of faults are: overvoltage, undervoltage, overcurrent, gate drive fault, motor overtemperature, and inverter overtemperature.
- **PI Gain:** The proportional and integral gains of the control loops of the motor controller must be adjustable for different motor applications.
- **Pulse Width Modulation:** The PWM strategy should be selected to have minimum harmonic content at the output of the inverter and the switching frequency be adjustable in the range of 10 kHz to about 20 kHz. The dead-band time of the PWM signals must also be adjustable.

In propulsion applications, generally the motor is operated in torque control mode. Propulsion motor determines the characteristics of the propulsion system and the controller. If the motor used is an induction motor, the torque of the motor is controlled using slip frequency control or vector control.

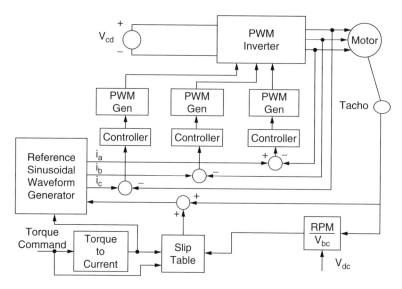

Figure 16.10 Sinusoidal PWM-slip frequency control of induction motor.

16.5.1 SLIP FREQUENCY CONTROL

There are several methods of controlling the slip frequency and the torque of the motor. Figure 16.10 shows a torque control scheme based on slip frequency look-up table and instantaneous current controllers. In the system shown, the induction motor is fed from a three-phase IGBT bridge inverter. The three-phase sinusoidal reference waveforms are generated in a reference waveform generator. The instantaneous values of the motor currents are subtracted from the reference currents, and the error waveforms are fed to proportional and integral controllers. The output of the PI controllers are also sinusoidal waveforms and are compared with a high-frequency triangular waveform to generate the PWM signals to control the switching time of the power devices. The slip frequency values optimized for a range of speed and torque values are stored in a look-up table. These values can be calculated using the motor parameter or from the experimental data obtained using the dynamometer tests. The slip frequency signal from the look-up table is added to the rotor speed values to obtain the stator frequency. The stator current depends on the commanded torque value. The implementation of this system is relatively simple. But the limitations are that the system performance is affected by the variation in motor parameters and temperature. Additionally, the slip frequency control system has a slower response time and creates larger transients than vector control system.

16.5.2 VECTOR CONTROL OF PROPULSION MOTOR

Vector control of an induction machine based on a standard indirect vector control method is shown in Figure 16.11. Direct axis and quadrature axis reference currents are derived from torque command, DC voltage, and motor speed. Each level of torque command at a given speed corresponds to a pair of Iq and Id values in the look-up table. In the event that the DC bus voltage varies from its nominal value, look-up table values are also correspondingly changed. The three-phase measured motor currents are transformed into synchronously rotating reference currents and are compared with the reference currents Iqs and Ids, and the errors are processed in PI controllers. The outputs of the two PI

Power Electronics and Control for Hybrid and Fuel Cell Vehicles

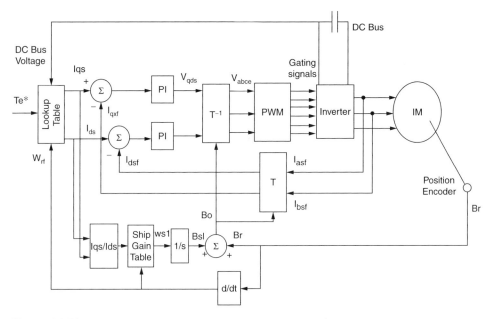

Figure 16.11 Vector control of induction machine in a propulsion system.

controllers are synchronous voltages Vqs and Vds, which are transformed into stationary reference frame currents by using the synchronous angle θ_e. Using these reference currents, the PWM unit generates the necessary gating signals to the inverter. The synchronous angle is obtained from the addition of the slip angle and the rotor angle from the encoder. The slip angle is obtained by integrating the slip frequency. The slip frequency ω_{sl} is calculated by using the steady state slip frequency equation,

$$\omega_{sl} = s \cdot \omega_e = \frac{1}{T_r} \cdot \frac{I_{qs}^*}{I_{ds}^*}$$

This calculation is done in the slip gain table of Figure 16.11. The slip gain itself is the inverse of the open loop rotor time constant T_r. The value of slip gain is affected by the variations in rotor temperature, speed, and saturation, causing the system to be improperly field oriented. Tuning the slip gain at different current level and speed level is one method of obtaining the correct slip gain. Other techniques of determining the slip gain are discussed in the literature [18,19].

16.5.3 Sensorless Operation

In the area of propulsion and industrial motor control technologies, methods to eliminate the speed/position sensors, inverter current sensors, and so on have been under investigation for several years. The speed sensorless control strategies discussed in the literature are [20,21]:

- Slip frequency calculation method
- Speed estimation using state equations
- Estimation based on slot space harmonic voltages or third harmonic detection
- Flux estimation and flux vector control

- Rotor flux orientation
- Stator flux orientation
- Direct control of flux and torque
- Fuzzy logic based direct self control
- Observer based speed sensorless control (Luenberger, etc.)
- Model reference adaptive systems
- Neural network-based sensorless control

The above technologies have not yet proven to be practical for automotive applications. Most of the speed sensorless control strategies depend on the parameters of the motor. The best speed sensorless strategy is the one that provides the speed information and speed regulation with an accuracy of 0.5% or better from zero speed to maximum speed under all operating conditions. The speed calculation should be independent of the parameter variations and saturation levels of the electric machine.

16.6 APU CONTROL SYSTEM IN SERIES HYBRID VEHICLES

In series hybrid vehicle applications, the output power of the generator driven by an engine is used to extend the range of the vehicle or provide a part of the propulsion power. The generator system requires a power and control electronics unit to convert mechanical power to required electrical power. This section describes a control system for the induction generator driven by a Stirling engine [22]. The generator control is based on indirect field orientation method, with an optimized flux profile to obtain highest possible efficiencies. The control system is designed to eliminate the mechanical controls generally required for controlling the Stirling engine power.

The Stirling engine provides almost a constant output torque over a wide speed range of operation. Hence, by properly regulating the speed of the engine, the output power of the generator can be regulated. The block diagram of the Stirling engine's induction generator power control scheme is shown in Figure 16.12. The generator speed, ω, which

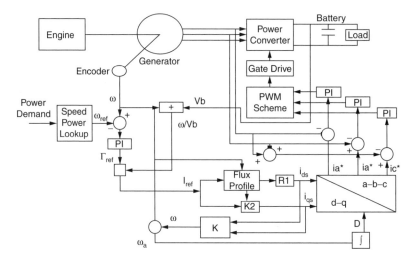

Figure 16.12 Stirling engine-driven induction generator APU control.

Power Electronics and Control for Hybrid and Fuel Cell Vehicles

is the same as the engine speed, is measured using an incremental encoder. The speed reference, ωref, as a function of the required power is obtained from a system controller lookup table. The speed reference is subtracted from the generator speed, and the speed error, after being processed through a PI controller, results in a signal proportional to the torque of the engine. This signal is then multiplied by the speed to obtain a signal proportional to the output power. Also, the DC voltage Vb, is measured at the load terminal and used in the generator control. The output power divided by the DC voltage results in a DC load current reference signal. This reference signal, in combination with the optimized machine flux profile and rotational speed, is used to obtain the d-q axes current reference signals for the generator control. The expression for the current reference is given below.

$$I_{(ref)} = \{(K_1 + K_2/s)\omega_{actual}(\omega_{actual} - \omega_{ref})\}/V_{dc}$$

where K_1 and K_2 are proportional and integral gains of the PI controller.

The flux profile is programmed as a function of speed and current references to operate the generator at maximum efficiencies at all operating points. The d-q axis currents are then transformed to the ABC reference frame currents. These currents are compared with the measured generator currents and the error is used in obtaining the PWM sinusoidal reference signals. These signals are then compared with a high-frequency triangular waveform to obtain the PWM signals, which are used to control the switching instants of the power devices in the power converter.

The generator control can be explained by referring to Figure 16.13. Assume that the APU system is operating at a power P_1. A step increase in system power demand to

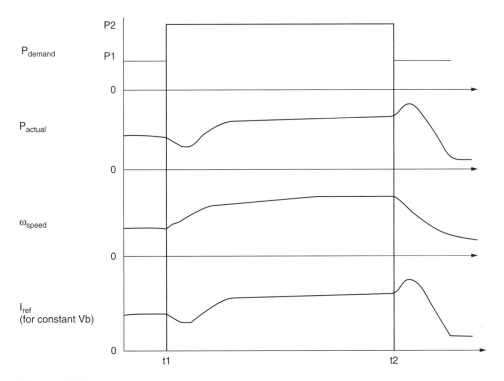

Figure 16.13 Step response for change in power command.

P_2 at instant t_1 would momentarily decrease the torque demand. This will reduce the current reference value, I_{ref}. When the current reference decreases, the load decreases, hence the speed of the engine increases toward the new value that would provide the required higher power. The higher power results in higher current and thus the current reference will also gradually increase to its new value dictated by the following relation. The APU system reaches the equilibrium point at the new power level P2.

$$P_1 = T_1 * \omega_1 = V_{dc1} * I_{dc1}$$

$$P_2 = T_2 * \omega_2 = V_{dc2} * I_{dc2}$$

T_1, T_2 are torque values at powers P_1, P_2 and speeds ω_1, ω_2, respectively.

A similar explanation can be given for a step decrease in the power demand. When the power demand is decreased at instant t_2, the torque demand momentarily increases. This increases the load current and hence the speed decreases. Because of the constant torque, decrease in speed decreases the output power and the system reaches the equilibrium point at a lower level. Thus, the power of the engine is controlled directly from the generator control system by adjusting the reference current.

16.7 FUEL CELL FOR APU APPLICATIONS [13,23,24]

The power required to feed the various electrical loads in an automobile is generally obtained using a belt-driven alternator driven by an internal combustion engine. The alternator can only produce power while the engine is running. The fuel cell can produce the on-board power even when the engine is not running and thus would eliminate the need for an alternator as shown in Figure 16.14. A fuel cell APU has higher efficiency and can be used in conventional or mild hybrid configurations and is not linked to a fully electric drivetrain. The engine-independent fuel cell auxiliary power unit could provide power for heating/cooling as well as the host of convenience items found in today's semi-trucks, thus avoiding the necessity for running the engine idle for a long time. This would considerably reduce the emissions from the trucks. High-temperature solid oxide fuel cells are particularly suitable for on-board power generation in automotive and truck applications because of the potential for internal reforming of more conventional petroleum fuels into hydrogen, eliminating the need for an external reformer. It has less stringent requirements for reformate quality and is less sensitive to contaminants such as sulfur.

Fuel cell APU could also be used to generate power at 42 V, and also meets the challenges of providing higher power to the increasing electrical loads such as electric air-conditioning and X-by-wire system. A dual 42 V/14 V architecture using an alternator is shown in Figure 16.15. In this architecture, a generator feeds a 42 V bus having 42 V loads and a battery. A DC-DC converter connects this bus to the conventional 14 V bus having 12 V loads and a 12 V battery. The architecture for a dual-voltage electrical system containing a fuel cell power source is shown in Figure 16.16. The alternator of Figure 16.15 is directly replaced by the fuel cell, and a new box, labeled "Power Conditioning Unit," is added. The functions of the power-conditioning unit are to make the fuel cell stack output voltage compatible with the 42 V standards, to protect the fuel cell from overload and short-circuit at the output, and to prevent current from flowing back into the fuel cell stack, which would damage the cell. The power conditioner needs to be controlled to achieve the optimum operation of the fuel cell unit and to achieve the highest efficiency. Generally, the power conditioner operates in the buck mode to obtain 42 V at the output.

Power Electronics and Control for Hybrid and Fuel Cell Vehicles

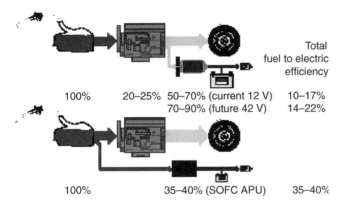

Figure 16.14 Comparison of fuel cell-based APU with conventional system.

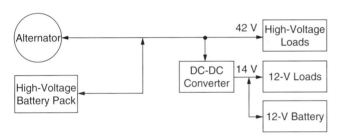

Figure 16.15 Dual-voltage system with generator as power source.

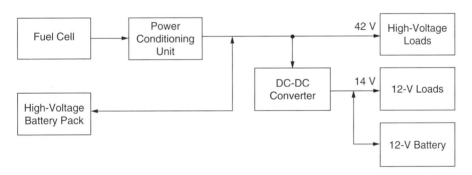

Figure 16.16 Dual-voltage system with fuel cell as power source.

For a certain voltage range, it may be more efficient to operate the power conditioner in linear mode or completely short the power conditioner and allow the output to follow the input, still meeting the 42 V specification requirements. Due to the overload conditions, if the fuel cell voltage falls below about 36 V, the power conditioner may need to be turned off to protect the fuel cell. In addition, the current controller has to control the reformate flow into the fuel cell to limit its output voltage.

A detailed electrical architecture with a typical fuel cell APU is shown in Figure 16.17. In this system, the fuel cell output has to power the 42 V and 14 V loads of the vehicles, and also the accessory loads of the fuel cell unit. The stack voltage can vary in

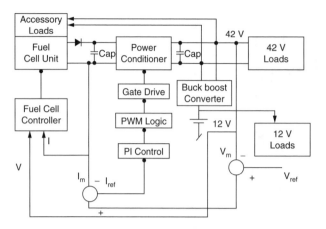

Figure 16.17 Fuel cell APU powering 42 V and 14 V loads.

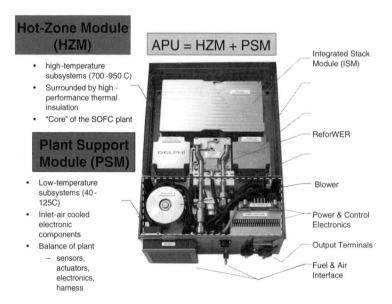

Figure 16.18 Delphi's Gen 2 solid oxide fuel cell-based APU (courtesy Delphi Corporation).

the range of 60 V to 36 V from no load to full load. The power conditioner maintains the output voltage of the fuel cell stack to the desired voltage range. A 42 V/14 V buck-boost converter is used for charging the vehicle 12 V, which powers all the 14 V accessories. The system is first started by boosting the 14 V to 42 V to power the accessory loads.

One of the major challenges in the SOFC APU is the packaging of all the electronics inside the APU unit. As can be seen from Figure 16.18 [24], the blower motor, power electronics, and the control unit are very close to the hot-zone, which is at a temperature of about 700–950°C. The plant support module that consists of all the electric accessory loads and power converters needs to be maintained at a temperature of less than 125°C. The thermal insulation between the hot-zone and the plant support module is a very critical factor in the development of the SOFC-based APU.

REFERENCES

1. J. Botti and C. Miller, Powertrains of the Future: Reducing the Impact of Transportation on the Environment, SAE 1999 World Congress, March 1–4, 1999, Detroit, MI.
2. J. Walters, H. Husted, and K. Rajashekara, Comparative Study of Hybrid Powertrain Strategies, SAE Future Transportation Technology Conference, August 20–22, 2001, Costa Mesa, CA.
3. Toyota Presentation. www.worldenergy.org/wec-geis/global/downloads/Nzconf/10c.pdf.
4. *Toyota Hybrid System THS II – Hybrid Synergy Drive*, Toyota Publications, 2004.
5. G. Simopulos, J. MacBain, et al., Fuel Economy Improvements in an SUV Equipped with an Integrated Starter Generator, SAE International Truck and Bus Meeting, November 12–14, 2001, Chicago, IL.
6. K. Rajashekara, Power Electronics for the Future of Automotive Industry, PCIM Europe, May 14–26, 2002.
7. G. Cameron, Integrated Starter Generator Systems, IEEE Workshop on Power Electronics in Transportation, December 5–6, 2000, Novi, MI, pp. 57–73.
8. R. Davis and T. Kizer, Energy Management in Daimler Chrysler's PNGV Concept Vehicle, SAE Convergence Paper No. 2000-01-C062, Warrendale, PA, 2000.
9. P. Duhr, P. Farah, and L. Schoester, *Stop/Start Function: The Claw Pole Machine, a Good Alternative to the ISG*, Haus der Technik, Munich, Germany, December 2000.
10. J. Sakano and K. Mashino, 42 V Side Mount Motor Generator System with a High Current Power Module and a Driver IC, MIT Consortium Meeting, Los Angeles, CA, January 31, 2002.
11. K. Rajashekara, Propulsion System Strategies for Fuel Cell Vehicles, SAE 2000 World Congress, March 6–9, 2000, Detroit, MI.
12. J. Tachtler, T. Dietsch, and G. Goetz, Fuel Cell Power Unit – Innovation for the Electric Supply of Passenger Cars, SAE 2000 World Congress, March 6–9, 2000, Detroit, MI. Paper No. 2000-01-0374, March 6–9, 2000.
13. J. Zizelman, S. Shaffer, and S. Mukerjee, Solid Oxide Fuel Cell Auxiliary Power Unit — A Development Update, SAE 2002 World Congress, 2002-01-0411, Detroit, MI.
14. K. Rajashekara, Propulsion System Issues in Electric and Hybrid Vehicle Applications, International Power Electronics Conference, IPEC-95, Yokohama, Japan, 1995, pp. 95–98.
15. K. Rajashekara, Evaluation of Power Devices for Electric Propulsion Systems, IEEE Workshop on Power Electronics in Transportation, October 22–23, 1992, Dearborn, MI.
16. www.nrel.org/vehiclesandfuels/powerelectronics/about.html.
17. L. Blair, FreedomCAR and Fuel Cells: Are They the Future?" SAE Fuel Cell TOPTEC, April 9, 2002, Dearborn, MI.
18. J.M.D. Murphy and F.G. Turnbull, *Power Electronic Control of AC Motors*, Pergamon Press, 1988.
19. D.W. Novotny and T.A. Lipo, *Vector Control and Dynamics of AC Drives,* Clarendon Press, 1996.
20. K. Rajashekara, A. Kawamura, and K. Matsuse, *Sensorless Control of AC Motor Drives,* IEEE Press, 1996.
21. D.O. Neacsu and K. Rajashekara, Comparative Analysis of Torque-Controlled IM Drives with Application in Electric and Hybrid Vehicles, *IEEE Trans. on Power Electronics,* Vol. 16, March 2001, pp. 240–247.
22. K. Rajashekara, R. Shah, and S. McMullen, Control System for a Stirling Engine Driven Generator in a Hybrid Electric Vehicle, EVS-15 Symposium, Brussels, Belgium, October 1998.
23. H. Husted, "Dual-Voltage Electrical System with a Fuel Cell Power Unit, SAE Future Transportation Technology Conference, August 21–23, 2000, Costa Mesa, CA.
24. J. Zizelman, Development Update on Delphi's Solid Oxide Fuel Cell System: From Gasoline to Electric Power, Fourth SECA Meeting, Seattle, WA, April 2003.
25. T. Matsumoto, N. Watanabe, et al., Development of Fuel-Cell Hybrid Vehicle, SAE 2002 World Congress, Paper No. 2002-01-0096.

Part IV

Automotive Motor Drives

17

Brushed-DC Electric Machinery for Automotive Applications

Babak Fahimi
University of Missouri-Rolla, Rolla, Missouri

Electric motors have played a crucial role in the evolution of the automotive industry. Existing trends in more electrification of automobiles indicate a further increase in deployment of electromechanical energy devices in coming years. Due to historical, technical, and economical incentives DC-brushed machines have been the favorite choice for numerous automotive applications ranging from starters to auxiliary devices. Ease of control, capital investment, and relatively low cost of manufacturing compared to other energy conversion devices are among the main reasons to justify the substantial use of DC-brushed machines, as advanced motor drive technologies emerge. Although maintenance and durability are still considered as main impeding factors, an impressive compactness and relative high efficiency seems to be of higher significance in the automotive industry. Introduction of power electronics into automotive products over the past two decades has further paved the road for high-grade performance and flexibility in four-quadrant applications. However, it must be mentioned that DC-brushed motor drives are primarily employed for smaller size motors; hence, the design practices should be done in the context of the application to maintain engineering and commercial sense. This necessitates an investigation of drive performance in the presence of the high temperatures that are typical in automotive applications. The present chapter provides an overview of the fundamentals, magnetic design, and control practices for DC-brushed motor drives. Due attention is given to permanent magnet DC-brushed motor drives, as they represent the dominant magnetic configuration used in most automotive DC-brushed motors.

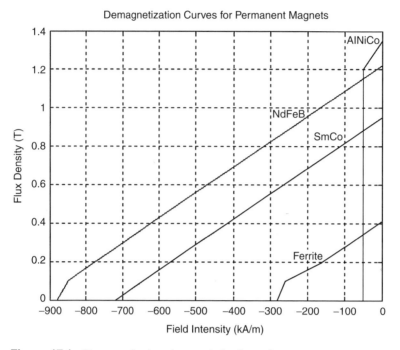

Figure 17.1 Demagnetization characteristics for various permanent magnets.

17.1 FUNDAMENTALS OF OPERATION

17.1.1 Introduction

Permanent magnet motor drives are being increasingly used in today's automotive industry. This is mainly due to an increase in energy density of permanent magnets. NdFeB magnets set a vivid example of this trend by offering an impressive energy density of about 300 kJ/m^3. In addition, highly efficient performance, more compact geometry, lower weight, and shorter electrical time constants are among other attractive attributes of these technologies.

Permanent magnets are described by their demagnetization characteristic. This is a portion of the hysteresis cycle located in the second quadrant of the flux density vs. field intensity (B-H) plane. Figure 17.1 shows demagnetization characteristics of various magnets at rated temperature (20°C).

As can be observed, the demagnetization curve of a permanent magnet can be fully characterized by the following parameters:

1. Remanent magnetism (B_r), where characteristic crosses the B-axis
2. Coercive field intensity (H_c), where characteristic crosses the H-axis
3. Curvature connecting these two points, mostly one or two linear segments

In fact, the second linear region that is close to the H-axis is considered as an unstable region. Therefore a proper operating point will be located in the first linear part, which is expressed as:

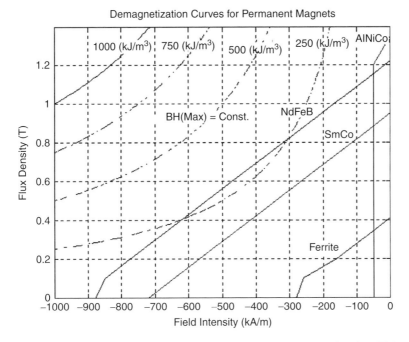

Figure 17.2 Constant energy loci and demagnetization curves of various PM.

$$B = B_r + \mu_0 \mu_r H \qquad (17.1)$$

where μ_r and μ_0 stand for relative and air permeability, respectively. It must be noted that in permanent magnets μ_r is very close to 1 (relative permeability of the air). Accordingly, the energy density of a permanent magnet (assuming a linear characteristic) can be computed as:

$$E = B.H \Rightarrow E_{Max} = \frac{B_r^2}{4\mu_0 \mu_r} [\frac{kJ}{m^3}] \qquad (17.2)$$

Figure 17.2 depicts the constant energy loci along with demagnetization characteristics of various permanent magnets.

Addressing different types of permanent magnet in the above graphs, it must be mentioned that ferrite magnets represent a low-cost solution while offering a limited flux density. AlNiCo magnets, which are more expensive as compared to ferrite magnets, demonstrate a very high remanent magnetism. This, however, is undermined by very limited coercive field intensity. The most expensive, SmCo magnets, represent quality in every aspect including high remanent magnetism, large coercive field intensity, and a fully linear demagnetization characteristic. They also are known as highly stable in the presence of high temperature variation. Finally, the NdFeB magnets demonstrate very high energy density at room temperature. Furthermore, the cost associated with NdFeB is much less than SmCo magnets. However, coercive field intensity in NdFeB magnets is highly sensitive to temperature changes. This, in turn, results in an inadequate performance at high temperatures.

Table 17.1 Summary of Attributes for Various Permanent Magnets

Parameters at 20°C	Ferrite	AlNiCo	SmCo5	NdFeB
Remanent magnetism (B_r[T])	0.38 ... 0.42	0.61 ... 1.35	0.85 ... 1.0	1.0 ... 1.23
Coercive field intensity (H_c[KA/m])	390 ... 280	59 ... 50	1000 ... 1200	1600 ... 960
Max. energy density (BH_{max}[kJ/m^3])	28 ... 34	13 ... 62	140 ... 200	195 ... 280
Temperature coefficient (K_B[%/C])	−0.2 ... −0.23	−0.02	−0.04 ... −0.05	−0.11 ... −0.13
Temperature coefficient (K_H[%/C])	+0.4 ... +0.22	+0.03 ... −0.07	−0.25	−0.6 ... −0.8
Reversible permeability (μ_{rev})	1.05	< 5	1.05	< 1.2
Density [ρ/(kg/dm^3)]	4.6 ... 4.9	6.7 ... 7.3	8.1 ... 8.3	7.3 ... 7.4
Specific electric resistance [$\rho_{CL}/\mu\Omega$cm]	10^{12} ... 10^{16}	40 ... 70	50 ... 60	140
Cost [%]	10 ... 15	40 ... 60	600 ... 800	200 ... 300

In general, the main objectives in selecting a permanent magnet for motor drive application can be summarized as:

- High energy density
- A linear demagnetization characteristic in the entire vicinity of the second quadrant (B-H plane)
- High stability with respect to temperature
- High specific resistance to mitigate eddy currents
- Durability against corrosion and demagnetization
- Low cost

Although achieving all these attributes in a single magnet is not possible, proper design can help us to optimize the performance of the drive in the context of the application. Table 17.1 summarizes the magnetic characteristics of four families of the permanent magnets.

Comparing various properties of permanent magnets, the following observation can be made:

- A rise in temperature will reduce the remanent magnetism in all magnet types. On a percentage basis, ferrite magnets seem to be the most sensitive magnets, while SmCo demonstrates the least sensitivity.
- Coercive field intensity portrays different behavior for various materials. While an increase in temperature results in significant decrease of coercive field in NdFeB, an opposite response is seen in ferrite magnets. Overall, SmCo offers the least sensitivity to temperature.

- SmCo is the heaviest and the most expensive alternative among all candidates. It also presents one of the lowest specific resistances, which translates to high eddy current losses.
- AlNiCo offers the highest remanent magnetism. This, however, is mainly undermined by a very limited coercive field intensity and extremely high conductivity.
- The reversible permeability in most cases is close to 1.

We will use these observations to predict the performance of brushed DC-motor drives in the presence of high temperature variations.

17.1.2 Torque Production in Brushed DC-Motor Drives

Electromagnetic torque in brushed DC-motor drives may be viewed as a product of interaction between two magnetic fields constructed by armature (rotor winding) and field (permanent magnet on the stator). Airgap, as a media in which electromechanical energy conversion takes place, has an essential role in this process. Therefore, it is important to calculate the components of the electromagnetic fields in the airgap. Figure 17.3 depicts a cross-view of a portion of machine cut and rolled as shown.

Assuming a linear form for demagnetization characteristic of permanent magnet, the field component generated by permanent magnets in the airgap can be expressed as:

$$B = \frac{B_r}{1+(\sigma/h_M)} \qquad (17.3)$$

where σ, h_M, and B_r denote airgap length, depth of the magnet, and remanent magnetism, respectively. This equation shows that a large B_r can result in relaxing of manufacturing cost by either allowing larger airgaps or smaller magnets. The armature component of the electromagnetic field in the airgap, on the other hand, is generated by the current in the stator phases and is approximated as:

$$H = \frac{2mwI}{\pi D} \qquad (17.4)$$

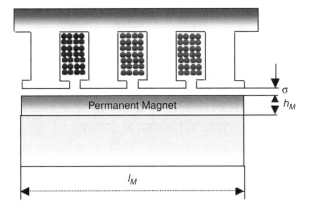

Figure 17.3 Cross-view of a segment of BLDC machine cut and rolled.

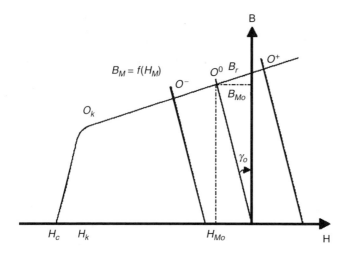

Figure 17.4 Magnetic behavior in PM machine.

where m, w, I, and D stand for number of phases, number of winding, phase current, and airgap outer diameter (rotor OD + airgap length). The force density in the airgap is given by:

$$f = \frac{1}{2\pi} 2 \int_{\pi/6}^{5\pi/6} H \cdot B \, d\beta = \frac{2}{3} H \cdot B \qquad (17.5)$$

Equation 17.3 and Equation 17.5 show that any change in remanent magnetism of permanent magnets impacts the torque productivity of the machine in a linear fashion. In addition, variation of field intensity in the magnet due to armature current can be considered another determining factor in evaluating the performance of the machine in the presence of parameter variations caused by temperature and so on. Figure 17.4 shows a graphical interpretation of the electromagnetic variations occuring in surface-mount magnets used in brushed DC machines.

Point $O°$ represents the no-load operating point as specified by the design and demagnetization curve of the permanent magnet. In selecting the operating point, specific criteria may be used. For instance, as shown in Figure 17.5, to minimize the size of used SmCo magnet (the most expensive type), the operating point is located such that maximum energy is achieved. However, in the case of ferrite magnet one should try to locate the operating point close to the highest remanent magnetism, thereby obtaining the highest possible flux density in the airgap at the expense of an increased magnet volume. The latter will be tolerated by low cost of the ferrite magnets.

Once the no-load operating point is selected, one can include the effects of armature magnetomotive force (mmf). This has been graphically shown in Figure 17.4 and is expressed analytically as follows:

$$B_M = B_{M0} \pm \Delta B_M \cong \frac{B_r}{1+(\sigma/h_<)} \pm \frac{\mu_0}{2\sigma} \frac{\Theta_A}{1+(h_M/\sigma)}$$

$$H_M = H_{M0} \pm \Delta H_M \cong \frac{-B_r}{\mu_0(1+h_M/\sigma)} \pm \frac{1}{2\sigma} \frac{\Theta_A}{1+(h_M/\sigma)} \qquad (17.6)$$

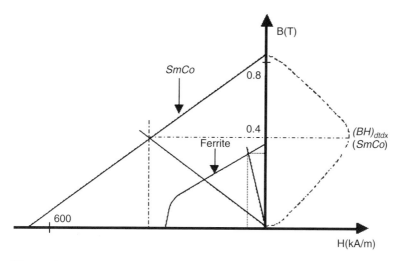

Figure 17.5 Selection of operating point using SmCo and ferrite magnets.

where Θ_A stands for magnetomotive force of the armature (stator winding). As can be seen, due to variations in armature current the operating point will fluctuate along demagnetization characteristics, traveling between A+ (field enforcement) and A− (field weakening). It is therefore crucial to avoid entering a region beyond knee point (O_k). Indeed, forcing the operating point to beyond this point will result in unstable operation of the magnetic circuit. As a result, design of the machine should take into account operation in linear region, even in the presence of parameter variations.

The above discussion shows that while remanent magnetism has a direct impact on airgap flux density, the coercive field intensity will have a significant impact on armature mmf. Given the sensitivity of a particular magnet, these facts can be effectively used in determining h_M/σ, and D_R (rotor outer diameter) and winding configuration respectively.

17.1.3 Impact of Temperature on Performance of a BLDC Drive

Temperature variations is an inevitable attribute of many industrial applications such as automotive products. In fact, in some automotive applications a temperature range of −40 to 150°C is reported. This, in turn, alters the properties of permanent magnets used in brushed DC-motor drives. Although flexibility in locating some of automotive motor drives can potentially avoid an extreme case of temperature variation, investigation of temperature impact on machine operation seems to be necessary. Figure 17.6 and Figure 17.7 show demagnetization curves for NdFeB and ferrite magnets at various temperatures. As can be seen, a rise in temperature reduces the remanent magnetism in both cases. However, a severe reduction in coercive field in NdFeB poses a real limitation for armature currents at high temperatures. This will clearly undermine the practicality of using NdFeB magnets at very high temperatures. It must be noted that a modest range of temperature, below 40°C, can be appropriate for employing the NdFeB. The ferrite magnets, on the other hand, depict a reverse behavior at high temperatures; they tend to have larger coercive fields. This, however, at very low temperatures may cause a problem. According to Table 17.1, SmCo permanent magnets demonstrate a superior performance in terms of temperature stability. However, this comes with a significantly higher cost. Therefore, if a modest range of temperatures can be accommodated, employment of NdFeB magnets is recommended.

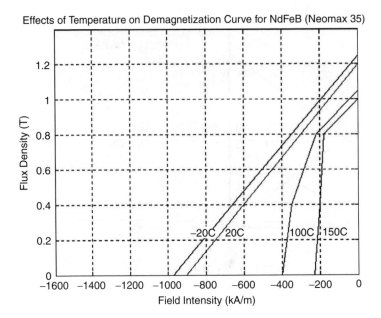

Figure 17.6 Thermal behavior of NdFeB magnets.

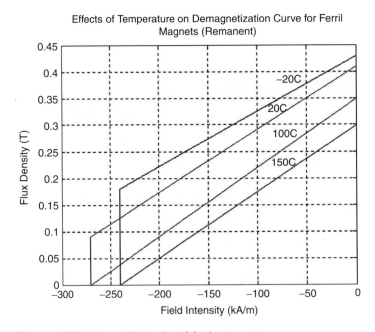

Figure 17.7 Thermal behavior of ferrite magnets.

In order to quantify the effects of temperature change in brushed DC-motor drives, the linear (stable) part of demagnetization characteristics is expressed as follows:

$$B = f(H) \cong B_r + \mu_0 \mu_r H$$

$$B_r \approx B_{r0}(1 + k_B T/100) \tag{17.7}$$

$$H_K \approx H_{k0}(1 + k_H T/100)$$

The no-load operating points are given by Equation 17.6 as follows:

$$B_{M0} \cong \frac{B_r}{1 + (\sigma/h_M)} = \frac{(B_{r0}(1 + k_B T/100))}{1 + (\sigma/h_M)} \tag{17.8}$$

A change in the temperature, therefore, directly shifts the location of operating point. This will affect the allowed armature current as the maximum coercive field has also been changed. This is expressed as:

$$\Delta H_M \leq H_{M0} - H_k = \frac{-B_r}{\mu_0(1 + h_M/\sigma)} - H_k$$

$$= \frac{-(B_{r0}(1 + k_B T/100))}{\mu_0(1 + h_M/\sigma)} - (H_{k0}(1 + k_H T/100)) \tag{17.9}$$

By combining Equation 17.5, Equation 17.8, and Equation 17.9 one can say that an increase in temperature will reduce remanent magnetism and, hence, will reduce the torque productivity of the machine by a factor of:

$$\frac{\partial B}{\partial T} = \left(\frac{(k_B)}{1 + (\sigma/h_M)}\right) \frac{B_{r0}}{100} \tag{17.10}$$

where T stands for temperature in degrees Celsius. Table 17.1 provides a range of k_B values for various permanent magnets. At the same time, the permitted armature field to maintain a stable operation is limited by:

$$\frac{\partial \Delta H_M}{\partial T} = \left(\frac{-k_B B_{r0}}{\mu_0(1 + h_M/\sigma)} - k_H H_{K0}\right) \frac{1}{100} \tag{17.11}$$

It must be noted that a more strict limit on the armature excitation will further impact the torque generation of the brushed DC-motor drive. Table 17.1 provides a range of k_H values for various permanent magnets. In order to illustrate the effects of temperature the following example is given. Table 17.2 shows three various designs (two-pole design is assumed) using ferrite, SmCo, and NdFeB permanent magnets.

Table 17.2 Example Designs for Illustration of Temperature Effects

Type	σ (mm)	h_M (mm)	D (mm)
Ferrite	0.5	4.5	50
SmCo	0.5	1	50
NdFeB	0.5	2	50

Table 17.3 Operational Condition at 20°C

Type	B_r [T] @ 20°C	H_A [kA/m] @ 20°C	Stack length (mm)	Torque (N-m)
Ferrite	0.42	30	165	5
SmCo	1	30	96	5
NdFeB	1.23	30	64	5

Once the airgap flux density and armature mmf are known the following equation can be used to compute electromagnetic torque:

$$T = \frac{\pi}{3} HBD^2 l \qquad (17.12)$$

Using Equation 17.12, we have selected the stack length such that a total torque of 5 N-m is achieved. Table 17.3 shows the results of torque calculation.

In the next step, using the above method, operating points and the electromagnetic torque at −20°, 20°, 100°, and 150°C are computed and listed in Table 17.4. Please note that the stack length of machines is kept constant in this exercise.

Table 17.4 Effects of Temperature on Operational Characteristics of BLDC

Type		B_r (T)	H_k (kA/m)	B_{airgap} (T)	H_M (kA/m)	Torque (N-m)
Ferrite	−20°	0.46	−246	0.414	−36	5.36
	20°	0.42	−280	0.378	−33	5
	100°	0.34	−246	0.306	−27	3.57
	150°	0.29	−280	0.261	−23	2.6
SmCo	−20°	1.02	−1320	0.68	−270	5.12
	20°	1	−1200	0.667	−265	5
	100°	0.96	−960	0.64	−254	4.83
	150°	0.94	−810	0.627	−250	4.73
NdFeB	−20°	1.29	−980	1.032	−205	5.18
	20°	1.23	−900	0.984	−196	5
	100°	1.1	−220	0.88	−175	4.42
	150°	1.02	−190	0.816	−162	3.82

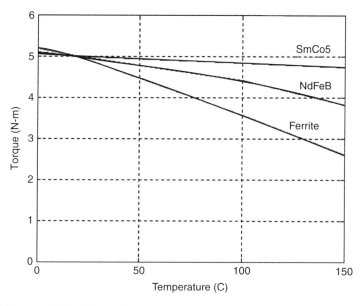

Figure 17.8 Effects of temperature on electromagnetic torque in brushed DC-motor drives.

The results of this study are graphically shown in Figure 17.8.

As can be seen, reduction in remanent magnetism, due to temperature variations, is the main cause for reduction in torque. At high temperatures, despite the relatively low armature mmf, low coercive/knee point field intensity also contributes to this reduction. It is therefore expected that a higher armature mmf would have resulted in yet higher reduction in torque under this design. It must also be noted that SmCo depicts the most stable operation; however, with a reasonable range of temperatures, employment of NeFeB magnets seems to be the most appropriate choice. NdFeB magnets not only are better economic choices, they also offer higher specific resistance at a lower weight compared to SmCo magnets. This in turn minimizes the effects of high-frequency eddy current losses caused by slot harmonics.

17.2 SERIES CONNECTED DC-MOTOR DRIVES

Due to its application in starters, series connected DC-machines are introduced in this section. Interestingly, series connected DC-machines were the primary choice for electric propulsion in early versions of electric vehicles. This is attributed to their natural torque vs. speed characteristic, which would allow for a compact motor size. This novel property of series connected DC-machines was recently reinvented by many researchers, who recommended motors with extended constant power region as optimal choice for electric propulsion. Under steady-state conditions when vital parameters of a DC machine are at quasi-static conditions, one can analytically extract the output characteristic of the series connected DC-motors.

It is important to note that due to commutation of the armature windings, the magnetic fields created by the field and armature windings are orthogonal; hence, there exists an inherent decoupling between their respective magnetomotive forces. Furthermore, unlike rotating field machinery the magnetic fields of armature and field windings are stationary with respect to each other. One can classify the DC-machines in accordance with their respective winding configurations. The commonly used architectures are:

- Separately excited DC-machines
- Parallel or shunt connected DC-machines
- Series connected DC-machines
- Compound (series + parallel) windings DC-machines

In order to obtain the output characteristics of DC-machines one needs to express the electromagnetic torque and induced voltages in the armature (located on the rotor); in the absence of magnetic saturation these equations can be expressed as follows:

$$T = L_{Af} i_A i_f$$
$$E = L_{Af} i_f \omega_r \tag{17.13}$$

in which L_{Af}, i_A, i_f, and ω_r stand for mutual inductance between armature and field windings, armature current, field current, and rotor speed, respectively. Furthermore, the voltage equations for the armature and field circuits can be used to simplify the above equations:

$$V_A = r_A I_A + E = r_A I_A + L_{Af} I_f \omega_r$$
$$V_f = r_f I_f \tag{17.14}$$

The use of capital letters in Equation 17.14 indicates the steady-state values for the armature and field currents and voltages. By combining Equation 17.14 and Equation 17.15 one can analytically express the torque vs. speed characteristic in a DC-machine as:

$$V_A = r_A I_A + L_{Af} \frac{V_f}{r_f} \omega_r$$

$$\Rightarrow I_A = \frac{V_A - L_{Af} \frac{V_f}{r_f} \omega_r}{r_A} \tag{17.15}$$

$$\Rightarrow T = L_{Af} \left(\frac{V_A - L_{Af} \frac{V_f}{r_f} \omega_r}{r_A} \right) \left(\frac{V_f}{r_f} \right)$$

Given the winding configuration one can use Equation 17.15 to obtain the output characteristic of the DC-machine in interest. In particular for series connected DC-machines one can make the following simplifications:

$$V_t = V_f + V_A$$

$$I_A = I_f$$

$$\Rightarrow I_A = \frac{V_t}{(r_A + r_f) + L_{Af}\omega_r} \qquad (17.16)$$

$$\Rightarrow T = L_{Af}\left(\frac{V_t}{(r_A + r_f) + L_{Af}\omega_r}\right)^2$$

where V_t represents the terminal voltage of the machine. A quick inspection of Equation 17.16 indicates that at standstill, a series connected DC-motor exhibits its maximum torque density. Moreover, as speed increases the generated electromagnetic torque will decline. These attributes optimally suit an application such as the starter in automobiles.

18

Induction Motor Drives

Khaled Nigim
University of Waterloo, Waterloo, Ontario, Canada

18.1 INTRODUCTION

Induction motors play an important part in many different types of domestic and industrial processing machinery. The popularity of the motor is due to its ruggedness and operational reliability. The rotational speed of the induction motor ω_m, as well as its developed torque T_d, may be controlled by several techniques to meet the load requirements [1]. The induction motor has two main parts; a stationary part called the stator and a rotating part called the rotor, in which the mechanical load is connected. The stator has windings slotted into a segmented iron frame to reduce eddy current and magnetic losses. The windings establish a rotating magnetic field necessary to circulate current, by induction, into rotor windings. The stator winding absorbs the electric power needed to balance the mechanical torque. The rotating part carries the rotor windings.

Two types of rotor are manufactured. The first type is a wound rotor in which the rotor windings are again slotted into a segmented iron frame and connected together at one end externally. The second type is simply a shorted aluminum bar. The bars are housed in a slotted iron frame. This is the most common type due to its rigidity and construction simplicity and is known as the squirrel cage induction machine. Squirrel cage induction motors are widely used and are the subject of our analysis.

The developed torque in the induction motor is dependent on the supply voltage (V_s) and the internal circuitry elements representing the motor. Current induced in the rotor winding rotating with the shaft is dependent on the slip. The slip is the difference between the rotor induced voltage frequency f and the synchronous reference frequency f_s. A practical equivalent circuit that is normally adopted for analyzing the motor speed characteristics at steady state is represented in Figure 18.1. The values of the impedances shown are referenced to the supply side. In the equivalent circuit, the impedance of the

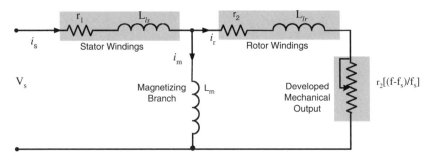

Figure 18.1 Steady-state induction motor per phase equivalent circuit (all values refer to the stator side).

stationary part is represented by a series element Z1 ($Z_1 = r_1 + j\omega_s L_{ls}$), constituting the windings influencing the rotating magnetic flux. The magnetic flux coupling the two parts is represented by magnetic inductive reactance X_m ($X_{1m} = j\omega_s L_m$). The rotating part impedance is represented by Z_2 ($Z_1 = r_2 + j\omega_s L_{lr}$), constituting the windings influencing the produced torque. In the equivalent circuit, the motor developed torque is represented by an equivalent variable resistor, given by the product of the rotor windings resistor and the normalized ratio between the supply and rotor frequencies. Quantitatively, this is given by

$$r_2 \cdot \left(\frac{f - f_s}{f_s} \right).$$

In squirrel cage induction motors, the motor shaft rotational speed depends directly on the supply frequency (f_s), resulting in the magnetic field rotating at synchronous rotational speed of $2\pi f_s$ rad/s. In addition, the speed is inversely proportional to the number of the magnetic poles (p) created by the stationary stator winding when energized by the AC power source. How the windings are connected together determines the number of poles. The most important requirement when dealing with adjustable speed drives (ASDs) is to study the performance of the motor under various voltages, current, magnetic flux, and frequency control strategies. Motor speed and torque are interrelated in the induction machine. Increasing the applied load torque (T_L) entails more demand on the motor to develop electromechanical torque (T_d). The increase in the load demand reduces the speed and vice versa. However, using today's power electronics and microprocessor technologies, both speed and torque can be controlled independently to meet the different types of loads that industry requires.

18.2 TORQUE AND SPEED CONTROL OF INDUCTION MOTOR

There are various methods to control the induction motor torque and speed, varying in complexity, performance, and cost [2]. One method of speed control is to change the physical connection of the stator windings, to increase or decrease the number of poles that those windings are creating. Stator windings embedded in the stationary part of the motor are responsible for developing the rotating flux and how it is distributed. This method is known as pole changing and is suitable for applications in which two speed settings are sufficient.

A second simple, yet practical, method is by varying the supply voltage by reducing the magnitude supplied to the motor. This technique provides marginal speed change at the expense of losing the developed torque as the voltage is reduced. Reduced voltage method is used in many small induction motors coupled to fans and air blowers. The same technique is also used as a way to reduce the starting current and, hence, provide safe starting of the motor.

A third method is incorporated to control the speed and torque independently by varying the supply voltage and frequency using switching power electronic devices known as inverters [3]. The switching patterns of the device are controlled through switching logic, known as pulse width modulation (PWM) or space vector modulation (SVM), in association with control circuitry. They are manipulated by either linear- or microprocessor-based control topology in a way to achieve the desired developed motor torque to meet the load characteristics with the highest energy efficiency and minimum losses. The controller could also be used to manipulate the motor to maintain constant torque or power at various operating speeds. The application of power electronic switching devices introduces greater flexibility in controlling the motor speed and torque for a wide variety of applications, including that of traction system in locomotives and electric vehicle.

18.3 BASICS OF POWER ELECTRONICS CONTROL IN INDUCTION MOTORS

Adjustable speed electric drives consist of power electronic switching device assemblage (converters) interfacing the electric power source and the motor. ASDs are manufactured in variable ratings from 1 hp to more than 1000 hp and are considered the working horse of today's industries.

Controlling the voltage and frequency supplied to the motor is the heart of the induction motor's speed and torque control. Voltage and frequency can be varied independently or dependently using the pulse width modulation technique, which controls the switching of the inverter feeding the motor, thus producing variable load at different speed settings. Direct torque and magnetic field control can also be achieved by controlling the space vector of the field and other variable parameters. Direct switching of the inverter feeding the motor through space vector modulation topology results in better dynamic performance. Vector-based induction motor drives are popular. The basic theoretical basis of interfacing the motor with the inverter switching states has been clearly presented in many textbooks [4–7].

Induction motor variable speed drives can be powered by either an AC or DC power source. For AC power-sourced drives, shown in Figure 18.2(a), the converter is made of two parts. The first part, simply called a converter, converts the AC into DC supply. The converter can be controlled to produce variable DC output voltage using controlled switching devices such as thyristors or transistors or simply constructed by a diode rectifier bridge. The second part, known as the inverter, inverts the DC voltage source into variable or constant voltage magnitude and a controllable frequency AC source. The generated frequency is carried out using switching devices again like transistors. The quality of the reproduced variable AC source is a main key element in the performance of the drive. In the DC or battery-sourced drives, only the inverter part is needed to convert the DC source into variable AC source, as shown in Figure 18.2(b).

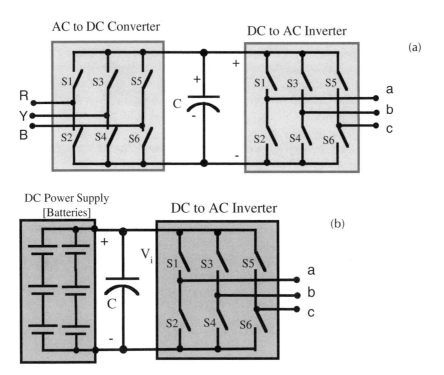

Figure 18.2 (a) AC-DC-AC conversion (voltage source inverter); (b) DC-AC conversion.

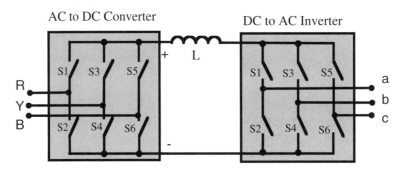

Figure 18.3 Current source inverter.

The converter and the inverter are connected together through the DC link. If the DC link voltage is maintained almost constant by connecting capacitors across the DC link, the assemblage is then identified as a voltage source inverter, shown in Figure 18.2. If the DC link current is maintained almost constant through the connection of DC choke, the converter assemblage is then identified as a current source inverter, as shown in Figure 18.3.

18.4 INDUCTION MOTOR VSD OPERATING MODES

ASDs can be controlled under different control strategies that enable the motor to operate in two distinct torque-speed regions, mainly the constant torque-speed and the constant

Induction Motor Drives

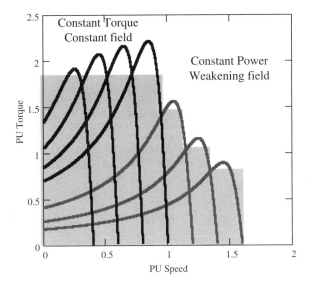

Figure 18.4 Torque-speed characteristics under v/f control.

power-speed regions as shown in Figure 18.4. The speed torque relationship of the induction motor can be found using the steady-state equivalent circuit shown. The developed torque that controls the behavior of the motor under different regions is given by Equation 18.1.

$$T_d = \frac{3p}{2\pi} \frac{A}{\left[\left(\frac{r_1}{f}+A\right)^2 + B^2\right]} \left(\frac{V_s}{f}\right)^2 \quad Nm \quad (18.1)$$

where

$$A = \frac{2\pi r_2}{\omega_s} \frac{(\omega_s L_m)^2}{\left[r_2^2 + \omega_s^2(L_{lr}+L_m)^2\right]} \quad \text{and} \quad B = 2\pi L_{ls} + \frac{\left[r_2^2 + \omega_s^2 L_{lr}(L_{lr}+L_m)\right]}{\left[r_2^2 + \omega_s^2(L_{lr}+L_m)^2\right]}.$$

For constant torque operation, the voltage to frequency (V_s/f) is maintained almost constant from starting to the maximum rated synchronous speed. To avoid saturation at low speed, the inverter voltage is boosted to compensate for the stator winding's ohmic losses and to establish the required starting torque at low frequency. For the constant power operation, the DC link voltage cannot be increased. However, the switching devices are switched at higher frequencies, resulting in lower torque as the speed increases beyond the motor rated speed.

Selection of an induction motor and its control scheme depends on the applied load characteristics and how the system is to be controlled. Loads can be classified according to inertia or torque-speed characteristics or by the load control dynamic. An electric vehicle is considered to be of the high inertia load that needs to be controlled during the acceleration and deceleration and be able to maintain constant speed during cruising. Moreover, electric vehicles are rapidly started and stopped. Therefore, the drive needs to be designed ultimately for driving the motor in the four operating quadrants.

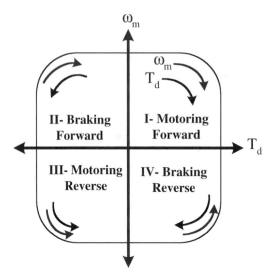

Figure 18.5 Induction motor VSD: four operating quadrants in the (ω_m,T_d) plane.

The concept of operating quadrants plays an important role in the theory and practice of ASDs. Both the torque developed in the motor T_d, and rotational speed ω_m, of the rotor can assume two polarities: one for motoring, in which the product of the torque and the speed produce positive mechanical power (i.e., power is drawn from the source), and another for generating or braking (i.e., power is fed into the source).

Many of today's ASDs have the capability to operate in both polarities of the motor speed. The operation with the two polarities results in four possible operating modes, as shown in Figure 18.5. Operating in the four quadrants, the drive is motoring forward in the first quadrant, braking in the second quadrant, reversing in the third quadrant, and reverse braking in the fourth quadrant; hence:

1. Mode I is for motoring in the positive torque-speed and is used to drive the vehicle in forward directions.
2. Mode III is for motoring in the negative torque-speed and is used to drive the vehicle in reverse directions.
3. Mode II is the braking mode in the positive torque-speed and is used for electronic slowing down before the mechanical brake is applied.
4. Mode IV is also a braking mode in the negative torque-speed and has the same function of mode II but when the vehicle is reversing.

Electric locomotives and electric vehicles have a provision allowing electric braking to relieve the mechanical load from wearing rapidly. However, in many small, low-cost, yet efficient electric vehicles such as the electric golf car, the braking is dependent on the mechanical system. Therefore, the electric drive is only allowed to operate in the motoring forward or reverse modes. In the case of four-quadrant operation, the battery source must be isolated every time the electronic braking is initiated to protect the battery from damage. The battery is then shortly replaced by a resistor to dissipate the energy rapidly as the motor is braking.

18.5 FUNDAMENTALS OF SCALAR AND VECTOR CONTROL FOR INDUCTION MOTORS

The developed voltage and current by the inverter in ASDs are the main controlling elements that dictate how the motor meets the load demand. In the steady state, the voltage v, current i, and magnetic flux linkage λ are defined by their magnitude and frequency (angular velocity). If the magnitude of parameters are observed and adjusted, the control technique is classified as a scalar control. It is usually adopted in low-performance ASDs. For high-performance ASD systems, vector control is adapted, in which the control variables consider the observation of the instantaneous values of positions of stator voltage v_s, current i_s, and the magnetic flux linkage λ space vectors [8]. Vector techniques are used to adjust the instantaneous values, thus permitting high dynamic performance of the drive. Vector control can be implemented in many different ways. The well-known techniques are field-oriented control (FOC) [9] and direct torque control (DTC) [10], which will be discussed briefly in the following sections. An update on the different techniques will also be outlined.

18.5.1 SCALAR CONTROL

18.5.1.1 Open Loop Scalar Control

Induction motor speed and torque can be controlled using inverters by regulating the voltage and frequency of the inverter AC output. To maintain constant torque at wide speed range, the voltage-to-frequency ratio is kept almost constant, to maintain constant air gap flux. This can be achieved in several ways. One of the simplest ways is to control the inverter switching to generate semisinusoidal voltage waveform with its associated frequency. Figure 18.6(a) shows a typical open loop torque-speed control. The voltage command V_s^* is generated directly from the frequency reference signal through a volt/Hertz gain constant (K_0). In steady state, the motor air gap flux is approximately related to the supply voltage and to the shaft rotational speed, thus maintaining a reasonably constant developed torque at different speed settings as given by Equation 18.1. At steady-state operation, if the load torque is increased, the slip will increase and a balance will be maintained between the developed torque and the applied load; thus, the system will be adequate for many applications that are not severely affected by minor speed deviations. At low speed, the voltage is boosted to compensate for the ohmic drop of the stator windings resistor by incorporating a reference voltage that limits the minimum voltage signal at low-speed settings to V_o.

18.5.1.2 Closed Loop Scalar Control

For more accuracy in controlling the speed and limiting the motor current, the drive needs to incorporate sensing devices to detect the motor current and the rotational rotor speed in order to control the torque or the speed, respectively. Both the motor current and speed must be observed to achieve reliable operation. Figure 18.6(b) shows a closed loop speed control in which the speed of the rotating shaft is observed. The difference between the actual speed and the reference is used to regulate the inverter voltage and frequency through a current limit control to avoid running the motor at different torque-speed regions. Again, at low speed, the voltage is boosted to compensate for the ohmic drop of the stator windings.

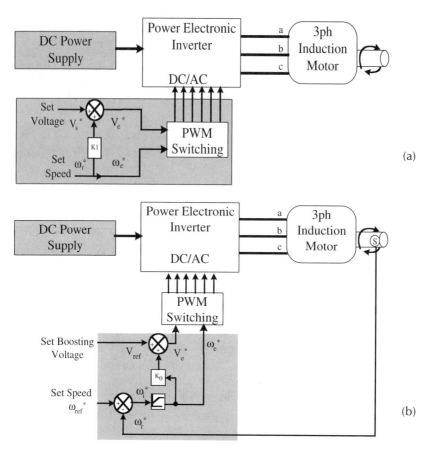

Figure 18.6 (a) Open loop (constant v/f) scalar control; (b) closed loop (constant v/f) speed scalar control.

18.5.2 Fundamentals of Field-Oriented Control (Vector Control) in Induction Motors

18.5.2.1 Field-Oriented Control

Field-oriented control is based on resolving the instantaneous line input motor currents into two components; flux and torque, producing current components. FOC motor controllers are essentially current-controlled systems. In this way, it is expected that the motor will produce controllable torque similar to the DC separately excited electric motor. In DC separately excited motors, the torque produced is a function of the magnetic field flux linkage λ_f and armature current component I_a (as given by Equation 18.2, in which K_t is a constant dependent on the motor size) and how the armature windings are connected.

$$T_d = K_t \lambda_f I_a \quad Nm \tag{18.2}$$

For the induction motor, a similar torque equation can be introduced using the revolving reference frame transformation. Induction motor space vectors in the stator reference frame are revolving, so that their d and q components are AC signals. A dynamic dq-to-DQ transformation allows conversion of those signals to the DC quantities.

Induction Motor Drives

The dq-to-DQ transformation introduces a revolving frame reference in which, in the steady state, space vectors appear as stationary. Therefore, if the flux linkage or current components of the stator and rotor can be transferred to the DQ frame, then it is possible to generate a torque equation that mimics the DC motor torque equation. The torque developed in an induction motor is proportional to the product of magnitudes of space vectors of two selected motor variables (two currents, two fluxes, or a current and flux) and the sine of the angle between these two vectors. The transformation from dq to DQ reference frame allows the generation of the torque equation represented by the interaction of DQ components.

The steps needed to evolve the torque equation using a series of static and dynamic transformation of the variables components are:

1. The instantaneous measured motor variable, say, the current supplied by the inverter (i_a, i_b, and i_c) is transferred to the dq axes components (i_d and i_q) using the transformation matrix as presented by Equation 18.3.

$$\begin{bmatrix} i_d \\ i_q \end{bmatrix} = \begin{bmatrix} 1 & -\frac{1}{2} & -\frac{1}{2} \\ 0 & \frac{\sqrt{3}}{2} & -\frac{\sqrt{3}}{2} \end{bmatrix} \begin{bmatrix} i_a \\ i_b \\ i_c \end{bmatrix} \quad (18.3)$$

2. The transformation between the static current components in the dq frame to the dynamic current components in the DQ frame (i_D and i_Q) is carried out using the transformation matrix given in Equation 18.4.

$$\begin{bmatrix} i_D \\ i_Q \end{bmatrix} = \begin{bmatrix} \cos(\omega_e t) & \sin(\omega_e t) \\ -\sin(\omega_e t) & \cos(\omega_e t) \end{bmatrix} \begin{bmatrix} i_d \\ i_q \end{bmatrix} \quad (18.4)$$

where ω_e is the angular velocity in the arbitrary revolving reference frame. Therefore, if a current vector revolves in the stator frame with the same revolving angular velocity ω_e, it is then possible to treat that particular vector component as DC quantities given by its DQ quantities.

3. The torque is then presented by the interaction of the stator current space vectors, the rotor flux linkage space vector λ_r, and, say, stator current i. The DQ component of the flux linkage is derived in the same manner as the current space vector using the transformation matrix in Equation 18.3 and Equation 18.4, respectively.

4. Based on the dynamic model and equations derived for the induction motor found in many textbooks [2–5] and summarized in the Appendix at the end of this chapter, the torque produced can be presented by Equation 18.5.

$$T_d = \left(\frac{2p}{3} \frac{L_m}{JL_{lr}} \right) (i_{QS} \lambda_{DR} - i_{DS} \lambda_{QR}) \quad N\,m \quad (18.5)$$

where i_{QS} and i_{DS} are the DQ components of the stator current, while λ_{DR} and λ_{QR} are the DQ components of the rotor flux linkage. Aligning the D axis of the revolving reference frame with the rotor flux vector λ_r in which quadrate flux linkage component is 0 will result in an equation format similar to the DC motor torque and is given by Equation 18.6.

$$T_d = \left(\frac{2p}{3}\frac{L_m}{JL_{lr}}\right)(i_{QS}\lambda_{DR}) \quad N\,m \tag{18.6}$$

where p is the number of pair poles, L_m is the magnetizing reactance, and L_{lr} is the sum rotor leakage and magnetizing reactance.

A similar torque equation can be obtained aligning the D axis with the stator air gap flux vector and is given by Equation 18.7.

$$T_d = \left(\frac{2p}{3J}\right)(i_{QS}\lambda_{DS}) \quad N\,m \tag{18.7}$$

The knowledge of the instantaneous position and magnitude of the flux space vector, with which the revolving reference frame is aligned, constitutes the necessary requirement for proper field orientation. Identification of the flux space vector can be based on direct measurements, resulting in direct field orientation or estimation from other measured variables. A direct FOC control scheme is shown in Figure 18.7 for the case when the rotor flux vector λ_r is aligned with the reference frame. Information regarding the flux vector components is detected using sensors embedded in the stator in close proximity of the air gap, which calls for a special arrangement to be made to the induction motor.

Figure 18.8 shows the indirect FOC in which the rotor flux vector is estimated by measuring the rotor angular displacement using a shaft position digital encoder and manipulating a series of transformed flux and current equations based on the motor

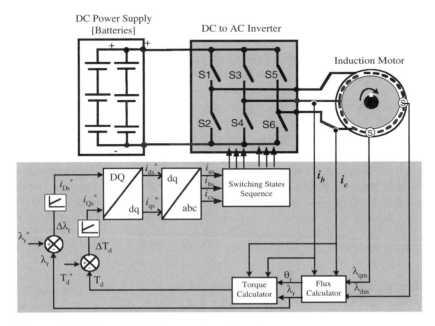

Figure 18.7 Direct FOC with rotor flux orientation.

Induction Motor Drives

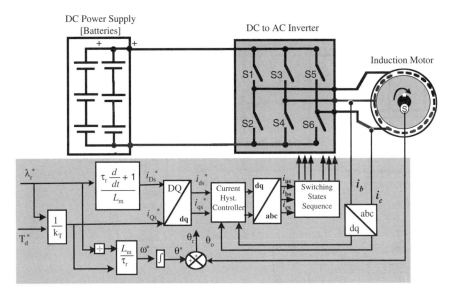

Figure 18.8 Indirect FOC with rotor flux orientation.

parameters including the rotor inertia τ_r ($\tau_r = L_2/r_2$). For this type, motor parameters must be accurately estimated, which entails the inclusion of a thermal correction factor as the motor is running. Other indirect schemes do not incorporate speed loop and provide successful control, as reported in References 10 and 11.

In summary, FOC control systems produce reference values of currents that are based on both torque and flux set demand, which ultimately control the sequence of inverter switching states. For reliable operation, both the rotor velocity and motor parameters including the rotor inertia should be accurately measured and estimated periodically. Many of today's drives incorporate on-line parameter adjustment as the parameter values change due to temperature. FOC, therefore, incorporates vector-switching techniques that are based on specific properties of the motor, resulting in a more accurate torque and speed control.

18.5.2.2 Direct Torque Control

Direct torque control is achieved by developing the following basic functions:

1. A motor model that estimates the actual torque, stator flux, and shaft speed by measuring the motor currents. Motor parameters are calculated frequently. Corrections for temperature and saturation effects can also be made. The motor parameters are established by an identification run, which is made during commissioning.
2. Measuring the DC voltage supplying the inverter and generating voltage and current vectors. Using the machine dynamic model, the flux and torque are estimated.
3. Mapping the information on the state of the power switches to select the optimum switching strategy.

Direct torque control method directly controls the stator flux and the torque based on the instantaneous errors in the flux and torque [12]. The error signal is a result of a

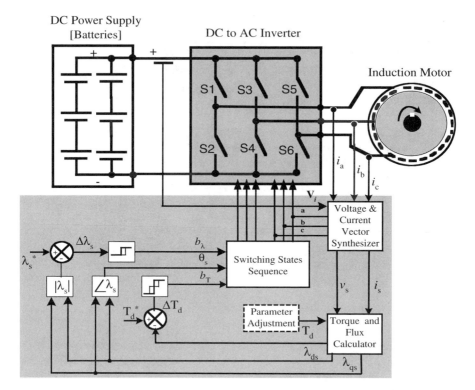

Figure 18.9 Direct torque control DTC.

two-level hysteresis controller. The magnitude of the stator flux is normally kept constant and the motor torque is controlled by means of the angle between the stator and the rotor flux. The switching strategy is based on space vector modulation technique. Adaptation of SVM results in reduced torque pulsation and improves the inverter current profile [13]. There are six voltage vectors and two different kinds of zero voltage vectors available in the two-level voltage source inverter configurations. The switching states can be selected from a look-up table as the variables are changing. In many drives, to achieve higher pullout torque and higher dynamic performance during flux changes, the rotor flux instead of the stator flux is assumed as reference.

A block diagram of a convectional DTC scheme for the induction motor is shown in Figure 18.9. Space vectors v_s and i_s are manipulated in the vector synthesizer block shown in the diagram to estimate the required torque and flux components [14].

To estimate the space vector v_s, the DC inverter voltage V_i is measured. V_i is the battery bank voltage and is assumed constant. V_i value is then used to generate the stator voltage vectors (v_a, v_b, v_c) using the information from the inverter binary switching states (a, b, c). Each state will have either 1 "the device is fully on" or 0 "the device is fully off." The implementation of a transformation matrix for the three-phase fully controlled transistor bridge is presented in Equation 18.8.

$$\begin{bmatrix} v_a \\ v_b \\ v_c \end{bmatrix} = \frac{V_i}{3} \begin{bmatrix} 2 & -1 & -1 \\ -1 & 2 & -1 \\ -1 & -1 & 2 \end{bmatrix} \begin{bmatrix} a \\ b \\ c \end{bmatrix} \quad (18.8)$$

Induction Motor Drives

The space vector v_s ($v_s = v_{ds} + jv_{qs}$) is then estimated using the abc-dq transformation matrix presented in Equation 18.9.

$$\begin{bmatrix} v_{ds} \\ v_{qs} \end{bmatrix} = \begin{bmatrix} 1 & -\frac{1}{2} & -\frac{1}{2} \\ 0 & \frac{\sqrt{3}}{2} & -\frac{\sqrt{3}}{2} \end{bmatrix} \begin{bmatrix} v_a \\ v_b \\ v_c \end{bmatrix} \quad (18.9)$$

Similarly, the stator current space vector i_s estimated from the instantaneous inverter line current d and q components through using Equation 18.9 and is given by Equation 18.10.

$$i_s = i_d + ji_q \quad (18.10)$$

where

$$i_d = \frac{3}{2} i_a$$

and

$$i_q = \frac{\sqrt{3}}{2} i_a - \sqrt{3} i_c.$$

Based on v_s and i_s, the stator flux vector λ_s, and developed torque T_d, are calculated. The magnitude, λ_s, of the stator flux is compared in the flux control loop with reference value λ_s^* and T_d is compared with the reference torque T_s^* in the control loop.

The flux and torque errors, $\Delta\lambda_s$ and ΔT_d, are applied to respective bang-bang controllers. The controller would have proper designed characteristics to meet the requirement of safe operation and allow four-quadrant operation of the drive. The output of the flux controller's output signal b_λ, can assume the values of 0 and 1, while torque signal b_T can assume values of −1, 0, and 1.

Selection of the inverter states is based on both values and on the sector of vector plane in which the stator flux vector $\Delta\lambda_s$ is currently located as well as on the direction of rotation of the motor. A look-up table based on stator vector reference is generated for the different possible non-zero binary switching states for the clockwise and counter-clockwise rotation is shown in Table 18.1.

18.6 INDUCTION MOTOR DRIVES FOR ELECTRIC VEHICLES

The basic requirements for electric vehicles are to provide high torque at zero speed, to allow high-speed cruising (field weakening), and to conserve battery energy with the possibility of operating at slightly variable DC battery voltage level.

The EV drive is configured to respond to a torque demand set by the driver in a similar setup as in the standard passenger car. The accelerator position provides a torque demand as a fraction of the maximum available torque. Similarly, the first portion of the brake pedal travel is used to derive a regenerative torque demand; the remaining pedal travel brings in a set of standard mechanical brakes.

Table 18.1 Inverter Switching States for Direct Torque Control [4]

	Counterclockwise Rotation						Clockwise Rotation					
b_λ	1			0			1			0		
b_T	1	0	−1	1	0	−1	1	0	−1	1	0	−1
	abc switching states			abc switching states			abc switching states			abc switching states		
Sector 1	6	7	5	2	0	1	5	7	6	1	0	2
Sector 2	2	0	4	3	7	5	1	0	4	3	7	6
Sector 3	3	7	6	1	0	4	3	7	5	2	0	4
Sector 4	1	0	2	5	7	6	2	0	1	6	7	5
Sector 5	5	7	3	4	0	2	6	7	3	4	0	1
Sector 6	4	0	1	6	7	3	4	0	2	5	7	3

Binary/state: 1 = 100, 2 = 010, 3 = 110, 4 = 001, 5 = 101, and 6 = 011. States 0 and 7 both = 000

Several reported EV prototypes are equipped with vector control drives that operate under the rotor field-oriented control topology [14,15] or direct flux and torque control [16]. A typical EV indirect FOC control is shown in Figure 18.10. However, as the control is dependent on the accuracy of parameter estimation, an on-line parameter estimation module is normally incorporated as part of the drive enhancement [17]. The decoupling of the torque and flux cannot be accurately achieved if the motor parameters are not adjusted for saturation and the effect of temperature changes on the motor parameters (namely, the rotor inertia and the rotor inductive reactance). Figure 18.11 shows the block diagram that allows on-line adjustment of the motor parameters using the model reference

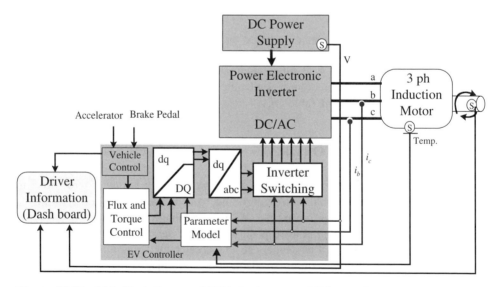

Figure 18.10 Main block diagram of EV induction motor DTC control.

Induction Motor Drives

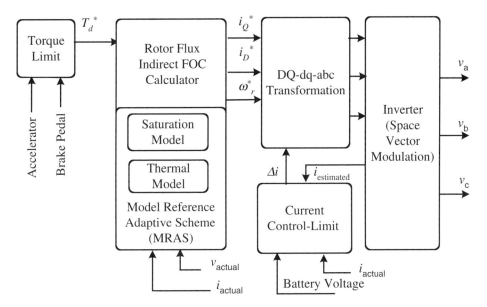

Figure 18.11 EV induction motor controller incorporating on-line parameter adjustment.

adaptive system (MRAS), in which the measured parameters are continuously compared with reference values from command quantities and the difference is used to generate new value of the parameters.

18.7 CONCLUSION

Today there are number of different control schemes for accurate torque control. Field-oriented control and direct torque control are the dominant control strategies for controlling the speed and torque in induction motors. DTC continues to improve as it incorporates space vector modulation and minimal torque ripple control strategies. FOC has a low dynamic performance because the current regulator and parameter detuning causes high torque and flux magnitude errors. DTC has a fast torque response, is relatively simple, and requires no speed or position encoder. Incorporating DTC with space vector modulation provides low steady-state torque ripple and current distortion that is equal to FOC and is less sensitive to parameter variation as in FOC. However, the control algorithm is very complex compared to other control schemes. It requires a relatively fast processor to implement at the desired switching frequency.

The current trend in induction motor control is to incorporate a neuro-fuzzy control strategy in order to achieve high-performance decoupled flux and torque control and improve the operation at very low speed [18]. Many of today's drives are sensorless and incorporate several control topologies for switching the inverter. The inverter is indirectly controlled based on torque and flux control errors using a space vector PWM switching topology and sliding mode control [19]. The overall control strategy is to meet the various load requirements from the motor. In many EV tested models, DTC, which incorporates on-line parameters adjustment, provides reliable torque and speed control of the vehicle.

REFERENCES

1. M. El-Hawary. *Principles of Electric Machines with Power Electronics Applications, Second Edition.* IEEE Press and John Wiley & Sons, Inc., New York, 2002.
2. B. K. Bose. *Power Electronics and AC Drives.* Prentice Hall, Inc., New York, 1986.
3. N. Mohan, T. M. Undeland, and W. P. Robbins. *Power Electronics, Converters, Applications, and Design.* John Wiley & Sons, Inc., New York, 2003.
4. A. M. Trzynadlowski. *Control of Induction Motors.* Academic Press, San Diego, CA, 2001.
5. M. P. Kazmierkowski, R. Krishnan, and F. Blaabjerg. *Control in Power Electronics.* Academic Press, San Diego, CA, 2002.
6. R. Krishnan. *Electric Motor Drives, Modeling, Analysis, and Control.* Prentice Hall, Inc., New York, 2001.
7. P. C. Sen. "Electric motor drives and control — Past, present and future." *IECON '88. Proceedings,* Industrial Electronics Society. Vol. 3, pp. 534–544, October 1988.
8. D. Casadei, G. Grandi, and G. Serra. "Study and implementation of a simplified and efficient digital vector controller for induction motors." Sixth International Conference on Electrical Machines and Drives (Conf. Publ. No. 76), pp. 196–201, September 1993.
9. D. Casadei, G. Grandi, and G. Serra. "Rotor flux oriented torque-control of induction machines based on stator flux vector control." Fifth European Conference on Power Electronics and Applications. Vol. 5, pp. 67–72, September 1993.
10. C. Lascu, L. Boldea, and F. Blaabjerand. "A modified direct torque control for induction motor sensorless drive." *IEEE Transactions on Industry Applications.* Vol. 36, pp. 122–130, January–February 2000.
11. S. A. Shirsavar and M. D. McCulloch. "Speed sensorless vector control of induction motors with parameter estimation." Sixth International Conference on Power Electronics and Variable Speed Drives (Conf. Publ. No. 429), pp. 267–272, September 1996.
12. J. R. G. Schofield. "Variable speed drives using induction motors and direct torque control." IEEE Colloquium on Vector Control Revisited, pp. 5/1–5/7, February 1998.
13. H. W. Van der Broeck, H. Ch. Skudelny, and G. Stanke. Analysis and realization of a pulse width modulator based on voltage space vectors. *IEEE Transactions Industry Application.* Vol. 24, pp. 142–150, 1988.
14. M. Schroedl, D. Hennerbichler; and T. M. Wolban. "Induction motor drive for electric vehicles without speed and position sensors." Fifth European Conference on Power Electronics and Applications, Vol. 5, pp. 271–275, September 1993.
15. S. Wall. "Vector control: a practical approach to electric vehicles." IEEE Colloquium on Vector Control and Direct Torque Control of Induction Motors, pp. 5/1–5/7, October 1995.
16. D. Casadei, G. Serra, and A. Tani. "Direct flux and torque control of induction machine for electric vehicle applications." Seventh International Conference on Electrical Machines and Drives (Conf. Publ. No. 412), pp. 349–353, September 1995.
17. J. Faiz, S. H. Hossieni, M. Ghaneei, A. Keyhani, and A. Proca. "Direct torque control of induction motors for electric propulsion systems." *Electric Power Systems Research,* Vol. 51, pp. 95–101, 1999.
18. P. Z. Grabowski, M. P. Kazmierkowski, B. K. Bose, and F. Blaaberg. "A simple direct torque neuro-fuzzy control of PWM inverter fed induction motor drive." *IEEE Transactions Industry Electronics,* Vol. 47, No. 4, pp. 863–870, August 2000.
19. C. Lascu and A. M. Trzynadlowski. "Combining the principles of sliding mode, direct torque control, and space vector modulation in a high-performance sensorless AC drive." 37th IAS Annual Meeting, *Conference Record of the Industry Applications Conference,* Vol. 3, pp. 2073–2079, October 2002.

APPENDIX: INDUCTION MOTOR MODEL IN THE STATIONARY FRAME

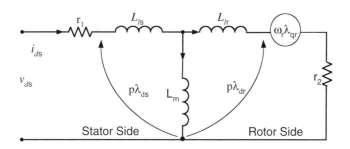

a) Direct axis—Stationary reference model

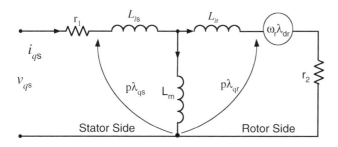

b) Quadrature axis—Stationary reference model

Figure 18.12 Induction motor models in the stationary frame.

The equations that describe the model are based on the equivalent circuits shown in Figure 18.12(a) and (b).

1. $\dfrac{d\omega_r}{dt} = p\dfrac{L_m}{JL_r}[\lambda_{dr} i_{qs} - \lambda_{qr} i_{ds}] - \dfrac{B}{J}\omega_r - \dfrac{T_L}{J}$

2. $\dfrac{d\lambda_{qr}}{dt} = p\omega_r \lambda_{dr} - \tau_r \lambda_{qr} + \tau_r L_m i_{qr}$

3. $\dfrac{d\lambda_{dr}}{dt} = -p\omega_r \lambda_{qr} - \tau_r \lambda_{dr} + \tau_r L_m i_{dr}$

4. $\dfrac{di_{qs}}{dt} = -\dfrac{L_m}{\sigma L_{ls} L_{lr}} p\omega_r \lambda_{dr} + \dfrac{L_m}{\sigma L_{ls} L_{lr}} \tau_r \lambda_{qr} - \left(\dfrac{L_m}{\sigma L_{ls} L_{lr}^2} + \dfrac{r_1}{\sigma L_{ls}}\right) i_{qs} + \dfrac{1}{\sigma L_{ls}} v_{qs}\mathfrak{M}$

5. $\dfrac{di_{ds}}{dt} = \dfrac{L_m}{\sigma L_{ls} L_{lr}} p\omega_r \lambda_{qr} + \dfrac{L_m}{\sigma L_{ls} L_{lr}} \tau_r \lambda_{dr} - \left(\dfrac{L_m}{\sigma L_{ls} L_{lr}^2} + \dfrac{r_1}{\sigma L_{ls}}\right) i_{ds} + \dfrac{1}{\sigma L_{ls}} v_{ds}$

6. $T_d = p \dfrac{L_m}{JL_{lr}} [\lambda_{dr} i_{qs} - \lambda_{qr} i_{ds}]$

where

$$\sigma = 1 - \dfrac{L_m^2}{L_{ls} L_{lr}}$$

and

ω_r = Rotor angular velocity
p = Number of poles
J = Rotor inertia
B = Friction coefficient
T_L = Applied torque

19

DSP-Based Implementation of Vector Control of Induction Motor Drives

Hossein Salehfar
University of North Dakota, Grand Forks, North Dakota

19.1 INTRODUCTION

Due to advances in solid-state power electronic devices and microprocessors, various inverter control techniques are becoming more popular in AC motor drive applications. Pulse width modulation (PWM)-based drives have been used extensively, and they can control both the frequency and the magnitude of the voltage applied to motors. Recently, using vector control concepts a new digital modulation technique known as space vector pulse width modulation (SVPWM) has become very popular. Compared to other modulation techniques and in terms of lower total harmonic distortions (THDs), torque ripples, switching losses, higher output-to-input voltage ratios, and so on, SVPWM-based drives offer superior performance and control.

This chapter presents and discusses a digital signal processor (DSP)-based implementation of SVPWM inverters to control three-phase induction machines. Several excellent references on vector control theory and the SVPWM technique already exist [1–3]. However, to make the chapter self-contained some of the basics of the vector control and SVPWM methods are discussed in the second section. Experimental measurements and results are presented in the third section. Some concluding remarks are given in the last section of the chapter.

19.2 SPACE VECTOR CONTROL

Given a three-phase balanced voltage set as shown in Figure 19.1, the sum of the phase voltages $V_a(t)$, $V_b(t)$, and $V_c(t)$ at any point in time is 0:

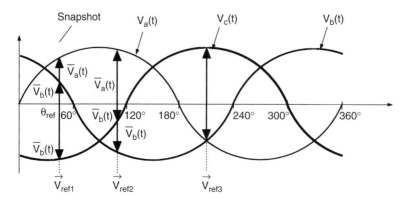

Figure 19.1 Three-phase voltage set.

$$V_a(t) + V_b(t) + V_c(t) = 0 \quad (19.1)$$

Let the following time-dependent voltage vector be defined as:

$$\vec{V}_{abc}(t) = \vec{V}_a(t) + \vec{V}_b(t) + \vec{V}_c(t) \quad (19.2)$$

where

$$\vec{V}_a(t) = 1 V_a(t)$$
$$\vec{V}_b(t) = V_b(t) e^{j120°} \quad (19.3)$$
$$\vec{V}_c(t) = V_c(t) e^{j240°}$$

A voltage space vector $\vec{V}$ is then defined as:

$$\vec{V} \equiv \tfrac{2}{3} \vec{V}_{abc}(t) = \tfrac{2}{3}\left[V_a(t) + \vec{a} V_b(t) + \vec{a}^2 V_c(t) \right] \quad (19.4)$$

where $\vec{a} = e^{j120°}$ and $\vec{a}^2 = e^{j240°}$. Using phase voltages, the construction of a voltage space vector at a given point in time is illustrated in Figure 19.2. Figure 19.3 shows the orthogonal projection of the voltage space vector onto the imaginary-β and the real-α axes in the complex plane. This projection is called the three-phase to two-phase direct transformation [18]. Using the direct transformation, the voltage vector can be written in the complex plane as:

$$\vec{V}_{abc}(t) = V_\alpha + j V_\beta \quad (19.5)$$

The opposite of direct transformation is the reverse vector transformation, as illustrated in Figure 19.2. The reverse vector transformation can be achieved by projecting $\tfrac{2}{3} \vec{V}_{abc}(t)$ onto the a, b, and c axes [4]. Combining Equations 19.4 and 19.5, the voltage space vector can be related to the complex plane as:

$$\vec{V} = \tfrac{2}{3} \vec{V}_{abc}(t) = \tfrac{2}{3}(V_\alpha + j V_\beta) = \tfrac{2}{3}\left[1 V_a(t) + \vec{a} V_b(t) + \vec{a}^2 V_c(t) \right] \quad (19.6)$$

DSP-Based Implementation of Vector Control of Induction Motor Drives 407

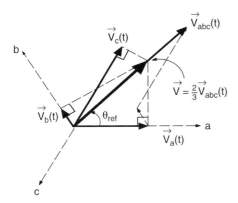

Figure 19.2 Construction of voltage space vector.

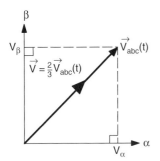

Figure 19.3 Direct transformation.

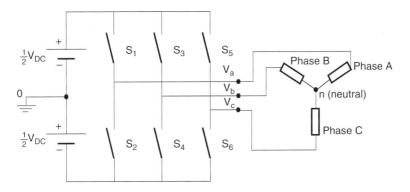

Figure 19.4 Three-phase inverter with load.

A three-phase inverter with a Y connected load is shown in Figure 19.4. In this figure V_{a0}, V_{b0}, and V_{c0} are the inverter output voltages with respect to the return terminal of the DC source marked as "0". These voltages are called the pole voltages or inverter phase voltages. And V_{an}, V_{bn}, and V_{cn} are the load phase voltages (the inverter output voltages) with respect to n. Each switching circuit configuration generates three independent pole voltages, V_{a0}, V_{b0}, and V_{c0}. There are eight possible switching configurations,

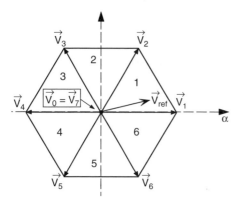

Figure 19.5 Space vectors and rotating reference vector.

called the operating states or the inverter states. The two possible pole voltage values that can be produced at any time are $-\frac{1}{2}V_{DC}$ and $+\frac{1}{2}V_{DC}$. For example, when switches S_1, S_4, and S_5 are closed, the corresponding pole voltages of the inverter state are:

$$V_{a0} = \tfrac{1}{2}V_{DC}$$
$$V_{b0} = -\tfrac{1}{2}V_{DC} \qquad (19.7)$$
$$V_{c0} = \tfrac{1}{2}V_{DC}$$

Using this procedure, the inverter state in Equation 19.7 is represented by the notation (101) and the corresponding space vector is denoted as $\vec{V}_{101}$ or $\vec{V}_6$. All the possible pole (phase) voltages created by the eight inverter states can be transformed into eight corresponding space vectors by plugging in the pole voltage values of each inverter state into Equation 19.6. The resultant eight space vectors are denoted as $\vec{V}_{000}$, $\vec{V}_{100}$, $\vec{V}_{110}$, $\vec{V}_{010}$, $\vec{V}_{011}$, $\vec{V}_{001}$, $\vec{V}_{101}$, $\vec{V}_{111}$ or $\vec{v}_0$, $\vec{v}_1$, $\vec{v}_2$, $\vec{v}_3$, $\vec{v}_4$, $\vec{v}_5$, $\vec{v}_6$, $\vec{v}_7$. Space vectors $\vec{v}_0$ and $\vec{v}_7$ are the null-state vectors and they apply zero voltages across the load. Similarly, $\vec{v}_1$, $\vec{v}_2$, $\vec{v}_3$, $\vec{v}_4$, $\vec{v}_5$, $\vec{v}_6$. are the active-state vectors that apply non-zero voltages to the load.

Figure 19.5 shows all the eight possible space vectors in the complex plane. The space vectors take six distinct angles as commanded by the inverter active-states, forming a hexagon with the two null vectors at the axis of the hexagon. The hexagon represents the range of all realizable space vectors [5]. In a SVPWM process, it is possible to realize any arbitrary and desired space vector that lays within each sector of the hexagon.

The SVPWM method generates the appropriate time-based space vectors so that any desired voltage vector $\vec{V}_{ref}$ can be approximated within a fundamental time period. For example, consider the voltage $\vec{V}_{ref}$ located in sector 1 of the hexagon of Figure 19.5. To approximate this $\vec{V}_{ref}$, the space vectors ($\vec{v}_0$, $\vec{v}_1$, $\vec{v}_2$, $\vec{v}_7$) are applied in sequence such that $\vec{v}_0$ is applied for a time period T_0, $\vec{V}_1$ for a time period T_1, $\vec{V}_2$ for a time period T_2, and $\vec{V}_7$ for a time period T_7 such that

$$\vec{V}_{ref} \cdot T_{PWM} = \vec{V}_0 \cdot T_0 + \vec{V}_1 \cdot T_1 + \vec{V}_2 \cdot T_2 + \vec{V}_7 \cdot T_7 \qquad (19.8)$$

The above sequence consists of four space vectors, two of which are the zero-state vectors. Figure 19.6 shows the four inverter states approximating this $\vec{V}_{ref}$. Figure 19.6

DSP-Based Implementation of Vector Control of Induction Motor Drives

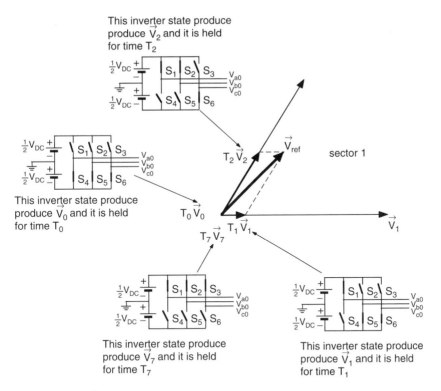

Figure 19.6 $\vec{V}_{ref}$ approximation.

also shows the projections of $\vec{V}_{ref}$ onto the adjacent space vectors. These projections can be used to determine the respective "on" times, T_1 and T_2 and the null times, T_0 and T_7, from the following equation 19.5.

$$T_{PWM} = T_0 + T_1 + T_2 + T_7$$
$$T_{null} = T_0 + T_7 \tag{19.9}$$

The inverter state time durations can be calculated as follows for various values of index n:

$$\begin{bmatrix} T_n \\ T_{n+1} \end{bmatrix} = \frac{\sqrt{3}T_{PWM}|\vec{V}_{ref}|}{2V_{DC}} \begin{bmatrix} \sin(n60°) & -\cos(n60°) \\ -\sin[(n-1)60°] & \cos[(n-1)60°] \end{bmatrix} \begin{bmatrix} \cos\omega_{ref}t \\ \sin\omega_{ref}t \end{bmatrix} \tag{19.10}$$

$$\theta_{ref} = \omega_{ref}t = \tan^{-1}\left(\frac{V_\beta}{V_\alpha}\right) \tag{19.11}$$

Figure 19.7 shows the production of some of the state vector signals as $\vec{V}_{ref}$ rotates in time.

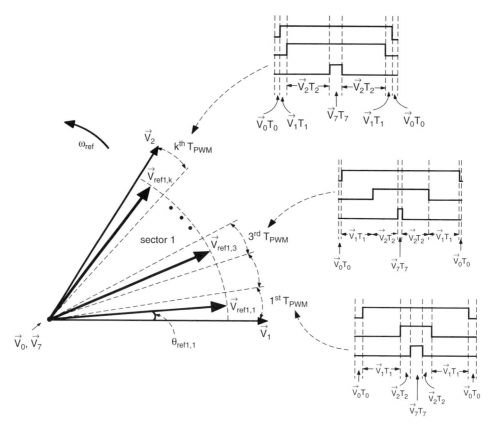

Figure 19.7 Production of state vector signals.

19.3 EXPERIMENTAL RESULTS

The experimental system consists of the following hardware and software items:

1. TI Digital Motor Controller (DMC) drive platform
2. TI eZdsp DSK platform
3. Three-phase induction motor
4. PC with TI Code Composer (CC) and VisSim Rapid Prototyper
5. Low/high pass filters
6. Variac
7. Isolation transformer

Prior to applying the SVPWM signals to the gates of the power transistors, the DSP-based SVPWM signals were analyzed to verify the theory. Figure 19.8 shows one of the experimental setups to view the PWM-phase "a" output signal at the interface connector (P2). Next, other line-to-ground and line-to-line SVPWM waveforms were observed with and without low pass filtering. Six space vector PWM signals are distributed through connector P2 to turn on and off the power transistors on the Digital Motor Controller board. The three-phase signals produced are PWM1 (SVPWM signal-phase "a"), PWM3

DSP-Based Implementation of Vector Control of Induction Motor Drives

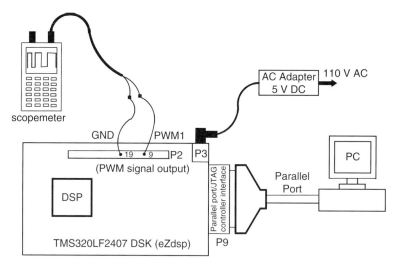

Figure 19.8 eZdsp setup to observe PWM signals.

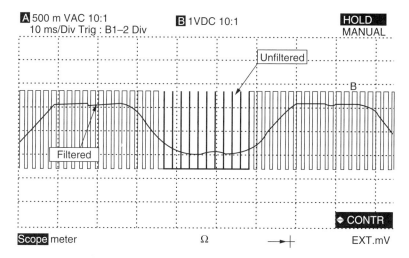

Figure 19.9 Filtered (200 Hz cut-off) and unfiltered PWM1 signal.

(SVPWM signal-phase "b"), and PWM5 (SVPWM signal-phase "c"), whereas PWM2, PWM4, and PWM6 are complementary SVPWM signals. Figure 19.9 shows the PWM1 signal superimposed on the low-pass-filtered (200 Hz cut-off frequency) PWM1 signal. The low-pass-filtered PWM1 signal is referred to as the duty ratio of the PWM1 signal.

It can be seen from Figure 19.9 that the phase "a" voltage is not purely sinusoidal, due to the presence of harmonics of triplen orders [6]. The triplen harmonics occur at 49.8 Hz, 99.6 Hz, 149.4 Hz , ... and so on. Since the PWM1 duty ratio waveform was low-pass filtered with a cut-off frequency of 200 Hz, the triplen harmonics are present up to 200 Hz. Figure 19.10 shows the fundamental (30 Hz cut-off), the filtered (200 Hz cut-off), and the unfiltered PWM1 signal.

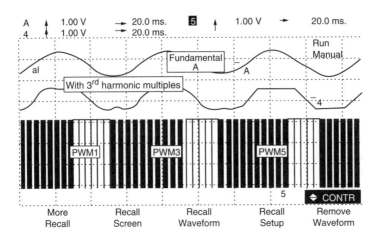

Figure 19.10 Fundamental, filtered, and unfiltered PWM1.

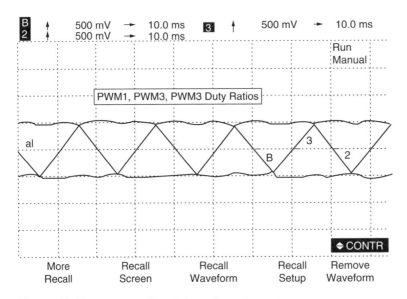

Figure 19.11 Low-pass-filtered three-phase duty ratios.

Figure 19.11 shows the three-phase duty ratio waveforms (PWM1, PWM3, and PWM5 signals). Keeping the cut-off frequency at 200 Hz excluded the switching frequency (2 kHz) harmonics and its multiples.

A fundamental frequency line-voltage waveform is obtained by taking the difference between PWM1 and PWM3 waveforms, as shown in Figure 19.12. The triplen order harmonics cancel each other, leaving a nearly pure sine wave. However, the high switching frequency harmonics do not cancel this way; rather, they are removed by the motor's inductances. The motor winding's inductances act as a low pass filter and remove the high frequency switching components, leaving the triplen order harmonics in the phase voltage waveforms. These harmonics do not affect the line-voltages [6].

DSP-Based Implementation of Vector Control of Induction Motor Drives

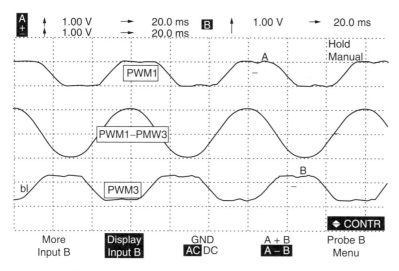

Figure 19.12 Duty ratio and line-voltage waveforms.

19.4 CONCLUSIONS

The SVPWM technique is relatively simple and more efficient than other conventional PWM control methods, such as the sine wave PWM technique. In SVPWM method, the voltage vector $\vec{V}_{ref}$, the rotating vector, is the desired three-phase output voltage of the inverter, mapped to the complex plane. When the desired waveform is a balanced three-phase voltage waveform, then $\vec{V}_{ref}$ becomes a rotating vector with a frequency and amplitude equal to that of the desired three-phase voltage waveform. The SVPWM method provides a more efficient use of the DC supply voltage compared to other PWM methods. With the SVPWM technique, the maximum inverter output voltage is 90.6% of the inverter capability, compared to the sine wave PWM, which is only 78.5% efficient. With SVPWM method (compared to sine wave PWM method) the percentage of the inverter capability is increased by 15.4%.

REFERENCES

[1] D.W. Novotny and T. A. Lipo, *Vector Control and Dynamics of AC Drives*, Oxford University Press, Great Britain, 1997.
[2] B.K. Bose, *Modern Power Electronics and AC Drives*, Prentice Hall, NJ, 2002.
[3] N. Mohan, *Electric Drives, An Integrated Approach,* MNPERE, Minneapolis, MN, 2000.
[4] S. Buso, "Digital Control of Three-Phase DC/AC Converters: Space Vector Modulation," University of Padova, September 1999.
[5] Texas Instruments, "Implementing Space Vector Modulation, Application Notes," 2001.
[6] P. Sanitua, W. Sangchai, T. Wiangtong, and P. Petchatuporn, "FPGA-Based IC Design for 3-Phase PWM Inverter with Optimized Space Vector Modulation Schemes," Mahanakorn University of Technology, Bangkok, Thailand.

20

Switched Reluctance Motor Drives

Babak Fahimi and Chris Edrington
University of Missouri-Rolla, Rolla, Missouri

20.1 INTRODUCTION

Switched reluctance motor (SRM) drives have received considerable attention over the past few years in the ever-growing market of adjustable speed motor drives (ASMDs). A rugged structure and modular magnetic configuration along with relatively high-grade performance indices are among the advantages of the SRM drive. In addition, the absence of magnetic sources (i.e., windings or permanent magnets) on the rotor make SRM relatively easy to cool and insensitive to high temperatures (see Figure 20.1). The latter is of prime interest in automotive applications, which demand operation under harsh ambient conditions.

As a singly excited synchronous machine, SRM generates its electromagnetic torque solely based on the principle of reluctance torque. Most electromechanical energy conversion devices are formed by the combination of two magnetic fields. The source for these magnetic fields is furnished by either a controllable electromagnet or by the virtue of using a permanent magnet. Consequently, the attraction or the repletion force between these magnetic fields originates the dominant part of the torque. In SRM, the tendency of a polarized rotor pole to get in alignment with an excited stator pole is the only source of torque generation. An optimal performance is achieved by proper positioning of the stator excitation with respect to the rotor position. Therefore, sensing of the rotor position becomes an integral part of the control in SRM drives.

A unipolar power inverter is usually used to supply electric power to SRM. Generation of the targeted current profile is performed using either a hysteresis or PWM-type current controller. Although a square-shaped current pulse is commonly used for excitation of SRM at low speeds, optimal current profiles are sometimes used to mitigate the undesirable effects such as excessive torque undulation and audible noise.

Figure 20.1 Rotor and stator of an 8/6 SRM.

A SRM drive portrays an outstanding example of the advanced motor drives in which the burden is shifted from complicated geometries into development of sophisticated control algorithms, a quality that is supported by the existing trend in development of high-grade yet cost-effective DSP-based controllers and fast semiconductor switches.

The present chapter explores the magnetic and power electronics fundamentals of the SRM drive. As optimal torque control is the major differentiating factor among various parts of electric drives, we will give a detailed explanation of the torque generation process and optimal torque control in four quadrants. Following this, development of speed controllers for SRM drives will be discussed.

20.2 HISTORICAL BACKGROUND

Switched reluctance machines are a modern version of the "electromagnetic engine," first invented in 1835 in U.K. by Taylor and Davidson [1]. This system was comprised of seven salient rotor poles and four electromagnets fixed on the stator. The commutation of the phases was performed using a mechanical commutator installed on the rotor. In 1838, this system was further improved and incorporated in the first electric locomotive of the world by Davidson and his colleagues.

The main practical shortcoming of this electromechanical motion device was the turn-off process. As one may imagine, the mechanical commutation of the coils occurs at the point where the magnetic field exhibits its maximum strength. This was due to the fact that sudden disconnection of a highly inductive circuit results in a strong spark on the mechanical commutator. This may explain why newspaper *Edinburgh Witness* published the following release on Davidson's electric locomotive: "One curious phenomenon connected with the motion of this new and ingenious instrument was the extent and brilliancy of the repeated electric flashes which accompanied the action of the machinery." The absence of a practical solution for commutation was in fact the main reason for the failure of switched reluctance machine on its early days. By the end of the 1960s and early 1970s the concept of SRM was reintroduced. This was at the same time that power electronics converters were making their entrance to the field of electro-mechanics. The introduction of power electronics made it possible to recover the stored magnetic energy to the source through novel switching strategies.

The beginning of this new era for SRMs came by the patents of Bedford in 1972 [2,3] as well as the article by Professor Nasar in 1969 [4]. The invention of Bedford was

Switched Reluctance Motor Drives

very similar to the conventional switched reluctance motors of today, whereas Nasar's machine was in the form of a disc rotor machine. In the 1980s the development of these new machines were seriously pursued in the U.K. and the U.S. as a result the number of publications and patents on SRMs witnessed a very strong increase. The state of the art research and development in switched reluctance machines over the past few years, and after almost 30 years of research, is mostly focused at optimization of performance, generation, and mitigation of noise and torque ripple. It is widely believed that once practical answers for these areas of research have been found switched reluctance machines can appear as strong contenders in the adjustable speed drives market.

20.3 FUNDAMENTALS OF OPERATION

Switched reluctance machines can operate as a motor or generator. To explain the torque generation process, one may investigate the mechanism of the electromechanical energy conversion. As shown in Figure 20.3, in order to establish a reluctance torque, a stator phase is excited at unaligned position, viz., a position at which a pair of stator and rotor poles exhibit their largest airgap length. By magnetizing the stator pole, the closest rotor pole will be magnetically polarized and forms an attraction force. The tangential component of this force substantiates an electromagnetic torque, which tends to reduce the airgap length. The shape of the current is usually controlled such that a maximum torque per ampere is generated. As the rotor approaches full alignment (also known as aligned position), the radial component of the attraction force becomes dominant, and the tangential component of the force reduces to zero. Therefore it makes economic sense to remove the current before aligned position. The shaded area in Figure 20.3 depicts the converted energy into mechanical form whereas the area denoted by "R" demonstrates the magnetic energy that has not been converted into useful work. Notably, the ratio between mechanical work and total converted energy into magnetic form is an indication of power quality in SRM drives.

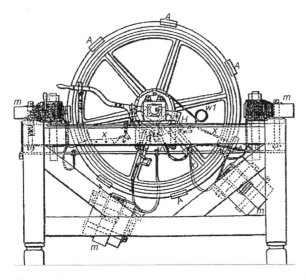

Figure 20.2 Basic construction of an electromagnetic engine.

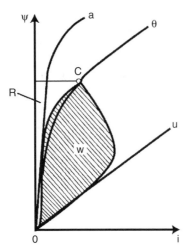

Figure 20.3 Electromechanical energy conversion in SRM.

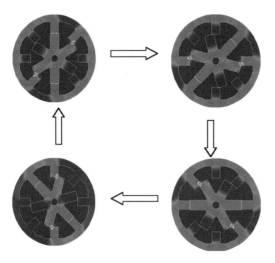

Figure 20.4 Illustration of short vs. long flux path for an 8/6 SRM.

In order to obtain motoring action a stator phase is excited when rotor is moving from unaligned position towards aligned position. Similarly by exciting a stator phase when rotor is moving from aligned towards unaligned position, a generating action will be achieved. By sequential excitation of the stator phases a continuous motion can be achieved. Figure 20.4 illustrates the distribution of the magnetic field during commutations in an 8/6 SRM drive. Notably, direction of the rotation is opposite to that of the stator excitation. As can be observed, a short flux path in the back-iron occurs in each electrical cycle. This, in turn, may cause asymmetry in the torque production process.

Proper synchronization of the stator excitation with the rotor position is a key step in development of the optimal control in SRM drives. Since magnetic characteristics of the SRM, such as phase inductance or phase flux linkage, portray a one-to-one correspondence with the rotor position, they may directly be used for control purposes. In either

Switched Reluctance Motor Drives

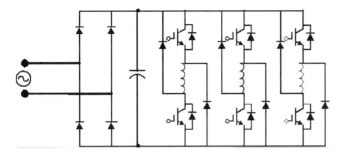

Figure 20.5 An asymmetric bridge with the front end rectifier for a 3-ϕ SRM drive.

case direct or indirect detection of the rotor position forms an integral part of the control in the SRM drives.

The asymmetric bridge shown in Figure 20.5 is the commonly used power electronics inverter that is employed for SRM drives. This topology offers a unipolar architecture that allows for a satisfactory operation in SRM drives. If both switches are closed, the available DC link voltage is applied to the coil. By opening the switches, the negative DC link voltage will be applied to the coil terminals and freewheeling diodes will guarantee a continuous current in the windings of the motor. Obviously, by keeping one of the switches closed while the other one is open, the respective freewheeling diode will provide a short-circuited path for the current. This topology can be effectively used to implement PWM-based or hysteresis-based current regulation as demanded by control objective. However, one should notice that at high speeds the induced motional voltage in the coils is dominant and does not allow an effective control on the current waveform. Therefore, current regulation is an issue that is related to low-speed mode of operation. During generation, the mechanical energy supplied by the prime mover will be converted into an electrical form manifested by the induced motional voltage. Unlike motoring mode of operation, this voltage acts as a voltage source that intensifies the increase of the current in the stator phase, thereby resulting in generation of electricity.

In addition to the above classic converter, over the past two decades alternative topologies have been suggested for the control of SRM drives. Figure 20.6 illustrates a few viable solutions that have been reported in the literature.

It must be noted that a number of new topologies, such as soft-switched and improved C-dumped converters, have been recently developed; these have not been included in Figure 20.6.

An ideal converter must satisfy the following:

- Low switches per phase ratio
- Ability to supply and control a commanded current independently and precisely
- Flexibility in adapting to any number of phases (odd or even)
- Low VA rating for a given rating of the drive
- Robustness and reliability
- Good efficiency
- Ability to operate in all four quadrants effectively
- Less torque ripple and noise

The most commonly used converter types are the classic half-bridge-type converter and the split phase-type converter. Classic-type converter is the most flexible type converter but it requires more switches. The split phase-type converter requires an even number of

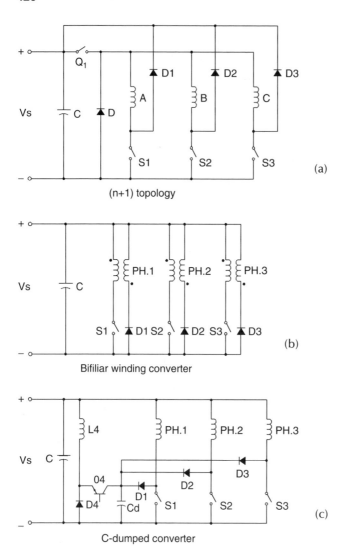

Figure 20.6 Various SRM converter topologies.

phases and has a high active device rating; hence, it is suitable only for low-voltage and low-power applications. Another type of converter suitable for star-connected-type SRM is the dual-decay-type converter, with flexibility of control in freewheeling mode and reduced device ratings. Most of the converters developed recently are aimed at reducing the number of switches and are found to be more application specific.

Due to its nonlinear behavior, the dynamics of SRM drives undergo a significant change as the operating point of the drive changes in torque vs. speed plane. What follows is a summary of the dynamics in SRM drives over the entire speed range.

Each phase of the SRM can ideally be considered as a decoupled magnetic circuit whose dynamic is given by the phase voltage differential equation:

$$V = Ri + \frac{\partial \psi}{\partial \theta_r}\omega + \frac{\partial \psi}{\partial i}\frac{di}{dt} \qquad (20.1)$$

Switched Reluctance Motor Drives

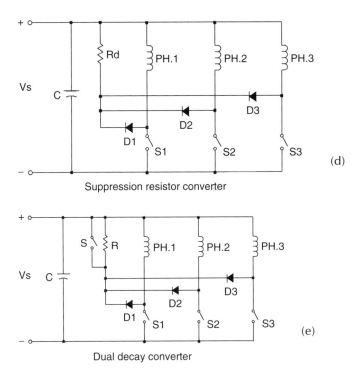

Figure 20.6 (continued)

in which V, R, ψ, θ_r, and ω denote phase voltage, stator phase resistance, flux linkage, rotor position, and angular speed, respectively. It is also important to note that the flux linkage is described as:

$$\psi(i,\theta_r) = L(i,\theta_r)i \tag{20.2}$$

where $L(i,\theta_r)$ represents the gross self-inductance of the stator phase winding and i stands for phase current. The dependency of L upon phase current manifests the nonlinear effects of saturation in iron. By combining Equation 20.1 and Equation 20.2, the differential equation governing the dynamics of a phase winding in SRM drive, in the absence of mutual effects, is expressed as:

$$V = \left(R + \omega \frac{\partial L}{\partial \theta_r}\right)i + \left(L(\theta_r) + i\frac{\partial L}{\partial i}\right)\frac{di}{dt} \tag{20.3a}$$

or equivalently:

$$V = R_{eq}i + L_{eq}\frac{di}{dt}$$

$$R < R_{eq} < R + \omega \frac{L_{max} - L_{min}}{\min(a_s, a_r)} \tag{20.3b}$$

$$L_{min} < L_{eq} < L_{max}$$

Table 20.1 Dynamic Behavior of SRM in Various Operating Regions

Region	Phase Voltage Equation
Region I	$V = Ri + \left(L^* + i\dfrac{dL^*}{di}\right)\dfrac{di}{dt}$
Region II	$V = Ri + \left(L(\theta_r) + i\dfrac{dL}{di}\right)\dfrac{di}{dt}$
Region III	$V = \left(R + \omega\dfrac{dL}{d\theta_r}\right)i + \left(L(\theta_r) + i\dfrac{dL}{di}\right)\dfrac{di}{dt}$
Region IV	$V = \left(R + \omega\dfrac{dL}{d\theta_r}\right)i + L(\theta_r)\dfrac{di}{dt}$
Region V	$V = \omega\dfrac{dL}{d\theta_r}i + L(\theta_r)\dfrac{di}{dt} + \sum M_j \dfrac{di_j}{dt}$

where a_s and a_r denote rotor and stator pole arcs, respectively. Due to the spatial distribution of the magnetic field and saturation effects, coefficients of this equation represent a time variant and nonlinear function. Furthermore, the electrical frequency of excitation for consecutive phases of SRM is given by:

$$f_e[Hz] = \frac{\omega[rad/\sec]N_r}{2\pi} \quad (20.4)$$

where N_r denotes number of rotor poles. This shows that the available time for computation in control reduces as the speed of the drive increases. Table 20.1 summarizes the dynamic behavior of SRM drives in various operational regions as shown in Figure 20.7.

Table 20.1 indicates that the dynamics of SRM undergo a significant change as the speed of the drive increases. This stems from minor role of motional back-EMF at very low speed, unsaturated operation along with significant contribution of mutual inductances at very high speed, and nonlinear effects of saturation in constant torque region (region III). This can be interpreted as a highly dynamic system with a variable structure. Consequently, any successful sensorless method needs to take these variations into account. It must also be noted that concepts such as low speed and high speed introduced here reflect relative measures of speed that are entirely defined by the dynamic behavior of the SRM drive. In other words, depending on magnetic design and operational conditions of SRM drive, a specific speed may fall into low- or high-speed categories. The low-speed region reflects a range of speed within which the induced voltage is less than the available DC link voltage. The high-speed region, in contrast, is a range of speeds where the motional back-EMF exceeds the DC link voltage (at least for a portion of the electrical cycle).

In practice, the ability to shape and regulate the current pulse is a representation of operating in low-speed regime. In high-speed regime, on the other hand, the shape of the current pulse is dictated by the motional back-EMF and is referred to as single pulse mode

Switched Reluctance Motor Drives

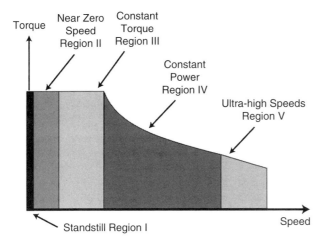

Figure 20.7 Various operational regions in SRM.

of operation. Due to essential role of motional back-EMF in forming various modes of operation, tracking motional back-EMF is of great importance to our proposed generalized sensorless methodology.

In order to understand the impact of the motional back-EMF, interdependence between operating point and motional back-EMF needs to be explored. Figure 20.4 shows a typical torque vs. speed characteristic of an SRM drive. We have further identified two operating points that represent high torque-low speed and low torque-high speed conditions. It is also important to note that the nonlinear effects of saturation are significant after phase current exceeds a certain threshold. A horizontal line in Figure 20.4 represents this boundary. Saturation in SRM drives tends to reduce the aligned position inductance, whereas the unaligned position is almost unchanged. The motional back-EMF and average electromagnetic torque in an SRM drive are approximated as:

$$E = \omega\, i \frac{\partial L}{\partial \theta_r} \cong \left(\frac{\pi}{N_r}\right) \omega\, i \{L_a(i) - L_u\}$$

$$T_{ave} = \int_{\theta_U}^{\theta_A} \int_0^i i' \frac{\partial L}{\partial \theta_r} di' \cong \frac{N_r}{\pi} \int_0^i i' \{L_a(i) - L_u\}\, di' \tag{20.5}$$

where the subscripts in L_a and L_u refer to aligned and unaligned inductances, respectively, and N_r stands for the number of rotor poles. Accordingly, the incremental variations in motional back-EMF and reference (maximum) current at every operating point are given by:

$$\Delta E = \left\{\frac{N_r}{\pi} i(L_a(i) - L_u)\right\} \Delta\omega + \left\{\frac{\omega}{i} + \frac{1}{i(L_a(i) - L_u)} \frac{dL_a}{di}\right\} \Delta T_{ave}$$

$$\Delta i = \left\{\frac{\pi}{(iL_a(i) - L_u i)N_r}\right\} \Delta T_{ave} \tag{20.6}$$

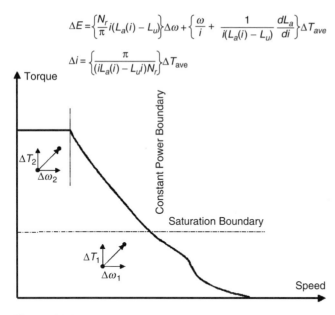

Figure 20.8 Effects of operating point on motional back-EMF.

As can be seen from Equation 20.6, saturation increases the percentage of rise in reference current as one enters into the saturated region. In other words, one needs more current to generate the same increase in demanded torque as compared to the low torque region. The motional back-EMF, on the other hand, is affected by the speed and torque. The impact of change in speed on the induced voltage seems to be linear in either a saturated or unsaturated region. Saturation and speed of operation, however, influence the impact of the torque component. At a constant speed, saturation decreases the impact of torque on motional back-EMF. Operational speed, however, acts as a multiplier to the torque component resulting in a high impact in induced voltage. Consequently, comparing the two operating points in Figure 20.8, for the same changes in speed and torque, a much higher jump in induced voltage of the first operating point is expected while the second operating point will result in a higher increase in the reference current. This example shows how differently SRM characteristics would react to the same set of changes in the operating point.

20.4 FUNDAMENTALS OF CONTROL IN SRM DRIVES

Control of electromagnetic torque is the main differentiating factor between various types of adjustable speed motor drives. In switched reluctance motor drives, electromagnetic torque is primarily controlled by tuning the commutation instants and the profile of phase current. Figure 20.9 depicts the basics of commutation in SRM drives. As can be seen, by proper positioning of the current pulse one can obtain positive (motoring) or negative (generating) types of operation.

Induced motional voltage and electromagnetic torque generated by SRM drive can be expressed in terms of co-energy as follows:

Switched Reluctance Motor Drives

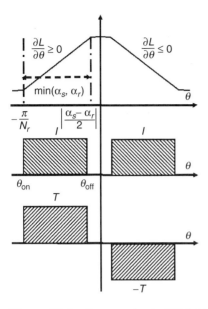

Figure 20.9 Commutation in SRM drives.

$$E = \frac{\partial^2 W_c}{\partial \theta \partial i} \omega \approx \frac{dL(\theta)}{d\theta} i\omega$$

$$T = \frac{\partial W_c}{\partial \theta} \approx \frac{1}{2} \frac{dL(\theta)}{d\theta} i^2$$

(20.7)

where W_c, L, θ, i, and ω stand for co-energy, phase inductance, rotor position, phase current, and angular speed, respectively. It must be noted that the nonlinear effects of magnetic saturation are neglected here. It is evident that a positive torque is achieved only if the current pulse is positioned in a region with an increasing inductance profile. Similarly, a generating mode of operation is achieved when excitation is positioned in a region with a decreasing inductance profile. In order to enhance the productivity of the SRM drive, the commutation instants, i.e., θ_{on}, θ_{off}, need to be tuned as a function of the angular speed and phase current. To fulfill this goal, optimization of the torque per ampere is a meaningful objective. Therefore, exciting the phase when inductance has a flat shape should be avoided. At the same time, the phase current needs to be removed well before the aligned position to avoid generation of the negative torque.

20.4.1 Open Loop Control Strategy for Torque

By proper selection of the controlled variables, viz., commutation instants and reference current, an open loop control strategy for SRM drives can be designed. The open loop control strategy comprises the following steps:

1. Detection of the initial rotor position.
2. Computation of the commutation thresholds in accordance with the sign of torque, current level, and speed.
3. Monitoring of the rotor position and selection of the active phases.
4. A control strategy for regulation of the phase current at low speeds.

What follows is an in-depth explanation of each step.

20.4.1.1 Detection of the Initial Rotor Position

The main task at standstill is to detect the most proper phase for initial excitation. Once this is established, according to the direction of rotation, a sequence of excitation between stator phases will be put in place. The major difficulty in using commercially available encoders is that they do not provide a position reference. The easiest way for startup process is then to align one of the stator phases. This can be achieved by exciting an arbitrary stator phase with an adequate current for a short period of time. Once at aligned position, given the geometry of the machine, a reference initial position can be established. As can be observed, this method requires an initial movement by the rotor, which may not be permitted in some applications. In those cases incorporation of a sensorless scheme at standstill is sought. Although explanation of sensorless control strategies for detection of the rotor position is beyond the extent of this chapter, due to its critical role, detection of rotor position at standstill is explained here.

To achieve this goal, a series of voltage pulses with fixed and sufficiently short duration is submitted to all phases. By consequent comparison between the magnitudes of the resulting peak currents, the most appropriate phase for conduction is selected. Figure 20.10 shows a set of normalized inductance profiles for a 12/8 SRM drive.

Notably, a full electrical period is divided into six separate regions according to the magnitude of the inductances. Due to the absence of the induced voltage and small amplitude of currents, one can prove that the following relationship will hold for the magnitude of measured currents:

$$I_{ABC} = \frac{V_{Bus} \Delta T}{L_{ABC}} \quad (20.8)$$

where ΔT, V_{Bus}, and L_{ABC} stand for duration of pulses, DC link voltage, and phase inductances, respectively. Table 20.2 summarizes the detection process for a 12/8 SRM drive.

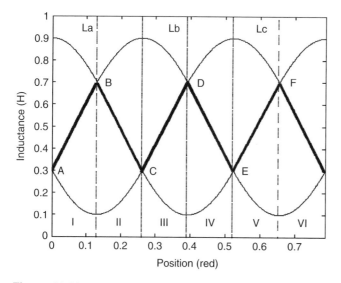

Figure 20.10 Assignment of various regions according to inductances in a 12/8 SRM drive.

Switched Reluctance Motor Drives

Table 20.2 Detection of the Best Phase to Excite at Standstill

Region	Condition	Rotor Angle (mech)
I	$I_A < I_B < I_C$	$0 < \theta^* < 7.5°$
II	$I_B < I_A < I_C$	$7.5° < \theta^* < 15°$
III	$I_B < I_C < I_A$	$15° < \theta^* < 22.5°$
IV	$I_C < I_B < I_A$	$22.5° < \theta^* < 30°$
V	$I_C < I_A < I_B$	$30° < \theta^* < 37.5°$
VI	$I_A < I_C < I_B$	$37.5° < \theta^* < 45°$

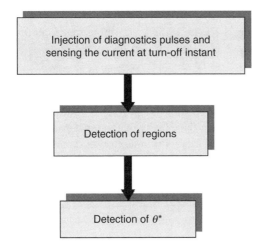

Figure 20.11 Detection of rotor position at standstill.

Once the range of position is detected, proper phase for starting can be easily detected. Furthermore, in each region there exists a phase that offers a linear inductance characteristic. This phase can be used for computation of the rotor position (using Equation 20.8).

The flow-chart in Figure 20.11 summarizes the detection process at standstill.

20.4.1.2 Computation of the Commutation Thresholds

In the next step the commutation angles for each phase should be computed and stored in the program. If the commutation angles are fixed, then the computation of the thresholds are relatively straightforward. It must be noted that within each electrical cycle, every phase should be excited once. In addition, in a symmetric machine, phases are shifted by:

$$\Delta\theta = \frac{(N_s - N_r)360°}{N_s} \text{[Electrical]} \qquad (20.9)$$

where N_s and N_r stand for number of rotor and stator poles. Given a reference for rotor position such as aligned position of phase-A, one can compute and store the commutation

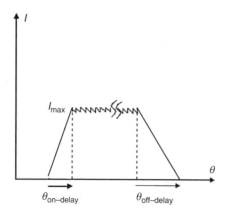

Figure 20.12 A typical current pulse at low speeds.

instants for each phase. The commutation thresholds are usually converted into a proper scale so they can be compared with the content of a counter that tracks the number of incoming pulses from the position sensor.

If an optimal performance of the machine is targeted, effects of rotational speed and current must be taken into account. Figure 20.12 shows a typical current pulse for an SRM drive. To achieve an optimal control, the delay angles during the turn-on and turn-off process need to be taken into account. By neglecting the effects of motional back-EMF in the neighborhood of commutation (this is a valid assumption as turn-on and turn-off instants occur close to unaligned and aligned position, respectively) one can calculate the delay angles as:

$$\theta_{\text{on-delay}} = \frac{\omega L_u}{r} \ln\left(\frac{V}{V - rI_{\max}}\right)$$

$$\theta_{\text{off-delay}} \approx \theta_{\text{on-delay}} \left(\frac{L_a(I_{\max})}{L_u}\right)$$

(20.10)

where L_u, L_a, ω, V, and r denote unaligned inductance, aligned inductance, angular speed, bus voltage, and stator phase resistance, respectively. The dependency of the aligned position inductance upon maximum phase current is an indication of the nonlinear effects of saturation that need to be taken into account. As the speed and level of current increases, one needs to adopt the commutation angles using Equation 20.10. As can be seen, dependency upon the angular speed is linear, while the dependency upon the maximum phase current has a nonlinear format.

20.4.1.3 Monitoring of the Rotor Position and Selection of the Active Phases

Once the previous steps are done, one can start with the main control tasks, namely enforcing the conduction band and regulation of the current. The block diagram in Figure 20.13 shows the structure used in a typical program, which forms the basic control of the SRM drive. As the first task in the interrupt routine the current value of the rotor position will be compared to the commutation thresholds and the phases, which should be or will be identified. In the next step the current in active phases (an active phase is referred to as a phase that has been turned on) will be regulated.

Switched Reluctance Motor Drives

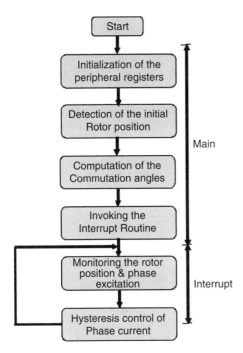

Figure 20.13 Block diagram of the basic control in SRM drives.

20.4.1.4 A Control Strategy for Regulation of the Phase Current at Low Speeds

At low speeds, where the induced motional voltage is small, a method for control of the phase current is necessary. In the absence of such routines, the phase current will increase exponentially and will damage the semiconductor devices and motor windings. Hysteresis and pulse width modulation types of control are commonly used for regulation of phase current at low speeds. At higher speeds, presence of a significantly large induced voltage limits the growth of the phase current, and therefore there is no need for such regulation schemes. The profile of the regulated current depends on the control objective. In most applications a flat-topped (square-shaped) current pulse will be used. Figure 20.14 shows a regulated current waveform along with the gate pulse that is recorded at low-speed region.

In order to conduct the hysteresis type of control currents in active phases are to be sensed. Once the phase current is sampled it needs to be converted into digital form. This can be done using the on-chip analog-to-digital converter in most processors. State-of-the-art on-chip A/D converters offer a conversion time of about 200 ηsec with a 12-bit resolution. The control rules for a classic two-switch-per-phase inverter (see Figure 20.15) are given below:

1. If $I_{min} \geq I$ then both switches are on. This results in applying the bus voltage across the coil terminals.
2. If $I_{max} \leq I$ then both switches are turned off. This results in applying the negative bus voltage across the coil terminals.
3. If $I_{min} \leq I \leq I_{max}$ there is no need to make any changes in the status of the switches; i.e., if the switches are on they remain on and if they are off they remain off.

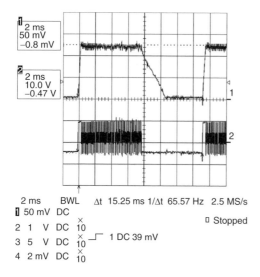

Figure 20.14 Phase current waveform and the gating signal without optimization; reference current = 5.5 A; conduction angle = 180° (electrical); operating speed = 980 rpm; output power = 120 W.

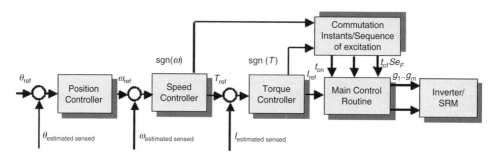

Figure 20.15 Cascaded control configuration for an SRM drive system.

Therefore, by performing a comparison between the sampled current and current limits one can develop a hysteresis control strategy. Since the check on the current occurs during each interrupt service, the period of interrupt should be small enough to allow for a tight regulation.

Once the basic operation of the SRM drive is established, one can design and develop closed loop forms of the control. In the following sections closed loop torque and speed control routines in SRM drive will be discussed. This includes a four-quadrant operation of the drive.

20.5 CLOSED LOOP TORQUE CONTROL OF THE SRM DRIVE

As SRM technology starts to emerge in the form of a viable candidate for industrial applications, the significance of reliable operation under closed-loop torque, speed, and position control for this class of adjustable speed drives increases. Figure 20.15 depicts a typical cascaded control configuration for SRM drives. The main control block is responsible

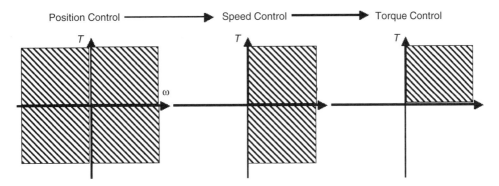

Figure 20.16 Minimum requirement of an adjustable speed drive for performing torque, speed, and position control.

for generation of gate signals for power switches. It also needs to perform current regulation and phase commutation. In order to perform these tasks, it then would require reference current, commutation instants, and sequence of excitation. The torque controller provides the reference current, while the information regarding the commutation is obtained from a separate block that coordinates the motoring, generating, and direction of rotation as demanded by various types of controllers. Feedback information is generated using either estimators or transducers.

Depending on the application, an adjustable speed motor drive may operate in various quadrants of the torque vs. speed plane. For instance, in a water pump application, where control of the output pressure is targeted, torque control in one quadrant is sufficient, whereas in an integrated starter/alternator a four-quadrant operation is necessary. Figure 20.16 shows the minimum requirement of an adjustable speed motor drive for performing torque, speed, and position control tasks. As can be seen, a speed controller may issue positive (motoring) or negative (generating) torque commands to regulate the speed. In a similar way, a position controller will ask for positive (clockwise) and negative (counterclockwise) speed commands. Accommodation of such commands will span all four quadrants of operation in the torque vs. speed plane. As a result, four-quadrant operation is a necessity for many applications in which positioning of the rotor is an objective. In order to achieve four-quadrant operation in SRM drives, the direction of rotation in airgap field needs to be altered. In addition, to generate negative torque (generation mode), the conduction band of the phase should be located in a region with negative inductance slope.

Figure 20.17 depicts a general block diagram of the closed loop torque control system. The main modules in this figure are:

- An estimator for the average/instantaneous electromagnetic torque
- A feed-forward function for fast and convergent tracking of the commanded torque
- A computational block to determine commutation instants according to the sign of demanded torque and magnitude of the phase current

The estimator for average/instantaneous electromagnetic torque is designed based on Equation 20.7 and an analytical expression of the inductance profile. This incorporates an analytical model of the phase inductance/flux linkage as shown below:

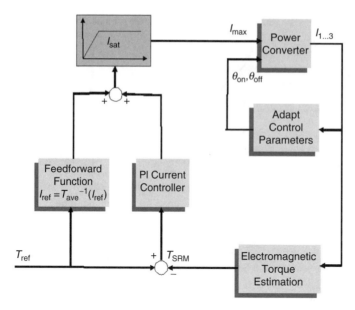

Figure 20.17 General block diagram of the torque control system.

$$L(i,\theta) = L_0(i) + L_1(i)\cos N_r\theta + L_2(i)\cos 2N_r\theta \qquad (20.11)$$

where L_0, L_1, and L_2 represent polynomials that reflect the nonlinear effects of saturation. Moreover, the inverse mapping of the torque estimator is used to form a feed-forward function. In the absence of the torque sensor/estimator this feed-forward function can be effectively used to perform an open loop control of the torque. The use of a feed-forward controller accelerates the convergence of the overall torque tracking. The partial mismatch between reference and estimated torque is then compensated via a PI controller. It must be noted that introduction of the measured torque into the control system requires an additional analog-to-digital conversion. Figure 20.18 shows a comparison between estimated and measured torque in a 12/8 SRM drive at steady state when responding to a periodic ramp function in closed loop. The average torque estimator shows good accuracy. The existing 0.4 N-m average error is due to the fact that iron and stray losses are not included in the torque estimator. In order to perform this test, a permanent magnet drive acting as an active load was set in a speed control loop running in the same direction at 800 rpm.

As mentioned earlier, operation in all four quadrants of the torque vs. speed plane is a requirement for many applications. This is to operate SR as a motor or generator during clockwise and counterclockwise motions. Given the symmetric shape of the inductance profile with respect to aligned position, one can expect that for a given conduction band and at a constant speed, current waveforms during motoring and generating will be the mirror image of each other. However, one should note that during generation the motional back-EMF acts as a voltage source, resulting in an increase of phase current even after a phase is shut down. This may cause some complications in terms of stability at high speeds. In order to alter the direction of rotation, the only necessary step is to change the sequence of excitation. Notably, the sequence of excitation among stator phases is opposite to the direction of rotation. The transition between two modes needs to be quick and smooth. Upon receipt of a command requesting a change in direction, the excited

Switched Reluctance Motor Drives

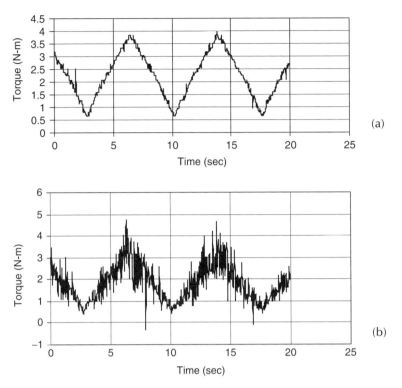

Figure 20.18 Comparison between measured and estimated torque: (a) estimated average torque; (b) measured torque.

phase needs to be turned off to avoid generation of additional torque. Simultaneously, regenerative braking should be performed. This requires detection of a phase in which the inductance profile has a negative slope. The operation in generating mode continues until speed decays to zero (or a tolerable near-zero speed). At this time, all the phases will be cleared and a new sequence of excitation can be implemented. Speed reversal during generating is not a usual case, since direction of the rotation is dictated by the prime mover. In case the speed reversal is initiated by the prime mover, the SR controller needs to be notified. Otherwise, a mechanism for detection of direction of rotation should be in place. Such mechanism would detect any unexpected change of mode, i.e., motoring to generating.

20.6 CLOSED LOOP SPEED CONTROL OF THE SRM DRIVE

As the next step in development of a high-grade SRM drive, speed control is explained. As shown in Figure 20.19, a cascaded type of control can be used to perform closed loop speed control. The speed can be sensed using the position information provided by the encoder to the microcontroller. Due to the fact that an SRM drive is a synchronous machine, one may choose the electrical frequency of the excitation for control purposes. The relationship between mechanical and electrical speeds is given below:

$$\omega_e = N_r \omega_m \quad (20.12)$$

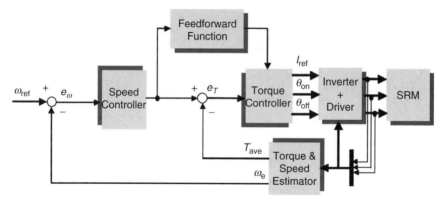

Figure 20.19 Closed loop speed control of an SRM drive.

The ultimate success in performing a tightly regulated speed control relies upon the performance of the inner torque control system, as depicted in Figure 20.19. It is recommended that a feed-forward function be used to mitigate the initial transients in issuing commands into the torque control system.

20.7 INDUSTRIAL APPLICATIONS: VEHICULAR COOLANT SYSTEM

Typical ambient temperatures within the engine compartment are in the neighborhood of 125°C [4]. However, working temperatures of coolants can exceed this temperature, resulting in operating conditions at a much higher temperature. This is due primarily to the heat transfer from the thermal contact between the coolant/pump-housing/electric motor system. Thus, it is evident that a machine capable of withstanding these harsh environmental conditions is required.

Furthermore, as performance enhancements are made with respect to automotive engine designs, coolant system requirements become more demanding. The current trend among automotive engine designers is to reduce the cylinder bore wall thickness and reduce coolant jacket/channel size. Reduction in wall thickness results in an increased heat transfer between the cylinder wall and the coolant jacket, which requires a greater coolant flow rate in order to remove the heat. Reduction in the coolant channel thickness in turn causes increased pressure within the coolant system. Increased flow rate and pressure in the coolant system translates to high speed and torque requirements for the electric actuator.

In addition to the aforementioned mechanical modifications, use of more lightweight materials such as aluminum to replace cast iron are being considered for cylinder heads and intakes. This results in an overall vehicle weight reduction, which helps to improve fuel efficiency. However, since aluminum has a thermal conductivity of 202 W/mK compared to cast iron (which has a thermal conductivity of 58 W/mK), additional load will be placed on the coolant system in order to remove the heat.

Environmental concerns are the impetus behind development of low-emission and fuel-efficient internal combustion engines (ICEs). Thus, there is room for considerable improvement for a large portion of vehicles in the market. Reduction of vehicle weight through use of lighter materials helps increase fuel efficiency. Greater modularity of engine components, incorporation of electronics into electromechanical devices, known as mechatronics, and elimination of the unibelt construction are also furtherances in increasing fuel efficiency and emission reduction.

Furthermore, the trend toward more electric vehicles has emphasized a need to move to a 42 V system. Moving to a 42 V system is necessary, since the increase in power consumption by the vehicle's electrical system creates current levels that are too high for conventional 12 V systems.

Thus, in light of the aforementioned trends it is necessary to design and construct an electrical machine that is capable of efficient 42 V operation, while meeting the requirements of high speed and torque demanded by the coolant system. In addition, the machine should be designed such that other necessary control components such as current sensors, inverters, and encoders can be incorporated into its overall structure or, if possible, eliminated (i.e., a greater focus on mechatronics).

There are several aspects of the SRM that make it one of the best overall candidates for this application. Although SRMs suffer from torque ripple and acoustic noise, these are not potential problems that prohibit its use for this application.

One of the primary characteristics of the SRM, which makes it a viable solution for the pump application, is its ruggedness and invariance to rotor temperatures. In an SRM, all the excitation requirements are given to the stator, which in turn sets up a magnetic dipole between stator and rotor poles. The resulting tendency is to reduce the airgap reluctance by moving the rotor pole toward an aligned position with an excited stator pole. This operational characteristic is much different than the electromechanical energy conversion that takes place in induction machines (IMs) or permanent magnet machines (PMs), which rely on rotor windings or magnets on the rotor in order to establish proper magnetomotive force (mmf) in the air-gap. Flux sources on the rotor typically have characteristics, such as winding resistance or the flux density of the permanent magnets, that are strongly affected by temperature. As stated before there will be a considerable amount of thermal coupling and, therefore, heat transfer into the rotor. Additionally, heat generated due to iron loss in the core of the SRM is principally in the stator. Thus, one additional benefit to the stator bearing the burden of excitation and losses is that it is easily cooled, which can be a potential problem to IM and PM machines whose control characteristics are functions of temperature and whose rotor structures, in the case of IMs, are major sources of heat generation with the machine. Furthermore, SRMs typically have a low-cost construction due to the absence of windings and permanent magnets on the rotor structure. This is of prime interest for automotive products.

The SRM is known to be capable of high-speed operation over a wide constant power region. This aspect of the SRM offers a large amount of flexibility, allowing a controller to tune the drive to achieve the appropriate coolant flow rate to attain maximum efficiency of the coolant system while the drive operates under constant power. Since high-speed operation, typically > 10,000 rpm, is necessary, the SRM certainly fulfills this criteria naturally, as opposed to a PM machine, which must have rotor modifications made that degrade performance/cost in order to operate in this range of speed.

Fault tolerance is also an important issue and desirable feature in automotive applications. Typically, when a component fails it is advantageous to have a certain amount of redundancy built into the system such that the vehicle can still operate until maintenance can be performed. However, in order to yield this redundancy, cost becomes exorbitant. In the case of the SRM, the machine is naturally fault tolerant. AC machines are generally comprised of sinusoidal or concentrated windings with star or delta connections. Thus, electromechanical energy conversion is interdependent upon proper excitation. Conversely, the SRM is a discrete wound machine and thus phase windings are independent of each other; so in a multiphase machine if one phase fails the machine can still operate at a somewhat degraded performance until repairs can be conducted. Additionally, as seen previously, the inverter topology used for an SRM protects it from serious electrical faults such as shoot-through, adding another degree of fault tolerance to the coolant drive.

In design of a drive system, such as the coolant pump drive, one of the main objectives is to achieve high performance at low cost. Therefore, elimination of sensors is a high priority. Normally, drive systems require an encoder/resolver in order to feedback speed and rotor angle information to the control unit. Unfortunately, encoders and resolvers that can operate under the harsh conditions of the engine compartment, at the required speed, are expensive. Less expensive encoders or resolvers are available but typically have poor resolution. However, it has been shown that sensorless control methodologies can be quite effective for the SRM at high-speed and super-high-speed ranges.

One of the primary failure modes in any electrical machine is the seizure of the bearing(s). Case studies have shown that induction machines operating under conditions similar to that of the engine compartment have an approximate life span of approximately 10,000 hours. Bearing wear, usually in the radial direction, will effectively change the airgap such that the magnetic characteristics of the machine deviate from nominal conditions. The SRM is no different in this aspect; however, the performance quality of the machine can be maintained over its life span by utilizing an auto-calibration technique. This technique takes advantage of the unique geometrical structure of the SRM, mechanical system stiffness, and high-speed DSP technology, allowing the magnetic characteristics of the SRM to be updated such that controller tuning can be performed that yields the highest performance of the drive system possible.

REFERENCES

[1] Taylor, W.H. "Obtaining motive power," Patent number 8255, England and Wales, 1839.
[2] Bedford, B.D. "Compatible brushless reluctance motors and controlled switch circuits," U.S. Patent number 3679953, 1972.
[3] Bedford, B.D. "Compatible permanent magnet or reluctance brushless motors and controlled switch circuits," U.S. Patent number 3678352, 1972.
[4] Nasar, S.A. "DC switch reluctance motor," *Proceedings of the IEE*, Vol. 116, No. 6, 1969.

21

Noise and Vibration in SRMs

William Cai
Remy International, Inc., Anderson, Indiana

Pragasen Pillay
Clarkson University, Potsdam, New York

21.1 INTRODUCTION

The advantages of switched reluctance machines (SRMs) over induction and PM machines include higher reliability, higher operating temperatures, and speeds and ease of manufacturing. Two disadvantages, which have received considerable attention recently, are torque ripple and acoustic noise. With respect to the latter, solutions may be application specific. For example, it is quite permissible to produce specialized designs for the automotive industry, because of the large volumes involved. The machines are also face-mounted in automotive applications rather than foot-mounted in industrial applications, opening doors for vibration reduction. Frameless designs are also possible, giving rise to different vibration and acoustic noise characteristics than designs with frames, particularly with ribs.

The primary cause of vibration and acoustic noise is stator ovalization, excited by normal attractive magnetic forces. The stator deformation corresponds to the second order mode shape of the stator yoke for diametrically opposite excitation, with significant noise emitted at natural resonant frequencies of the stator assembly [1,2]. This phenomenon can be observed in two-phase 4/2, three-phase 6/4, four-phase 8/6, five-phase 10/8, and six-phase 12/10 pole combinations. However, there are many other combinations of the stator and rotor poles. The lowest order mode shape, which can be excited by the principal components of the electromagnetic forces, need not correspond to the 2nd order mode shape or oval deformation in those cases. For example, the 4th order mode shape is the lowest mode shape that can be excited by a three-phase 12/8 pole SRM, and the dominant acoustic noise source will correspond to a quasi-square deformation, instead of the oval deformation. Obviously, a high stiffness stator can help reduce vibration deformation, and low vibration and noise can be achieved furthermore by mismatching the waveforms and frequencies of the excitation force with the stator mode shapes and resonant frequencies of the SRM. This chapter considers machine designs with 12 poles and tries to determine a more reasonable design for the low vibration.

The resonant frequencies and mode shapes of the SRM can be obtained by analytical calculation [2–5], numerical computation (finite element method and so on) [2,3,5–8], or experimental techniques [1,2,4,5,7]. The experimental method can be performed only after manufacturing a prototype motor. Therefore, there is advantage in using numerical methods for the computation of resonant frequencies and mode shapes during motor design, since the simplifications in the analytical model may cause unacceptable errors. The conventional round stator stack has been analyzed at depth, including the fillets at pole root [7], the shapes and sizes of the pole [8], the length and thickness of the yoke [2,3], and various frames [2,5]. In this chapter, the effects of different shapes of the stator laminations and frame are examined, which is a new contribution to low-vibration motor design. Five different geometrical topologies of stator stacks are considered.

21.2 NUMERICAL MODELS OF SRM STATOR MODAL ANALYSIS

The stator of a traditional SRM with 12/8 poles is used as a prototype to investigate modal vibrations, as shown in Model I of Figure 21.1 (I). Based on the same copper, stator pole height and utilization of lamination steel, four other outlines of the stator lamination stacks are derived. These stator configurations can be applied to SRMs with two, three, four, and six phases although a two-phase SRM can only be used for unique direction drives. The geometrical topologies and the corresponding finite element models are shown in Figure 21.1. Two stator stacks have the same polygonal outlines and different pole positions. The poles are arranged under polygonal vertexes in Model II in Figure 21.1 (II) while the poles in Model III are aligned with the sides of the polygonal lamination sheets, as shown in Figure 21.1 (III). For lamination stacks with square outline, the difference in Models IV and V lies in four of the poles aligned at the corners as opposed to the sides of the square, as shown in Figure 21.1 (IV) and (V), respectively. The fillets with the same radius are located at the pole roots in all five models of the SRM stator since it has been shown that they increase the stiffness of the stator.

21.3 FINITE ELEMENT RESULTS OF THE STATOR MODAL ANALYSIS

The structural finite element analysis is performed for five stator stacks shown in Figure 21.1 and for the stacks plus smooth frames with a thickness of 10 mm. Although a 3D finite element analysis is carried out, only selected in-plane flexural mode shapes are given in this chapter, as shown in Figure 21.2. The in-plane flexural modes play an important role in electromagnetic vibrations, while the bending, torsional, and out-of-plane modes contribute to vibrations under unbalanced or fault situations. If a smooth frictional frame is added to the lamination stack, their stiffness and modal frequencies increase.

Observing the computed results, the first six mode shapes are related to rigid motion of the lamination stacks in all models; i.e., three displacements along the x-, y-, and z-axis

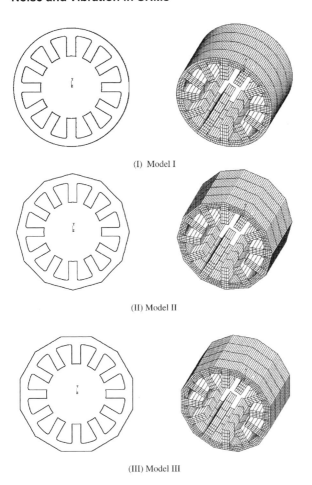

Figure 21.1 Configurations and numerical models of the SRM lamination stack with 12 poles.

and three rotations around the three axes. The resonant frequencies in all five models increase with the order of the mode shapes, except for the 0th order mode. In Models I, II, and III of Figure 21.1, the mode shapes appear in pairs, which have the same shape and frequency but a 90° phase difference from each other, such as the mode shapes from the 2nd order to the 5th order in these three models. This phenomenon occurs only four times (i.e., the 3rd and 5th order modes as well as the two high-frequency modes) up to 30,000 Hz for the in-plane flexural modes in Models IV and V of Figure 21.1. Some modes of the same order have different resonant frequencies in Models IV and V. For example, the 2nd order mode in Model IV has frequencies of 2485 Hz and 3556 Hz, as shown in Figure 21.2 (a4) and (a'4), respectively. For most modes of the same order, the resonant frequencies of Models II to V are higher than the modal frequencies of Model I of traditional circular laminations, except for rows (g), (h) and (k) in the Table 21.1. This becomes more evident for the square modes IV and V, which means higher stiffness for these two models. The pole deformation of the same order modes in different models can show differences from each other. For instance, the amplitudes of the 2nd order mode lie right behind the poles in Figure 21.2 (a1), (a3), and (a4), while the amplitudes of the 2nd order mode in Models II and V lie behind the centers of neighboring poles, as shown in

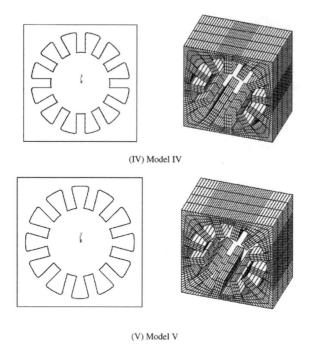

(IV) Model IV

(V) Model V

Figure 21.1 (continued)

Figure 21.2 (a2) and (a5). Attention should be paid to this during vibration analysis and measurement as well as low vibration design.

21.4 DESIGN SELECTION OF LOW VIBRATION SRMS

There are several guidelines to select low vibration designs: First, avoid the mode shapes that are similar to the waveforms of electromagnetic excitations. Second, mismatch the modal frequencies with the frequencies of possible excitation, or push the resonant frequency over the audible range if possible. Third, the stator stack has a relatively higher resonant frequency (high stiffness) at its lowest order mode shape, so that a small vibration amplitude results under the excitation of the same electromagnetic force amplitude. Finally, the stator assembly has enough motion constraints. Based on these rules, several design discussions are given as follows:

1. Using the stators for a three- or two-phase motor, instead of a six-phase motor, the order of the main components of the radial force waveform will increase from the 2nd order to the 4th order (for three phases, 12/8 poles) or 6th order (for two phases). The increase in the resonant frequencies with the order of the mode shape means that the stiffness of the stator increase and the vibrations will be reduced as the consequence. There is also a larger range of excitation operational frequency before excitation of resonance. For magnetic forces with the same amplitude, the 6th order waveform and 4th order waveform will produce a lower vibration than the 2nd order waveforms.

Noise and Vibration in SRMs

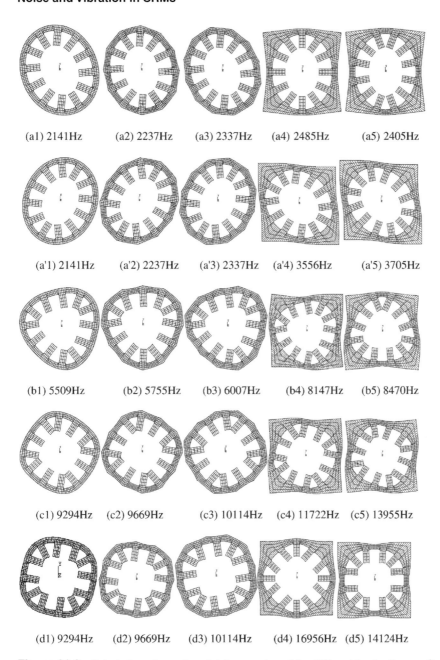

(a1) 2141Hz   (a2) 2237Hz   (a3) 2337Hz   (a4) 2485Hz   (a5) 2405Hz

(a'1) 2141Hz   (a'2) 2237Hz   (a'3) 2337Hz   (a'4) 3556Hz   (a'5) 3705Hz

(b1) 5509Hz   (b2) 5755Hz   (b3) 6007Hz   (b4) 8147Hz   (b5) 8470Hz

(c1) 9294Hz   (c2) 9669Hz   (c3) 10114Hz   (c4) 11722Hz   (c5) 13955Hz

(d1) 9294Hz   (d2) 9669Hz   (d3) 10114Hz   (d4) 16956Hz   (d5) 14124Hz

Figure 21.2 Selected modes and modal frequencies of the different lamination configurations.

2. If the five-stator lamination stacks are used for a six-phase SRM with 12/10 poles, the dominant mode shape will be the 2nd order. It seems the normal square stack has the highest stiffness among the five candidates. If all 10 of the 2nd order modes (Figure 21.2 (a_i) and (a'_i), i = 1 to 5) are carefully examined, the mode shapes in Figure 21.2 (a2), (a'4), and (a5) can only be excited under overlapping operation of two phases. In this case the excitation forces turn out to be small,

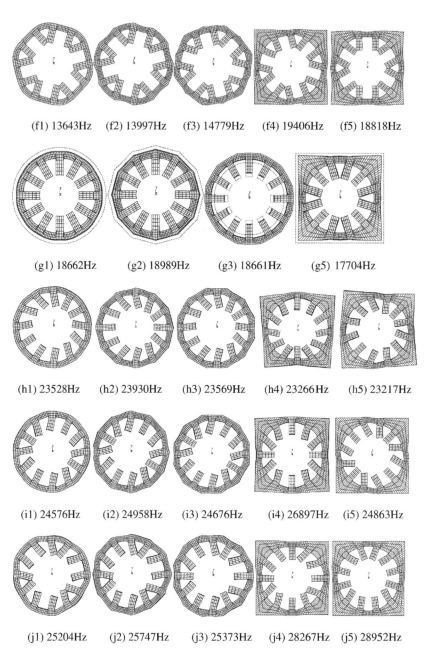

(f1) 13643Hz   (f2) 13997Hz   (f3) 14779Hz   (f4) 19406Hz   (f5) 18818Hz

(g1) 18662Hz   (g2) 18989Hz   (g3) 18661Hz   (g5) 17704Hz

(h1) 23528Hz   (h2) 23930Hz   (h3) 23569Hz   (h4) 23266Hz   (h5) 23217Hz

(i1) 24576Hz   (i2) 24958Hz   (i3) 24676Hz   (i4) 26897Hz   (i5) 24863Hz

(j1) 25204Hz   (j2) 25747Hz   (j3) 25373Hz   (j4) 28267Hz   (j5) 28952Hz

Figure 21.2 (continued)

especially for the leading phase. Thus, the dominant excitation force with the lowest order, while turning off the conducting phase, may correspond to the other 7 modes in the first two rows of Figure 21.2. Of these 7 modes, the sequence with ascending resonant frequencies is from Models I to V. The selection of a low vibration six-phase SRM should in turn be from Models V to I.

Noise and Vibration in SRMs

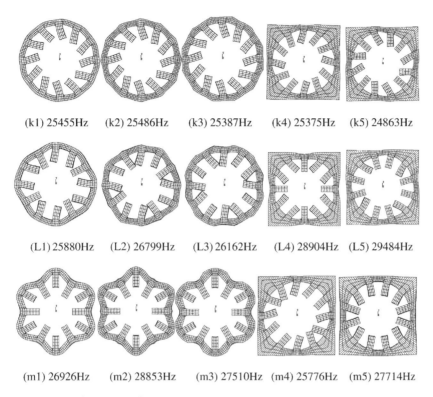

(k1) 25455Hz (k2) 25486Hz (k3) 25387Hz (k4) 25375Hz (k5) 24863Hz

(L1) 25880Hz (L2) 26799Hz (L3) 26162Hz (L4) 28904Hz (L5) 29484Hz

(m1) 26926Hz (m2) 28853Hz (m3) 27510Hz (m4) 25776Hz (m5) 27714Hz

Figure 21.2 (continued)

3. If the stator stacks are used for a three-phase 12/10 SRM with short flux-paths, the superior sequence for low vibration design should be Models IV, V, III, II, and I, according to 2nd order mode frequencies.
4. If the stator stacks are used for a three-phase SRM with 12/8 poles, the lowest order waveforms of the magnetic forces correspond to the modes in rows (c) and (d) of Figure 21.2. The best choice is still between the two stacks with the square outline, i.e., Models IV and V of Figure 21.1. The lowest frequency of Model V corresponding to the quasi-4th or 4th order mode is 13,955 Hz, which is the highest among the five candidates. Compared to the two low order mode shapes of Model V, there are three mode shapes related to the quasi-4th and 4th order waveforms in Model IV, which may cause difficulties in noise control. Thus, Model V is a good low vibration design. In fact, the high order modes in row (L) of Figure 21.2 can also be excited under high-speed operation, and their frequencies may fall into the audible range (i.e., lower than 20,000 Hz) after mounting the phase windings, especially for the first three models. However, the main concern in low vibration design is the low order modes, i.e., quasi-4th and 4th order modes. Comparisons of the deformation transition of the quasi-4th and 4th order modes of Models IV and V in Figure 21.1 are given in Figure 21.3 and Figure 21.4, which show the low order mode shapes corresponding to a three-phase 10/8 pole SRM.
5. If these stator stacks are used for two-phase SRMs, the 6th order mode in row (f) of Figure 21.2 may play an important role since the resonant frequencies of the

Table 21.1 Comparison of the Modal Frequencies among Different Stator Shapes

Mode Reference No. in Figure 21.2	Resonant Frequencies (Hz) and Relative Variations (%)								
	Model I	Model II		Model III		Model IV		Model V	
	Hz	Hz	%	Hz	%	Hz	%	Hz	%
(a) [2nd]	2141	2237	4.48	2337	9.15	2485	16.07	2405	12.33
(a′) [2nd]	2141	2237	4.48	2337	9.15	3556	66.09	3705	73.05
(b) [3rd]*	5509	5755	4.47	6007	9.04	8147	47.89	8470	53.75
(c) Quasi-4th	9294	9669	4.03	10,114	8.82	11,722	26.12	13,955	50.15
(d) [4th]	N/A	N/A	N/A	N/A	N/A	16,956 17,915	N/A	14,124	N/A
(e) [5th]*	12,413	12,804	3.15	13,471	8.52	18,635	50.12	17,700	42.59
(f) [6th]	13,643	13,997	2.59	14,779	8.33	19,406	42.24	18,818	37.93
(g) [0th]	18,662	18,989	1.75	18,611	−0.27	N/A	N/A	17,704	−5.13
(h)	23,528	23,930	1.71	23,569	0.17	23,266	−1.11	23,217	−1.32
(i)	24,576	24,958	1.55	24,676	0.41	25,776	4.88	24,863	1.17
(j)	25,204	25,747	2.15	25,373	0.67	28,267	12.15	N/A	N/A
(k) [tangential]	25,455	25,486	0.12	25,387	−0.27	25,375	−0.31	24,863	−2.33
(l)	25,880	26,799	3.55	26,162	1.09	28,904	11.68	N/A	N/A
(m)	26,573	28,108	5.78	27,035	1.74	N/A	N/A	N/A	N/A
(n)	26,926	28,853	7.16	27,510	2.17	N/A	N/A	29,484	9.50

Note: Relative variations are based on the traditional round SRM lamination stack.

* Modes appear in pair.

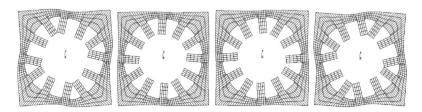

(a) Mode at 11722Hz (sub=13) in Model IV

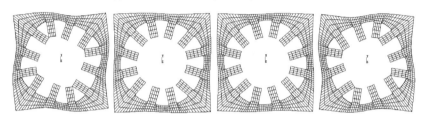

(b) Mode at 13955Hz (sub=15) in Model V

Figure 21.3 Deformation transition comparison of square modes in Models IV and V.

Noise and Vibration in SRMs

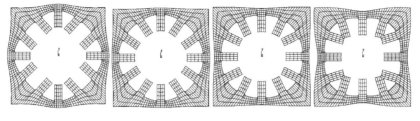

(a) Mode at 16956Hz (sub=19) in Model IV

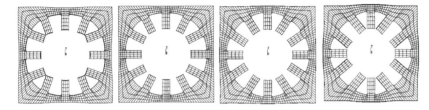

(a') Mode at 17915Hz (sub=20) in Model IV

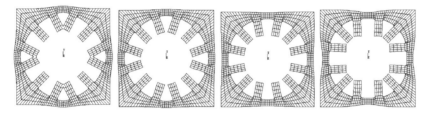

(b) Mode at 14124HZ (sub=16) in Model V

Figure 21.4 Deformation transition comparison of the fourth-order modes in Model IV and V.

mode shapes in row (m) are beyond the audible range. In this case, the superior sequence for low vibration design is Models IV, V, III, II, and I of Figure 21.1 in terms of the 6th order modal frequencies.

6. There is no 0th order mode shape (i.e., expansion/contraction motion) in Model IV, which benefits low vibration operation. The rest of the models with ascending resonant frequencies of this mode are Models III, I, II, and V of Figure 21.1.
7. For the modes related to tangential deformation shown in row (k) of Figure 21.2, the first four models have good stiffness, but Model V shows lower stiffness.
8. If active noise control techniques are used to reduce monotone acoustic noise, selecting a square shape lamination should be avoided, since it breaks the resonant frequency of the same order mode into two different frequencies.

21.5 THE EFFECTS OF A SMOOTH FRAME ON THE RESONANT FREQUENCIES

If the lamination stack is installed in a frame with a friction fit, the selected results of the low order mode shapes are given in Table 21.2. Because the frame material is steel, which

Table 21.2 Comparison of the Modal Frequencies among Different Stator Shapes plus Frame

Mode Reference No. in Figure 21.2	Resonant Frequencies (Hz) and Relative Variations (%)								
	Model I	Model II		Model III		Model IV		Model V	
	Hz	Hz	%	Hz	%	Hz	%	Hz	%
(a) [2nd]	3466	3583	3.38	3632	4.79	3773	8.86	3741	7.93
(a′) [2nd]	3466	3583	3.38	3632	4.79	4752	37.10	4814	38.89
(b) [3rd]*	8799	9082	3.22	9207	4.64	11,060	25.70	11,188	27.15
(c) Quasi-4th	14,341	14,722	2.66	14,964	4.34	16,079	12.12	17,629	22.93
(d) [4th]	N/A	N/A	N/A	N/A	N/A	19,825	N/A	17,714	N/A
(e) [5th]*	18,069	18,404	1.85	18,753	3.79	21,457	18.75	21,030	16.39
(f) [6th]	19,222	19,512	1.51	19,890	3.48	22,111	15.03	21,832	13.58
(g) [0th]	19,899	19,970	0.36	19,835	−0.32	18,369	−7.69	18,510	−6.98

Note: Relative variations are based on the traditional round SRM lamination stack with smooth frame.

* Modes appear in pair.

has similar mechanical properties to lamination steel, the effects of adding a frame can be treated approximately as the thickening of the yoke. In Table 21.2, the variation in percentage refers to an increase of the resonant frequencies of the other four modes, compared to the traditional round stator shown in Model I of Figure 21.1. For most mode shapes, Models II to V have higher modal frequencies than Model I, except for the 0th order mode shape. The increase is more evident in Models IV and V, which means they have higher stiffness. The superior sequence for selection of low vibration design among the five models has not been changed after adding the smooth frame of 10 mm. However, the variation for the models with a frame becomes smaller than the vibration in the five lamination models. This demonstrates that Model I is more sensitive to frame structure or yoke thickness. The square-shaped Models IV and V are still the better choices for low vibration design.

To further compare the variation of the resonant frequencies of the models with and without frame, the frequencies corresponding to selected mode shapes are given in Table 21.3. The resonant frequencies and stiffness of the first two models are more sensitive to frame or yoke thickness, especially for Model I of Figure 21.1. For instance, the resonant frequencies of the square modes (or quasi-4th order modes) of Models I and II increase 54.3% and 52.3%, respectively, due to the frame, while the frequencies of similar modes in Models IV and V increase only 37.17% and 26.33% after addition of a frame. Another observation is that the frame causes a larger increase of the modal frequencies for the mode shapes of lower order. For example, the 2nd order mode frequency in Model II increases 60.17% (from 2337 Hz to 3583 Hz) due to the addition of a 10 mm frame, while the resonant frequency of the 6th order mode of the same model increases only 39.4% (from 13,997 Hz to 19,512 Hz) after adding the frame.

Although Models IV and V show low vibration characteristics without increasing the utilization of the materials, they are heavier and have unsymmetrical magnetic paths. But this tradeoff results in reduced SRM vibrations.

Noise and Vibration in SRMs

Table 21.3 Variation of the Mode Frequencies after Adding the Frame

Mode Shapes (Ref. No. in Figure 21.2)	Resonant Frequencies (Hz) and Relative Variations (%)							
	(a), 2nd	(a′), 2nd	(b), 3rd	(c)	(d), 4th	(e), 5th	(f), 6th	(g), 0th
Model I								
Stack	2141	2141	5509	9294	N/A	12,413	13,643	18,662
+ frame	3466	3466	8799	14,341	N/A	18,069	19,222	19,899
Increase (%)	61.89	61.89	59.72	54.30	N/A	45.57	40.89	6.63
Model II								
Stack	2237	2237	5755	9669	N/A	12,804	13,997	18,989
+ frame	3583	3583	9082	14,722	N/A	18,404	19,512	19,970
Increase (%)	60.17	60.17	57.81	52.26	N/A	43.74	39.40	5.17
Model III								
Stack	2337	2337	6007	10,114	N/A	13471	14,779	18,611
+ frame	3632	3632	9207	14,964	N/A	18,753	19,890	19,835
Increase (%)	55.41	55.41	53.27	47.95	N/A	39.21	34.58	6.58
Model IV								
Stack	2485	3556	8147	11,722	16,956	18,635	19,406	N/A
+ frame	3773	4752	11,060	16,079	19,825	21,457	22,111	18,369
Increase (%)	51.83	33.63	35.76	37.17	16.92	15.14	13.94	N/A
Model V								
Stack	2405	3705	8470	13,955	14,124	17,700	18,818	17,704
+ frame	3741	4814	11,188	17,629	17,714	21,030	21,832	18,510
Increase (%)	55.55	29.93	32.09	26.33	25.42	18.81	16.02	4.55

Note: Increase = (frame-stack)/stack × 100%

21.6 CONCLUSIONS

This chapter has examined the design of SRMs for automotive applications. The modal analysis of five different stator lamination stacks with different shapes has been performed by a 3D structural finite element method. The effects of a smooth frame were examined for each case. The resonant frequencies and mode shapes of five candidate models are compared with each other. Criteria for low vibration designs are discussed for SRMs of different phase numbers. This new contribution will benefit the design selection of low vibration and low noise SRMs in automotive applications.

REFERENCES

[1] D.E. Cameron, J.H. Lang, S.D. Umans. The origin and reduction of acoustic noise in variable-reluctance motors, *Conference Record of the 1989 IEEE Industry Applications Conference,* 24th IAS Annual Meeting, San Diego, CA, October 1989, pp. 108–115. Also see D.E. Cameron, J.H. Lang, S.D. Umans. The origin and reduction of acoustic noise in doubly salient variable-reluctance motors, *IEEE Trans. on Industry Applications,* Vol. 28, No. 6, November/December, 1992, pp. 1250–1255.

[2] P. Pillay, W. Cai. An investigation into vibration in switched reluctance motor. *Conference Record of the 1998 IEEE Industrial Application Society*, 33rd IAS Annual Meeting, Vol. 1, St. Louis, MO, Oct. 12–16, 1998.

[3] R.S. Colby, F. Mottier, T.J.E. Miller. Vibration modes and acoustic noise in a four-phase switched reluctance motor. *Conference Record of the 1995 IEEE Industrial Application Society,* 30th IAS Annual Meeting, Vol. 1, Orlando, FL, Oct. 8–12, 1995, pp. 441–447.

[4] C. Yongxiao, W. Jianhua, H. Jun. Analytical calculation of natural frequencies of stator of switched reluctance motor. Eighth International Conference on Electrical Machines and Drives (EMD'97, IEEE Conference. Publ. No.444), Cambridge, U.K., September 1–3, 1997, pp. 81–85.

[5] W. Cai, P. Pillay. Resonance frequencies and mode shapes of switched reluctance motor, The 1999 IEEE Biennial International Electric Machines and Drives Conference, Seattle, WA (IEMDC'99), May 9–12, 1999, pp. 44–47.

[6] F. Camus, M. Gabsi, B. Multon. Prediction des vibrations du stator d'une machine a reluctance variable en fonction du courant absorbes, *Journal de Physique III: Application Physics, Materials Science and Fluids Plasma Instruments,* No. 7, France, 1997, pp. 387–404.

[7] M. Besbes, C. Picod, F. Camus, M. Gabsi. Influence of stator geometry upon vibratory behaviour and electromagnetic performances of switched reluctance motors, *IEEE Proceedings on Electric Power Applications*, Vol. 145, No. 5, September 1998, pp. 462–468.

[8] B. Fahimi, G. Suresh, M. Ehsani. Design considerations of switched reluctance motors: Vibration and control issues, *Conference Record of the 1999 IEEE Industry Applications Conference 34th Annual Meeting*, Vol. 4, Phoenix, AZ, October 3–7, 1999, pp. 2259–2266.

22

Modeling and Parameter Identification of Electric Machines

Ali Keyhani, Wenzhe Lu, and Bogdan Proca
Ohio State University, Columbus, Ohio

NOMENCLATURE

($\hat{}$): Estimate of ($\cdot$)
$[\ ]^T$: Transpose of $[\ ]$
$E[\cdot]$: The operation of taking the expected value of $[\cdot]$
$w(\cdot)$: Process noise sequence
$v(\cdot)$: Measurement noise sequence
$X(\cdot)$: State vector
$Y(\cdot)$: Measured output vector in the presence of noise
Q: Covariance of the process noise sequence
R_0: Covariance of the measurement noise sequence
$R(\cdot)$: Covariance of the state vector
$e(\cdot)$: Estimation error, $e(k) = Y(k) - \hat{Y}(k)$
exp: The exponential operator
det: Determinant
$\hat{Y}(k|k-1)$: The estimated value of $Y(k)$ at time instant k given the data up to $k-1$
$U(\cdot)$: Input vector
$\theta(\cdot)$: Parameter vector

22.1 INTRODUCTION

Modeling the dynamical properties of a system is an important step in analysis and design of control systems. Modeling often results in a parametric model of the system that contains several unknown parameters. Experimental data are needed to estimate the unknown parameters.

Electric machines are now widely used in electric/hybrid vehicles. Identifying appropriate model structures of these machines and estimating the parameters of the models has become an important part of the automotive control design.

Generally, the parameter estimation from test data can be done in frequency-domain or time-domain. Since noise, which may cause problems to parameter estimation, is an inherent part of the test data we will first study the effects of noise on frequency-domain parameter estimation. To examine this issue, we will study the identification of synchronous machine parameters from noise-corrupted measurements. Then, we will show how the time-domain maximum likelihood technique can be used to remove the effect of noise from estimated parameters. The models and the procedures to identify the parameters of synchronous, induction, and switched reluctance machines using experimental data will be presented.

22.2 CASE STUDY: THE EFFECTS OF NOISE ON FREQUENCY-DOMAIN PARAMETER ESTIMATION OF SYNCHRONOUS MACHINE

22.2.1 Problem Description

A solid-rotor machine consists essentially of an infinite number of rotor circuits. However, in practice, only a three-rotor-winding or a two-rotor-winding model is used in estimating machine parameters from test data. Experience gained in modeling of many machines shows that neither the second nor the third order model structure can be an exact mathematical representation of a machine.

In estimating the parameters of a system, one question needs to be answered: If the assumed model structure is correct, then can one obtain a unique estimate of the parameters from noise-corrupted frequency response data? The answer to this question cannot be found from measurements, since the measurements are made on a machine with a complex, high order rotor circuit, with unknown structure and unknown parameters.

If one assumes a model structure and then proceeds with estimating its parameters from actual measurements, then the structural error and the effect of noise in the measurements will result in inaccurate parameters. Therefore, it will not be clear whether the discrepancy between the simulated model response and the measured response is due to the effect of noise on the parameters, inadequacy of the assumed model structure, or both. Therefore, the structural identification problem and the parameter estimation problem should be studied separately. There is a need to show that the measurements noise will not corrupt the estimated parameters when the parameters of an assumed structure are estimated from the frequency response measurements.

In this section, a third order machine model with known parameters is simulated, and then the data are noise-corrupted using a known noise distribution. The objective is to estimate the parameters of this model from the noise-corrupted data and evaluate the estimated parameters by comparison with the known parameters.

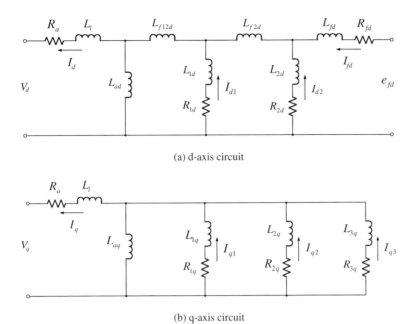

Figure 22.1 SSFR3 equivalent circuit structures.

22.2.2 PARAMETERS ESTIMATION TECHNIQUE

In the literature the second order model of synchronous machine is referred to as SSFR2 and the third order model as SSFR3. These notations will be used in this section. It is generally assumed that the synchronous machine d-axis and q-axis circuit structures can be represented by the SSFR3 or the SSFR2 models. The SSFR3 model is shown in Figure 22.1. The SSFR2 model structure can be obtained from the SSFR3 model by reducing the number of rotor body circuits from two to one and also assuming that L_{f2d}, which reflects the leakage flux effect, is zero. The standard circuit model structure can be obtained by assuming that L_{f12d} is also negligible.

22.3.2.1 Estimation of D-Axis Parameters from the Time Constants

The transfer functions of the d-axis SSFR3 equivalent circuits are:

$$L_d(s) = K_d \frac{(1+T_1s)(1+T_2s)(1+T_3s)}{(1+T_4s)(1+T_5s)(1+T_6s)} \quad (22.1)$$

$$sG(s) = G_d \frac{(1+T_7s)(1+T_8s)}{(1+T_4s)(1+T_5s)(1+T_6s)} \quad (22.2)$$

Using an assumed value of armature resistance, R_a, the $L_d(s)$ is calculated from the operational impedance, $Z_d(s) = -V_d(s)/I_d(s)$, and $sG(s)$ is calculated from $I_{fd}(s)/I_d(s)$ when the field is short-circuited.

The equations that relate the circuit parameters to the time constants can be obtained from Equation 22.1 and Equation 22.2. These equations are described in terms of the

Table 22.1 Definitions for D-Axis Circuit Unknowns and Knowns

Unknown Circuit Parameters	Unknown Vector $\bar{x}$	Known Constants	Known Vector $\bar{y}$
L_{ad}	x_1	L_1	y_0
L_{f12d}	x_2	K_d	y_1
R_{1d}	x_3	$T_1 T_2 T_3$	y_2
L_{1d}	x_4	$T_1 T_2 + T_1 T_3 + T_2 T_3$	y_3
L_{f2d}	x_5	$T_1 + T_2 + T_3$	y_4
R_{2d}	x_6	$T_4 T_5 T_6$	y_5
L_{2d}	x_7	$T_4 T_5 + T_4 T_6 + T_5 T_6$	y_6
R_{fd}	x_8	$T_4 + T_5 + T_6$	y_7
L_{fd}	x_9	G_d	y_8
—		$T_7 T_8$	y_9
—		$T_7 + T_8$	y_{10}

unknown vector $\bar{x}$ (i.e., the circuit parameters) and the known vector $\bar{y}$, as defined in Table 22.1.

The vector $\bar{y}$ is estimated from the measured frequency response data of transfer functions. The time constants are estimated by using a curve-fitting technique described in References 15, 16, and 20. The functional form of the vector $\bar{y}$ that relates to the circuit parameters (i.e., the vector $\bar{x}$) can be derived using MACSYMA [21], a computer-aided symbolic processor. These relationships are complex and nonlinear, and can be written as:

$$f_i(\bar{x}) = \bar{y}_i + g_i(\bar{x}, \bar{y}) + \zeta_i = 0 \tag{22.3}$$

where $i = 1, \ldots, 10$.

Details of these equations are given in Appendix A. In general, these 10 equations are nonlinear in nature and are not consistent with each other. This is due to the noise ζ imbedded in vector $\bar{y}$. Because of the nonlinearity of these equations, a closed form solution for vector $\bar{x}$ may not be possible, and a numerical technique such as Newton-Raphson method may have to be used to solve these equations iteratively. Moreover, these are a redundant set of equations, 10 equations with 9 unknown parameters. Because of the inconsistency of these equations, multiple solutions will be obtained depending on which equation is ignored.

If the measured frequency response data are noise free (i.e., $\zeta_i = 0$, $i = 1, \ldots, 10$), then Equation A.1 through Equation A.10, given in Appendix A, would be consistent, and a unique solution will be obtained regardless of which equation is ignored.

The set of nonlinear equations, $F(\bar{x}) = [f_1(\bar{x}), f_2(\bar{x}), \ldots f_{10}(\bar{x})] = 0$, can be solved by updating $\bar{x}$ as:

$$\bar{x}^{K+1} = \bar{x}^K + \Delta \bar{x}^K, \quad K = 0, 1, 2 \ldots \tag{22.4}$$

where

Modeling and Parameter Identification of Electric Machines

$$\Delta \bar{x}^K = -\left[\frac{\partial F}{\partial x}\right]^{-1}_{\bar{x}=\bar{x}^K} F(\bar{x}^K) \tag{22.5}$$

until the residuals are smaller than a predetermined error ε (i.e., $\left|f_i(\bar{x}^K)\right| \leq \varepsilon$).

The Newton-Raphson solution is formulated by discarding one equation from the set described by Equation A.1 through Equation A.10. This is necessary because there are only 9 independent equations out of the 10 equations. Since, for noisy data, these 10 equations are inconsistent, a multiple solution set is obtained, with the solution depending on which equation is ignored.

Before the iterative approximation can be carried out, a good initial estimate of the unknown vector $\bar{x}$ is essential for convergence to a solution. In this study, the initialization of the unknown vector $\bar{x}$ is performed by using the method developed by Umans et al. [15]. In his method, Equation A.8 is discarded and the remaining nine equations are solved for the nine parameters.

22.3.2.2 Estimation of Q-Axis Parameters

The q-axis transfer function of the SSFR3 equivalent circuit can be written as:

$$L_q(s) = K_q \frac{(1+T_1's)(1+T_2's)(1+T_3's)}{(1+T_4's)(1+T_5's)(1+T_6's)} \tag{22.6}$$

Using an assumed value of armature resistance, R_a, the $L_q(s)$ is calculated from the operational impedance, $Z_q(s) = -V_q(s)/I_q(s)$. The q-axis parameters can be determined from a consistent set of linear equations (see Reference 15), which relate the $L_q(s)$ transfer function's time constants to the equivalent circuit parameters.

22.2.3 STUDY PROCESS

For the purpose of this study, synthetic frequency response data were created using the Monticello generating unit SSFR3 model parameters derived by Dandeno and Poray [11]. The frequency response data so developed were then corrupted with a uniformly distributed noise of zero mean and varying degrees of signal-to-noise (S/N) ratios. the following relationships were used in creating the noise-corrupted data:

$$\tilde{Z}_d(s) = Z_d(s) + \eta_1 \tag{22.7}$$

$$\tilde{Z}_q(s) = Z_q(s) + \eta_2 \tag{22.8}$$

$$s\tilde{G}(s) = sG(s) + \eta_3 \tag{22.9}$$

where $\tilde{Z}_d(s)$, $\tilde{Z}_q(s)$, and $s\tilde{G}(s)$ represent the noise-corrupted data, and η_1, η_2, and η_3 represent noise.

The noise-corrupted $\tilde{L}_d(s)$ and $\tilde{L}_q(s)$ data were then developed using the following relationships:

$$\tilde{L}_d(s) = \frac{\tilde{Z}_d(s) - R_a}{s} \tag{22.10}$$

$$\tilde{L}_q(s) = \frac{\tilde{Z}_q(s) - R_a}{s} \quad (22.11)$$

where R_a is the armature resistance originally used in creating the synthetic data.

Following this, the required d-q axes transfer functions were computed based on the nonlinear least square curve-fitting techniques developed by Marquardt [20]. Both magnitude and phase angle data were used in estimating the time constants. Monticello generator parameters, corresponding to the SSFR3 model structures, were then recalculated using the Newton-Raphson method discussed earlier. The same model structure was retained so that any discrepancy observed in the recalculated values of the machines parameters could be specifically ascribed to the noise introduced in the synthetic data.

22.2.4 Analysis of Results

For the purpose of evaluating the effect of noise on estimated parameters of the Monticello machine used as the study machine, various uniformly distributed noise sequences were used with zero mean and with signal-to-noise ratios varying from 3100:1 to 250:1, where S/N = $(\Sigma(\text{signal})^2/(\text{noise})^2)^{1/2}$.

To assess the appropriate level of S/N ratio that should be considered in the study, an effort was made to roughly estimate the level of S/N ratio normally achievable in an SSFR field test. For this purpose, noise-corrupted synthetic data of the Monticello generator with S/N ratios ranging from 3100:1 to 250:1 were plotted and compared with the corresponding data acquired during the August 1984 test on Rockport Unit #1. This is a 1300 MW cross-compounded unit owned and operated by the American Electric Power Company.

Figure 22.2 shows the $L_d(s)$ magnitude and phase plots of the Rockport field test data and the noise-corrupted synthetic data of the Monticello generator with S/N ratio of 3100:1. The two sets of plots are similar, showing similar noise effects. Plots corresponding to S/N ratio of 250:1 were found to be too noisy, but because of space constrains are not included in this section. However, to evaluate the full impact of measurement noise, some results pertaining to such noisy data are also provided in this section.

22.2.4.1 D-Axis Parameter Estimation

Table 22.2 shows estimated values of the d-axis transfer function time constants of the Monticello machine, corresponding to various degrees of S/N ratios.

Results obtained indicate that because of the noise in the synthetic data, an error is introduced in the estimated values of the transfer function time constants. Moreover, the magnitude of the error increases significantly as the S/N ratio deteriorates (i.e., noise level is higher) from 3100:1 to 250:1.

The transfer function time constants corresponding to S/N ratio of 3100:1 were then used to estimate the d-axis machine parameters. For this purpose, the nonlinear set of equations (A.1–A.10) was solved using the Newton-Raphson method. As indicated earlier, these are a redundant and inconsistent set of equations with the number of unknowns being one less than the number of equations. Therefore, to obtain the solution, one of the equations has to be discarded. However, the authors of this chapter feel that there are no obvious reasons for discarding any particular equation. Therefore, in this study, an effort was made to solve the subsets of equations obtained by discarding one equation at a time.

Results presented in Table 22.3 indicate that by using this approach, four solution sets are obtained even when the S/N ratio is as high as 3100:1. In this case, the same

Modeling and Parameter Identification of Electric Machines

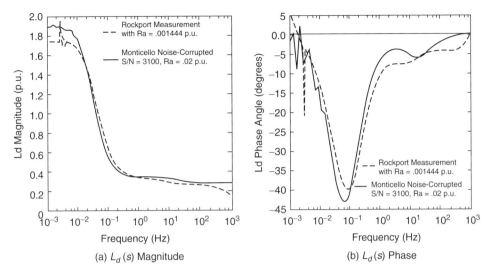

Figure 22.2 $L_d(s)$ magnitude and phase plots of field test and synthetic noise-corrupted data.

Table 22.2 Estimated Values of D-Axis Transfer Function Time Constants with $R_a = 0.02$ p.u.

TFT*	Original Values with $R_a = 0.02$	Estimated Values Obtained with $R_a = 0.02$					
		S/N Ratio 3100:1		S/N Ratio 500:1		S/N Ratio 250:1	
		Est.	% Error	Est.	% Error	Est.	% Error
K_d	1.9	1.9025	−0.13	2.0009	−5.31	2.2419	−18.0
T_1	1.184	1.1842	−0.01	1.1655	1.56	1.0645	10.09
T_2	0.00772	0.00772	0.05	0.00759	1.63	0.00779	−0.92
T_3	0.0026	0.00259	0.40	0.00245	5.72	0.00224	13.71
T_4	6.601	6.6121	−0.03	6.9155	−4.62	7.3178	−10.70
T_5	0.00951	0.00949	0.07	0.00917	3.42	0.00882	7.10
T_6	0.0026	0.0026	0.0	0.00248	4.47	0.00235	9.43
T_7	0.00453	0.00452	0.22	0.00421	6.56	0.00389	13.78
T_8	0.00041	0.00041	−0.01	0.0004	0.96	0.0004	1.88
C_d	5.5308	5.5387	−0.15	5.7649	−4.24	6.0636	−9.64

* TFT = transfer function time constants.

solution set was obtained when Equations A.2–A.7 were discarded one at a time; no solution could be obtained when Equation A.1 was discarded.

A study of Table 22.3 indicates that some of the parameters in these solution sets differ by as much as 130%. In particular, the value of the generator field winding inductance (L_{fd}) in solution set 2 differs from its original value by as much as 42%. This means that for the same value of R_{fd}, the d-axis transient short circuit time constant will differ significantly from the original value.

Table 22.3 Estimated Values of D-Axis Machine Parameters with S/N Ratio 3100:1 and $R_a = 0.02$ p.u.

Machine Parameters in p.u.	Original Values	Estimated Values			
		Set 1* with EQD** of A.2–A.7	Set 2 with EQD of A.8	Set 3 with EQD of A.9	Set 4 with EQD of A.10
L_{ad}	1.691	1.69352	1.69352	1.69352	1.69352
L_{f12d}	0.0093	0.00953	0.0103	0.00378	−0.00301
R_{1d}	0.067	0.06684	0.0671	0.07326	0.08013
L_{1d}	0.1144	0.11377	0.11422	0.13152	0.15096
L_{f2d}	0.1287	0.12826	0.12374	0.13323	0.14065
R_{2d}	0.0092	0.0091	0.01316	0.00899	0.00889
L_{2d}	0.0014	0.00139	0.00201	0.00054	0.00123
R_{fd}	0.00081	0.00081	0.00079	0.00081	0.00081
L_{fd}	0.0087	0.00864	0.01237	0.00942	0.00854
Res***	—	0.78	57.9	−0.15	0.17

* Solution obtained by discarding any one of the Equations A.2 through A.7.
** EQD = equation number discarded.
*** Res = residual of the discarded equation.

An effort was also made to estimate the accuracy of each of these solution sets. For this purpose, frequency response data were created corresponding to each of the four solution sets. Data so obtained were compared with the noise-corrupted synthetic data of the Monticello machine. Accuracy was measured in terms of the mean error and the RMS error, which are defined as follows:

$$\text{Mean error} = \frac{1}{n}\sum_{K=1}^{n} e_{(K)}$$

$$\text{RMS error} = \sqrt{\frac{1}{n}\sum_{K=1}^{n} e_{(K)}^2}$$

where
n = Number of data points
$e_{(K)}$ = (Value of the noise-corrupted synthetic data at the Kth frequency) − (Value of the created data corresponding to a particular solution set at the Kth frequency)

Results presented in Table 22.4 indicate that each of the solution sets is quite accurate; mean error and RMS error of each solution set are quite small.

Therefore, it may be observed that by using frequency analytical techniques, multiple solution sets are obtained with each of the solution sets being quite accurate. However,

Modeling and Parameter Identification of Electric Machines

Table 22.4 D-Axis Mean and RMS Errors of Estimated Frequency Response with S/N Ratio of 3100:1 and $R_a = 0.02$ p.u.

Set No.	Mean Error		RMS Error	
	LD Magnitude	sG Magnitude	LD Magnitude	sG Magnitude
1	0.52E-05	0.35E-04	0.59E-02	0.22E-03
2	0.26E-04	0.14E-01	0.59E-02	0.16E-01
3	0.26E-04	−0.10E-02	0.59E-02	0.82E-02
4	0.26E-04	0.47E-02	0.59E-02	0.91E-02

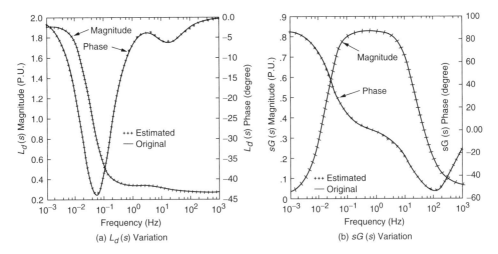

Figure 22.3 Monticello synthetic noise-corrupted and estimated variation of $L_d(s)$ and $sG(s)$ with frequency.

estimated values of some of the machine parameters may differ significantly from the corresponding values in the other solution sets. In view of this, it may sometimes be difficult to choose a solution from the multiple solution sets.

Figure 22.3 shows the noise-corrupted synthetic data plots of $L_d(s)$ and $sG(s)$ generated from the original values of the Monticello machine parameters given in Table 22.3. These plots are superimposed over the frequency response data generated from the estimated values of the machine parameters shown in solution set 1 in Table 22.3. The two sets of plots overlap each other almost completely. This confirms a high level of accuracy can be shown to exist for the other solution sets.

For each of the transfer function time constant sets shown in Table 22.2, corresponding d-axis machine parameters were estimated. The machine parameters obtained by discarding Equation A.8 are presented in Table 22.5. It may be noted that Equation A.8 corresponds to the equation discarded by Umans et al. in Reference 16.

The purpose of obtaining these parameters by discarding only one particular equation is to specifically study the impact of noise in the test data while circumventing the multiple solution set issue.

Table 22.5 Estimated Values of D-Axis Machine Parameters with $R_a = 0.02$ p.u. and Discarding Equation A.8

Machine Parameters in p.u.	Original Values with $R_a = 0.02$	Estimated Values Obtained with $R_a = 0.02$ p.u.		
		S/N Ratio 3100:1	S/N Ratio 500:1	S/N Ratio 250:1
L_{ad}	1.691	1.69352	1.79189	2.03293
L_{f12d}	0.0093	0.01030	0.02429	0.06845
R_{1d}	0.067	0.06710	0.06624	0.07578
L_{1d}	0.1144	0.11422	0.10521	0.11107
L_{f2d}	0.1287	0.12374	0.07407	−0.00502
R_{2d}	0.0092	0.01316	0.03978	0.03237
L_{2d}	0.0014	0.00201	0.00603	0.00486
R_{fd}	0.00081	0.00079	0.00076	0.00081
L_{fd}	0.0087	0.01237	0.04327	0.06467
Res*	0.0	57.9	176	245

* Res = residual of the discarded equation.

A study of Table 22.5 indicates that as the S/N ratio deteriorates, estimated values of some of the machine parameters vary significantly. In particular, for the case of the S/N ratio of 250:1, values of L_{ad}, L_{f2d}, and L_{fd} become unrealistic. This is primarily because noise in the test data introduces error in the estimated values of the transfer function time constants. This error is then amplified during the process of estimating machine parameters from the subset of Equations A.1–A.10.

During this study, an effort was also made to assess sensitivity of the estimated values of the machine parameters, to the error in the value of armature resistance R_a used in deriving the operational inductance $L_d(s)$ from the operational impedance $Z_d(s)$ data.

The value of R_a is generally calculated from the low-frequency asymptote of the $Z_d(s)$ or $Z_q(s)$ data (i.e., $R_a = \text{Lim}_{s \to 0} Z_d(s)$). However, experience shows that the data resolution is very poor in the low-frequency range. Therefore, calculated value of R_a is bound to have a certain degree of error. This will be true to some extent when R_a is measured directly with the help of a sensitive bridge circuit.

In view of the above fact, the machine parameters were estimated for two sets of values of R_a, i.e., 0.02 p.u. and 0.0201 p.u. The corresponding results obtained are presented in Table 22.5 and Table 22.6.

A study of Table 22.6 shows that when the value of R_a is 0.0201 p.u. instead of 0.02 p.u., the machine parameters estimated become unrealistic even when then S/N ratio is as high as 3100:1. L_{ad}, which can be measured quite accurately with the help of a number of well-established testing procedures, is approximately 60% higher than the original value used for generating the synthetic data. The value of L_{fd} is negative, which is totally unrealistic. Similarly, a negative value of R_{2d} cannot be justified.

Therefore, these results clearly show that estimated values of the machine parameters are very sensitive to the value of R_a. Even a 0.5% error in R_a could result in unrealistic estimation of the machine parameters.

Table 22.6 Estimated Values of D-Axis Machine Parameters with $R_a = 0.0201$ p.u. and Discarding Equation A.8

Machine Parameters in p.u.	Original Values with $R_a = 0.02$	Estimated Values Obtained with $R_a = 0.0201$ p.u.		
		S/N Ratio 3100:1	S/N Ratio 500:1	S/N Ratio 250:1
L_{ad}	1.691	2.69676	2.73873	2.85214
L_{f12d}	0.0093	0.22544	0.23764	0.24937
R_{1d}	0.067	0.12913	0.14289	0.16692
L_{1d}	0.1144	0.1778	0.19490	0.21758
L_{f2d}	0.1287	−0.08238	−0.08943	−0.09757
R_{2d}	0.0092	−0.01695	−0.1894	−0.02134
L_{2d}	0.0014	−0.0025	−0.00281	−0.00315
R_{fd}	0.00081	0.009	0.00091	0.00093
L_{fd}	0.0087	−0.05903	−0.06577	−0.07234
Res*	0.0	571	569	582

* Res = residual of the discarded equation.

It may be noted that noise, inherent in the test data, is a random process and cannot be removed. In practice, one can only remove unwanted signals, which are not part of the system to be modeled, by using filters (hardware or software filters). Therefore, to minimize the effect of noise, the analytical technique used should be robust and should not be affected by noise.

A study of the q-axis parameters estimation gives similar results as above.

22.2.5 Conclusions

Based on the results of this study, it is concluded that:

1. Noise, which is inherently present in the field test data, has significant impact on the synchronous machine parameters estimated from the SSFR test data using curve-fitting techniques.
2. Multiple solution sets for the machine parameters are obtained depending upon the equation ignored from the set of relevant equations. In some cases the solution may not even converge.
3. Estimated values of the machine parameters are very sensitive to the values of armature resistance used in the data analysis. Even a 0.5% error in the value of armature resistance could result in unrealistic estimation of the machine parameters.
4. A technique should be developed that provides a unique physically realizable machine model even when the test data are noise-corrupted. This problem is studied in the next section.

22.3 MAXIMUM LIKELIHOOD ESTIMATION OF SOLID-ROTOR SYNCHRONOUS MACHINE PARAMETERS

22.3.1 INTRODUCTION

In the previous section, it was shown that multiple parameter sets will be obtained when the transfer functions of a solid-rotor synchronous machine are estimated from noise-corrupted, frequency-domain data, and then, the machine parameters are computed from the estimated machine transfer function's time constants. Moreover, the estimated machine parameters are very sensitive to the value of the armature resistance used in the study.

In this section, a time-domain identification technique is used to estimate machine parameters. The objective is to show that the multiple solution set problem encountered in the frequency response technique can be eliminated if the time-domain estimation data are generated from the d-q axis transfer functions estimated for the SSFR test data. The maximum likelihood (ML) estimation technique is then used to estimate the machine parameters.

The ML identification method has been applied to the parameter estimation of many engineering problems [25–30]. It has been established that the ML algorithm has the advantage of computing consistent parameter estimates from noise-corrupted data. This means that the estimate will converge to the true parameter values as the number of observations goes to infinity [29–30]. This is not the case for the least-square estimators, which are commonly used in power system applications.

22.3.2 STANDSTILL SYNCHRONOUS MACHINE MODEL FOR TIME-DOMAIN PARAMETER ESTIMATION

22.3.2.1 D-Axis Model

Assuming that the d-axis rotor body can be represented by two damper windings (i.e., SSFR3), the standstill discrete d-axis model of a round rotor machine is given by [19]:

$$X(k+1) = A_d(\theta_d)X(k) + B_d(\theta_d)U(k) + w(k) \quad (22.12)$$

$$Y(k+1) = C_d X(k+1) + v(k+1) \quad (22.13)$$

where

$$C_d = \begin{bmatrix} 1 & 0 & 0 & 0 \\ 0 & 0 & 0 & 1 \end{bmatrix}$$

$$X = \begin{bmatrix} i_d & i_{1d} & i_{2d} & i_{fd} \end{bmatrix}^T$$

$$Y = \begin{bmatrix} i_d & i_{fd} \end{bmatrix}, \quad U = \begin{bmatrix} v_d \end{bmatrix}$$

$$\theta_d = \begin{bmatrix} L_{ad}, L_{f12d}, R_{1d}, L_{1d}, L_{f2d}, R_{2d}, L_{2d}, R_{fd}, L_{fd} \end{bmatrix}$$

In addition, $w(\cdot)$ and $v(\cdot)$ denote the process noise and measurement noise respectively. It is assumed that

Modeling and Parameter Identification of Electric Machines

$$E[w] = 0, \quad Q = E[ww^T] \tag{22.14}$$

$$E[v] = 0, \quad R_0 = E[vv^T] \tag{22.15}$$

$$E[X[0]] = [0], \quad P_0 = E[XX^T] \tag{22.16}$$

22.3.2.2 Q-Axis Model

Assuming that the q-axis rotor body can be represented by three damper windings, the standstill discrete q-axis model is given by:

$$X(k+1) = A_q(\theta_q)X(k) + B_q(\theta_q)U(k) + w(k) \tag{22.17}$$

$$Y(k+1) = C_q X(k+1) + v(k+1) \tag{22.18}$$

where

$$C_d = \begin{bmatrix} 1 & 0 & 0 & 0 \end{bmatrix}$$

$$X = \begin{bmatrix} i_q & i_{1q} & i_{2q} & i_{3q} \end{bmatrix}^T$$

$$Y = \begin{bmatrix} i_q \end{bmatrix}, \qquad U = \begin{bmatrix} v_q \end{bmatrix}$$

$$\theta_q = \begin{bmatrix} L_{aq}, R_{1q}, L_{1q}, R_{2q}, L_{2q}, R_{3q}, L_{3q} \end{bmatrix}$$

The initial value of the state and statistics of measurement noise are described by Equation 22.14 through Equation 22.16. The computations of $A_d(\theta_d), B_d(\theta_d), A_q(\theta_q), B_q(\theta_q)$, from the continuous time-domain representation are described in Reference 19, and the explicit parameterization in terms of θ_d and θ_q is shown in Appendix B.

In this study, the effect of noise on parameter estimation is studied by using the simulated noisy data of a known model structure for parameter identification. The identification problem is to estimate the parameter vector θ_d and θ_q from a record of the time-domain sequence of i_d, i_{fd}, i_q, v_d and v_q.

22.3.3 Effect of Noise on the Process and the Measurement

Figure 22.4 shows the block diagram of the effect of noise on the process and the measurements. The model, which mathematically describes the process, is subjected to the deterministic input at each time instant k. Nature also subjects the process to a random input sequence $w(\cdot)$. The sequence $w(\cdot)$ is designated as the process noise sequence. It is assumed to be Gaussian with zero mean and covariance matrix $Q(\cdot)$. The covariance matrix Q gives a measure of the intensity of the process noise on the model. A high value of the covariance matrix Q corresponds to a noisy process. The reason for introducing the measurement noise sequence $v(\cdot)$ is that in physical problems, the measurements are inherently

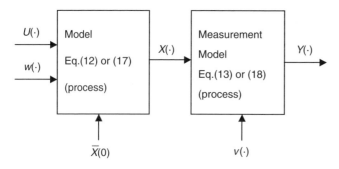

Figure 22.4 Block diagram representation of measurement and process noise.

subjected to errors. The signal conditioning equipment and sensors introduce measurement noise, which is random. The measurement errors $v(\cdot)$ are assumed to be independent and Gaussian with zero mean value and a known covariance matrix R_0. It is further assumed that the sequence $w(\cdot)$, $v(\cdot)$ $v(\cdot)$, and $X(0)$ are independent.

Let us denote the variance of $v_1(\cdot)$ and $v_2(\cdot)$ by $\sigma_{v_1}^2$ and $\sigma_{v_2}^2$. Also let $v_1(\cdot)$ and $v_2(\cdot)$ represent the measurement noise of i_d and i_{fd}. The assumption that $v_1(\cdot)$ and $v_2(\cdot)$ are independent ensures that measurement of i_d will not introduce additional uncertainty (i.e., measurement noise) in the measurement of i_{fd}. This assumption is not completely true. For example, the use of shunt resistances for current measurements will introduce its own uncertainty in the process variables to be measured. In this book, however, it is assumed that measurement errors are independent; therefore the covariance $R(\cdot)$ is a diagonal matrix, and the diagonal elements represent the variance of the measurement errors. Note that the standard deviations of measurement errors represent the percentage errors associated with the sensors. The accuracy of the sensors may be known from manufacturer data or from carefully controlled experiments on the sensors themselves.

The initial covariance R_0 is constructed from the knowledge of sensor errors, and it represents a measure of the prior confidence in the sensors to produce accurate measurements. Strictly speaking, two experiments performed on the same process will not result in identical measurements. Therefore, the covariance of the estimation error is calculated as part of Kalman filter [26–30] for estimating the machine states and then the parameters. The covariance of the estimation error is defined as

$$R(k) = COV\big(e(k), e(k)\big)$$
$$e(k) = Y(k) - \hat{Y}(k)$$
(22.19)

22.3.4 Maximum Likelihood Parameter Estimation

Consider the system described by the linear difference Equation 22.12 and Equation 22.13 or Equation 22.17 and Equation 22.18. To apply the maximum likelihood method, the first step is to specify the likelihood function [26–30]. The likelihood function $L(\theta)$, where θ represents θ_d or θ_q, is defined as

$$L(\theta) = \prod_{k=1}^{N}\left[\frac{1}{\sqrt{(2\pi)^m \det(R(k))}} \exp\left(-\frac{1}{2} e(k)^T R(k)^{-1} e(k)\right)\right]$$
(22.20)

where $e(\cdot)$, $R(\cdot)$, N, and m denotes the estimation error, the covariance of the estimation error (see Equation 22.19), the number of data points, and the dimension of Y, respectively.

Maximizing $L(\theta)$ is equivalent to minimizing its negative log function, which is defined as:

$$V(\theta) = -\log L(\theta)$$

$$V(\theta) = \frac{1}{2}\sum_{k=1}^{N}\left[e(k)^T R(k)^{-1} e(k)\right] + \frac{1}{2}\sum_{k=1}^{N} \log \det(R(k)) + \frac{1}{2}mN \log(2\pi) \qquad (22.21)$$

The vector θ (i.e., θ_d or θ_q) can be computed iteratively using Newton's approach [27,31]:

$$H\Delta\theta + G = 0$$
$$\theta_{new} = \theta_{old} + \Delta\theta \qquad (22.22)$$

where H and G are the Hessian matrix and the gradient vector of $V(\theta)$. They are defined by:

$$H = \frac{\partial^2 V(\theta)}{\partial \theta^2} \qquad G = \frac{\partial V(\theta)}{\partial \theta} \qquad (22.23)$$

The H and G matrices are calculated using the numerical finite difference method as described in References 24 and 30.

To start iterative approximation of θ, the covariance of estimation error $R(k)$ (see Equation 22.19) is obtained using the Kalman filter theory [25–29]. The steps are as follows:

1. Initial conditions: The initial value of the state is set equal to zero. The initial covariance state matrix P_0 is assumed to be a diagonal matrix with large positive numbers. Furthermore, assume an initial set of parameter vector θ.
2. Using the initial values of the parameters vector θ, compute the matrices A, B, and C for d or q-axis.
3. Compute estimate $\hat{Y}(k|k-1)$ from $\hat{X}(k|k-1)$:

$$\hat{Y}(k|k-1) = C\hat{X}(k|k-1) \qquad (22.24)$$

4. Compute the estimation error of $Y(k)$:

$$e(k) = Y(k) - \hat{Y}(k|k-1) \qquad (22.25)$$

5. Compute the estimation error covariance matrix $R(k)$:

$$R(k) = R_0 + C \cdot P(k|k-1) \cdot C^T \qquad (22.26)$$

6. Compute the Kalman gain matrix:

$$K(k) = P(k|k-1) \cdot C^T \cdot R(k)^{-1} \qquad (22.27)$$

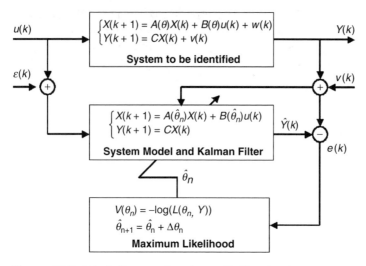

Figure 22.5 Block diagram of maximum likelihood estimation.

7. Compute the state estimation covariance matrix at instant k and $k + 1$:

$$P(k|k) = P(k|k-1) - K(k) \cdot C \cdot P(k|k-1)$$
$$P(k+1|k) = A(\theta) \cdot P(k|k) \cdot A^T(\theta) + Q \qquad (22.28)$$

8. Compute the state at instant k and $k + 1$:

$$\hat{X}(k|k) = \hat{X}(k|k-1) + K(k) \cdot e(k)$$
$$\hat{X}(k+1|k) = A(\theta) \cdot \hat{X}(k|k) + B(\theta) \cdot U(k) \qquad (22.29)$$

9. Solve Equation 22.22 for $\Delta\theta$ and compute the new θ such that

$$\theta_{min} \leq \theta_{new} \leq \theta_{max} \qquad (22.30)$$

10. Repeat steps 2 through 9 until $V(\theta)$ is minimized.

The above mechanism for maximum likelihood estimation is illustrated in Figure 22.5. A model of the system is excited with the same input as the real system. The error between the estimated output and the measured output is used to adjust the model parameters to minimize the error (maximize the likelihood). This process is repeated until the cost function $V(\theta)$ is minimized.

22.3.5 Estimation Procedure Using SSFR Test Data

The machine parameters estimated from the frequency-domain data are very sensitive to the value of the armature resistance used in deriving the operational inductance L_d and L_q

Modeling and Parameter Identification of Electric Machines

data [22]. However, the values of L_{ad} and L_{aq} provided by manufacturers are quite accurate. In our estimation of rotor body circuit parameters, the manufacturer's specified values of L_{ad} and L_{aq} will be utilized to decide the appropriate value for armature resistance. The θ_{min} and θ_{max} (e.g., $L_{ad,min}$ and $L_{ad,max}$) are selected within ±15% of the values supplied by the manufacturers. The rotor body circuit parameters are constrained to be greater than zero. The steps are as follows:

1. Estimate the value of R_a:

$$R_a = \lim_{\omega \to 0} Z_d(j\omega)$$

2. Compute the operational inductance L_d and L_q data using the measured frequency response data or Z_d and Z_q.
3. Fit very high order transfer function to $L_d(s)$ and $L_q(s)$ data of step 2 and $sG(s)$ transfer function [22].
4. Compute the step-response of the transfer function of step 3.
5. Use the *ML* identification technique to estimate the machine parameters as a constraint minimization problem.

As indicated in the previous section, the transfer function's time constants are very sensitive to the value of R_a. If the initial value of R_a is not accurately estimated, then this may not give a minimum value of $V(\theta)$. Therefore, it is suggested that the constraint *ML* identification process be repeated for another value of R_a until $V(\theta)$ is minimized.

This problem of iterating with different values of R_a occurs only when the time-domain data are generated from the SSFR test data. If the time-domain data are directly measured, all parameters can be estimated without going through the above iterative procedure.

It may be noted that accurate results will be obtained by this approach, because in step 3 high order transfer functions can be fitted to the $L_d(s)$, $L_q(s)$, and $sG(s)$ data. In the classical frequency response technique, only the third- or second-order transfer functions are used. Therefore, the estimated transfer functions will not accurately represent the SSFR data in the subtransient region [11]. Furthermore, as indicated in the previous section, multiple parameter sets will be obtained when the machine parameters are estimated using the classical SSFR technique. Note that the use of higher order transfer functions, in the classical SSFR technique, will result in a larger set of nonlinear and inconsistent equations (see Appendix A), which cannot be accurately solved for the machine parameters. However, the proposed approach can be used to obtain an accurate and unique estimation of machine parameters.

22.3.6 Results

The machine parameters are estimated from the time-domain data, which are computed by using the estimated transfer functions. The transfer functions are estimated from the noise-corrupted SSFR data as described in previous section. The signal-to-noise ratio of 3100:1 (where S/N = $[\Sigma(signal)^2/(noise)^2]^{1/2}$) is used in this study. The noise sequence used is a normally distributed random variable with zero mean.

To obtain a record of the time-domain data, the step response of the estimated d-q axis transfer functions [1] is computed. The input step voltage is defined as

$$V_d(t) = \begin{cases} 1 & t \leq 0.025s \\ 0 & t > 0.025s \end{cases}$$
$$V_q(t) = \begin{cases} 1 & t \leq 0.06s \\ 0 & t > 0.06s \end{cases} \quad (22.31)$$

The input signals to be applied to the transfer functions can be chosen without any constraints. The step voltage signals are used because of their rich frequency contents. This input signal excited all the estimated transfer function modes. The time-domain data of $i_d(\cdot)$, $i_{fd}(\cdot)$, and $i_q(\cdot)$ are computed using the input step defined by Equation 22.31. The variance of the process noise is assumed to be negligible, and its signal-to-noise ratio is very high. The signal-to-noise ratio of measurement noise is 3100:1. This measurement noise was introduced in the SSFR data.

The ML estimation is used to estimate the machine parameters. The results are given in Table 22.7 for d-axis, and in Table 22.8 for q-axis. The parameters L_{ad}, L_{aq}, and R_{fd} were initialized at 15% below their original values. Since the ranges of these variables are known, the initial values of these parameters are quite reasonable. However, since *a priori*

Table 22.7 Maximum Likelihood Identification of D-Axis Parameters

SSFR3 Parameter p.u.	Initial Values	Estimated Values	Original Values	Percent Errors
L_{ad}	1.4400E-0	1.6933E-0	1.691E-0	0.1
L_{f12d}	1.0000E-1	9.7909E-3	9.3E-3	5.0
R_{1d}	1.0000E-1	6.6717E-2	6.7E-2	0.5
L_{1d}	1.0000E-1	1.1355E-1	1.144E-1	0.8
L_{f2d}	1.0000E-1	1.2801E-1	1.287E-1	0.5
R_{2d}	1.0000E-1	9.0980E-3	9.15E-3	0.5
L_{2d}	1.0000E-1	1.3919E-3	1.400E-3	0.5
R_{fd}	7.0000E-4	8.1093E-4	8.11E-4	0.0
L_{fd}	1.0000E-2	8.6397E-3	8.70E-3	0.7

Table 22.8 Maximum Likelihood Identification of Q-Axis Parameters

SSFR3 Parameter p.u.	Initial Values	Estimated Values	Original Values	Percent Errors
L_{aq}	1.4400E-0	1.6328E-0	1.627E-0	0.4
R_{1q}	1.0000E-1	1.0619E-2	1.06E-2	0.0
L_{1q}	1.0000E-1	1.9245E-0	1.918E-0	0.4
R_{2q}	1.0000E-1	1.3004E-1	1.293E-1	0.5
L_{2q}	1.0000E-1	1.2584E-1	1.247E-1	0.9
R_{3q}	1.0000E-1	2.1050E-2	2.10E-2	0.2
L_{3q}	1.0000E-1	3.8318E-1	3.816E-1	0.4

Modeling and Parameter Identification of Electric Machines

knowledge of the rotor body circuits' parameters is much less precise, they are initialized arbitrarily.

The estimated parameters can be evaluated by comparing the mean and RMS errors of the $i_d(t)$, $i_{fd}(t)$, and $i_q(t)$ for both methods, namely, the SSFR method as stated in previous section and the ML method.

It is established [29,30] that the ML method gives unique solution even when the data are noise-corrupted. The unique estimated values of the d-q axis parameters obtained in this study are given in Tables 22.7 and 22.8.

Figure 22.6 through Figure 22.8 show that the time-domain simulated responses of $i_d(t)$, $i_{fd}(t)$, and $i_q(t)$ respectively as step voltage are used as input in the d- and q-axis models of the Monticello machine. In each figure, three sets of responses are plotted. One

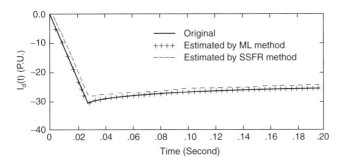

Figure 22.6 Monticello GS, original and estimated $i_d(t)$.

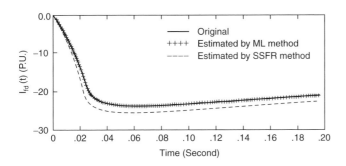

Figure 22.7 Monticello GS, original and estimated $i_{fd}(t)$.

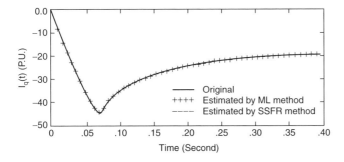

Figure 22.8 Monticello GS, original and estimated $i_q(t)$.

Table 22.9 Mean and RMS Error Comparisons of D-Axis and Q-Axis Time-Domain Response

Method	Parameter Set	Mean Errors		RMS Errors	
		$i_d(t)$	$i_{fd}(t)$	$i_d(t)$	$i_{fd}(t)$
D-axis					
SSFR [1]	1	9.5E-4	6.6E-3	1.0E-3	7.0E-3
	2	1.5E-1	1.0E-0	1.6E-1	1.0E-0
	3	6.9E-3	6.0E-3	2.1E-3	1.8E-2
	4	1.1E-2	6.9E-2	2.3E-2	1.5E-1
ML	1	8.0E-4	5.5E-7	4.1E-5	6.1E-6

Method	Mean Error $i_q(t)$	RMS Errors $i_q(t)$
Q-axis		
SSFR	4.8E-8	1.2E-7
ML	4.8E-8	1.2E-7

of the responses corresponds to the original set of parameters of the Monticello machine, estimated in Reference 11. The remaining two sets of responses correspond to the parameters estimated by the ML and SSFR methods.

A study of these figures and Table 22.9 shows that for the same set of synthetic SSFR data, the d-axis results obtained by using the ML method are more accurate; and the q-axis results are the same for both methods. However, it can be shown that if an actual measured set of SSFR test data is used, the q-axis results of the ML method will also be more accurate. This is because, with the ML method, transfer functions higher than third order will represent the measured data more closely. Using higher order transfer functions will provide more accurate results both in the d- and q-axis parameters. It should be noted that the synthetic SSFR data used in this study were created from the SSFR3 model provided in Reference 11. Therefore, the transfer functions estimated to represent the noise-corrupted synthetic data could only be of third order.

The proposed approach can also be applied directly to measured standstill time test data. The standstill test data can be obtained by closing a suitable DC source across the stator windings while the field winding is short-circuited [6] so as to introduce a step change in voltage. The rotor position and stator connections are the same as SSFR testing procedures described in Reference 6 and Reference 11. The ML technique can be used to estimate the armature, field, and rotor body parameters directly from the measured standstill test data.

22.4 MODELING AND PARAMETER IDENTIFICATION OF INDUCTION MACHINES

Induction motors are used in automotive applications, either as stand-alone propulsion systems (electric vehicles) or in combination with an internal combustion engine (hybrid

Modeling and Parameter Identification of Electric Machines

electric vehicles). Accurate knowledge of the induction motor model and its parameters is critical when field orientation techniques are used. The induction motor parameters vary with the operating conditions, as is the case with all electric motors. The inductances tend to saturate at high flux levels, and the resistances tend to increase as an effect of heating and skin effect. Temperature can have a large span of values, load can vary anywhere from no-load to full load, and flux levels can change as commanded by an efficiency optimization algorithm. It could then be expected that the model parameters also vary considerably.

22.4.1 Model Identification

Although there are many models to describe induction motors, some are highly complex and not suitable to be used in control. The authors will only concentrate on the models that can be used in induction motor control. Also, since modern induction motor control is field oriented, d-q models will be analyzed. An excellent presentation on available model types can be found in Reference 34. The classical induction motor model (used in most control schemes) has identical d- and q-axis circuits, as shown in Figure 22.9. Since the classical model is a fourth-order system with six elements of storage (inductances) the model can be reduced to a simpler model without any loss of information [34].

The notations represent:

v_{ds}, v_{qs}: stator voltages in stationary reference frame
i_{ds}, i_{qs}: currents in stationary reference frame
λ_{dr}, λ_{qr}: rotor fluxes in stationary reference frame
L_l, L_m: magnetizing and leakage inductance (*r* for rotor, *s* for stator)
R_s, R_r: stator, rotor resistance
ω_r: synchronous and mechanical frequency (*rad/s*).

The transformation combines the leakage inductances in a single inductance. This schematic is preferred for control applications and is called the Γ model (the classical model is denoted as the T model). Depending on whether a stator flux or rotor flux controller is sought the leakage inductance can be placed in the stator or in the rotor. The

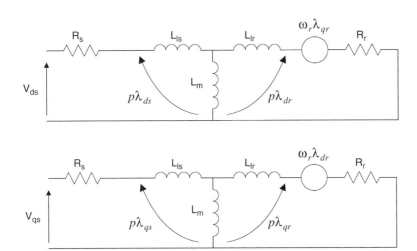

Figure 22.9 Equivalent circuits in d-q stationary.

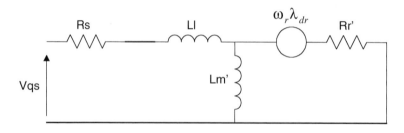

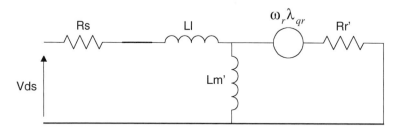

Figure 22.10 Reduced equivalent circuits in d-q stationary for RFO control.

transformation is meant to have L_m, be equal either to L_s or L_m of the classical model. Figure 22.10 shows the reduced model for rotor flux-oriented (RFO) control.

Although there are more complicated models used in performance analysis, transient stability, and short-circuit studies, their complexity (expressed in the number of differential equations used in the model) makes them unattractive for control purposes. The known variations of the classical model are derived by allowing parameter variations and by representing core losses. Although all parameters are known to vary with the operating conditions, the effect of the variation of the leakage inductances is usually neglected. The magnetizing inductance is shown to vary as a function of the magnetizing current, rotor flux, or input voltage. The stator and rotor resistances are mainly affected by the rotor temperature and skin effect. In steady-state models, core losses are typically represented as a resistance in parallel with the magnetizing inductance. However, by doing so the order of the model increases by two and adversely affects the control task. In literature, there are two trends to avoid this problem. One consists in adding the core loss resistance in parallel with the rotor resistance. The other [34] adds an R-L branch in both d and q axes and supplies it with the voltage created by the rotor flux of the corresponding axis. Then the differential terms associated with this branch are neglected to maintain the order of the system. A third method consists in adding the core loss resistance in series with the magnetizing inductance.

Figure 22.11 shows the induction motor model used in this work in stationary reference frame. The core loss branch is added to account for both stator and rotor core losses. Since the core loss resistance is much larger than the rotor resistance, it will be neglected in this part of modeling. The following basic equations of induction machine can be derived:

$$\frac{d\omega_r}{dt} = \mu\left(\lambda_{dr} i_{qs} - \lambda_{qr} i_{ds}\right) - \frac{B}{J}\omega_r - \frac{T_L}{J} \qquad (22.32)$$

Modeling and Parameter Identification of Electric Machines

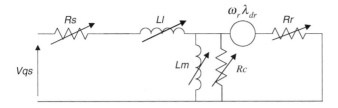

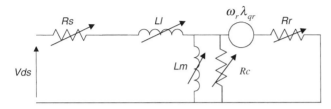

Figure 22.11 Induction motor models in stationary reference frame.

$$\frac{d\lambda_{qr}}{dt} = n_p\omega_r\lambda_{dr} - \eta\lambda_{qr} + \eta L_m i_{qr} \tag{22.33}$$

$$\frac{d\lambda_{dr}}{dt} = -n_p\omega_r\lambda_{qr} - \eta\lambda_{dr} + \eta L_m i_{dr} \tag{22.34}$$

$$\frac{di_{qs}}{dt} = -\beta n_p\omega_r\lambda_{dr} + \eta\beta\lambda_{qr} - \gamma i_{qs} + \frac{1}{\sigma L_s}v_{qs} \tag{22.35}$$

$$\frac{di_{ds}}{dt} = \beta n_p\omega_r\lambda_{qr} + \eta\beta\lambda_{dr} - \gamma i_{ds} + \frac{1}{\sigma L_s}v_{ds} \tag{22.36}$$

where:

$\eta \equiv \frac{1}{T_R} = \frac{R_r}{L_m}$: inverse of the rotor time constant

$\sigma \equiv 1 - \frac{L_m}{L_s}$: leakage coefficient

$\beta \equiv \frac{1}{L_l}$: inverse of leakage inductance

$\gamma \equiv \frac{R_s + R_r}{L_l}$: inverse of the stator time constant

$L_s = L_l + L_m$: stator inductance

n_p: number of poles pairs

The electromagnetic torque expressed in terms of the state variables is:

$$T_e = \mu J\left(\lambda_{dr}i_{qs} - \lambda_{qr}i_{ds}\right) \tag{22.37}$$

where:

$\mu \equiv \dfrac{n_p}{J}$: constant

J : inertia of the rotor

T_e: electromagnetic torque $(N \cdot m)$

In synchronously rotating reference frame, the motor equations can be expressed as:

$$\frac{d\omega_r}{dt} = \mu \cdot \lambda_r \cdot i_{qs}^e - \frac{B}{J}\omega_r - \frac{T_L}{J} \tag{22.38}$$

$$\frac{d\lambda_r^e}{dt} = -\eta \cdot \lambda_r + \eta \cdot L_m \cdot i_{ds}^e \tag{22.39}$$

$$\frac{di_{qs}^e}{dt} = -\gamma i_{qs}^e - \beta n_p \omega_r \lambda_r - n_p \omega_r i_{ds}^e - \eta L_m \frac{i_{qs}^e i_{ds}^e}{\lambda_r} + \frac{1}{\sigma L_s} v_{qs}^e \tag{22.40}$$

$$\frac{di_{ds}^e}{dt} = -\gamma i_{ds}^e + \eta \beta \lambda_r + n_p \omega_r i_{qs}^e + \eta L_m \frac{{i_{qs}^e}^2}{\lambda_{dr}} + \frac{1}{\sigma L_s} v_{ds}^e \tag{22.41}$$

$$\frac{d\theta_e}{dt} = n_p \omega_r + \eta L_m \frac{i_{qs}^e}{\lambda_r} \tag{22.42}$$

where:

v_{ds}^e, v_{qs}^e: stator voltage in synchronous reference frame

i_{ds}^e, i_{qs}^e: currents in synchronous reference frame

λ_r: rotor flux in synchronous reference frame

and the expression for torque is given by:

$$T_e = \mu J \cdot \lambda_r \cdot i_{qs}^e \tag{22.43}$$

22.4.2 Parameter Estimation

There are many parameter estimation techniques for the induction motor. Depending on the type of tests performed on the motor, the testing methods could be classified as:

Off-site methods, which test the motor separately from its application site [36–42]. The motor is tested individually, in the sense that it is not necessarily connected to the load it is going to drive or in the industrial setup it is going to operate in. The most common such tests are the no-load test and the locked-rotor test. The advantage of the above methods is their simplicity. However, these tests usually represent poorly the real operating conditions of the machines (for example, they lack the effect of PWM switching on the machine parameters).

Modeling and Parameter Identification of Electric Machines

On-site and off-line methods, which are performed with the motor already connected in the industrial setup and supplied by its power converter [35,43–47]. These tests are usually meant to allow the tuning of the controller parameters to the unknown motor it supplies and are also known as self-commissioning. As they are convenient for the controller manufacturer (one control program could work for different motors), they usually are less precise than the individual tests.

On-line methods: Some parameters are estimated while the motor is running on-site [48–52,55]. These methods are concerned usually with rotor parameters (L_m and R_r or the time constant, T_r) and assume that the other parameters are known. These methods usually perform well only for a good initial value of the parameter to be determined and for relatively small variations (within 10%).

The purpose of this section is the development of an induction motor model with parameters that vary as a function of operating conditions. The development is on-site and off-line. While stator resistance is measured through simple DC test, the leakage inductance, the magnetizing inductance, and the rotor resistance are estimated from transient data using a constrained optimization algorithm. Through a sensitivity analysis study, for each operating condition, the parameters to which the output error is less sensitive are eliminated. The parameters are estimated under all operating conditions and mapped to them (e.g., analytical functions relating parameters to operating conditions are created). A correlation analysis is used to isolate the operating conditions that have most influence on each parameter. A core loss resistance models core losses. This resistance is estimated using a power approach and Artificial Neural Networks. No additional hardware is necessary. The same power converter and DSP board that controls the motor in the industrial setting is used to generate the signals necessary to model the motor. Therefore, phenomenon related to operation (for example, PWM effects) is captured in modeling.

22.4.2.1 Estimation of Stator Resistance

The estimation of the stator resistance was carried out through a DC test, as shown in Figure 22.12. The resistance can be calculated as:

$$R_S = \frac{2}{3} \cdot \frac{V_a - V_c}{I_c} \quad (22.44)$$

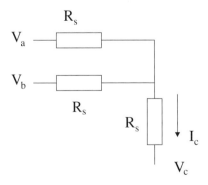

Figure 22.12 Circuit for stator resistance.

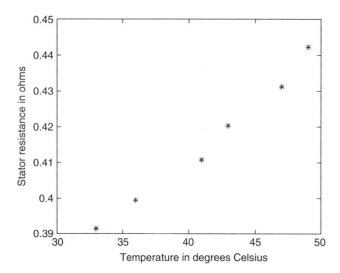

Figure 22.13 Stator resistance as a function of temperature.

To capture the effect of temperature on the stator resistance, the following sequence of testing was used:

At each test, the motor was run with an increased load.
The stator resistance test was performed immediately after the motor stopped.

The temperature of the stator winding was also measured. The temperature dependency of the stator resistance is shown in the Figure 22.13.

22.4.2.2 Estimation of L_l, L_m, and R_r

Transient data was used to determine L_m, L_l, and R_r. The data consisted of small disturbance in the steady-state operation of the IM by stepping the supply voltage with 10%. The tests encompass a wide variation of frequency, supply voltage and load:

- The frequency was varied from 30 to 80 Hz in steps of 10 Hz.
- The supply voltage was varied from 10 to 100% of the rated voltage value in steps of 10% for each frequency.
- The load was varied from no-load to maximum load in eight steps.

A total of 290 data files were obtained. The estimation was performed using a constrained optimization method available in Matlab ("constr"). Figure 22.14 shows the block diagram of the estimation procedure.

The induction motor model can be expressed in state space form as:

$$\dot{X} = AX + BU \qquad (22.45)$$

and the output equation is:

$$Y = C \cdot X \qquad (22.46)$$

Modeling and Parameter Identification of Electric Machines

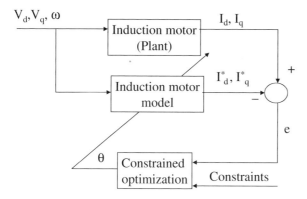

Figure 22.14 Estimation block diagram.

where

$$X = \begin{bmatrix} i_{qs} & i_{ds} & \lambda_{qr} & \lambda_{dr} \end{bmatrix}^T \qquad (22.47)$$

$$A = \begin{bmatrix} -\dfrac{R_s + R_r}{L_l} & 0 & \dfrac{R_r}{L_l L_m} & -\dfrac{\omega_r}{L_l} \\ 0 & -\dfrac{R_s + R_r}{L_l} & \dfrac{\omega_r}{L_l} & \dfrac{R_r}{L_l L_m} \\ -R_r & 0 & -\dfrac{R_r}{L_m} & \omega_r \\ 0 & R_r & -\omega_r & -\dfrac{R_r}{L_m} \end{bmatrix} \qquad (22.48)$$

$$C = \begin{bmatrix} 1 & 0 & 0 & 0 \\ 0 & 1 & 0 & 0 \end{bmatrix} \qquad (22.49)$$

$$\text{and } U = \begin{bmatrix} V_q \\ V_d \end{bmatrix}$$

The initial conditions for the model were established as:

$$X = \begin{bmatrix} i_{qs(0)} & i_{ds(0)} & \hat{\lambda}_{qr} & \hat{\lambda}_{rq} \end{bmatrix}^T \qquad (22.50)$$

The error between model and measurements was calculated as:

$$e = \sqrt{\sum_{k=1}^{N} \left(\hat{i}_{ds(k)} - i_{ds(k)} \right)^2 + \left(\hat{i}_{qs(k)} - i_{qs(k)} \right)^2} \qquad (22.51)$$

where ^ = estimated values

The constrained optimization function is used to minimize the error function by modifying the parameter vector, θ:

$$\theta = \begin{bmatrix} L_m & R_r & L_l & \hat{\lambda}_{qr(0)} & \hat{\lambda}_{dr(0)} \end{bmatrix} \qquad (22.52)$$

The values of stator resistance were based on temperature measurements. The initial values of the fluxes are not normally included in the parameter vector, since they can be calculated from the initial conditions of the currents at steady state. However, these currents are noise corrupted and their measurement error will propagate into the calculation of the initial values of the flux. Furthermore, since flux equations have a large time constant, the initial condition error would influence the flux observation over the entire transient measurement (the self-correction of an otherwise convergent flux observer [57] will not have the time to correct the initial condition error) and will yield erroneous parameter estimates.

The authors observed that the parameter vector modification increased the rate of convergence of the algorithm. Constraints on R_r, L_l, and L_m were imposed as 10% of the rated value for the lower bound and 300% for the upper bound. For $\lambda_{d(0)}$ and $\lambda_{d(0)}$ the constraints were imposed as ± 200% of the saturation value (0.5 Wb).

22.4.3 SENSITIVITY ANALYSIS

Since an output error estimation method is used, there is no theoretical guarantee that the parameters will converge to their actual values. Therefore, it is necessary to study the effect of each parameter on the total error. It is obvious that those parameters with little effect on the total error will be more prone to estimation errors than parameters that affect it more. For any data point, the error can be expressed as:

$$E(t) = I \cdot \sin(\omega t + \varphi) - \hat{I} \cdot \sin(\omega t + \hat{\varphi}) \qquad (22.53)$$

At steady state, the squared error per period is:

$$e^2 = \frac{1}{T} \int_0^T E^2(t) dt = \frac{1}{2} \left(I^2 + \hat{I}^2 - 2\hat{I} \cdot I \cos(\varphi - \hat{\varphi}) \right) \qquad (22.54)$$

The sensitivity of the squared error to a parameter (y) can be expressed as:

$$S_y^e = \frac{e^2(\hat{I}(y + \Delta y), I(y))}{\Delta y} \cdot \frac{y}{I} \qquad (22.55)$$

For the proposed model, the steady-state current (complex form) can be expressed as:

$$\vec{I} = V \cdot \frac{\frac{R_r}{s} + jX_m}{-X_l X_m + \frac{R_s R_r}{s} + j\left(\frac{R_r}{s}(X_l + X_m) + R_s X_m\right)} \qquad (22.56)$$

and I and φ are the module and phase angle of $\vec{I}$.

The sensitivity analysis was conducted for a slip ranging from 0 to 10% (larger values of s are unobtainable at steady state) and a frequency from 20 to 100 Hz. The rated

Modeling and Parameter Identification of Electric Machines

values of the parameters were used. Figure 22.15 shows a comparison of sensitivity for R_r, L_m, and L_l at 60 Hz. Figure 22.16 through Figure 22.18 represent the sensitivity of each individual parameter for different frequencies and slips. It can be seen that the sensitivity of the error to L_l or R_r is low at low slip. Large errors can be introduced at low slip since their effect on the error is small. A limit of 2% on the slip was imposed on the slip values. The L_l and R_r estimates below this value are discarded. For large values of the slip the sensitivity of the error to L_m decreases to 0. L_m estimates for slip values larger than 2% were discarded.

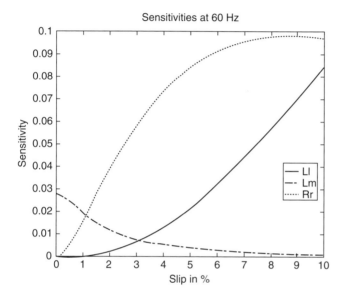

Figure 22.15 Sensitivity of error to parameters as a function of slip at 60 Hz.

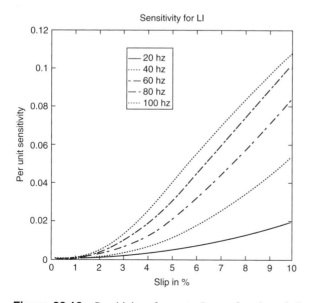

Figure 22.16 Sensitivity of error to L_l as a function of slip at different frequencies.

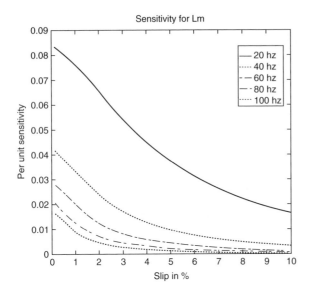

Figure 22.17 Sensitivity of error to L_m as a function of slip at different frequencies.

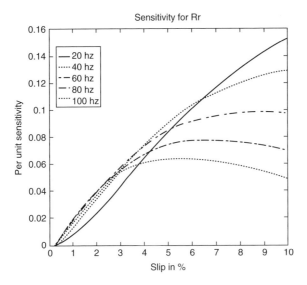

Figure 22.18 Sensitivity of error to R_r as a function of slip at different frequencies.

Observation

The concept of sensitivity of the currents (error) to the parameters can be extended to more classical induction motor tests: In the no-load test, only L_m is estimated, whereas in the locked rotor test R_r and L_l are estimated.

22.4.4 Parameter Mapping to Operating Conditions

The model proposed here is dependent on the operating conditions. Up to this point, the parameters of the motor were estimated for various operating conditions. The purpose of

Modeling and Parameter Identification of Electric Machines

Table 22.10 Correlation between Parameters and Operating Conditions

Parameter	i_{ds}^e	i_{qs}^e	Is	Ws
L_m	−0.9287	0.0167	−0.6652	0.3473
R_r	0.1543	0.8061	0.8023	0.5485
L_l	−0.3212	−0.4264	−0.7135	0.0177

this section is to find the relation of the parameters to the operating conditions in a form that allows for use in a control environment. However, in order to be able to define an operating condition or to relate (map) a parameter to a condition, a correlation analysis is necessary. This establishes the "strong" and "weak" dependencies of parameters to operating variables. The variables for the correlation study are selected intuitively as:

i_{ds}^e, i_{qs}^e : the stator currents in synchronous reference frame
Is : the stator current (peak value)
w_s: slip frequency

It could be argued that temperature is also a factor in this mapping. However, since the only temperature measurement available was the stator temperature (and was used for stator resistance calculation), it was not used in this correlation study. The correlation between two variables (in this case one variable is a parameter [y] and the other an operating condition variable [x]) can be defined as:

$$C_{x,y} = \frac{\frac{1}{N-1}\sum_{k=1}^{N}\left((x_k - \bar{x})(y_k - \bar{y})\right)}{\sigma_x \sigma_y} \qquad (22.57)$$

where $\bar{x}, \bar{y}$ are the mean of x and y, respectively, and σ_x, σ_y are their standard deviations. Table 22.10 shows the results of the correlation:

Mapping consists of expressing the parameters of the motor as analytical functions of the operating conditions. The selection of the variables describing the operating conditions is based on the correlation study.

22.4.4.1 Magnetizing Inductance, L_m

A strong correlation was observed between L_m and i_{ds}^e. L_m clearly saturates with an increase in i_{ds}^e. A second-order polynomial was used to represent the dependency of L_m to i_{ds}^e in the saturated region.

$$L_m(i_{ds}^e) = k_1 \cdot i_{ds}^{e\,2} + k_2 \cdot i_{ds}^e + k_3 \qquad (22.58)$$

Figure 22.19 shows a comparison between the polynomial and the results of the estimation.

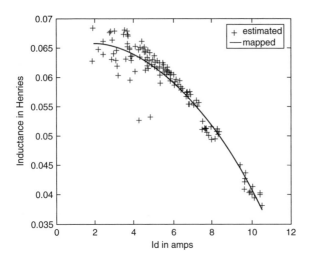

Figure 22.19 L_m as function of i_{ds}^e.

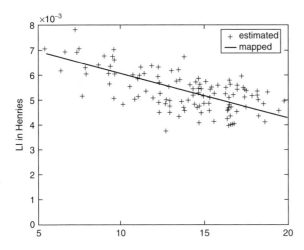

Figure 22.20 L_l as function of Is.

22.4.4.2 Leakage Inductance, L_l

A strong correlation was also observed between L_l and Is. L_l saturates with an increase in Is. A linear approximation was used to represent the dependency of L_l to Is and is shown in Figure 22.20.

$$L_l(I_s) = k_4 \cdot I_s + k_5 \qquad (22.59)$$

22.4.4.3 Rotor Resistance, R_r

It can be safely assumed that the rotor resistance varies as a function of two factors: slip frequency (through skin effect) and rotor temperature (immeasurable). However, Table 22.10 shows a correlation between R_r and ws but also i_{qs}^e. The correlation is shown in

Modeling and Parameter Identification of Electric Machines

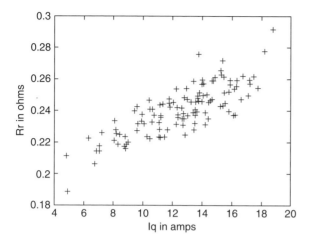

Figure 22.21 R_r as function of i_{qs}^e.

Figure 22.21. The correlation is due to the fact that both slip frequency and temperature are proportional to i_{qs}^e. It was observed that the $R_r(i_{qs}^e)$ correlation holds only if the motor runs for a few minutes at a certain operating condition, to allow for temperature to reach a steady state. A sudden variation in i_{qs}^e would not determine a sudden change in R_r if slip frequency remains constant, since temperature does not change as fast. Therefore, the $R_r(i_{qs}^e)$ relation can only be used at steady state.

In order to establish the influence of slip frequency on R_r, a test similar to a locked rotor was used. The difference was that the rotor was not mechanically locked, but the voltages were small enough that the rotor would not move. The frequency was varied between 5 and 120 Hz (1 Hz increments in the 5–10 Hz region and 10 Hz increments for the rest). Prior to each series of tests, the motor was run under a loading condition (no load, medium load, and full load) to ensure heating of the rotor. A temperature sensor was mounted on the stator. This sensor was used for an indication when temperature has reached a steady state (for each loading condition). Figure 22.22 shows the results of the R_r

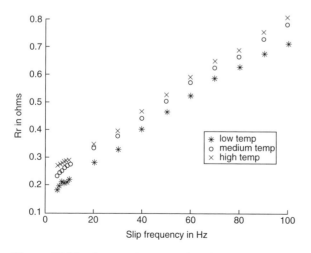

Figure 22.22 Rotor resistance as function of slip frequency for different temperatures.

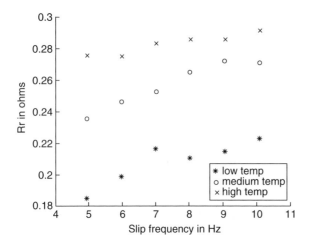

Figure 22.23 Rotor resistance as function of slip frequency for lower frequencies.

estimation as a function of slip frequency (for locked rotor, equal to stator frequency). Figure 22.23 shows just the 5–10 Hz region, which is of more interest for us, since slip rarely exceeds this range.

Observation 1. Since rotor frequency (slip frequency) influences the values of rotor resistance, the locked rotor test must be carried out at a low frequency if the rated value of rotor resistance is sought. This is particularly important for squirrel-cage motors in which skin effect is present. For example, for the motor used in this research, a locked rotor test at rated frequency would yield a value of rotor resistance approximately 3 times higher than real. Since rotor temperature measurements are hardly possible, a precise off-line mapping of rotor resistance to operating conditions is impossible. However, due to the linearity (within the range of interest) of the relation between rotor resistance and slip frequency, an on-line observer can be developed.

The observation is based on the assumptions that the rotor temperature varies much slower than the other variables (current, speed, and so on) and that steady-state operating conditions exist (e.g., the motor is not in continuous transient). The rotor resistance dependency to slip frequency and rotor temperature can be expressed as:

$$R_r(\omega_s, T) = R_1(T) + k_6 \cdot \omega_s \tag{22.60}$$

in which $R_1(T)$ is the influence of temperature (unknown). The coefficient k_6 (influence of slip frequency) can be estimated off-line from the locked rotor tests measurements. At each operating condition (steady state), the value of rotor resistance and slip frequency can be estimated with an observer, as shown in the next section. Then for each loading condition (temperature):

$$\hat{R}_1(T) = \hat{R}_r(\omega_s, T) - k_6 \cdot \hat{\omega}_s \tag{22.61}$$

Assuming that temperature changes slowly, at each instant of time, knowing the slip frequency allows for the determination of rotor resistance. Each time a steady-state condition is detected, $R_1(T)$ is re-evaluated and rotor resistance calculated as function of slip frequency.

Observation 2. It can be argued that since rotor resistance is estimated, there is no need in determining $R_1(T)$. This is true while the motor operates at steady state. However, for efficiency optimization it is important to predict the variation of rotor resistance prior to a new steady-state condition. For this case it is important to have the value of $R_1(T)$ and predict the variation of R_r based on slip frequency.

22.4.5 CORE LOSS ESTIMATION

One should note that since the slip is non-zero for the no-load test and R_r is already known, R_c could be theoretically calculated from the parallel resistance of R_r and R_c. However, even for most precise speed encoders, the error in calculating a slip approaching zero could translate in an order of magnitude of error when calculating R_r/s (s being the slip $s = \omega_s/\omega_e$).

A power-based approach is used for calculating the core resistance.

22.4.5.1 Calculation of Rotor Losses at Frequencies of Interest

Use the no-load tests and calculate the rotor power losses for each data set:

$$P_{rotor} = V_s I_s \cos(\varphi) - R_s I_s^2 \tag{22.62}$$

A plot of these losses is shown in Figure 22.24 for various frequencies. The losses increase with both the frequency and the rotor flux.

22.4.5.2 Calculation of Friction and Windage Losses Using ANN

Since core losses are zero when flux is zero, the intersection of the power curves with the vertical axis determines the friction and windage losses for a specific frequency. To find the friction and windage losses for all frequencies, an ANN was used to map the rotor losses to frequency and flux. Multilayer feed-forward neural networks have often been used in system identification studies. These networks consist of a number of basic computational units called processing elements connected together to form multiple layers. A

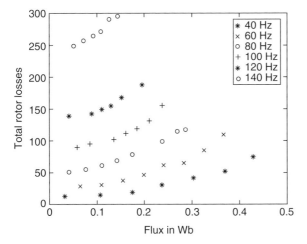

Figure 22.24 Rotor power losses for no-load test.

typical processing element forms a weighted sum of its inputs and passes the result through a nonlinear transformation (also called transfer function) to the output. The transfer function may also be linear, in which case the weighted sum is propagated directly to the output path. The ANN used in this research consists of processing elements arranged in three distinct layers. Data presented at the network input layer are processed and propagated, through a hidden layer, to the output layer. Training a network is the process of iteratively modifying the strengths (weights) of the connecting links between processing elements as patterns of inputs and corresponding desired outputs are presented to the network.

In this work, the mathematical relationship between the input and output patterns can be described as:

$$P_{rotor\_losses} = N_d(\lambda, \omega_e) \tag{22.63}$$

where N_d is a nonlinear neural network mapping to be established. The ANN used in this study is shown in Figure 22.25 and consists of two processing elements in the input layer, corresponding to each variable. A single processing element in the output layer corresponds to the losses being modeled. The number of elements in the hidden layer is arbitrarily chosen depending on the complexity of the mapping to be learned. A hyperbolic tangent (tanh) transfer function is used in all hidden layer elements, while all elements in the input layer and output layer have linear (1:1) transformations.

The back-propagation algorithm is used to train the neural network such that the sum squared error, E, between actual network outputs, O, and corresponding desired outputs, ζ, is minimized over all training patterns μ

$$E = \sum_{\mu}[\zeta_\mu - O_\mu]^2 \tag{22.64}$$

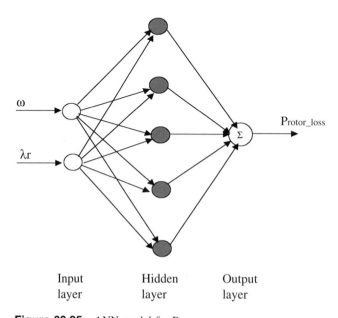

Figure 22.25 ANN model for P_{rotor_losses}.

Modeling and Parameter Identification of Electric Machines

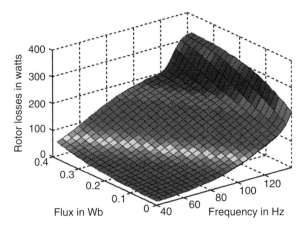

Figure 22.26 Mapping of rotor losses using ANN.

After estimating the nonlinear mapping N_d in terms of the neural network, the network output P_{rotor_losses} is computed from the 2 × 1 input vector P according to the following equation:

$$P_{rotor\_losses} = W_2 \cdot \tanh(W_1 \cdot P + B_1) + B_2 \tag{22.65}$$

W_2 denotes the matrix of connecting weights from the hidden layer to the output layer. W_1 is the weight matrix from the input layer to the hidden layer. If there are m processing elements in the hidden layer, W_2 is of size $1 \times m$, and W_1 is of size $m \times 1$. Bias terms B_2 and B_1 are used as connection weights from an input with a constant value of 1. B_2 and B_1 denote the 1×1 and $m \times 1$ bias vectors from the bias to the output layer, and from the bias to the hidden layer, respectively. The task of training is to determine the matrices W_1, W_2, and bias vectors B_1, B_2. The training patterns for the neural network models are composed of the no-load test data. Each data set is a vector of λ, ω_e and P_{rotor_losses}. The results of the mapping are shown in Figure 22.26. Friction and windage losses can be calculated for ANN at zero flux.

22.4.5.3 Calculation of Core Losses

Core losses for each frequency and flux can be determined by subtracting the friction and windage losses and the resistive rotor losses from the rotor losses.

$$P_{core}(\omega, \lambda_r) = P_{rotor\_losses}(\omega, \lambda_r) - P_{f\&w}(\omega_r) - R_r I_{r2}^2 \tag{22.66}$$

where

$$P_{f\&w} = R_r \frac{1-s}{s} I_{r2}^2 = P_{rotor\_losses}(\omega, 0) = N_d(\omega, 0) \tag{22.67}$$

Since

$$P_{f\&w} = R_r \frac{1-s}{s} I_{r2}^2 \gg R_r I_{r2}^2,$$

the last term of the previous equation is neglected.

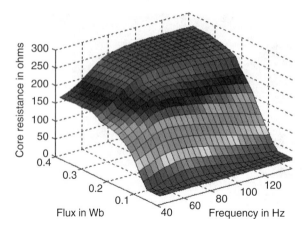

Figure 22.27 Rotor core losses as function of flux and frequency.

22.4.5.4 Calculation of Core Resistance

For each data point, calculate the core loss resistance as:

$$R_c(\omega_e, \lambda_r) = \frac{P^2_{rotor\_losses}(\omega_e, \lambda_r)}{I_r^2 \cdot P_{core}(\omega_e, \lambda_r)} \quad (22.68)$$

Map the core loss resistance to flux and frequency using ANN. The procedure is similar to the rotor loss mapping. Figure 22.27 presents the results of the mapping.

22.4.6 Model Validation

22.4.6.1 Steady-State Power Input

In order to validate the model at steady state, tests encompassing the entire range of operation of the induction motor were used. The frequency of the motor was varied from 30 to 70 Hz. The supply voltage was varied from 10 to 100% of rated. For each voltage and frequency entry, the load was varied from zero to maximum value. For all data sets, input power was measured and compared to the input power calculated using measured voltage and speed and the model. Figure 22.28 shows the results of the comparison.

22.4.6.2 Dynamic

The model was used to predict the transient performance after an input voltage disturbance. Figure 22.29 through Figure 22.31 present the results in terms of the stationary reference frame currents. The measured and estimated currents are represented on the same graph (measured = solid line and estimated = dotted line).

A second type of tests consisted in transient behavior when starting the motor. The start-up currents (measured and simulated) are shown in Figure 22.32 through Figure 22.33 in synchronous reference frame. Figure 22.32 represents the results when variable parameters were used, whereas Figure 22.33 represents the results when rated (fixed) parameters were used.

Modeling and Parameter Identification of Electric Machines

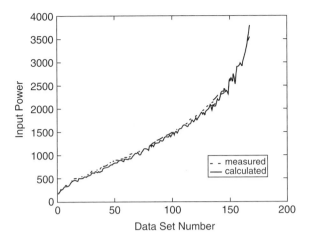

Figure 22.28 Measured and calculated input power.

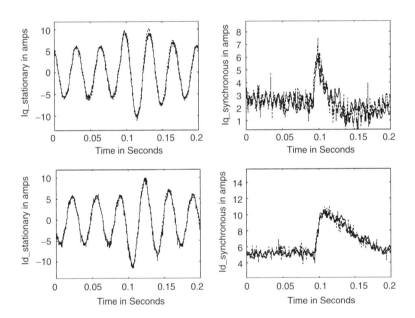

Figure 22.29 Model validation for large disturbance test at low frequency.

22.4.7 Conclusions

A systematic procedure for induction motor modeling was developed in this chapter. The model includes the effects of inductance saturation (both for magnetizing and leakage inductance) and the effects of the core losses. It is also shown that there is a variation of rotor resistance as a function of slip frequency. The leakage inductance, magnetizing inductance, and rotor resistance are estimated from transient data information using a constrained optimization method. Sensitivity analysis is employed to show that error sensitivity to parameters varies as a function of slip. The analysis eliminates parameters

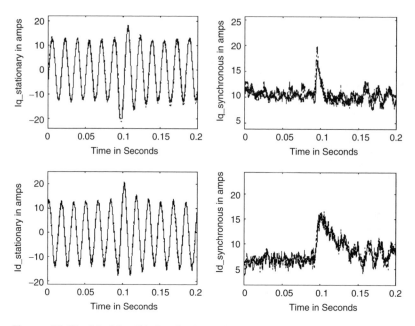

Figure 22.30 Model validation for large disturbance test at medium frequency.

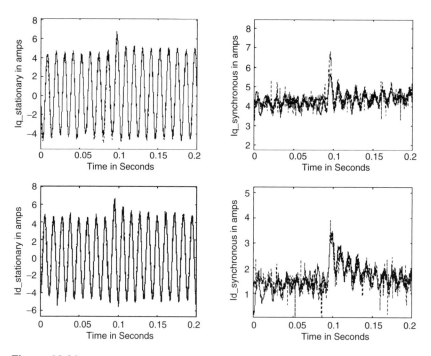

Figure 22.31 Model validation for large disturbance test at high frequency.

that yield low sensitivity. Analytical functions are used to map the parameters to operating conditions. Core losses are estimated using a power approach. ANN are used to map the total rotor losses (iron losses, friction, and windage losses) to flux and frequency. The core

Modeling and Parameter Identification of Electric Machines

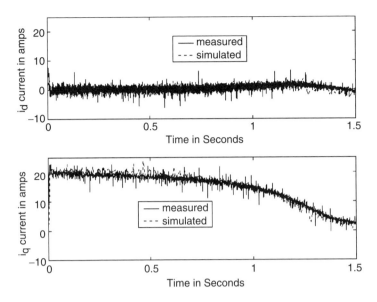

Figure 22.32 Transient response for start-up from zero speed.

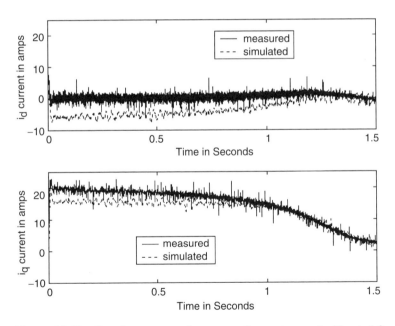

Figure 22.33 Transient response for start-up from zero speed with rated fixed parameters.

losses are obtained by subtracting the rotor losses at zero flux (generated by the ANN) from the rotor loss surface. The model is validated using tests covering various operating conditions. For steady-state validation, the model is shown to correctly predict the power input of the motor. For dynamic validation, input voltage disturbance tests and start-from-zero tests were employed. The model correctly predicted both tests.

22.5 MODELING AND PARAMETER IDENTIFICATION OF SWITCHED RELUCTANCE MACHINES

22.5.1 INTRODUCTION

Switched reluctance machines (SRMs) have undergone rapid development in hybrid electric vehicles and other automotive applications over the last two decades. This is mainly due to the various advantages of SRMs over the other electric machines, such as simple and robust construction and fault-tolerant performance.

In most of these applications, speed and torque control are necessary. To obtain high-quality control, a proper model of SRM is often needed. At the same time, to increase reliability and reduce cost, sensorless controllers (without rotor position/speed sensors) are preferred. With the rapid progress in microprocessors (DSPs), million instructions per second (MIPS)-intensive control techniques such as sliding mode observers and controllers [57] become more and more promising. An accurate nonlinear model of the SRM is essential to realize such control algorithms.

The nonlinear nature of SRM and high saturation of phase winding during operation makes the modeling of SRM a challenging work. The flux linkage and phase inductance of SRM change with both the rotor position and the phase current. Therefore the nonlinear model of SRM must be identified as a function of the phase current and rotor position. Two main models of SRM have been suggested in the literature: the flux model [58] and the inductance model [59]. In the latter, "the position dependency of the phase inductance is represented by a limited number of Fourier series terms and the nonlinear variation of the inductance with current is expressed by means of polynomial functions" [59]. This model can describe the nonlinearity of SRM inductance quite well. We will use this model here.

Once a model is selected, how to identify the parameters in the model becomes an important issue. Finite element analysis can provide a model that will be subjected to substantial variation after the machine is constructed with manufacturing tolerances. Therefore, the model and parameters need to be identified from test data. As a first step, the machine model can be estimated from standstill test using output error estimation (OE) or maximum likelihood estimation (MLE) techniques. This method has already been applied successfully to identify the model and parameters of induction and synchronous machines [60,61].

Furthermore, during on-line operation, the model structures and parameters of SRMs may differ from the standstill ones because of saturation and losses, especially at high current. To model this effect, a damper winding may be added into the model structure, which is in parallel with the magnetizing winding. The magnetizing current and damper current are highly nonlinear functions of phase voltage, rotor position, and rotor speed. They are not measurable during operation and are hard to be expressed with analytical functions. Neural network mapping are usually good choices for such tasks [62–64]. A two-layer recurrent neural network has been adopted here to estimate these two currents, which takes the phase voltage, phase current, rotor position, and rotor speed as inputs. When the damper current is estimated and damper voltage is computed, the damper parameters can be identified using output error or maximum likelihood estimation techniques.

In this section, the procedures to identify an 8/6 SRM parameters from standstill test data are presented after an introduction to the inductance model of SRM. Then a two-layer recurrent neural network is constructed, trained, and applied to identify the damper parameters of SRM from operating data. Model validation through on-line test is also given, which proves the applicability of the proposed methods.

Modeling and Parameter Identification of Electric Machines

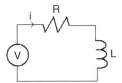

Figure 22.34 Inductance model of SRM at standstill.

22.5.2 Inductance Model of SRM at Standstill

The inductance model of switched reluctance motor is shown in Figure 22.34. The phase winding is modeled as a resistance (R) in series of an inductance (L).

Since the phase inductance changes periodically with the rotor position angle, it can be expressed as a Fourier series with respect to rotor position angle θ:

$$L(\theta, i) = \sum_{k=0}^{m} L_k(i) \cos k N_r \theta \tag{22.69}$$

where Nr is the number of rotor poles and m is the number of terms included in the Fourier series. High-order terms (HOTs) can be omitted without bringing significant errors.

In Reference 59, the authors suggest using the first three terms of the Fourier series (m = 3), but more terms can be added to meet accuracy requirements.

To determine the coefficients $L_k(i)$ in the Fourier series, we need to know the inductances at several specific positions. Use $L_\theta(i)$ to represent the inductance at position θ, which is a function of phase current i and can be approximated by a polynomial:

$$L_\theta(i) = \sum_{n=0}^{k} a_{\theta,n} i^n \tag{22.70}$$

where k is the order of the polynomial and $a_{\theta,n}$ are the coefficients of the polynomial. In our research, $k = 5$.

For an 8/6 machine, Nr = 6. When θ = 0 is chosen at the aligned position of phase A, then θ = 30 is the unaligned position of phase A. Usually the inductance at unaligned position can be treated as a constant.

$$L_{30^0} = const \tag{22.71}$$

22.5.2.1 Three-Term Inductance Model

If three terms are used in the Fourier series (m = 3), then we can compute the three coefficients L_0, L_1, and L_2 from L_{0^0} (aligned position), L_{15^0} (midway position), and L_{30^0} (unaligned position). Since

$$\begin{bmatrix} L_{0^0} \\ L_{15^0} \\ L_{30^0} \end{bmatrix} = \begin{bmatrix} 1 & 1 & 1 \\ 1 & \cos(6*15^0) & \cos(12*15^0) \\ 1 & \cos(6*30^0) & \cos(12*30^0) \end{bmatrix} \begin{bmatrix} L_0 \\ L_1 \\ L_2 \end{bmatrix} = \begin{bmatrix} 1 & 1 & 1 \\ 1 & 0 & -1 \\ 1 & -1 & 1 \end{bmatrix} \begin{bmatrix} L_0 \\ L_1 \\ L_2 \end{bmatrix} \tag{22.72}$$

so

$$\begin{bmatrix} L_0 \\ L_1 \\ L_2 \end{bmatrix} = \begin{bmatrix} 1 & 1 & 1 \\ 1 & 0 & -1 \\ 1 & -1 & 1 \end{bmatrix}^{-1} \begin{bmatrix} L_{0^0} \\ L_{15^0} \\ L_{30^0} \end{bmatrix} = \begin{bmatrix} 1/4 & 1/2 & 1/4 \\ 1/2 & 0 & -1/2 \\ 1/4 & -1/2 & 1/4 \end{bmatrix} \begin{bmatrix} L_{0^0} \\ L_{15^0} \\ L_{30^0} \end{bmatrix}. \quad (22.73)$$

Or in separate form:

$$L_0 = \frac{1}{2}\left[\frac{1}{2}(L_{0^0} + L_{30^0}) + L_{15^0}\right],$$

$$L_1 = \frac{1}{2}(L_{0^0} - L_{30^0}), \quad (22.74)$$

$$L_2 = \frac{1}{2}\left[\frac{1}{2}(L_{0^0} + L_{30^0}) - L_{15^0}\right].$$

22.5.2.2 Four-Term Inductance Model

If four terms are used in the Fourier series ($m = 4$), then we can compute the four coefficients L_0, L_1, L_2, and L_3 from L_{0^0} (aligned position), L_{10^0}, L_{20^0}, and L_{30^0} (unaligned position). Since

$$\begin{bmatrix} L_{0^0} \\ L_{10^0} \\ L_{20^0} \\ L_{30^0} \end{bmatrix} = \begin{bmatrix} 1 & 1 & 1 & 1 \\ 1 & \cos(6*10^0) & \cos(12*10^0) & \cos(18*10^0) \\ 1 & \cos(6*20^0) & \cos(12*20^0) & \cos(18*20^0) \\ 1 & \cos(6*30^0) & \cos(12*30^0) & \cos(18*30^0) \end{bmatrix} \begin{bmatrix} L_0 \\ L_1 \\ L_2 \\ L_3 \end{bmatrix}, \quad (22.75)$$

so

$$\begin{bmatrix} L_0 \\ L_1 \\ L_2 \\ L_3 \end{bmatrix} = \begin{bmatrix} 1/6 & 1/3 & 1/3 & 1/6 \\ 1/3 & 1/3 & -1/3 & -1/3 \\ 1/3 & -1/3 & -1/3 & 1/3 \\ 1/6 & -1/3 & 1/3 & -1/6 \end{bmatrix} \begin{bmatrix} L_{0^0} \\ L_{10^0} \\ L_{20^0} \\ L_{30^0} \end{bmatrix} \quad (22.76)$$

Or in separate form:

$$L_0 = \frac{1}{3}\left[\frac{1}{2}(L_{0^0} + L_{30^0}) + (L_{10^0} + L_{20^0})\right],$$

$$L_1 = \frac{1}{3}(L_{0^0} + L_{10^0} - L_{20^0} - L_{30^0}), \quad (22.77)$$

$$L_2 = \frac{1}{3}(L_{0^0} - L_{10^0} - L_{20^0} + L_{30^0})$$

$$L_3 = \frac{1}{3}\left[\frac{1}{2}(L_{0^0} - L_{30^0}) - (L_{10^0} - L_{20^0})\right].$$

22.5.2.3 Five-Term Inductance Model

If five terms are used in the Fourier series ($m = 5$), then we can compute the five coefficients L_0, L_1, L_2, L_3, and L_4 from L_{0^0} (aligned position), $L_{7.5^0}$, L_{15^0} (midway position), $L_{22.5^0}$, and L_{30^0} (aligned position). Since

$$\begin{bmatrix} L_{0^0} \\ L_{7.5^0} \\ L_{15^0} \\ L_{22.5^0} \\ L_{30^0} \end{bmatrix} = \begin{bmatrix} 1 & 1 & 1 & 1 & 1 \\ 1 & \cos(6*7.5^0) & \cos(12*7.5^0) & \cos(18*7.5^0) & \cos(24*7.5^0) \\ 1 & \cos(6*15^0) & \cos(12*15^0) & \cos(18*15^0) & \cos(24*15^0) \\ 1 & \cos(6*22.5^0) & \cos(12*22.5^0) & \cos(18*22.5^0) & \cos(24*22.5^0) \\ 1 & \cos(6*30^0) & \cos(12*30^0) & \cos(18*30^0) & \cos(24*30^0) \end{bmatrix} \begin{bmatrix} L_0 \\ L_1 \\ L_2 \\ L_3 \\ L_4 \end{bmatrix},$$

(22.78)

so

$$\begin{bmatrix} L_0 \\ L_1 \\ L_2 \\ L_3 \\ L_4 \end{bmatrix} = \begin{bmatrix} 1/8 & 1/4 & 1/4 & 1/4 & 1/8 \\ 1/4 & \sqrt{2}/4 & 0 & -\sqrt{2}/4 & 1/4 \\ 1/4 & 0 & -1/2 & 0 & 1/4 \\ 1/4 & -\sqrt{2}/4 & 0 & \sqrt{2}/4 & 1/4 \\ 1/8 & -1/4 & 1/4 & -1/4 & 1/8 \end{bmatrix} \begin{bmatrix} L_{0^0} \\ L_{7.5^0} \\ L_{15^0} \\ L_{22.5^0} \\ L_{30^0} \end{bmatrix}. \quad (22.79)$$

Or in separate form:

$$L_0 = \frac{1}{4}\left[\frac{1}{2}(L_{0^0} + L_{30^0}) + (L_{7.5^0} + L_{15^0} + L_{22.5^0})\right],$$

$$L_1 = \frac{1}{4}\left[(L_{0^0} + L_{30^0}) + \sqrt{2}(L_{7.5^0} - L_{22.5^0})\right]$$

$$L_2 = \frac{1}{4}(L_{0^0} - 2L_{15^0} + L_{30^0}), \quad (22.80)$$

$$L_3 = \frac{1}{4}\left[(L_{0^0} - L_{30^0}) - \sqrt{2}(L_{7.5^0} - L_{22.5^0})\right]$$

$$L_4 = \frac{1}{4}\left[\frac{1}{2}(L_{0^0} + L_{30^0}) - (L_{7.5^0} - L_{15^0} + L_{22.5^0})\right].$$

22.5.2.4 Voltages and Torque Computation

Besides inductance, the phase winding also contains resistance (R). A simple model structure of phase winding is shown in Figure 22.34.

Based on the inductance model described above, the phase voltage equations can be formed and the electromagnetic torque can be computed from the partial derivative of magnetic co-energy with respect to rotor angle θ. They are listed here:

$$\begin{aligned} V &= R \cdot i + L\frac{di}{dt} + i\frac{dL}{dt} \\ &= R \cdot i + L\frac{di}{dt} + i\left(\frac{\partial L}{\partial \theta}\omega + \frac{\partial L}{\partial i}\frac{di}{dt}\right) \end{aligned}, \quad (22.81)$$

where

$$\frac{\partial L}{\partial i} = \sum_{k=0}^{m} \frac{\partial L_k(i)}{\partial i} \cos kN_r\theta \quad (22.82)$$

$$\frac{\partial L}{\partial \theta} = -\sum_{k=0}^{m} L_k(i)kN_r \sin kN_r\theta . \quad (22.83)$$

And

$$T = \frac{\partial W_c(\theta,i)}{\partial \theta} = \frac{\partial\left\{\int [L(\theta,i)i]di\right\}}{\partial \theta} = \frac{\partial\left\{\int \sum_{k=0}^{m}\left[L_k(i)\cos(kN_r\theta)i\right]di\right\}}{\partial \theta}$$

$$= -\sum_{k=0}^{m}\left\{kN_r \sin(kN_r\theta)\int [L_k(i)i]di\right\} \quad (22.84)$$

22.5.3 Parameter Identification from Standstill Test Data

22.5.5.3.1 Standstill Test Configuration

The basic idea of a standstill test is to apply a short voltage pulse to the phase winding with the rotor blocked, record the current generated in the winding, and then use maximum likelihood estimation to estimate the resistances and inductances of the winding. By performing this test at different current levels, the relationship between inductance and current can be curve-fitted with polynomials.

The experimental setup is shown in Figure 22.35. An 8/6 SRM is used in this test. Before testing, the motor is rotated to a specific position (with one of the phase windings aligned, unaligned, or at other positions) and blocked. A DSP system (dSPACE DS1103 controller board) is used to generate the gating signal to a power converter to apply appropriate voltage pulses to that winding. The voltage and current at the winding is sampled and recorded. Later on, the test data is used to identify the winding parameters.

When all the test data are collected, MLE is used to identify the winding parameters (R and L). The results of identification are validated with the test data: the voltages

Modeling and Parameter Identification of Electric Machines

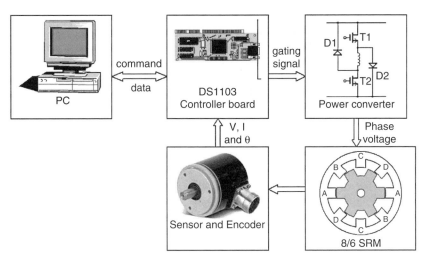

Figure 22.35 Experimental setup.

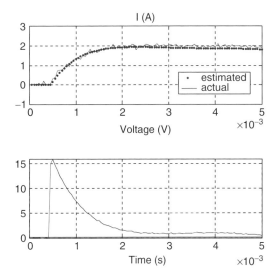

Figure 22.36 Validation of model for standstill test.

measured at standstill test are applied to the SRM model and the current responses are compared with the measured currents. From Figure 22.36, the estimated current (dotted curve) matches the measured current (solid curve) very well. This proves that the estimated parameters are quite satisfactory.

22.5.3.2 Standstill Test Results

The motor used in this test is an 8/6 SRM. Tests are performed at several specific positions for current between 0 and 50 A. The inductance estimation and curve-fitting results at aligned, midway, and unaligned position are shown in Figure 22.37 through Figure 22.39 (results are obtained using Matlab/Simulink).

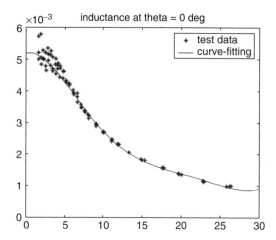

Figure 22.37 Standstill test results for inductance at 0°.

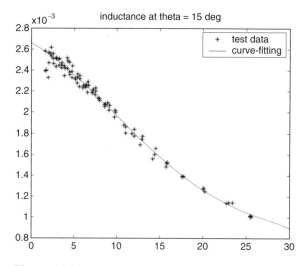

Figure 22.38 Standstill test results for inductance at 15°.

The results show that the inductance at unaligned position doesn't change much with the phase current and can be treated as a constant. The inductances at midway and aligned position decrease when current increases due to saturation.

A 3D plot of inductance shown in Figure 22.40 depicts the profile of inductance vs. rotor position and phase current.

At theta = 0 and 60°, phase A is at its aligned positions and has the highest value of inductance. It decreases when the phase current increases. At theta = 30°, phase A is at its unaligned position and has lowest value of inductance. The inductance here keeps nearly constant when the phase current changes.

In Figure 22.41, the flux linkage vs. rotor position and phase current based on the estimated inductance model is shown. The saturation of phase winding at high currents is clearly represented. At aligned position, the winding is highly saturated at rated current.

Modeling and Parameter Identification of Electric Machines

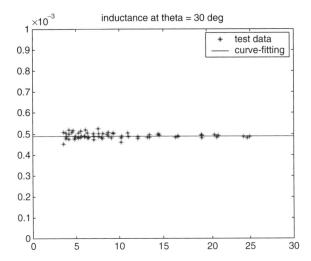

Figure 22.39 Standstill test results for inductance at 30°.

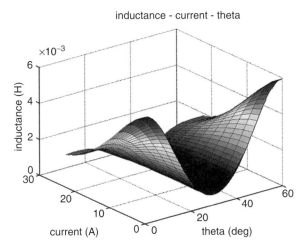

Figure 22.40 Standstill test result: nonlinear phase inductance.

22.5.4 Inductance Model of SRM for On-Line Operation

For on-line operation case, especially under high load, the losses become significant. There are no windings on the rotor of SRMs. But similar to synchronous machines, there will be circulating currents flowing in the rotor body, making it work as a damper winding. Considering this, the model structure may be modified as shown in Figure 22.42, with R_d and L_d added to represent the losses on the rotor.

The phase voltage equations can be written as:

$$\begin{bmatrix} L & -L_d \\ 0 & L_d \end{bmatrix} \cdot \begin{bmatrix} \dot{i}_1 \\ \dot{i}_2 \end{bmatrix} = \begin{bmatrix} 0 & R_d \\ -R & -R-R_d \end{bmatrix} \cdot \begin{bmatrix} i_1 \\ i_2 \end{bmatrix} + \begin{bmatrix} 0 \\ 1 \end{bmatrix} V, \qquad (22.85)$$

where i_1 and i_2 are the magnetizing current and damper current.

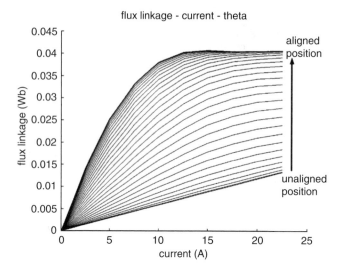

Figure 22.41 Flux linkage at different currents and different rotor positions.

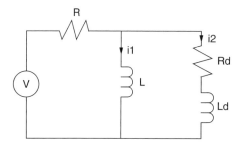

Figure 22.42 Model structure of SRM under saturation.

It can be rewritten in state space form as:

$$\dot{X} = AX + BU$$
$$Y = CX + DU \qquad (22.86)$$

where

$$X = [i_1 \quad i_2], \ U = [V],$$

$$A = \begin{bmatrix} L & -L_d \\ 0 & L_d \end{bmatrix}^{-1} \cdot \begin{bmatrix} 0 & R_d \\ -R & -R-R_d \end{bmatrix},$$

$$B = \begin{bmatrix} L & -L_d \\ 0 & L_d \end{bmatrix}^{-1} \cdot \begin{bmatrix} 0 \\ 1 \end{bmatrix},$$

Modeling and Parameter Identification of Electric Machines

$$Y = i_1 + i_2,$$

$$C = [1 \quad 1], \text{ and } D = 0.$$

The torque can be computed as follows (notice that L is the magnetizing winding):

$$T = \frac{\partial W_c(\theta, i_1)}{\partial \theta} = \frac{\partial \left\{ \int [L(\theta, i_1) i_1] di_1 \right\}}{\partial \theta}$$

$$= \frac{\partial \left\{ \int \sum_{k=0}^{m} [L_k(i_1) \cos(kN_r \theta) i_1] di_1 \right\}}{\partial \theta} \quad (22.87)$$

$$= -\sum_{k=1}^{m} \left\{ kN_r \sin(kN_r \theta) \int [L_k(i_1) i_1] di_1 \right\}$$

During operation, we can easily measure phase voltage V and phase current $i = i_1 + i_2$. But we cannot measure the magnetizing current (i_1) and the damper winding current (i_2). Let's assume that the phase parameters R and L obtained from standstill test data are accurate enough for low-current case. And we want to attribute all the errors at high-current case to damper parameters. If we can estimate the exciting i_1 during on-line operation, then it will be very easy to estimate the damper parameters. This is described in later sections.

22.5.5 Two-Layer Recurrent Neural Network for Damper Current Estimation

22.5.5.1 Structure of Two-Layer Recurrent Neural Network

During on-line operation, there will be motional back-EMF in the phase winding. So the exciting current i_1 will be affected by:

Phase voltage V
Phase current i
Rotor position θ
Rotor speed ω

To map the relationship between i_1 and V, i, θ, ω, the neural network structure shown in Figure 22.43 is used. It is a two-layer recurrent neural network.
 The first layer is the input layer. The inputs of the network are V, i, θ, ω (with possible delays). One of the outputs, the current i, is also fed back to the input layer to form a recurrent neural network.
 The second layer is the output layer. The outputs are i (used as training objective) and i_1.
 A hyperbolic tangent sigmoid transfer function — *tansig()* — is chosen to be the activation function of the input layer, which gives the following relationship between its inputs and outputs:

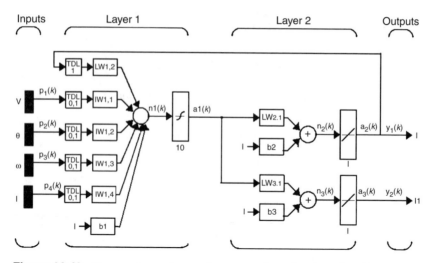

Figure 22.43 Recurrent neural network structure for estimation of exciting current.

$$n_1 = \sum_{i=1}^{4} IW_{1,i} \cdot p_i + LW_{1,2} \cdot y_1 + b_1$$

$$a_1 = tansig(n_1) = \frac{2}{1+e^{-2n_1}} - 1 \quad (22.88)$$

A pure linear function is chosen to be the activation of the output layers, which gives:

$$n_2 = LW_{2,1} \cdot a_1 + b_2$$

$$y_1 = a_2 = purelin(n_2) = n_2 \quad (22.89)$$

$$n_3 = LW_{3,1} \cdot a_1 + b_3$$

$$y_2 = a_3 = purelin(n_3) = n_3 . \quad (22.90)$$

After the neural network is trained with simulation data (using parameters obtained from the standstill test), it can be used to estimate exciting current during on-line operation. When i_1 is estimated, the damper current can be computed as

$$i_2 = i - i_1 , \quad (22.91)$$

and the damper voltage can be computed as

$$V_2 = V - i \cdot R , \quad (22.92)$$

then the damper resistance R_d and inductance L_d can be identified using output error or maximum likelihood estimation.

Modeling and Parameter Identification of Electric Machines

22.5.5.2 Training of Neural Network

The data used for training is generated from a simulation of the SRM model obtained from a standstill test.

First, from standstill test results, we can estimate the winding parameters (R and L) and damper parameters (R_d and L_d). The R_d and L_d from standstill test data may not be accurate enough for an on-line model, but it can be used as initial values that will be improved later.

Second, build an SRM model with the above parameters and simulate the motor with hysteresis current control and speed control. The operating data under different reference currents and different rotor speeds is collected and sent to neural network for training.

Third, when training is done, use the trained ANN model to estimate the magnetizing current (i_1) from on-line operating data. Compute damper voltage and current according to Equation 22.91 and Equation 22.92. Then estimate R_d and L_d from the computed V_2 and i_2 using output error estimation. This R_d and L_d can be treated as improved values of standstill test results.

Repeat the above procedures until R_d and L_d are accurate enough to represent on-line operation (it means that the simulation data matches the measurements well).

In our research, the neural network can map the exciting current from V, i, θ, ω very well after training of 200 epochs.

22.5.6 Estimation Results and Model Validation

The parameters for damper winding are successfully estimated from operating data by using the neural network mapping described above.

To test the validity of the parameters obtained from the above test, a simple on-line test has been performed. In this test, the motor is accelerated with a fixed reference current of 20 Å. All the operating data such as phase voltages, currents, rotor position, and rotor speed are measured. Then the phase voltages are fed to an SRM model, running in Simulink, which has the same rotor position and speed as the real motor. All the phase currents are estimated from the Simulink mode and compared with the measured currents. In Figure 22.44, the phase current responses are shown. The dashed curve is the voltage applied to phase winding, the solid curve is the measured current, and the dotted curve is the estimated current. An enlarged view of the curves for phase A is shown in Figure 22.45. It is clear that the estimation approximates to the measurement quite well.

To compare an on-line model with a standstill one, we compute the covariance of the errors between the estimated phase currents and the measured currents. The average covariance for a standstill model is 0.9127, while that for an on-line model is 0.6885. This means that the on-line model gives much better estimation of operating phase currents.

22.5.7 Conclusions

During online operation, the exciting current i_1 changes with phase voltage V, rotor position θ, and rotor speed ω. The relationship between them is highly nonlinear and cannot be easily expressed by any analytical equation. The neural network can provide very good mapping if trained correctly. This makes it a good choice for such a task.

Once the NN is trained, it can estimate the exciting current from inputs very quickly, without solving any differential equations necessary in conventional methods. So, it can be used for on-line parameter identification with no computational difficulties. This method has been successfully applied to synchronous machines and induction machines [62,64,65]; it can be applied to SRMs, as well.

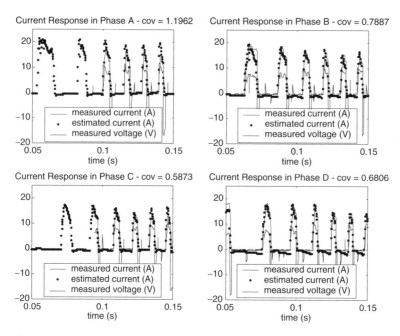

Figure 22.44 Validation of model with on-line operating data.

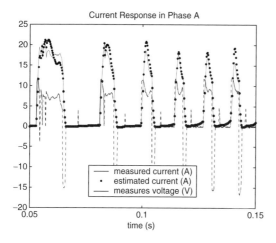

Figure 22.45 Validation of model with on-line operating data (Phase A).

This section presents the idea and procedure to use artificial neural network to help identify the resistance and nonlinear inductance of SRM winding from operating data. First the resistance and inductance of the magnetizing winding are identified from standstill test data. Then a two-layer recurrent neural network is set up and trained with simulation data based on a standstill model. By applying this neural network to on-line operating data, the magnetizing current can be estimated and the damper current can be computed. Then the parameters of the damper winding can be identified using maximum likelihood estimation. Tests performed on a 50 Å 8/6 SRM show satisfactory results of this method.

REFERENCES

1. A. Keyhani, S. Hao, G. Dayal, Maximum likelihood estimation of solid-rotor synchronous machine parameters from SSFR test data, IEEE/PES Winter Meeting, New York, 1989.
2. Y.N. Yu, H.A.M. Moussa, Experimental determination of exact equivalent circuit parameters of synchronous machines, *IEEE Transactions on Power App. Systems,* Vol. PAS-90, No. 6, November/December 1971, pp. 2555–2560.
3. M.V.K. Chari, S.H. Minnich, R.P. Schulz, Improvement in accuracy of prediction of electrical machine constants, and generator models for subsynchronous resonance conditions, EPRI Project RP-1288 and RP-1513, *Report #EL-3359,* Vol. 1, 2, and 3, 1984, 1987.
4. S.H. Minnich, R.P. Schulz, D.H. Baker, R.G. Farmer, D.K. Sharma, J.H. Fish, Saturation functions for synchronous generators from finite elements, *IEEE Transactions on Energy Conversion,* Vol. EC-2, No. 4, December 1987, pp. 680–692.
5. J.J. Sanchez-Gasca, C.J. Bridenbaugh, C.E.J. Bowler, J.S. Edmonds, Trajectory sensitivity-based identification of synchronous generator and excitation system parameters, IEEE/PES Winter Meeting, New York, 1988.
6. E.S. Bose, K.C. Balda, R.G. Harley, R.C. Beck, Time-domain identification of synchronous machine parameters from simple standstill tests, IEEE/PES Winter Meeting, New York, 1988.
7. J.C. Balda, M.F. Hadingham, R.E. Fairbairnm R.C. Harley, E. Eitelberg, Measurement of frequency response method — Part I: Theory, *IEEE Transactions on Energy Conversion,* Vol. EC-2, No. 4, December 1987, pp. 646–651.
8. J.C. Balda, M.F. Hadingham, R.E. Fairbairnm R.C. Harley, E. Eitelberg, Measurement of frequency response method — Part II: Measured results, *IEEE Transactions on Energy Conversion,* Vol. EC-2, No. 4, December 1987, pp. 652–657.
9. F.P. deMello, L.N. Hannett, Determination of synchronous machine electrical characteristics by test, *IEEE Transactions on Power App. Systems,* Vol. PAS-102, December 1983, pp. 3810–3815.
10. IEEE Committee report, Current usage and suggested practices in power system stability simulations for synchronous machines, *IEEE Transactions on Energy Conversion,* Vol. EC-1, No. 1, March 1986, pp. 77–93.
11. P.L. Dandeno, A.T. Poray, Development of detailed turbogenerator equivalent circuits from standstill frequency response measurements, *IEEE Transactions on Power App. Systems,* Vol. PAS-100, No. 4, April 1981, pp. 1646–1655.
12. P.L. Dandeno, P. Kundur, A.T. Poray, M. Coultes, Validation of turbogenerator stability models by comparison with power system tests, *IEEE Transactions on Power App. Systems,* Vol. PAS-100, April 1981, pp. 1637–1645.
13. M.E. Coultes, W. Watson, Synchronous machine models by standstill frequency response tests, *IEEE Transactions on Power App. Systems, Vol.* PAS-100, No. 4, 1981, pp. 1480–1489.
14. IEEE standard procedure for obtaining synchronous machine parameters by standstill frequency response testing, *IEEE Standards,* 115A-1987.
15. S.D. Umans, J.A. Mallick, G.L. Wilson, Modeling of solid rotor turbogenerators — Part I: Theory and techniques, *IEEE Transactions on Power App. Systems,* Vol. PAS-97, No. 1, 1978, pp. 269–277.
16. S.D. Umans, J.A. Mallick, G.L. Wilson, Modeling of solid rotor turbogenerators — Part II: Example of model derivation and use in digital simulation, *IEEE Transactions on Power App. Systems,* Vol. PAS-97, No. 1, 1978, pp. 278–291.
17. S.D. Umans, Modeling of solid rotor turbogenerators, Ph.D. thesis, Massachusetts Institute of Technology, Cambridge, MA, January 1976.
18. I.M. Canay, *Extended Synchronous Machine Model for the Calculation of Transient Processes and Stability, Electric Machine and Electromechanics: An International Quarterly,* 1977, pp. 137–150.
19. A. Keyhani, S.M. Miri, Observers for tracking of synchronous machine parameters and detection of incipient faults, *IEEE Transactions on Energy Conversion,* Vol. EC-1, No. 2, June 1986, pp. 184–192.

20. D. Marquardt, An algorithm for least-square estimation of nonlinear parameters, *Journal of Soc. Indust. and Appl. Math. II*, 1963, pp. 431–441.
21. Symbolics Inc., MACSYMA, Computer-Aided Mathematics Group of Symbolics, Cambridge, MA.
22. A. Keyhani, S. Hao, G. Dayal, The effect of noise on frequency-domain parameter estimation of synchronous machine models, IEEE/PES Winter Meeting, New York, 1989.
23. L.X. Le, W.J. Wilson, Synchronous machine parameter identification: A time domain approach, *IEEE Transactions on Energy Conversion*, Vol. 3, No. 2, June 1988, pp. 241–247.
24. N. Jaleeli, M.S. Bourawi, J.H. Fish III, A quasilinearization-based algorithm for identification of transient and subtransient parameters of synchronous machines, *IEEE Transactions on Power Systems*, Vol. PWRS-1, No. 3, August 1986, pp. 46–52.
25. M. Namba, T. Nishiwaki, S. Yokokawa, K. Ohtsuka, Y. Ueki, Identification of parameters for power system stability analysis using Kalman filter, *IEEE Transactions on App. Systems*, Vol. PAS-100, July 1981, pp. 3304–3311.
26. K.J. Astrom, Maximum likelihood and prediction error methods, *Automatica*, Vol. 16, 1980, pp. 551–574.
27. N.K. Gupta, R.K. Mehra, Computational aspects of maximum likelihood estimation and reduction in sensitivity function calculations, *IEEE Transactions on Automatic Control*, VOL. AC-19, No. 6, December 1974, pp. 774–783.
28. G.C. Goodwin, R.L. Payne, *Dynamic System Identification*, Academic Press, New York, 1977.
29. K.J. Astrom and T. Soderstrom, Uniqueness of the maximum likelihood estimates of the parameters of an ARMA model, *IEEE Transactions on Automatic Control*, Vol. 19, 1974, pp. 769–773.
30. T. Bohlin, On the maximum likelihood method of identification, *IBM Journal on Research and Development*, Vol. 4, No. 1, 1970, pp. 41–51.
31. J.E. Dennis, R.B. Schnabel, *Numerical Methods for Unconstrained Optimization and Nonlinear Equations*, Prentice Hall, Inc., Englewood Cliffs, NJ, 1983.
32. R.P. Schulz, Synchronous Machine Modeling, IEEE 1975 Symposium, "Adequacy and Philosophy of Modeling: Dynamic System Performance," *IEEE Monograph 75* CH0970-4 PWR, pp. 24–28.
33. W.B. Jackson, E.L. Winchester, Direct- and quadrature-axis equivalent circuits for solid-rotor turbine generators, *IEEE Transactions on Power App. Systems*, Vol. PAS-88, No. 7, July 1969, pp. 1121–1136.
34. G.R. Slemon, Modeling of induction machines for electric drives, *IEEE Transactions on Industry Applications*, Vol. 25, No. 6, 1989, pp. 1126–1131.
35. N.R. Klaes, Parameter identification of an induction machine with regard to dependencies on saturation, *IEEE Transactions on Industry Applications*, Vol. 29, No. 6, 1993, pp. 1135–1140.
36. S.I. Moon, A. Keyhani, S. Pillutla, Nonlinear neural-network modeling of an induction machine, *IEEE Transactions on Control Systems Technology*, Vol. 7, No. 2, 1999, pp. 203–211.
37. J.A. de Kock, F.S. van der Merwe, H.J. Vermeulen, Induction motor parameter estimation through an output error technique, *IEEE Transactions on Energy Conversion*, Vol. 9, No. 1, 1994, pp. 69–76.
38. A.M.N. Lima, C.B. Jacobina, E.B.F. de Souza, Nonlinear parameter estimation of steady-state induction machine models, *IEEE Transactions on Industrial Electronics*, Vol. 44, No. 3, 1997, pp. 390–397.
39. L.A. de Souza Ribeiro, C.B. Jacobina, A.M. Nogueira Lima, Linear parameter estimation for induction machines considering the operating conditions, *IEEE Transactions on Power Electronics*, Vol. 14, No. 1, 1999, pp. 62–73.
40. P. Pillay, R. Nolan, T. Haque, Application of genetic algorithms to motor parameter determination for transient torque calculations, *IEEE Transactions on Industry Applications*, Vol. 33, No. 5, 1997, pp. 1273–1282.

41. E. Mendes, A. Razek, A simple model for core losses and magnetic saturation in induction machines adapted for direct stator flux orientation control, IEEE Conference Publication, Vol. 399, 1994, pp. 192–197.
42. S. Ansuj, F. Shokooh, R. Schinzinger, Parameter estimation for induction machines based on sensitivity analysis, *IEEE Transactions on Industry Applications,* Vol. 25, No. 6, 1989, pp. 1035–1040.
43. J. Stephan, M. Bodson, J. Chiasson, Real-time estimation of the parameters and fluxes of induction motors, *IEEE Transactions on Industry Applications,* Vol. 30, No. 3, 1994, pp. 746–759.
44. X. Xu, D.W. Novotny, Implementation of direct stator flux orientation control on a versatile DSP-based system, *IEEE Transactions on Industry Applications,* Vol. 27, No. 4, 1991, pp. 694–700.
45. J.K. Seok, S.I. Moon, S.K. Sul, Induction machine parameter identification using PWM inverter at standstill, *IEEE Transactions on Energy Conversion,* Vol. 12, No. 2, 1997, pp. 127–132.
46. C. Wang, D.W. Novotny, T.A. Lipo, An automated rotor time constant measurement system for indirect field-oriented drives, *IEEE Transactions on Industry Applications,* Vol. 24, No. 1, 1988, pp. 151–159.
47. Y.N. Lin, C.L. Chen, Automatic IM parameter measurement under sensorless field-oriented control, *IEEE Transactions on Industrial Electronics,* Vol. 46, No. 1, 1999, pp. 111–118.
48. S.K. Sul, A novel technique of rotor resistance estimation considering variation of mutual inductance, *IEEE Transactions on Industry Applications,* Vol. 25, No. 4, 1989, pp. 578–587.
49. T. Noguchi, S. Kondo, I. Takahashi, Field-oriented control of an induction motor with robust on-line tuning of its parameters, *IEEE Transactions on Industry Applications,* Vol. 33, No. 1, 1997, pp. 35–42.
50. L.C. Zai, C.L. DeMarco, T.A. Lipo, An extended Kalman filter approach to rotor time constant measurement in PWM induction motor drives, *IEEE Transactions on Industry Applications,* Vol. 28, No. 1, 1992, pp. 96–104.
51. D.J. Atkinson, P.P. Acarnley, J.W. Finch, Observers for induction motor state and parameter estimation, *IEEE Transactions on Industry Applications,* Vol. 27, No. 6, 1991, pp. 1119–1127.
52. S. Wade, M.W. Dunnigan, B.W. Williams, New method of rotor resistance estimation for vector-controlled induction machines, *IEEE Transactions on Industrial Electronics,* Vol. 44, No. 2, 1997, pp. 247–257.
53. K. Matsuse, T. Yoshizumi, S. Katsuta, S. Taniguchi, High response flux control of direct-field-oriented induction motor with high efficiency taking core loss into account, *IEEE Transactions on Industry Applications,* Vol. 35, No. 1, 1999, pp. 62–69.
54. B. Robyns, P.A. Sente, H.A. Buyse, F. Labrique, Influence of digital current control strategy on the sensitivity to electrical parameter uncertainties of induction motor indirect field-oriented control, *IEEE Transactions on Power Electronics,* Vol. 14, No. 4, 1999, pp. 690–699.
55. K. Akatsu, A. Kawamura, On-line rotor resistance estimation using the transient state under the speed sensorless control of induction motor, *IEEE Transactions on Power Electronics,* Vol. 15, No. 3, 2000, pp. 553–560.
56. S.J. Chapman, *Electric Machinery Fundamentals,* 2nd Edition, New York: McGraw-Hill, 1991.
57. V. Utkin, J. Guldner, J. Shi, *Sliding Mode control in Electromechanical Systems,* Taylor & Francis, London, 1999.
58. S. Mir, I. Husain, M.E. Elbuluk, Switched reluctance motor modeling with on-line parameter identification. *IEEE Transactions on Industry Applications,* Vol. 34, No. 4, July/August 1998.
59. B. Fahimi, G. Suresh, J. Mahdavi, M. Ehsani, A new approach to model switched reluctance motor drive application to dynamic performance prediction, control, and design, Power Electronics Specialists Conference, Vol. 2, 1998.
60. L. Xu, E. Ruckstadter, Direct modeling of switched reluctance machine by coupled field-circuit method, *IEEE Transactions on Energy Conversion,* Vol. 10, No. 3, September 1995.

61. S.I. Moon, A. Keyhani, Estimation of induction machine parameters from standstill time-domain data, *IEEE Transactions on Industry Applications,* Vol. 30, No. 6, November/December 1994.
62. A. Keyhani, H. Tsai, T. Leksan, Maximum likelihood estimation of synchronous machine parameters from standstill time response data. IEEE/PES Winter Meeting, Columbus, OH, January 31–February 5, 1993.
63. K.M. Passino, Intelligent Control: Biomimicry for Optimization, Adaptation, and Decision-Making, in *Computer Control and Automation Textbook*, The Ohio State University, March 2001.
64. S. Pillutla, A. Keyhani, Neural network-based modeling of round rotor synchronous generator rotor body parameters from operating data, *IEEE Transactions on Energy Conversion,* 1998.
65. S. Pillutla, A. Keyhani, Neural network-based saturation model of round rotor synchronous generator rotor, *IEEE Transactions on Energy Conversion,* 1998.
66. S.S. Ramamurthy, R.M. Schupbach, J.C. Balda, Artificial neural networks-based models for the multiple excited switched reluctance motor. APEC 2001.
67. B. Fahimi, G. Suresh, J. Mahdavi, M. Ehsani. A new approach to model switched reluctance motor drive application to dynamic performance prediction, control, and design, Power Electronics Specialists Conference, Vol. 2, 1998.
68. W. Lu, A. Keyhani, A. Fardoun, Neural network-based modeling and parameter identification of switched reluctance motors, *IEEE Transactions on Energy Conversion,* Vol. 18 No. 2, June 2003, pp. 284–290.
69. H.B. Karayaka, A. Keyhani, G.T. Heydt, B.L. Agrawal, D.A. Selin, Synchronous generator model identification and parameter estimation from operating data, *IEEE Transactions on Energy Conversion,* Vol. 18, No. 1, March 2003, pp. 121–126.
70. W. Lu, A. Keyhani, H. Klode, A.B. Proca, Modeling and parameter identification of switched reluctance motors from operating data using neural networks, IEEE International Electric Machines and Drives Conference (IEMDC'03), Vol. 3, 1–4 June, 2003, pp. 1709–1713.
71. A.B. Proca, A. Keyhani, Induction motor parameter identification from operating data for electric drive applications, *Proceedings of 18th Digital Avionics Systems Conference,* Vol. 2, 24–29 Oct. 1999, pp. 8.C.2-1–8.C.2-6.
72. A.B. Proca, A. Keyhani, Identification of variable frequency induction motor models from operating data, *IEEE Transactions on Energy Conversion,* Vol. 17, No. 1, March 2002, pp. 24–1.
73. A. El-Serafi, A. Abdallah, M. El-Sherbiny, E. Badawy, Experimental study of the saturation and the cross-magnetizing phenomenon in saturated synchronous machines, *IEEE Transactions on Energy Conversion,* Vol. EC-3, December 1988, pp. 815–823.
74. F. De Mello, L. Hannett, Representation of saturation in synchronous machines, *IEEE Transactions on Power Systems,* Vol. 1, November 1986, pp. 8–14.
75. S. Minnich, R. Schulz, D. Baker, D. Sharma, R. Farmer, J. Fish, Saturation functions for synchronous generators from finite elements, *IEEE Transactions on Energy Conversion,* Vol. 2, December 1987, pp. 680–687.
76. H. Tsai, A. Keyhani, J.A. Demcko, D.A. Selin, Development of a neural network saturation model for synchronous generator analysis, *IEEE Transactions on Energy Conversion,* Vol. 10, No. 4, December 1995, pp. 617–624.
77. L. Xu, Z. Zhao, J. Jiang, On-line estimation of variable parameters of synchronous machines using a novel adaptive algorithm — estimation and experimental verification, *IEEE Transactions on Energy Conversion,* September 1997, Vol. 12, No. 3, pp. 200–210.
78. H. Tsai, A. Keyhani, J.A. Demcko, R.G. Farmer, On-line synchronous machine parameter estimation from small disturbance operating data, *IEEE Transactions on Energy Conversion,* Vol. 10, No. 1, March 1995, pp. 25–36.
79. K.S. Narendra, K. Parthasarathy, Identification and control of dynamical systems using neural networks, *IEEE Transactions on Neural Networks,* Vol. 1, pp. 4–27, 1990.

80. I.M. Canay, Causes of discrepancies on calculation of rotor quantities and exact equivalent diagrams of the synchronous machine, *IEEE Transactions on Power Apparatus and Systems,* Vol. PAS-88, No. 7, July 1969, pp. 1114–1120.
81. J.L. Kirtley Jr., On turbine-generator rotor equivalent circuit structures for empirical modeling of turbine generators, *IEEE Transactions on Power Systems,* Vol. PWRS-9(1), 1994, pp. 269–271.
82. I. Kamwa, P. Viarouge, J. Dickinson, Identification of generalized models of synchronous machines from time-domain tests, *IEEE Proceedings C,* 138 (6), November 1991, pp. 485–498.
83. S. Salon, Obtaining synchronous machine parameters from test, Symposium on Synchronous Machine Modeling for Power Systems Studies. Paper No. 83THO101-6-PWR. Available from IEEE Service Center, Piscataway, NJ.
84. S.R. Chaudhary, S. Ahmed-Zaid, N.A. Demerdash, An artificial neural network model for the identification of saturated turbogenerator parameters based on a coupled finite-element/state-space computational algorithm, *IEEE Transactions on Energy Conversion,* Vol. 10, December 1995, pp. 625–633.
85. J.A. Demcko, J.P. Chrysty, Self-calibrating power angle instrument, EPRI GS-6475, Vol. 2, Research Project 2591-1 Final Report, EPRI, August 1989.
86. M.A. Arjona, D.C. Macdonald, A new lumped steady-state synchronous machine model derived from finite element analysis, *IEEE Transactions on Energy Conversion,* Vol. 14, No. 1 March 1999, pp. 1–7.
87. H. Karayaka, A. Keyhani, B. Agrawal, D. Selin, G. Heydt, Identification of armature, field, and saturated parameters of a large steam turbine-generator from operating data, *IEEE Transactions on Energy Conversion,*.
88. M. Hagan, M.B. Menhaj, Training feed-forward networks with the Marquardt algorithm, *IEEE Transactions on Neural Networks,* Vol. 5, No. 6, November 1994, pp. 989–993.
89. H. Karayaka, A. Keyhani, B. Agrawal, D. Selin, G. Heydt, Methodology development for estimation of armature circuit and field winding parameters of large utility generators, *IEEE Transactions on Energy Conversion,* Vol. 14, No. 4, December 1999, pp. 901–908.
90. A. Derdiyok, H. Rehman, M.K. Guven, N. Inanc, L. Xu, A robust sliding mode observer for speed estimation of induction machine, *APEC* 2001, Vol. 1, pp. 413–418.

APPENDIX A

The complex nonlinear equations that relate the d-axis SSFR3 model parameters with the time constants of the corresponding transfer function can be concisely written as:

$$f_i(\bar{x}) = \bar{y}_i + g_i(\bar{x}, \bar{y}) + \zeta_i = 0$$

where $i = 1, \ldots, 10$.

Vector $\bar{x}$ and $\bar{y}$ are defined in Table 22.1. ζ_i represents the noise associated with each element y_i. This is because noise is inherently present in the test data.

The above set of nonlinear equations can be expanded as:

$$f_1(\bar{x}) = y_1 - y_0 + x_1 + \zeta_1 = 0 \tag{A.1}$$

$$\begin{aligned}
f_2(\bar{x}) = y_2 - [&(x_9 + x_5 + x_2 + x_1)x_4 x_7 y_0 \\
&+ (x_2 + x_1)x_9 x_7 + (x_2 + x_1)x_5 x_7 y_0 \\
&+ (x_5 + x_2 + x_1)x_9 x_4 y_0 \\
&+ (x_2 + x_1)x_5 x_9 y_0 + (x_9 + x_5 + x_2)x_1 x_4 x_7 \\
&+ (x_2 x_9 + x_2 x_5)x_1 x_7 + (x_5 + x_2)x_1 x_9 x_4 \\
&+ x_1 x_2 x_5 x_9]/[(y_0 + x_1)x_3 x_6 x_8] \\
&+ \zeta_2 = 0
\end{aligned} \tag{A.2}$$

$$\begin{aligned}
f_3(\bar{x}) = y_3 - [&x_6(x_9 + x_5 + x_2 + x_1)x_4 y_0 \\
&+ x_6(x_2 + x_1)x_9 y_0 + x_6(x_2 + x_1)x_5 y_0 \\
&+ x_6(x_9 + x_5 + x_2)x_1 x_4 + x_6 x_2 x_9 x_1 \\
&+ x_6 x_2 x_1 x_5 + x_3(x_9 + x_5 + x_2 + x_1)x_7 y_0 \\
&+ x_3(x_5 + x_2 + x_1)x_9 y_0 \\
&+ x_3(x_9 + x_5 + x_2)x_1 x_7 + x_3(x_5 + x_2)x_1 x_9 \\
&+ x_8(x_4 + x_2 + x_1)x_7 y_0 \\
&+ x_8(x_5 + x_2 + x_1)x_4 y_0 + x_8(x_2 + x_1)x_5 y_0 \\
&+ x_8(x_4 + x_2)x_1 x_7 + x_8(x_5 + x_2)x_1 x_4 \\
&+ x_8 x_1 x_2 x_5]/[(y_0 + x_1)x_3 x_6 x_8] \\
&+ \zeta_3 = 0
\end{aligned} \tag{A.3}$$

Modeling and Parameter Identification of Electric Machines

$$f_4(\bar{x}) = y_4 - [x_3x_6(x_9+x_5+x_2+x_1)y_0$$
$$+ x_3x_6x_1x_9 + x_3x_6x_1(x_5+x_2)$$
$$+ x_8x_6(x_4+x_2+x_1)y_0 + x_8x_6x_1(x_4+x_2)$$
$$+ x_3x_8(x_7+x_5+x_2+x_1)y_0$$
$$+ x_3x_8x_1(x_7+x_5+x_2)]/[(y_0+x_1)x_3x_6x_8]$$
$$+ \zeta_4 = 0 \qquad (A.4)$$

$$f_5(\bar{x}) = y_5 - [x_4x_7(x_9+x_5+x_2+x_1)$$
$$+ x_7x_9(x_2+x_1) + x_7x_5(x_1+x_2)$$
$$+ x_7x_4(x_5+x_2+x_1)$$
$$+ x_9x_5(x_1+x_2)]/[x_3x_6x_8]$$
$$+ \zeta_5 = 0 \qquad (A.5)$$

$$f_6(\bar{x}) = y_6 - [x_6(x_9+x_5+x_2+x_1)x_4$$
$$+ x_6(x_2+x_1)x_9 + x_5x_6(x_2+x_1)$$
$$+ x_3(x_9+x_5+x_2+x_1)x_7$$
$$+ x_3x_9(x_5+x_2+x_1)$$
$$+ x_8(x_4+x_2+x_1)x_7$$
$$+ x_8(x_5+x_2+x_1)x_4$$
$$+ x_8x_5(x_1+x_2)]/[x_3x_6x_8]$$
$$+ \zeta_6 = 0 \qquad (A.6)$$

$$f_7(\bar{x}) = y_7 - [x_3x_6(x_9+x_5+x_2+x_1)$$
$$+ x_8x_6(x_4+x_2+x_1)$$
$$+ x_3x_8(x_7+x_5+x_2+x_1)]/[x_3x_6x_8]$$
$$+ \zeta_7 = 0 \qquad (A.7)$$

$$f_8(\bar{x}) = y_8 - x_1/x_8 + \zeta_8 = 0 \qquad (A.8)$$

$$f_9(\bar{x}) = y_9 - x_4x_7/(x_3x_6) + \zeta_9 = 0 \qquad (A.9)$$

$$f_{10}(\bar{x}) = y_{10} - x_4/x_3 - x_7/x_6 + \zeta_{10} = 0 \qquad (A.10)$$

APPENDIX B

The differential equations of the standstill d- and q-axis circuit model (see Figure 22.B1), assuming the field winding is short-circuited, can be written as:

D-axis:

$$v_d = -R_a i_d + p\lambda_d \quad (B.1)$$

$$0 = R_{1d} i_{1d} + p\lambda_{1d} \quad (B.2)$$

$$0 = R_{2d} i_{2d} + p\lambda_{2d} \quad (B.3)$$

$$0 = R_{fd} i_{fd} + p\lambda_{fd} \quad (B.4)$$

where p: d/dt

$$\lambda_d = -(L_l + L_{ad})i_d + L_{ad}(i_{fd} + i_{1d} + i_{2d}) \quad (B.5)$$

$$\lambda_{1d} = (L_{ad} + L_{f12d} + L_{1d})i_{1d} + L_{ad}(i_{2d} + i_{fd} - i_d) + L_{f12d}(i_{2d} + i_{fd}) \quad (B.6)$$

$$\lambda_{2d} = (L_{ad} + L_{f12d} + L_{f2d} + L_{2d})i_{2d} + L_{ad}(i_{1d} + i_{fd} - i_d) + L_{f12d}(i_{1d} + i_{fd}) + L_{f2d}i_{fd} \quad (B.7)$$

$$\lambda_{fd} = (L_{ad} + L_{f12d} + L_{f2d} + L_{fd})i_{fd} + L_{ad}(i_{1d} + i_{2d} - i_d) + L_{f12d}(i_{1d} + i_{2d}) + L_{f2d}i_{2d} \quad (B.8)$$

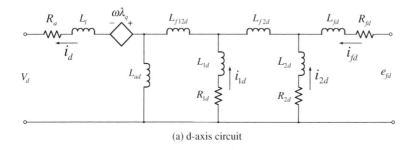

(a) d-axis circuit

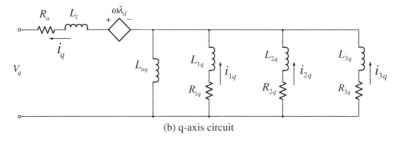

(b) q-axis circuit

Figure 22.B1 SSFR3 model structures.

Modeling and Parameter Identification of Electric Machines

Defining:

$$X = \begin{bmatrix} i_d & i_{1d} & i_{2d} & i_{fd} \end{bmatrix}^T$$

$$U = \begin{bmatrix} v_d \end{bmatrix}$$

Equation B.1 through Equation B.8 can be written as the following:

$$F_d(\theta_d)\dot{X} = G_d(\theta_d)X + DU \qquad (B.9)$$

where $\dot{X} = dX/dt$

$$D = \begin{bmatrix} 1 & 0 & 0 & 0 \end{bmatrix}^T$$

$$\theta_d = \begin{bmatrix} L_{ad}, L_{f12d}, R_{1d}, L_{1d}, L_{f2d}, R_{2d}, L_{2d}, R_{fd}, L_{fd} \end{bmatrix}$$
$$= \begin{bmatrix} \theta_1, \theta_2, \theta_3, \theta_4, \theta_5, \theta_6, \theta_7, \theta_8, \theta_9 \end{bmatrix}$$

$$F_d(\theta_d) = \begin{bmatrix} -L_l - \theta_1 & \theta_1 & \theta_1 & \theta_1 \\ -\theta_1 & \theta_1 + \theta_2 + \theta_4 & \theta_1 + \theta_2 & \theta_1 + \theta_2 \\ -\theta_1 & \theta_1 + \theta_2 & \theta_1 + \theta_2 + \theta_5 + \theta_7 & \theta_1 + \theta_2 + \theta_5 \\ -\theta_1 & \theta_1 + \theta_2 & \theta_1 + \theta_2 + \theta_5 & \theta_1 + \theta_2 + \theta_5 + \theta_9 \end{bmatrix}$$

$$G_d(\theta_d) = \begin{bmatrix} R_a & 0 & 0 & 0 \\ 0 & -\theta_3 & 0 & 0 \\ 0 & 0 & -\theta_5 & 0 \\ 0 & 0 & 0 & -\theta_8 \end{bmatrix}$$

Now Equation B.9 can be rewritten as:

$$\begin{aligned} \dot{X} &= F_d^{-1}(\theta_d)G_d(\theta_d)X + F_d^{-1}(\theta_d)DU \\ &= \bar{A}_d X + \bar{B}_d U \end{aligned} \qquad (B.10)$$

where

$$\bar{A}_d = F_d^{-1}(\theta_d)G_d(\theta_d)$$
$$\bar{B}_d = F_d^{-1}(\theta_d)D$$

The discrete dynamic representation of Equation B.10 is

$$X(k+1) = A_d(\theta_d)X(k) + B_d(\theta_d)U(k) \qquad (B.11)$$

The derivation of discrete system matrices, $A_d(\theta_d)$ and $B_d(\theta_d)$, from continuous system matrices, $\bar{A}_d(\theta_d)$ and $\bar{B}_d(\theta_d)$, is given in Reference 19.

Q-axis:

$$v_q = -R_a i_q + p\lambda_q \tag{B.12}$$

$$0 = R_{1q} i_{1q} + p\lambda_{1q} \tag{B.13}$$

$$0 = R_{2q} i_{2q} + p\lambda_{2q} \tag{B.14}$$

$$0 = R_{3q} i_{3q} + p\lambda_{3q} \tag{B.15}$$

where $p: d/dt$

$$\lambda_q = -(L_l + L_{aq})i_q + L_{aq}(i_{1q} + i_{2q} + i_{3q}) \tag{B.16}$$

$$\lambda_{1q} = (L_{1q} + L_{aq})i_{1q} + L_{aq}(i_{2q} + i_{3q} - i_q) \tag{B.17}$$

$$\lambda_{2q} = (L_{2q} + L_{aq})i_{2q} + L_{aq}(i_{1q} + i_{3q} - i_q) \tag{B.18}$$

$$\lambda_{3q} = (L_{3q} + L_{aq})i_{3q} + L_{aq}(i_{1q} + i_{2q} - i_q) \tag{B.19}$$

$$X = \begin{bmatrix} i_q & i_{1q} & i_{2q} & i_{3q} \end{bmatrix}^T$$

$$U = \begin{bmatrix} v_q \end{bmatrix}$$

Equation B.12 through Equation B.19 can be put into the form of

$$F_q(\theta_q)\dot{X} = G_q(\theta_q)X + DU \tag{B.20}$$

where $\dot{X} = dX/dt$

$$\theta_q = \begin{bmatrix} L_{aq}, R_{1q}, L_{1q}, R_{2q}, L_{2q}, R_{3q}, L_{3q} \end{bmatrix}$$
$$= \begin{bmatrix} \theta_1, \theta_2, \theta_3, \theta_4, \theta_5, \theta_6, \theta_7 \end{bmatrix}$$

$$F_q(\theta_q) = \begin{bmatrix} -L_l - \theta_1 & \theta_1 & \theta_1 & \theta_1 \\ -\theta_1 & \theta_1 + \theta_3 & \theta_1 & \theta_1 \\ -\theta_1 & \theta_1 & \theta_1 + \theta_5 & \theta_1 \\ -\theta_1 & \theta_1 & \theta_1 & \theta_1 + \theta_7 \end{bmatrix}$$

$$G_q(\theta_q) = \begin{bmatrix} R_a & 0 & 0 & 0 \\ 0 & -\theta_2 & 0 & 0 \\ 0 & 0 & -\theta_4 & 0 \\ 0 & 0 & 0 & -\theta_6 \end{bmatrix}$$

Now Equation B.20 can be rewritten as:

$$\begin{aligned}\dot{X} &= F_q^{-1}(\theta_q)G_q(\theta_q)X + F_q^{-1}(\theta_q)DU \\ &= \bar{A}_q X + \bar{B}_q U\end{aligned} \quad (B.21)$$

where

$$\bar{A}_q = F_q^{-1}(\theta_q)G_q(\theta_q)$$

$$\bar{B}_q = F_q^{-1}(\theta_q)D$$

The discrete dynamic representation of Equation B.21 is

$$X(k+1) = A_q(\theta_q)X(k) + B_q(\theta_q)U(k) \quad (B.22)$$

The derivation of discrete system matrices, $A_q(\theta_q)$ and $B_q(\theta_q)$, from continuous system matrices, $\bar{A}_q(\theta_q)$ and $\bar{B}_q(\theta_q)$, is given in Reference 19.

23

Brushless DC Drives

James P. Johnson
Caterpillar Inc., Washington, Illinois

This chapter's subject is brushless DC motors and drives. The brushless DC (BLDC) motor operates in such a way as to emulate the operation of a shunt DC motor with brushes, or a permanent-magnet DC motor. This is possible by turning the DC motor inside-out, using an AC stator consisting of the armature windings and a field-producing rotor. As will be further discussed in this chapter, the AC imposed on the stator windings in the BLDC must be switched in synchronism with the rotor position in order to maintain field-orientation, or optimal stator current and rotor flux interaction, as the commutator, brushes, and windings act to provide in a DC machine. This is only possible with switching electronics, which is a relatively new development, and thus the BLDC is only beginning to emerge in our infrastructure. BLDC motor control is now one of the fastest growing areas of motion control in the world today, and is expected to continue to proliferate as the advantages of the BLDC become more widely known, and fuel costs continue to increase.

23.1 BLDC FUNDAMENTALS

Brushless DC has found many definitions in the literature. The NEMA standard, "Motion/Position Control Motors and Controls," defines a "Brushless DC Motor" as follows: "A brushless DC motor is a rotating self-synchronous machine with a permanent magnet rotor and with known rotor shaft positions for electronic commutation. A motor meets this definition whether the drive electronics are integral with the motor or separate from it."

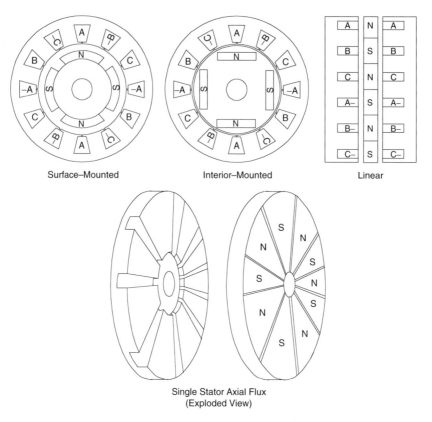

Figure 23.1 BLDC configurations.

Several types of electrical machines and drives have been referred to as BLDC, including:

1. Permanent magnet synchronous machines (PMSMs)
2. BLDC trapezoidal back-electromotive force (back-EMF) surface-mounted magnet machines,
3. BLDC sinusoidal back-EMF surface-mounted magnet machines
4. BLDC interior-magnet machines
5. The BLDC motor and drive combination
6. Axial-flux BLDC

Sketches of the more common types are shown in Figure 23.1. The PMSM has a sinusoidal back-EMF and windings such as are found in other AC machines that usually are not full, or nearly full-pitch, concentrated windings, as are found in many BLDC machines. The surface-mounted magnet BLDC back-EMF waveshape is often dependent on the magnetic orientation of the magnets. A common method of obtaining a sinusoidal back-EMF is by a parallel orientation of the magnetization in the magnet. Trapezoidal back-EMFs use a radially oriented magnetization. The most common type of BLDC is the four-pole, surface-mounted magnet machine with a quasi-trapezoidal back-EMF waveform.

Brushless DC Drives

23.2 CONTROL PRINCIPLES AND STRATEGIES

The common self-synchronous BLDC inverter and drive configuration is shown in Figure 23.2. The drive shown is the more commonly used voltage source inverter (VSI), as opposed to the current source inverter (CSI). The VSI is more widely used due to cost and weight properties, dynamic performance capabilities, and ease of controlling, which surpass those of the CSI [1]. The weight and cost differences between the two drives is the difference between capacitors for the DC link of the VSI vs. a heavy inductor between the rectifier and inverter in the CSI. The VSI outperforms the CSI in dynamic response capability, as the large inductor that acts to make the rectifier of the CSI act as a constant current source requires greater commutation overlap angles, prevents fast current build-up in the motor windings inhibiting high-speed servo operations, and may increase the need and size of snubber components in the drive. The desirable property of constant current control, and thus constant torque control of the CSI can be approximately obtained in the VSI by using a hysteresis-type current control in an inner current control loop.

The term "self-synchronous" refers to the drive circuitry's requirement of the knowledge of the instantaneous rotor position in order to determine the correct gate firing sequence needed to properly align stator phase current pulses and machine phase back-EMF.

Figure 23.3 is a block diagram of a classical position and speed control scheme for BLDC. If only speed control is desired, the "position controller" and position feedback circuitry may be eliminated. Usually, both position and velocity feedback transducers are required in high-performance position controllers. Just a position sensor, without a velocity sensor, would require differentiating the position signal, which would tend to amplify noise in an analog system; however, in digital systems this usually is not a problem. The position sensor, or some means of gathering position information, is required in position and speed control of BLDC. Many high-performance applications include current feedback for torque control [1]. At the minimum, a DC bus current feedback is required to protect

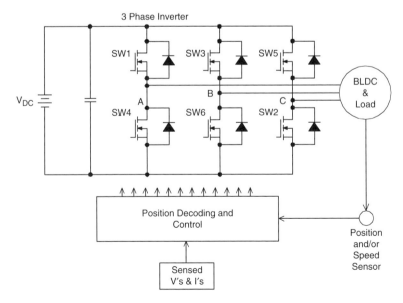

Figure 23.2 Basic BLDC drive.

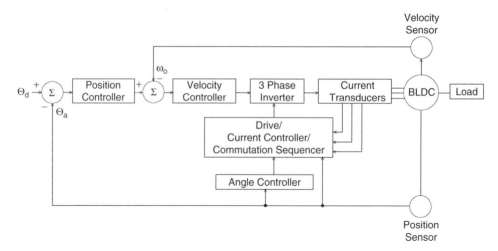

Figure 23.3 Block diagram of a classical speed and position control for a BLDC.

the drive and machine from overcurrents. The use of an inner closed-loop current control provides the action of a very fast CSI, without the need of the DC link inductor, and is referred to as a current-regulated VSI (CRVSI) [1]. The DC voltage regulation in the drive may also be performed by a controlled rectifier acting as the DC power source, or in the inverter by applying PWM to either both the upper and lower switches, which are simultaneously turned on, or to only the lower or only the upper switches.

Using PWM with only the lower switches or only the upper switches reduces switching losses, as opposed to switching both the upper and lower. However, if utilizing advance angle techniques, both switches should be turned on and off, as paths may exist through the switch in a phase leg that remains closed and the freewheeling diode of a different phase leg, which could introduce currents producing negative torque. Not utilizing a "chopping" switch to regulate the DC bus voltage eliminates a switch from the drive, but with the DC regulating switch, there is only one power semiconductor device that experiences the higher carrier frequency switching losses of the PWM. Utilizing a controlled rectifier to vary the DC bus voltage requires extra control measures and increases switching losses, initial drive cost, and line power factor control complexity. When the input to the drive is supplied by the utility, an inductor is usually included after the rectifier to reduce the utility current harmonic content. The inductor acts with DC link capacitance, forming a low-pass LC or proportional–integral (PI) (CLC) filter that has a cutoff frequency low enough to block the PWM carrier frequency at a minimum, and lower frequency components, if applicable, such as in a variable-speed drive.

The DC link capacitance provides the high-frequency ripple currents of the inverter, while the inductor blocks the higher frequency, allowing the average current to pass. If the drive is supplied by a DC source, a filter may also be used to reduce EMI from passing to the source. If PWM is not utilized, current control alone can be effective in providing high-performance torque control without varying the DC bus. The controller blocks "position controller" and "velocity controller" in Figure 23.3 may be any type of classical controller such as a proportional–integral controller, or a more advanced controller. The "current controller and commutation sequencer" provides the properly sequenced gating signals to the "three-phase inverter," while comparing sensed currents to a reference to maintain a current control by hysteresis (current chopping) or with a voltage source

Brushless DC Drives

(PWM)-type current control. Hysteresis current control could be a constant frequency hysteresis control, band hysteresis, or level hysteresis. The current control could be used to create sinusoidal current waveforms, limit peaks, or square-wave current waveforms, especially at lower frequency when operating in the torque-limited region of the machines capability curve. Using position information, the commutation sequencer causes the inverter to "electronically commutate," acting as the mechanical commutator of a DC machine [2].

A detailed guide to switching was described in Reference 3, "The commutation angle associated with a brushless motor is normally set so that the motor will commutate around the peak of the torque angle curve. Considering a three-phase motor, connected in Delta or Wye, commutation occurs at electrical angles which are plus or minus 30° (electrical) from the peaks of the torque-angle curves. When the motor position moves beyond the peaks by an amount equal to 30° (electrical), then the commutation sensors cause the stator phase excitation to switch to move the motor suddenly to –30° relative to the peak of the next torque-angle curve." The torque curves can be obtained by either electrically exciting a line-line connection and then forcing the rotor to turn while measuring the torque, or by forcing the shaft, loading the windings, and measuring the torque vs. rotor position [3]. The actual shape of these curves for a trapezoidal back-EMF machine should be trapezoidal. However, due to winding configurations, localized saturations, bulk saturations, and flux leakages, the back-EMF and torque-angle curves of trapezoidal machines are more nearly sinusoids with flattened peaks [4].

The "position sensor" is usually either a three-element Hall-effect sensor or an optical encoder. The "angle controller" is another option that allows phase shifting (advancing) of the current pulses with respect to rotor position such as to allow the current pulse to build up more fully before current pulse/phase back-EMF alignment, which can increase the speed range of the motor. Advancing of the angle is required due to the electrical time constant of the windings. A given amount of time is needed for a current pulse to build up. At higher speeds there is less time available for this current build-up before it is required that the current pulse be aligned with the back-EMF. A problem with this type of operation is that the drive may become "soft," as in field-weakening operations of DC machines. A "soft" drive is one that has speed/load torque characteristics such that the speed change is greater for a given load change as compared to a normal stiff drive. In Reference 5, it was concluded that the reactive power demanded during angle advancing operation is increased, considering the system to be a sinusoidal drive (permanent-magnet synchronous motor) corresponding to considering only the fundamental of a quasi-square-wave drive current and trapezoidal back-EMF voltage waveforms.

23.3 TORQUE PRODUCTION

Figure 23.4 shows a cutaway view of a three-phase, 4-pole, 12-slot, full-pitch, surface-mounted permanent-magnet, trapezoidal back-EMF BLDC, equivalent circuit, and associated waveforms. The voltages seen in this diagram, V_{ab}, V_{bc}, and V_{ca}, are the line-to-line back-EMF, which are the result of the permanent-magnet flux crossing the airgap in a radial direction, cutting the coils of the stator at a rate proportional to the rotor speed. The waveforms V_{an}, V_{bn}, and V_{cn} are the line-to-motor neutral back-EMF, or phase back-EMF voltages, which are represented by voltage sources in the equivalent circuit of the motor. The coils of the stator are positioned in the standard three-phase full-pitch, concentrated arrangement, and thus the phase trapezoidal back-EMF waveforms are displaced by 120

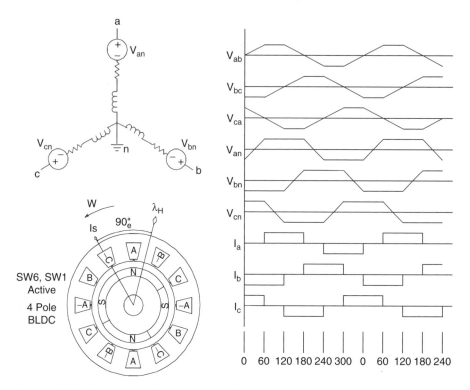

Figure 23.4 Three-phase, 4-pole cutaway BLDC, equivalent circuit, and associated waveforms.

degrees electrical (120°e). The current pulse generation technique shown in Figure 23.4 is a "120°e on, 60°e off" type, meaning each phase current is flowing for two thirds of an electrical 360°e period, 120°e positively and 120°e negatively. Between the "on" times of a phase are 60°e periods when the phase is "off," allowing the designation of the phase during this period as the "silent phase." The "silent phase" period is typically used in "sensorless" control of BLDC as a period in which the back-EMF is observed in order to determine rotor position. Another switching algorithm possibility includes changing the "dwell" of the current pulses; i.e., changing the "on" time of the pulses. The dwell can be theoretically increased to 180°e; however, in an actual circuit with inductances, a commutation lag may exist and some margin must be reserved for commutation overlap (usually under 15°e). It was determined in Reference 6 that by systematically increasing the dwell angle from 120°e at low speeds to 180°e at high speeds, the maximum torque is attained at all speeds.

To drive the motor with maximum torque/ampere it is desired that the line current pulses be overlapped by the line-neutral back-EMF voltages of the particular phase. This allows a maximum torque output by the fundamental physical principle of torque generation, i.e., Torque = Total Force × Moment Arm, where the force is produced by the interaction of the flux produced by the rotor magnets and the current in the stator coil sides. From the Lorentz force equation,

$$\text{Force}_{1\,\text{coil side}} = \int_L N(I \times B)\,dl$$

Brushless DC Drives

where

N = number of turns per slot per phase
I = current in the coil
B = the flux density vector
L = length of the coil side

At any given time, two phases are excited with DC current. With current flowing in the same direction, a radially magnetized magnet of a given polarity that is circumferentially wide enough to overlap two adjacent slots results in forces in the coil sides of both the slots that adds to form the total force of a pole. The total force is then the sum of the forces of all the poles.

For example, in a BLDC with radially magnetized magnets, full-pitched windings, the number of stator slots = (# of phases) × (# of poles), two phases simultaneously active with square-wave excitation, and the magnets' circumferencial distance approximately equal to the pole arc, the torque is given as follows [7]:

$$\text{Torque} = N_P \cdot N_t \cdot N_{spp} \cdot P \cdot I \cdot B_g \cdot L \cdot R,$$

where

N_p = number of active phases (two in a three-phase system)
N_t = number of turns per slot per phase
N_{spp} = number of slots per pole per phase
P = number of magnet poles
I = DC current magnitude
B_g = radial flux density in the gap due to the magnets
L = length of the core where both the stator and rotor overlap
R = outer radius of rotor (moment arm)

The most accurate static torque profile for a specific machine geometry, prior to the fabrication of the motor, is determined by using a finite element software package that uses numerical methods. Finite element methods are utilized on modern computers and are known to require lengthy periods of time to obtain solutions with a greater degree of accuracy than hand calculations with required approximations. Finite element methods for BLDC are discussed later in this chapter.

23.4 ADVANTAGES AND DISADVANTAGES

The objective of this section is to clearly discuss the advantages, disadvantages, and fundamentals of operation of the permanent-magnet brushless DC motor. Also presented are some of the more important performance characteristics.

The permanent-magnet BLDC (PM-BLDC) motor offers automobile and aerospace industry manufacturer's many advantages:

1. Low noise. There are no mechanical brushes or slip-rings required in the PM-BLDC. This eliminates virtually all mechanical noise except that of the bearings, coupling, and load. Electromagnetically, the commutation frequency depends on the speed of the motor and the number of poles of the motor, being related by $\omega_e = P/2\, \omega_m$, where ω_e is the electrical (commutation) frequency (per phase)

in electrical rad/s, ω_m is the mechanical rotor frequency in mechanical rad/s, and P is the number of poles in the machine.

The actual on/off switching of each individual semiconductor device in the driving inverter, required for commutation, occurs at the electrical frequency rate; i.e., each switch is turned on and off once per electrical cycle. As pulse width modulation (PWM) is often used to control the duty cycle of the bus voltage applied to the inverter, the carrier frequency of the PWM must also be considered. In low-power systems, < 10 kW, the PWM carrier is usually chosen to be higher than audible frequencies, i.e., > 15 kHz, and thus no noise from the PWM is produced by the carrier. The harmonics due to variations of the duty cycle, if operating at a high carrier frequency, are modulated around the higher carrier frequency and are thus inaudible as well. Even for baseband harmonics, magnetostriction in the BLDC generates negligible noise, as the BLDC is not normally operated in a highly saturated state, which would induce magnetostriction – (creating noise as in switched-reluctance machines that operate in saturation).

2. High efficiency. PM-BLDCs have been shown to be the highest efficiency machines available today [1,6]. In 1985, in Reference 8, a study revealed that due to the higher efficiency of the integral horsepower PM-BLDC, an annual electrical energy savings of 2.8×10^{10} kWh would be possible by replacing the presently used integral horsepower induction motors with PM motors. This was based on a total 6.6×10^{11} kWh used by induction motors. The savings would be equivalent to 20 million barrels of oil saved per year, or $7 billion initial reduction in electrical generating capacity requirements, plus an annual $2 billion per year ongoing fuel savings at the utilities. A similar study in Reference 9 showed that not only was there an energy savings, but the savings in energy costs may compensate for the higher initial cost of the PM-BLDC in less than a year of operation, the economic advantage increasing with higher ratings. The higher efficiency of PM-BLDC can primarily be accredited to the presence of the permanent magnet field that provides a nearly constant and continuous magnetic field and consumes no electrical power. Another important feature of magnets is their longevity: Under proper operating conditions magnets have extremely small remagnetizing coefficients [7]; i.e., permanent magnets tend to keep their magnetic properties over long periods of time. Another efficiency factor is that there is no additional torque requirement for friction caused by brushes or slip-rings, as in DC and some AC machines.

3. Reduction of excitation requirements. As stated above, permanent magnets provide a constant field which increases the efficiency by reducing the need for the creation of an electromagnetic field as required in most other machines.

4. Low maintenance and greater longevity. No brushes and no slip-rings are required and thus the lifetime of the motor is based solely on insulation, bearing, and magnet life.

5. Ease of control. Continuing advancements in control and inverter semiconductor packaging have reduced the technical requirements in design and fabrication of BLDC drives. Many semiconductor manufacturers produce control integrated circuits (ICs) specifically for BLDC control, enabling the development of an extremely inexpensive one-chip drive controller. Recently, integration of power semiconductor devices with gate drive circuitry in ICs specifically designed to meet motor drive inverter requirements are becoming available, reducing the overall cost of system development and initial cost of the drive.

6. Lighter, more compact construction. Aerospace and automotive applications demand lighter, more compact devices to increase fuel efficiency and reduce fuel storage requirements. Recently, higher energy density magnets, samarium cobalt and neodymium iron boron, have come into use to provide increased power density machines for use in these applications.
7. The ease in cooling. As the armature (stator) windings are on the outside of the motor, the PM-BLDC has the inherent property of being thermally manageable. It is well known that the majority of losses occurring in a PM-BLDC will be in the windings as I^2R losses [10]. As the windings are on the outside of the machine, heat may freely pass out of the machine via its outer surface. This offers advantages over the DC machine, in which heat may be trapped on the rotor (armature).

As in all modern machines, there are some inherent disadvantages:

1. Cost of permanent magnets. Costs of the higher energy density magnets prohibits their use in applications where initial cost is more a concern than obtaining a system with all of the advantages previously listed. Typically, ceramics are the least expensive and have the lowest energy density. Neodymium iron boron magnets have the highest energy density products, at about 3 times the cost of ceramics. Samarium cobalt magnets have energy densities comparable to neodymium iron boron magnets at around 6 times the cost of ceramics [11]. The cost is primarily a function of the availability of the raw materials used to fabricate the magnets; however, as discussed in a later section, considerations other than cost, e.g., thermal, may necessitate the use of a particular type of magnetic material in certain applications.
2. The possibility of demagnetization of the permanent magnets. Extreme care must be taken in utilizing permanent magnets, as high levels of demagnetizing force applied to the magnets or subjecting the magnets to high temperatures can cause demagnetization of the magnets.
3. The dangerous nature of large magnets. Permanent magnet machines have rarely been utilized in larger drives, i.e., > 100 Hp. One reason for this is the difficulty in handling the larger permanent magnets, and the danger presented by large permanent magnets. It has been reported that attempts to build large permanent magnet machines have resulted in injuries due to metallic objects flying across a room toward the magnets. Also, to implace magnets that are premagnetized, and larger in size, requires power machinery, with control capabilities of relatively high precision, to avoid damaging the magnets. Safer and more practical *in situ* magnetization requires that either a specially designed coil system that can be utilized around and in the assembled machine, or additional windings in-built in the machine, provide the capability to magnetize the permanent magnets. Either method adds considerable cost to the system.

23.5 TORQUE RIPPLE

Torque ripple can be a problem in BLDC drives. There are a variety of reasons for torque ripple in BLDC drives:

Commutation torque ripple is due to the switching on and off of the phases at 60°e intervals. This ripple is at a frequency that is 6 times the electrical frequency. The reason for this type of ripple is that as one phase turns off (commutates), while another turns on, the rise and fall rates of the respective phase currents are not equivalent, and thus the torque generated by the two currents during commutation does not instantaneously add to the value of torque of one fully excited phase, which would allow a smooth torque over the commutation interval. In Reference 16, a ramping current control with a speed-dependent slope was used for both the phases (the phase turning on and the one commutating), which required the torque, due to the two ramping current waveforms, to equal that due to a single fully on current level. It was reported in Reference 6 that a closed-loop control can eliminate the commutation torque pulsations (6 times electrical frequency), provided that the frequency of these pulsations is lower than the closed-loop speed bandwidth.

Cogging torque is apparent at low speeds and is due to the natural attraction and repulsion of the stator teeth by the magnets. Some may prefer to think in terms of reluctance torque; i.e., cogging torque is due to the flux of the magnets attempting to seek out the least reluctance path by moving the rotor or stator toward magnet-tooth alignments [6]. A typical reluctance torque profile is a function of the angular position, the rotor within the slotted stator, and is periodic with a period of $\int$ pole pitch [12]. Cogging torque is overcome by various machine design methods. The most commonly used machine design method to reduce the effects of cogging torque includes skewing either the magnets or the stator slots by one slot pitch; however, this increases the distortion of the back-EMF waveform, and may also increase axial forces.

Distortions in the trapezoidal back-EMF can also cause torque ripple. These distortions may be due to leakage flux paths between magnets, and thus a rounding of the corners of the back-EMF waveform [1]. A variety of other reasons exist as well, such as local saturations in the pole shoes, armature reaction effects due to strong stator currents and thus fields produced by the stator effecting the rotor field, partial and localized demagnetizations of the permanent magnets, and inherent machine-specific characteristics due to winding configurations, geometry, and design. The distortion of the phase back-EMF, which causes it to have less than a 120°e crest (flat peaks), prohibits the possibility of full phase back-EMF/current alignment for 120°e (discontinuous [13]) operation. Without full alignment over 120°e for each phase current, zero torque ripple cannot be achieved in a CRVSI BLDC drive without special control techniques [14].

Ripple may also exist due to the hysteresis current control frequency, or at the PWM carrier frequency rate due to the chopping of the current waveform.

Harmonics due to machine slotting and winding configurations coupled with quasi-square-wave current excitation may also cause torque components other than the fundamental.

In general, torque ripple is most noticeable at lower speeds and when operating the machine with heavy loads. Torque ripple usually manifests itself in speed ripple, which is more readily measured. Speed ripple also depends on the coupling between the machine and the load, and the mechanical (dynamic) characteristics of the machine and load. A highly compliant (with a lack of torsional stiffness) coupling or load will more readily suffer from torque pulsations in the form of speed pulsations and mechanical oscillatory behavior, as will a system with less rotational inertia. The general solution is a more solid, stiff coupling between machine and load where possible. The load is whatever it is, and thus, unless additional stiffness, damping, or increased inertia can be afforded in the system, the mechanical load and coupling dictate a certain amount of the inherent susceptibility of the system to torque ripple.

23.6 DESIGN CONSIDERATIONS

The magnetic circuit, the use of finite element analysis, a set of geometrically intensive calculations, and recursive design and redesign enable the conception of BLDC machines. Typically, a first design is made with calculations and the magnetic circuit. Following, a more time-consuming "fat-trimming" process may be undertaken with the aid of finite element analysis. The fat-trimming process consists of reducing the machine size as much as possible without compromising the required design specifications. This is possible as the flux patterns obtained in finite element enable the user to determine saturated and unsaturated conditions in the motor's materials at any desired operating point and rotor position. A simplified, one-pass design of a surface-mounted trapezoidal back-EMF machine considering the magnetic metal to have an infinite permeability could easily be developed in Matlab code. The equations and method can be found in Reference 15. A simple computer machine design program would typically provide a quick first-try solution, which could then be modeled, analyzed, and modified with the aid of a finite element package.

23.7 FINITE ELEMENT ANALYSIS AND DESIGN CONSIDERATIONS FOR BLDC

To obtain a static torque profile, a plot of flux vs. rotor position, or some other graphical representation or data for a particular machine, a geometrical drawing of the machine must first be entered into the finite element package. Each portion of the machine is assigned appropriate material properties; e.g., magnets are assigned a B_r, H_c, μ_{recoil}, and perhaps other magnet parameters, a direction or directional function assigning their flux orientations, and magnetic material permeabilities, and polarizations if anisotropic. Current density (or total current) is assigned to the areas of the copper material, where it is desired that current flow. Some packages offer parametric features that allow parts of the motor geometry or material property to rotate, move, or vary in value by some amount for each finite element solution. Typically, many solutions may be found for the machine with certain parameters changed to determine the effect of the variation of position or material property on the performance of the machine. As an example, a torque profile for a BLDC would be obtained by rotating the rotor structure at small angular increments and finding a solution for the torque on the rotor at each angular increment, while the appropriate phase currents are maintained active.

Other factors in machine design that finite element analysis can determine include levels of saturation in the magnetic materials, flux density levels throughout the machine, thermal properties and possible hot spots, and in vibrational analysis studying the effects of eddy currents and magnetostriction. In design, by observing the flux densities over a parametric solution, the determination of what materials in the machine are underutilized, thereby allowing the material to be made thinner, such as the back iron thickness if the flux density is too low during peak torque conditions. Also, the amount of magnet circumferential and radial length can be found optimally after initial hand calculations and computer program have found first-try figures. More comprehensive finite element packages have the ability to do mechanical as well as electromagnetic, thermal, electrostatic, and eddy current analysis, and even connect electrical circuitry and mechanical loads to a finite element model to verify functionality and analyze the electrical waveforms and mechanical response when excitation is provided the machine. This allows the user to not only design for the electromagnetic constraints, but to ensure mechanical specifications are met.

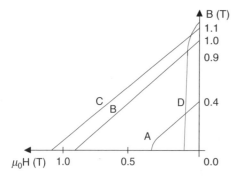

Figure 23.5 Typical demagnetization characteristics of the most common permanent magnet materials: A. ceramic, B. samarium cobalt, C. neodymium iron boron, D. Alnico.

23.8 PERMANENT MAGNETS

Permanent magnets are essential ingredients to the BLDC. In the 1950s, the first machines that were considered BLDC were PMSM using Alnico magnets [1]. A diagram of typical demagnetization characteristics of the most common types of permanent magnets is shown in Figure 23.5. The graphs of this figure are actually only the PM characteristics in the second quadrant of the B-μ_0H plane. The intersection of a characteristic with the μ_0H-axis is at a value of H that is called the coercivity, H_c, of the magnetic material. The B-axis intercept is termed the remanence, B_r. The characteristic in the second quadrant, above the knee, has a slope equal to the relative permeability of the magnet. The coercivity is defined as the amount of magnetizing force required to reduce the magnet's flux density to zero. This is a demagnetizing force and is thus a negative value, showing opposition to the inherent property of flux production in the magnet. The remanence of a magnet is considered that flux density available at the flux-producing surfaces of the magnet when no external magnetizing force is present. This is the flux density in the magnet when it is "keepered," or when a material with infinite permeability is used to provide a flux path from one end of the magnet to the other [7].

Magnets usually come in one of two forms: sintered or cast (bonded) [11]. Although a wide range of magnetic properties are available in both types of magnets, the sintered type are usually preferred in electrical machines, as their properties are more capable of higher performance operation [15]. The characteristic of the permanent magnet through all four quadrants of the B-μ_0H plane is actually a large hysteresis curve. Normally, in machine design and drive considerations, only the second quadrant portion is considered, as this is where desirable operating points are located. When designing a BLDC machine, a magnetic circuit representing an area of the machine equivalent to one pole must be developed that includes MMF sources due to the magnets and the stator ampere-turns. Other components of this magnetic equivalent circuit include the permeances, or reluctances of all of the portions of magnetic material, i.e., stator and rotor back iron, stator teeth (or poles), pole shoes, and portions of nonmagnetic material of the machine, i.e., the airgap, slots, windings, and magnets.

The total permeance of the magnetic circuit at any instant, as seen by the magnets acting as a flux source, determines the instantaneous operating point of the magnets, as shown in Figure 23.6. The absolute value of the slope of the load line is determined by machine dimensions. A "permeance coefficient" (*PC*) was defined in Reference 7. The

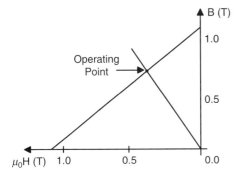

Figure 23.6 Demagnetization characteristic showing load line.

load line's slope is the absolute value of $PC \times \mu_0$. The simplest expression given in Reference 7 relates the magnet flux density, B_M, the remanence flux density, B_r, and the recoil permeability, μ_{rec}, to the PC as

$$\frac{B_M}{B_r} = \frac{PC}{PC + \mu_{rec}},$$

and is a measure of how far down the demagnetization curve the machine operates on an open circuit. The additional field due to armature excitation causes the load line to shift to the left with increased armature excitation. In Figure 23.6, the load line rotates counterclockwise around the origin for an increase in airgap length. In any variation of the operating point that places the operating point closer to or past the knee, in moving the operating point from right to left, the return path for a reversed variation will follow the slope of recoil permeability. Thus, variations that take the operating point to the left past the knee should especially be avoided as this has the capability of causing a large irreversible demagnetization, depending on the magnet demagnetization characteristic.

The BLDC is designed such that, during normal operation, the static permeance characteristic (static load line) is chosen to intercept the magnet's demagnetization characteristic in the second quadrant, usually with a static load line slope of around −4 [15]. The higher the intersection of the static load line on the magnet characteristic in the second quadrant, the higher the permeance of the machine's magnetic circuit without current applied. As current is applied to the windings, the amount of reluctance in the magnetic material varies according to the levels of saturation (local and bulk). This causes the load line to vary from the static case. If the load line moves down the second quadrant portion of the curve toward the x-axis, and back again toward the y-axis along the straight portion of the characteristic, the operating point of the magnet remains on the characteristic. If the current is such that the flux produced in the magnets, due to the current, causes heavy saturation in the machine, the permeance decreases and the load line tends to rotate toward alignment with the negative x-axis of the B-μ_0H plane.

Special care must be taken in design to ensure that the maximum coercive force presented to the machine does not cause the load line to intersect the demagnetization characteristic below the knee. Included in this must be the consideration of the variation of the position of the demagnetization curve for variations in temperature. Due to the negative temperature dependencies of the coercivity of the higher-energy density magnets, such as samarium cobalt and neodymium iron boron magnets, the coercivity of the demagnetization characteristic tends to decrease in amplitude, while the knee tends to move higher on the characteristic with increasing temperature, making it easier to demagnetize the machine at higher temperatures. In the case of ferrite PMs, which have a positive coercivity temperature coefficient, the magnets are more susceptible at lower temperatures to demagnetization, as the knee moves lower in the second or third quadrant with increasing temperature [16]. In all cases, however, the remanence temperature coefficient is negative, and thus with increasing temperature, the characteristic always moves downward vertically. The temperature effects are reversible unless the Curie temperature is reached and slightly exceeded in the case of ceramics, or in the case of rare-earth magnets at temperatures somewhat lower than their Curie temperature when metallurgical changes take place in the magnets [7]. Curie temperatures of the most commonly used magnets in BLDC, ferrites, samarium cobalt, and neodymium iron boron, are approximately 450, 700 to 800, and 310°C, respectively.

Magnets are often given ratings based on a "maximum energy-density." In Reference 7, the maximum energy-product or maximum energy-density is considered the product of the B and $\mu_0 H$ values, which form curves in the second quadrant. The rating for a particular magnet is that $B \times \mu_0 H$ curve that the demagnetization curve of the magnet is just tangent to.

23.9 BLDC SIMULATION MODEL

This section will present equations that can be used in a computer program simulation model for a BLDC motor and drive. The drive considered uses a "120°e on, 60°e off" or discontinuous-type switching with quasi-square-wave excitation currents. The concept of this simulation model can be modified to include continuous square-wave operation or "180°e on, 180°e off" type, or sinusoidal excitation or back-EMF.

In order to determine what equations are required to perform a computer program simulation of a BLDC motor and drive, the topology and modes of the drive must be examined. The diagrams of the motor and drive model presented are shown in Figure 23.2 and Figure 23.4. The total switch and winding resistance have been lumped into a single R value for each phase. Mutual and leakage inductances have been included in the total winding inductance, L. We first examine two modes in this drive. The first mode is when an upper and a lower switch conduct. The second mode is during a commutation, when one switch is turning off, one is turning on, and the other remains on.

Referring to Figure 23.2 and Figure 23.4, first consider the case of switches 1 and 6 beginning to conduct, while switch 2 is commutating (turning off). The line-line voltage equations are

$$V_{AB} = V_{DC} = i_a R + L \frac{di_a}{dt} + e_a - e_b - L \frac{di_b}{dt} - i_b R \qquad (23.1)$$

$$V_{BC} = 0 = i_b R + L \frac{di_b}{dt} + e_b - e_c - L \frac{di_c}{dt} - i_c R \qquad (23.2)$$

$$V_{CA} = -V_{DC} = i_c R + L \frac{di_c}{dt} + e_c - e_a - L \frac{di_a}{dt} - i_a R \qquad (23.3)$$

Brushless DC Drives

Knowing

$$i_a + i_b + i_c = 0 \tag{23.4}$$

$$i_a = -(i_b + i_c) \tag{23.5}$$

$$i_b = -(i_a + i_c) \tag{23.6}$$

$$i_c = -(i_a + i_b) \tag{23.7}$$

From Equation 23.1,

$$i_a R + L\frac{di_a}{dt} = i_b R + L\frac{di_b}{dt} + V_{DC} + e_b - e_a \tag{23.8}$$

From Equation 23.2,

$$i_b R + L\frac{di_b}{dt} = i_c R + L\frac{di_c}{dt} + e_c - e_b \tag{23.9}$$

From Equation 23.3,

$$i_c R + L\frac{di_c}{dt} = i_a R + L\frac{di_a}{dt} - V_{DC} + e_a - e_c \tag{23.10}$$

Substituting Equation 23.6 into Equation 23.8,

$$2\left(i_a R + L\frac{di_a}{dt}\right) = -i_c R - L\frac{di_c}{dt} + V_{DC} + e_b - e_a \tag{23.11}$$

Substituting Equation 23.10 into Equation 23.11,

$$\frac{di_a}{dt} + \frac{R}{L}i_a = \frac{2V_{DC} - 2e_a + e_b + e_c}{3L} \tag{23.12}$$

Substituting Equation 23.7 into Equation 23.9,

$$2\left(i_b R + L\frac{di_b}{dt}\right) = -i_a R - L\frac{di_a}{dt} + V_{DC} + e_c - e_b \tag{23.13}$$

Substituting Equation 23.8 into Equation 23.13,

$$\frac{di_b}{dt} + \frac{R}{L}i_b + = \frac{-V_{DC} + e_a - 2e_b + e_c}{3L} \tag{23.14}$$

Substituting Equation 23.5 into Equation 23.10,

$$2\left(i_c R + L\frac{di_c}{dt}\right) = -i_b R - L\frac{di_b}{dt} - V_{DC} + e_a - e_c \tag{23.15}$$

Substituting Equation 23.9 into Equation 23.15, and rearranging,

$$\frac{di_c}{dt} + \frac{R}{L}i_c = \frac{-V_{DC} + e_a + e_b - 2e_c}{3L} \tag{23.16}$$

Equation 23.12, Equation 23.14, and Equation 23.16 are the differential equations representing the three-phase currents, which are valid until the commutating phase current goes to zero, i.e., in this case $i_c = 0$. After $i_c = 0$ until the next commutation sequence, the voltage equations become

$$V_{AB} = V_{DC} = i_a R + L\frac{di_a}{dt} + e_a - e_b - L\frac{di_b}{dt} - i_b R \tag{23.17}$$

$$V_{BC} = 0 = i_b R + L\frac{di_b}{dt} + e_b - e_c \tag{23.18}$$

$$V_{CA} = 0 = +e_c - e_a - L\frac{di_a}{dt} - i_a R \tag{23.19}$$

where Equation 23.20 has been substituted into Equation 23.1, Equation 23.2, and Equation 23.3.

$$i_c = 0$$
$$\frac{di_c}{dt} = 0 \tag{23.20}$$

From Equation 23.17,

$$i_a R + L\frac{di_a}{dt} = i_b R + L\frac{di_b}{dt} + V_{DC} + e_b - e_a \tag{23.21}$$

Now, $i_a = -i_b$, so substituting for i_b in Equation 23.21 produces, after some manipulation,

$$\frac{di_a}{dt} + \frac{R}{L}i_a = \frac{V_{DC} - e_a + e_b}{2L} \tag{23.22}$$

which is true after $i_c = 0$. Although left in for completeness, during this mode, Equation 23.18 and Equation 23.19 are invalid, as V_{BC} and V_{CA} are indeterminate.

Working through the next sequence, the results can be generalized. Next, switches 5 and 6 are beginning to conduct while switch 1 is commutating.

Before $i_a = 0$,

$$V_{AB} = V_{DC} = i_a R + L\frac{di_a}{dt} + e_a - e_b - L\frac{di_b}{dt} - i_b R \tag{23.23}$$

$$V_{BC} = -V_{DC} = i_b R + L\frac{di_b}{dt} + e_b - e_c - L\frac{di_c}{dt} - i_c R \tag{23.24}$$

Brushless DC Drives

$$V_{CA} = 0 = i_c R + L\frac{di_c}{dt} + e_c - e_a - L\frac{di_a}{dt} - i_a R \tag{23.25}$$

From Equation 23.23,

$$i_a R + L\frac{di_a}{dt} = i_b R + L\frac{di_b}{dt} + V_{DC} + e_b - e_a \tag{23.26}$$

From Equation 23.24,

$$i_b R + L\frac{di_b}{dt} = i_c R + L\frac{di_c}{dt} + e_c - e_b \tag{23.27}$$

From Equation 23.25,

$$i_c R + L\frac{di_c}{dt} = i_a R + L\frac{di_a}{dt} - V_{DC} + e_a - e_c \tag{23.28}$$

Substituting Equation 23.6 into Equation 23.26,

$$2\left(i_a R + L\frac{di_a}{dt}\right) = -i_c R - L\frac{di_c}{dt} + V_{DC} + e_b - e_a \tag{23.29}$$

Substituting Equation 23.28 into Equation 23.29,

$$\frac{di_a}{dt} + \frac{R}{L}i_a = \frac{V_{DC} - 2e_a + e_b + e_c}{3L} \tag{23.30}$$

Substituting Equation 23.7 into Equation 23.27,

$$2\left(i_b R + L\frac{di_b}{dt}\right) = -i_a R - L\frac{di_a}{dt} - V_{DC} + e_c - e_b \tag{23.31}$$

Substituting Equation 23.26 into Equation 23.31,

$$\frac{di_b}{dt} + \frac{R}{L}i_b + = \frac{-2V_{DC} + e_a - 2e_b + e_c}{3L} \tag{23.32}$$

Substituting Equation 23.5 into Equation 23.28,

$$2\left(i_c R + L\frac{di_c}{dt}\right) = -i_b R - L\frac{di_b}{dt} + e_a - e_c \tag{23.33}$$

Substituting Equation 23.27 into 23.33, and rearranging,

$$\frac{di_c}{dt} + \frac{R}{L}i_c = \frac{V_{DC} + e_a + e_b - 2e_c}{3L} \tag{23.34}$$

After $i_a = 0$, $i_b = -i_c$, and $i_a = di_a/dt = 0$, and therefore the only valid voltage equation is Equation 23.24, which when rearranged yields the differential equation

$$\frac{di_b}{dt} + \frac{R}{L}i_b = \frac{-V_{DC} - e_b + e_c}{2L} \tag{23.35}$$

The first set of equations, when switches 1 and 6 were starting to conduct and switch 2 was commutating, represent the situation when a lower switch is commutating. The second set, when 5 and 6 are starting to conduct and switch 1 is commutating, represent the situation when an upper switch is commutating. In both cases the three differential equations are completely correlated, except the sign of V. Generalizing, W is the phase that stays on, X is the phase that is just coming on, and Y is the phase turning off. Also V is positive for a commutating switch in the upper half of the converter and negative V for a commutating switch in the lower half; i.e., switches 1, 3, and 5 are upper, and 4, 6, and 2 are lower switches. The generalized equations are then as follows.

Before $i_Y = 0$,

$$\frac{di_W}{dt} + \frac{R}{L}i_W + = \frac{-2V_{DC} + e_Y - 2e_W + e_X}{3L} \tag{23.36}$$

$$\frac{di_Y}{dt} + \frac{R}{L}i_Y + = \frac{V_{DC} + e_X - 2e_Y + e_W}{3L} \tag{23.37}$$

$$\frac{di_X}{dt} + \frac{R}{L}i_X + = \frac{V_{DC} + e_W - 2e_X + e_Y}{3L} \tag{23.38}$$

After $i_Y = 0$,

$$\frac{di_W}{dt} + \frac{R}{L}i_W + = \frac{-V_{DC} - e_W + e_X}{2L} \tag{23.39}$$

Toward a computer program simulation model, many options exist in the method of solution of the Equation 23.36 through Equation 23.39. The most commonly chosen methods are numerical integration and off-line analytical solutions. The off-line analytical solutions can be performed in the LaPlace domain, with the time domain equations for the three currents resulting. The solutions are given below.

Considering Equation 23.36, the LaPlace domain equation is written,

$$I_W(s)R + L[sI_W(s) - i(0)] = \frac{-2V_{DC} + e_Y - 2e_W + e_X}{3s} \tag{23.40a}$$

manipulating,

$$I_W(s) = \frac{-2V_{DC} + e_Y - 2e_W + e_X}{3s(R + sL)} + \frac{Li(0)}{R + sL} \tag{23.40b}$$

which provides the time domain solution,

Brushless DC Drives

$$i_W(t) = \frac{-2V_{DC} + e_Y - 2e_W + e_X}{3R}\left(1 - e^{-\frac{R}{L}t}\right) + i_W(0)\, e^{-\frac{R}{L}t} \tag{23.41}$$

Also for the other two phase currents,

$$i_Y(t) = \frac{V_{DC} + e_X - 2e_Y + e_W}{3R}\left(1 - e^{-\frac{R}{L}t}\right) + i_Y(0)\, e^{-\frac{R}{L}t} \tag{23.42}$$

$$i_X(t) = \frac{V_{DC} + e_W - 2e_X + e_Y}{3R}\left(1 - e^{-\frac{R}{L}t}\right) + i_X(0)\, e^{-\frac{R}{L}t} \tag{23.43}$$

Equation 23.41, Equation 23.42, and Equation 23.43 are the time domain solutions before $i_Y = 0$.

After $i_Y = 0$, the solution is

$$i_W(t) = \frac{-V_{DC} + e_W + e_X}{2R}\left(1 - e^{-\frac{R}{L}t}\right) + i_W(t = t_1)\, e^{-\frac{R}{L}t} \tag{23.41}$$

where t_1 is the initial time of the second mode, i.e., when $i_Y = 0$. The subtle difficulty in using these equations is the time, t, in the exponentials in each of these equations. At the start of each rise or fall of a particular phase current, the t of the exponential is $t = 0$.

The second method of solving the phase current differential equations utilizes numerical integration. A trapezoidal integration routine can be found in Reference 17. It consists of putting the first-order linear differential equation to be solved into a state space form of the type

$$\dot{x} = Ax + Bu \tag{23.42}$$

where

    x = the phase current vector
    A = the system matrix
    B = the input coupling matrix [18]
    u = the voltage vector

An example of putting an equation into this form with Equation 23.36 is as follows:

$$\dot{I}_W = \left[-\frac{R}{L}\right] I_W + \left[\frac{1}{3L}\right]\left[-2V_{DC} + e_Y - 2e_W + e_X\right] \tag{23.43}$$

where a capital I was used rather than a small i to avoid confusion with the dot denoting the derivative of i. This is easily extended to include all three phase currents. If small enough time steps are taken in the simulation integration of these equations, a continuous-time domain solution for Equation 23.42 is of the form:

$$x(t) = Mx(t - \Delta t) + N[u(t - \Delta t) + u(t)] \tag{23.44}$$

where

$$M = [\mathbf{I} - \Delta t\, \mathbf{A}]^{-1}\, [\mathbf{I} + \Delta t\, \mathbf{A}]$$

$$N = [\mathbf{I} - \Delta t\, \mathbf{A}]^{-1}\, (\Delta t)\, \mathbf{B}$$

from Reference 17, where
    $\mathbf{I}$ = the identity matrix
    Δt = the time step duration

In the BLDC motor and drive, which operates self-synchronously, the steps of the simulation include

1. Loading in initial conditions
2. Calculating rotor position and determining which switches should be on and which if any are commutating
3. Calculating or using a look-up table to obtain the back-EMF at the particular rotor position
4. Calculation of the currents
5. Calculation of the torque using

$$T_e = \frac{P}{\omega}(i_a e_a + i_b e_b + i_c e_c) \tag{23.45}$$

where
    T_e is the electromagnetic torque,
    P is the number of poles,
    ω is the excitation frequency,
and the current and voltage values are their instantaneous values. The current values are found from the numerical integration, while the voltage values are found from a time domain function or look-up table.

6. Calculation of speed using

$$J\dot{\omega} + B\omega + T_L + T_F = T_e \tag{23.46}$$

where
    J is the total system rotational inertia
    ω is the speed
    B is the damping constant of the motor
    T_L is the load torque
    T_F are other frictional torques present due to coupling
    T_e is the electromagnetic torque
This equation is also solved using a numerical integration.

7. Calculation of whatever else is required by the control and application of the control or increment a counter for the control if the control loop period is longer than the time step period
8. Repeat for next time step

Brushless DC Drives

23.10 SENSORLESS

Sensorless BLDC motor control refers to the operation of the BLDC without its usually required rotor position sensor. The study of sensorless control, and the use of sensorless methods, is found only infrequently in today's industrial applications; however, it is the method of BLDC control of the future.

Sensorless means fewer parts, i.e., the elimination of optical encoder, Hall-effect sensors, resolver, small-signal lines to the position transducer, and associated decoding circuitry. This means savings in manufacturing costs, which is of primary importance in today's industry, as well as increased reliability and durability. High-performance BLDC drives require highly accurate rotor position information in order to know when to commutate the phases to optimize performance, i.e., maintain an optimal torque per amp, optimal machine or drive efficiency, or maximum torque output. Many academic and industrial studies in the area of sensorless BLDC control in high-performance drives have utilized some type of microprocessor to implement computationally intensive algorithms to estimate rotor position and optimize one of these figures of merit. Usually this microprocessor is in the form of a microcontroller, digital signal processor (DSP), or DSP controller. Thus, the technical evolution of microprocessors is also of interest to the motion control engineer.

In the early 1980s, the first microcomputers were being used by the general public. The first were Apples, Radio Shack's TRS-80s, and other game-type systems that could be connected to a television. These ran at clock speeds in the low megahertz and had little memory. The technology and basic concepts of microprocessors was new then and difficult to fathom for those who were not intimately familiar with microprocessor technology. Writing and debugging even a short program could take hours or days, and recording data usually was done on magnetic tape rather than on hard or floppy disks. CD-ROM wasn't even thought of yet, except by an elite few. Today, software makes computers easy. The microcomputer industry continues to reduce the size and cost, while increasing the speed and capabilities of processors. The present state-of-the-art in microcomputers realizes clock speeds in the gigahertz range. Memory for computers and associated peripherals for computers are now inexpensive. Thus, the trends are smaller, faster, less expensive, and easier to use.

Utilizing microprocessors in our future sensorless BLDC control systems holds great promise. The one argument against using a microprocessor-based control for a sensorless BLDC drive is that the microprocessor and chip-set may be more expensive than the existing sensors and controller. The justification of the microprocessor-based system is in its use in larger, more complex, and perhaps multimotor systems. In these systems, the microprocessor's speed can be utilized to provide control to all of the various drives, relays, and other controls and switches, while also providing a means of analyzing, processing, and recording feedback information, monitoring interlocks, and providing an overall supervisory system control. Thus, the sensorless control of the BLDC motors in the system is just part of the control provided by the "master" (microprocessor-based) controller. The cost of the fastest microprocessor systems is still a problem; however, the cost of faster microprocessor-based systems continues to rapidly decline as technology advances.

REFERENCES

1. T.M. Jahns, "Motion Control with Permanent-Magnet AC Machines," *Proc. IEEE*, Vol. 82, pp. 1241–1252, August 1994.
2. T.M. Jahns, "Torque Production in Permanent-Magnet Synchronous Motor Drives with Rectangular Current Excitation," *IEEE Trans. on Industry Applications*, Vol. IA-20, pp. 803–813, July/August 1984.
3. C.K. Taft and T.J. Harned, "The Dynamic Characteristics of a Three Phase Brushless DC Motor," *Proc. of the 14th Annual Symposium on Incremental Motion Control Systems and Devices*, pp. 51–62, June 1985.
4. C.K. Taft and R. Gauthier, "Brushless Motor Torque Speed Curves," *Proc. of the 14th Annual Symposium on Incremental Motion Control Systems and Devices*, pp. 73–89, June 1985.
5. W. Leonhard, *Control of Electrical Drives*, New York: Springer-Verlag, 1996.
6. T.J.E. Miller and R. Rabinovici, "Back-EMF Waveforms and Core Losses in Brushless DC Motors," *IEEE Proc. on Electronic Power Applications*, Vol. 141, pp. 144–154, May 1994.
7. T.J.E. Miller, *Brushless Permanent-Magnet and Reluctance Motor Drives*, New York: Oxford University Press, 1989.
8. E. Richter, T.J.E. Miller, T. W. Neumann, and T. L. Hudson, "The Ferrite Permanent Magnet AC Motor — A Technical and Economical Assessment," *IEEE Trans. on Industry Applications*, Vol. IA-21, pp. 644–650, May/June 1985.
9. G.R. Slemon, "On the Design of High-Performance Surface-Mounted PM Motors," *IEEE Trans. on Industry Applications*, Vol. 30, pp. 134–140, January/February 1994.
10. E.K. Persson, "Brushless Low-Inertia Motors," *IEEE Conf. on Small Electric Machines*, pp. 19–22, March 1976.
11. S.A. Nasar, I. Boldea, and L.E. Unnewehr, *Permanent Magnet, Reluctance, and Self-Synchronous Motors*, Boca Raton, FL: CRC Press, 1993.
12. W. Radziwill, "Brushless DC Machines for Space Applications," *IEEE Conf. on Small Electric Machines*, pp. 23–26, March 1976.
13. P.C. Krause, O. Wasynczuk, and S.D. Sudhoff, *Analysis of Electric Machinery*, New York: IEEE, Inc., 1995.
14. Y. Sozer and D.A. Torrey, "Adaptive Torque Ripple Control of Permanent Magnet Brushless DC Motors," *Conf. Proc. of IEEE Applied Power Electronics Conf.*, pp. 86–92, 1998.
15. D.C. Hanselman, *Brushless Permanent-Magnet Motor Design*, New York: McGraw-Hill, Inc., 1994.
16. Z.J. Liu, D. Howe, P.H. Mellor, and M.K. Jenkins, "Coupled Thermal and Electromagnetic Analysis of a Permanent Magnet Brushless DC Servo Motor," *IEEE Sixth Int'l. Conf. on Electrical Machines and Drives*, pp. 631–635, 1993.
17. N. Mohan, T.M. Undeland, and W.P. Robbins, *Power Electronics — Converters, Applications, and Design, Second Edition*, New York: John Wiley & Sons, Inc., 1989.
18. M.S. Santini, A.R. Stubberud, and G.H. Hostetter, *Digital Control System Design*, New York: Saunders College Publishing, 1994.

24

Testing of Electric Motors and Controllers for Electric and Hybrid Electric Vehicles

Sung Chul Oh
Korea University of Technology and Education, Chungnam, South Korea

24.1 INTRODUCTION

Testing of electric motors and controllers for electric vehicles is essential for commercialization. Unfortunately, since commercialization of EV is somewhat delayed due to various reasons, setting up the standard test procedure for the motor and controller doesn't receive wide attention. But since various types of motors and controllers are adopted for the electric vehicle and hybrid, a test procedure that is unique compared with industrial application for the electric vehicle should be identified first. In this chapter, various standardization issues regarding the motor and controller are covered. Different approaches regarding standardization of components between car manufacturers and electric component manufacturers are compared.

    For motor/controller standards for electric vehicles, test methods for selected items are analyzed based on existing standards. Most related standards mainly focus on the minimum safety requirements, and detailed test methods are not covered. But based on test methods for the power electronics equipment, test items applicable to the electric vehicle use are proposed. Test conditions and test items for the combined motor and controller are proposed.

    Proposed test methods are reviewed by a corresponding industrial expert group. Based on this draft, several tests were carried out. Some practical aspects of testing procedure depending on different load absorbing system such as M-G set, eddy current type engine dynamometer, and AC dynamometer are analyzed. Finally, for the driving cycle test without installing a motor and controller in a vehicle, the concept of hardware in the loop (HIL) is introduced. Based on this concept, driving cycle test was performed.

Motor and controller performance in the vehicle environment can be checked without installing an actual motor and controller in vehicle. Some test results are shown in a later section.

24.2 CURRENT STATUS OF STANDARDIZATION OF ELECTRIC VEHICLES

24.2.1 ELECTRIC VEHICLES AND STANDARDIZATION [1]

The technologies involved in electric vehicles are moving very quickly, particularly in the field of batteries, power electronics, and drive systems. Generic standards to ensure safety of persons, to measure performances, and to ensure compatibility will continue to be developed as the technology advances.

With standardization of the electric road vehicle becoming an important issue, the question arises which body would be responsible for these standards. The problem is less straightforward than it looks: The electric vehicle, which introduces electric traction technology in a road vehicle environment, represents in fact mixed technology. On the one hand, the electric vehicle is a road vehicle, the standardization competence for which is the province of ISO. On the other hand, the electric vehicle is a piece of electrical equipment, the standardization competence for which falls under the wings of the International Electrotechnical Commission (IEC).

The difference is even more stressed by the constitution of the technical committee working groups in the two organization; in the International Organization for Standardization (ISO), there is a strong input from vehicle manufacturers, while in IEC many of the delegated experts are electricians. Furthermore, there is a fundamentally different approach taken toward the concept of standardization in the automotive and the electrotechnical world. There is a different "standard culture," the origin of which can be traced back in history.

In the car manufacturing world, standardization is not so widespread: Manufacturers desire to develop their own technical solutions, which in fact make their product unique. Standardization for road vehicles is limited to issues covered by regulation (safety, environment impact) and to areas where interchangeability of components is important. For components like combustion engines, for example, there are very few standards.

In the automotive industry, in fact, most manufacturers were responsible for the manufacturing of all components (e.g., the combustion engine) for a certain vehicle. This made the need for overall standardization much more stringent. Also, the individual customer is unlikely to require strict compliance to standards. Safety regulations, however, may be enforced by government.

In the electric world, there is a much longer tradition for standardization (the IEC was founded as early as 1906, when the electrotechnical industry was at its very beginning) and a stronger tendency to standardize all and everything; furthermore, standards are more looked upon as being a legally binding document. Electric motors are covered by extensive standards covering their construction and testing. Even the color code of wires is standardized.

In the electrotechnical industry, in fact, the role of specialist component manufacturers acting as suppliers to equipment manufacturers has always been more important. Furthermore, the customers of the electrotechnical industry are more likely to be powerful corporations (e.g., railway companies) that tend to enforce very strict specifications on the equipment they order or purchase; hence, the need for more elaborate standards to ensure the compliance of the equipment.

24.2.2 STANDARDIZATION BODIES ACTIVE IN THE FIELD

The organizations involved with electric vehicle standardization world-wide include the following:

24.2.2.1 The International Electrotechnical Commission

Inside IEC, the Technical Committee 69 (TC69) is dedicated to electric vehicles. IEC TC 69 (Electric Road Vehicles and Electric Industrial Trucks) was established in the 1970s, at a time when environmental concerns, potential oil shortages, and available technology made the prospect of electric vehicles attractive. After a slow start, a number of standards and reports were published in the 1980s dealing with chargers wiring and electric drive systems. More recently, the increasing awareness of environmental matters, coupled with direct legislation to promote zero emission vehicles, has prompted direct involvement at all levels by the vehicle industry.

The following working groups are active inside TC 69:

WG2: Motors and Controllers
WG3: Batteries
WG4: Infrastructure

All these working groups are constituted of experts who are designated by their national committees.

24.2.2.2 The International Organization for Standardization

Inside ISO, Technical Committee 22 is responsible for road vehicles; its Sub-Committee 21 (ISO TC22 SC21) is dedicated to electric road vehicles. The following working groups are active:

WG 1: Vehicle operation conditions, vehicle safety, and energy storage installation
WG 2: Terminology: Definitions and methods of measurement of vehicle performance and of energy consumption. The committee is constituted also of experts designated by national standardization organizations.

24.2.2.3 Other Regional Organizations

In the European Union, harmonized standards on the European level are being developed by:

- CEN, the European Commission for Standardization. Inside CEN, TC 301 is responsible for electric road vehicles. Its active working groups are the following:
  - WG1: Measurement of performances
  - WG4: Liaison and dialogue between vehicle and charging station
  - WG5: Safety: Other aspects
- CENELEC, the European Commission for Standardization in the field of electrotechnics.

Inside CENELEC, TC 69X is responsible for electric vehicles.

Its working groups are the following:

WG1: Charging design and operation
WG2: Charging environmental aspect
WG3: Safety

At the national level, activities are performed by the national standardization or electrotechnical committees, for example:

- In the U.S., the Society of Automotive Engineers (SAE) has issued a number of technical documents concerning electric vehicles.
- In Japan, the Japanese Electric Vehicle Association (JEVA) is concerned with electric vehicle standardization.

24.2.3 STANDARDIZATION OF VEHICLE COMPONENTS

In the field of standardization of traction components like motors and controllers, the clear opposition between the electrotechnical and the automotive industries mentioned above is obvious. The division of standardization labor between the two main concerned bodies, IEC and ISO, is the central issue in this field.

Inside IEC TC 69, this subject is the responsibility of Working Group 2. This WG was active in the 1980s, producing several documents regarding cables, instrumentation, motors, and controllers. It has been dormant for a number of years, but is now revitalized. Its first commitment is the preparation of a standard considering electric vehicle motors and controllers. This document merges and revises the documents IEC 785 and IEC 786.

The rapid evolution of electric vehicle technology makes it desirable to consider a revision, taking into account the latest developments. The concept of integrated electric drives makes a merge of the motor and controller parts of the document desirable. The "new work item proposal" regarding this document has been approved; however, there was a strong opposition from a number of car manufacturers that, according to their traditions, deemed this standardization work unnecessary.

The new document should be an answer, however, to the needs of ISO, which has requested clear standards including characteristics, specifications, and testing procedures for electric vehicle components. Activities on this document will resume with the outcome of the IEC/ISO agreement.

24.2.4 STANDARDIZATION ACTIVITIES IN JAPAN [2]

In 1987, Japan established Japanese Electric Vehicle Standard (JEVS) Z103-87, concerning the range test of EVs. Test conditions of this JEVS had differed from that of actual vehicle usage in driving conditions, and the value of its deviation has a tendency to increase due to the improvement of road and vehicle. The range test has been modified corresponding to ISO/CD 8714 (Reference Energy Consumption) and the change in the Japanese Industrial Standard (JIS) on ICEVs. Namely, the range test at cruising speed has been deleted to prevent users from being confused by two results obtained by two different test procedures. In the modified range test, range is evaluated by only the 10.15-mode driving cycle test (10-mode is allowed for testing those vehicles designed for town use only).

ISO/CD8715 (Vehicle Specification) and ISO/CD8714 (Reference Energy Consumption) are to be conducted in sequence. Discussions on energy consumption, climbing ability, maximum speed, and acceleration test came up, referring to the ISO/CDs mentioned above, but the resultant JEVSs are slightly different from the ISO/CDs in content. Due to Japanese regulations, the Japanese must choose the 10.15-mode driving cycle instead of the ECE R15-mode.

Test procedures according to a scheduled sequence are not included in these JEVSs, so that each test can be conducted independently. This decision was made to avoid useless repetition of tests when part of a sequential test fails. In connection with this decision, the general condition of battery SOC is assessed by 15 minutes' driving at a speed of 80% of the 30-minute maximum speed. Estimation by calculation and tests using a chassis dynamometer is allowed for the maximum climbing ability test.

The Z107-88 combined test of motor and controller enables a motor and controller combination test to be conducted by simulating the operating conditions similar to the on-board motor and controller. The results are:

- Output power and efficiency
- Operating temperature
- Operation of interlock function for safety
- Operation of regenerative brake and its efficiency

24.2.4.1   Z108-1994: Measurement of Range and Energy Consumption
(at Charger Input)

Energy consumption is defined in the following equation:

Consumption (kWh/km) = (Recharged energy after range test)/(Range covered)

24.2.4.2   Z109-1995: Acceleration Measurement Test

The test shall be conducted after 15 minutes running at a speed 80% of the 30-minute maximum speed (battery pre-conditioning).

24.2.4.3   Z110-1995: Test Method for Maximum Cruising Speed

Test procedure to measure the 30-minute maximum speed and practical maximum speed (1 km maximum speed) as defined by the ISO standard.

24.2.4.4   Z111-1995: Measurement for Reference Energy Consumption
(at Battery Output)

Energy and distance data is obtained in five test cycles following two cycles for warming up. Z108 will be replaced with this standard if capacity estimation of the battery by dynamic discharge is put into practical use.

24.2.4.5   Z901-1995: Electric Vehicle: Standard Form of Specifications
(Form of Main Specifications)

This defines items and formats of EV specification. This defines two different levels (mandatory specification format and optional specification format).

24.2.4.6   Z112-1996: Electric Vehicle: Standard Measurement of Hill
Climbing Ability

The existing climbing test has been modified to harmonize with ISO/DIS 8715. (This standard allows the results to be estimated by calculation.)

24.2.4.7   E701-1994: Combined Power Measurement of Motor and Controller

This describes the basic power characteristics test procedure necessary for motor/controller designers to obtain data for vehicle designers. Although industrial motors are designed for long-time operation at steady load conditions, EV motors can cope with a wide range of loads. So, E701 and E702 determine the maximum short-duration power and the maximum power available for more than 1 hour. The time for short-duration power can be chosen in pre-determined values (30 seconds; 1 minute; and 3, 5, 10, and 15 minutes), referring to the condition in which the EV is used. In every measurement, the load should be set so that the temperatures of the motor and controller reach their equilibrium temperature not exceeding their permissible temperatures.

24.2.4.8   E702-1994: Power Measurement of Motors Equivalent
             to On-Board Application

This standard defines the torque-speed characteristics of a motor as a part of the vehicle characteristics. This test procedure is necessary for vehicle designers to evaluate vehicle characteristics (for users).

24.2.4.9   Japanese Standards Concerning Vehicle Performance
             and Energy Economy

Concerning the performance of EVs, such as maximum cruising speed, acceleration, and hill climbing ability, sufficient result data with enough repeatability can be obtained if proper battery conditions can be obtained. However, the measurement of energy economy is affected by various factors, such as by the driving conditions, driving schedule, and vehicle conditions, or by the recharging procedure such as the stopping condition of recharging and the temperature of the batteries. To discuss in detail the energy economy test procedure, following existing JEVSs are reviewed by the corresponding sub-committee.

> Z108-1994 Measurement for range and energy consumption (at charger input)
> Consumption (kWh/km) = (Recharged energy after range test)/(Range covered)
> Z111-1995 Measurement for reference energy consumption (at battery output)
> Energy and distance data is obtained by five test cycles following two cycles for
>   warming up.
> Standardization

24.3 TEST PROCEDURE USING M-G SET [3]

24.3.1 ELECTRIC MOTOR

Due to the rapid development of power switching devices, control of electric motors evolves from constant-speed application to variable-speed application. A conventional DC motor is replaced by an induction motor and permanent magnet motor with advanced controller. Standardization on base speed and maximum speed is needed. Some suggested items such as motor output and speed are excerpted from the standard for inverter motors. Many items should be modified to consider control characteristic of electric vehicles. For cooling, because the motor is operating in a relatively low-speed region, force cooling is recommended. Most cases, a speed detector such as an encoder would be installed as a standard accessory. If a motor with the same design is produced, the motor equivalent circuit parameter should be controlled to get the same performance. Use of a power supply that can provide maximum output would be recommended. Cooling should simulate actual driving conditions, but that is quite difficult.

24.3.2 CONTROLLER

Minimum safety requirements for the user are proposed. Minimum requirement value will be updated based on advice of experts. Especially, limitation values for EMI/EMC will be modified.

Test method for the electric performance of the controller is suggested. Electrical items can be measured with this standard. A combined test procedure for motor and controller is suggested. The definition of short time rating and cycle test should be determined.

The following opinions from experts are received.

1. More detailed definition is needed for the coverage of suggested standard (system controller and motor controller).
2. Precise definition of terms used in standard should be established.
3. Setting of minimum requirements should be careful. Expert group should review it.
4. Combined test would be more focused.
5. For harmonic test, more frequency bands should be included.
6. In some items, maximum and minimum values should be limited at the same time.
7. More detailed test methods for output and temperature tests should be proposed.

After discussion, further standard activity will focus on development of combined test for motor and controller.

24.3.3 APPLICATION OF TEST PROCEDURE

A performance test for the electric motor and controller is carried out based on the proposed draft. A general test for the motor such as size, weight, wire resistance, withstand/insulation voltage and temperature test, speed-torque test, noise test, and overspeed test is carried out. Since test procedures for noise and overspeed are not established yet, more research is needed. A dedicated controller is needed for the motor speed-torque test and temperature test. Load consists of motor-generator set. DC generator is driven by test motor, and resistance load connected to generator is controlled to get required power. During temperature test, main focus was how the temperature at various points of motor and controller will saturate within recommended value. For the output power test, main focus was checking variation of output due to load variation.

24.3.4 ANALYSIS OF TEST ITEMS FOR THE TYPE TEST

The most important test items for the motor and controller are temperature rise and output power test. Especially for the temperature rise test, selection of measurement point and variation of load (continuous, pattern, and overload) are necessary. In this section, the test procedure for the type test of sample motor controller is analyzed.

24.3.4.1 Motor Test

M-G set was used as a load absorbing for the motor test. Load variation was done by changing the combination of resistor bank. Power analyzer and strain gauge are used for the measurement. First continuous output power test was performed. At 100% rated power, time and temperature rise until temperature of motor wire saturated, and efficiency are measured.

100% Rated Load Test. When load is varied, characteristics at rated power at each motor speed (1500, 3000, 4500, 6000, 7500 rpm) were measured. Following items were measured: motor voltage, motor current, motor output power, motor power factor, and motor efficiency.

Motor Test at 200% Output Power. Same items as in 100% ouput power test were measured.

Temperature Rise Test. Temperature rise was measured while motor was driven by controller. Main purpose of temperature rise test was to check if motor temperature reached to temperature limited by its insulation level and to measure the elapsed time when the motor temperature reaches safety temperature. When motor is operated with inverter, temperature rise is higher than operation with conventional AC, because more harmonics are involved. Measurement points would be ambient temperature, motor frame, motor wire end ring, motor bearing, and inverter heat sink. Mainly, motor winding temperature should be measured.

24.3.4.2 Controller Test (Controller Only)

1. Switch spike voltage
2. Temperature rise test

Measurement points would be heat sink, SPS, transformer, DC link capacitor, internal temperature, and ambient temperature. Test condition would be no load, rated ouput power, duty cycle, and fan lock operation.

24.4 TEST PROCEDURE USING EDDY CURRENT-TYPE ENGINE DYNAMOMETER

24.4.1 Test Strategy

When M-G set was used as a load system for the output power and temperature test, we experienced the following problems.

1. Continuous load control is difficult.
2. Because resistor bank was used to control load, active load control was impossible.
3. Since engineering sample of motor/controller was used, selection of test item is limited.

The following strategy was set up for the test.

1. Test motor and controller will be commercially available product that is used in KATECH EV development.
2. Engine dynamometer will be used to control load actively.
3. Lead-acid battery pack will be used as a power supply for the motor/controller test.

Table 24.1 Comparison of Test Items [1]

Test Item	M-G set	Engine Dyno
Output power test	Ratio of rated power test	Ratio of rated power (Torque, Speed varied)
Temperature rise test	Continuous rated power	Constant speed
		Continuous rated power
		Maximum power
		Duty cycle operation
Electrical performance test		Output current harmonic
Regenerative braking test		Load variation

24.4.2 Test Procedure

A comparison of test items with M-G set and engine dynamometer is shown in Table 24.1. The test procedure was as follows:

1. Output power test: Rated continuous power (18 kW) is set as 100%, when output powers are 25, 50, 75, 100, 150, and 200%; speeds are 1250, 2500, 3750, and 5000 rpm; battery voltage, current, controller output current, and voltage and motor speed and torque are measured. Based on measurement data, overall efficiency at each operating point is calculated.
2. Temperature rise test: Temperature at each measuring point on motor and controller is measured at each operating point. Most controllers have overtemperature trip function, which in this case measures the elapsed time until controller is tripped.
   a. Constant speed operation: Based on engine test item, 40 km/h constant speed is simulated (throttle opening: 56%, torque: 30 Nm, motor speed: 1500 rpm; output power: 4.7 kW). Temperature rise at each point is measured.
   b. Maximum output power operation: Due to limitation of output torque of engine dynamo, motor is operated at 28.2 kW (156%) and elapsed controller trip time is measured.
   c. Continuous rated power operation: At rated power (rated torque and speed) of 18 kW temperature rise was measured.
   d. Duty cycle operation: To simulate vehicle operation, load is varied to 156% in 30 sec, 100% in 1 min, 0% in 30 sec, and the whole cycle is repeated 10 times and temperature rise was measured.
   e. Electrical performance test: Harmonic analysis is done on motor current.
   f. Regenerative braking: While motor is operating with the accelerator opening of 25, 50, 75, and 100%, accelerator pedal is suddenly released and charging current flowing into battery is measured.

24.4.3 Discussion on Test Procedure

1. As a main power supply, a lead-acid battery (60 Ah × 20 each) was used. But even after full charging, the battery pack couldn't supply sufficient power for

maximum power and continuous power test. Use of large DC power supply is recommended.
2. Output power test and temperature rise test are the core of the motor/controller test, if the motor is not installed in vehicle. Especially during the temperature rise test, the cooling method should simulate actual driving conditions. During the test, a cooling fan is used to cool down the radiator. Detailed cooling condition should be specified in the test report.
3. Data acquisition for electric parameter such as voltage and current should be upgraded.
4. Standardization on test condition of constant speed test and duty cycle test is necessary. For the temperature rise test, test method in case of abnormal cooling should be specified.
5. Standardization on testing condition of regenerative braking test (opening of accelerator, percentage load, and braking time) is necessary.

24.5 TEST PROCEDURE USING AC DYNAMOMETER [4]

The standard reviewed for adoption as a Korean Standard (KS) is shown in Table 24.2. The test was carried out according to the standard. Adoption as Korean Standard is still pending.

24.5.1 TEST STRATEGY

When an eddy current-type engine dynamometer was used as a load absorbing system, the following problems were encountered:

1. Active load control was impossible.
2. Detailed test method for regenerative braking was not established.
3. Automatic data acquisition system was not established.

The following test strategy is suggested:

1. Commercially available motor and controller will be used as test motor and controller.
2. Vector-controlled induction motor drive will be used as a dynamometer for the active load control
3. During powering and regenerating, power supply should either supply or absorb power to and from motor.

Block diagram of test setup is shown in Figure 24.1.

Table 24.2 List of Proposed Standards

No.	Title
G7-MC01	Test procedure for electric motor for electric vehicle
G7-MC02	General test procedure for controller for electric vehicle
G7-MC03	Combined test for motor and controller for electric vehicle

Testing of Electric Motors and Controllers for Electric and Hybrid Electric Vehicles 547

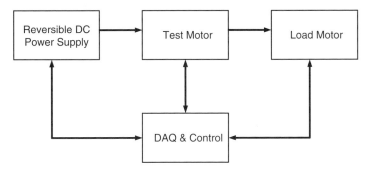

Figure 24.1 Block diagram of test setup.

Table 24.3 Comparison of Test Items [2]

Test Item	M-G Set	Engine Dyno	AC Dyno
Output power test	Ratio of rated power Continuous rated power	Torque, speed varied Constant speed Continuous rated speed	Torque, speed varied
Temperature rise test		Maximum power Duty cycle	Continuous rated power Maximum power Duty cycle
Electrical performance test		Harmonic analysis on motor current Load change test	
Regenerative braking			Continuous four-quadrant operation

24.5.2 TEST ITEMS

A comparison of the test item with a different load absorbing system is shown in Table 24.3. The definition of four-quadrant operation and motor/load mode is shown in Table 24.4.

24.5.3 TEST PROCEDURE

1. Maximum torque: Dynamometer speed is kept constant and increase torque command until measured torque saturates. Saturated torque is defined as maximum torque.
2. Maximum speed: Increase dynamometer speed until measured torque reaches 0. That speed is defined as maximum speed.
3. Output power DC input voltage and current and motor speed and torque at each operating point and efficiency are calculated.

Table 24.4 Definition of Four-Quadrant Operation and Motor/Load Mode

Torque: POSITIVE Rotation: REVERSE Motor mode: REGEN : 2nd Quad Motor direction: REVERSE Load direction: REVERSE	Torque: POSITIVE Rotation: FORWARD Motor mode: THROTTLE : 1st Quad Motor direction: FORWARD Load direction: FORWARD
Torque: NEGATIVE Rotation: REVERSE Motor mode: THROTTLE : 3rd Quad Motor direction: REVERSE Load direction: REVERSE	Torque: NEGATIVE Rotation: FORWARD Motor mode: REGEN : 4th Quad Motor direction: FORWARD Load direction: FORWARD

4. Temperature rise test: At given operating point, temperature at each measuring point of motor controller is measured.
5. Regenerative braking: System efficiency at each four-quadrant operation (positive/negative torque, positive/negative speed) is measured.

24.6 TESTING OF ELECTRIC MOTOR/CONTROLLER IN VEHICLE ENVIRONMENT

24.6.1 CONCEPT OF HARDWARE IN THE LOOP

Hardware in the loop (HIL) requires dynamic models that simulate the remainder of the system under evaluation. The system must include all significant interactions that exist within a particular HEV. The HIL simulation monitors the actual hardware measurements to drive the simulation and commands the dynamometer to respond based upon the rest of the simulated system. Outputs from the tested system typically include sensor information (measurements). Input signals to the test system include the commands from an operator.

The HIL concept can be applied to a wide variety of vehicle system studies. For instance, in order to test a drive train in its operational environment, by using the HIL concept, we do not have to develop a drive train to go into a vehicle; we can do it in a test setup with more instrumentation.

24.6.2 HIL APPLICATION TO MOTOR/CONTROLLER

Test setup was developed to study the motor with gap control using HIL. Post-transmission HEV with four-axial flux motors shown in Figure 24.2 was used for the HIL control. In addition, the block diagram of the test setup is shown in Figure 24.3. The induction motor was controlled in speed mode based on the vehicle simulator. Since post-transmission HEV is simulated, induction (load) motor speed is proportional to the vehicle speed, which does not change quickly. Many control strategy parameters can be altered with HIL results. Vehicle performance is simulated with the vehicle simulator. Actual axial flux motor and induction motor are controlled based on the vehicle simulation results. Interface between the vehicle simulator and hardware is realized by dSPACE real-time processor.

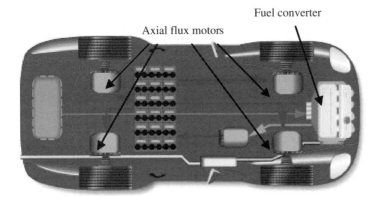

Figure 24.2 Series HEV with four axial flux motors.

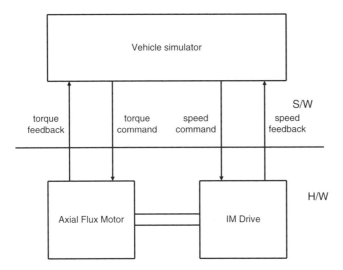

Figure 24.3 Block diagram of the test setup.

In the vehicle simulator, vehicle parameters such as weight, drag resistance, and wheel radius are stored. In real applications, driver behavior such as applying accelerator or brake and gearshift should be considered in a vehicle simulator. However, in this study, only the vehicle speed regulator is used to control the vehicle. The vehicle simulator consists of a PI speed controller. Reference speed from the specified driving cycle is compared with the measured speed. The difference goes through the PI controller, and torque command is generated. Torque command inputs to the axial motor controller. In this case, the test motor simulates one wheel of the four-wheel-drive vehicle. Based on the measured torque, vehicle acceleration and speed are calculated in the vehicle simulator. Furthermore, the vehicle simulator controls the induction motor based on the calculated speed. The induction motor is controlled in speed mode, and the axial flux motor is operated in torque mode. The induction motor simulates the actual speed of the vehicle. Performance of the test motor during the drive cycle can be measured and compared with the simulation results in the HIL setup.

24.6.3 Test Description

The vehicle simulator allows simultaneous control of the axial flux motor and the simulated vehicle. It is then possible to perform the vehicle drive cycle test. The induction motor drive was used to propel the simulated vehicle on a Japan 10 drive cycle. The test was performed with a variable gap operation. Airgap was controlled by the servomotor installed on the lead screw on the shaft of the axial flux motor.

24.6.4 Test Results

The vehicle powered by four axial flux motor wheels successfully followed a Japan 10 drive cycle. For the experiment, only one motor is needed to simulate the vehicle performance. To the vehicle simulator, 4 times the measured torque was fed back. Figure 24.4 shows how the vehicle speed follows a Japan 10 drive cycle.

In addition, for the quantitative analysis, motor characteristics such as torque, speed, input voltage, and current are measured. Figure 24.5 shows the motor characteristics. Figure 24.6 shows the axial flux motor torque command and the response during the same Japan 10 drive cycle test.

Measured axial flux motor torque shows vibration, which accompanies acoustic noise during the test acceleration phases at low speed. The vehicle inertia fortunately smoothed a lot of the speed ripples on the motor shaft. The noise was identified as a result of the vibration at low speed and high motor torque regions. Figure 24.7 shows the airgap command and measured airgap. One of the advantages of using HIL is that both simulation results from the vehicle simulator and actual hardware measurements can be compared at the same time.

Based on the measurement, overall motor efficiency during driving mode is analyzed. Based on the measured results, overall motor efficiency can be calculated. Overall efficiency is defined as follows.

$$\text{Overall efficiency} = \frac{\int p_{out}.dt}{\int P_{in}.dt} \qquad (24.1)$$

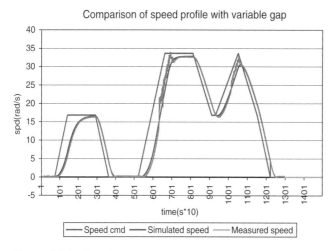

Figure 24.4 Speed command and measured speed.

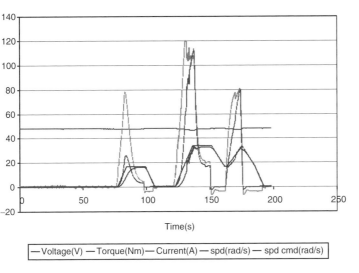

Figure 24.5 Motor characteristics.

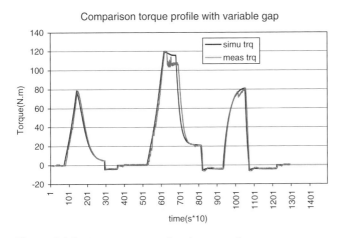

Figure 24.6 Torque command and measured torque.

Overall efficiencies during Japan 10 drive cycle and Japan 15 drive cycle are shown in Table 24.5.

As shown in Table 24.5, simulation and experimental results are quite matched. During the high-speed driving cycle such as Japan 15 drive cycle, motor has better efficiency. Power consumption by the gap actuator is not considered in the simulation. There would be some arguments about the definition of the overall efficiency. Since motor has different characteristics during motoring and regeneration modes of operation, in some references, motor efficiency is defined in motoring and regeneration modes separately. In this case, overall efficiency can be defined as weighted average of the motoring and regenerating efficiencies.

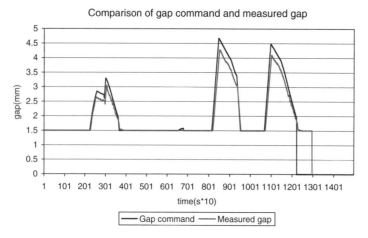

Figure 24.7 Gap command and measured gap.

Table 24.5 Comparison of Overall Efficiency

Cycle	Constant Gap		Variable Gap	
	Simulation	Experiment	Simulation	Experiment
Japan 10	61.6%	60.3%	60.9%	60.9%
Japan 15	77.9%	77.4%	77.1%	77.2%

24.7 CONCLUSION

In this chapter, many aspects of testing of motor and controller for the electric vehicle were reviewed. For the past 4 years, test procedures for the motor and controller for electric vehicles have been developed. Many types of motors and controllers were tested and test results were used to update test setup and test procedure. Proposed test procedure will be adopted as a Korean Standard. It will contribute the commercialization of EVs.

The current status of standardization was also reviewed. Many practical problems applying existing standards to the testing of motors and controllers were investigated, and finally a test procedure for the driving cycle with HIL is proposed.

Suitability of axial flux motors for vehicular propulsion systems has been investigated. Axial flux motor characteristics during drive cycles have been simulated with a vehicle simulator. Motor model for the simulation was developed based on the efficiency map of the motor in different operating conditions. Actual motor characteristics of the variable airgap axial flux motor during different drive cycles have also been measured through the hardware-in-the-loop concept. Motor speed follows the speed command during the variable gap operation. Feasibility of using variable gap operation has been verified as well. Extended speed operating range and improved efficiency have been presented as the advantages of the variable gap operation. Other types of HEV drive train configurations can be tested with HIL without changing the hardware setup. Only software modifications are needed to test other configurations. Other components such as battery and new control strategies can also be tested.

REFERENCES

1. P. Van den Bossche. The need for standards in an emerging market — The case of the electric vehicle. *Proceedings of Electric Vehicle Symposium,* Los Angeles, CA, 1997.
2. K. Shimizu. Status of Japanese EV Standardization Activities in '97. *Proceedings of Electric Vehicle Symposium*, Los Angeles, CA, 1997.
3. G.B. Chung and S.C. Oh. Standardization activity on EV motor, controller and charger in Korea. *Proceedings of Electric Vehicle Symposium.* Pusan, 2002.
4. S.C. Oh, J. Kern, T. Bohn, A. Rousseau, and M. Pasquier. Axial flux variable gap motor: application in vehicle systems, SAE World Congress and Exhibition, Paper No. 2002-01-1088, Detroit, MI, 2002.

Part V
Other Automotive Applications

25

Integrated Starter Alternator

William Cai
Remy International, Inc., Anderson, Indiana

An integrated starter generator (ISG) is an electric subsystem in which the functions of the starting engine and the generating electric power are performed by one electric machine on-board the vehicle, instead of two separated electric machines in a traditional automotive vehicle. Since AC generators combined with rectifiers/bridges are used to produce electric power, charging batteries and running on-board electric loads in all modern vehicles, the generator on-board the vehicle is generally called as alternator. The ISG is accordingly named integrated starter alternator (ISA). The ISA can be connected with gasoline or diesel engines either through the crankshaft directly or the belt system indirectly. The ISA is sometimes called belt-driven alternator starter, or BAS (order change of starter and alternator in the phrase is for simplifying pronunciation only) if it is connected with the engine shaft through a belt system like the existing alternator. In crank shaft-driven ISA, the rotor of an ISA machine is sometimes served as the flywheel, providing additional mechanical damping function, so that the system can be called integrated starter alternator damper (ISAD).

The starter and generator are two key components in the electric energy system on-board vehicles. The function of a starter is to crank engine. That is, the engine is fired by the ignition system when the starter drives the engine to certain speed, such as 120 to 400 rpm. Until recently, most starters were either series DC motor or permanent magnet (PM) DC motor with or without soft magnetic shunt. However, the function of a generator is to convert mechanical energy to electric energy for charging the battery and running on-board electric loads, such as lights, electric heater, and electric motors. The generator has experienced an evolution from DC generator to AC alternator because the commutation system, especially the brushes, limited the life of a DC generating system; the invention of electronic rectifier/bridge and regulating devices enables the AC generator to provide low fluctuated voltage. The early preliminary concept of the ISG can be dated back to the 1930s after the starter was invented in 1911. Automotive engineers noticed the reciprocal

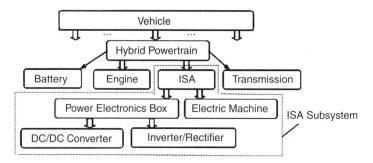

Figure 25.1 ISA subsystem in vehicle system.

principle of electromechanical energy conversion and tried to use one DC electric machine to crank the engine and generate electric power. However, the ISG egg was not hatched to practical ISG chicken because of the difference of specifications between starter motor and generator and the complexity of the system control at the ages without high-developed power electronics.

Besides the function of cranking engine and generating electric power, the ISA system also provides similar functions to hybrid electric vehicles (HEVs), such as driving/launching assistance to engine during vehicle acceleration, braking regeneration during deceleration of the vehicle, and so on. Obviously, these functions are far beyond what the name of ISA suggests. Therefore, ISA vehicles are sometimes called mild hybrid electric vehicles, or HEV-mild for short.

25.1 ISA SUBSYSTEM IN VEHICLE SYSTEMS

Generally, a vehicle has eight vehicle-level systems; the powertrain system is one of them. A traditional powertrain system is composed of gasoline/diesel engine, energy storage battery/batteries, mechanical transmission, powertrain control, and so on. If an ISA system is added to the powertrain-level subsystem, a traditional powertrain system becomes a hybrid powertrain system. The ISA electric drive subsystem consists of an electric machine and power electronics box. The power electronics box is composed of an inverter/rectifier and a bidirectional DC-DC converter. The ISA subsystem is outlined within the dotted line loop shown in Figure 25.1.

25.2 POWERTRAIN COUPLING ARCHITECTURE

One end of a typical internal combustion (IC) engine, either diesel or gasoline engine, is connected to the transmission through the engine shaft while the other end of the engine is connected to a pulley assembly through a belt, shown in Figure 25.2. The crankshaft of the engine with the help of a belt tensioner drives power steering, compressor of the air conditioner, coolant/water pump, and alternator in a conventional pulley system. How to connect the ISA machine to the conventional engine and where the ISA machine is to be mounted become critical issues in the mild hybrid electric system. Different powertrain coupling configurations are described and the related discussions are addressed in the following paragraphs.

Integrated Starter Alternator

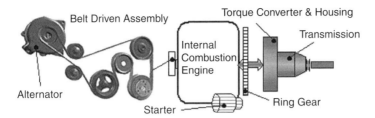

Figure 25.2 Typical hardware configuration in a conventional powertrain.

Compared to the powertrain system in a fully hybrid electric vehicle, the requirements for reconfiguration of the powertrain hardware are compromised in the mild hybrid electric vehicles. That is, an ISA subsystem can be added to a traditional powertrain system with minimum or minor modifications to its hardware configuration, because the system does not require the ISA to perform fully driving function continuously, but launching assistance temporarily only. The ISA machine can be mounted on the engine crankshaft at the transmission end of the engine in direct drive configuration. The shaft of the ISA machine can also be connected to the engine shaft through belt, gear, or chain at either belt pulley end or the transmission end of the engine in the indirect drive configuration, in which the ISA machine is normally mounted on the engine block and offset from the centerline of the engine crankshaft. In general, the vehicle mounted with the ISA subsystem belongs to the category of the parallel hybrid electric vehicles-mild. See Chapter 2 and Chapter 3 for the definition of parallel and series hybrid systems.

25.2.1 Crankshaft-Mounted ISA Configuration

In the conventional powertrain configuration shown in Figure 25.2, a starter offsetting the engine centerline is mounted on the engine block at the end facing the transmission, while an alternator is located at the belt assembly end of the engine block. During cranking time, the starter drives the IC engine shaft through a ring gear and cranks the engine at cranking speed. Once the engine is cranked, the alternator is driven by the engine shaft through the belt. Obviously the function of either starter or alternator is performed through the engine shaft.

If a new electric machine is mounted directly on the crankshaft, the conventional alternator and starter as well as the mass damper–flywheel including the ring gear can be eliminated and their functions can be replaced by the new electric machine. This is the concept of the initial crankshaft-mounted integrated starter alternator. A typical architecture of the crankshaft-mounted ISA subsystem is shown in Figure 25.3. The ISA machine is mounted between IC engine block and transmission. In order to minimize the modification to the mechanical components of the powertrain and drivetrain, the room occupied by the ISA machine is made available through assembling the downsized torque converter inside the rotor hub. To ensure the initial cranking of the engine in the winter season, a super auto-capacitor might be required in parallel with 36 V battery, although it is not given in Figure 25.3. The relative locations of the ISA machine and drivetrain components [1] are shown in Figure 25.4.

An obvious advantage is that the IC engine and the ISA machine are parallel and are connected to the drivetrain directly. Besides the common functions of the ISA, such as cranking engine, generating electric power, and stop/go at urban operation, the crankshaft-mounted ISA can smooth the engine vibration and driveline shake [2]; it can also act as an engine assistance to improve acceleration and up-hill climbing ability, and even

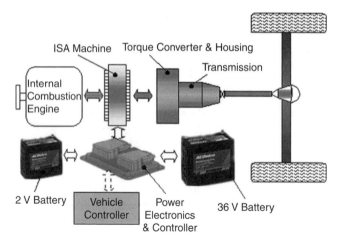

Figure 25.3 Configuration of the crankshaft-mounted ISA.

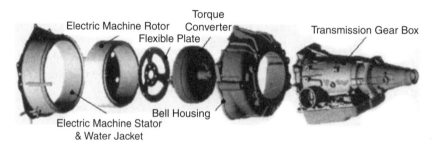

Figure 25.4 The location of the ISA machine.

drive the vehicle independently like the motor in the fully hybrid electric vehicle. The ISA machine can also operate at generating state to power on-board and off-board three-phase or single-phase AC loads if an additional DC-AC inverter and plugs are added to the system. In heavy-duty vehicle applications, the ISA machine can be disengaged from the powerful IC engine and driveline by adding clutches at the IC engine end and transmission end of the ISA machine, and driven by an additional small engine. In the case, the ISA machine and the small engine perform as an auxiliary power unit (APU).

Due to the small available space between the IC engine and the transmission, a shorter and thicker ISA machine is required for packaging reasons. The system configuration like this may prevent from adding cooling fans at either side of the electric machine due to insufficient cooling surfaces and high percentage of the used/heated air recirculation. Liquid cooling is a favorite style although forced duct air-cooling might be an option for the crankshaft-mounted ISA subsystem. The details of thermal issues will be addressed later in this chapter.

25.2.2 Offset-Mounted ISA Configuration

Three mechanical drives with different offset package options are compared. They are gear drive at the transmission side, chain drive at the transmission side, and belt drive at the accessory side of the IC engine [3].

Integrated Starter Alternator

Gear drive. The ISA machine is mounted at the location of the conventional starter. The gear of the ISA machine is engaged with an engine shaft gear, which is similar to the relationship between starter pinion gear and ring gear. The gear ratio can be fixed at 2.5–4:1 for the single stage gear drive. A challenging issue is that the pitch-line velocity increases to 4–8 m/s when the engine is running at near redline speed, such as 6000 rpm, compared to the velocity of 0.5 m/s for common commercial gears. Therefore, the much higher material hardness of the steel used for both pinion and ring gears is required in the ISA system with the single gear drive, due to the higher pitch-line velocity, which increases the high material and machining cost. A two-stage gear drive can solve the problem. However, the package issue may cause the requirements for relocating/modifying the transmission or the engine case.

Chain drive. The ISA machine can be located nearby the conventional starter position, while the chain drive might be located inside the transmission case. To extend the wear life of the chain drive, forced oil lubrication rather than the conventional bath lubrication is required at engine speeds over 1500 rpm, which requires redesigning the existing lubricating system. With this drive, it becomes possible to add active engine torque smooth function. Compared to the belt drive, the chain width can be reduced without compromising fatigue life so that the system will be more compact. However, adding chain drive in the transmission case may require redesigning it besides the limited space issue there. The loud noise, except for silent chains, is a disadvantage for the chain drive in the offset-mounted ISA system.

Belt drive. The ISA machine is mounted, offsetting the crankshaft, at the accessory side of the engine block. It can be either driven by a separate belt or included in the existing belt system. The conventional starter may still be required for initial cold engine cranking due to the limitation of the belt tension, wearing issues, and power rating.

A wider belt is required only for the ISA machine, and the rest can be kept with the original narrow belt unchanged if the separate belt drive is introduced to the system. However, an extension of the engine shaft length is required to accommodate the additional belt, which needs more space and adds extra load to the crankshaft bearing due to longer shaft overhang.

Adding the ISA machine to the existing belt system is the configuration used by most vehicle manufacturers. To accommodate the increased load due to introducing the ISA machine, the belt width and pulley width of all accessories has to be increased, besides the addition of an idler and so on. The difference between a traditional belt drive system and a new belt-driven ISA system is shown in Figure 25.5. In the conventional belt system, the engine crankshaft drives an alternator, power steering pump, air conditioner compressor, water pump, and other accessories (not shown in Figure 25.5). In the belt-driven ISA

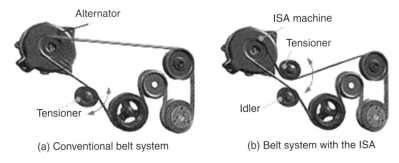

Figure 25.5 Belt system with the ISA [4] vs. conventional belt system.

system, the ISA machine is located at the place of the alternator of a conventional engine. The belt used in the ISA system can be either ribbed-V like a timing belt or a poly-V belt. To transfer higher torque for cranking or restarting the IC engine by the ISA machine, a fixed idler replaces the belt tensioner of the conventional engine, while the conventional belt tensioner is relocated as shown in Figure 25.5. The belt tensioner can be controlled either mechanically or electrically. Generally, the conventional starter shown in Figure 25.2 may still be required for initial cold engine cranking, while the ISA machine acts as motor and rotates the engine crankshaft via the belt to restart the engine or warm start during stop/go operation mode in the system with the belt-driven ISA. This may help lower the system requirements for power package, cable, harness, and electronic component ratings if a small ISA machine can satisfy with all the rest specifications except for cold cranking, especially at temperatures of −29° to −40°C in the winter season. Of course, the conventional starter can be eliminated if the on-board or off-board electric loads need a larger ISA machine to generate enough electric power; the large ISA machine with its corresponding power electronic drive and belt drive can perform the initial cold cranking at motoring operation mode. In the powertrain system with ISA subsystem, whether a conventional starter remains there depends on system power specifications and new system cost targets.

The powertrain architecture containing ISA subsystem can offer stop/start, braking regeneration, and torque assistance at launch besides electric power generation. The offset belt driven ISA has many advantages, such as low cost, low noise, relatively flexible package, and no lubrication, although it cannot damp the engine vibration effectively. To perform the stop/start function without compromising the customer's comfort and cooling system capability during engine stop, a clutch between engine crankshafts and drive/driven pulley is required. With help of the clutch, the pulley at crankshaft is disengaged from the engine shaft, and the ISA machine powered by the battery-stored energy acts as motor and keeps accessories, such as water pump, power steering pump, and air conditioner compressors, running. An electric oil pump is also required to keep the transmission oil flow at the engine stop and to prevent the transmission from overheating. A typical belt-driven ISA subsystem is shown in Figure 25.6.

The ISA machine can be cooled by either conventional air-cooling or the liquid cooling since it is located at the same location of the conventional alternator. The choice of cooling styles depends on cooling requirement and system packaging. However, the power electronics box is normally cooled by liquid/water since the ambient temperature of 105 to 125°C under the hood causes much difficulty in the air-cooling system. Therefore, it is essential to keep the water pump running during stop/start in the thermal design of different ISA systems.

25.3 FEATURES AND PERFORMANCES OF THE ISA SYSTEM

The goal to add an ISA subsystem to the conventional powertrain system lies in not only replacing the existing starter and alternator on-board the vehicle, but also targeting many new features and performance improvements on the vehicle energy system. In some cases, the latter is more attractive to the customer than the former. Actually, a belt-driven ISA subsystem with a small power package still requires a conventional starter for initial cold engine cranking. Compared to the conventional powertrain system, the features of the ISA subsystem and the effects of the ISA subsystem on the performance of the vehicle powertrain and the entire vehicle are described in detail as follows.

Integrated Starter Alternator

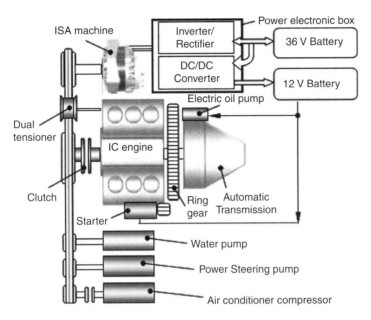

Figure 25.6 Powertrain with a belt-driven ISA subsystem.

25.3.1 STATE OF THE ART

A brief summary of the state of the art of the energy system on-board conventional vehicles is helpful for understanding the features and performance of the ISA subsystem in the new generation vehicles. For a conventional vehicle, the starter pulls the engine from its static state to cranking speed (normally 80–200 rpm, varied from engine to engine). Once the engine is cranked, the task of the starter is finished and the starter is on standby for the next starting. The cranked/ignited engine continues accelerating up to the idle speed, such as 500 to 800 rpm. When the engine runs near idle speed, the alternator, a multiphase AC electric machine (most of them are three-phase AC claw pole Lundell machine), starts to generate electric power all the way from idle to the engine redline speed of 3500 to 6500 rpm. A hydraulic drive braking system is responsible for deceleration or vehicle stop through mechanical friction force while the engine runs at idle speed during urban rush hour or other traffic stop, in which the fuel is consumed and poison emission is produced. The power steering, if any, is driven by hydraulic system, too. The vibration and oscillation of the engine shaft is damped by a dual mass flywheel, i.e., a mass damper and a ring gear carrier.

Starter. This is a series DC motor or a permanent magnet DC motor with a solenoid engagement system. It is offset-mounted on the engine block near the transmission side. The starter via its pinion gear drives the ring gear on the engine crankshaft and cranks the engine at about 80–200 rpm. The cranked engine begins consuming fuel and emitting exhaust pollution, especially at low engine speeds. Generally, the highest current during engine cranking occurs at stall point, i.e., starter stall current. The current at the engine cranking point is defined as the cranking current, which is slightly lower than half of the stall current for a reasonable cranking system design. So the components and harness as well as cables in the cranking circuit have to be able to conduct the current from the stall to cranking current, at least for tens of milliseconds. For instance, the cranking circuit carries the stall current of 550 ADC to the cranking current of 270 ADC for about 20 to 30 milliseconds during the cranking of a 0.9 to 1.2 L gasoline engine in a 14 V system. The stall DC current can reach 3500 A in the cranking circuit of a 15 to 16 L engine. If the temperature drops to $-29°$ to $-40°C$ in a cold winter, the breakaway torque driving the engine from static to initial movement will challenge the capability of the initial cranking of a starter. Due to meshing of the pinion and ring gear, and moving of DC machine brushes on the surface of the commutator, a loud noise can be heard during engine cranking. Please also notice that it is impossible to use a conventional brush-type DC motor for frequent stop/starts since the wearing of brushes prevents the cranking life of a starter from exceeding 40,000 cranking times.

Alternator. This is a multiphase synchronous machine with passive diode rectifier(s) and outputs DC current at a constant voltage level (either 14 V or 28 V at DC bus) via a field current regulator. Almost all of the conventional alternators are Lundell type or claw pole type. Although the most powerful alternator with the help of liquid cooling can output up to 4.5 to 5.5 kW, most of Lundell alternators can only output about the maximum power of 2.8 kW. The average efficiency of all existing Lundell alternators is about 50%, although the peak efficiency can reach 68% at certain speeds and under light load conditions for some so-called "high-efficiency" alternators. Among others, the low-cost material, compact size, and diode rectifier cause the alternator's low efficiency. The conventional alternator is normally mounted at the accessory side of the engine and driven by the engine crankshaft via a belt system at the drive ratio of 2.3:3.8, which adds extra power loss up to 3%. At least one belt tensioner is required to accommodate the pulley load variation. The magnetic and mechanical vibration produces not only acoustic noise but also high unit failure rate.

Flywheel damper. This is a mass disc and ring gear carrier, which adds extra weight and cost to the vehicle. The flywheel can only passively damp the vibration from the powertrain and drivetrain within a small frequency range. Introducing a flywheel to the system causes additional fuel consumption.

Voltage of vehicle electrical system. This refers the voltage across the terminals of the electric energy storage device (i.e., battery/batteries in the conventional vehicle) when it is being charged. A 12 V battery consists of six 2 V battery cells in series, while 12 V is the open circuit voltage of the battery. However, the voltage of each battery cell increases to about 2.33 V when the battery is being charged, so the voltage of a six-cell battery becomes 14 V at charging in the system. So the system voltage is 14 V, which the electrical designers have to deal with, when a 12 V battery is employed in the system. Most North American customers use a 12 V battery, while Asian and European customers may use a 24 V battery. Therefore, the system voltage is 14 V in North America, while it may be 28 V in Asia and Europe on-board the conventional vehicle. Generally, the alternator is driven by the engine and starts to charge the battery once the engine runs, so the system voltage is sometimes called "engine-on" voltage. It should be noticed that the battery voltage is about 12 V or lower if it is used to power electric loads, such as the starter, at engine-off. The low-voltage system requires the high current to run some powerful loads, which increases either system losses or system cost, even both.

25.3.2 Features of the ISA Subsystem

25.3.2.1 Initial Cranking and Stop/Start

The brush type Lundell machine has been used in the ISA system [7]. The contacts between brushes and slip rings are located at the field side, while the armature windings in the main working circuit are without contact. Most of the ISA systems are composed of the machine without mechanical contact, and solid-state power electronics devices are used. This can increase the cranking times dramatically due to non-wearing fashion, which adds the stop/start function to the vehicle.

Initial cranking or cold cranking is of the essential specifications in the modern vehicle. The engine is required to be cranked from still state to normal running under all severe environments, especially at low temperature, such as −40°C for commercial applications, even −50°C for military applications. The ISA machine has to produce a so-called breakaway torque in order to overcome the static engine torque and accelerate the engine to cranking speed. This torque is ranged at 1.15 to 1.5 times (even higher than this) of cranking torque, depending on the ambient temperature, the grade of engine oil, and the type of internal combustion engine. This is the highest torque that the ISA machine has to provide during the initial cranking period. In other words, the initial cranking needs to establish the highest torque by ISA machine. As common knowledge, the ISA machine size is governed by its output torque, which means that the initial cranking requires a machine large enough. In spite of this, the duration from breakaway to engine being cranked is about 0.09 to 0.35 sec each crank, which allows the ISA machine to run at its peak torque, instead of continuous torque. For the light-duty and heavy-duty application, the electric loads of the on-board and off-board vehicle demand a larger machine to generate the required power. It is possible that this big machine can meet the needs of the initial cranking specifications. Generally the engine cranking speed decreases with dropping of ambient temperature. It is possible for a powerful ISA machine to crank the engine at a higher speed even near the idle speed, which reduces the exhaust emissions dramatically. However, increasing machine size solely for the initial cold cranking might not be the best investment if the other functions, like powering electric loads on-board and off-board vehicles or performing frequent stop/start, do not required a larger machine. The conventional starter and gear assembly are kept unchanged [5,8] in this case. In a belt-driven ISA system like this, the conventional starter is used for the initial cold cranking of the engine, while the ISA machine is used to crank the engine during stop/start.

On the other hand, the crankshaft mounted ISA can perform the initial cold cranking while it meets the needs of the high electric power requirement of the vehicle. With help of the powerful ISA machine, furthermore, the engine can be cranked at a high engine speed, for example, the cranking speed near engine idle speed. With cranking speed increase, the fuel consumption and the tail emission pollution are improved significantly. Because of the direct connection between the engine and the ISA machine, the efficiency of the energy transfer is much higher than the belt-driven ISA system.

Stop/start or auto-start is believed the most important of the ISA functions. It is a main contributor to fuel saving in the vehicle with an ISA system. It is a mistake to define the stop/start as the state of entire vehicle. Actually, stop/start refers to only the engine operation state, i.e., the engine-off and restart. The engine is shut down after a specified delay, while the vehicle is running at downhill, deceleration, and braking states and traffic stops, and then the ISA machine cranks the engine automatically once the engine is required to drive the vehicle or to output mechanical power driving the ISA machine as generator. The stop/start requirements are combined with both a torque specification and a power specification over a given speed range. This event requires accelerating the engine quickly and smoothly to a given speed as near engine idle speed as possible in order to reduce the fuel consumption and pollution emission. To accomplish the stop/start function, the ISA machine has to meet the minimum torque and shaft power specifications, i.e., the ISA machine runs at constant torque range until a given speed, and then runs at a constant power or reduced power range. The vehicle level computer, with help of the ISA subsystem controller, controls the stop/start function.

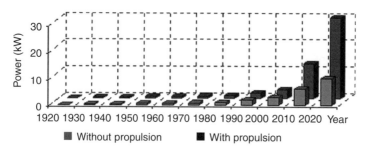

Figure 25.7 Peak electric power of average passenger vehicle.

It is a concern whether the acceleration performance is compromised after adding the stop/start function. The vehicle acceleration tests [10,11] show that the acceleration time from 0 to 50 km/h (or 110 km/h) is reduced, although the time from control signal input to the initial movement is longer for a vehicle with the ISA subsystem, compared to the conventional vehicles with the same engine size. The vehicle with the ISA system has a shorter passing acceleration time from 100 km/h to 120 km/h, too.

25.3.2.2 High-Efficient Large-Power Generation

Based on the research data, the average peak electric power required by the passenger vehicle has experienced a steady increase for over 90 years and will continuously increase in the future. As shown in Figure 25.7, the electric power on-board the vehicle was only a couple hundred watts in 1930s. The growth rate of the electric power was 2% per year from 1940 to 1970, while the rate became 6% per year from 1970 to 1990. With the introduction of hybrid electric vehicle and fuel cell vehicle (FCV) propulsion, the vehicle electric power will be boosted at the rate of 8% per year while the electric power on the vehicle without electric propulsion may still increase steadily from 1990 to 2020.

With increase of the electric power requirement, the output power and the efficiency of the alternator become more and more important in the vehicle. The reason is that the performance of a large alternator plays a big role on the vehicle performance and fuel consumption, compared to a small alternator's effects. The belt-driven ISA system can output the electric power from 3 kW to 8 kW while the crankshaft-mounted ISA machine can output up to 15 kW continuously for the passenger or light duty vehicle, and it may increase further, depending on application requirements. Because the generating state is required to be continuous, the generating efficiency of the ISA system is paid more attention by the design engineers. A typical comparison between the ISA system and the conventional generating system is given in Figure 25.8. The highest efficiency of the existing generating system, based on the Lundell alternator, is about 65% at full load. Although a so-called high-efficiency alternator can improve efficiency by 4 to 6% and reach the peak of 68% at full load, it is still much lower than the efficiency of the ISA system operated at generating state. Compared with the efficiency curves in Figure 25.8, it can also be found that the dropping rate of the generating efficiency for the ISA is much slower than the conventional Lundell alternator. For example, the ISA machine can still run with the efficiency over 70% while the efficiency of a high-efficient alternator drops below 50% at high speed region.

Besides the careful design of the ISA components and system, the high efficiency of the system is achieved also through better materials, new electronic devices, and efficient cooling methods. With increase of requirements for the launching torque assistance and

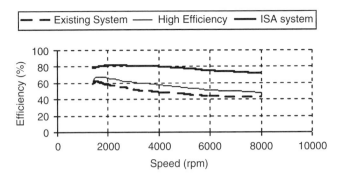

Figure 25.8 Comparison of the efficiency between existing generating system and ISA system.

propulsion, the ISA machine can be more powerful and high efficient. Actually it already becomes a real hybrid electric vehicle, which is beyond the scope coverage of this chapter (see Chapter 2).

25.3.2.3 Launching Torque Assistant

In addition to the IC engine, the ISA machine helps boost the power of the vehicle powertrain. A high-voltage battery, such as 36 V, serves as electric energy storage. Because of the incorporation between the IC engine and the ISA system, the powertrain can supply much higher driving torque for the vehicle's acceleration if the IC engine remains unchanged. It should be noticed here that boosting the powertrain overall torque dramatically would cause the necessity to check the mechanical stress and even to strengthen the driveline. However, this is not the goal for the ISA application. On the other hand, the integration of the IC engine and the ISA machine makes it possible for the IC engine to be downsized without compromising the acceleration capability. The fuel consumption and pollution are reduced because a smaller engine runs with higher efficiency at average combustion pressure levels. Although the vehicle with the ISA machine cannot perform as a powerful hybrid electric vehicle, the torque curve is still compensated by the ISA machine at motoring operation mode. The output torque of the powertrain with ISA machine is boosted and smoothed, compared to the torque curve of a conventional powertrain, shown in Figure 25.9. It can be found that the hybrid torque is raised with help of the ISA machine, which consequently compensates the IC engine torque and

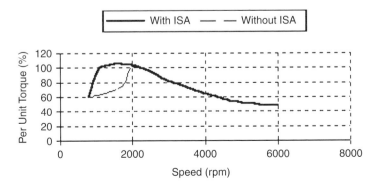

Figure 25.9 Launching torque assistance of the ISA machine.

reduces the acceleration/launching time. This is also the reason why a vehicle with the ISA system has the shorter acceleration time from engine-off to normal city driving speed (50 km/h) or to highway speed (110 km/h), in comparison with the conventional vehicle with running engine at the idle speed.

25.3.2.4 Braking Energy Regeneration

In the conventional brake system, the hydraulic system drives the static brake shoe against the rotor, and friction force, which retards the vehicle movement, is produced between them. During the braking procedure, the dynamic energy stored in the vehicle is converted into thermal energy and emitted into ambient air. The wasted energy ranges up to several hundred kilowatts. In the powertrain with the ISA machine, fuel to the IC engine is cut off and the vehicle is driven by the drive wheels during deceleration. The engine shaft transfers the mechanical energy from drive wheels to the ISA rotor indirectly or directly, and the ISA machine is operated at regenerating state by controlling the stator field vector or the current pulse, depending on machine types. The contactless braking torque produced by the ISA machine at regenerating state has the same function as the conventional contact brake assembly, and the wasted energy during the conventional braking is recovered by the ISA machine. A large amount of the braking energy is converted into electric energy and stored in the high-voltage battery for reuse. The regenerating function can achieve the reduction of the fuel consumption significantly. A system like this could potentially replace the costly and heavy retarder.

25.3.2.5 Low Loss and Cost via High System Voltage

Raising system voltage can reduce cable and harness size at the same power loss, or can operate on-board/off-board much higher power loads with the existing electric energy network. The 42 V system voltage or 36 V battery is proposed by an automotive consortium at MIT and used by some main automotive manufacturers, which has the maximum system voltage of 60 V during road dump or at other transient operations.

The objective of introducing 42 V system is to meet the needs of the high-power loads like the ISA machine and electric air conditioner. The machine windings in the low-voltage system require more parallel paths per phase, and it is difficult to arrange the AC bus bars in the limited end region. To handle the same electric power, the high current in the low-voltage system requires large cable size, especially the shielded three-phase AC cables, which will cost as much as the machine cost. More parallel MOS-FETs to conduct high current are also costly in the low-voltage system. It is suggested that the high-voltage system be considered with a high priority, as long as the machine is required to perform initial cold cranking and other motoring function for the ISA system in the high electric power application.

However, considering traditional 12 V loads like light systems and motors running power doors, power windows, and windshield wipers, the 42 V and 14 V will be used simultaneously for on-board vehicles for a long time. The dual-voltage system will reduce the investment for replacing the existing 14 V electrical system and parts while introducing the 42 V system to accommodate the electric power increase. Actually, some electric loads like light bulbs and the conventional starter take more benefits in 14 V systems than in 42 V systems. There is a debate over whether a 12 V battery is required in the dual-voltage system. Although the 12 V loads can be run indirectly by a 36 V battery via a DC bulk converter, it is preferable to run those loads by a 12 V battery, since keeping a DC converter active may increase system loss and raise safety issues. Existing 42 V systems and 42 V systems under development use dual batteries, i.e., 36 V and 12 V batteries. However, it might be possible to equip only 36 V batteries on-board in the future.

For the heavy-duty applications, the system voltage could be up to DC 340 V in order to run the ISA machine at the reasonable current level, for example, less than 500 A/phase.

25.3.2.6 Active Damping Oscillation and Absorbing Vibration

As mentioned in Section 25.3.1, the conventional flywheel dual-mass damper is costly, and the passive damping is effective only within a limited frequency range. The ISA machine can provide the electromagnetic damping actively, which is more effective than the flywheel mass damper, and the effective damping speed range (or frequency) is much wider. The damping is accomplished by the ISA machine torque, which is out of phase of the engine torque ripple produced by gas and inertia. When the crankshaft driven by additional positive torque ripple is rotating faster than the stator magnetic field, the ISA machine operates at generating state and produces braking torque against the increase of the engine speed. The electric energy generated by the ISA machine is temporarily stored in an intermediate capacitor. If the engine runs slower than the armature magnetic field due to negative torque ripple, the ISA machine operates at motoring state and produces driving torque to accelerate the speed of the IC engine.

The ISA machine at motoring operation state will absorb the electric energy from the capacitor, a temporary energy storage component. Figure 25.10 shows the comparison of the engine speed oscillation between the powertrain system with and without the ISA active damping activated. It is found that the engine speed variation is reduced/damped dramatically at the average speed of 1150 rpm. The torsion vibration of the third-order engine speed is reduced by about 80%. Compared to the flywheel mechanical damper, the ISA machine active damping system also has the benefit to lower fuel consumption with help of the intermediate capacitor.

A big challenge for an ISA system to perform frequent stop/starts is the issue of the engine vibration at stopping procedure [12]. In the conventional vehicle, the engine speed drops sharply in area I of Figure 25.11, since the creep torque driving the wheels disappears suddenly due to the fuel cut-off. The dynamic balance maintained by the creep torque, brake force, and suspension expansion is lost, which causes a longitudinal vibration of

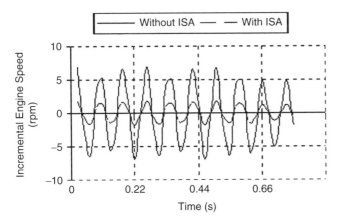

Figure 25.10 Engine speed oscillation within four rotations at 1150 rpm.

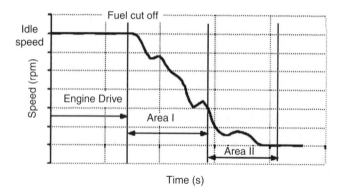

Figure 25.11 Stopping behavior of a conventional engine.

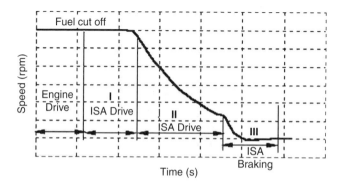

Figure 25.12 Stopping behavior of the engine with ISA.

the vehicle. When the speed drops to area II, the torque ripple due to the compression and expansion with piston movement inside the engine cylinder produces the rolling vibration to the engine mounting system, i.e., engine speed oscillation. Generally the rolling vibration occurs at or below a half of the engine idle speed.

The ISA machine makes it possible for both the longitudinal vehicle vibration and the engine rolling vibration to be suppressed by stopping procedure control. Figure 25.12 shows the control steps for vibration reduction with help of the ISA machine. First, once the engine is stopped via fuel cut-off, the ISA machine is first operated as a motoring state within the predetermined short time. In area I, the ISA machine drives the engine at almost the same engine idle speed, and then the throttle valve is completely closed to release the cylinder pressure. At the idle speed, the ISA machine continues driving the engine crankshaft for a given time as long as the negative braking pressure is insufficient. In this way, the rolling vibration is suppressed. Second, when the brake pressure reaches the required negative level, the ISA machine begins to reduce its driving torque gradually, and the crankshaft speed decreases smoothly. And last, the engine speed drops within mounting resonant speed. The torque ripple can be reduced if the ISA machine produces a braking torque when the piston in the cylinder is accelerated in its expansion cycle. The control for braking begins by operating the ISA machine at generating state before the crankshaft stops rotation [12].

25.3.2.7 Cylinder Shutoff

The consumption map shows that the specific fuel consumption of an engine, i.e., the fuel consumption per unit output power, is lower at near full load than that in the partial load range. For example, the fuel saving is about 45% at idle speed when three cylinders of a six-cylinder engine are deactivated [13]. The improvement in fuel consumption is achieved by having the rest of the cylinders run in the effective mean pressure range to provide the required overall power while the partial cylinders are shut off. The advantage of saving fuel and lowering emissions is compromised by some disadvantages, such as losing comfort due to vibration system detuned at one or more cylinder shutoff in the conventional vehicle. The ISA machine, especially the crankshaft-mounted ISA machine, can damp or reduce the vibrations caused by the cylinder shutoff dramatically, and maintain the comfort index (noise, structure-borne sound) of the vehicle at the level with original engine. As a result, the ISA makes the cylinder shutoff strategy for fuel saving and emission reducing possible without compromising the vehicle comfort.

25.3.2.8 Power APU and Other Electric Loads

Many U.S. states have prohibited heavy-duty trucks from running idle in parking lots or highway rest areas because of the large amount of emissions. The electric loads (such as electric appliances, audio/video devices, air-conditioners, and electric heaters) required by truck drivers overnight on-board the vehicle have to be powered by an auxiliary power unit. The APU is a generator unit operated by a small engine to reduce the emissions. The ISA machine can be detached by a clutch from powertrain/drivetrain systems and serves as the APU generator system. In this way, the conventional APU is replaced by the ISA system. This is an extended application for the ISA system in heavy-duty transportation areas.

25.4 COMPONENTS IN THE ISA SUBSYSTEM [7]

The ISA subsystem, either crankshaft-mounted style or belt-driven style, is connected to the engine (or transmission) mechanically and the energy storage components (battery/batteries) electrically. Figure 25.2 and Figure 25.3 show the different mechanical installations. The options of the electrical architectures and the key components in the ISA subsystem will be discussed in detail. The life of the ISA system should be targeted at operation of 8,000 to 10,000 hours, or at 15 years or 150,000 miles, whichever comes first. The ISA subsystem has to fully function at ambient temperature of –40 to 125°C.

The system voltage with the ISA subsystem can be selected from 14 V (or 28 V in some Asian and European applications), 42 V, dual levels of 14 V (or 28 V) and 42 V, and dual levels of 14/28 V and high voltage (such as 340 V). Among choices of the system voltage levels, the existing 14 V or 28 V always pops into designers' minds first. As a simple background, the high current has to be carried by the harness, cable, and electronic devices at low-voltage system (12 or 24 V) when the large power is transferred or applied. With the power increase, the current in 14/28 V system may become unreasonably high. The 14/28 V voltage level is not recommended in the ISA system with high-power requirements, such as above 2–3 kW, because of system cost and efficiency. As mentioned in Section 25.2.3.5, the 14 V and 28 V voltages are used in the electrical system on-board conventional vehicles, and many electric loads are still rated at 14/28 V level in the future

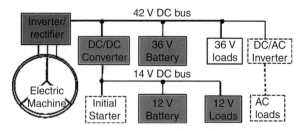

Figure 25.13 The electrical diagram of a typical ISA subsystem.

due to component life and manufacturing investment. Therefore, the single 42 V or 340 V voltage is not practical, and the dual-voltage system, i.e., 14/28 V and 42 V or higher, is always required if high voltage (42 V or 340 V) is preferred for running high-power loads. In practical applications, the dual-voltage system with 14/28 V and 42 V is used in the ISA systems of passenger cars and light duty trucks with power requirements below 20 kW, while the dual-voltage system with 14/28 V and high voltage (> 42 V) is used for heavy-duty trucks and transit buses. The different dual-voltage configurations of 14 V and 42 V, as examples, are discussed in this section.

25.4.1 Electric Machine with Dual-Voltage Output

The AC electric machine operating as generator provides two voltage levels, i.e., 14 V and 42 VDC, which is done through armature winding design and separate power electronic converters/bridges. This system does not need a DC-DC converter for dual-voltage outputs. However, the machine operates under unbalanced condition between 14 V and 42 V with loading variation. Furthermore, besides the high cost of the converters it is difficult to harmonize the two voltage levels when the machine runs as motoring. So the machine with dual-voltage levels does not fit the ISA application well, although it can be used for electric power generation.

25.4.2 36 V Battery with 12 V Intermediate Terminal

The partial 36 V battery cells serve as 12 V output, while the entire battery provides 36 VDC power. That is, the battery has three terminals — "+36," "+12," and "–" — connected to 42 V and 14 VDC buses as well as ground, respectively. There is only one battery in the system, so the cost will be low. However, the common cells shared by 14 V and 42 V overrun, compared to the rest of the battery cells. This will result in short battery life. It is not recommended for the ISA application.

25.4.3 Typical ISA Electrical System

A typical ISA subsystem consists of an electric machine, a power electronic inverter/rectifier, a DC-DC converter, and an optional starter for initial cold cranking, shown in Figure 25.13.

The electric machine can be operated under either motor state for creeping or launching assistance or generator state for charging batteries or powering on-board/off-board electric loads. It also provides the torque to crank the engine at initial engine cranking or at recranking after engine stop. The power electronic inverter/rectifier runs as an inverter that converts the DC power to AC power and drives the machine operating under motor state, and runs as a rectifier that converts the AC power to DC power at generating state. The DC-DC converter is used to convert the high DC bus voltage (such as 42 V) to low DC bus voltage (such as 14 V) or vice versa. The conventional starter is an optional component and is used only for initial cold engine cranking, instead of engine cranking at the stop/start, when the torque of the electric machine in the ISA system is not high enough for initial cold cranking. The initial cranking starter is a conventional 12 V starter, instead of a 36 V starter, which will save the investment of the 36 V starter development and avoid some drawbacks of the 36 V starter. Generally, the generating power of the electric machine is higher than the power generated by a conventional alternator, so an optional DC-AC inverter can be added into the ISA electrical system to run the AC loads. The 36 V battery and DC loads are connected to the 42 VDC bus, while the 12 V battery and DC loads are connected to 14 VDC bus.

The 12 V battery can be eliminated if a signal is sent to the DC-DC converter at the requirement of 12 V loads, even during engine shutting down. Of course, the complicated control of the DC-DC converter and monitoring the requirement of 12 V power/loads should be introduced to the ISA system without a 12 V battery, and the 14 VDC bus voltage ripples will increase due to absence of the 12 V battery. Actually the 12 V loads, such as bulbs, lighting systems, and different 12 VDC motors, can work properly when powered by a 42 V pulse width modulation (PWM) chopper [14].

25.4.4 Multifunction Inverter with a Neutral Inductor

A multifunction inverter and an inductor connected at the neutral point of electric machine are introduced [15] into the ISA electrical system to replace the conventional inverter/rectifier and DC-DC converter of Figure 25.13, shown in Figure 25.14. Through modulating duty cycles of the voltage pulse, the 14 V voltage can be provided at the inductor output terminal. The rest of the system component configuration, such as energy storage batteries, electric loads, and optional initial starter, is the same as that in the conventional ISA electrical system in Figure 25.13.

The advantage of this system lies in cost reduction, because a simple inductor replaces a complicated DC-DC converter. The disadvantage of the system, unlike the system with a bidirectional DC-DC converter, is that the electric energy stored in a 12 V battery cannot be boosted to high-voltage level to power the machine at motoring operation. The principle of the multifunction inverter will be addressed in Section 25.4.7.3.

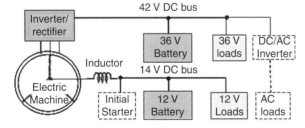

Figure 25.14 The electrical diagram of an ISA subsystem with multifunction inverter.

25.4.5 Electric Machine

The electric machine is a key component in the ISA system and operates at three states, i.e., motoring, generating, and regenerating. The entire system efficiency mainly depends on the machine's efficiency. At motoring operation, it converts 36 V electrical energy into mechanical energy, which performs the following functions: initial cold key crank to static engine, auto-crank at stop/start, launch assist during engine acceleration, creep assist during hill climbing, fuel cutoff smoothing during deceleration, vibration smoothing during stop, and so on. At generating operation, it converts mechanical energy of the engine into 42 VDC electrical energy powering 36 V loads or charging 36 V batteries. At regenerating operation, it converts the kinetic energy of the vehicle into 42 VDC electrical energy while offering braking function. Ideally the electric machine, through an inverter, runs a DC bus with the maximum voltage of 36 V under motoring state while through a rectifier it should provide the DC bus a charging voltage up to 42 V.

In practice, the available 30–36 V battery voltage at motoring and the required 36 to 42 V charging voltage at generating are the main challenge during electric machine design. Besides the dilemma from voltage specifications, the requirement of a large speed range and the high temperature of cooling media bring several new critical issues to the ISA machine. The ambient temperature ranges from −40 to 125°C, which can be used as the specification of an air-cooled machine. If the machine is cooled by liquid, the available engine coolant has up to 120–130°C, while the available transmission oil temperature will be 135 to 150°C, unless a separate liquid-cooling loop is introduced. The speed of the electric machine runs from 0 to 6,000 rpm for the crankshaft-mounted ISA system, while the maximum operation speed of the machine runs as high as 13,800–19,200 rpm for the belt-driven ISA system with the belt transfer ratio of 2.3–3.2.

25.4.5.1 Specifications of the ISA Electric Machine

The electric machine of the ISA system does not require reverse rotation, so its mechanical specifications are limited within the first and second quadrants in the plane of torque vs. speed. The ISA machine runs in the first quadrant with positive torque and speed under motoring state, while it runs in the second quadrant with negative torque and positive speed under generating and regenerating brake state. In the first quadrant, the machine provides a large torque within a short time interval at low speed and supplies driving torque (with constant power characteristic) within a longer time interval at higher speed range. In the second quadrant, the machine produces a large braking torque and feeding the energy back to DC bus within a short time interval, while it is required to generate electric power and charge batteries as well as run electric loads continuously. The ratio of the high to low speed requires higher than 7 in most applications, and even over 10 for some applications. The following specifications of the ISA machine provide maximum acceleration and reduce fuel consumption: high peak torque, low rotor moment of inertia, high efficiency, low mass, and small size. Low torque ripple and noise, high reliability, and durability are also demanded. Furthermore, the capital investment, the system performance, and the operation cost have to be fully considered and harmonized.

Motoring Requirements. Unlike the electric machine on-board a fully hybrid vehicle, the machine in the ISA system of a mild hybrid vehicle requires a shorter time running under motor state. Therefore, its motoring specifications are mainly defined as short-term or intermittent operation schedule, although the continuous performance of the machine at motoring state is demanded. This allows the motor to provide more torque/power within a limited duration. the following specifications are normally required for the ISA machine running as a motor.

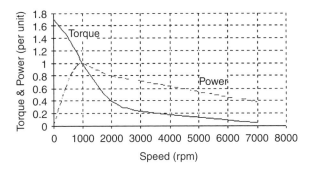

Figure 25.15 Torque and power requirement at motoring of a belt-driven ISA machine.

Start torque or lock rotor torque. If a conventional starter is not installed, the ISA machine is required for initial cold cranking of the engine. Then the electric machine has to provide a breakaway torque that overcomes the engine static torque and rotates the engine from 0 to 10 to 20 rpm, especially in a bitter cold winter when the temperature drops down to −29 to −50°C. Normally, the breakaway torque is about 1.5 to 1.8 times the minimum cranking torque or the torque corresponding to the maximum motoring power, which depends on the ambient temperature, the grade of the lubricating oil, and the engine itself. As an example in Figure 25.15, the mechanical characteristic of the ISA machine shows the breakaway torque requirement (1.68 in per unit value) of a typical internal combustion engine. Obviously, the start torque or lock rotor torque at the lowest frequency has to be higher than the breakaway torque. Fortunately, the maximum peak torque is required only within a short time, such as 0.08 to 0.20 sec, and the torque requirement can gradually be reduced to 1.2 to 1.3 per unit value once the engine is accelerated above 10 to 20 rpm. The duration of the cycle is normally up to 1 to 3 sec. The procedure is repeated after a proper interval rest.

Automatic crank at stop/start. Restarting the engine after a stop at a traffic light or traffic jam or a low battery charging state being monitored in city driving is a main function of the ISA system. The automatic crank is also called warm crank, since both the engine oil and the transmission oil are at warm state, such as 20 to 60°C. Besides cranking the engine quickly, the ISA machine must be capable of assisting the engine to accelerate it up to 800 to 1200 engine rpm within a very short time, such as 0.3 to 0.7 sec, in order to maintain or improve the acceleration performance (from stop to 20 to 35 mph) of the conventional vehicle without an ISA system. For the quick response and frequent stop/start, the moment of inertia for the ISA machine rotor should be as low as possible, especially for the belt-driven ISA. The electric machine must be capable of providing the maximum torque or the maximum power, whichever results in a lower torque, from a static engine to 800 to 1200 engine rpm. The torque at motoring state of automatic crank should maintain within a required time and be capable of repeating after a short rest.

Launch assist torque or power boost. An extra torque added to the top of the engine torque, shown in Figure 25.9, can improve the acceleration performance of a vehicle, which is equivalent to an elevation of the engine power from smaller to higher level. The launch assist torque should be available after key controlled initial engine start, automatic engine crank, or normal engine driving. It could be specified from engine cranking speed (i.e., 100 to 400 engine rpm) to highway driving engine speed (i.e., 1800 to 2500 engine rpm) or even up to engine redline speed (i.e., 5000 to 6000 rpm) if applicable. The duration requirement of the launch assist torque could be from 3 to 5 sec up to 15 sec. Some crankshaft-mounted ISA machines provide the driving power continuously, like specifications in a fully hybrid electric vehicle.

Torque smooth and fuel cutoff vibration reduction. The drivetrain disturbances and deceleration fuel cutoff perturbations on engine torque at low-speed range shall be reduced dramatically by an ISA machine. The controller of the ISA machine should be capable of switching between motoring and generating states quickly and providing torque and vibration damping effectively at idle/low engine speed and up to 2000 to 3000 engine rpm.

In the 42 VDC electrical system the motoring performance specifications should be met even at a lower voltage level of 30 V to 33 VDC, although the maximum available battery voltage is 36 V at the DC input of the electric machine drive. If a 14 VDC electrical system is used for the ISA system on-board the vehicle, the motoring operations of the machine have to be fully functional at 10 to 11 VDC voltage, in spite of the battery voltage of 12 V. The practical low available voltage is caused by battery charge state and ambient temperature, as well as internal resistance and so on. If a three-phase induction machine is used for the ISA machine and the space-vector control is introduced for its control, it must meet all motoring specifications at the minimum available line-to-line voltage of 21 to 23 VAC and 7 to 8 VAC in rms values for the 42 VDC and the 14 VDC systems, respectively.

Generating Specifications. Once the engine is cranked and accelerated to idle speed, the ISA machine has to be capable of converting mechanical power to electrical power to charge batteries and operate electric loads. The normal generating operation is required for the ISA machine from the idle speed up to the redline speed of the engine. Furthermore, the ISA machine shall be capable of not only providing braking torque but also feeding back the mechanical power to the electrical system during braking.

The generated voltage at the rectifier output of DC bus requires at least 36 VDC, and may run up to 48 VDC, but should not be over 52 V for a normal 42 V electrical system on-board vehicle. The DC output under generating condition of the ISA machine shall be at least 12 VDC, may run up to 16 VDC, but should not be over 18 V for a 14 V ISA system. If the ISA machine is a three-phase induction machine controlled by synchronizing rectification, the minimum generated line-to-line voltage should be 27 VAC and 9 VAC in rms values for the 42 V and 14 V electrical system, respectively. In addition to the voltage requirement, the ISA machine has to properly function at the ambient temperature of up to 125°C for air-cooling style and the coolant temperature of up to 135 and 150°C for the liquid-cooling style by the engine coolant and transmission oil, respectively.

Furthermore, the rotor of the machine has to tolerate the mechanical stress without damaging at engine redline speed, such as up to 6,500 rpm for the crankshaft-mounted ISA and up to 21,000 rpm for the belt-driven ISA. All electric power required by electric vehicle loads on- or off-board is supplied by the ISA machine under generating condition, so the generating performances of the machine, including power output and efficiency, are critical specifications. Generally, the output power at idle engine speed shall be larger than 35 to 60% of the maximum continuous output. Within a short time, such as 1 to 3 minutes, about 1.3 to 1.4 per unit power is required for rush charge. Figure 25.16 shows a typical specification for power generation of a belt-driven ISA, in which less than 1,750 to 2,000 rpm corresponds to idle engine speed of 550 to 700 rpm while the 6,000 to 7,000 rpm motor speed corresponds to the engine speed of 1,900 to 2,200 rpm at highway driving. When the motoring speed is higher than 12,000 rpm, i.e., engine speed > 4,000 rpm, the generated power can be reduced since it is beyond the normal driving speed. It will lead to either machine oversize or performance reduction at low-speed range if too much performance is required at the speed near engine redline speed. Actually, the fuel economy and other vehicle performance do not get much benefit when the ISA machine

Integrated Starter Alternator

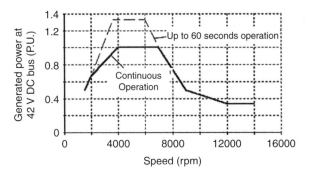

Figure 25.16 Generating power requirement at 42 VDC output of a belt-driven ISA.

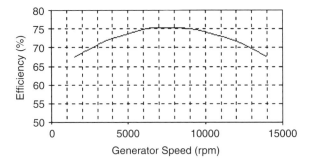

Figure 25.17 Efficiency requirement at 42 VDC output and power > 0.2 per unit.

provides the same output power and efficiency at the redline speed as that at normal drive speed.

Compared to conventional automotive Lundell alternators with diode rectifier/bridge, the ISA system consisting of optimal designed machine and MOSFET converter should provide more efficient electromechanical energy conversion. A typical efficiency curve of a belt-driven ISA machine shall be achieved, shown in Figure 25.17, when the electrical output power is larger than 20% rated power at 42 VDC bus. When the power output is more than rated power, for instance, the short time requirement shown by the dashed line in Figure 25.16, the machine efficiency is allowed to be slightly lower. When cooling fans are used for the ISA machine cooling, its efficiency at high-speed range will be compromised due to higher fan losses.

25.4.5.2 Types of ISA Electric Machines

This section summarizes the machine types used in ISA machine development. The DC machine has been eliminated from the candidate list of the ISA machine because of the issues from its commutator. With development of the power electronics and microprocessors, the frequency variable inverter/converter makes multiphase AC machines dominate ISA applications. They are claw pole Lundell machine, induction machine, switched reluctance machine, permanent magnet machines including sinusoidal PM machines, and trapezoidal PM machines, as well as PM reluctance machines.

Claw Pole Lundell Machine. Based on the operation principle, the Lundell machine belongs to the synchronous machine category. Its stator armature is the same as

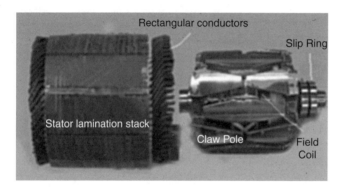

Figure 25.18 Brush-type Lundell machine with dual field coils.

other multiphase AC machines, while its rotor consists of claw pole segments, rotor cores, field coils, and slip rings. The DC current through two slip rings is exerted on the circular field coil. The excited field coil produces axial flux in the rotor core, and the flux lines circulate along the rotor end disc, claw pole, airgap, stator lamination stack, another adjacent rotor pole, and then back to the rotor core. The airgap flux can be adjusted through the controlling field current. The Lundell machines, either brush type or brushless type, are widely used as alternators in the automotive industry.

The power range in applications is up to 2.5 kW, although Remy International has produced a 5 kW Lundell alternator since the 1950s. The power limitation is caused by high rotor leakage flux between claw poles. With the increase of the axial length, the magnetomotive force (MMF) drops of rotor poles become unreasonably higher, which avoids any effort to raise the output power through increasing airgap flux. If two conventional Lundell-type rotors are attached together, as shown in Figure 25.18, the attempt to extend the effective length of the stator and rotor becomes possible. This technology can extend the power range of the Lundell machine up to 5 to 6 kW.

To reduce resistance loss and increase output, especially at low-speed range, the rectangular conductors are used for stator windings with high slot fill ratio (SFR), although the conventional round-wire wound stator windings are the well-established technology and cost less. Here the slot fill ratio is defined as the ratio of the total cross-section areas of all bare copper in a slot to the cross-section area of the entire slot. Obviously the higher the SFR, the lower the phase resistance. On the other hand, the slot leakage flux may cause AC resistance increase if the thickness of the conductor in the slot height direction is too high. The so-called skin effect reduces the effective cross-section area of a conductor, especially at high frequency or speed range. To reduce the skin effect, the thickness of a conductor in a slot must be limited to the proper thinness level, which requires increasing the number of conductors in the slot depth direction. The following formula can help estimate the reasonable thickness of a conductor in slots:

$$h = \frac{1.32}{\sqrt{m}} \sqrt{\frac{b_s}{b_c} \frac{1}{\pi \sigma f \mu}} \qquad (25.1)$$

where b_s is the slot width; b_c is the conductor width in the slot width direction; σ is the electric conductivity of the rectangular conductor at a given temperature and the copper conductivity is about 1.72×10^{-7} $(\Omega \cdot m)^{-1}$ at room temperature of 20°C; f is the frequency of current carried by the conductor in the slot; μ is the permeability of the conductor and

Integrated Starter Alternator

slot space, which is $\mu_0 = 4\pi \times 10^{-7}$ H/m for copper and space filled with air; m is the number of conductor layers in the slot depth direction.

According to the above formula, the proper thickness of a conductor is the function of temperature and operation speed. In practice, the thickness of the conductor should be chosen at the speed and temperature of the machine often running.

Remy, Nippon Denso, Valeo, Visteon, and Mitsubishi have presented many patents and manufacturing reports about how to manufacture the stator windings with rectangular segment conductors, but that is beyond the scope of this book.

Besides the established manufacturing technology, the Lundell machine enables the flux weakening at high speed by simply controlling the field current for the constant power operation. At the low-speed range, the field current produces the maximum magnetic field in the airgap, which results in a high torque to crank the engine or run in EV mode. The technology, running a Lundell machine as an alternator, is well established. However, the Lundell machine running at motoring state requires a position sensor to feed the accurate rotor position back to control loop because the stator magnetic field produced by three-phase PWM modulation has to be strictly synchronized with the rotor magnetic field. The field current control is on the top of three-phase inverter control, which makes the control system more complicated.

Permanent Magnet Machines. The permanent magnet machines are paid more and more attention in the automotive industry. The airgap magnetic field is produced by permanent magnet (PM), unlike what is produced by field coils in the wound field synchronous machines, such as Lundell machines. The main advantage of PM machines is their high efficiency due to the absence of the field coil losses. From the efficiency point of view, the PMAC machine with 8 to 12 poles is preferable [16]. The encased rotor is preferred, which can prevent permanent magnets from iron dregs in the cooling air or corrosion of the NdFeB if applicable.

Based on the structure of the PMAC machines, they can be categorized as surface-mounted PM type and interior PM type. The PMs are attached on the rotor surface in the surface-mounted PMAC machine (SPM) shown in Figure 25.19(a), while PMs are imbedded inside the rotor in the interior PMAC machine (IPM) shown in Figure 25.22 and Figure 25.23a. The equivalent airgap in either d- or q-axis direction is very large in the SPM since the thickness of the PM ($\mu_r \approx 1$) is added to the airgap length. So its d-axis inductance L_d is approximately equal to its q-axis inductance L_q, or there is no significant salient effect in the SPM because of the negligible difference between d- and q-axis reluctance. Furthermore, the magnetizing inductance is very small due to the large equivalent

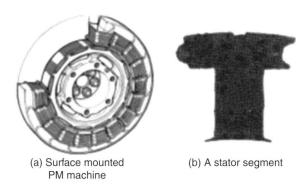

(a) Surface mounted PM machine

(b) A stator segment

Figure 25.19 Permanent magnet brushless DC machine.

airgap in the surface-mounted PMAC machine. On the other hand, the interior PMAC machine displays a high magnetizing inductance, and the salient pole effect (i.e., $L_d \neq L_q$) is displayed in the IPM machine because of the significant difference of the reluctance in its d- and q-axis magnetic paths.

According to the waveform of the electromotive force (EMF) and current in the operation, PMAC machines can be classified as the trapezoidal PMAC (or switched PMAC or brushless DC PM machine) and the sinusoidal PMAC machine.

The trapezoidal PMAC is operated with 120° trapezoidal current pulses. To turn on and off the switches of the inverter/converter (in Section 25.4.6.1) properly, a position sensor is required to feed back the rotor position signal. Its stator three-phase windings are placed in the slots within 60° phase belt and all windings are in full pitch (the span of two coil sides is equal to a pole pitch). The three-phase windings can be connected into either Y-connection or Δ-connection in the trapezoidal PMAC machine. The rotor pole arc occupied by a PM is 180° if stator windings are in Y-connection, while the rotor pole arc shall be 120° if stator windings are connected into Δ-connection [17]. The trapezoidal PMAC machine with either Y- or Δ-connection provides slightly different torque density [16]. Most of the trapezoidal PMAC machines have the structure of surface-mounted PMs.

The sinusoidal PMAC machine is operated with 120° sinusoidal currents. A position sensor installed on the rotor and a power electronic bridge is required to operate the machine. Three-phase stator windings can be the same as those in the conventional synchronous machines, in which the distribution and the short pitch of the windings are applied to reduce the space harmonics. The concentrated windings are possible, in which each coil is installed on one stator tooth. Figure 25.19 shows a sinusoidal PMAC machine with 12 rotor poles and 18 stator teeth [18]. To reduce the stator resistance losses through higher slot fill ratio, the stator teeth are segmented, as shown in Figure 25.19(b), and each coil is wound on the segmented body after adding unsymmetrical insulation to the segment. Finally, all the segments with wound coils are assembled together. The design of the PMAC machines like this is described in Reference 19. The rotor of a sinusoidal PMAC machine can be either the interior PM type or the surface-mounted PM type with the pole arc of 110 to 130° and the parallel magnetization.

Unlike other industrial applications, the ISA machines require extremely wide speed variable ratio besides the very high cranking torque. It is difficult for a trapezoidal PMAC or brushless DC machine with surface-mounted PMs to perform constant power operation, because the flux produced by PM is almost constant, and a large demagnetizing component of the stator current is demanded for flux weakening. The high stator current causes high stator loss. The trapezoidal PMAC machine can operate at higher than its base speed by controlling the current leading angle and extending the conduction angle from 120 to 180° [20]. However, the speed ratio can only be extended to 2~3 with the expense of the high current and torque ripple. In the pure traction application, the surface-mounted PMAC machine without flux weakening capability can be so designed that the power supply voltage is balanced by the EMF produced in the windings at the maximum speed; i.e., the rated speed is chosen at the maximum speed. The machine runs at the reduced voltage level below the rated speed, while the stator current decreases above the rated speed. The stator resistance loss decreases above the rated speed, which benefits the EV application. The voltage exerted on the inverter switches will be still within their voltage ratings even at losing DC power supply (or battery). Unfortunately, the stator windings, designed according to the maximum speed, cannot be used for the ISA machine because the ISA machine must generate the voltage high enough to charge the battery from idle speed. Therefore, the trapezoidal PMAC machine with surface-mounted magnets and windings designed at the maximum speed does not meet the generating specification of the ISA

Integrated Starter Alternator

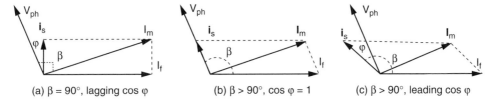

Figure 25.20 Vector diagram of nonsalient sinusoidal PMAC machine.

machine, unless a DC-DC boost converter is added between the electronic bridge and battery.

Extending constant power operation requires flux weakening in the sinusoidal PMAC machine. As discussed in the above paragraph, the surface-mounted PMAC machine has constant flux characteristic, low magnetizing inductance L_m, and negligible salient pole effect ($L_{ad} \approx L_{aq} \approx L_m$). A limited flux weakening can be performed by increasing the leading angle β of the stator current I_s. The torque component of the stator current, I_q, is proportional to $\sin\beta$ (i.e., $I_q = I_s\sin\beta$) while the flux-weakening component of the stator current, I_d, is proportional to $\cos\beta$ (i.e., $I_d = I_s\cos\beta$), as shown in Figure 25.20. The demagnetizing flux linkage is $\psi_d = I_d L_m$. A very high I_d is required to perform flux weakening because the magnetizing inductance L_m is very low in the sinusoidal PMAC machine with surface-mounted PMs. Actually, the stator resistance loss increases due to the high current. The equivalent magnetizing current is 2 to 5 times the stator current in the surface-mounted PMAC machine, compared to the magnetizing current of 20 to 50% stator current in induction machines. The surface-mounted PMAC machines are difficult to fulfill the constant power operation by flux weakening, especially at high speed and light load. Therefore, the surface-mounted PMAC machine, either trapezoidal or sinusoidal type, does not fit the applications of the ISA machine with constant power specification of wide speed variable ratio.

The surface-mounted PMAC machines are not suitable to operate under constant power specification because too high a stator current is required for flux weakening. Then two questions are very interesting: (1) Besides other specifications, how can you know if a sinusoidal PMAC machine, existing or under development, meets the constant power specifications given the current rating of the machine and inverter? (2) What kinds of the sinusoidal PMAC machines can meet power specifications of the ISA at wide speed range?

To answer the first question, the current limitation circle and the voltage limitation circle/ellipse are introduced to explore the state restrictions at high-speed regime of the sinusoidal PMAC machines. As a basic concept, the surface-mounted sinusoidal PMAC machine is set as an example, although it is not the candidate for ISA applications. From the system design point of view, the stator current of the ISA machine is limited by the thermal capability of the machine and the current rating of the inverter switches. The current limitation of inverter switches can be converted into the stator current limitation I_{max} in the $i_d \sim i_q$ plane, i.e.:

$$i_s^2 = i_d^2 + i_q^2 \leq I_{max}^2 \tag{25.2}$$

where i_s, i_d, i_q are the stator phase current, its d-axis component, and its q-axis component, respectively.

This stator current limitation can be expressed by a circle with radius of I_{max} and the center at the origin of the $I_d - I_q$ plane, shown with a solid line in Figure 25.21. This

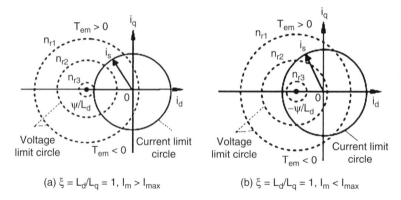

Figure 25.21 Current limit circle and voltage limit circle of nonsalient PMAC machine ($n_{r1} < n_{r2} < n_{r3}$).

means that the current regulator can produce any stator current whose vector arrow is on or within the current limitation circle. On the other hand, the DC bus voltage, depending on the battery output voltage, limits the available stator voltage U_{max}. Neglected the effect of stator resistance, the maximum available stator voltage is governed by the equation as follows [21,22]:

$$i_q^2 + \left(\frac{L_d}{L_q}\right)^2 \left(i_d + \frac{\Psi_f}{L_d}\right)^2 \leq \left(\frac{30 U_{max}}{\pi p L_q} \frac{1}{n_r}\right)^2 \tag{25.3}$$

where L_d, L_q are the d- and q-axis inductance (H), respectively; p is the number of pole pairs; n_r is the rotor speed (rpm); Ψ_f is the stator PM flux linkage ($V \cdot s$); and U_{max} is the limitation of the phase voltage.

There is no salient pole effect in the surface-mounted PMAC machine, i.e., $L_d = L_q$. Equation 25.3 represents a circle with radius of

$$\frac{30 U_{max}}{\pi p L_q} \frac{1}{n_r}$$

and center at point ($-\Psi_f/L_d$, 0) in the $i_d - i_q$ plane, i.e., the voltage limitation circle shown in dashed line in the Figure 25.21. It can be proved that the Ψ_f/L_d represents the phase current I_m under three-phase short-circuit condition if the stator resistance is negligible. The radius of the voltage limitation circle is proportional to the maximum available voltage U_{max} and inversely proportional to the rotation speed n_r of the machine shaft. To meet the voltage limitation, the stator current vector can be at any point within the voltage circle without considering the current limitation circle. When both current limitation and voltage limitation are taken into account, the stator current has to be within the intersection area of the two circles.

In the first and second quadrants of the d-q plane in Figure 25.21, the PMAC machine runs at motoring state with positive torque, i.e., $T_{em} > 0$. In the third and fourth quadrants, it runs at generating or brake regenerating state with negative torque, i.e., $T_{em} < 0$. The stator current in the second and third quadrants has the flux weakening effect. With speed increase, the voltage limitation circle contracts towards its origin at $-\Psi_f/L_d$ on the i_d-axis.

Integrated Starter Alternator

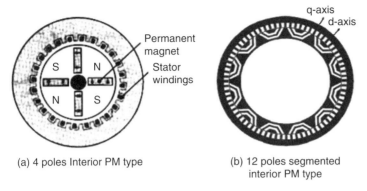

(a) 4 poles Interior PM type

(b) 12 poles segmented interior PM type

Figure 25.22 Interior PMAC machines ($L_q > L_d$).

The intersection part between the current and voltage limitation circles becomes smaller and smaller with the speed increase ($n_{r3} < n_{r2} < n_{r1}$ in Figure 25.21), which results in a low I_q and low torque, consequently. When n = n_{r3} in Figure 25.21(a), there only exists one common point of the current and voltage limitation circles. At this speed, the stator current is only used to limit the stator voltage. At this speed, there is no electromechanical energy conversion. However, the flux weakening operation can not be a speed limitation when the origin of the voltage limitation circle is within the current limitation circle, if the reduced power generation is allowed, shown in Figure 25.21(b). The location of the origin of the voltage limitation circle depends on the three-phase short-circuit current, $I_m = \psi_f/L_d$. Generally I_m is very high, and an overdesigned inverter or machine is not preferred. So protecting the inverter and machine from short-circuit becomes very important for the surface-mounted PMAC, since the origin of the voltage limitation circle is far from the current limitation circle. Actually, the surface-mounted sinusoidal PMAC machine is not recommended for the wide speed range applications in the ISA.

Generally, the trapezoidal PMAC machine and the sinusoidal PMAC machine with surface-mounted PM are not suitable for the ISA applications required for wide speed operation by a flux weakening method.

Now let us answer the question of what kind of sinusoidal PMAC machine can meet the wide speed and the constant voltage requirements under both motoring and generating operation for the ISA application. If the sinusoidal PMAC machine combines with characteristics of the synchronous reluctance machine, it can operates at wide speed range by a flux weakening method. When the PMs with pole arc of 110 to 130° (in electrical degrees) are aligned with d-axis and inlaid into the rotor body [16], the reluctance difference is produced between the d- and q-axis. The d-axis inductance L_d is low due to the low permeability of the PM, while the high q-axis inductance L_q comes from the small airgap, i.e., $L_d < L_q$. To produce the similar salient-pole reluctance effect, the one-layer PMs or more-layer segmented PMs can also be imbedded inside the rotor body, shown in Figure 25.22(a) and (b), respectively. The magnetizing direction of the PMs in the segmented arrangement of Figure 25.22(b) has d-axis orientation. The machines with the imbedded PMs are called interior PM machine.

On the other hand, the nonmagnetic barriers or axial laminated structures can be added between the imbedded PMs, shown in Figure 25.23(a), which exhibits the conventional salient pole effect (i.e., $L_d > L_q$) because the q-axis reluctance is higher than d-axis reluctance. In the structure shown in Figure 25.23(a), the PMs represented by a solid block inside the rotor are magnetized in the tangent direction. A more direct idea is to combine

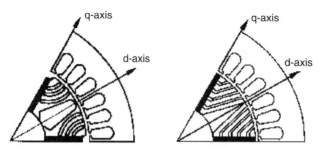

(a) Interior PM machine with magnetic barriers in q-axis

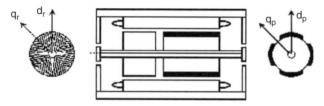

(b) Hybrid rotor with two parts: surface mounted PM & reluctance

Figure 25.23 Different PMAC machines with $L_d > L_q$.

the PMAC machine and the reluctance machine, i.e., a hybrid rotor with two parts [21]. The rotor consists of two blocks in the axial direction: One is the conventional surface-mounted part and the other is the axial laminated part, shown in Figure 25.23(b). The combination of the two rotor parts displays the characteristic of the sinusoidal salient PMAC machine. The machine in Figure 25.23(b) has the salient pole effect of $L_d > L_q$. However, this salient pole effect can be changed into $L_d < L_q$ if the two rotor parts are assembled after one of them is rotated by 90°.

Generally, the sinusoidal interior PMAC machines, in either Figure 25.22 or Figure 25.23, display high armature reaction inductance and salient pole effect. By proper design, the flux weakening can be performed to meet the requirement of the wide speed range and the constant voltage operation in ISA applications. The electromagnetic torque of the sinusoidal IPM can be written as follows:

$$T_{em} = \frac{3}{2} p \left[\psi_f i_q + i_d i_q (L_d - L_q) \right] \tag{25.4}$$

where all variables have the same meanings as those in Equation 25.2 and Equation 25.3.

The first term $\psi_f I_q$ corresponds to the primary or main component of electromagnetic torque, while the second term $i_d i_q (L_d - L_q)$ corresponds to the reluctance torque component. The reluctance torque will be null if there is no salient pole effect or $L_d = L_q$. The voltage equation of the sinusoidal IPM machine can be written in phasor style:

$$V_{ph} = E_f + rI_s + j\omega L_d I_d + j\omega L_q I_q \tag{25.5}$$

where ω is the electric angular speed (rad/s); V_{ph} is the phase voltage (V); and r is the stator phase resistance (Ω). At a given speed, Equation 25.5 can be expressed by the phasor diagram shown in Figure 25.24. Figure 25.24(a) shows the stator current is in the first quadrant (i.e., $0 < \beta < 90°$, $I_d > 0$, and $I_q > 0$), while Figure 25.24(b) shows the stator

Integrated Starter Alternator

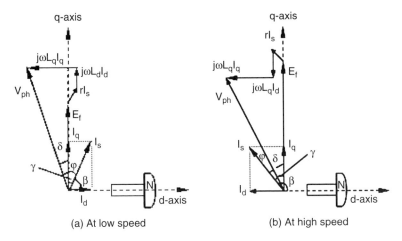

Figure 25.24 Phasor diagram for the sinusoidal IPM machine running at motoring state.

current is in the second quadrant (i.e., $90° < \beta < 180°$, $I_d < 0$, and $I_q > 0$), in which β is the current phase angle. When the stator current phasor is in the third and fourth quadrants, the machine runs at generating state. When the stator resistance r is negligible, the electromagnetic torque, either at motoring or generating state, can also be expressed as:

$$T_{em} = \frac{m}{p}\left(\frac{30}{\pi n_r}\right)^2 \left(\frac{E_f V_{ph}}{L_d}\sin\delta + \frac{V_{ph}^2}{2}\frac{L_d - L_q}{L_d L_q}\sin 2\delta\right) \quad (25.6)$$

where E_f is the phase EMF produced solely by PM field (rms, V); m is the phase number; p is the number of the pole pairs; n_r is the rotor shaft speed (rpm); δ is the angle between E_f and V_{ph}, i.e., the power angle or torque angle. δ is defined as positive when V_{ph} leads E_f, and δ is defined as negative when V_{ph} lags E_f. In Figure 25.24, φ is the angle between phase voltage V_{ph} and current I_s, and $\cos\varphi$ is the power factor; β is the phase angle of the current I_s referenced to d-axis, and $I_s\cos\beta = I_d$, $I_s\sin\beta = I_q$. γ is the phase angle of the current I_s referenced to q-axis; ω is the synchronous angular frequency ($\omega = 2\pi f$, rad/s).

The relationship between the electromagnetic torque T_{em} and the power angle δ in Equation 25.6 can be drawn in Figure 25.25. The IPM machines in Figure 25.22 display the salient pole effect of $L_q > L_d$, and their torque vs. power angle can be shown in Figure 25.25(a). The resultant torque consists of main component ($\propto \sin\delta$) and reluctance component ($\propto \sin 2\delta$). Compared to a surface-mounted PM AC machine, the IPM machine provides an additional reluctance torque component, which moves the maximum torque from the position $\delta_{max} = 90°$ to $\delta_{max} > 90°$. The stable operation regime of the IPM machine is $0° < \delta < \delta_{max}$ at motoring state and $-|\delta_{max}| < \delta < 0°$ at generating state. On the other hand, if the IPM machines in Figure 25.23 display the salient pole effect of $L_d > L_q$, their torque vs. power angle can be shown in Figure 25.25(b). In this case, the additional reluctance torque causes the maximum resultant torque to move from position $\delta_{max} = 90°$ to $\delta_{max} < 90°$. The stable operation regime of the sinusoidal IPM machine with $L_d > L_q$ is $0° < \delta < \delta_{max}$ at motoring state and $-|\delta_{max}| < \delta < 0°$ at generating state.

A main reason to choose the sinusoidal IPM machine as the candidate of the ISA application is its flux weakening potential. A so-called critical EMF that corresponds to the critical state of the stator current between flux aiding and flux weakening is defined to determine the state of the machine, i.e.,

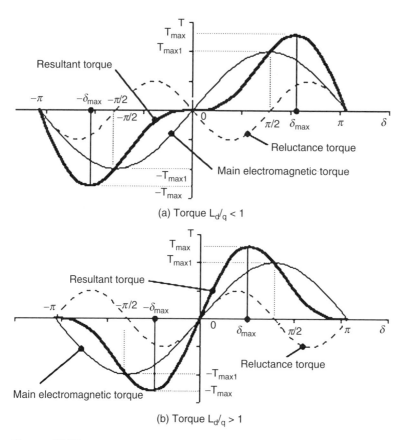

Figure 25.25 Relationship between torque and power angle of interior PM machine.

$$E_c = \sqrt{V_{ph}^2 - (\omega L_q I_s)^2} - I_s r \tag{25.7}$$

Actually, the critical state separating flux aiding and weakening corresponds to zero d-axis current (i.e., $I_d = 0$), at which all the stator current produces torque, i.e., $I_q = I_s$. The machine is running in flux aiding state if the no-load EMF E_f produced by PM is higher than the critical EMF E_c, while it is running in flux weakening state if $E_f < E_c$. However, it can be proved that the current locus at the maximum torque/current control goes along the asymptote at 45° lines in the second and third quadrants of d-q plane [16]. At given current limitation, the maximum torque is achieved by controlling the phase angle γ between 45° and 90° or β = 135~180°.

Figure 25.24(a) gives the phasor diagram under flux aiding state, while Figure 25.24(b) displays the phasor relationship at flux weakening state. The incremental part of the no-load EMF with speed increase is offset by the back-EMF component $\omega L_d I_d$ in Figure 25.24(b), and the terminal voltage is maintained at a similar level at high speed. During flux weakening operation, the high d-axis armature reaction inductance L_{ad} (= $L_d - L_1$) requires a relatively low d-axis current to produce certain flux weakening effect. L_1 is the stator phase leakage inductance. The sinusoidal IPM machine with high L_{ad} benefits its constant power operation at wide speed regime by flux weakening. However, the IPM machine with $L_q > L_d$ displays other advantages besides its easy manufacturing. A salient

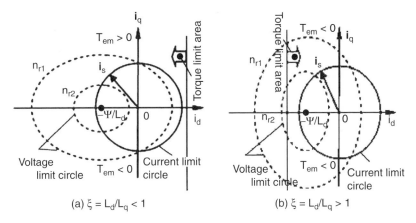

Figure 25.26 Current limit circle, voltage limit ellipse, and torque limit area of IPM machine ($n_{r1} < n_{r2}$).

ratio is defined as $\xi = L_d/L_q$, whose value describes different salient effects. The surface-mounted PMAC machine has no salient effect and $\xi = 1$. The $\xi < 1$ represents that the reluctance in d-axis is higher than that in q-axis, and the machines in Figure 25.22 belong to this category. The $\xi > 1$ represents the machine with characteristic of $L_d > L_q$, and the examples are shown in Figure 25.23. The comparison and analysis on the IPM machine with different salient ratios have been done [21–24]. In order to understand better the effect of the different salient ratio on the sinusoidal IPM machines in the ISA applications, the current limitation circle and the voltage limitation ellipse are drawn in Figure 25.26 first. The inverter volt-ampere ratings, the maximum speed, and the related parameters are highlighted as follows.

The current limitation circle in the sinusoidal IPM machine is the same as that in the surface-mounted PM machine. The stator current phasor must be inside the current limit circle to prevent it from damaging the inverter and overheating the machine, which has low limitations, i.e., $i_s < I_{max}$.

Unlike the surface-mounted PM machine, the reluctance torque in the IPM machine adds one more limitation to the operation area in the d-q plane. Based on Equation 25.4, the d-axis current i_d has to meet the torque limitation of $[\psi_f + i_d (L_d - L_q)] \geq 0$ if motoring operation with $I_q > 0$ or generating operation with $I_q < 0$. The torque limit area can be further expressed by the following half-plane equation

$$\left. \begin{array}{ll} i_d \geq -\Psi_f/(L_d - L_q) & (L_d > L_q) \\ i_d \leq -\Psi_f/(L_d - L_q) & (L_d < L_q) \end{array} \right\} \quad (25.8)$$

That is, if adjusting $i_q > 0$ for motoring and $i_q < 0$ for generating, Equation 25.8 displays the right-half plane at the critical torque line of $i_d = -\psi_f/(L_d - L_q)$ for $L_d > L_q$ or the left-half plane at the critical torque line of $i_d = -\psi_f/(L_d - L_q)$ for $L_d < L_q$. The critical torque line is located at the second and third quadrants for $L_d > L_q$ in Figure 25.26(a), while it lies in the first and fourth quadrants for $L_d < L_q$ in Figure 25.26(b). If the parameters are so designed that $|\psi_f/(L_d - L_q)| \geq I_{max}$, then the torque critical line is outside the current limit circle. In this case, the torque limit in Equation 25.8 can be neglected because the current limit is much stricter than the torque limit. Otherwise, the intersection area of Equation 25.2 and Equation 25.8 is the potential operation area without voltage limitation.

The voltage limitation circles of the surface-mounted PM machine become the voltage limitation ellipses in the IPM machines. Based on Equation 25.3, the ellipse center is located on the d-axis, and its distance from the origin of the d-q current plane is approximately equal to the three-phase short-circuit current ($I_{sc} = -\psi_f/L_d$), i.e., at point ($-\psi_f/L_d$, 0). The voltage ellipse center may fall inside or outside the current limit circle, depending on the parameters ψ_f and L_d of the machine. The machine can run for constant power or reduced power until infinite speed by flux weakening if $\psi_f/L_d < I_{max}$. Actually, this design also guarantees no inverter damage at a three-phase short-circuit of the machine, which is the preferable machine design [16]. However, the risk for this design consideration is possible demagnetization of the PM when the d-component of the stator current is higher than the short-circuit current $|I_{sc}|$ (but still within the current limit circle) at deep flux weakening operation. The PM materials (such as NdFeB or SmCo) with property curve extending to the third quadrant in B–H plane become necessary [16]. If the center of the voltage limit ellipse is outside the current limit circle, the maximum speed is the speed at which the voltage limit ellipse is tangent to the current limit circle for all machines of $L_q > L_d$ and the machines of $L_d > L_q$ satisfying torque limitation. One axis of the ellipse aligns with horizontal d-axis, while the other is parallel with the vertical q-axis. The lengths of the horizontal and vertical axes of the voltage limit ellipse are

$$\frac{30 U_{max}}{\pi p L_d} \frac{1}{n_r} \text{ and } \frac{30 U_{max}}{\pi p L_q} \frac{1}{n_r},$$

respectively. The size of the ellipse is inversely proportional to the speed n_r at given voltage limitation, and $n_{r2} > n_{r1}$ in Figure 25.26. The horizontal axis is longer at $L_d < L_q$, as shown in Figure 25.26(a), while the vertical axis is longer at $L_d > L_q$, as shown in Figure 25.26(b). The current arrow has to be within the voltage limitation ellipse corresponding to this speed.

For a given torque, stator current locus in $i_d - i_q$ coordinate is hyperbola whose asymptotes are the i_d-axis and the critical torque line. During IPM control, the stator phase currents must be limited within the intersection area of the current limit circle, the voltage limit ellipse at given speed, and the area bounded by torque hyperbola toward its focus side, shown in Figure 25.26.

It has been presented that the sinusoidal IPM machines with moderate saliency $\xi = L_d/L_q > 1$ in Figure 25.23(b) shows lower rated volt-ampere ratings and higher performance during flux weakening [21]. The variation of the current phase angle, with variation of the speed or power, is more significant than that in machines with $\xi < 1$, which allows a low-resolution position sensor to be used for the IPM machines with $\xi > 1$. However, the comparison of advantages and disadvantages between machines with $\xi > 1$ and with $\xi < 1$ is still under discussion [20–24]. A practical design of an IPM machine with $\xi < 1$ for the ISA application is given [25] to meet the requirement of EMF, cold cranking torque, and generating power at high speed.

In summary, the sinusoidal IPM machines are good candidates for the ISA application. The surface-mounted PM machines, either trapezoidal or sinusoidal type, do not fit the requirements for the large speed range operation by flux weakening, unless an extra DC-DC converter is added between the inverter and 42 VDC bus.

Induction Machines. Most induction machines in ISA applications are three-phase machines and connected with DC bus through a power electronic inverter/rectifier. An induction machine consists of a normal three-phase stator and a cage or wound-winding rotor. The wound rotor is seldom used as an ISA machine rotor because it belongs to brush type structure, which requires three slip rings and three stationary brushes. A typical

Integrated Starter Alternator

(a) Stator   (b) Cast rotor

Figure 25.27 Induction machine.

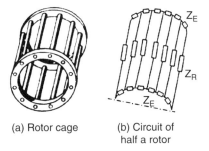

(a) Rotor cage   (b) Circuit of half a rotor

Figure 25.28 Rotor cage and its equivalent impedance network of an induction machine.

crankshaft-mounted induction machine for ISA application is shown in Figure 25.27. The stator windings can be wound by round wire or made by square conductors. The L-shaped rectangular copper bar winding shown in Figure 25.27(a) can be used to increase slot fill ratio and decrease end-turn length because of the limited available space between IC engine and transmission [10,26]. Unlike the field current in synchronous machine, the rotor current of induction machine is induced by the rotation magnetic field in the airgap. The laminated core is required to carry the rotor cage, and the rotor cage can be cast with copper or aluminum, as in Figure 25.27(b). For prototypes, copper or aluminum bars can first be inserted into rotor slots, and then welded with end rings. A rotor cage after removing the rotor core is shown in Figure 25.28(a), and the rotor equivalent impedance circuit net is shown in Figure 25.28(b), where Z_R and Z_E are the impedance of a rotor bar and an end-ring segment, respectively.

Like stator windings in sinusoidal PMAC machines and synchronous machines, they are preferred to dominate the fundamental magnetomotive force and depress the low-order harmonic MMF, especially the 5th, 7th, 11th and 13th harmonics in three-phase windings. The traditional technology, such as the distribution and short pitch (y ≈ 5/6 pole pitch), is effective for harmonic reduction, while the full pitch concentrated windings are avoided in induction machine. The windings with high slot fill ratio can help compact the machine size and reduce stator resistance losses as well as improve the performance of the machine, especially at low-speed range. To increase the reliability, the wire insulation with inverter grade can reduce the stator failure rate because all ISA machines interact with power electronic control.

When three-phase sinusoidal currents with 120° phase shift conduct in the three-phase symmetrical windings with phase difference of 120° in space, a sinusoidal rotating magnetic field is produced in the airgap. The magnetizing component of the stator current is about 20 to 50% of the rated stator current, but it may reach above 60% at low speed for an ISA machine at generating state. This is especially true for the crankshaft-mounted ISA machine requiring larger airgap (0.6 to 0.8 mm) because of the fluctuation of the engine shaft. The airgap flux density is about 0.7 to 1.0 Tesla for ISA low-speed range. The airgap magnetic field rotates at the same electric angular speed ω_s as the stator current, which is called the synchronous angular speed/frequency. Changing the stator current frequency can change the rotation speed of the magnetic field in the airgap. Changing the conduction sequence of three-phase currents or solely switching any two stator terminals can reverse the rotation direction of the magnetic field in the airgap, which is used to change the rotation direction of the machine or produce braking torque in conventional applications. The rotor currents are induced in the rotor cage because of the rotation magnetic field in the airgap. The interaction between the rotor current and airgap magnetic field produces the magnetic torque. The magnetic torque drives the rotor rotation at motoring state and contradicts the rotor rotation at generating state. The machine operates at motoring state if the magnetic field runs faster than the rotor ($\omega_s > \omega_r$, ω_r – rotor electric angular speed). The machine operates at generating state if the magnetic field runs slower than the rotor speed ($\omega_s < \omega_r$). However, there is no electromechanical energy conversion if the rotor runs at the same speed of the magnetic field ($\omega_s = \omega_r$), since no rotor current is induced if there is no relative movement between rotor bars and the magnetic field. The speed difference, i.e., $\Delta\omega = \omega_s - \omega_r$, is a key parameter for induction machine, and the speed difference in per unit is defined as slip and given as follows

$$s = \Delta\omega/\omega_s = (\omega_s - \omega_r)/\omega_s \qquad (25.9)$$

where $\omega_s = 2\pi f$ in rad/s, and f is the stator current frequency (Hz). The synchronous electric angular speed ω_s can be expressed by the synchronous mechanical speed n_s in rpm, i.e., $n_s = 2\pi\omega_s/p = 60f/p$, where p is the number of machine pole pairs. It can be proved that $\Delta\omega$ is equal to the angular frequency of the rotor current (rad/s), i.e., $\Delta\omega = 2\pi f_r$, where f_r is the rotor current frequency (Hz). The rotor speed in rpm can be written as $n_r = n_s(1-s)$.

At a given stator current frequency, the operation state of an induction machine can be expressed by the slip as the braking state by reversing conduction sequence at $s > 1$ (or $n_r < 0$), the motoring state at $0 < s \leq 1$ (or $n_s > n_r > 0$), the generating state at $s < 0$ (or $n_r > n_s$), and no energy conversion at $s = 0$ (or $n_r = n_s$). The electromagnetic torque T_{em} of the induction machine can be written as the function of the slip s at a given phase voltage and stator frequency as well as electrical parameters of the machine, such as stator and rotor resistances, leakage inductances, and magnetizing inductance. Figure 25.29 shows curves of torque vs. rotor speed when the induction machine runs for ISA application. Each thin line in Figure 25.29 represents a torque-slip curve at a given stator frequency. The solid thick curve represents the motor operating at constant flux/torque control and the constant power or reduced power control by flux weakening. The dashed thick line represents the electric power generation from the idle through the maximum speed.

At the beginning of engine cranking, the induction machine produces 1.5 to 1.6 times the cranking torque to overcome the breakaway torque of the engine, which requires the start torque of the machine reaching or exceeding the breakaway torque at the lowest start frequency f_{min} of the stator current. At the given f_{min}, the stator voltage required to drive the engine from static state can be derived from E/f = constant (here E is the back-EMF) or even higher than this, and the voltage is higher than that from V_{ph}/f = constant

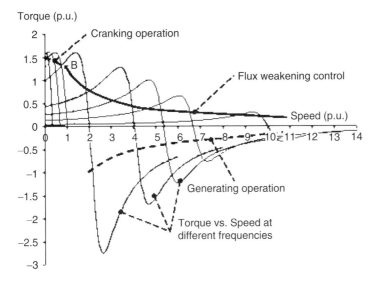

Figure 25.29 Torque-speed characteristics of an induction machine for ISA application.

during conventional constant torque/flux control because the stator resistance has more effect than the short-circuit reactance at low frequency. Once the engine starts rotating to 10 to 30 rpm, the required ISA machine torque can be gradually reduced from the engine static point to cranking speed, although the constant high torque benefits the high cranking speed. However, the declined torque requirement helps design the ISA machine more compact.

The induction machine is allowed to run at the slip values corresponding to the maximum torques with frequency variation during cranking procedure because the cranking procedure is very short, for example, 0.2 to 3 sec, although the slip at the maximum torque causes low power factor and high machine losses. Running on the outline of torque-speed curves can only be set as a short time schedule, while the speed for the continuous operation at each given inverter output frequency has to fall into the torque curve segment between peak torque and the zero torque at synchronous speed, i.e., at the right side of the peak torque. At each individual frequency point, the maximum torque decreases with the stator resistance while the slip corresponding to the maximum torque (i.e., the critical slip s_m) is proportional to rotor resistance. It can also be proved that the rotor resistance loss is proportional to slip at given stator frequency. This means low stator resistance benefits to torque capability while low rotor resistance helps the thermal capability of the ISA induction machine. This is the reason why rectangular stator windings and low-cost copper rotors are attractive. The maximum torque is proportional to the square of the ratio between voltage and frequency at the high frequency range. Therefore, the power capability of the induction machine decreases with speed increase at constant voltage operation by flux weakening (especially for ISA electric power generation), and the constant power operation can only be valid within the speed variable ratio of 3~5. Actually, existing publications unveil the unstable phenomenon at the high-speed end of the ISA induction machine [27]. Fortunately, it is not necessary for the ISA machine to run at engine redline speed without compromising the power output.

The induction machines fit the ISA application. However, its rotor losses should be properly controlled. This is especially true for the machine with liquid-cooled stator and without rotor cooling fan. The rotor fan blades can be cast together with rotor rings for the belt-driven ISA. However, the rotor fan does not function well for the crankshaft-mounted ISA because of the high percentage of hot air recirculation within the narrow space. To the fan cooling induction machine, the continuous power should be lowered to 80% at 20% rated speed. Forced-air cooling does not have this limitation. The pole number of the induction machine for ISA application is about 8–16. Higher pole number benefits to high torque design, but the power factor of the induction machine with high pole number is relatively low and the required carrying frequency for inverter switches is high. For example, the power factor of a 16-pole 8 kW induction machine may be lower than that of an 8-pole machine by 10 to 15%. Low-power factor requires high inverter ratings and high stator losses due to the high reactive power. The rotor speed signal must be fed back to the power inverter/rectifier in the ISA application. However, the speed signal sensor, which can be derived from rotor position signal, is much less expensive, compared to the position sensor required for sinusoidal IPM machine. The issues related to sensors and their applications are addressed in Chapter 10.

Switched Reluctance Machines (SRM). The SRM has double saliency structure. The cross-section of a three-phase 12/8 pole SRM, with 8 rotor poles and 12 stator poles, is shown in Figure 25.30. A concentrated coil (not shown in Figure 25.30) is wound on a stator pole. Each phase among the three phases consists of four coils wound on the four stator poles on the two perpendicular radial lines. For example, phase A consists of four coils on the stator poles A, A′, A″ and A‴ in Figure 25.30; so do phases B and C. The four coils can be connected into four paths with all four in parallel, two paths with two in series and then in parallel, and one path with all four in series, depending on the design specification. The polarity of four poles in each phase may be NNSS or NSNS. The NSNS arrangement is preferred because it provides low flux density in its magnetic circuits, which provides higher varying rate of inductance, i.e., higher $L(\theta,i)/\theta$. The phases are connected with the power electronic inverter (see Section 25.4.6.2). The conduction sequence of the three phases is based on the rotor position provided by a position sensor mounted on the rotor shaft.

To understand the principle of the SRM, two rotor positions (i.e., alignment and nonalignment) are defined first. When the stator pole center in a phase and the rotor pole center are aligned with each other, for example, the position of phase A in Figure 25.30(a) and phase B in Figure 25.30(b), the positions are in alignment. Otherwise, they are in nonalignment if the stator pole center is aligned to the span center between two rotor poles. The stator phase inductance vs. rotor position angle is shown in Figure 25.31. The stator phase inductance reaches its peak value at alignment position and becomes the

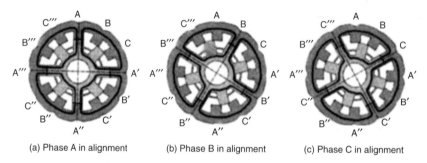

(a) Phase A in alignment   (b) Phase B in alignment   (c) Phase C in alignment

Figure 25.30 Three-phase switched reluctance machine with 12/8 poles.

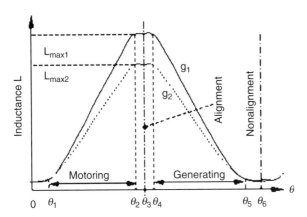

Figure 25.31 Inductance vs. rotor position angle of the SRM.

minimum at nonalignment position. The electromagnetic torque trends to pull the rotor to alignment position. The torque can be given as follows without magnetic saturation:

$$T_{em} = \frac{1}{2} i^2 \frac{dL(\theta)}{d\theta} \qquad (25.10)$$

where i is the stator phase current; L is the phase inductance; and θ is the rotor position angle.

In Figure 25.31, the torque is positive when $dL/d\theta > 0$ from θ_1 to θ_2, and the SRM runs as a motor. The phase B is excited at the rotor position shown in Figure 25.30(a), then phase C conducts at rotor position shown in Figure 25.30(b), and phase A conducts at rotor position of Figure 25.30(c), and then phase B again, and so on. In this way, the rotor is pulled and rotates in counterclockwise direction. The rotor rotates 15° during every phase conduction, and 24 switching shifts are required for one revolution. The machine rotates in a clockwise direction if the conduction sequence in Figure 25.30(a), (b), (c) is C-A-B, then C again, and so on. However, the rotor rotation direction is independent of the current direction, based on Equation 25.10. The phase is switched on at offset alignment position, but switched off at or before the alignment position when the SRM runs at motoring state. The SRM output power can mainly be controlled by adjusting switching-on position while the vibration and noise reduction can be controlled by adjusting switching-off angle and smoothing the dropping voltage [28,29]. The torque is negative when $dL/d\theta < 0$ from θ_4 to θ_5, and the SRM runs as generator. When the engine drives the SRM rotating in the counterclockwise direction, the SRM generates electric power if the conduction sequence in Figure 25.30(a), (b), and (c) is A-B-C, then A, and so on. At generating state, the phase is switched on at alignment position, and switched off at offset alignment position. With increase of the SRM airgap, the maximum inductance and the $dL/d\theta$ decrease, For instance, $L_{max2} < L_{max1}$ due to airgap $g_2 > g_1$ in Figure 25.31, and the torque decreases at the given phase current with a large airgap. A small airgap benefits to high torque density of the SRM. Therefore, the SRM power is compromised when it is used in the crankshaft-mounted ISA application due to larger airgap.

The main losses in the SRM exist in the stator windings. The iron losses in the stator and rotor are relatively low at low speed, while they increase dramatically at high speed. Fortunately, the iron volume used in the SRM is not so much. The salient rotor has the fan effects, and an extra rotor fan can be avoided. Its efficiency is similar to high-efficient induction machine. The low rotor moment of inertia and no rotor winding benefit to high-speed operation. The SRM is suitable for wide speed variable ratio, and the development of the SRM for aircraft ISA application targets the speed up to 52,000 rpm [30]. The sensorless control technology is also introduced to the SRM for ISA applications [31]. The SRM for ISA application has been extended to heavy-duty trucks [9].

The torque ripple, vibration, and acoustic noise are the main disadvantages of the SRM. The torque ripple can be reduced by controlling the current waveform and the switching angles including conduction angle, switching on and off position, and so on. The radial magnetic force is the main reason for the SRM noise, especially when the force frequency is near the stator resonant frequency and with the same mode shape [28]. The acoustic noise can be suppressed by proper design (such as, three-phase 12/8 poles is better than three-phase 6/4 poles) and smooth voltage at switching-off as well as performing active noise control technology [29]. Chapter 21 addresses the vibration and noise issues of the SRM in detail.

25.4.5.3 Application Comparison of ISA Electric Machines

The electric machines used for ISA applications are claw pole Lundell synchronous machine, induction machine, sinusoidal PMAC machine, and switched reluctance machine. Table 25.1 compares the practical properties for the above machines by qualitative analysis. The trapezoidal PMAC or brushless DC PM machine with surface-mounted PMs is also included in the table, although it does not fit the specifications of the ISA for wide speed variable ratio by flux weakening. A PM machine has a higher efficiency than an induction machine and the SRM. The price paid for the simple structure and control of the SRM is its high torque ripple and acoustic noise. The Lundell machine can easily run at high speed by adjusting its field current, but output power is limited due to its structure, and its control is complicated because of the additional field control. The efficiency and capability of the PM machine are compromised during constant voltage operation, especially in the light-load and high-speed range, while the SRM can run very fast and even be used as aerospace ISA at speeds up to 60,000 rpm or higher. There are many on-the-shelf drives and technologies for induction machine control, and the resolution requirement to its speed sensor is relatively low. Except for induction machines and trapezoidal PM machines, all other machines require high-resolution position sensors, although sensorless technology is under development.

Table 25.1 Comparison of Different Machines for ISA Applications

Electric Machine Types	Claw Pole Lundell	Trapezoidal PMAC	Sinusoidal PMAC	Induction Machine	Switched Reluctance
Efficiency and compactness		✓	✓		
Low torque ripple and noise	✓		✓	✓	
Easy close loop control		✓			✓
Fewer control sensors		✓			
Wide speed range	✓		✓	✓	✓
High power application		✓	✓	✓	✓

25.4.6 DC-AC INVERTER AND AC-DC RECTIFIER

The electric energy is stored in a DC battery on-board the vehicle. Any electromechanical energy conversion through an AC electric machine or a switched reluctance machine requires a power electronic bridge. The electronic bridge inverts the DC-to-AC power supplying to the machine under motoring operation while it rectifies the AC-to-DC power charging the battery or running electric loads under generating operation of the ISA machine. Both inverter and rectifier function have to be fully considered during the power electronic bridge design. Running as a motor, the ISA machine requires constant torque operation during engine cranking and constant power operation during driving assistance. Running as a generator, the ISA system requires providing DC power with constant voltage, such as 42 V, in the machine speed range corresponding to the engine speed from idle to redline. Besides the mechanical requirement, the inverter/rectifier has to fully function at ambient temperature of –40 to 105°C and partially function at ambient temperature of 105 to 125°C.

The following paragraphs discuss three-phase inverter/rectifier for three-phase AC machines and switched reluctance machines.

25.4.6.1 Configuration of Three-Phase Converter

In the electric drives of the ISA machine, the voltage source inverter (VSI) is much more popular than the current source inverter (CSI) due to different factors. With the same power rating of the inverter, the capacitor (if any, but can be reduced or eliminated because of the existence of the battery on-board the vehicle) in the VSI is cheaper and smaller than the inductor in the CSI. In addition, the new electronic devices without reverse blocking capability are more suitable for VSI application. Figure 25.32 shows a topological configuration of the three-phase inverter/rectifier (VSI type) consisting of a capacitor, C, and six electronic switches, S_1–S_6. The MOSFET and IGBT are the most commonly used power electronic switches in the ISA system, although there can be others, like BJT, GTO, and MCT. The MOSFET has the highest switching frequency in spite of its low controllable capacity. Its resistance at conduction is proportional to 2.6 times its blocking voltage, so the MOSFET is more suitable for low-voltage application. The symbol and the rating [16] of a typical MOSFET are shown in Figure 25.33(a). The low switching losses and low gate drive power are the advantages of the IGBT. However, a high voltage variation rate dv/dt occurs during fast switching, which requires paying much more attention to electromagnetic interference (EMI) of the connection cables and the special machine insulation against dv/dt stress. The rating and the symbol of a typical IGBT are shown in Figure 25.33(b). It is preferable to use MOSFET in the 42 V system, while the IGBT can be used

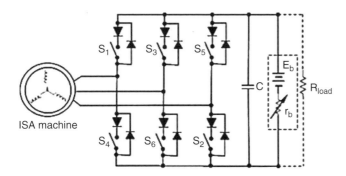

Figure 25.32 Inverter/rectifier diagram.

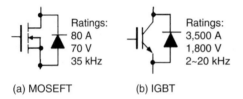

(a) MOSEFT (b) IGBT

Figure 25.33 Electronic components used in an ISA system.

in the high-voltage system in full hybrid power train, such as 600 VDC in an Allison electric transmission.

In the CSI for trapezoidal PMAC, the parallel capacitor in Figure 25.32 is replaced by a series inductor between the battery and power switching bridge. It is also possible to insert a current regulator between the capacitor and the power switching bridge. Every switch conducts 120° (electrical degrees), and two switches are on during each time interval. The switching sequence of the six switches is S_6S_1, S_1S_2, S_2S_3, S_3S_4, S_4S_5, S_5S_6, and then back to S_6S_1 again. Although three-phase currents are regulated, only one DC current sensor is required because the regulated DC bus current conducts in two-series phase windings (if Y-connection) at any instant. However, the current regulation and the electronic switching can be performed in the VSI by PWM [16].

In the VSI, the relationship between DC voltage and AC line-to-line rms voltage varies, depending on the modulation strategies. There are three modulation strategies for the control of output three-phase AC voltage, i.e., six step control and sinusoidal modulation as well as space vector modulation.

In the six-step or rectangular waveform control, three switches are on and off in each time interval. The switching sequence of the six switches in Figure 25.32 is $S_5S_6S_1$, $S_6S_1S_2$, $S_1S_2S_3$, $S_2S_3S_4$, $S_3S_4S_5$, $S_4S_5S_6$, and then back to $S_5S_6S_1$ again. The rms value of the AC line-to-line fundamental voltage can be given as:

$$U_{LL(rms)} = \frac{\sqrt{6}}{\pi} U_{DC} \approx 0.78 U_{DC} \qquad (25.11)$$

The magnitude of the fundamental phase voltage with six-step control is $(2/\pi)U_{DC}$, shown in Figure 25.34. A high AC voltage can be achieved at the given DC voltage, which is the main advantage of this control method. The higher voltage leads to lower resistance losses, since the current is relatively low due to high voltage at given power output. However, the low-order harmonics, such as 5th and 7th, appear in the voltage waveforms, which cause high stray losses of the machine.

When the variation of the switching duty cycle is sinusoidal, an approximate sinusoidal phase current can be obtained. This modulation strategy is called sinusoidal PWM. With sinusoidal PWM strategy, the rms value of the AC line-to-line fundamental voltage can be obtained as follows:

$$U_{LL(rms)} = \sqrt{\frac{3}{8}} U_{DC} \approx 0.61 U_{DC} \qquad (25.12)$$

The magnitude of AC fundamental phase voltage is $0.5U_{DC}$ with sinusoidal PWM modulation, shown also in Figure 25.34. Lower harmonic voltage and current is the main advantages, which helps reduce the stray losses of the machine. On the other hand, its AC

Integrated Starter Alternator

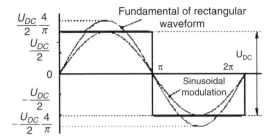

Figure 25.34 Fundamental phase voltage under rectangular and sinusoidal voltage controls.

output voltage is lower than that in six-step control, which will increase the resistance losses due to higher AC current for the same output power requirement. Compared to the six-step control, furthermore, the chopping frequency is high in the sinusoidal modulation, which causes higher inverter losses.

To increase the output AC voltage at a given DC bus voltage, the space vector modulation [16] can be used without introducing much harmonic components. With space vector PWM control, the rms value of the AC line-to-line fundamental voltage can be achieved as follows:

$$U_{LL(rms)} = \frac{U_{DC}}{\sqrt{2}} \approx 0.707 U_{DC} \qquad (25.13)$$

The relationship between space vector PWM control and sinusoidal PWM modulation is drawn in Figure 25.35. Actually the AC output voltage increases by 15% in the space vector PWM, compared to the sinusoidal PWM. In practical application of the ISA control, the space vector PWM is the popular strategy, while the six-step modulation is introduced to expand the operation to high speed. Generally, changing the control strategy between space vector control and six-step control can increase the speed varying ratio of the ISA machine. Please notice the practically available AC voltage is lower than the voltage given in Equation 25.11, Equation 25.12, and Equation 25.13 because there are voltage drops across the power electronic switches.

In Figure 25.32, besides the switch components there exists a reverse parallel diode at each bridge arm. Those diodes are called feedback or freewheeling diodes. They provide

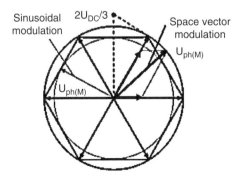

Figure 25.35 Voltage relationship between space vector control and sinusoidal control.

an alternative path for the machine current when the main power switch is off at motoring state. When the ISA machine runs at generating or regenerating state, the diodes provide the main active power path for battery charging, while the switchable components supply reactive power to build airgap flux for an induction machine or to weaken the airgap flux for constant power operation of a permanent magnet machine. In the ISA system, the diodes are the main power channel at generating state and behave much beyond the freewheeling in the conventional industry drives.

25.4.6.2 Inverter Configuration of the SRM

Figure 25.36 shows a classic converter for three-phase switched reluctance machine of Figure 25.30. E_b and r_b in the dashed line box represent a real energy storage battery. Three windings represent the three phases, i.e., phases A, B, and C. Each phase is connected with two MOSFETs in series. Two diodes at each phase provide alternative paths for the current when the main switch(es) is off. It should be noticed that there are many other main circuit topologies beyond the circuit in Figure 25.36, and the power switch components could be any type in spite of the shown MOSFET in Figure 25.36. When the SRM runs under motoring state, the circuit of each phase provides three loops, i.e., positive, zero, and negative loops.

1. *Positive loop:* When the two main switches in upper and lower arms are on and the machine runs at the positive slope range of phase inductance, the DC bus voltage is exerted fully across the phase windings, and $I_{AU} = I_{phA} = I_{AL} > 0$, while $I_{DU} = I_{DL} = 0$. The machine produces positive torque operating at motoring state if it runs at positive slope of the inductance variation. The current from DC bus is provided to the working phase at this state.
2. *Zero loop:* When one switch at either upper or lower arm is off while the other is on, there is no external voltage exerted across the phase winding, and the phase winding is in the zero voltage loop. For example, if the switch of the upper arm is on while the lower arm switch is off, $I_{AU} = I_{phA} = I_{DU}$ while $I_{DL} = 0$. This operation state ensures the phase current decreases at a slow rate. The current decays in the zero voltage loop.
3. *Negative loop:* When two main switches are off at the same time, the DC bus voltage is exerted across the phase winding inversely, and $I_{DL} = I_{phA} = I_{DU}$ while $I_{AU} = I_{AL} = 0$. The phase current decreases rapidly at the negative loop. This is also the regeneration state, which feeds back the power to the DC bus.

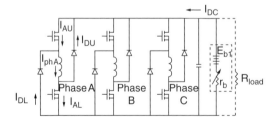

Figure 25.36 Classical three-phase SRM converter with two switches per phase.

Integrated Starter Alternator

25.4.7 DC-to-DC Converter

As discussed in Section 25.3.2.5, it is necessary to provide both the 14 V/28 VDC electric power and the 42 VDC or high-voltage power in the dual-voltage system on-board the vehicle. A bidirectional DC-to-DC converter or at least a unidirectional DC-to-DC buck converter is demanded for the ISA electrical system. The DC-to-DC converter shall have the minimum function as follows: (a) charge a 36 V (or other higher voltage battery bank if applicable) or 12 V (or 24 V) battery in a dual-battery system; (b) charge a 36 V battery from a 12 V battery that is either in the same vehicle or in another vehicle for a jump-start; (c) supply the power to all 12 V electric loads. As mentioned in the previous paragraphs, all the electronic components and devices have to fully function at ambient temperature between –40 and 105°C and can operate at a derated output level at ambient temperature over 105 and up to 135°C, which is independent of cooling styles.

The principle of the conventional DC-to-DC converter has been discussed in Chapter 11. The following paragraphs will be limited to discuss the specifications of the DC-to-DC converter in the ISA system. Furthermore, a multifunction inverter with an inductor connected to the neutral point of the three-phase electric machine is introduced, which can replace the buck inverter from 42 V to 14 V besides its inverter function.

25.4.7.1 Buck Mode of the DC-to-DC Converter

When the DC-to-DC converter converts high voltage into low voltage, it works at buck mode. In the 42 V system, the output is 14 VDC, while the input is 42 VDC. The buck output variation at the 14 V side should be limited between 12 V and 16 V when the input voltage is between 30 V and 48 V at the 42 V side. It is preferable to output the voltage > 8 V at half an input voltage. The buck mode shall be enabled during engine cranking and running. The response to all commands including initial default setting and disabling signals is required as soon as possible, such as within 0.2–0.5 sec. The continuous output power depends on the vehicle requirements, but most of automotive application requests the capability of at least 1.5–2 kW. The voltage ripple of the buck output for 14 VDC should not be over 0.5 V peak-to-peak value. The protection from undervoltage (such as 8 V) and overvoltage (such as 18 V) should fully function. The 14 VDC output must be able to suppress transient overvoltage at full load dump like lost output, battery connection broken, disconnected ground path, or load switch/change, in which the transient voltage should not be over 2.5 times output voltage, i.e., 35 V. The efficiency of the 14 V output at buck mode requires higher than 90% at 20% of the rated load or higher while the minimum efficiency should not be lower than 60% below 6% of rated load.

25.4.7.2 Boost Mode of the DC-to-DC Converter

The boost mode is referred to convert the low DC voltage to a higher voltage level during the engine cranking for the dual-battery system or during jump-start by another vehicle's battery. For example, the input is 14 V and the output is 42 VDC. The variation of the 42 V output should be limited between 30 V and 46 V when the input voltage is between 9 V and 16 V. To prevent the 12 V battery from overly discharging, however, the boost operation can be terminated when the input voltage is below 11 V. Besides the quick response to command signals, the maximum ripple of output voltage must be less than 1.8 V peak to peak without battery connection. The boost mode can tolerate different load dumps at full load, and the output voltage cannot exceed 56–58 V. However, the converter should not be damaged during a short time of the negative input voltage shock due to wrong connection of the 12 V battery or the overvoltage shock (but below 26 V) at jump-start. At the same level of the rated load, the efficiency of 42 V output under boost mode operation allows 3 to 5% lower than the efficiency at 14 V output under buck mode operation because the boost mode brings more switching and component losses than those at buck mode.

25.4.7.3 Multifunction Inverter

Besides driving the electric machine in the ISA system, a multifunction inverter can replace the buck DC-to-DC converter, which provides a cost reduction solution for the ISA system. The entire system diagram is shown in Figure 25.14 in which the conventional DC-to-DC converter is replaced by the existing inverter and a slightly modified ISA machine. To describe the principle of the multifunction inverter, the system topology is redrawn in details, as shown in Figure 25.37. The 36 V battery is simplified into EMF E_b in series with an equivalent internal resistance r_b, which is the function of the discharging current and the state of charge (SOC) of the battery, and so does the 12 V battery. The simplified battery model is given in Section 25.5.1. There is no difference in the circuit topology on 42 V side from a conventional inverter. The 14 V power source is taken from the neutral point of the ISA machine with Y-connection. An inductor L is located between the neutral point and the 14 V output terminal to reduce the current ripple, in which the inductance L depends on the requirement to the low current ripple. Compared with a conventional DC-to-DC converter, the additional inductor consists of fewer parts and costs much less.

The inverter works at synchronous rectification like the buck DC-to-DC converter. To explain how the inverter performs the function like a conventional DC-to-DC converter, only one phase in the circuit configuration is drawn out and shown in Figure 25.38. Switching the upper and lower arm MOSFETs alternatively enables the output DC voltage to be controlled. The output voltage can be written as [15]:

$$V_0 = [T_{on}/(T_{on} - T_{off})]V_{in} \qquad (25.14)$$

where V_o is the output voltage; V_{in} is the input voltage; T_{on} is the time interval in which the upper arm MOSFET is on while the lower arm MOSFET is off; and T_{off} is the time interval in which the upper arm MOSFET is off while the lower arm MOSFET is on.

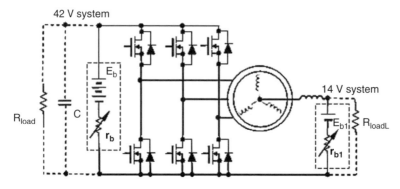

Figure 25.37 System topology of the multifunction inverter.

Integrated Starter Alternator

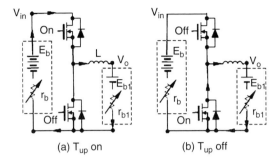

(a) T_{up} on (b) T_{up} off

Figure 25.38 Buck mode operation of a single phase.

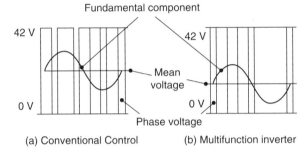

(a) Conventional Control   (b) Multifunction inverter

Figure 25.39 Switching comparison.

From Equation 25.14, the output DC voltage is proportional to the duty cycle of the upper arm MOSFET. The three-phase system can be treated as the expansion of a single phase while their symmetry enables the equal power provided by each phase. At the same time, the three phases' AC power is provided by pulse width modulation control. The bucked DC output voltage, as an additional output, does not affect the ISA machine output, because magnetic fields produced in the airgap by the DC components cancel each other for three symmetrical phase windings. Thus, a system can provide a DC output voltage while it performs as a normal three-phase AC inverter.

For a conventional ISA machine control in Figure 25.39(a), the DC output voltage V_o at neutral point is 21 V if the input voltage is 42 VDC, because the switching-on time of the upper arm and lower arm MOSFETs is the same on average. To output 14 VDC at 42 VDC input, based on Equation 25.14, the PWM switching-on time of the upper arm MOSFET is set to about half of the switching-on time of the lower arm MOSFET, i.e., $T_{on} = 0.5 T_{off}$, shown in Figure 25.39(b). In practice, the targeted DC output is obtained, according to the state of the battery, through a proportional and integrated controller [15].

It should be noticed that the system efficiency of the multifunction inverter becomes lower when the 42 V power requirement is higher than the 14 V power requirement, compared to the conventional ISA system with a DC-to-DC converter. The improvement on control algorithm at higher 42 V power demanding range is required in this system. A big inductor is required to reduce the current ripple, which is another disadvantage of the multifunction inverter.

25.5 ISA SYSTEM ISSUES

The ISA system is a part of the powertrain system. There are many interfaces between the ISA and other systems, mechanically or electrically. At first it has to be mounted on-board the vehicle, and its machine shaft has to interact with the engine crankshaft directly or indirectly in order to exchange mechanical energy between the ISA machine and the engine. Second, the electric terminals of the ISA system must be connected to the vehicle power system to exchange electric energy with the energy storage system and to power electric loads. Third, ISA power electronic controller has to interact with vehicle controller through wire-type or wireless communication. Finally, the ISA cooling system may be inserted into one of the liquid circuit loops on-board the vehicle, except it is cooled separately. However, this section discusses only a part of these issues.

25.5.1 Energy Storage and ISA System

Generally, both the energy storage system and the ISA system are the subsystems of the powertrain, shown in Figure 25.1. However, the ISA system cannot be developed independently, since many of its performances depend on the capability of the energy storage and discharge. Most vehicle manufacturers demand that suppliers provide the ISA and energy storage system at the same time or at least develop them together.

One battery or a group of 12 V batteries on-board the vehicle serve as energy storage in the conventional powertrain system. In the ISA system, one 42 V battery with high-power capacity is added to meet the needs of the ISA machine and high power loads, while one 12 V battery takes care of the conventional 12 V loads. The available battery voltage U_b is critical to the motoring performance of the ISA machine, especially to the initial cold engine cranking and launching assistance at high speed. The battery at discharging can be modeled as:

$$U_b = E_b - r_b \cdot I_b \qquad (25.15)$$

where U_b is the battery output voltage (V); E_b is the open circuit voltage of the battery (V); r_b is the equivalent internal resistance of the battery; and I_b is the discharging current of the battery.

The state of charge is used to describe the ampere-hour capacity, defined as:

$$SOC = \left(1 - \int_0^t i_b dt / C_N \right) \cdot 100\% \qquad (25.16)$$

where C_N is the normal ampere-hour capacity of the battery (Ah).

The open circuit voltage E_b of a battery depends on the SOC. The open circuit voltage of a 12 V battery is given in Figure 25.40 [32]. The open circuit voltage increases with SOC. The internal resistance of a battery is the function of the SOC and ambient temperature. Based on further test results, the equivalent internal resistances of a battery are given in Figure 25.41. The equivalent battery internal resistance increases with decreasing SOC and dropping temperature. According to Equation 25.15 and Figure 25.40 and Figure 25.41, the output voltage of a 12 V battery at the full SOC and the temperature of 20°C (as well as the discharging current of 200 A) is 12.28 V, while it drops to 11.3 V at 60% SOC, temperature of −20°C, and the current of 200 A.

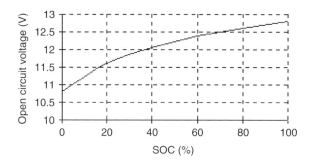

Figure 25.40 Open circuit voltage vs. SOC of a 12 V battery.

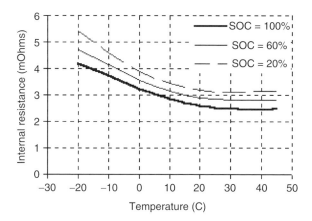

Figure 25.41 Internal resistance vs. SOC of a 12 V battery.

At different discharging currents, curves of the battery voltage vs. the discharging time are measured and given in Figure 25.42. This test is done at 100% SOC and temperature of −29°C. It is found that the battery internal resistance varies with discharging current at a given SOC and temperature, so the internal resistance r_b in Equation 25.15 is an equivalent average value. With extension of the discharging time, the battery voltage drops further even at the constant discharging current. The reason is that the battery SOC drops with discharging time according to Equation 25.16.

Furthermore, the maximum output power for a battery is achieved only if the equivalent resistance at the input terminals of the real loads is equal to the battery internal resistance. Although it is fully considered during the ISA system design, the inductance effect of the ISA machine and the resistance variation due to operation conditions cause the maximum available battery power to never be fully used. The sufficient battery power margin must be left.

The available voltage for a 36 V battery may be as low as 30 V at ambient temperature of −40°C, 60% SOC, and 200 A load current. This is the worst case for an ISA machine to meet the specifications of the engine cranking since the voltage drops by nearly 18%. If the ISA machine is designed to meet the low-voltage motoring requirements, then it is difficult to produce 42 V charging voltage under generating state at low-speed range. Thus, minimizing the battery voltage drop challenges the ISA system design. The system voltage

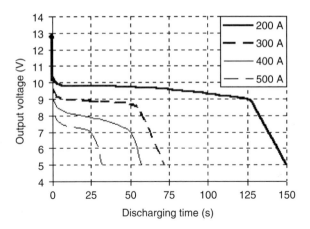

Figure 25.42 Battery voltage vs. time at different discharging currents (100% SOC and –29°C).

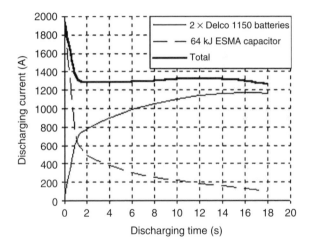

Figure 25.43 Discharging current of the batteries and capacitor in parallel.

issues have to be solved with the power and torque issue of the ISA machine at the same time, because the isolated ISA power system is different from the drive system for other industry applications at infinite power system grid.

As basic knowledge, the ultracapacitor or auto-capacitor can maintain the DC bus voltage constant for a certain amount of time. A 64 kJ capacitor is connected to a group of two 12 V batteries in parallel. The recorded discharging currents show that the total available DC bus current increases dramatically with help of the auto capacitor, shown in Figure 25.43. The test also shows that the drop of the available DC bus voltage is reduced significantly at the first few seconds, which provides enough time for engine cranking. The tested system is for heavy-duty applications, but the concept can be used in the ISA energy storage system and help the ISA machine during initial cold cranking, especially in the bitter cold winter season.

In the 42 V ISA system, an auto-capacitor with proper power rating can be connected in parallel with the 36 V battery, which enables the ISA machine not only to perform initial cold cranking in the bitter cold winter season, but also to provide higher driving torque during launch assistance. The details and other applications of the auto-capacitor are addressed in Chapter 7.

25.5.2 ISA Cooling Styles

The cooling styles and performances govern the performance and the package size as well as system efficiency or fuel economy of the ISA. Based on the cooling media, the cooling types can be divided into air and liquid cooling. Unlike the conventional alternator, the cooling of the power electronic box must be fully considered besides the electric machine, because a single integrated package of the machine and the power electronic drives becomes impossible in ISA applications. Most power electronic devices can only tolerate up to 190°C, while the machine insulation can tolerate up to 230°C. The objective of the cooling is to reduce the temperature rise; to protect the insulation, conductors, bearings, and power electronic devices from damaging; or simply to preserve the life of the ISA system. The statistical data show that every 10°C increase above the rated temperature of the insulation reduces the insulation life by 50%. The warranty analysis confirms that the alternators with high temperature rise have high stator failure rate. The winding resistance losses are high at low speed, while the bearing and core losses are critical at high speed, especially for a belt-driven ISA machine. For the power electronics box, the device damages very quickly if its junction temperature is beyond the temperature limitation. The heat is removed by means of three methods: conduction, radiation, and convention, based on the thermal principle. The thermal theory is beyond the scope of this book.

25.5.2.1 Air Cooling

The highest ambient temperature under the hood is 125°C for air cooling. The internal or external fans can be mounted on the machine shaft for self-cooling if the housing is open. The self-cooling machine should be designed according to the continuous power at low speed. The electric machine requires a downgrading operation if the operation speed is below the rated speed, such as operating with 80% rated power at 20% rated speed [16]. The cooling air can also be sent to the encased housing machine through a duct and forced to circulate by an exterior mechanism, i.e., forced cooling style. If the power electronics box is included in a single cooling loop, the cool air must pass it first, and then goes to the machine because of the thermal limitation difference between them. For the forced air-cooling machine, the continuous power rating of the machine depends on the maximum loss of the machine and has nothing to do with speed, because the cooling system is independent of the machine speed. In some ISA systems, the machine is cooled by shaft-mounted fans, while the power electronics box is cooled by liquid.

It should be mentioned that the self-cooling fan cannot be used for the crankshaft-mounted ISA because the narrow space between engine and transmission causes too much hot air at outlets to feed back to the inlets and results in a lower cooling system efficiency. Actually the belt-driven ISA machine with self-air-cooling also brings 15 to 30% of the hot air back to the cold air inlet, which was confirmed by the cooling system in the conventional alternator. The maximum rate of removing heat by natural convection and radiation is 775 W/m^2 with temperature rise 40°C, based on application experience [33,34]. With forced-air convection the rate increases to about 3100 W/m^2 at the same temperature rise. The two limitation values can be used to roughly calibrate the air-cooling design. Painting the rough casting metal surface black can triple the heat radiation of the polished metal surface.

Table 25.2 Thermal Property of Material

Materials	Air	Water	Engine Oil
Mass density, kg/m^3	1.21	1000	888
Specific heat, J/(kg·°K)	1005	4186	1880
Thermal conductivity, W/(m·°K)	0.026	0.580	0.150

25.5.2.2 Liquid Cooling

Compared to air, liquid like water and engine oil has higher specific heat and thermal conductivity, shown in Table 25.2. The high thermal capacity and the better thermal conduction of the liquid can improve the performance and capability of the cooling system, or remove more heat from the ISA system efficiently. The liquid-cooled ISA system can be more compact at a given power requirement or more efficient at a given package size.

As a rule of thumb, the maximum rate of the heating rejection increases to about 6,200 W/m^2 or higher at temperature rise of 40°C by direct liquid cooling. The rate limits the allowable generated heat per unit volume to about 610,000 W/m^3, which is twice as high as the loss limitation in forced air-cooling. Therefore, the current density of machine windings can increase proportionally in the liquid cooling system. The typical current density values with different cooling styles are given in Table 25.3 for ISA machine design reference [34].

It should be noticed that a low rotor current density might benefit the crankshaft-mounted induction machine for ISA applications, because no room is available for a rotor fan even if the stator is cooled by liquid. A previous EV manufacturer has experienced burnt rotors. The oil directly cooled stator windings has been reported in the induction machine for hybrid transit bus application.

There are three kinds of fluid available on-board the vehicle: engine oil, transmission oil, and engine coolant. The typical temperatures of them at inlet are 150, 135, and 125°C, respectively. Generally the power electronic box is cooled by either engine coolant or a separate cooling loop because of junction temperature requirements of the electronic devices. The power electronic devices are cooled by liquid through their heating sinks, while the components are sealed inside the box. The cool liquid passes the box first, then the electric machine if a single cooling loop is applied for the ISA system. Either engine oil or transmission fluid can be used for direct machine winding cooling, but it is seldom used for power electronic box cooling. A hollow stator housing called a "water jacket" conducts the coolant in the tangent direction around the stator, which is the typical ISA

Table 25.3 Current Density of ISA Machines

Cooling Method	Current Density (A/mm^2)
Self fan cooling	7.8~10.9
Forced air cooling	14~15.5
Liquid cooling	23.2~31

machine cooling style. It should be mentioned that an additional motor is required to run the transmission fluid during engine stop at stop/start operation to prevent the transmission from overheating.

Thermal management is a complicated issue because of its 3D distribution and many variable heating dissipation coefficients. The difficulty does not come from 3D finite element method itself, but from the accuracy of the input data, especially thermal coefficients and material properties. Intensive tests and measurements are always required even after carefully detailed design and finite element analysis.

25.5.3 OTHER ISSUES

The ISA is a relatively new but a transitional system to fully hybrid electric vehicles, compared to conventional vehicles. However, it has been paid much attention since the first commercial vehicle launched in 2002, although the concept and development can be dated back to the 1930s. Many other system issues are under exploration and discussion; for instance, integration of the ISA with powertrain and drivetrain, control and communication between an ISA subsystem and vehicle CPU through CAN, mechanical stress and fatigue, vibration and acoustic noise, electrical and electronic devices, and so on.

25.6 SUMMARY

The goal of development and production for the ISA system is much beyond the original objectives like replacement of starter and alternator, fuel economy, and emission reduction. Vehicles with the ISA system are expected to provide higher performance and more convenience as well as better customer comfort.

The ISA system is a subsystem of the powertrain system on-board the vehicle. It interacts with drivetrain, energy storage, vehicle controller, and communication systems. Its basic components are an electric machine and power electronic inverter/rectifier. The unidirectional or bidirectional DC-to-DC converter is added for high-power applications with high DC bus voltage, such as 42 V for passenger car and 340 V for heavy-duty application. The bidirectional DC converter provides jump-start function at low or dead 36 V battery. A multifunction inverter can perform both inverter/rectifier and unidirectional DC-to-DC converter. An extensional DC-to-AC inverter may be included in the ISA system to operate other on-board or off-board electric loads. The traditional starter may be remained for initial cold cranking or eliminated, depending on the power package of the ISA machine.

The ISA machine runs at motoring state for hot or cold engine cranking, launch assistance, creep drive, and so on. It runs as a generator to charge the battery and operate the on-board or off-board electric loads, while the ISA machine also runs at regenerating state and feeds the braking power back to the battery during braking deceleration of the vehicle. The ISA machine performs active damping function for vibration reduction at a variety of the IC engine states. The engine can be shut down at a traffic jam or for urban traffic lights, and automatically restarted right away at the driver's demand or the low-battery power state. Because of the high available electric power from the ISA, some conventional hydraulic equipment could potentially be replaced by electric motors, such as power steering, air conditioner, catalyst converter, and brakes.

Based on the mounting style, the ISA system can mainly be categorized as crank-shaft-mounted ISA or belt-driven ISA, although other styles are possible. The electric machine for ISA application requires running at high speed range, such as 0 to 6,000 rpm for a crankshaft-mounted machine and 0 to 21,000 rpm for a belt-driven ISA. The ISA is an isolated power system, compared to other industry applications. In some speed ranges, the low available DC bus voltage (29 to 30 VDC for 36 V battery) for motoring operation at the worst condition and the high DC bus voltage requirement for charging the battery (42 V in the 42 V electric system) are the critical challenges to the ISA engineers. A parallel auto-capacitor can release the low-voltage burden at motoring state of the ISA machine; however, it adds system cost. AC electric machines are the main candidates for the ISA application. Induction machine, sinusoidal interior permanent magnet machine, claw pole Lundell synchronous machine, switched reluctance machine, and other hybrid machines are successfully developed for the ISA systems. However, the surface-mounted permanent magnet machines, either brushless DC type or sinusoidal type, are not suitable for the ISA system because they cannot run within large speed ranges by flux weakening control to meet the constant voltage control requirements, especially at generating state. Different control strategies for a variety of machines in the ISA applications are outlined in this chapter. Some system issues, such as energy storage and system cooling, are also discussed.

The objective of this section is to provide a guideline to ISA engineers and other automotive engineers as well as people who are interested in the mild hybrid electric vehicles. Many important issues are still under discussion, since the development and commercial production of ISA systems are relatively new. With ISA applications, the technology and experience will benefit not only the ISA developers and customers, but also the entire generation of the hybrid electric vehicles and their users.

REFERENCES

1. D. Evans, M. Polom, S. Poulos, K. Van Maanen, T. Zarger. Powertrain architecture and controls integration for GM's hybrid full-size pickup truck. SAE 2003 World Congress, Detroit, MI, March 3–5, 2003, Paper 2003-01-0085.
2. K.P. Zeyen, T. Pels. ISAD – a computer controlled integrated-starter-alternator-damper system – a new approach to energy engineering. SAE Conference for Future Transportation Technologies, San Diego, CA, August 5–8, 1997, Paper 972660.
3. R. Henry, B. Lequesne, S. Chen. Belt-driven starter-generator for future 42 V systems. SAE 2001 World Congress, Detroit, MI, March 5–8, 2001, Paper 2001-01-0728.
4. Visteon Corporation. Innovations: electromechanical drive system. http://www.visteon.com/about/features/2001/082201.shtml, August 22, 2001.
5. M. O-hori, K. Tanei, Y. Yano, K. Hayama, S. Sako, T. Ezoe. Engine starting system development by belt drive mechanism. SAE 2001 World Congress, Detroit, MI, March 4–7, 2002, Paper 2002-01-1086.
6. J. Kassakian, J. Miller, N. Traub. Automotive electronics power up. *IEEE Spectrum,* May 2000, pp. 34–39.
7. H. Kusumi, K. Yagi, Y. Ny, S. Abo, H. Sato, S. Furuta, M. Morikawa. 42 V power control system for mild hybrid vehicle (MHV). SAE 2001 World Congress, Detroit, MI, March 4–7, 2002, Paper 2002-01-0519.
8. G. Witzenburg. GM Prepares to serve up hybrids in three distinct flavors. *Automotive Industries,* February 2003, pp. 43–44.
9. M. Algrain, W. Lane, D. Orr. A case study in the electrification of class-8 trucks. 2003 IEEE International Electric Machines and Drives Conference (IEMDC'2003), Madison, WI, June 1–4, 2003, pp. 647–655.

10. D. Evans, K.V. Maanen. Electric machine powertrain integration for GM's hybrid full-size pickup truck. SAE 2003 World Congress, Detroit, MI, March 3–6, 2003, Paper 2003-01-0084.
11. D. Polletta, T. Louckes, A. Severinsky. Fuel economy and performance impact of hybrid drive systems in light-duty trucks, vans and SUVs. SAE 2001 World Congress, Detroit, MI, March 5–8, 2001, Paper 2001-01-2826.
12. T. Teratani, K. Kuramochi, H. Nakao, T. Tachibana, K. Yagi, S. Abou. Development of Toyota mild hybrid system (THS-M) with 42 V power net. 2003 IEEE International Electric Machines and Drives Conference (IEMDC'2003), Madison, WI, June 1–4, 2003, pp. 3–10.
13. F. Hartig, K. Hockel, S. Friedmann. Die BMW Zylinderabschaltung. *ATZ*, Vol. 83, No. 2, 1981.
14. J. Berryhill, K. Li, J. Ward. Summary of 42 V PWM testing results. SAE 2002 World Congress, Detroit, MI, March 4–7, 2002, Paper 2002-01-0518.
15. Y. Kusaka, K. Tsuji. Novel power conversion system for cost reduction in vehicles with 42 V/14 V power supply. SAE 2003 World Congress, Detroit, MI, March 3–6, 2003, Paper 2003-01-0307.
16. B.K. Bose. *Power Electronics and Variable Frequency Drives: Technology and Applications*. New York: John Wiley and Sons Inc., July 1996.
17. T.J.M. Miller. *Brushless Permanent-Magnet and Reluctance Motor Drives*. Oxford, U.K.: Oxford Science Press, 1989.
18. H. Ogawa, M. Matsuki, T. Eguchi. Development of a powertrain for the hybrid automobile – the Civic hybrid. SAE 2003 World Congress, Detroit, MI, March 3–5, 2003, Paper 2003-01-0083.
19. D. Hanselman. *Brushless Permanent Magnet Motor Design,* 2nd ed. University of Maine, The Writers Collective, March 2003.
20. T.M. Jahns. Torque production in permanent magnet synchronous motor drives with rectangular current excitation. *IEEE Transactions on Industry Applications,* Vol. 20, No. 4, July/August 1984, pp. 803–813.
21. N. Bianchi, S. Bolognani, B.J. Chalmers. Salient-rotor PM synchronous motors for an extended flux-wakening operation range, *IEEE Transactions on Industry Applications,* Vol. 36, No. 4, July/August 2000, pp. 1118–1125.
22. T. M. Jahns. Flux weakening regime operation of an interior PM synchronous motor drive. *IEEE Transactions on Industry Applications,* Vol. 23, No. 4, July/August 2000, pp. 681–689.
23. N. Bianchi, S. Bolognani. Parameters and volt-ampere ratings of a synchronous motor drive for flux weakening applications. *IEEE Transactions on Power Electronics,* Vol. 12, No. 5, September 1997, pp. 895–903.
24. W.L. Soong, T.J.E. Miller. Practical field-weakening performance of the five classes of brushless synchronous AC motor drive. *Proceedings of IEEE European Power Electronics Conference,* Brighton, U.K., 1993, pp. 303-310.
25. B.H. Bae, S.K. Sul. Practical design criteria of interior permanent magnet synchronous motor for 42 V integrated starter-generator. 2003 IEEE International Electric Machines and Drives Conference (IEMDC'2003), Madison, WI, June 1–4, 2003, pp. 656–662.
26. F. Kozlowski, H. Hengstenberger. Compact stator alternator systems with high efficiency for 42 V and 12 V. SAE 2001 World Congress, Detroit, MI, March 4–7, 2002, Paper 2002-01-0522.
27. S. Chen, B. Lequesne, R.R. Henry, Y. Xue, J.J. Ronning. Design and testing of a belt-driven induction starter-generator. *IEEE Transactions on Industry Applications,* Vol. 38, No. 6, November/December 2002, pp. 1525–1533.
28. W. Cai, P. Pillay. Resonance frequencies and mode shapes of switched reluctance motor. *IEEE Transactions on Energy Conversion,* Vol. 16, No. 1, March 2001, pp. 43–48.
29. W. Cai, P. Pillay, A. Omekanda. Low Vibration Design SRMs for Automotive Applications Using Modal Analysis, *IEEE Transactions on Industry Applications,* Vol. 39, No. 4, July/August, 2003, pp. 971–977.

30. C.A. Ferreria, S.R. Jones, W.S. Heglund, W.D. Jones. Detailed design of a 30 kW switched reluctance starter/generator system for a gas turbine engine application. *IEEE Transactions on Industry Applications,* Vol. 31, No. 3, May/June 1995, pp. 553–561.
31. S.R. Jones, B.T. Drager. Sensorless switched reluctance starter/generator performance. *IEEE Industry Applications Magazine,* November/December 1997, pp. 33–42.
32. F.C. Sun, H.W. He, X.L. Yu. Power battery modeling based on the characteristic field theory. SAE 2003 World Congress, Detroit, MI, March 3–5, 2003, Paper 2003-01-0091.
33. J.R. Hendershot, T.J.E. Miller. *Design of Brushless Permanent-Magnets Motors. Switched Reluctance Motors and Their Control.* Hillsboro, OH: Magna Physics Publishing, 1994, pp. 15-1–15-24.
34. T.J.M. Miller. *Switched Reluctance Motors and Their Control.* Hillsboro, OH: Magna Physics Publishing, 1993, pp. 115–134.

26

Fault Tolerant Adjustable Speed Motor Drives for Automotive Applications

Babak Fahimi
University of Missouri-Rolla, Rolla, Missouri

26.1 INTRODUCTION

Adjustable speed motor drives, and in particular vector-controlled AC motor drives, are being increasingly used as mission-critical components in a variety of high-performance automotive applications [1]. However, proper operation of the closed-loop control system depends on reliable operation of the electromechanical actuator, power electronic power processor, control circuitry, and feedback sensors. In the event of sensor failures, it is desirable that the AC adjustable speed motor system continues to operate, even if under a diminished performance capacity. This is especially important in the high-impact automotive applications where even limp-back operation is preferred over no operation.

In this chapter an adjustable speed induction motor drive has been chosen to illustrate the concept of fault tolerance. Furthermore, failures in a sensory part of the drive system have been targeted for this presentation. The concepts presented in this chapter can be easily modified to cope with specific characteristics of permanent magnet synchronous motor drives.

Control techniques for induction motor drives are well treated in the literature [2]. For high performance Induction motor drive systems over the entire speed range, indirect field oriented current controllers (IFOCs) are commonly used. These controllers typically incorporate rotor position feedback and motor current feedback [3]. Sensorless vector controllers eliminate the need for position feedback but operate poorly at very low speeds [4]. Scalar control methods have been used when rotor flux orientation cannot be maintained. However, they do not allow for decoupled torque and flux control [5]. Volts/hertz open loop induction motor control can be used without feedback sensors. This, however, results in slow dynamics and sluggish response. Furthermore, operation of volts/hertz is characterized by large transient behavior in the motor currents and torque [6]. Much discussion in the literature has focused on choosing between the various control alternatives and deciding upon which one to implement subject to the requirements of a particular application.

An example of a fault tolerant control strategy in the event of failure in sensors can be explained as follows. Vector-oriented control is preferred to satisfy high-performance applications. It is used when position and current sensors are available. This gives excellent dynamic operation over the entire speed range. In the event of position sensor failure, the sensorless controller is activated. High-performance vector control can be maintained above low speeds. Outside of its stability region, the controller reorganizes to a scalar controller, where magnitude of the currents is controlled but vector orientation is lost. In the event that the current sensors also fail, the control system transitions to volts/hertz control. This gives the poorest dynamic operation but allows the system to continue operating. The transition criteria are continually re-evaluated in the event that a failure recovery occurs, and it then becomes possible to return to a more desirable level of control authority.

26.1.1 Self-Organizing Controllers

The high-impact nature of applications such as electric propulsion systems in vehicles necessitates a fault tolerant performance. This can be realized in a flexible controller architecture, which squeezes maximum performance in the event of a failure in any part of the induction motor drive. To achieve this goal, a reorganizing controller will adopt the best control methodology according to available feedback and operational hardware. The reorganizing controller comprises two parts, namely failure detection and fallback strategy. While the first part monitors the status of system components such as sensors, motors, and inverters, the latter will engage the most appropriate control strategy based on a hierarchical order. A graphical summary of the reorganizing controller for IM drive systems is shown in Figure 26.1. The reorganizing action is highlighted in this figure.

As can be seen, a vector represents the operational status of each sensor. The constituents of the status vector are updated via a failure detection system. A healthy operation of a specific component corresponds to 1, while 0 demonstrates a faulty operation. To give an example, let's focus on the status of sensors.

$$\begin{pmatrix} S_1 \\ S_2 \\ S_3 \\ S_4 \end{pmatrix} = \begin{pmatrix} S(\text{Position Sensor}) \\ S(\text{Flux Estimator}) \\ S(\text{Current Sensor \#1}) \\ S(\text{Current Sensor \#2}) \end{pmatrix} \qquad (26.1)$$

Depending upon the functionality of sensors, the most appropriate control strategy will be employed. A summary of the various control strategies along with required sensors are shown in Figure 26.2. According to this classification a fault tree for various sensor failures in induction motor drives has been developed and is shown in Figure 26.3.

In Figure 26.3, the three switches S1, S2, and S3 indicate the status of position, voltage, and current sensors, respectively. If all of the sensors are in a satisfactory condition, an indirect vector control of the IM drive is applied. Upon detecting a failure in position sensor, observer-based vector control will be incorporated. Due to a limited speed range or failure in the voltage measurement system a transition to current control mode is performed. A loss of current sensor will leave us with the option of volts/hertz control. A novel reconstruction of the phase currents, which will be explained later, uses a DC-link

Fault Tolerant Adjustable Speed Motor Drives for Automotive Applications

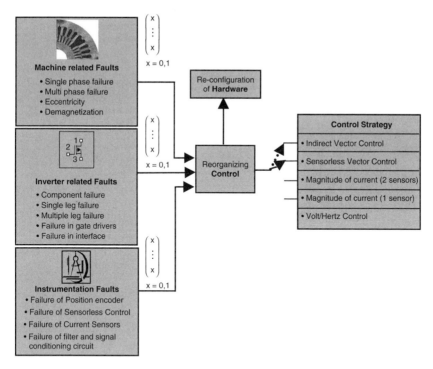

Figure 26.1 Graphical illustration of reorganizing controller.

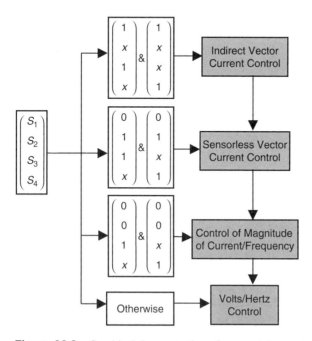

Figure 26.2 Graphical demonstration of reorganizing control in the event of sensor loss.

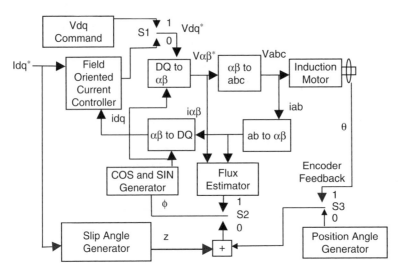

Figure 26.3 Fault tree for various sensor failures in IM drive system.

current sensor, which can be integrated with the existing hierarchy to further strengthen the maneuverability of the proposed fault tolerance control strategy. Notably, these transitions are not necessarily made in the proposed order. In addition, once a sensor or a faulty part of the system recovers, the detection unit will switch back to the best control possible. This is a clear interpretation of "resilience" in control.

26.1.1.1 Hierarchy of Control Methods in Induction Motor Drives

Quality of performance in induction motor drives depends strongly upon their control strategy. Indirect vector control method has proven to offer the best transient and steady-state performance over the entire speed range. The main principle in implementing this technique is to match the excitation of the stator with the angle of the rotor flux [2]. This can be interpreted mathematically as:

$$\frac{d\phi}{dt} = P\omega + \frac{M_{dq} i_q}{\tau_r \lambda_r}$$

$$\frac{d\lambda_r}{dt} = \frac{-\lambda_r}{\tau_r} + \frac{M_{dq}}{\tau_r} i_d$$

(26.2)

where λ_r is the magnitude of the rotor flux, ϕ is the angle of the rotor flux, i_d and i_q are direct and quadrature axes currents respectively, P is the number of pole pairs, τ_r is the rotor time constant, M_{dq} is the magnetizing inductance, and ω is the rotor velocity. Assuming that the magnitude of rotor flux has reached its steady state, one can compute the rotor flux angle using the first equation in (26.2). A successful implementation of this method, however, relies on the availability of high-accuracy position and current measurements. In addition, sophisticated adaptive and identification techniques need to be integrated into the control algorithm to compensate for the nonlinear and time varying nature of the induction motor drives [4–6]. These effects are mainly due to the dependency of the rotor time constant upon temperature rise and saturation effects. The high number of sensors in this method indicates considerable vulnerability to sensor failures. Inherent unreliability of discrete position sensors has thus resulted in development of a range of position sensorless techniques.

Estimation of rotor flux components is the main building block in the majority of the position sensorless techniques developed for induction motor drives. In fact, normalized rotor flux components are directly used as trigonometric functions of rotor flux angle.

$$\sin(\phi_r) = \frac{\lambda_{\beta r}}{\lambda_r}$$
$$\cos(\phi_r) = \frac{\lambda_{\alpha r}}{\lambda_r}$$
(26.3)

This assumption, however, largely depends on the accuracy of the flux estimation scheme. The existing tradeoff between mitigation of measurement noise, limited sampling time, and speed range prohibits an accurate performance over the entire speed range. Indeed, the performance of the induction motor drive will be substantially deteriorated outside the targeted speed range. This, along with the need for external voltage sensors in the event of imperfection in inverter operation, has called for yet another alternative that can substitute the sensorless technique as required by the considerable variation in the speed of the internal combustion engine. The very next best control methodology is scalar control of current magnitude.

Having lost the access to rotor flux angle, an emulated rotor flux position can be used to control the induction motor drive. This internal angle generator maintains the very last known electrical velocity of the rotor flux and continues to control the magnitude of the stator current in the new dq frame. Although the new dq frame does not necessarily match the actual dq frame of reference, it can effectively serve the purpose of the scalar current control. As a direct consequence of scalar control, a load-dependent torque production along with a deteriorated dynamic performance is expected. It must be noted that despite a reduction in the quality of performance, this method has the potential to be used along with an incipient pullout detection method, which can ultimately replace the existing indirect or sensorless control techniques. Evidently, implementation of scalar control of current magnitude is contingent on the availability of current sensors. In the event of failure in current sensing, scalar control in terms of volts/hertz should be engaged.

The volts/hertz control is considered as the last step in the hierarchical classification of control strategies for induction motor drives. Although a significant deterioration of transient performance and a loss in torque/ampere is expected, this method depicts one of the most direct methods for operating a voltage source inverter. There is an apparent compromise between simplicity and high-grade operation of induction motor drives, which is symbolically shown in Figure 26.4. From Figure 26.4 one can conclude that for each operating condition under vector control there is at least one equivalent scalar control that would provide the same level of performance.

26.1.1.2 Smooth Transition between Various Control Methods

In preparation for developing a smooth transition from sensorless to encoder-based IFOC and vice versa, one should develop a signature detection scheme that will identify the loss of encoder. Detection of anomalies in the performance of the encoder is an integral part of the reorganizing control. By online analyzing of feedback (position and current) information, detection systems should identify any faulty operation of the discrete position encoder.

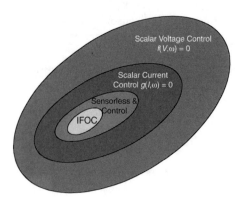

Figure 26.4 Classification of various control strategies in induction motor drives.

Furthermore, this algorithm should be fast enough to avoid a substantial deterioration of the performance upon occurrence of the sensor loss. Therefore, design of a signature detection subsystem can be considered as a threefold problem requiring existence, uniqueness, and fast detectability. The same philosophy will be applied in design of detection modules for the remaining parts of the system.

Assuming an optical encoder, one can choose the first- and second-order statistical characteristics (mean and standard deviation) of position data for our investigation. This would automatically ensure the existence and uniqueness of the signature. The fast detection of the targeted signature was tested on the experimental setup, and satisfactory results were obtained. A drastic change in the average speed over a short period of time (much shorter than the mechanical time constant of system) stands for a mechanical or electronic breakdown of the encoder system. However, in a more realistic situation, mechanical slipping of encoder can also cause inaccurate position information that will finally contribute to a malfunction of field-oriented control. Unlike the previous case, this type of anomaly will not cause an immediate change in the average speed. However, a substantial change in the moving standard deviation of the system will be observed due to missing encoder pulses.

This has led us to perform a simulation study on an induction motor drive system in which the effectiveness of the detection system in the event of such failures is examined. Figure 26.5 depicts the sequence of events in this experiment. At t = 0.1 sec, 1 pulse per each group of 20 pulses obtained from the encoder was masked. This was effectively integrated to our control algorithm. As a result, an increase in the moving standard deviation of the system is observed. At t = 0.18 sec, a failure in the encoder was detected. Consequently, control reconfigures itself to sensorless while the encoder loss persists. At t = 0.45 sec, the induced anomaly was removed. This was reflected in the detected signature (standard deviation), and a reconfiguration command to the encoder-based system was issued at t = 0.51 sec.

This system will serve as a building block in our reorganizing control. A further extension of this concept will be employed to detect anomalies in sensorless control, current magnitude control, machine related failures, and inverter-related faults.

A smooth and bump-free transition between encoder-based and sensorless vector control of an induction machine will significantly enhance the reliability and durability of the drive system, among other advantages. Our study of the various reconfiguration scenarios has resulted in the following observations:

Fault Tolerant Adjustable Speed Motor Drives for Automotive Applications

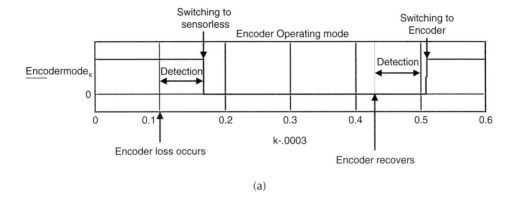

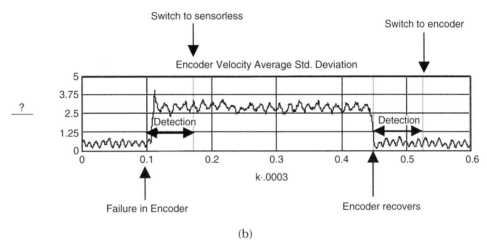

Figure 26.5 Sequence of events along with the detected signature during the validation test: (a) sequence of events in the validation test; (b) detected standard deviation during the test.

- While running the IM drive system from an encoder-based vector control, the estimated rotor angle (ϕ_e) follows the calculated rotor angle by the encoder, i.e.:

$$\theta_e(\text{Encoder}) + \frac{1}{\tau_r}\frac{i_q}{i_d} \approx \phi_e(\text{Flux Estimator}) \qquad (26.4)$$

where τ_r, i_q, i_d stand for rotor time constant, torque, and field components, respectively.
- Due to the imperfection of digital filtering, some phase shift between the estimated and actual rotor angle might occur. It was observed, while simulating the IM drive system using an encoder-based vector control, the phase shift will not increase. In other words, the estimated rotor flux angle will be locked with the calculated rotor flux angle obtained from encoder-based control.
- While running the IM drive by estimated rotor flux angle obtained from rotor flux estimation, an initial mismatch between two angles will accumulate in the course of time. This would result in a substantial drift between estimated and calculated rotor angles.

The latter mismatch between estimated and calculated rotor flux angle will result in a sudden change of electromagnetic torque. This in turn results in an uncomfortable and damaging bump during the switchover from sensorless to encoder-based system.

It is important to note that while running the system from an encoder, the calculated and estimated rotor flux speeds are tied together via an internal feedback. In other words, any mismatch between these two speeds will be interpreted as a perturbation to a closed loop stable system and will be rejected. This will, then, avoid any accumulation of error between two angles in the course of time. However, while running the system from sensorless control, the calculated speed is not in the feedback loop. Therefore, any difference between estimated and calculated speed will eventually result in a drift between the calculated and estimated angles.

To demonstrate the impact of mismatch between calculated and estimated rotor angles, a case study is represented in Figure 26.6. In this simulation study, the IM drive system is running under sensorless vector control at t = 1.5 sec when a switching command from sensorless to encoder-based control is issued. Due to a considerable difference

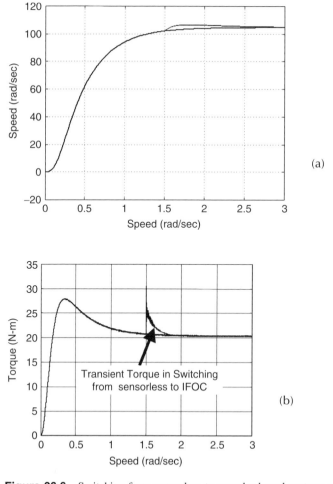

Figure 26.6 Switching from sensorless to encoder-based system; red stands for sensorless operation and blue stands for encoder-based system. At t = 1.5 sec, switching to encoder-based system occurs: (a) speed; (b) electromagnetic torque.

between calculated and estimated rotor angles, an immediate increase in the torque occurs. This contributes to a transient overspeed, as shown in Figure 26.6(a). In practice this phenomena will be also be recognized by a noise and shock to the system. It must be noted that, if not properly compensated, this would have an adverse effect on the durability of the IM drive system. To avoid undesirable consequences of this behavior a smooth transition technique from sensorless to encoder-based system was designed and implemented in the control algorithm of the IM drive.

The main idea in development of a smooth transition between sensorless and encoder-based system is to compensate for the existing difference between estimated and computed rotor flux angles. To cope with this objective, a monitoring system will identify the calculated and estimated rotor flux angles. Accordingly, the initial value of the computed slip angle will be modified such that both estimated as well as computed rotor flux angles match at the instant of crossover. This will then guarantee a smooth transition between two sensorless and indirect vector control methods.

26.1.1.3 Reconstruction of the Phase Currents

Modern drive systems often require fast and accurate closed-loop control of electromagnetic torque, and consequently motor phase current. A common phase current regulation strategy is to achieve a closed-loop current control using delta hysteresis regulation (DHR). Within DHR, switch-level control is achieved by comparing a commanded current to a measured current at fixed sampling intervals. This method has been proven to be effective in controlling both magnitude and phase of inverter output current. Typically, DHR implementation requires three- or at least two-phase current sensors, and is typically realized using analog techniques.

However, phase current information does not have to depend upon the phase current sensors. An alternative is to measure the DC-link current and use the switching state information of the inverter to reconstruct the phase currents. Several researchers have investigated current reconstruction strategies.

A block diagram of a current-regulated induction motor drive is shown in Figure 26.7. For the system shown, a delta hysteresis regulation scheme is used to control phase currents. Traditionally, the regulation is realized using analog techniques. Specifically, the inputs to the DHR are the commanded currents from the motor controller and the information of motor phase currents. The output of the DHR is a six-channel switching signal.

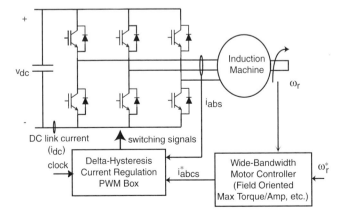

Figure 26.7 Fully protected inverter-motor drive system using DHR.

Table 26.1 Relationship between Switching States, DC-Link Current, and Actual Phase Currents

Switching States #	S_c	S_b	S_a	i_{dc}
0	0	0	0	0
1	0	0	1	i_a
2	0	1	0	i_b
3	0	1	1	$-i_a$
4	1	0	0	i_a
5	1	0	1	$-i_b$
6	1	1	0	i_a
7	1	1	1	0

An accurate method to obtain full information of output phase currents is to measure them directly using three current sensors (or at least two). Additionally, most drives use an additional DC-link sensor for overcurrent protection.

An alternative to obtain the phase current information considered in this research is to reconstruct phase current using measured i_{dc} and knowledge of the switching states. For the drive shown in Figure 26.1, the switching state and correlation to i_{dc} is detailed in Table 26.1. It is noted that in state 0 or 7, all of either the upper switches or lower switches are turned on, thus i_{dc} is always 0 and doesn't represent any phase current. This should be avoided as described in the following section. In all other states, i_{dc} represents one of the phase currents.

26.2 DIGITAL DELTA HYSTERESIS REGULATION

Based upon the relationship between i_{dc} and the three-phase currents, a current regulation has been designed for accurate reconstruction and control of the phase currents using a single DC-link current sensor. In contrast to the traditional DHR, this method is realized through digital techniques. Herein it is referred to as digital delta hysteresis regulation (DDHR). A block diagram of DDHR is shown in Figure 26.8. Within the block diagram, $i^*_{abcs}(k + 1)$ represents commanded phase currents; $i_{abcsre}(k)$ represents reconstructed phase currents; and $i_{dc}(k)$ is the measured DC-link current. The vector $S_{abc}(k)$ represents the present switching state, $\hat{S}_{abc}(k)$ is the switching state one obtains by comparing commanded current to reconstructed current, and $\hat{S}_{abc}(k + 1)$ is the actual switching command to be sent to the gate drive at the next clock cycle.

The block diagram in Figure 26.8 shows that the DDHR is composed of two parts. The lower portion is the current reconstruction, where the phase current is determined as close as possible from the DC-link current and switching states. The upper part is the Switching State Generator that establishes the next switching state based upon commanded phase currents, reconstructed phase currents, and present switching states.

Fault Tolerant Adjustable Speed Motor Drives for Automotive Applications

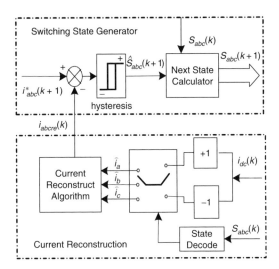

Figure 26.8 Digital delta-hysteresis current regulation.

26.2.1 CURRENT RECONSTRUCTION ALGORITHM FOR DDHR

Based upon Table 26.1, line currents can be reconstructed using knowledge of the DC-link current and switching states. There is a reconstruction error that results from the fact that at each instant of time only one phase current is updated directly from measurement. Therefore, high-operation frequency is preferred, as the phase currents are more frequently updated. As power electronic device DSP technology is improved, it becomes possible to operate at higher switching frequencies. This also helps to improve control bandwidth and decrease low-frequency harmonics.

However, this does not mean that the higher the switching frequency, the better the current reconstruction. Specifically, high-frequency operation raises the following problems for line currents reconstruction:

1. High dv/dt Excites Parasitic Resonances that Create Electromagnetic Noise on the DC-Link Sensor — Due to high dv/dt of IGBT switching, there is electromagnetic noise coupled to DC-link current. Adding snubbers is a way to mitigate this problem. However, the noise cannot be absolutely eliminated. For the system studied herein, noise was present in frequencies above 50 kHz when the system operated at a switching frequency around 15 kHz. Figure 26.9 shows the experimental measurements of the noise coupled in DC-link current. Noise in the DC-link current is transferred to line current reconstruction and presents error if it is not mitigated.

2. DC-Link Loop and Parasitic Inductance Forbids Link Current to Change Instantaneously — The phase current attributed to the DC-link current is expected to change instantaneously. For example, when switching state S_{cba} changes from 1 to 2, i_{dc} in theory has an instantaneous change from i_{as} to i_{bs}. However, due to DC side loop and parasitic inductance, the phase current attributed to i_{dc} cannot change instantaneously. The slow change loop will cause a significant error to the reconstructed current, particularly when switching occurs at a high frequency. Figure 26.10 shows a measured i_{dc} over a small time-scale. As shown,

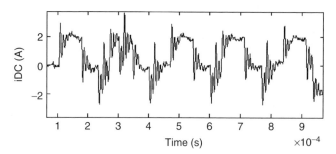

Figure 26.9 EMI/EMC noise coupled in DC-link current.

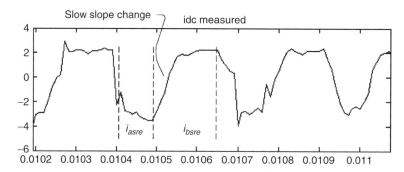

Figure 26.10 Effect of loop and parasitic inductance on DC-link current.

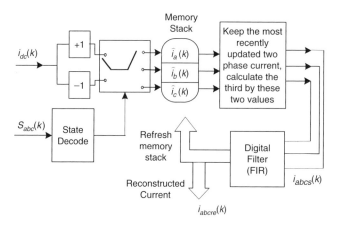

Figure 26.11 Block diagram of current reconstruction.

when the switching states changes, the relatively slow response of i_{dc} will yield errors in the reconstructed current.

A block diagram of the current reconstruction is shown in Figure 26.11. At each clock cycle, i_{dc} is measured. Based upon the present switching state, the observer updates a single phase current in the memory stack using the DC-link current. Since only one phase current is updated, it is important to select the most recently updated two phase currents and use them to calculate the third.

In Figure 26.11, $i'_{abcs}(k)$ represents predicted values of the reconstructed currents. These consist of the two most recently updated phase currents and the calculated third phase current. In theory, these can be used to represent the reconstructed currents sent to the switching state generator. However, it has been found that to improve the accuracy of the approach, an important step implemented is to apply a digital filter to the predicted currents to create a set of reconstructed currents.

REFERENCES

[1] Miller, J. M., P. J. McCleer, and J. H. Lang, "Starter-alternator for hybrid electric vehicle: comparison of induction and variable reluctance machines and drives," *Proceeding of the 33rd IEEE Industry Application Society Annual Meeting*, October 1998, pp. 513–523.

[2] Leonhard, W., *Control of Electrical Drives*, Berlin: Springer-Verlag, 1985.

[3] Finch, J. W. and D. J. Atkinson, "Scalar to vector: General principles of modern induction motor control," *Proceeding of the IEE Conf. PEVD (London)*, July 1990, pp. 364–369.

[4] Jansen, P. L., R. D. Lorenz, and D. W. Novotny, "Observer-based direct field orientation: Analysis and comparison of alternative methods," *IEEE Transactions on Industry Applications*, Vol. 30, No. 4, July/August 1994, pp. 945–953.

[5] Bose, B. K., "Scalar decoupled control of induction motor," *IEEE Transactions on Industry Applications*, Vol. IA-20, No. 1, January/February 1984, pp. 216–225.

27

Automotive Steering Systems

Tomy Sebastian, Mohammad S. Islam, and Sayeed Mir
Delphi Corporation, Saginaw, Michigan

27.1 INTRODUCTION

The steering system in an automobile converts the driver's rotational input at the steering wheel or hand wheel into a change in the steering angle of the vehicle's road wheels to control the direction of motion. Effort is required to turn the steering wheel due to the presence of friction between the tires and the road surface. In earlier steering systems, the driver provided the required torque to steer the vehicle. For heavier vehicles, the driver is unable to provide sufficient torque to steer the vehicle. In such cases an additional mechanism is needed to assist the driver. Hydraulically assisted power steering was introduced around the 1950s. Developments in power and control electronics and in electric machines led to the development of electrically assisted steering. This chapter will discuss various types of steering mechanisms.

27.2 STEERING SYSTEM

Based on the operating principle, steering systems can be classified as shown in Figure 27.1. The following sections will discuss these systems in more detail. In all these systems, the objective is to move the road wheels by a certain angle for a given angle rotation of the hand wheel. The ratio of the road wheel angle rotation to the steering angle rotation is normally referred to as the steer ratio.

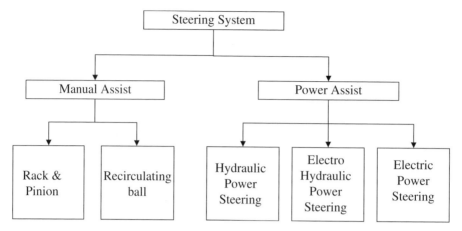

Figure 27.1 Classification of steering systems.

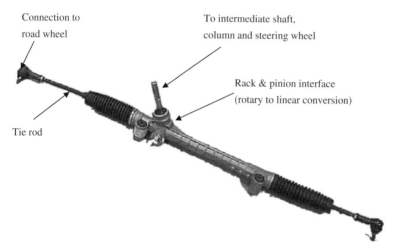

Figure 27.2 A manual rack-and-pinion arrangement for a steering system (courtesy Delphi Corporation).

27.2.1 MANUAL STEERING

A simple way of controlling the road wheel angle from the steering wheel is by means of a rack-and-pinion gear mechanism, shown in Figure 27.2. In this case the driver provides all the energy required for the angular movement of the road wheels. The rack-and-pinion mechanism converts the rotational input of the driver to a linear motion of the rack. This conversion also allows reducing the driver effort to turn the steering wheel.

The required driver torque is a function of the rack force and the rack-and-pinion ratio called *C-factor* (sometimes *C-factor* is also referred to as steering ratio). This factor is defined as the distance traveled by the rack in mm when the steering wheel is rotated for 360 degrees.

Thus, the relationship between the steering wheel speed and torque to the rack speed and force can be written as:

Automotive Steering Systems

$$T_s = \frac{F_r C}{2000\pi\,\eta_{rp}} \; N\text{-}m \qquad (27.1)$$

$$v_r = \frac{\omega_s C}{2000\pi} \; m/s \qquad (27.2)$$

where T_s is the steering wheel torque in *N-m*, ω_s is the steering wheel speed in *rad/s*, F_r is the rack force in *N*, v_r is the rack velocity in *m/s*, and η_{rp} is the rack-and-pinion efficiency [1]. For heavier vehicles, the steering effort torque will be large; therefore, these systems are only used in smaller vehicles.

Another mechanism to convert the driver torque into road wheel rotation is by means of a recirculating ball arrangement, also known as an integral gear mechanism [1,2]. Since this system is not currently used with electric power steering, this is outside the scope of this book.

27.2.2 Hydraulically Assisted Steering

Traditionally, the power assist is obtained by hydraulic means. Figure 27.3 shows the schematic of such a system. In this case, a hydraulic pump is driven from the vehicle engine through a belt and pulley arrangement. The high-pressure fluid is used to move a piston in the steering gear assembly to assist the driver. The direction of movement is controlled by a valve mechanism.

The hydraulically assisted system has an assist characteristic, which is independent of the vehicle speed. It is preferred to have higher assist (or lower driver effort) at low

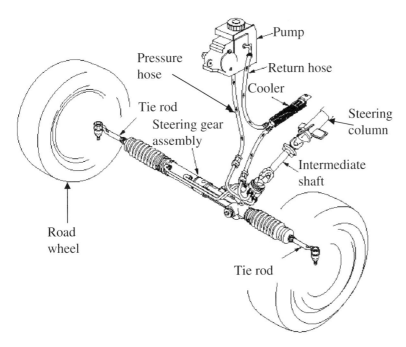

Figure 27.3 The schematic of a hydraulically assisted steering system.

speed and parking conditions, and lower assist at higher vehicle speed situations such as on a highway. The variable speed-effort characteristic is obtained by using electronically controlled valve mechanisms or by electromagnetically controlled systems such as Magna Steer™.

The hydraulically assisted steering system provides exceptionally good steering feel characteristics, though it has several disadvantages:

1. The continuously running pump creates constant power loss, thus increasing the fuel consumption of the vehicle.
2. At the end of life of the vehicle, one has to deal with the hydraulic fluid and the hoses.
3. Tuning of the vehicle steering characteristics is complicated and time consuming.
4. Assembly of the system in the vehicle is time consuming due to the large number of components to be assembled.
5. Packaging is difficult, as engine accessories are needed for coupling the pump to the engine.
6. As the power assist is dependent on the engine speed, if the engine is stalled, the power assist also will be lost (engine dependency).

27.2.3 Electrohydraulic Power Steering

Some of these disadvantages could be overcome by running the pump from an electric motor. Such a system as shown in Figure 27.4 is often referred to as an electrohydraulic power steering system. This system addresses the issues 4 through 6 mentioned above. It also provides reduced fuel consumption, as the pump speed is independent of the engine speed and the speed could be controlled to reduce the losses. These systems draw a continuous current of the order of few amperes from the battery, mostly to support the hydraulic losses. During the steering maneuver, depending on the vehicle size, the peak current drawn may be as high as 100 amperes from a 12 V power system.

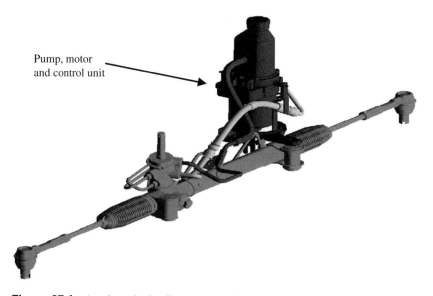

Figure 27.4 An electrohydraulic power steering system.

Automotive Steering Systems

Figure 27.5 Mechanization of an electric power steering system (courtesy Delphi Corporation).

27.2.4 ELECTRIC POWER STEERING

An electrically assisted steering system addresses most of the disadvantages of the hydraulic system, though it brings in some new issues and challenges. In an electric power steering system, the assist to the driver is provided by an electric motor. The base system is very similar to the manual rack-and-pinion arrangement explained with Figure 27.2. An electric motor with a gear reduction mechanism is coupled to the main steering path to provide the assist. In Figure 27.5, the assist mechanism is coupled to the steering path at the column (column assist). The assist could also be provided at the pinion (pinion assist) or even at the rack (rack assist).

The required motor torque can be written as:

$$T_m = \frac{(T_s - T_d)}{\eta n} \tag{27.3}$$

where T_d is the driver input torque, η is the efficiency of conversion, and n is the gear ratio between the column and the motor. T_s is the total load as in Equation 27.1.

Also, the motor speed ω_m and the steering wheel speed ω_s are related by:

$$\omega_m = n\omega_s \tag{27.4}$$

A curve showing the assist torque vs. steering wheel input torque is shown in Figure 27.6. Due to the shape of this curve, it is sometimes also known as the "bathtub" curve. This curve is for a particular steering wheel speed and vehicle speed. The torque-speed requirements of the motor are derived from the "bathtub" curve using the gear ratio of the mechanical gear and the efficiencies of other system components. Figure 27.7 shows the typical power flow and losses in a typical electric power steering system. The power source for an automotive application is usually the battery with a nominal voltage of 12 V. The maximum current draw allowed from the battery is usually about 75 to 100 A, depending on the vehicle type and the manufacturer. This automatically places a limit on the input power (= 12 × 75 = 900 W peak power). Based on this input power limitation, the designer has to allocate the efficiencies to the system components in order to get the

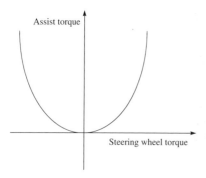

Figure 27.6 The assist torque vs. the steering wheel torque of a typical electric power steering system.

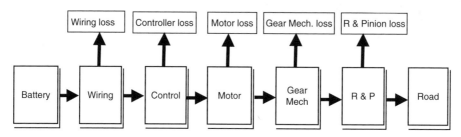

Figure 27.7 Typical power flow and losses in an electric power steering system.

required output power, which can vary from 250 W to 550 W, depending on the steering loads, which in turn depend on the gross vehicle weight (GVW). Usually the overall system cost and efficiency is balanced with tradeoffs between the different components (i.e., gear reduction mechanism, electronic controller, motor, and wiring harness).

The electric power steering eliminates the hydraulic fluid. This is an on-demand system, since the motor does not take in any power if the driver does not steer. This results in an improved fuel economy to the vehicle. When the vehicle is not in a steering maneuver, the system takes very little current to support the control electronics. In a steering maneuver, the battery current could reach up to 100 A. Heavier vehicles will require higher current. This limits the usage of electric power steering to smaller vehicles, as the available battery current is limited. Also due to the engine-independent nature, the steering assist will be provided even with the engine stalled as long as the battery power is present. It also reduces the engine accessories, making it simpler.

Another advantage of the electric power steering system is in the ease of tunability of the steering feel. The assist as a function of the vehicle speed can be easily programmed. As the assist characteristics can be programmed using software, one does not have to change the mechanical components, as in a hydraulic system, to tune the steering characteristics. This reduces the tuning time substantially compared to a hydraulic system.

27.3 ADVANCED STEERING SYSTEMS

Steering systems have moved from just providing driver assist, to also providing added comfort and enhanced vehicle stability. Recently, four-wheel steering and other systems that enabled better vehicle controllability were introduced in the market.

Figure 27.8 Four-wheel steering under low vehicle speed condition, where the front and rear wheels are in phase opposition (courtesy Delphi Corporation).

27.3.1 FOUR-WHEEL STEERING

The introduction of electric power steering also allowed other modes of vehicle control. The addition of steering capability to the rear wheels provides the vehicle with much more control options. The four-wheel steering introduced in trucks and sport utility vehicles allows for better vehicle maneuverability. The advantages are obvious when the vehicle is used for trailering and during parking. At low vehicle speeds, the rear wheels are in phase opposition to the front wheels and at higher vehicle speeds they are in phase with each other. Figure 27.8 shows a low-speed maneuver of a four-wheel steering system where the front and rear wheels are in phase opposition.

27.3.2 FUTURE-GENERATION STEERING SYSTEMS

Future steering systems will integrate other functions such as braking, throttle, and suspension to improve the vehicle stability and control. To accomplish this, there has to be some level of decoupling between the driver and the road wheels. Active front steering systems use a differential arrangement so as to be able to control the road wheels either by the driver or by the motor.

Steer-by-wire systems give complete mechanical decoupling of the driver to the road wheel by eliminating all mechanical linkages between the steering wheel and the road wheel. The sensors and control along with the drive motors precisely position the road wheels at the desired position. Such systems will require fault tolerant communication and control schemes. These systems will require actuators to actuate the road wheels and to provide feedback to the driver.

REFERENCES

[1] J.W. Post, E.H. Law, "Procedure for the characterization of friction in automobile power steering systems," SAE International Congress and Exposition, Detroit, MI, February 26–29, 1996, Document Order Number SP-1136.

[2] R.K. Jurgen, *Electronic Steering and Suspension Systems*, Society of Automotive Engineers Inc.

[3] C.P. Cho, R.H. Johnston, "Electric motors in vehicle applications," *Proceedings of the IEEE International Vehicle Electronics Conference (IVEC'99)*, Changchun, China, 6–9 September 1999.
[4] N. Iwama, Y. Inaguma, K. Asano, T. Mori, Y. Hayashi, "Independent rear wheel control by electric motors," 23rd FISITA Congress, Torino, Italy, 1990.
[5] E.A. Bretz, "By-wire cars turn the corner," *IEEE Spectrum*, April 2001, pp. 68–73.

28

Current Intensive Motor Drives: A New Challenge for Modern Vehicular Technology

Babak Fahimi
University of Missouri-Rolla, Rolla, Missouri

28.1 BACKGROUND

With the vehicular industry turning into a primary market for electromechanical energy conversion devices and power electronics base products, there is little doubt that the existing 12 V electric power system in conventional automobiles can no longer satisfy the ever-growing demand of more electrification. The main problem with this is the fact that increased electric power consumption results in a current intensive condition. This is a problem known and avoided by the majority of experts in conventional power system. The old rationale for avoidance of current intensive scenarios stems from a simple fact that current intensive systems tend to be less efficient due to their high levels of copper losses. This phenomenon, however, appears to have additional consequences in electric power system of the automobiles.

A quick inspection of the existing trend in electrification of automobiles indicates that in the near-term future excessive levels of current intensiveness will be exhibited. The predictions as illustrated in Figure 28.1 are so dramatic that researchers over the past decade have seriously weighed the option of increasing the operational voltage to 42 V in order to relax the growing dimensions of the above problem while avoiding substantial additional costs associated with the new high-voltage insulation requirements for 60 V and above. Although this sounds compatible with the existing needs and seems to be a good intermediate solution for the near future, it has not been yet implemented and has already faced new set of demands as automotive industry evolves with an exemplary pace. It must be noted that although current intensiveness is a major factor in the need for a

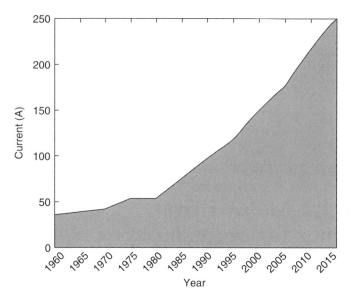

Figure 28.1 Anticipated current requirements in a 12 V conventional car.

new power system for cars, the drastic increase in the complexity of wiring, need for higher generation units, and enhancement of the existing power system architecture in terms of survivability and efficiency have played a crucial role in this regard.

As depicted in Figure 28.1 the excessive levels of current in the next decade will subject a 42 V power net to the same set of problems. From this perspective, investigation of current intensive energy conversion devices deserves more attention.

Figure 28.2 illustrates the existing architecture for the electric power system of automobiles along with a possible topology proposed for the 42 V PowerNet. As can be observed, integration of the starter motor by introducing a bidirectional power electronic converter, dual-voltage system to segregate between current intensive loads and other low-power, low-voltage type of loads are among the most profound differences between these two topologies.

Inclusion of electric propulsion can further intensify the existing problem in which a medium-size adjustable speed motor drive will provide the entire or part of the propelling force in the vehicle. Although conservation of energy, environmental impact, and even political consequences of electric and hybrid electric vehicles are undeniable, a successful implementation of these concepts is subject to the feasibility of the development and control of current intensive motor drives. In fact the outstanding issues related to design and control of current intensive motor/generator drives may explain the selection of dual-voltage systems in most existing electric drives.

28.2 MAGNETIC DESIGN OF CURRENT INTENSIVE MOTOR DRIVES

At first glance, the problem of magnetic design in current intensive motor drives appears to be trivial. Figure 28.3 shows a typical torque vs. speed characteristic of an adjustable speed motor drive. As can be noted this characteristic can be partitioned into three major operational regions. These regions are:

Current Intensive Motor Drives: A New Challenge for Modern Vehicular Technology 635

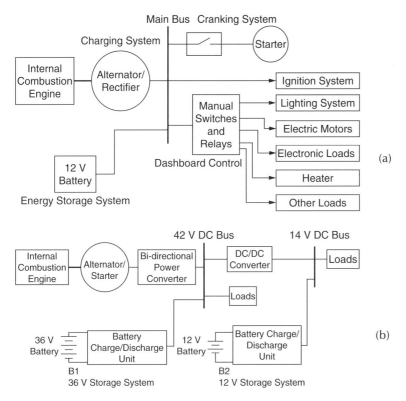

Figure 28.2 Transition from the existing 12 V into a dual-voltage system: (a) simplified architecture for the existing power system of the conventional cars; (b) a possible architecture for the 42 V automotive electric power distribution system.

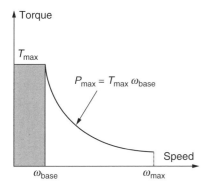

Figure 28.3 Torque vs. speed characteristic of an adjustable speed motor drive.

- Constant torque region, which starts at standstill and continues up to the base speed. Base speed is normally referred to as the speed at which the motional back-EMF induced in a phase of the machine exceeds the available part of the input voltage acting on the respective phase. In other words, base speed is when the unconditional control over the current waveform will start to diminish.

- Constant power region, which starts at or about the base speed and continues up to the beginning of the natural region of the machine. In this region, by proper control of the excitation a constant output power can be maintained. The speed range during which the above condition is satisfied is referred to as the speed range of the adjustable speed drive. Investigation of several independent researches indicates that an optimal choice for traction in electric vehicles should have a substantially wide speed range (3 to 4 times that from the base speed). Among various types of electric motor drives, reluctance type machines (switched reluctance, interior permanent magnet, and variable reluctance) represent the widest speed range, while surface-mounted permanent magnet machines represent the shortest speed range in their so-called field weakening region.
- Natural region, which starts at the end of the constant power and ends at the breakdown speed of the motor, beyond which the electromechanical actuator is not able to generate an effective electromagnetic torque. This region does not have major practical usage and in most parts is avoided by design and control algorithms.

Depending on the function sought from the electromechanical energy conversion device, various parts of the torque vs. speed characteristic may have a profound impact on the overall performance of an automotive application. For instance, for direct power steering assist, constant torque region plays a central role, whereas for an electric coolant pump constant power region is of particular significance.

In the process of transferring a voltage intensive design into a current intensive motor drive, the operational characteristic of the actuator should remain unaltered. In other words the maximum torque, base speed, and maximum speed of the motor should not change. Common sense dictates that maintaining a constant ampere-turn in the windings of the stator will exactly satisfy this requirement. In order to shed some light on this statement let's consider the case of a multiphase switched reluctance motor (SRM) drive. Figure 28.4 illustrates the inductance profile of a multiphase SRM drive. In this figure α_s, α_r, and N_r denote stator pole arc, rotor pole arc, and number of rotor poles, respectively. The dependency between the inductance value and rotor position indicates the substantial saliencies in the magnetic configuration of the machine. Due to its singly excited architecture and the absence of magnetic sources on the rotor, dynamics of the SRM drive can be entirely described by the behavior of its stator current/flux linkage.

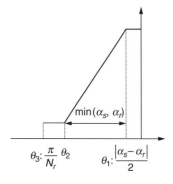

Figure 28.4 Inductance profile of an SRM drive.

In order to maintain a constant level of magnetomotive force (mmf) the number of turns and gauge of the conductors should be adjusted such that the targeted current with equal or less copper losses can circulate through the stator winding while generating the same level of mmf. The base speed of an SRM can be approximated using the following expression:

$$\omega_{base} \cong \left(\frac{V_{bus}}{I_{max}} - r\right)\left(\frac{\partial L(\theta, I_{max})}{\partial \theta}\right)^{-1} \quad (28.1)$$

where L, θ, r, I_{max}, and V_{bus} represents phase inductance, rotor position, stator phase resistance, maximum phase current and available DC link voltage, respectively. In an ideal case one can assume that if the shape and average distribution of the current in stator coils have not changed then one can claim that the following relationship exists between the old and new inductances, representing the current intensive and voltage intensive cases.

$$\frac{\partial L_1/\partial \theta}{\partial L_2/\partial \theta} = \frac{r_1}{r_2} = \frac{N_1^2}{N_2^2} = \left(\frac{V_1}{V_2}\right)^2 = \left(\frac{i_2}{i_1}\right)^2$$

$$\frac{N_1}{N_2} = k \quad (28.2)$$

A quick inspection of Equation 28.2 would indicate that under ideal conditions the base speed and maximum torque of the machine would not change if the ratio of the number of turns is selected as the ratio between the two voltage levels. However, as expected, in practice there are changes due to physical distribution of the current in the stator windings that would impact the distribution of the magnetic flux in the airgap and hence the electromagnetic torque developed by the SRM. In addition, distribution of the current would have a profound impact on the thermal response of the actuator and its cooling mechanism. In special cases high-frequency response of the motor influenced by the skin effects will also alter the resistance of the stator coils. Intensified levels of electromagnetic emission in current intensive motor drives can also be considered as a factor that would require further investigation. As can be seen, although under ideal conditions magnetic design of the machine can be easily modified by tuning the number of turns, the secondary issues imposed by operational conditions can drastically influence the overall performance of the drive.

28.3 STABILITY CONSIDERATIONS IN MULTICONVERTER SYSTEMS

Continuous increase of the power electronic-based energy processing units in vehicles manifests the occurrence of a multiconverter system. These multiconverter systems are subject to new stability issues, among which negative impedance instability has attracted considerable attention over the past few years. In order to address this problem let us consider the system shown in Figure 28.5. Assuming that some of the load converters supply electric power to adjustable speed motor drives that in part are of considerable size and sink significant power, one can describe a scenario for negative impedance instability. The adjustable speed motor drives operating at a constant speed, whose electromagnetic torque is tightly regulated, provide a constant mechanical power in respect to the switching

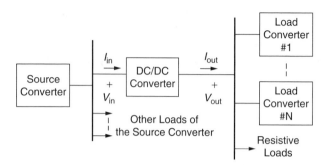

Figure 28.5 A typical multiconverter architecture for automotive applications.

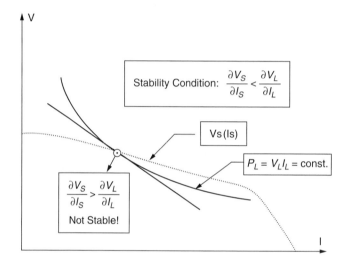

Figure 28.6 Graphical representation of the constant power instability.

frequency of the DC/DC converter. This means that within a few switching periods the mechanical output power can be kept constant. Operating at a fixed efficiency one can claim that the input electric power taken by the adjustable speed motor drive is a constant value. This results in a rather profound stability crisis in the system, for which the selection of operating voltage and current plays a crucial role. In order to understand the problem, the output characteristic of the main DC/DC converter along with the required input electric power by the adjustable speed motor drive are shown in Figure 28.6.

As can be observed, due to the difference between the relative slopes of the source and load characteristics in voltage vs. current plane, the first operating point represents an unstable condition. This means due to an unavoidable disturbance the state of the system will adjourn the equilibrium state and will never return. This is a classic definition of instability among system experts. To create a more realistic condition, one needs to consider the small signal equivalent circuit of this representative system in which a dependent current source represents the effects of constant power load. Professor Emadi from Illinois Institute of Technology has shown that the transfer function between the input and output voltages in a DC/DC buck converter supplying the motor will exhibit unstable mode depending upon the selection of system parameters, including the operational voltage of the adjustable speed motor drive.

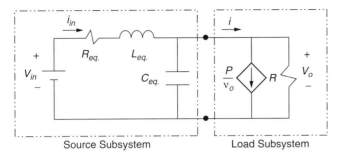

Figure 28.7 Equivalent averaged model of a DC/DC converter and a constant power load.

Figure 28.7 illustrates the equivalent system comprised of an averaged model of a DC/DC front-end converter along with that of a constant power system with a finite efficiency (losses in the machine are represented by a parallel resistance). The condition for stability in this system is:

$$\text{Stability condition} \Rightarrow P < \frac{V_o^2}{R} + \frac{R_{eq.} C_{eq.}}{L_{eq.}} V_o^2 \qquad (28.3)$$

It can be seen that reduction of voltage with respect to the power (current intensiveness) can have an adverse impact on the stability of the system. This concept can be further extended to cover the condition for a combination of constant power and constant voltage loads.

28.4 ENERGY TRANSFER

An effective transfer of energy can be viewed as a significant challenge in the electric power system of a vehicle. It is anticipated that there will be close to 1600 circuits in the conventional car of the future. Transfer of energy in a typical adjustable speed motor drive comprises several steps that are summarized in Figure 28.8. In each step of the way, there is a drop of the voltage and there is a loss of the electric power. In a current intensive system the tolerance on the voltage drop is limited. However, there are engineering and cost thresholds on minimization of the voltage drop that can result in critical reduction of the operational voltage in a current intensive scenario. The copper losses and silicon losses are similarly subject to higher levels as compared to their voltage intensive counterparts. This issue scan mount into serious design problems that need to be answered before a feasible current intensive motor drive can be put in place.

As far as the transfer of active power is concerned, design of the wiring system needs to be inspected. Figure 28.9 shows a typical example for a simple power system

Figure 28.8 Various components in an adjustable speed motor drive for automotive applications.

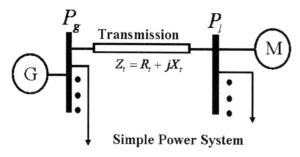

Figure 28.9 Transfer of active and reactive power in a simple power system for automotive applications.

that one may find in an automotive system. The active and reactive powers can be expressed using the following equations:

$$\text{Active power:} \quad P_g = \sum R_t \|I_g\|^2 + \sum P_l$$
$$\text{Reactive power:} \quad Q_g = \sum X_t \|I_g\|^2 + \sum Q_l \quad (28.4)$$

In order to maximize the level of transferred active power to the loads, an impedance matching is necessary. This can be done either using a proper selection of the cable parameters or using power electronics techniques. In either case it must be noted that significant saving in terms of overall efficiency of the transmission system and effective delivery of the power can be achieved.

28.5 IMPACT ON CONTROL

Advanced adjustable speed motor drives are devised with monitoring systems that would require a precise estimation of the machine parameters. Placement of adjustable speed motor drives in critical path and high impact applications such as those in the starter/alternator or power steering creates additional demand in terms of functionality and, above all, enhanced survivability. In order to achieve fault tolerance incorporating advanced techniques such position sensorless algorithms are sought. These methods are, however, highly dependent upon an accurate estimation of the machine parameters such as rotor time constant in induction motor drives, stator phase inductance in reluctance type machines, and self-inductance in permanent magnet motors, to name a few. The accuracy of inductance and resistance values in a current intensive motor drive is usually less than that of an equivalent voltage intensive motor drive. This implies a limitation on the available range and sensitivity of the targeted parameter over the entire operational regime. This in turn undermines the practicality of the respective control technique. Therefore, many advanced technologies that are well established may face implementation issues that eventually will impede the overall effectiveness of the motor drive system. This is an ongoing area of research, and as the quest for employment of current intensive motor drives continues, the need for modified or alternative control techniques would be more apparent.

29

Power Electronics Applications in Vehicle and Passenger Safety

D.M.G. Preethichandra
Kyushu Institute of Technology, Kitakyushu, Japan

Saman Kumara Halgamuge
University of Melbourne, Melbourne, Australia

29.1 INTRODUCTION

Vehicle safety has been drastically improved during the last couple of decades, and most of the safety systems are electronically controlled. Whenever the mechanical power needs to be electronically controlled, the obvious answer is a power electronic unit interfacing the control module and the actuator. Therefore, all safety devices in modern vehicles are associated with relevant power electronic circuits. In this chapter we will investigate the involvement of power electronics in important safety devices individually.

29.2 POWER ELECTRONICS IN VEHICLE SAFETY

Implementation of comprehensive safety measures in vehicles is a complicated task to achieve and has to be addressed from numerous different perspectives including road accidents, engine malfunctioning, user mistakes, and theft. Therefore, a dedicated power electronic system for each task of concern is generally implemented in modern vehicles. As a result, a common control system for the automobile power electronics subsystems was a long-standing demand in the industry and had been addressed by all major manufacturers over the last couple of decades. With the development of manufacturer-specific intelligent control systems, the service industry seemed to be running into a more complicated situation. Then the need for standardizing of such control systems was treated as a major issue. In a similar way that Hewlett-Packard's instrument networking standard (HPIB) became the default industry standard with IEEE recognition as the IEEE488 (or GPIB) bus, the controller area network (CAN) developed by Robert Bosch GMBH in the late 1980s (on the request of Mercedes) became the default industry standard for automobile industry.

29.2.1 THE CAN BUS USED TO NETWORK VEHICLE POWER ELECTRONIC MODULES

The CAN was first used in the high-end vehicles of Mercedes, and soon followed by BMW, Porsche, and Jaguar. Then the acceptance by the other European manufacturers such as Volkswagen, Fiat, and Renault paved the road for the CAN bus to become the industry standard. Today the CAN bus works as the universal backbone of all automobile power electronics systems.

CAN uses only three of the seven-layer Open System Interconnection (OSI) model of the International Organization for Standardization (ISO). They are the Application, Data Link, and Physical layers [1]. It has neither master nor slave devices on the system, and the bus arbitration totally depends on the message priority level defined by the sender. As stated earlier, CAN is a multimaster network. It uses Carrier Sense Multiple Access/Collision Detection with Arbitration on Message Priority (CSMA/CD+AMP). Before sending a message the CAN node checks if the bus is busy. It also uses collision detection. In these ways it is similar to Ethernet. However, when an Ethernet network detects a collision, both sending nodes stop transmitting. They then wait for a random period of time before trying to send again. This makes Ethernet networks very sensitive to high load in the bus. The CAN protocol solves this problem with a very clever principle of message priority-based bus arbitration.

Predominantly, there are two standards for the CAN derived by the ISO, the High Speed CAN and the Low Speed CAN [2,3]. Figure 29.1 and Figure 29.2 show the Low and High Speed CAN bus configurations. Usually a twisted-pair cable with 33–55 twists per meter is used for High Speed CAN bus in automobiles. The data transfer rate (baud rate or number of bits per second) is dependent of the bus length. Table 29.1 shows the recommended baud rates for various bus lengths. One might wonder why bus lengths over

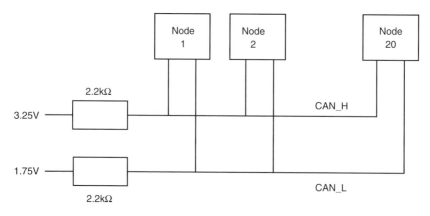

Figure 29.1 The low-speed CAN bus configuration.

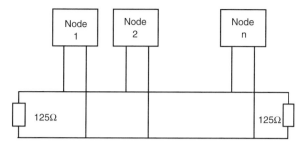

Figure 29.2 The high-speed CAN bus configuration.

Table 29.1 CAN Bus Lengths and Their Recommended Maximum Baud Rates

Bus Length	Maximum Baud Rate
< 40 m	1 Mbps
< 100 m	500 kbps
< 200 m	250 kbps
< 500 m	125 kbps
< 1000 m	40 kbps
< 6000 m	10 kbps

100 m are defined if it is for automobile internal use. The reason is the CAN bus is widely being used in production line and building automation systems where the effective bus length may be a couple of kilometers.

The CAN bus has two lines and each line has two possible states, namely dominant and recessive. Any device on the network is to be connected onto these two lines, referred to as high-wire (CAN_H) and low wire (CAN_L). The bus logic is very much similar to the wired — AND, where the recessive bits are overwritten by the dominant bits. Mostly, but not essentially, the recessive bits are equivalent to logic 1 and dominant bits are equivalent to logic 0. The actual voltages for each state in the wires are different from one standard to the other. Table 29.2 shows a summary of these voltages.

Table 29.2 CAN Bus Voltages

| Standard | Bus Status | Voltage | |
		CAN_H	CAN_L
ISO 11519	Recessive	1.75 V	3.25 V
(low speed CAN)	Dominant	4.0 V	1.0 V
ISO 11898	Recessive	2.5 V	1.5 V
(high speed CAN)	Dominant	3.5 V	1.5 V

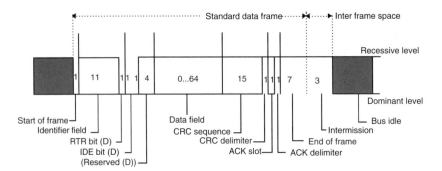

Figure 29.3 Standard CAN data frame.

As long as there is no node sending any data to the CAN bus, it remains in the recessive state. Before any data is put on the bus, each device checks the bus status. Whenever the bus recognizes any data transfer request by any node, it goes to the dominant state. Any CAN message has an identifier of 11 bits (or 29 bits for extended CAN), which has the priority level. If there are two nodes requesting the bus at the same time, then the winner will be selected purely on the priority level of the message. Figure 29.3 shows the standard CAN data frame. The unsuccessful candidate will try again in the next free time slot until it get a chance to transmit.

29.2.2 Engine Safety Systems

Engine safety is basically achieved by sensing every vital engine parameter concerned, and subsequent processing of information leading to adjusting the relevant devices appropriately or alarming the driver. As more and more electronic systems becoming available in newer vehicles, fully or partially replacing the old mechanical components, the vulnerability of failure is higher compared to the fully mechanical systems with basic features. Therefore, these systems are continuously being monitored and adjusted.

The valve timing is one of the most important factors to guarantee safe and environmental-friendly operation of an engine. Figure 29.4 shows the standard valve timing diagrams of an old engine and a variable valve timing (VVT) engine. In the mechanical valve control systems, the valve timings were constant relative to the crank angle. As commonly known, valves activate the breathing of the engine. The timing of breathing, that is, the timing of air intake and exhaust, is controlled by the shape and phase angle of cams. To obtain optimal power and minimal exhaust, engines need to have different valve timing at different speeds. When the engine revolution increases, the effective duration of intake and exhaust strokes decreases, so that fresh air becomes not fast enough to enter the combustion chamber, while the exhaust becomes not fast enough to leave the combustion chamber. Therefore, the best solution is to open the inlet valves a little earlier and close the exhaust valves a little later. In other words, the overlapping between intake period and exhaust period should be increased as revolution increases.

Different automobile manufacturers use their own methods for obtaining the optimum condition by changing the valve timing. Honda introduced its Valve Timing Electronic Control (VTEC) in the late 1980s into its cars and soon followed by all the major manufacturers. This includes Toyota VVT-I and VVTL-I, Mitsubishi MIVEC, Nissan Neo VVL, BMW Double Vanos, Porsche Varicam Plus, and Rover VVC. The most sophisticated method among them at that time was Rover's Variable Valve Control (VVC), introduced

Power Electronics Applications in Vehicle and Passenger Safety

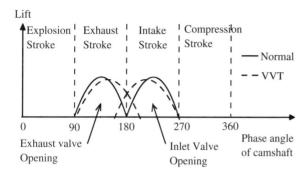

Figure 29.4 Valve timing of a four-stroke engine.

Figure 29.5 Phasers used for cam-phasing (photo courtesy of Ford Motor Company [Australia] Ltd.).

in 1995. This system has individual power electronic module to control each single valve in the engine. The advantage is these valves can be controlled very smoothly over the full-speed and -load range. This was then introduced to most of the high-end vehicles by the majority of car manufacturers.

The most common and economical technique is the cam-changing and cam-phasing VVT system first used by Toyota and Porsche, which has now become the most popular in the family car range, used by many manufacturers.

Let us see how the cam-changing and cam-phasing system works. Variable cam timing is achieved through actuators called phasers fitted to the camshaft. In V engines there are two phasers coupled by a chain for synchronization. Figure 29.5 shows the phasers of a V engine used in a Ford Falcon. Each phaser is associated with an oil control valve, which controls the phaser's relative angular position. Figure 29.6 shows these oil control valves from the engine side.

Figure 29.7 shows the cross-section of the oil control valve. The solenoid is controlled by a pulse width modulated (PWM) signal from the powertrain control module (PCM). The linear motion of the solenoid core is proportional to the duty cycle (or to mark to space ratio) of the PWM signal. At the start and in idling condition the actuator is in the fully advanced position under the spring pressure, where the duty cycle is 0% (off). The supply port is fed by high-pressure oil from the engine oil pump and the advance port acts as the output port. This will rotate the phaser fully in advance direction.

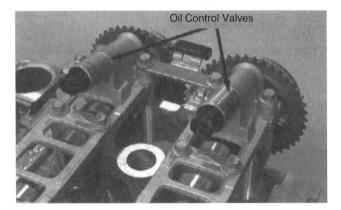

Figure 29.6 Oil control valves (photo courtesy of Ford Motor Company [Australia] Ltd.).

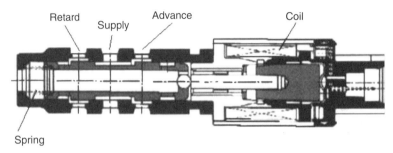

Figure 29.7 Oil control valve (diagram courtesy of Ford Motor Company [Australia] Ltd.).

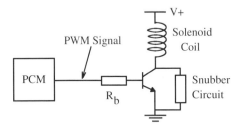

Figure 29.8 Schematic of power electronic circuit controlling the solenoid.

Figure 29.8 shows the schematic diagram of the power electronic circuit that controls the oil control valve. The signal to the solenoid is a pulse width modulated square wave signal, so that the power transistor in series with the solenoid coil is switched between on- and off-states accordingly. Even though the power transistor can be switched on or off instantly, the current flow through the coil cannot undergo a sudden change. Therefore, a snubber circuit is essential to protect the transistor from the back-EMF generated during the switching time. Then the current will decay down with a time constant governed by the snubber circuit. When this happens at a higher frequency, the average current flow

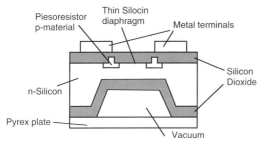

Figure 29.9 T-MAP sensor.

through the solenoid is proportional to the duty cycle (the ratio of conducting period to the total period) of the control signal.

With the increasing engine speed (above 1000 rpm), the variable valve timing is operational. The PCM generates the solenoid control signal in relation to the sensor signals, such as engine speed sensor, manifold pressure (MAP) sensor, and throttle position sensor (TPS). When the solenoid is actuated, it starts moving leftwards until the shrunk spring force is enough to balance it. With this action the advance port size is reduced, hence moving the oil pressure to the phaser in the advance direction. The result is the phaser stars rotating in the retard direction. With a higher duty cycle signal from the PCM, the core moves further leftwards, closing the advance port completely, and opens the retard port accordingly, moving the rotation of the phaser further in the retard direction. In the drive mode the duty cycle can vary from 20 to 80%. When turning the engine off, the actuators will return to the fully advanced (locked) position.

The timing of valves should be exactly on the right position to optimize the performance and should not go out from a certain limit to prevent the piston from being crashed onto the valves. This is the most crucial engine safety issue facing the designers. This was achieved by sensing the actual cam position and calculating the cam angle error (the difference between the actual and the desired). In a V engine the cam angle error for each cam is calculated separately, and the PCM can control each camshaft individually and adjust the position to match the single desired cam angle.

Engine temperature, manifold pressure, and oil condition are some of the important factors to be monitored for safe engine operation and performance optimization. The intake air temperature and manifold pressure are sometimes measured by a combined sensor called temperature and manifold pressure (T-MAP) sensor. The internal structure of a T-MAP sensor is shown in Figure 29.9. The piezoresistive type diaphragm provides a linear analog signal corresponding to the manifold pressure and the P-N junction provides a signal on intake mass air temperature. The relationship is governed by the following equation [4]:

$$I_0 = K_1 T^2 \varepsilon^{-V_{GO}/V_T}$$

where

I_0 = the reverse saturation current
K_1 = a constant dependent of temperature
T = the temperature of junction in Kelvin
V_{GO} = forbidden gap-voltage
V_T = the volt equivalent of temperature

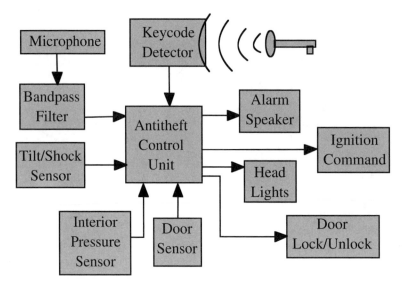

Figure 29.10 Block diagram of an antitheft alarm system.

29.2.3 ANTITHEFT ALARM SYSTEMS

Vehicle theft is a big problem faced by owners, and so the demand for intelligent antitheft systems is never declining. There are two types of antitheft systems available today. One is for vehicles manufactured without such systems and the other one is integrated to the vehicle control system and communicates with the powertrain control module and body electronics module (BEM) via the CAN bus. All these systems have different features and provide different degree of security.

Figure 29.10 shows the schematic diagram of a sophisticated antitheft system. The keyless entry switch sends an encrypted code to the keycode detector, and if and only if it matches with the vehicle's code, the control unit sends the relevant commands to the door locking system and ignition system (if remote ignition is available). Any shock or tilt beyond a permissible limit, unauthorized door opening, or a sudden change in the interior pressure will send a signal to the control unit and the alarm and the blinking headlights will be activated. Shock sensor consists of a vertical central electrode in a cylindrical container and another electrode placed on the base around the electrode 1, as shown in Figure 29.11. The electrodes are in contact through a metal ball and in case of

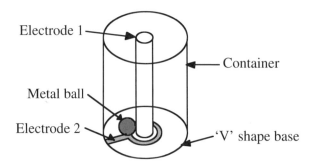

Figure 29.11 Shock sensor construction.

Power Electronics Applications in Vehicle and Passenger Safety

Figure 29.12 Vehicle antitheft security system (photo courtesy of Ford Motor Company [Australia] Ltd.).

a shock, the metal ball will move outwards, resulting in a broken connection between electrodes 1 and 2.

The microphone is employed to detect a glass breaking sound, and the band-pass filter cascaded has a very narrow bandwidth adjusted to the frequency of a sound from breaking glass. This prevents unauthorized entry by breaking the windows and prevents the system from responding to other noises.

More advanced systems that come with newer vehicles are capable of communicating with the BEM and PCM, and they have a special transponder integrated to the key and a transceiver mounted on the key-hole assembly, as shown in Figure 29.12. These systems have all the above discussed features and are capable of validating the physical key in the key-hole.

This transponder-key is a specially encoded ignition key, and each transponder contains a unique digital identification code. The engine immobilization system uses this code to validate the ignition key. Usually the transceiver communicates directly with the body electronics module and thereafter communications with the powertrain control module is via the CAN bus. Only the owner's key can start the vehicle, and for any other key, the BEM sends an immobilizing signal to the PCM. These keys do not require external power, so that the lifetime is that of the vehicle.

29.2.4 ADAPTIVE CRUISE CONTROL (ACC)

Keeping the speed at a constant value in highway driving is very important, not only to avoid overspeeding, but also to make the driver less stressful. All the major automobile research groups have addressed this need in the past, and as a result, modern vehicles come with built-in adaptive cruise control systems. The main features of these systems are that they operate at speeds higher than the city speed limits (usually 40 kmph), take over the engine control on the driver's request, and instantly hand over the engine control system to the driver if the brake or the accelerator is pressed. For example, if the ACC switch is pressed at a speed of 95 kmph, then it will adjust the engine control system to keep the vehicle at 95 kmph unless there is a slow vehicle detected in front.

Mitsubishi and Nissan proposed their ACC systems in 1997 and 1999, respectively, with laser-guided systems and radar-guided systems. In Europe, Mercedes S-Class came with an ACC system called Distronic in 1999, and this feature was soon available in other makes [5]. The distance measurement is associated with sensing systems of different kinds; Japanese manufacturers prefer the Lidar systems, while European manufacturers prefer radar systems.

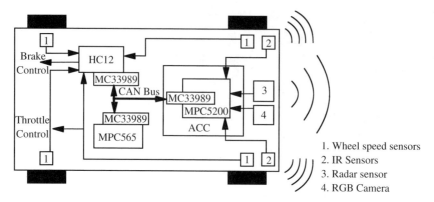

Figure 29.13 Block diagram of ACC system.

The block diagram shown in Figure 29.13 shows a complete ACC system proposed by Motorola [6] (all the numbers starting with the letter M or H are chip numbers). The ACC control processor (MPC5200) gets the front vehicle distance from the radar sensor, near-distance disturbances from the IR sensors, visual information on front vision from the RGB camera, and individual wheel speeds from wheel speed sensors (via the HC12 chip and CAN bus). Then it processes this information and sends a command to the PCM via CAN bus to activate the throttle valve or brakes as necessary.

Automatic maneuvering is a feature expected to appear in the next generation of vehicles, and there is a lot of research work going on all around the globe in this area [7]. Some systems use differential global positioning systems (DGPS), while the others rely on visionary information. However, the problem is the processing speed to keep up with the vehicle speed. In the recent past a group of Japanese researchers had successfully developed a low-speed automatic steering system [8] for a snow remover. They employed DGPS in one of their designs, and the other one depends on road sensors that detect preburied radio wave or magnetic line markers along the road in regular intervals. This lane tracking method cannot be used on normal roads but the DGPS system is a viable solution.

29.2.5 REVERSE SENSING AND PARKING SYSTEM

Reverse sensing systems provide an enormous support to the driver, especially in a compact environment. Figure 29.14 shows a typical setup of the reverse sensors and their detecting zones. The length of this lobe is determined considering the vehicle length and geometry. The 180 cm shown in Figure 29.14 is a typical lobe length for a standard family sedan-type vehicle. Whenever any of these sensors detect an obstacle within its lobe, the control module alarms the driver.

More sophisticated systems provide automatic maneuvering during the parking phase, supported by additional sensors detecting the distances to curbs or other obstacles in lateral and front directions. Toyota Prius 2003 model is one of the first commercially available vehicles with such an automatic parking system.

29.3 POWER ELECTRONICS IN PASSENGER SAFETY

Passenger safety is one of the most important things when selecting a vehicle. Modern vehicles are facilitated with highly sophisticated and intelligent safety devices to provide

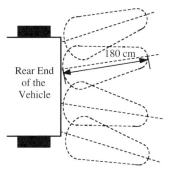

Figure 29.14 Reverse sensing system of a vehicle.

protection in case of a crash or an unexpected situation. These were first developed as mechanical systems, and little by little the electronics took control. As a result, most of the safety devices today are microprocessor controlled and activated by means of power electronics. Safety of vehicles is being continuously improved, and many research work can be found on real-world tests on vehicle safety [9–13].

29.3.1 Seatbelt Control Systems

The seatbelt plays a great role in passenger safety in case of a severe crash. The mechanical seatbelt system has a belt webbing wound on a spool connected to a spiral spring. When you pull and place the belt in its lock position, the spiralled spring is loaded, tightening you towards the seat gently. The ends of this spool have two wheels with teeth and form a ratchet mechanism when a sudden deceleration is detected. In case of a crash, it detects a huge deceleration and therefore the ratchet mechanism doesn't allow the belt to release. Thus, the passenger is tightened to the seat safely.

However, any initial slack or the initial release of a seatbelt during the crash may cause some damage. The modern seatbelt systems with pretensioners overcome this situation by tensioning the seatbelt by a little in a crash. Figure 29.15 shows the electronics behind this pretensioner. Whenever the crash sensors detect a severe crash, the control module sends a firing signal to the actuating unit of the pretensioner — simply a fast-heating-up filament. This filament produces enough heat to ignite the chemical inside the pretensioner chamber, and then the extra gas generated by the chemical explosion pushes a rack with gear teeth on it. This rack engages with the gear teeth of the spool and winds it a little bit. The extra slack or initial release can be rewound in this manner. Though this looks like a long process, the whole thing takes only a couple of milliseconds.

There are some situations in which the seatbelt has to be a little slacked, depending on the degree of the impact, to reduce the force applied by the belt on the body of the

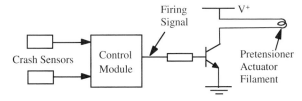

Figure 29.15 Schematic diagram of electronic actuator system of a seatbelt pretensioner.

passenger. This is achieved by introducing a device called a load limiter. Basic mechanical-type load limiters have a folded sewn segment, and on an excessive load the stitches holding the fold will be broken down and the segment will be unfolded. This will provide some extra belt length and reduce the force on passenger. However, this type can have only two stages: i.e., either fold or unfold. But the real-world situations are far more complex than what these two stages can address, and intelligent release of desired length is in demand. A large amount of research work is going on in this area, and passenger size, weight, and nature of the object (whether it is a passenger or a cargo) plays an active role in load limiter reaction.

29.3.2 Power Window Safety Systems

Manual or hand-operated windows are gradually disappearing in modern vehicles. Power windows are a great comfort to passengers but may cause severe damage to unattended children inside the vehicle if they handle it inappropriately. Therefore, the safety of power window operation is a vital issue in automobile passenger safety.

Generally, the power window works with a single DC motor lifting up or down the window glass on request. The control of a passenger window is basically done by double-pole three-way switches, as shown in Figure 29.16. These switches normally rest on the mid-position and return back to the rest position as soon as the finger is taken off the switch. Switch SW1 is the driver-side switch and SW3 is the passenger-side one where they are in series with the driver-side lock switch SW2. When the lock switch is in unlock position, pressing either SW1 or SW3 would rotate the motor clockwise or counterclockwise, dependent on which side of the switch is pressed.

In one type of commonly used window control, pressing the corresponding side performs switches both up and down operations. The danger with this type is if a child puts their head outside a half-opened window and steps onto the "up" side of the switch, it will drive the glass upwards, squeezing the child's neck. Identifying this problem, most of the manufacturers now have replaced these press-type switches with press for down operation and pull for up operation-type switches. However, some of the high-end vehicles come with obstacle monitoring systems integrated to the window power electronic circuits and former type of switches. There are couple of IR emitters and detectors mounted around the window frame to detect obstacles. The switch SW4 in Figure 29.16 is controlled by the control module, dependent on the above-mentioned sensor input signals. If there is an object across the window plane, it will open SW4, deactivating the power window operation.

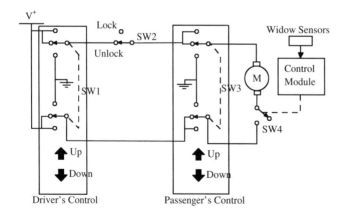

Figure 29.16 Circuit diagram of an intelligent power window system.

Power Electronics Applications in Vehicle and Passenger Safety

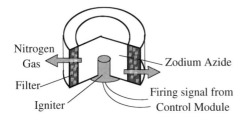

Figure 29.17 Inside of an airbag.

29.3.3 AIRBAGS

At present, airbags are a very common feature in automobile safety but their safety levels are often questioned by researchers [14–19]. First of all, we will see how the airbag system works. As shown in Figure 29.17, the airbag inflator consists of igniters surrounded by sodium azide and a filter between the sodium azide and the deflated airbag. Just like the seatbelt pretensioner, the airbag is activated by a signal from the control module once the crash sensors detect a crash. The activating circuit is similar to the circuit shown in Figure 29.15. Sodium azide is a very highly explosive material, which generates nitrogen gas when exploded. This nitrogen gas passes through the filter to the airbag and inflates it. Though this looks like a long procedure, it takes less than 40 ms to deploy the airbag from the moment a crash is detected, and it is sufficient to protect the occupants from a frontal impact. However, side airbags are much faster because the side impact time is much shorter compared to a frontal impact.

The deployment of airbag should not be the same for any passenger. Especially young children, pregnant women, and elderly passengers in the front seat require different levels of airbag inflations. Many research groups are addressing this, and weight and size of the passenger are the main parameters used in discrimination, but visual information processing results in a great advantage on identification of passenger category more accurately.

29.3.4 DRIVER ASSISTANCE SYSTEMS AND STRESS MONITORING

Driver assistance systems (DSSs) are being developed mainly based on vision and AI technology by most signal processing providers for vehicles. As such systems need to be integrated with the vehicle electrical system, power electronics must be used for the implementation.

A DSS using vision systems generally has four basic units:

1. Detection of objects from a camera mounted on-board the vehicle (e.g., traffic signs, pedestrians, other vehicles, the positioning of the driver, and the facial analysis)
2. Classification of such objects based on prior knowledge and expectation
3. Tracking of objects where it is required (e.g., traffic sign, another vehicle)
4. Decision making to support the driver by creating warnings and alarms such as, "Stop sign is getting closer," "You feel sleepy and tired," or even to interfere with the action of the driver

Such systems try to support the driver by observing the driver and passengers, so that warnings can be generated, if the driver looks tired, or to monitor airbags accurately at the last microsecond, if the car is involved in an accident.

Detection of stress and the fatigue level of the driver and passengers may mainly occur due to the depletion of oxygen and carbon monoxide poisoning. Carbon monoxide poisoning could happen due to the exhaust gases from the same vehicle or other vehicles on the road. For example, countries like Australia and Canada have the problem of suicide by feeding the vehicles exhaust gas to the vehicle with the windows closed. In such cases fatigue or suicide detection is possible using a proper device that checks the air quality of the cabin.

AI technologies have been used to develop DSS in the past [20–22]. Driving a vehicle with a trailer in the reverse direction is challenging. Even experienced drivers have to undertake a trial-and-error approach; i.e., if the trailer comes to a position where the angle between the longitudinal axes of the vehicle and the trailer cannot be increased even by maximum angle of the steering wheel in reverse driving, then the driver has to change the gear and drive forward in order to avoid further bending (critical angle). Since this is a nonlinear problem, a support system can be designed using fuzzy control in an efficient manner to overcome this situation. The expert knowledge of continuous reverse driving (i.e., without changing the gear) in such situations can be formulated using fuzzy rules. The real-time implementation of a hierarchically organized hybrid controller to handle this problem of a model truck (52.2 cm long, 18 cm wide, and 23.5 cm high) and trailer (67.8 cm long, 18.7 cm wide, and 20 cm high) is described in Reference 20.

Physical limits of driving dynamics are determined by friction between the tire and the road surface. According to the measurements of friction taken from different real road surfaces, the maximum value of the friction coefficient is ranging from a low value (on ice) to a high value (on well-textured dry asphalt). Therefore, automatic evaluation of the maximum possible friction coefficient is of importance for driving safety, especially in considering its use for antilock braking systems (ABS). Since this is closely related to the state of the road, identification of the state of the road surface can lead to the friction coefficient of the surface [21]. A Hall sensor transfers the deformation of the tire tread elements to the preprocessing unit, where the sensor signal is filtered. Training generates the AI system with experimental data drawn by driving in different known road surfaces.

29.4 CONCLUSIONS

The application of power electronics in vehicle and passenger safety attracted increased interest with the advancement of technology and the public awareness in safety systems for vehicles. The advent of by-wire technologies and wireless networking in vehicles also put forward new challenges into this area of research. This chapter described a few of the well-known applications of power electronics as well as a few applications still in the development phase. The advancement of sensors and actuators and high-speed signal processing will open new pathways to provide more and more sophisticated safety devices in the future vehicles.

Acknowledgments

The authors acknowledge Ford Motor Company (Australia) Ltd. and Mitsubishi Motors (Australia) Ltd. for providing technical details.

The second author acknowledges the Australian Research Council for partially funding some of the work described in this chapter.

REFERENCES

1. N. Liang and D. Popvic. The CAN Bus. In L. Vlacic, M. Parent, and F. Harashima (Eds.), *Intelligent Vehicle Systems*. SAE International, 2001, pp. 21–64.
2. International Standards Organization. *Road Vehicles — Interchange of Digital Information — Control Area Network (CAN) for High Speed Communications*. ISO11898:1993.
3. International Standards Organization. *Road Vehicles — Low Speed Serial Data Communications — Part 2 — Low Speed Control Area Network (CAN)*. ISO11519:1994.
4. J. Millman and C.C. Halkias. *Integrated Electronics*. Singapore: McGraw-Hill, 1987, pp. 726–762.
5. D. Maurel and S. Donikian. ACC Systems — Overview and Examples. In L. Vlacic, M. Parent, and F. Harashima (Eds.), *Intelligent Vehicle Systems*. SAE International, 2001, pp. 422–441.
6. http://e-www.motorola.com/webapp/sps/.
7. C.N. Xuan and Y. Youm. *Intelligent Online Driving System*. Korean Automatic Control Conference (KACC), Seoul, 2000.
8. H. Hirashita, T. Arai, and T. Yoshida. Automatic Steering System for Rotary Snow Removers. *Proceedings of International Symposium on Automation and Robotics in Construction* (19th ISARC), Maryland, 2002, pp. 443–448.
9. P. Thomas and A. Morris. *Real-World Accident Data — Coordinated Methodologies for Data Collection to Improve Vehicle and Road Safety*. 18th International Conference on the Enhanced Safety of Vehicles, Nagoya, 2003.
10. R. Cuerden, H. Lunt, A. Fails, and J. Hill. *On the Spot Crash Investigation in the U.K.: New Insights for Vehicle Research*. 18th International Conference on the Enhanced Safety of Vehicles, Nagoya, 2003.
11. P. Thomas and R. Frampton. *Crash Testing for Real-World Safety — What Are the Priorities for Casualty Reduction?* 18th International Conference on the Enhanced Safety of Vehicles, Nagoya, 2003.
12. B. Hurley and R. Welsh. *Analysis of Real-World Rollover Accidents in the U.K.* VDI, Germany, 2001.
13. P. Thomas and R. Frampton. *Real-World Crash Performance of Recent Model Cars — Next Step in Injury Prevention*. 18th International Conference on the Enhanced Safety of Vehicles, Nagoya, 2003.
14. A. Morris, R. Welsh, and A. Hassan. *Requirements for Crash Protection of Older Vehicle Drivers*. 47th Annual Conference of the Association for the Advancement of Automotive Medicine, Lisbon, 2003.
15. R. Welsh, A. Morris, and L. Clift. *The Effects on Height on Injury Outcome for Drivers of European Passenger Cars*. 47th Annual Conference of the Association for the Advancement of Automotive Medicine, Lisbon, 2003.
16. A. Kirk, R. Frampton, and P. Thomas. *An Evaluation of Airbag Benefits/Disbenifites in European Vehicles — A Combined Statistical and Case Study Approach*. IRCOBI Conference on Biomechanics of Impact, Munich, 2002.
17. A. Kirk and A. Morris. *Side Airbag Deployments in the U.K. — Initial Case Reviews*. 18th International Conference on the Enhanced Safety of Vehicles, Nagoya, 2003.
18. J.S. Barnes, A.P. Morris, and R.J. Frampton. Airbag Effectiveness in Real-World Crashes in Australia. *Proc., ImechE Conference*, Institute of Mechanical Engineers, London, 2002.
19. R. Frampton, R. Welsh, and A. Kirk. *Effectiveness of Airbag Restraints in Frontal Crashes — What European Field Studies Tell Us*. IRCOBI Conference on Biomechanics of Injury, Montpelier, VT, 2000, pp. 425–438.
20. S.K. Halgamuge, T.A. Runkler, and M. Glesner. *A Hierarchical Hybrid Fuzzy Controller for Realtime Reverse Driving Support of Vehicles with Long Trailers*. IEEE International Conference on Fuzzy Systems '94, Orlando, FL, pp. 1207–1210, 1994.

21. H.-J. Herpel, S.K. Halgamuge, M. Glesner, J. Stoecker, and S. Ernesti. *Fuzzy Logic Applied to Control and Data Analysis Problems in Automotive Applications.* International Symposium on Automotive Technology and Automation, pp. 225–233, Aachen, Germany, 1994.
22. W. Rattasiri and S.K. Halgamuge. Computationally Advantageous and Stable Hierarchical Fuzzy Systems for Active Suspensions. *IEEE Transactions on Industrial Electronics,* Vol. 50, No. 1, 2003.

30

Drive and Control System for Hybrid Electric Vehicles

Weng Keong Kevin Lim, Saman Kumara Halgamuge, and Harry Charles Watson
University of Melbourne, Melbourne, Australia

30.1 INTRODUCTION

Hybrid vehicles have multiple power sources that can be separately or simultaneously used to propel the vehicle. Different ways of integrating these power-producing components with electrical energy storage components allow for distinct types of hybrid electric vehicle (HEV) configurations. Generally, fuel energy may be converted by a number of distinct heat engines, such as an internal combustion engine (ICE), while an electrical motor (EM) derives its electrical energy from batteries, ultracapacitors, solar cells, fuel cells, or generators driven by heat engines or flywheels. To date, the most promising hybrid vehicle is the HEV driven by an ICE and an EM powered by on-board batteries. HEV solves many pure electric vehicles' problems and minimizes the shortcomings of conventional vehicles, while providing the benefits of both electric and conventional vehicles.

The major hybrid configurations (HCs) are series and parallel configurations, shown in Figure 30.1 and Figure 30.2, respectively. In series hybrids, an ICE powers a generator that either supplies electrical power to the EM or charges the batteries. The ICE does not mechanically drive the wheels directly. As for the parallel hybrid, an ICE supplies mechanical power directly to drive the wheels, with the EM directly coupled to the propulsion system. Parallel hybrids have the flexibility to propel a vehicle with pure ICE only, pure EM only, or a combination of both ICE and EM simultaneously, based on the Control Strategy (CS) settings. Currently there exists a third type of hybrid configuration that combines the best aspects of both series and parallel configurations, known as the series-parallel hybrid, as shown in Figure 30.3. This configuration allows the ICE to directly drive the wheels, but also has the ability to simultaneously charge the batteries and to power the EM through a generator.

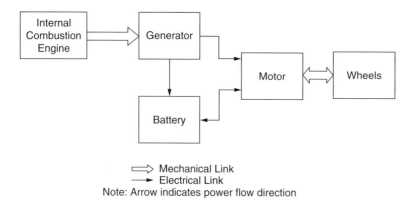

Figure 30.1 Series hybrid configuration.

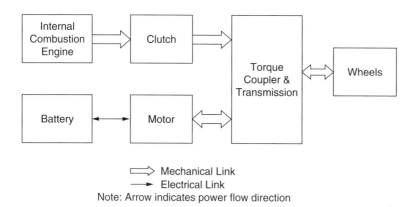

Figure 30.2 Parallel hybrid configuration.

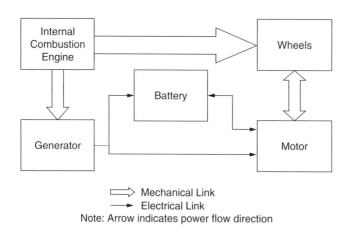

Figure 30.3 Series-parallel hybrid configuration.

Besides HC, HEV also makes a distinction between charge depleting hybrids and charge sustaining hybrids. Charge depleting hybrids allow their batteries to be depleted and do not have the ability to recharge them at the same rate as they are being discharged. This type of hybrid gives more priority to propelling the vehicle under pure electric mode most of the time and depletes its battery state of charge (SOC) to its minimum level within the allowable SOC operating range. When the SOC reaches the minimum level, the ICE will be triggered to drive the generator to recharge the battery back to its maximum SOC. When the SOC is high, the ICE is shut off and the vehicle is driven under pure electric mode until the SOC depletes to its minimum level again.

In charge sustaining hybrids, the ICE is adequately sized to meet the average power load, and whenever operated under the expected conditions, it will be able to maintain adequate reserve of energy in the batteries at all times. During high-power demand the battery provides the peak power, while in low-power periods the battery is recharged such that the SOC always maintains around the mid-level of the allowable SOC operating range. Whenever the SOC is high above the mid-level, more electrical power will be used to drive the vehicle, thus reducing fuel consumption and emission. If the SOC is far below the mid-level, ICE will be used to drive the generator to quickly recharge the SOC back to its mid-level.

When multiple power sources are used to drive an HEV, an effective power management and control algorithm is essential to control the flow of energy as well as to maintain a sufficient reserve of energy in the storage devices. The existence of a power management and control algorithm, which is better known as a Control Strategy, ensures that all power components operate together at optimal efficiency to achieve multiple design objectives like maximized fuel economy and minimized emission. Both the HC and Control Strategy must be designed together to achieve these objectives. Although HC might determine to some extent what type of CS should be implemented for a specific HEV, there is still a wide range of CS for each HC. These CSs manage the energy flow between all hybrid drivetrain components and optimize power generation as well as power conversion in individual components. Selecting the most suitable CS will have a significant impact on fuel economy, vehicle performance, and emissions.

30.2 CONTROL STRATEGY

The main objective of a hybrid electric vehicle is to maximize fuel economy while minimizing emissions. To achieve these objectives, the HEV must be carefully designed with a proper combination of hybrid hardware configuration and a well-developed power management algorithm. These power management algorithms are better known as Control Strategies that handle the proper flow of power and maintain adequate amounts of reserve energy within the on-board battery packs. In general, the internal combustion engine of an HEV is often undersized, and the CS will ensure that it always operates near or at its most efficient point. To overcome the vehicle's requirement during transients, an electric motor is adopted to provide the additional peak power to assist the ICE whenever necessary. Therefore, the role of the CS is to determine when and how much energy should be drawn from each power source to propel the vehicle while achieving the objectives of the HEV.

During deceleration or braking, the CS will also activate the regenerative braking mode. The spinning driveline is disengaged from the driving power sources and used to drive either an on-board generator or an EM as a generator, to produce electrical energy for recharging the batteries. The external friction force and the generator load will slow down the driveline and, hence, the vehicle; this is known as driveline braking. The remaining braking power will be supplied by the conventional brake pads acting on disc or drum to bring the vehicle to a stop at the desired pace. In this way some brake energy can be recovered to improve the overall energy efficiency of the vehicle. If the braking is abrupt then less brake energy will be recovered, as the friction braking will be handling more braking jobs than the driveline braking to stop the vehicle instantly. Hence, regenerative braking will recover more brake power whenever the vehicle is gradually decelerated and brought to a stop. The following will discuss some common Control Strategies that are adopted by most HEVs.

30.2.1 Thermostat Series Control Strategy

The Series Thermostat CS is a series hybrid configuration CS that uses the ICE along with a generator to power the EM to propel the vehicle and recharge the depleted batteries. The battery state of charge is allowed to fluctuate between the maximum and minimum set points, rather like a thermostat that maintains the temperature within the desired range. The principle of this CS is to deplete the battery to a very low SOC and then trigger the ICE to drive the generator to recharge the batteries while powering the EM. Once the batteries are fully recharged, the ICE is shut off again until such needs arise again. During deceleration, some brake energy is recovered to help recharge the battery through regenerative braking.

The aim of this CS is to propel the vehicle entirely under pure electrical mode as often as possible. This gives the advantage of setting the ICE to operate at one point of torque and speed that is most efficient and least polluting. It also prevents the ICE from handling transient loads where the highest level of emissions is usually produced. The EM that propels the vehicle under all driving conditions will handle the transient load. The ICE is set to run on only one fixed gear ratio that is either optimized for fuel economy or low emission when driving the generator. Moreover, another potential emission reduction may come from the application of an electrically heated exhaust catalyst converter. Since there is ample knowledge of when the ICE will be started, the exhaust catalyst converter can be electrically preheated to reduce cold-start emission [11].

The major disadvantage of this CS is that all of the ICE's power must be transmitted through the generator and then to the EM. Due to the inefficiency of these components, some energy will be lost during energy conversion from one form to another. This energy will usually not be lost if the ICE is directly used to mechanically drive the wheels as in conventional vehicles. Besides, this CS requires both an EM and a generator, which usually results in a more costly and heavier vehicle.

30.2.2 Series Power Follower Control Strategy

The Series Power Follower CS determines at what torque and speed the ICE should operate. Electrical power is generated, according to the given conditions of EM, batteries, ICE, and the power demanded by the vehicle. This CS is usually designed to maximize fuel economy, or minimize emission, or maximize battery life. The ICE may be turned off if the SOC gets too high, and turned on again if the power required reaches a certain threshold, or if the SOC hits the minimum level. This CS also incorporates regenerative braking to recycle some brake energy back into the batteries during vehicle deceleration.

When the ICE is on, its power output tends to follow the power required, accounting for losses in the generator so that the generator output power converges with the power requirement. Therefore, in some instances, the ICE output power may be adjusted by the SOC, which tends to bring the SOC back to the middle of its operating range, or just keep the SOC above some minimum value. At other times, the ICE's output power might be kept near to the power of the ICE at maximum efficiency, or allowed to change no faster than a prescribed rate. Thus, this CS changes according to the preset SOC conditions and the power required for propelling the vehicle.

In general, when the SOC is low and the power demand is less than the power of the ICE at its maximum efficiency, the generator is run at a power as close as possible to the ICE's most efficient operation point, without exceeding the system voltage power constraint. The batteries are charged as much as possible to keep the ICE's efficiency as high as possible while maintaining a mid-level SOC. When the power demand is less than the power at maximum efficiency of the ICE, but greater than the battery charge power, the generator is set to run at a power equivalent to the ICE's most efficient operation point. The required power is used to propel the vehicle while excess power generated is used to charge the batteries. During high-power demand, where the requested power is greater than the power at maximum efficiency of the ICE, the generator is set to run at a power equivalent to the ICE's most efficient operation point. Additional peak power is requested from the batteries to satisfy the total power requested. In city driving conditions, where the SOC is high, the ICE is shut off and the vehicle operates under pure electric mode as a zero emission vehicle.

The advantage of this CS is that the battery packs are relatively small and the SOC is always maintained around a mid-level. In general this allows the overall weight of the propulsion system to be lighter. The disadvantage is that the ICE is forced to operate at multiple points in its efficiency and emission maps to adjust for load changes. This causes the emission to increase as the operation of ICE moves away from its maximum efficiency point. However, changing the throttle slowly may compensate for this negative effect [6].

30.2.3 PARALLEL ICE ASSIST CONTROL STRATEGY

The Parallel Internal Combustion Engine Assist CS uses the EM and batteries as the main power source to propel the vehicle. This is a type of parallel hybrid configuration CS. The ICE is only activated during low SOC: accelerating, hill climbing, and high speed. When the ICE is operating it directly drives the wheels mechanically, while simultaneously spinning the motor mechanically as a generator to recharge the batteries. During deceleration, this CS allows for regenerative braking to recover the brake energy back into the batteries.

Under low-power and normal driving conditions, this CS will often drive the vehicle under pure electrical mode. Electrical power is derived from the battery packs to power the EM. The ICE will be operated to drive the vehicle whenever the batteries are depleted. During these driving conditions, where the SOC is low, the ICE will replace the EM to propel the vehicle mechanically, while using the EM as a generator to recharge the battery. When the batteries are fully recharged, the ICE will be shut off and the vehicle will be driven electrically again. Whenever high power is needed during acceleration or hill climbing, the ICE is used to assist the EM by providing additional peak power to fulfill the power demand. Therefore, the ICE is usually undersized and operates near full load conditions where it is most efficient.

The primary advantage of this CS is that the vehicle is driven electrically most of the time. Thus, it tends to reduce overall emissions and the amount of fuel used if the battery, EM, and ICE are properly sized. It also eliminates the need of an electrical starter, as the vehicle is always moving whenever the ICE is started up. The disadvantage is that emission levels may not be as good when the ICE is operating, because the ICE usually operates during high-load conditions where emission is the highest.

30.2.4 Parallel Electrical Assist Control Strategy

The Parallel Electrical Assist CS is the most commonly adopted CS for parallel configuration HEVs due to its robustness and simplicity. This CS has proven to be reliable, as both the Honda Insight and Hybrid Honda Civic adopted it as their primary CS. This CS uses the ICE to drive the vehicle, while the EM is often used for starting up the ICE and assisting the ICE during high-power demand. This concept allows the ICE to operate in a more efficient region to reduce fuel consumption, while keep emissions low by avoiding full-throttle conditions usually needed for acceleration and steep gradient. Regenerative braking is incorporated to help boost the energy efficiency, especially during urban driving conditions.

There are two distinct ways of using the EM to assist the ICE under this CS, as summarized in Figure 30.4 and Figure 30.5. The first method illustrated in Figure 30.4 uses the EM to start up the ICE and to assist the ICE during high-performance driving conditions. At low-power and normal driving conditions, the ICE is fully responsible for propelling the vehicle and recharging the batteries. Therefore, the ICE will be operating whenever the vehicle is travelling, except for idling stops, deceleration, and coasting downhill. The CS will maintain the ICE to operate within its most efficient region at all times.

Low Performance Driving Condition:	Engine only
Normal Driving Condition:	Engine only
High Performance Driving Condition:	Engine and Electrical Motor Assist
Engine Cut-off Condition	
Idling Stops, Deceleration, and Coasting Down Hill	

Figure 30.4 Parallel Electrical Assist Control Strategy (first method).

Low Performance Driving Condition:	Either Engine or Electrical Motor (Subjected to Conditions)
Normal Driving Condition:	Engine only
High Performance Driving Condition:	Engine and Electrical Motor Assist
Engine Cut-off Condition	
Engine Temperature greater or equal to 85 degree Celsius And SOC > SOC minimum And Low Vehicle Speed or Idling Stops or Deceleration or Coasting Down Hill	

Figure 30.5 Parallel Electrical Assist Control Strategy (second method).

Thus, at low-power driving conditions, the batteries will recharge as much as possible to maintain the ICE operating point as close as possible to its maximum efficiency point. Some excess energy is used to power the on-board electrical accessories load as well. Under normal driving conditions, ICE is operating at its maximum efficiency point, and most of the power is used to propel the vehicle, while the remaining power is used to charge the batteries. When additional power is needed during high-performance driving conditions, the EM will be activated to assist the ICE by providing additional peak power. Therefore, the ICE will be maintained at its maximum efficiency point while the EM supplies the additional peak power required. During deceleration, the CS switches to regenerative braking mode to recycle some braking power to recharge the batteries. Around 20% of the total energy consumed by the HEV is derived from regenerative braking, which improves fuel economy.

The second method of utilizing the EM to assist the ICE under this CS is summarized in Figure 30.5. During low-power driving conditions, the ICE produces more power than needed to drive the vehicle. This is due to the fact that the CS is always trying to maintain the ICE to operate near its maximum efficiency point. Since the excess energy produced by the ICE under low-power driving conditions is unable to be stored in the batteries whenever the SOC is high, some of this energy will be wasted, as not all of it can be used by the on-board electrical accessories load.

Hence, to prevent this unnecessary energy being lost during low-power driving conditions, the EM will be used to propel the vehicle instead of the ICE. The ICE will not always be operated whenever the vehicle is travelling. The vehicle travelling speed will determine when the ICE will be operated. The ICE will only be operated during normal and high-speed driving conditions. During low-speed driving, the CS will propel the vehicle under pure electric mode where only the EM is used to drive the vehicle. ICE will only be used to drive the vehicle under low-power driving conditions whenever the ICE temperature drops below a certain value or the SOC hits its minimum level. This is to ensure that the ICE's temperature is always kept warmed up, and to avoid cold starting the ICE when it is needed. As the vehicle's speed increases above a certain threshold, the ICE will be turned on to replace the EM in propelling the vehicle. At this stage most of the power produced by the ICE operating at its maximum efficiency point will be used to drive the vehicle, while the remaining power will be used to spin the EM as a generator to recharge the batteries and power the electrical accessories. At high-speed or high-power demand the EM will be powered by the batteries to assist the ICE to propel the vehicle by supplying the required peak power. Through this method, the CS will efficiently maintain the ICE to operate at its maximum efficiency point at all times. Regenerative braking is adopted to further improve overall energy efficiency.

Table 30.1 shows that the second method has a better fuel economy than the first method. The fuel economy improvement is much more significant, especially for urban drive cycles. The reason for this is that urban drive cycles involve low-speed travelling most of the time, which favors the second method. During low-speed travelling the second method is often propelling the vehicle under pure electrical mode and, hence, reducing fuel consumption and emission significantly. Besides, urban drive cycles involve many idling stops and deceleration, where regenerative braking and shutting off the ICE during idling helps improve the overall energy efficiency. The first method has higher fuel consumption due to the ICE producing more power than is needed during slow-speed travelling, as the CS tends to maintain the ICE at near its maximum efficiency point. As for highway drive cycles, the fuel economy improvement is not so significant when comparing both methods. However, the second method still maintains a slightly better fuel economy. In highway

Table 30.1 Comparison of Fuel Economy between the First and Second Method

Drive Cycle	First Method (L/100 km)	Second Method (L/100 km)
US FTP city cycle	10	7.6
US FTP highway cycle	7	6.8
European city cycle	9.9	8.8
Extra urban cycle	8.3	7.9
Australian urban cycle	11.3	9.6
Melbourne peak cycle	11.2	9
Australian truck highway cycle	8.1	8

Note: The above results are obtained from a simulation model developed using ADVISOR software by NREL. The test vehicle is based on a parallel hybrid model converted from an existing conventional Ford Falcon model. The above results meet the zero delta SOC requirement.

Table 30.2 Comparison of Performance between the First and Second Method

Performance	First Method	Second Method
0–96.6 km/h	8.2 s	8 s
64.4–96.6 km/h	3.2 s	3.1 s
0–137 km/h	14.7 s	14.4 s
Maximum acceleration	3.8 m/s^2	3.8 ms^2
Grade ability	15.5%	17.5%
Maximum speed	221.5 km/h	221.5 km/h

Note: Objective of grade test to achieve 12.5% grade ability while sustaining the vehicle's speed at 88.5 km/h for 82 sec with additional 1000 kg external load on vehicle test mass.

drive cycles, the vehicle will be travelling at high speed most of the time with no idling stops, thus there are fewer chances for the vehicle to be driven in pure electrical mode. The effect of shutting off the ICE during low speeds, as implemented in the second method, is not so relevant in the high-speed drive cycle, and hence the fuel economy improvement is not so significant. Table 30.2 shows that the second method has a slightly better performance than the first method. The reason is that the second method is slightly more efficient in controlling and maintaining the ICE to operate at its maximum efficiency point at all times.

30.2.5 Adaptive Control Strategy

The Adaptive CS adjusts its control behavior based on the current driving conditions that affect both the emissions and fuel consumption. Therefore, to determine the ideal operating point for the ICE and EM, the CS considers all possible ICE–EM torque pairs. This CS attempts to optimize the torque distribution between ICE and EM at each time step. For a given operating point, the CS calculates the possible fuel consumption and emitted emission, together with the energy needed to restore the SOC back to its initial level. The CS also takes into account user-defined and standard-based weightings of time-averaged fuel economy and emission performance to determine an overall impact function for selecting an ideal operating point. The CS then continuously selects the best operating point that gives the best possible fuel economy with minimum emission, while trying to restore the SOC back to its initial level. This CS gives the advantage of ensuring that the vehicle is always maintaining its optimal settings for optimizing fuel economy and emission reduction under all driving conditions. However, to implement this CS requires a controller with a very large computational memory and processing power that can handle high-speed and heavy computational demand for updating the control system within a split second.

This CS considers both fuel economy and emission in its choice of operating points at each time step. Over the valid range of operating torques, the CS normalizes all five competing metrics (i.e., energy used, hydrocarbons, carbon monoxide, nitrous oxides, and particulate matter emissions) by using user-defined and standard-based weighting of time-averaged fuel economy and emission performance to determine an overall impact function. Next the CS optimizes the instantaneous efficiency of the ICE, exhaust removal, EM, and batteries. It adjusts its behavior based on driving conditions such as ICE, EM, battery temperature, and amount of available regenerative braking. The amount of regenerative energy is calculated from time to time as the vehicle moves. User-defined fuel economy and emissions targets are taken into consideration as well. At each operating point the CS looks at the entire range of possible ICE–EM torque combinations to determine optimum operation point. The impact function is minimized, and performance is determined by the weighted sum of instantaneous fuel consumption [12]. The following describes the steps for implementing this CS, and the flow chart in [12] (i.e., Figure 5 of Reference 12) summarizes the entire CS algorithm:

1. Define the range of operating points, represented by the range of acceptable motor torques for the current torque request.
2. For each operating point, calculate the constituent factors for optimization:
   - Calculate the fuel energy that would be consumed by the ICE
   - Calculate the effective fuel energy that would be consumed by electromechanical energy conversion
   - Calculate the total energy that would be consumed by the vehicle
   - Calculate the emissions that would be produced by the engine
3. Normalize the constituent factors for each candidate operating points.
4. Apply user weighting to result from step 3.
5. Apply target performance weighting to result from step 4.
6. Compute overall impact function, a composite of results from steps 3–5, for all operating points. The minimum operating point calculated is the final operating point to be updated at each time step.

30.2.6 Fuzzy Logic Control Strategy

The Fuzzy Logic CS mimics human reasoning when interpreting inputs and outputs, and may be represented as a set of IF-THEN rules in a way that is readable by computers. It is very different from binary logic, which uses only two distinct states. For instance, when

Table 30.3 Fuzzy Logic Rule

1. If SOC is at Maximum level Then P_{gen} is 0 kW
2. If SOC is Normal and P_{driver} is Normal and ω_{EM} is Optimal Then P_{gen} is 10 kW
3. If SOC is Normal and ω_{EM} is not Optimal Then P_{gen} is 0 kW
4. If SOC is Low and P_{driver} is Normal and ω_{EM} is Low Then P_{gen} is 5 kW
5. If SOC is Low and P_{driver} is Normal and ω_{EM} is Not Low Then P_{gen} is 15 kW
6. If SOC is at Minimum level Then P_{gen} is $P_{gen,max}$
7. If SOC is at Minimum level Then Scale Factor is 0
8. If SOC is Not at Minimum level and P_{driver} is high Then P_{gen} is 0 kW
9. If SOC is Not at Minimum level Then Scale Factor is 1

Note: Adapted from Reference 9.

interpreting the speed of a vehicle as input, binary logic can only categorize the vehicle's speed input as either fast or slow. However, fuzzy logic uses continuously varying degrees of states or membership functions. Therefore, vehicle speed input can be extremely slow, slow, normal, fast, and extremely fast. The fuzzy logic controller assigns membership grades to variables. It may interpret an analogue speed of 120 km/h to be 75% fast and 30% extremely fast. Thus, the CS can vary the throttle to be wide open, medium open, or slightly open, depending on the preset rules. In automatic cruise control systems, if the vehicle's speed falls under part of the fast and the extremely fast categories, the final throttle opening will be adjusted according to a combination of some percentages of medium open and slightly open to slow down the vehicle.

Since driving conditions and vehicle loads are highly nonlinear and cannot be explicitly described when constructing a CS of an HEV, it is very difficult to determine when to control the EM to assist the driving torque or recharge the battery. Besides, a different driver, who has his or her own driving patterns, will yield different ways of handling the vehicle. Hence, if the CS were followed in a deterministic way, the objectives of optimizing fuel economy and emissions reduction as well as SOC balance may not be achieved. The Fuzzy Logic CS is useful in controlling nonlinear and uncertain systems such as HEV application, and it is immune to various vehicle load and road conditions [5].

In the Fuzzy Logic CS, fuzzification is the first step is to determine the set of rules (Table 30.3) and membership functions (Figure 30.6, Figure 30.7, and Figure 30.8) that allow the controller to have a fuzzy reasoning mechanism. The second step is to determine the degree of fulfillment for the antecedent of each rule using fuzzy logic operators. This degree of fulfillment determines to which degree the nth rule is valid. Once this degree of fulfillment is determined, it is used to modify the consequent rules accordingly. This is usually done by multiplying the degree of fulfillment with the consequent of nth rule. The final step is defuzzification, where the results of the inference step are combined into a single value. This is done by averaging the inference results weighted by the degree of fulfillment of the rules. Inference and aggregation can be combined into a single equation, where K is the number of rules, A is the degree of fulfillment, and C is the consequent of nth rule [4,9].

$$U_j = \sum_{n=1}^{k} A_n C_n \bigg/ \sum_{n=1}^{k} A_n \tag{30.1}$$

Drive and Control System for Hybrid Electric Vehicles

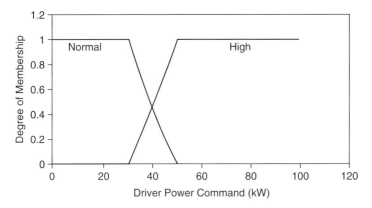

Figure 30.6 Driver power command membership function.

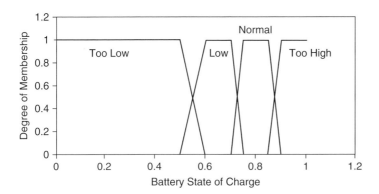

Figure 30.7 Battery state of charge membership function.

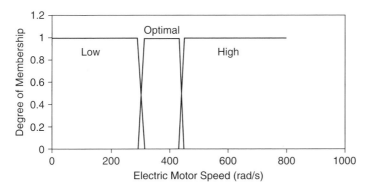

Figure 30.8 Electrical motor membership function.

A simplified block diagram of this Fuzzy Logic CS is depicted in Figure 30.9. The signals from the accelerator and brake pedals are normalized to a value between 0 and 1. Zero represents the pedal is not pressed, while one represents the pedal is fully pressed. The braking signal is then subtracted from the accelerating signal, thus giving the driving

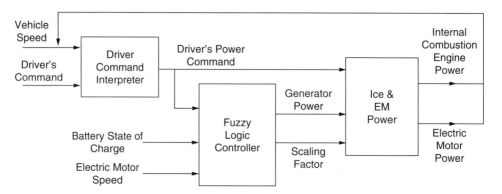

Figure 30.9 Fuzzy Logic Control Strategy schematic diagram.

inputs value between the range of negative one and positive one. The negative part of the driver input is sent to the brake controller, which computes the regenerative braking and friction braking power required to decelerate the vehicle. The positive part of the driver input is multiplied by the maximum available power at the current vehicle speed so that all power is available to the driver at all times. The maximum available power is the sum of maximum available ICE and EM power.

After driver power command is determined, the Fuzzy Logic CS calculates the optimal generator power for the EM when it is used to recharge the batteries and a scaling factor for the EM when it is used as a motor. The scaling factor falls between the range of zero and one. When the scaling factor is zero it indicates that the SOC is too low and the EM should not be used to drive the wheels, to prevent damaging the batteries. However, when the SOC is high the scaling factor equals one. Power demand, SOC, and EM speed are the main inputs to the Fuzzy Logic CS, as shown in Figure 30.9. Figure 30.6 presents the membership function for power demand. Fuzzy set Normal represents power range for normal or low-speed driving conditions. High represents power range for high acceleration and high speed. The power demand range, from 30 kW to 50 kW, is the transition between normal and high-power demand. Figure 30.7 presents the membership function for SOC. Fuzzy sets Too Low and Too High represent the ranges that the SOC should avoid. Fuzzy set Normal represents the ideal range where the SOC should be, and Low acts as a buffer range between Normal and Too Low fuzzy sets. Figure 30.8 presents the membership function for EM speed. Fuzzy set Optimal represents the optimal speed range. High and Low fuzzy sets represent EM speed too high and too low, respectively. The transition range is very narrow, as the EM efficiency usually drops rapidly when the speed is outside the optimal range.

From the rule base in Table 30.3, if the SOC is too high the desired generator power (P_{gen}) will be zero, to prevent overcharging the batteries. If SOC is normal the battery will only be charged when both the EM speed is optimal and drive power is normal. If SOC drops too low, the battery will be charged at a higher power level. This is to ensure a fast recharge of SOC to always maintain the SOC within its optimal operating range. If the SOC drops too low, the desired generator power is set to maximum available power and the scale factor is decreased to zero. This is to recharge the batteries as fast as possible to prevent damaging them. The final rule is to prevent the ICE from recharging the batteries whenever the power demand is high and the SOC is not too low. Under this condition the EM should be used to assist the ICE in providing peak power to ensure that the ICE is operating at its maximum efficiency point and its power level is not shifted outside its optimal range. Whenever the SOC is not too low, the scaling factor is set to one.

Finally, the CS determines the ICE and EM output power by using the power demand (P_{driver}), generator power (P_{gen}), and the scaling factor (S_f). The ICE power is the sum of power demand and generator power, as in the following equation.

$$P_{ICE} = P_{driver} + P_{gen} \tag{30.2}$$

The EM power is the reverse power of the generator, thus is negative of desired generator power.

$$P_{EM} = -P_{gen} \tag{30.3}$$

When the power demand is less than the threshold value, which is the scaling factor multiplied by 6 kW, the HEV will be driven under pure electric mode to prevent the ICE from operating outside its efficient range. Therefore, the ICE output power is set to zero, while the EM output power is set to a value equivalent to the total power demand. When the power demand is greater than the maximum ICE output power at current engine speed, the ICE output power is set to the maximum available power at current engine speed. For the EM, its output power is set to a value equivalent to the total power demand minus the maximum ICE output power at current engine speed. Whenever the EM is used as a motor to assist the ICE, which means the EM power is positive, it must be multiplied by the scaling factor to determine the peak power required from the EM.

30.3 POWER ELECTRONIC CONTROL SYSTEM AND STRATEGY

The main function of the electronic controller is to adjust control parameters for the smooth operation of the parts and select the optimum mode of operation under all driving conditions. In general, the basic power electronics components essential for HEV are converter and motor drive, DC-link brake resistor, DC-link main connector, battery charger, and board power supply. The converter and motor driver work together in converting and controlling the energy flow between the energy source and the electrical drive. The function of the DC-link brake resistor is to take over part of the energy that cannot be stored in the batteries when the SOC is too high. If this excess energy were not removed, it would damage the power electronics by raising the DC-link voltage. Besides, with DC-link brake resistors a constant brake torque can be maintained, even if the battery is already fully charged.

For safety reasons, the main connector acts as a buffer to separate the energy storage and the drive system. In case of failure the disconnection of drive system from energy storage has to be triggered automatically. Besides charging the batteries, the battery charger monitors the SOC and acts as a battery management system to prevent the damage of battery due to overload or deep discharge. The on-board power supply converts the energy from the main energy source to the conventional 12 V electrical energy supply consumed by the vehicle's electrical accessories [1].

The development of CS in HEV is to address the specialized needs of hybrid vehicles in power management, as there is always a tradeoff between energy efficiency and emissions [12]. In general, CSs developed for HEV are classified into three categories. The first type of CS involves control techniques like bang-bang control techniques, rule-based

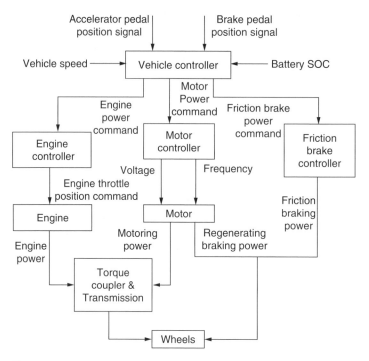

Figure 30.10 A general control schematic diagram of an HEV.

fuzzy logic, and load follower control techniques for control algorithm development. The second approach utilizes static optimization methods, where the optimization scheme determines the proper energy and power split between the two energy sources under steady-state operation. The last approach takes into account the dynamic nature of the system when performing the optimization. This optimization is done with respect to time rather than for a fixed point in time, thus dynamic optimization will be more accurate under transient conditions. However, this approach has extremely high computational costs and is sophisticated to implement.

The function of the hybrid controller in an HEV is to implement the CS to manage the distribution of desired propelling power from two different power sources under various driving conditions. Different CSs implemented by the hybrid controller would result in different fuel economy, performance, and emissions characteristics. However, all CSs have similar control modes that are targeted to handle all possible different driving conditions. Different CSs are only a matter of different combinations and sequences of utilizing these control modes in handling different driving conditions. Figure 30.10 shows an overall configuration of a general HEV hybrid controller that is capable of implementing the CS. The following are descriptions of all the control modes for handling all possible driving conditions.

The propelling mode is categorized into EM only mode, hybrid propelling mode, battery charging mode, and ICE only mode. When the accelerator is fully released, the value of accelerator position, C_a is zero. While the accelerator is fully depressed, the value of C_a is one. C_a value is within the range of zero and one. Therefore, the power output from the drive train is expressed as the following equation, where P_p is the power output at accelerator command and $P_{p\max}$ is the maximum power of the drivetrain, which is the sum of ICE and EM maximum output power.

$$P_p = C_a P_{pmax} \tag{30.4}$$

According to this power command and feedback parameters, such as vehicle speed, engine speed, motor speed, and battery SOC, the hybrid controller gives relevant commands to the ICE controller, EM controller, and brake controller to take necessary actions. The ICE output power P_e can be express in the following equation, where C_e ($0 <= C_e <= 1$) is ICE power command and P_{emax} is the maximum ICE power available. When C_e is zero the ICE throttle is fully closed, while when L_e is one the throttle is fully open.

$$P_e = C_e P_{emax} \tag{30.5}$$

The EM output power P_m is expressed in the following equation, where C_m ($-1 <= C_m <= 1$) is the motor power command and P_{mmax} is the maximum power of the motor. When C_m is less then zero the EM operates as a generator, and when C_m is greater than zero the motor functions in motoring mode.

Whenever the vehicle speed is less than the minimum speed of the engine, the CS will be operating in EM Only Mode. Under this mode the motor will be the only power source to propel the vehicle. Therefore, the final power output $P_p = P_m$ and $C_m = C_a$, while $C_e = 0$. Under Hybrid Propelling Mode, the EM will be used to assist the ICE in propelling the vehicle. The ICE will only operate at its maximum efficiency point and maintain the ICE output at its maximum available power. The EM will be used to provide the peak power to meet the total power demand. Therefore, $P_m = P_p - P_{emax}$ where $C_m = P_m/P_{mmax}$. When the power demand of the vehicle is less than the maximum ICE output power and the SOC is not full, the ICE will run at full throttle and the remaining power is used to recharge the batteries. Thus, the CS will be operating in Battery Charging Mode. Under this mode the $P_m = P_{emax} - P_p$, and $C_m = -P_m/P_{mmax}$. When the power demand is less than or equal to the maximum ICE power and the batteries are fully charged, the vehicle will be propelled under Engine Only Mode. Therefore, the EM will be shut off and $P_m = 0$ with $C_m = 0$. ICE output power will be equivalent to the vehicle power demand where $P_e = P_p$ with $C_e = P_e/P_{emax}$.

During Braking Mode the accelerator pedal will be fully released, and the brake pedal is depressed depending on how much braking is require to decelerate the vehicle or bring the vehicle to a stop. The fraction of maximum braking power C_b ($0 <= C_b <= 1$) is the signal from the brake pedal demanding brake power. When $C_b = 1$ the brake pedal is fully depressed, and when $C_b = 0$ the brake pedal is fully released. The total brake power demand P_b is expressed in the following equation, where P_{bmax} is the maximum brake power demand of the vehicle:

$$P_b = C_b P_{bmax} \tag{30.6}$$

The Braking Mode can be classified into two categories, which is the Pure Regenerative Braking Mode and Hybrid Braking Mode. Pure Regenerative Braking Mode only applies to driveline braking. However, Hybrid Braking Mode requires friction braking to assists driveline braking to decelerate and to stop the vehicle. In driveline braking, the electric brake power P_m is expressed in the following equation, where P_{mmax} is the maximum brake power supplied by the electrical machine (i.e., EM as it is also used as a generator in HEV) when driven by the driveline.

$$P_m = -C_m P_{mmax} \tag{30.7}$$

As for friction braking, the mechanical friction brake power, P_{fb} is expressed in the following equation, where C_{fb} ($0 <= C_{fb} <= 1$) is the fraction of the maximum mechanical friction brake power. $P_{fb\max}$ is the maximum mechanical friction brake power, which is the difference between the maximum brake power demand and the maximum electric regenerating brake force.

$$P_{fb} = C_{fb} P_{fb\max} \tag{30.8}$$

Under Pure Regenerative Braking Mode, vehicle brake power demand P_b is less than the maximum break power of driveline braking. All the needed braking power of the vehicle will be fully borne by the driveline braking when the regenerative braking takes place. Therefore, under this situation, the brake power demand can be expressed as $P_b = -P_m$ where $C_m = -P_m/P_{m\max}$. Whenever the vehicle brake power demand is greater than the maximum brake power of driveline braking, Hybrid Braking Mode takes place. The Regenerative Braking Mode should always be used first, for the purpose of energy saving. The friction braking must then be activated to assist the driveline braking by supplying the remaining brake power to decelerate the vehicle or bring it to a stop. The brake power supply by the mechanical friction brake system can be expressed as $P_{fb} = P_b - P_{m\max}$ where $C_{fb} = P_{fb}/P_{fb\max}$ [8].

In conclusion, the main task of the hybrid controller is to implement the CS and function as a logical control unit. Its main task is to send commands to lower level controllers such as the ICE throttle controller, motor controller, and brake controller. To ensure correct commands are given out to each component, the maximum capability of each component must be stored in the vehicle controller. This information is maximum ICE power curve, maximum EM motoring power and regenerating power, as well as maximum vehicle brake power demand. This information, together with the logic controller, will enable the hybrid controller to implement all sorts of CSs to manage the energy flow and conversion within an HEV.

30.4 CURRENT HEVS AND THEIR CONTROL STRATEGIES

An HEV's fuel economy and emissions depend on the way electrical energy stored in the batteries is substituted by the chemical energy of fuel, or on the way the ICE needs to provide extra energy to recharge the batteries. Therefore, a CS is needed for managing the energy flow and conversion efficiently. The CS implementation is handled by the electronic controller unit (ECU), where various factors like SOC, vehicle dynamics, and characteristics as well as parameters of various components are taken into consideration. The ECU then communicates with the lower level control unit of each specific component such as the motor controller (MC), engine management controller (EMC), and battery charger as well as converter, to control the HEV based on the selected CS. The following sections will discuss the CS adopted by the Honda Insight and Toyota Prius for their respective power management systems.

30.4.1 HONDA INSIGHT

Unlike the default parallel hybrid CS, the Honda Insight CS uses the direct split of the torque signal, where the controller directly commands the torque from the ICE and EM. In the default parallel hybrid CS, the ICE supplies all the demand torque, and only the unmet torque is drawn from the EM. In Insight's CS, there is a separate torque command sent to the EM directly. This is to ensure that the EM provides the correct amount of torque demanded by the CS. The torque coupler will sum up the torques from the ICE and EM, and send them to the transmission to drive the wheels.

Drive and Control System for Hybrid Electric Vehicles

Due to the hybrid architecture of the Honda Insight, it does not allow for pure electric driving. Whenever the vehicle is travelling the ICE will be operated, except for deceleration and coasting downhill. During low-speed and power driving, the ICE will operate as near to the ICE maximum efficiency point as possible by recharging the ICE as much as possible. Excess electrical energy is also used to power the on-board electrical accessories. Under normal driving conditions, the ICE will be operated at its maximum efficiency point, where all power will be used to drive the vehicle; the excess energy is then used to recharge the batteries. At high-speed and power driving conditions, the EM will be used to assist the ICE to achieve the requested power. Therefore, the ICE will be operated at all times whenever the vehicle is driving. During deceleration, regenerative braking is used to recover some brake energy back into the battery packs to improve energy efficiency.

Since the Honda Insight CS uses the direct split of the torque signal, the controller decides the torque contribution of the ICE and the EM. Based on the total torque request, and the characteristic of the motor, the controller decides the contribution of the EM. The remaining unmet torque request is supplied by the ICE, up to its maximum torque limit. If the two sources acting synergistically cannot meet the torque demand, the performance of the vehicle will degrade. In the Honda Insight, the battery pack SOC limits are set within a range of 20 to 80%. When the SOC is near the lower limits, the CS reduces the amount of electrical assist. As the SOC reaches 20% (i.e., the lower boundary limit), no assist is allowed, to prevent overdischarging and damaging the battery packs. Similarly, if the SOC reaches its upper bound limit (80%), all regeneration by the EM is stopped to prevent overcharging the battery and damaging the system [7,10].

30.4.2 Toyota Prius

Toyota's Prius combines the features of both a series and parallel HEV. It recharges its batteries primarily by using the ICE, in addition to regenerative braking. Prius uses a planetary gearbox for dividing the ICE output, plus a generator and a motor. This planetary gearbox also acts as a continuously variable transmission (CVT) system that enables the ICE to operate at its most efficient operating point at all times and eliminates the need of a clutch. Approximately 70% of the ICE torque is directed to the drive axle by means of a counter-gear and differential, through which the motor is also connected to the drive axle. The remaining 30% of ICE torque is used to drive the generator. According to the CS, the generator output may be directed either to recharge the battery packs or to power the EM for extra drive force. The generator also functions as a starter for the ICE, resulting in a quiet and seamless system.

The Prius hybrid architecture is a combination of series and parallel hybrid configuration, and hence it is much more flexible than the Insight's hybrid architecture. The Prius's architecture allows its CS to implement pure electrical drive mode when the power demand is below a certain threshold. This boosts the average efficiency of urban driving by increasing the efficiency by 40%. Thus, there will be significant improvement in fuel economy and emission reduction. The following is an overall summary of Prius's CS in handling all possible driving conditions [3,10].

- When ICE demand is low, such as starting, travelling at light load, or coasting downhill, the CS drives the Prius under pure electrical mode, where the on-board battery packs are used to power the EM.
- During normal travelling, the CS engages the ICE to drive the wheels and the generator to power the EM or recharge the battery, depending on the SOC.

- At full acceleration, the CS directs additional boost power from the battery to power the EM for assisting the ICE to achieve the power demand.
- During deceleration or braking, the CS turns the EM into a generator, recovering kinetic energy from the wheels into electricity to charge the battery.
- Whenever the car is idling or the power demand is low, the CS shuts off the ICE. The ICE is only operated when the battery needs charging or to power the air conditioning.

30.5 CONCLUSION

Hybrid CS and power management algorithms are essential for all hybrid vehicles to operate at their most efficient state. Without a CS, an HEV is just an ordinary vehicle with more than one power source hardwired together. Therefore, the vehicle will not have the intelligence to operate in its most efficient condition to achieve the goals of HEV. When applying a particular CS, there is always a tradeoff between fuel economy and emissions reduction. Thus, HEVs are specifically designed to operate in an optimum operation range. This range may be different if a given situation requires optimizing fuel economy or emission reduction.

Acknowledgment

The second author acknowledges the Rowden White Foundation in Australia for partially funding this work.

REFERENCES

1. A. Vezzini, K. Reichert. Power Electronics Layout in a Hybrid Electric or Electric Vehicle Drive System. IEEE workshop 0-7803-3292-X/96, 1996.
2. C. Lin, Z. Filipi, Y. Wang, L. Louca, H. Peng, D. Assanis, J. Stein. Integrated, Feed-Forward Hybrid Electric Vehicle Simulation in SIMULINK and its Use for Power Management Studies. *J SAE* 2001-01-1334, 2001.
3. D. Hermance, S. Sasaki. Hybrid electric vehicles take to the streets. *IEEE Spectrum*, 1998, p. 4852.
4. E. Cerruto, A. Consoli, A. Raciti, A. Testa. Fuzzy Logic Based Efficiency Improvement of an Urban Electric Vehicle. *J IEEE* 0-7803-1328-3/94, 1994.
5. E. Koo, H. Lee, S. Sul, J. Kim. Torque Control Strategy for a Parallel Hybrid Vehicle using Fuzzy Logic. *J IEEE* 0-7803-4943-1/98, 1998.
6. J. Marcinkoski, C. Whiteley, J. Goldman, R. Simons. 1999 University of Maryland FutureCar Design Report. *J SAE*, 1999.
7. K. Kelly, M. Zolot, G. Glinsky, A. Hieronymus. Test Result and Modeling of the Honda Insight using ADVISOR. *J IEEE* 2001-01-2537, 2001.
8. L. Chu, Q. Wang, Y. Li, Z. Ma, Z. Zhoa, D. Lui. Study of the Electronic Control Strategy for the Power Train of Hybrid Electric Vehicle. IEEE workshop 0-7803-5296-3/99, 1999.
9. N. Schouten, M. Salman, N. Kheir. Fuzzy Logic Control for Parallel Hybrid Vehicles. *J IEEE Trans.* 1063-6536/02, 2002.
10. R. Trigui, F. Badin, B. Jeanneret, F. Harel, G. Coquery, R. Lallemand, JP. Ousten, M. Castagne, M. Debest, E. Gittard, F. Vangraefshepe, V. Morel, L. Baghli, A. Rezzoug, J. Labbe, S. Biscaglia. Hybrid Light Duty Vehicle Evaluation Program. *J Automotive Technology* 1229-9138/012-02, 2002.

11. T. Ciccarelli, R. Toossi. Assessment of Hybrid Configuration and Control Strategy in Planning Future Metropolitan/Urban Transit System. Final Report, METRANS Transportation Center, California State University, Long Beach, CA, 2002.
12. V. Johnson, K. Wipke, D. Rausen. HEV Control Strategy for Real-Time Optimisation of Fuel Economy and Emission. *J SAE* 2000-01-1543, 2000.

31

Battery Technology for Automotive Applications

Dell A. Crouch
Delphi Corporation, Indianapolis, Indiana

31.1 INTRODUCTION

Batteries have been used in automobiles as energy storage and conversion devices since the early 1900s. The first usage was in electric cars. In 1911, Charles Kettering invented the electric starter, and it first appeared on the 1912 Cadillac [1]. This feature created an additional load for the electrical system — the battery had to be recharged for the next engine start. Adding a generator maintained the convenience advantage of an automobile powered by an internal combustion engine. Unlike an electric car, the battery did not have to be recharged for several hours from an external source of electricity.

Today, about 80% of the lead-acid battery industry is devoted to the production of so-called SLI batteries (Starting, Lighting, and Ignition). This abbreviation has become outdated, as new loads have been added to the electrical system. Many of these new loads are covered elsewhere in this book. In addition, new requirements for lower exhaust emissions and higher fuel economy are creating a market for hybrid electric vehicles (HEVs). These vehicles have two sources of energy for propulsion: an internal combustion engine and an electric motor. The combination of these two systems presents a new set of requirements for automotive batteries. This chapter deals with the effect of the present and future of the automotive electrical system on battery design, weight, volume, and performance. We will consider three types of battery chemistry: lead-acid, nickel-metal hydride (NiMH), and lithium ion (Li ion). Each has advantages and disadvantages. Interfaces between the electrical system and the battery will also be covered.

31.1.1 Battery Technology

Batteries are groups of electrochemical cells connected in series to meet the voltage requirements of a particular application. The voltage supplied by each cell is determined by its electrochemical reaction. The open circuit voltage of an electrochemical cell can be calculated from the energy of the reaction:

$$\Delta G = -nFE_0$$

where

ΔG = the energy of the reaction in Joules per mole (The free energy of a spontaneous reaction is negative by convention.)

n = the number of equivalents per mole for the reaction

F = the Faraday, 96,500 coulombs per equivalent

E_0 = the cell voltage when the electrolyte concentration is 1 equivalent per liter at standard temperature and pressure

It is also affected by temperature and by the concentration of the electrolyte. This relationship can be calculated using the Nernst equation:

$$E = E_0 + RT/nF \times \ln((([a-]\,[b+])/[gh])$$

where

E = the cell voltage at ambient temperature and present electrolyte concentration

T = the temperature in °K

R = the ideal gas constant, 8.3145 Joules/mole/°K

$[a-]$ = the concentration of the negative ions in the electrolyte

$[b+]$ = the concentration of the positive ions in the electrolyte

$[gh]$ = the concentration of the electrolyte solvent

(The actual concentration terms in these equations are called "activities." The distinction between activity and concentration is covered by any text in physical chemistry.)

Cell reactions and typical open circuit (no current flow) voltages for cells of various types are shown in Table 31.1. Discharge reactions are read from left to right. Charge reactions are read from right to left.

In batteries, current and reaction rate in grams/second are one and the same. One can be converted directly into the other. In all battery applications, we expect current to flow as soon as the external circuit is completed. Therefore, all discharge reactions are

Table 31.1 Cell Reactions and Open Circuit Voltages for Various Rechargeable Battery Cells

Battery Chemistry	Cell Reaction	Open Circuit Voltage Range (Volts per Cell)
Lead-acid	$Pb + PbO_2 + 2H_2SO4 \rightarrow 2PbSO_4 + 2H_2O$	2.06–2.15
Nickel-cadmium	$2Ni(OOH) + Cd + 2H_2O \rightarrow 2Ni(OH)_2 + Cd(OH)_2$	1.30
Nickel-metal hydride	$MH + Ni(OOH) \rightarrow M + Ni(OH)_2$	1.35
Lithium ion	$Li_xC + Li_{1-x}O_2 \rightarrow LiMO_2 + C$	3.8–4.0

Battery Technology for Automotive Applications

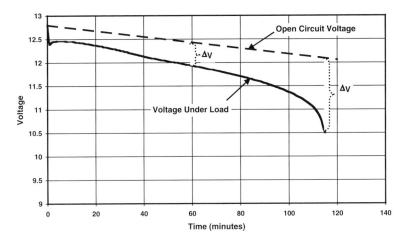

Figure 31.1 Discharge curve of automotive lead-acid battery.

spontaneous. In addition, the reaction kinetics must be fast enough to permit the required current flow. All of this must happen with a relatively small change in voltage. Voltage change is only partly dependent on internal resistance. The rest of the voltage change is nonlinear with respect to current.

$$\Delta V = IR + a\log I + b + RT/nF \ln(1 - (I/I_{limit}))$$

where

I = current
I_{limit} = the limiting current of the cell at the end of discharge or charge
V = voltage
a = slope of ΔV vs. log I curve
b = intercept of this curve

Figure 31.1 shows an automotive lead-acid battery being discharged at 25 A. The battery runs at this rate for 115 minutes to 10.5 V under load. Note that ΔV changes dramatically at the end of the discharge even though the current is the same throughout this test. The last part of the discharge curve is often referred to as the "knee" of the curve because of its shape. The increase in ΔV is caused by the fact that the battery is running out of reaction sites that will support a current of 25 A. The ΔV must increase continuously to drive the mass transport of sulfate ions from the electrolyte to the remaining reaction sites to maintain the current. At the very end of the discharge, the cell reaches limiting current density.

In contrast, charge reactions are forced because we are converting the active material back to a higher energy state, a higher entropy. There is no law of chemical kinetics that states that the reaction rate must be the same in both directions, discharge and charge. In fact, many commonly used battery systems will not recharge at all (for example, dry cells and alkaline manganese). In most battery systems, a battery with a given discharge performance will not recharge at the same rate with the same ΔV. Figure 31.2 shows the recharge of a lead-acid battery under modified constant current control. The charge begins at a constant predetermined current. When the voltage reaches a predetermined limit, the controlled variable changes from the battery current to the battery voltage. At this point, charge current decreases and approaches some equilibrium value asymptotically. Note that the change in ΔV still occurs at the end of the charge and for the same reason. There aren't enough reaction sites at the plates to support this charge current.

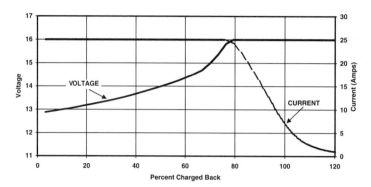

Figure 31.2 Charge curve for automotive lead-acid battery.

If the charge current had been maintained, the voltage curve would have shown an abrupt increase vs. time when the limiting charge current was reached. In lead-acid and nickel-metal hydride cells, the voltage increases until the next reaction, decomposition of water (gassing), begins. In lithium ion cells, this increase in voltage is not permissible: The electrolyte would be decomposed. So, charge control in these batteries occurs at cell level, not at battery level.

31.1.1.1 Valve Regulated Batteries

When some batteries are overcharged, they evolve gas by decomposing the water in the electrolyte. Standard "flooded" batteries have vents that allow these gases to escape. Valve regulated batteries incorporate a pressure relief valve that allows them to vent these gases only above some positive pressure (usually 1.5 to 7.5 psig). This valve prevents distortion of the case due to excess pressure. The technology is used in lead-acid, nickel-cadmium, and nickel-metal hydride batteries that are valve regulated. In all three cases, the oxygen evolved from the positive electrode on overcharge is recombined at the negative electrode. This recombination reaction acts as a slight discharge at the negative electrode and reduces the amount of hydrogen that is evolved. Therefore, it takes prolonged and severe overcharge to force the valve to open.

31.1.2 Present Automotive Battery Requirements

Battery requirements vary considerably as the electrical system becomes more extensive. Today's industry specifications cover requirements of the present 12 V electrical system [2]. The SAE, EN, and JIS specifications include:

- Performance: power for engine starting, energy storage for key-off drains, or charging system problems with the engine running
- Charging conditions: charge acceptance, charge voltage, and temperature
- Electrical interfaces: battery terminal dimensions and configurations
- Mechanical interfaces: case dimensions, terminal locations, maximum torque for making electrical connections, and battery mounting features (sometimes called "holddowns")

- Environmental requirements: temperature ranges for operation and storage, vibration requirements, and maximum tilt
- Specific tests required by certain battery chemistries: self-discharge ("stand" tests), gassing, and flame arrestor vent tests

31.1.2.1 Battery Performance Requirements

Performance requirements for any battery application can be looked upon as two components: the maximum power required and the maximum energy storage required. Generally, the peak power demand for an automotive battery occurs when the engine is started under cold conditions. The engine oil is viscous, the inertia of the engine has to be overcome and the engine operating temperature is far below its design point. The battery is also under less than ideal conditions. As chemical devices, batteries deliver less power when they are cold because the discharge reaction cannot run as fast as it would at normal temperatures. To enable the battery to operate at low temperatures and at high discharge currents, battery designers increase the area of reactants that are exposed to the current collectors and to the electrolyte. In general, this means putting more electrodes of the same polarity in parallel in the same cell. In lead-acid batteries, each plate is thinner and uses more of them to maintain the battery's overall dimensions and voltage while adding power capability.

The energy storage requirement is meant to cover the need to keep certain loads powered when the vehicle is parked and the engine is off. European requirements call for a battery "capacity" in ampere-hours that is equal to 20 hours the ampere rate required to discharge the battery in that amount of time. This is meant to represent the energy available to support key-off loads. To some degree, energy storage requirements are also meant to keep the engine running for some length of time when the charging system is not working properly. This was the original intent of the SAE reserve capacity requirement. Today, the 25 A called for in the reserve capacity test is no longer adequate. Since this requirement was proposed, automobiles have gone to electronic fuel injection, electric fuel pumps, high-energy ignition systems, rear window defoggers, and other devices that did not exist when the specification was written.

31.1.2.2 Battery Charging Requirements

In the absence of DC-DC converters, the battery charging voltage and the system bus voltage must be the same. The upper limits on system bus voltage come from the incandescent lamp filament requirements and from the power supplies for the various electronic devices on the vehicle. The battery requirements come from the state of charge of the battery, the battery chemistry, the number of cells in series, and the temperature. In general, when a battery is charged, an external source of electrical energy is connected to the battery. When charge current flows, the voltage of the battery is driven above its open circuit value. This forces the cell reaction to run backwards: The reaction products of the discharge reaction are converted back to their charged state. Each battery chemistry has a unique range of open circuit voltages, depending on the temperature and on the concentration of the electrolyte. So, the system engineer and the battery engineer jointly determine if it is possible for a battery with 1 to 4 V per cell to power the required loads and be charged within the voltage limits of the electrical system. The present system works well because the battery and the automotive electrical system evolved together.

31.1.2.3   Battery Termination Standards

The interface between the battery and the rest of the electrical system is the battery terminal. The ideal battery terminal has the following characteristics:

- Low resistance
- Remains tight even during temperature changes or vibration
- Can be disconnected and reconnected without damage to allow battery or cable replacement
- Polarity is readily recognized

New applications such as x-by-wire, emission and fuel economy devices, and HEVs are not covered by today's industry standards because they have not yet reached a stage of evolution where their requirements can be standardized. However, the direction of future battery requirements is becoming clearer.

31.2   FUTURE AUTOMOTIVE BATTERIES

The present 12 V electrical system has been growing rapidly for the last 50 years and more rapidly in the last 20 years. Today, we are approaching the practical limit for peak power requirements at 12 V. This is generally defined as 200 to 250 A at the bus voltage, about 2.5 to 3.5 kW. New power requirements for lower emissions, higher fuel economy, increased safety, and greater comfort and convenience are forcing the automotive industry to re-examine the question of system voltage. In Europe, an organization called Forum Bordnetz was formed in 1996 [3]. In the U.S., a consortium of industry members and the Massachusetts Institute of Technology was formed in 1997 [4]. Both of these organizations are intended to bring together automotive manufacturers, their suppliers, and experts from the academic world to propose standards for an automotive electrical system with a higher bus voltage. Both organizations settled on a system with a nominal bus voltage of 42 V. This level was chosen because it is near the highest voltage that can be easily tolerated by the human body. When we add range to this voltage to cover ripple and battery charging requirements at cold temperatures, we reach the voltage limits given by Underwriters Laboratories and others.

When the above requirements include more substantial propulsion loads, the power levels and, therefore, the voltage increase even more. Vehicles that are powered by a combination of an internal combustion engine and an electric motor are called hybrid electric vehicles. Figure 31.3 shows the approximate peak power requirement envelope based on a maximum current of 200–300 A at various voltages. The open boxes represent present and future automotive electrical system power limits for vehicles ranging from today's 12 V passenger car systems to 13-meter hybrid electric buses operating at 600 V or more. Despite the range of the size of these vehicles, this current limit works reasonably well.

This current limit is important for battery engineers because they can combine it with the duration of the power pulse to, in part, determine the size and weight of the individual battery cells. As the power and/or duration increase at a given voltage, the size of the battery increases. In higher voltage HEV systems, the trend has been to use batteries with relatively new chemistries: nickel-metal hydride and lithium ion [5]. The Ragone curves in Figure 31.4 and Figure 31.5 show the relative changes in mass and volume for

Battery Technology for Automotive Applications

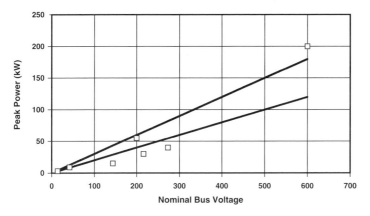

Figure 31.3 Peak power vs. bus voltage for two peak currents.

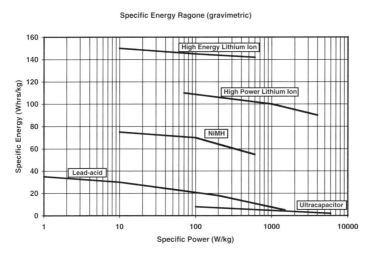

Figure 31.4 Specific energy Ragone plot for various battery chemistries.

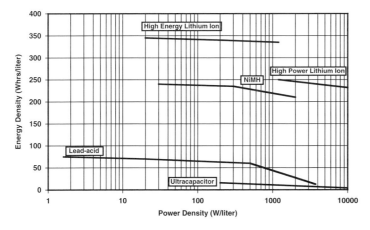

Figure 31.5 Energy density Ragone plot for various battery chemistries.

lead-acid, NiMH, and Li ion batteries. Since the curves are plotted as energy vs. power, the slope of any point on a curve is time, the duration of the discharge. Specific energy and specific power refer to energy and power per unit of mass. Energy density and power density refer to energy and power per unit volume. The ratio between the two is the apparent density of the battery. For example:

$$watts/kg \times kg/liter = watts/liter$$

where kg/liter is the apparent density of the battery.

The location of the various systems on these plots is obviously dependent on their chemical thermodynamics and kinetics. However, it is also dependent on battery engineering to application requirements. As can be seen, both specific energy and energy density decrease as power increases.

Part of this is due to losses in batteries as discussed earlier in Section 31.1. Another component is the loss of capacity per cell when the discharge current is increased. Ideally, the ampere-hour capacity of any cell would be the same regardless of discharge rate or temperature. The product of discharge time and discharge current would be a constant. The Peukert equation gives a means of fitting discharge performance data to a plot where capacity is not constant. The key variable is the value of the exponent n:

$$(I^n)t = C$$

where

C = a constant
I = the discharge current
t = the discharge time
n = a value greater than 1

In logarithmic form, the equation becomes:

$$n(\log(I)) + \log(t) = C$$

When the log of discharge time is plotted vs. the log of discharge current, n becomes the slope of the curve and it becomes negative, its value being less than -1. If n were

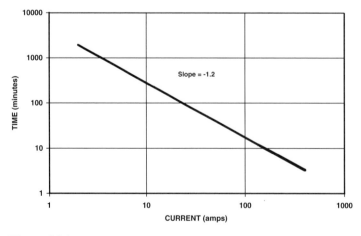

Figure 31.6 Peukert curve for automotive lead-acid battery.

Battery Technology for Automotive Applications

equal to −1, then the capacity of the battery would be the same for any discharge current. For lead-acid automotive batteries, the value of n ranges from −1.1 to −1.3 at room temperature. The value of n becomes more negative with decreasing temperature. The value of C decreases with decreasing temperature. For nickel-metal hydride and lithium ion, n ranges from −1.05 to −1.2.

Still another requirement for hybrid electric vehicle batteries is high charge acceptance. When the vehicle is rated for fuel economy or emissions, it is "driven" on a chassis dynamometer according to a specific schedule of speed vs. time. The various schedules used include acceleration, deceleration, constant speeds, and stops. Regenerative braking is used to charge the batteries in HEVs to increase overall vehicle efficiency. Since the mass of the vehicle can be very high, the charging power available during deceleration can be more than the batteries can accept. Under these conditions, the batteries reach the limiting charge current and either generate gas (lead-acid, nickel chemistries) or reach cell voltage limit (lithium ion). If battery charge acceptance can be increased, this additional stress on the batteries can be reduced.

31.3 COMBINATIONS OF BATTERIES AND ULTRACAPACITORS

If a battery is required to provide both long-term energy storage and short-duration pulses for engine starting or vehicle launch, its design is a compromise. The total electrode surface area must be increased by adding more electrodes in each cell. The increased current is distributed over a larger electrode area to maintain the battery voltage within the system requirements. If the pulse power requirement were to come from another device, the battery could use thicker, more durable electrodes and still deliver the required energy storage at lower rates.

One way to do this is to use an ultracapacitor to provide pulse power and a battery for energy storage only. The ultracapacitor can be recharged at some slower rate in preparation for the next pulse, or it can be recharged with regenerative braking. When the ultracapacitor is charged, the battery can be cycled over a wider range of SOC because the power for starting is already stored in the ultracapacitor. Clearly, there is a need for a more sophisticated charging system when we combine ultracapacitors and batteries. In all likelihood, the maximum charge voltage for the two devices will not be the same because their discharge characteristics are so different. Therefore, we are probably looking at some sort of DC-DC conversion or switching to control both devices on the same DC bus.

31.4 BATTERY MONITORING AND CHARGE CONTROL

The new applications of start-stop, hybrid electric drive, and totally electric drive require battery monitoring and charge control. Their starting or propulsion functions do not occur only when the vehicle is parked. They occur while the vehicle is on the road in traffic. For the last 10 to 15 years, the amount of work in this area has increased greatly. In a general sense, we can only measure four properties that reflect the battery's condition: voltage, current, temperature, and time. All other data such as power, energy, state of charge, and power capability is derived from these measurements and some model of the battery design. The charge control method is determined by the battery's ability to tolerate overcharge. In general, aqueous chemistries such as lead-acid, nickel-cadmium, and nickel-metal hydride have a gassing reaction that allows the battery to accept overcharge. Lithium-ion batteries, on the other hand, do not accept overcharge at all. The cells must have

Table 31.2 Charge Characteristics and Control Methods for Various Battery Chemistries

Battery Chemistry	Discharge Voltage Slope vs. Time	Tolerates Overcharge?	Charge Control Methods
Lead-acid	Related to state of charge	Yes — gassing	Battery level
Nickel-cadmium	Relatively little	Yes — gassing	Battery level
Nickel-metal hydride	Relatively little	Yes — gassing	Battery level
Lithium ion	Related to state of charge	No — charge current must stop at end of charge	Cell level

individual charge control with switching circuits to route charge current around a cell that is already fully charged.

Battery monitoring is done with several different technologies. The traditional ones are open circuit voltage and current-time integration. More recently, various methods based on AC impedance and conductance have been used to indicate the battery's state of charge and state of health. Battery state of charge is generally expressed as the percentage of fully charged capacity that remains in the battery:

$$SOC = (\text{ampere-hours remaining/fully charged ampere-hours}) \times 100$$

State of health is not as well defined mathematically. The intent is to show how much of the design life of the battery remains or to indicate that battery replacement is required soon. This requires mathematical models that are dependent on specific battery designs. If a replacement battery is not of the same design as the original, the accuracy of this software is reduced.

Table 31.2 shows a group of battery chemistries, their applicable characteristics, and the technologies used to control recharge. As the recharge approaches completion and charge current continues, cell voltages begin to increase until the battery voltage reaches the limit set by the charger. Since a battery is a group of cells connected in series, each cell sees the same current, but the cell voltages can vary. Therefore, the cells can reach their limits at different times. In aqueous systems, those cells produce gas by decomposing the electrolyte. In lithium-ion cells, this is not permissible, as noted above.

31.5 CONCLUSION

As this chapter is being written, the hybrid electric vehicle industry continues to increase. It now includes or will soon include passenger cars, light trucks, delivery trucks, and transit buses. The battery continues to be a key technology that determines a large part of the performance and durability of these hybrid electric vehicles. With continued development, they will provide more environmentally acceptable transportation in urban and suburban environments. They will also conserve petroleum energy resources.

REFERENCES

1. U.S. Patent 1,150,523, Engine-Starting Device, Charles F. Kettering, August 17, 1915.
2. SAE Surface Vehicle Standard J537, Storage Batteries, September 2000, Warrendale, PA.
3. Forum Vehicle Systems Electric Architecture, http://www.bornetzforum-42v.de/.
4. MIT/Industry Consortium on Advanced Automotive Electrical/Electronic Components and Systems, http://mit42v.mit.edu/.
5. United States Council for Automotive Research, http://www.uscar.org/consortia&teams/techteamhomepages/BATTERY.htm.

Index*

A

ABC, *see* United States Advanced Battery Consortium (US ABC)
AC-AC converters
    basics, 333, 345
    direct AC-AC converter, 336–345, *337–341, 344–345*
    indirect AC-AC converter, *334–335,* 334–336
    topologies, 334–345
Acceleration, 33, 222–223, *223*
Acceleration Measurement Test standard, 541
AC-DC rectifiers
    characteristics, 255, *256*
    circuit configuration, 255, *256*
    DC-link capacitor calculations, 260–263
    DC-link power calculations, 259
    diode type, 255–264
    dynamic breaking unit design, 263–264
    fire angle control scheme, 264–267, *266*
    input phase current analysis, 259
    integrated starter alternator, 595–598
    load conditions, 260–263
    operation mode, 264, *265*
    output current analysis, 259
    three-phase full-bridge type, 255–259, 268–271
    thyristor type, 264–271
    topology, 264, *265*
AC dynamometers, 546–548, *546–548*
AC motors, electric vehicles, 57
Adaptive control, HEVs, 664–665
Adaptive cruise control, 649–650, *650*
Adaptors, electric vehicles, 62
Advanced automotive steering systems, 630–631
Advanced Battery Consortium, *see* United States Advanced Battery Consortium (US ABC)
ADVISOR, 106
Airbags, *651,* 653, *653*
Air cooling, 605–607
Alternative architectures, conventional cars, 16–18

Alternator, ISA, 564
Aluminum-air batteries, 66
Ammeters, 59
Analytical *vs.* simulation methods, 251
Angle controller, 519
Antitheft alarm systems, 648–649, *648–649*
APU, *see* Auxiliary power unit (APU) control
Architectures
    combination, HEVs, 31–33
    conventional cars, 15–18
    optimal power management and distribution, *99,* 99–101
Audi company, 14
Automatic crank, electric machine, 575
Automotive control network protocols, 13–15, *14*
Automotive power semiconductor devices
    basics, 117–120, *118–119,* 156
    clamping devices, 120, 124, *125*
    diodes, 120–125
    emerging technologies, 150–154, *151–153*
    freewheeling devices, 120–124, *122, 124*
    high-voltage power switches, 139–148, *140–148*
    IGBTs, 139–148, *140–148*
    low-voltage load drivers, 125–139, *126–139*
    power integrated circuits, 148–150, *149–150*
    power losses, 154–156, *155*
    power MOSFETs, 125–139, *126–139*
    rectification devices, 120–121, *121*
    Schottky diodes, 125
    SiC devices, 150–154, *151–153*
    smart power devices, 148–150, *149–150*
    super-junction devices, 150–154, *151–153*
    thermal management, 154–156, *155*
    zener diodes, 120, 124, *125*
Automotive steering systems
    advanced systems, 630–631
    basics, 625
    classifications, 625, *626*
    electric power steering, *626,* 629–630, *629–630*
    electrohydraulic power steering, 628, *628*
    four-wheel steering, *631*

---

* Page numbers in italics indicate figures and tables.

future-generation systems, 631
hydraulically assisted steering, *627,* 627–628
manual steering, *626,* 626–627
Auto-start, ISA subsystem, 565
Auxiliaries, electric vehicles, 60–62
Auxiliary automotive motors control, 328–330
Auxiliary power unit (APU) control
hybrid and fuel cell vehicles, *364–365,* 364–368, *367–368*
series hybrid vehicle propulsion system, 348–349
Aware studies, 295–330
Axle architecture, 45–46

B

Balanced input impedances, *274,* 279–286, *281, 284–286*
Balda, Schupbach and, studies, 177, 182
Ballard Power Systems company, 71
Bansal studies, 55–73
Barron and Powers studies, 213
Batteries
charging, 39, 48
constraints, 103
conventional cars, 4, 6
cost, 64
electric vehicles, 63–67
integrated starter alternator, 572
SLI (starting, lighting, and ignition), 677
ultracapacitors, combination, 182–186, *183–184*
Battery technology, automotive applications
basics, 677–680, *678–680,* 686
charge control, 685–686, *686*
charging requirements, 681
future directions, 682, *683–684,* 684–685
monitoring, 685–686, *686*
performance requirements, 681
requirements, 680–682
termination standards, 682
ultracapacitors, 685
valve regulated batteries, 680
Belleville washers, 61
Belt drive, 561–562
Bibliographical survey, electric vehicles, 72–73
Bipolar junction transistor (BJT), 202–203, *204,* 299
BLDC, *see* Brushless DC (BLDC) drives
BMW company, 14–15, 644
Bode plots, 246
Body subsystems, MOSFETs, 126
Boost converters, 239–240, 599–600
Boost rectifiers
analysis, 275–279, *277*
balanced input impedances, 279–286
basics, 273, 293–294
control methods, 279–293
harmonics, 278–293
operating principles, 274–275, *274–275*
system description, 274–275, *274–275*

unbalanced input impedances, *287,* 287–292
unbalanced input voltages, 279–292, *287*
Bosch studies, 14
Braking energy regeneration, 568
Brushed-DC electric machinery
basics, 373
operating basics, *374–380,* 374–383, *382–383*
series connected DC-motor drives, 383–385
temperature impact, *376,* 379–383, *380, 382–383*
torque production, 377–379, *377–379*
Brushless DC (BLDC) drives
advantages and disadvantages, 521–523
basics, 515–516, *516*
control, *517–518,* 517–519
design considerations, 525
finite element analysis and design, 525
permanent magnets, *526–527,* 526–528
sensorless, 535
simulation model, *517, 520,* 528–534
torque, 519–521, *520,* 523–524
Brushless DC motors, electric vehicles, 57
Brush-type DC motors, voltage, 16
BTS640 (Infineon), 12, *12*
BTS441R (Infineon), 149–150, *150*
Buck-boost converters, *235,* 240–241, *241*
Buck converters, *234,* 237, 239, *240,* 599
Burke studies, 182
Byteflight, 15

C

Cable grommets, electric vehicles, 61
Cables, electric vehicles, 61
Cai studies, 437–447, 557–608
CAN, *see* Controller area network (CAN)
Canteli studies, 213–227
Capacitive acceleration sensors, 223, *223*
Case studies
electric machines, 450–459
game-theoretic strategy, 104–110, *111*
Cauer I model, *170,* 171–172
CDM, *see* Charged device model (CDM)
Cell balancing, ultracapacitors, *168–173,* 168–174
CEN/CENELEC, *see* European Commission for Standardization (CEN/CENELEC)
Chain drive, offset-mounted ISA configuration, 561
Chapman, Gouy and, studies, 167
Charge control, battery technology, 685–686, *686*
Charged device model (CDM), 196–200, *196–201*
Chargers, electric vehicles, 60
Charging, batteries, 16, 681
Charging system, conventional cars, 6
Chassis and safety subsystems, 126
Chemical detectors, 224
Chen studies, 97–112
Circuit breakers, 58
Circuit configuration, 255, *256*
Circuits, 255–256, *257*

Index

Civic, *see* Honda Civic hybrid
Clamping devices, 120, 124, *125*
Classifications, automotive steering systems, 625, *626*
Claw Pole Lundell Machine, 577–578, *see also* Lundell alternator/machine
Closed loop control, 430–433, *430–433*
Closed loop scalar control, 393, *394*
Cogging torque, 524
Cold cranking, 565
Combination architectures, 31–33, *32*
Combined Power Measurement of Motor and Controller standard, 542
Comfort subsystems, 126
Commonalities, DC-DC converters, 234, *234–235, 236–237, 238–239*
Commutation analysis, input inductance, 257–259, *258*
Commutation torque ripple, 524
Commutator motors, DC-AC inverters, *302*, 329, *329*
Components
    conventional cars, 6
    subsystem, ISA, *559–560*, 571–601
Compressed hydrogen, 23
Compressed ignited direct injected (CIDI) engines, 23–24
Configurations, HEVs, *26–28*, 26–31, *30*
Connectors, electric vehicles, 61
Constant power region, 638
Constant speed operation, 545
Constant torque region, 637
Continuous full load conditions, *260*, 260–261
Continuously variable transmission (CVT), 189–192
Continuous rated power operation, 545
Control
    brushless DC drives, *517–518,* 517–519
    DC-DC converters, 245–247, *246*
    impact, current intensive motor drives, 640
    switched reluctance motor drives, 424–430, *425–430*
Controller area network (CAN)
    conventional cars, 14
    power electronics, 101, 642–644, *642–644*
Controllers, 543–544
Control strategies
    conventional cars, 5
    DC-AC inverters, 310–321
    HEV drive and control systems, 659–669
    three-phase boost rectifiers, 279–293
Convenience subsystems, 126
Conventional cars
    alternative architectures, 16–18
    architectures, 15–18
    automotive control network protocols, 13–15, *14*
    basics, 3
    battery, 4, 6
    byteflight, 15
    charging system, 6
    components, 6
    controller area network, 14
    control strategy, 5

dual-voltage DC bus, 17–18, *17–18*
electrical distribution systems, 3–6, *5*
electric security, 15–16
fuses, 8–12, *9*
high frequency AC bus system, 16–18
load control, 13–15, *14*
local interconnect network, 14–15
management system, 6
motor starter system, 6
new architectures, 15–16
power bus topology, 5
protection devices, behavior comparison, 12, *12–13*
time triggered protocol, 15
voltage effect, 16
wiring system, 6–12, *7–8*
Converter circuits, essential, *247–248*
Converters, ultracapacitors, *176,* 177–182, *178–180, see also* specific type of converter
Coolant system, vehicular, 434–436
Cooling styles, 605–607
COOLMOS, 150
Core losses, induction machines, 483–486, *483–486*
Core resistance, induction machines, 486, *486*
Cosine wave crossing scheme, 267, *267*
Costs
    battery, electric vehicles, 64
    integrated starter alternator, 568–569
Crankshaft configuration, 559–560, *559–560*
Crankshaft-mounted integrated starter-generator system, 352–353, *352–353*
Crouch studies, 677–686
Crown, *see* Toyota Crown
Cruise, combination architecture, 33
Cúk converters, 169, 245
Current control techniques, 318–321, *319–321*
Current intensive motor drives
    basics, 633–634, *634–635*
    control impact, 640
    energy transfer, 639–640, *639–640*
    magnetic design, 634–637, *635–636*
    multiconverter systems, 637–639, *638–639*
    stability, 637–639, *638–639*
Current measurement, sensors, 218–220, *218–220*
Current ratings, power electronics requirements, 359
Current reconstruction algorithm, *620,* 621–623, *622*
Current source DC-AC inverters, 308–310, *309*
Current source inverter (CSI), 517–518, 595–596
Cycle life, electric vehicles, 64
Cycling and reliability requirements, 360
Cylinder shutoff, 571

D

Daimler-Chrysler company, 14, 71
Damper current estimation, 499–501, *500*
Damping oscillation, 569–570, *569–570*
Dandeno and Poray studies, 453

Davenport, Thomas, 55
Davidson, Taylor and, studies, 416
Davidson studies, 416
D-axis model and parameters
  frequency-domain parameter estimation, 451–459, *452, 455–459*
  solid-rotor synchronous machine parameters, 460–461
DC-AC conversion, 295–298, *296*
DC-AC inverters
  auxiliary automotive motors control, 328–330
  basics, 295, 298
  commutator motors, *302,* 329, *329*
  control techniques, 310–321
  current control techniques, 318–321, *319–321*
  current source type, 308–310, *309*
  DC-AC conversion, 295–298, *296*
  device stress, 325
  electromagnetic interference, 325
  hard switching effects, 325–326
  insulation effect, 325
  integrated starter alternator, 595–598
  machine bearing circuit, 325–326, *326*
  machine terminal over voltage, 326
  multilevel type, *315,* 321–324, *322, 324*
  resonant link DC converter, 327–328, *328*
  resonant type, 326–328
  single-phase type, *300–302,* 300–304, *305*
  soft-switching principle, 326–327, *327*
  switched field motors, 330, *330*
  switching loss, 325
  three-phase type, 304–308, *306–307, 309*
  types, 298
  voltage control techniques, 310–318, *311–317*
  voltage source type, *299,* 299–308
DC conventional supply (12 V), 4
DC-DC converters
  analytical methods, 251
  basics, 231–233, *232,* 250–251
  boost converters, 239–240
  buck-boost converters, *235,* 240–241, *241*
  buck type, *234,* 237, 239, *240*
  commonalities, 234, *234–235,* 236–237, *238–239*
  considerations, 252–253
  control, 245–247, *246*
  Cúk type, 245
  electric vehicles, 58
  electromagnetic interference, 252
  essential converter circuits, *247–248,* 247–250
  full-bridge type, 244, *244*
  half-bridge type, *243,* 243–244
  integrated starter alternator, 599–601
  isolated inverter driven type, 241–242
  loss calculations, 251
  power device selections, 251–252
  push-pull type, *242,* 242–242
  resonant type, 244–245
  simulation methods, 251
  types, 233–234
DC electric machinery, *see* Brushed-DC electric machinery
DC-link capacitor, 255–256, *257,* 260–263
DC-link power calculations, 259
Decel fuel shut off (DFSO), 24
Deep-cycle lead-acid batteries, 64–65
Design selection, noise and vibration, 440–443, *441–445,* 445
Device stress, DC-AC inverters, 325
Digital delta hysteresis regulation (DDHR), *620–622,* 620–623
Diode AC-DC rectifiers
  characteristics, 255, *256*
  circuit configuration, 255, *256*
  DC-link capacitor, 260–263
  DC-link power, 259
  dynamic breaking unit, 263–264
  input phase current analysis, 259
  load conditions, 260–263
  output current analysis, 259
  three-phase full-bridge type, 255–259
Diodes, 120–125, 202
Direct AC-AC converter
  basics, 336, *337–341, 344–345*
  forced-commutated cycloconverter, *339–341,* 339–345, *344–345*
  naturally commutated cycloconverter, 336–339, *337–338*
Direct torque control, 397–399, *398, 400*
Displacement, sensors, 223–224, *225*
Distribution systems, MOSFETs, 127
Distronic, 649
Double layer capacitance, electronic, 161–168
dq-DQ transformation, 394–395
d-q stationary frame, 283–286, *284–286*
Drive and control systems, hybrid electric vehicles
  adaptive control, 664–665
  basics, 657, *658,* 659, 674
  control strategy, 659–669
  fuzzy logic control, 665–669, *666–668*
  Honda Insight, 672–673
  parallel electrical assist control, *662,* 662–664, *664*
  parallel ICE assist control, 661
  power electronic control system, 669–672, *670*
  series power follower control, 660–661
  thermostat series control, 660
  Toyota Prius, 673–674
Driver assistance systems, 653–654
Drivetrains, hybrids
  basics, 37–38, *38*
  fuel cell-powered hybrids, 50–52, *51–52*
  parallel hybrids, 40–48, *41*
  parallel-series hybrids, 50, *50–51*
  selective coupling, 48–50, *49–51*
  series hybrids, 38–40, *39, 41*
  speed coupling, *46,* 46–48
  torque coupling, 40–46, *41*
DSP-based implementation vector control, 405
Dual-voltage DC bus, 17–18, *17–18*

Index

Dual-voltage output, 572
Dupont company, 163
Duty cycle operation, 545
Dynamic breaking unit design, 263–264
Dynamic characteristic requirements, 359–360
Dynamic model validation, 486, *487–489*
Dynamic resource allocation, 103

E

Early fuel shut-off (EFSO), 24
Eddy currents, sensor measurements, 224
Eddy current-type engine dynamometers, 544–546, *545*
EDLC, *see* Electronic double layer capacitance (EDLC)
Edrington studies, 415–436
Efficiency, propulsion motor control strategies, 361
Ehsani studies, 37–52
Electrical distribution systems, 3–6, *5*
Electrical load, 101
Electrical performance, testing, 545
Electrical systems, ISA, 564, *572,* 572–573
Electric fraction, 27
Electric loads, 571
Electric machines
  applications comparison, 594, *594*
  basics, 572, 574
  Claw Pole Lundell Machine, 577–578
  induction machines, 588–592
  permanent magnet machines, 579–588
  specifications, *567,* 574–577, *575, 577*
  switched reluctance machines, 592–594
  types, *577,* 577–594, *583–584*
Electric machines, modeling and parameter identification
  basics, 449–450
  case study, 450–459
  core loss estimation, 483–486, *483–486*
  damper current estimation, 499–501, *500*
  frequency-domain parameter estimation, 450–459
  identification of models, *469–471,* 469–472
  induction machines, 468–489
  mapping, 478–483, *479–482*
  maximum likelihood estimation, 460–468
  noise effects, 450–459, 461–462
  on-line operation, 497–499, *498*
  operating conditions, 478–483, *479–482*
  parameter estimation technique, *451–452,* 451–453, 472–476, *473–475*
  sensitivity analysis, 476–478, *477–478*
  solid-rotor synchronous machine parameters, 460–468
  SSFR test data, 464–465
  standstill model, 460–461, *491,* 491–496, *495–498*
  study process, 453–454
  switched reluctance machines, 490–502

time-domain parameter estimation, 460–461
two-layer recurrent neural network, 499–501, *500*
validation of models, 486, *487–489,* 501, *502*
Electric motors and controllers, testing
  AC dynamometer, 546–548, *546–548*
  application, test procedure, 543
  basics, 537–538, 552
  controller, 543
  eddy current-type engine dynamometer, 544–546, *545*
  electric motor, 542–543
  electric vehicles standardization, 538
  European Commission for Standardization (CEN/CENELEC), 539–540
  hardware in the loop, 548–549, *549*
  International Electrotechnical Commission (IEC), 539
  International Organization for Standardization (ISO), 539
  Japan, standardization activities, 540–542
  Japanese Electric Vehicle Association (JEVA), 540
  M-G set, 542–544
  Society of Automotive Engineers (SAE), 540
  standardization, 538–542
  testing, 542–544
  vehicle components standardization, 540
  vehicle environment, 548–551, *549–552*
Electric power generation, MOSFETs, 127
Electric power steering, *626, 629*–630, *629–630*
Electric security, conventional cars, 15–16
Electric Vehicle: Standard Form of Specifications (Form of Main Specifications) standard, 541
Electric Vehicle: Standard Measurement of Hill Climbing Ability standard, 541
Electric vehicles (EVs), *see also* Hybrid electric vehicles (HEVs)
  adaptors, 62
  auxiliaries, 60–62
  basics, 55–56
  batteries, 63–67
  bibliographical survey, 72–73
  circuit breakers, 58
  components, 57–59
  DC-DC converter, 58
  emissions performance, 68–69
  flywheels, 67
  fuel cell cars, 69–72
  fusible link, 59
  heat sink grease, 62
  hybrid electric vehicles, 56–57
  ignition key main contactor, 59
  induction motor drives, *400–401, 403,* 499–401
  instrumentation, 59–60
  motors, 57–58
  parallel hybrids, 57
  potbox, 59
  power storage, 63–68
  series hybrids, 57
  solar cars, 69

speed controller, 58
taper lock hub, 62
ultracapacitors, 67–68
vacuum power brake system, 62
Electric vehicles standardization, 538
Electrohydraulic power steering, 628, *628*
Electromagnetic interference (EMI)
   DC-AC inverters, 325
   DC-DC converters, 252
   lighting system, 18
Electronic control units, architecture, 214, 216–217, *216–217*
Electronic double layer capacitance (EDLC), 161–168, *162–167*
Electrostatic discharge (ESD) protection, electronics
   basics, 195
   failures and test models, 196–200, *196–201*
   MOSFETs, 129
   on-chip protection, 201–210, *202–211*
ELMOS company, 15
Emadi (Professor), 638
Emerging technologies, power semiconductor devices, 150–154, *151–153*
EMI, *see* Electromagnetic interference (EMI)
Emissions performance, 68–69
Energy, electric vehicles, 63–64
Energy density, 64
Energy recuperation, *see* Regenerative braking
Energy storage
   flywheels, 183–184
   integrated starter alternator, *558,* 602–605, *603–604*
   optimal power management and distribution, 100
Energy transfer, 639–640, *639–640*
Engine alone traction, 47
Engines
   cranking, 32
   power electronics safety, 644–647, *645–647*
   traction mode, 39
Equivalent circuits, three-phase full-bridge thyristor rectifier, 268, *268*
ESD, *see* Electrostatic discharge (ESD) protection, electronics
Essential converter circuits, *247–248,* 247–250
European Commission for Standardization (CEN/CENELEC), 539–540
Existence function, 296–297

F

Fahimi studies
   brushed-DC electric machinery, automotive applications, 373–385
   current intensive motor drives, 633–640
   fault tolerant adjustable speed motor drives, 611–623
   switched reluctance motor drives, 415–436
Failures and test models, 196–200, *196–201*

Falcon, *see* Ford Falcon
Farmer, Moses, 55
Fault detection, 361
Fault diagnosis and prognosis, 104
Fault tolerance, 435–436
Fault tolerant adjustable speed motor drives
   basics, 611–612
   current reconstruction algorithm, *620,* 621–623, *622*
   digital delta hysteresis regulation, *620–622,* 620–623
   induction motor drives, 614–615, *616*
   phase current reconstruction, *613,* 619–620, *619–620*
   self-organizing controllers, 612–620, *613–620*
   smooth transitions, 615–619, *617–618*
Faure, Camille, 55
Field-oriented control (FOC), 394–397, *396–397*
Finite element analysis and design, 438–440, *439–444,* 525
Fire angle control scheme, 264–267, *266*
Five-term inductance model, 493
Flexguard, electric vehicles, 62
Flywheels
   damper, ISA, 564
   electric vehicles, 67
   energy storage systems, 183–184
   hybrid vehicle applications, 192–193
   theory, 189–194, *190–191*
Forced-commutated cycloconverter, *339–341,* 339–345, *344–345*
Ford company, 71
Ford Falcon, 645
Foster II model, ultracapacitors, *170,* 171–172
Four-phase fully controlled bridge configuration, 27–28
Four-quadrant operation, 360
Four-term inductance model, 492–493
Four-wheel steering, *631*
Freewheeling devices, 120–124, *122, 124*
Frequency control techniques, three-phase inverters, 308
Frequency domain methods, 246–247
Frequency-domain parameter estimation
   basics, 450, 459
   d-axis parameters, 451–459, *452, 455–459*
   q-axis parameters, 453
   results analysis, 454–459, *455–459*
   study process, 453–454
   time-constraints, 451–453, *452*technique, *451,* 451–453
Friction, induction machines, 483–485, *484–485*
Fuel cell alone traction, 51
Fuel cell-powered hybrids, 50–52, *51–52*
Fuel cell vehicles
   Autonomy, 22
   auxiliary power unit, 366–368, *367–368*
   electric vehicles, 69–72
   historical perspectives, 22
   hybrid drivetrains, 51

Index

load transients, 31
power electronics and control, 354–358
power plant costs, 22–23
propulsion system, 355–358, *356–358*
sensorless operation, 363–364
slip frequency control, 362, *362*
vector control, 362–363, *363*
Fuel level, sensors, 224
Full-bridge converters, 244, *244*
Full-bridge inverters, 301–304, *302, 305*
Full load conditions, continuous, 260–261, *261*
Fuse blocks, electric vehicles, 61
Fuses, conventional cars, 8–12, *9*
Fusible link, 59
Future directions, battery technology, 682, *683–684*, 684–685
Future-generation automotive steering systems, 631
Fuzzy logic control, HEVs, 665–669, *666–668*

G

Game-theoretic approach, 107–110, *109–110*
Gao studies, 37–52
Garg studies, 97–112
Gauges, electric vehicles, 59
Gear drive, offset-mounted ISA configuration, 561
General Motors company, 71
Giral-Castillón studies, 3–18
Gordon, Andreas, 3
Gouy and Chapman studies, 167
Grid connected hybrids, 33–35, *35*

H

Half-bridge connections, 118
Half-bridge converters, *243*, 243–244
Half-bridge inverters, 300–301, *301*
Halgamuge studies, 641–654, 657–674
Hall effect
control, 519
sensors, 217–220, *219*, 224
Hard switching effects, 325–326
Hardware in the loop (HIL), 537, 548–549, *549*
Harmonics
three-phase boost rectifiers, unbalanced operation, 278–293, 303–304
torque ripple, 524
vector control, induction motor drives, 405, 411–412
HBM, *see* Human body model (HBM)
H-bridge circuits, 118
HCCI, *see* Homogeneous charge compression ignition (HCCI)
Heat shrink tube, 61–62
Heat sink grease, 62
Helmholtz distance and model, 167, 169

Henry, Joseph, 55
High-efficient large-power generation, 566–567, *566–567*
High frequency AC bus system, 16–18
High Intensity Discharge (HID) lamps
basics, 6
future directions, 18
voltage effects, 16
High system voltage, 568–569
High-voltage meters, 60
High-voltage power switches, 139–148, *140–148*
HIL, *see* Hardware in the loop (HIL)
Historical perspectives, 3–4, 416–417
Hoechst-Diafoil company, 163
Homogeneous charge compression ignition (HCCI), 22
Honda Civic hybrid, 24, 350
Honda company, 71, 644
Honda Insight, 672–673
Horizon lead-acid batteries, 65
Human body model (HBM), 196–200, *196–201*
Hybrid battery charging mode, 39
Hybrid drivetrains
basics, 37–38, *38*
fuel cell-powered hybrids, 50–52, *51–52*
parallel hybrids, 40–48, *41*
parallel-series hybrids, 50, *50–51*
selective coupling, 48–50, *49–51*
series hybrids, 38–40, *39, 41*
speed coupling, *46*, 46–48
torque coupling, 40–46, *41*
Hybrid electric vehicles (HEVs), *see also* Electric vehicles (EVs); Parallel hybrid electric vehicles; Series hybrid vehicles
basics, 21–26, *22–23, 25*
combination architectures, 31–33, *32*
electric vehicles, 56–57
flywheels, 192–193
grid connected hybrids, 33–35, *35*
power electronics and control, 347–353
propulsion system, 348–353, *349–350*
Hybrid electric vehicles (HEVs), drive and control system
adaptive control, 664–665
basics, 657, *658*, 659, 674
control strategy, 659–669
fuzzy logic control, 665–669, *666–668*
Honda Insight, 672–673
parallel electrical assist control, *662*, 662–664, *664*
parallel ICE assist control, 661
power electronic control system, 669–672, *670*
series power follower control, 660–661
thermostat series control, 660
Toyota Prius, 673–674
Hybrid mode, hybrid electric drivetrains, 39
Hybrid Synergy Drive (HSD) vehicle, 34–35, 351
Hybrid traction, 47, 51
Hydraulically assisted steering, *627*, 627–628
Hydrogen, 23, 70
Hysteresis, 318–319, *319*, 524

I

Identification Friend or Foe systems, 103
Identification of models, *469–471*, 469–472
Idle stop, 24
IEC, *see* International Electrotechnical Commission (IEC)
IGBTs, *see* Insulated gate bipolar transistors (IGBTs)
Ignition key main contactor, 59
Indirect AC-AC converter, *334–335*, 334–336
Induction machines
    basics, 468–469, 487–489, 588–592
    core loss estimation, 483–486, *483–486*
    identification of models, *469–471*, 469–472
    operating conditions, 478–483, *479–482*
    parameter estimation, 472–476
    sensitivity analysis, 476–478, *477–478*
    validation of models, 486, *487–489*
Induction motor drives
    basics, 387–388, *388*, 401
    electric vehicles, *400–401, 403*, 499–401
    fault tolerant adjustable speed motor drives, 614–615, *616*
    induction motor VSD operating modes, 390–392, *391–392*
    power electronics control, 389–390, *390*
    scalar control, 393, *394*
    speed control, 388–389
    torque, 388–389
    vector control, 393–399, *396–398, 400*
Induction motor drives, DSP-based implemented vector control
    basics, 405, 413
    experimental results, 410–412, *411–413*
    space vector control, 405–409, *406–410*
Induction motor VSD operating modes, 390–392, *391–392*
Industrial applications, switched reluctance motor drives, 434–436
Infineon, 11–12, 15, 149
Infrared detectors, 226
Initial cranking, 564–566
Input indicators, circuits, 255
Input inductance
    commutation analysis, 257–259, *258*
    selection, 271
    three-phase full-bridge thyristor rectifier, 268–270, *269*
Input phase current analysis, 259
Insight, *see* Honda Insight
Instrumentation, electric vehicles, 59–60
Insulated gate bipolar transistors (IGBTs), *see also* Automotive power semiconductor devices
    basics, 139–145, *140–145*
    converter interface, 180–181
    essential converter circuits, 248–249
    ignition type, 147–148, *147–148*
    inverters, 298–299
    matrix converter, 339
    power modules, *146*, 146–147
    series hybrid vehicle propulsion system, 349
    slip frequency control, 362
Insulation effect, DC-AC inverters, 325
Integral gains, propulsion motor control, 361
Integrated starter alternator (ISA)
    AC-DC rectifiers, 595–598
    basics, 557–558, 607–608
    battery, 36 V, 572
    braking energy regeneration, 568
    components, subsystem, *559–560*, 571–601
    cooling styles, 605–607
    costs, 568–569
    crankshaft configuration, 559–560, *559–560*
    cylinder shutoff, 571
    damping oscillation, 569–570, *569–570*
    DC-AC inverters, 595–598
    DC-DC converter, 599–601
    dual-voltage output, 572
    electrical system, *572*, 572–573
    electric loads, 571
    electric machine, 572, 574–594
    energy storage, *558*, 602–605, *603–604*
    features, 562–571
    high-efficient large-power generation, 566–567, *566–567*
    high system voltage, 568–569
    initial cranking, 564–566
    intermediate terminal (12 V), 572
    issues, 602–607
    launching torque assistant, *567*, 567–568
    losses, 568–569
    multifunction inverter, *572–573*, 573
    neutral inductor, *572–573*, 573
    offset-mounted configuration, 560–562, *561, 563*
    performance, 562–571
    power APU, 571
    powertrain coupling architecture, 558–562, *559*
    state of the art, 563–564
    stop/start, 564–566
    subsystem, 558, *558*
    typical electrical system, *572*, 572–573
    vibration, absorbing, *569–570*
Intermediate terminal (12 V), 572
Internal combustion engines (ICEs)
    automotive power semiconductor devices, 119
    drive and control systems, 657, 659
    historical perspectives, 2
    hybrid drivetrains, 37
    hydrogen, 24
    parallel assist control, 661–664
    series power follower control, 660–661
    thermostat series control, 660
International Electrotechnical Commission (IEC), 538–539
International Organization for Standardization (ISO), 538–539, 642
Inverters, 29, 598, *see also* specific type of inverter
ISA, *see* Integrated starter alternator (ISA)
Islam studies, 625–631

Index

Isolated inverter driven converters, 241–242
Issues, integrated starter alternator, 602–607

J

Japan, standardization activities, 540–542
Japanese Electric Vehicle Association (JEVA), 540
Johnson studies, 231–253, 515–535

K

Kazerani studies, 333–345
Kettering, Charles, 677
Keyhani studies, 449–502

L

Laser detectors, 226
Launching torque assistant, *567*, 567–568, 575
Leakage inductance, induction machines, 480, *480*
Lee studies, 255–271
Lidar systems, 649
Lighting systems, 18
Lim studies, 657–674
LIN, *see* Local interconnect network (LIN)
Linear current control, 321, *321*
Linear fire angle control scheme, 265, *266*
Lipo, Stankovic and, studies, 279
Liquid cooling, 606, *606*
Liquid hydrogen, 23, 70
Lithium-iron batteries, 66
Lithium metal sulfide batteries, 66
Lithium-polymer batteries, 66
Load conditions, AC-DC rectifiers, 260–263
Load control, conventional cars, 13–15, *14*
Local interconnect network (LIN), 14–15
Lock rotor torque, 575
Losses, 251, 568–569
Low-voltage load drivers, 125–139, *126–139*
Low-voltage meters, 60
Lundell alternator/machine, *see also* Claw Pole Lundell machine
  basics, 25
  CD/AC inverters, 295
  ISA subsystem, 564, 577–579
Lu studies, 449–502

M

Machine bearing circuit, 325–326, *326*
Machine model (MM), 196–200, *196–201*
Machine terminal over voltage, 326
Machine vision detectors, 226

Magnetic design, current intensive motor drives, 634–637, *635–636*
Magnetizing inductance, induction machines, 479, *480*
Magnetoresistive technology, 224
Maixé-Altés studies, 3–18
Management system, conventional cars, 6
Manual steering, *626*, 626–627
Mapping, electric machines, 478–483, *479–482*
Marquardt studies, 454
Martínez-Salamero studies, 3–18
Masrur studies, 97–112
Matlab modeling, 106
Matrix cycloconverter, *339–341*, 339–345, *344–345*
Maximum likelihood estimation, solid-rotor synchronous machine parameters
  basics, 460, 462–464, *464*
  d-axis model, 460–461
  estimation, 462–464
  noise effects, 461–462, *462*
  q-axis model, 461
  results analysis, 465–468, *466–468*
  SSFR test data, 464–465
  time-domain parameter estimation, 460–461
Maximum output power operation, 545
Maxwell Technologies, 161
MC33291L (Motorola), 148–149
MDmesh, 150
Measurement for Reference Energy Consumption (at Battery Output) standard, 541
Measurement of Range and Energy Consumption standard, 541
Mechanical transmission losses, 29
Mercedes company, 649
Metal oxide semiconductor field effect transistors (MOSFETs), *see also* Automotive power semiconductor devices
  basics, 125–128, *126–128*
  characteristics, 129–139, *130–139*
  essential converter circuits, 248–249
  inverters, 298–299
  multifunction inverter, 600–601
  smart power switches, 11
  three-phase converters, 595
Miller capacitance, 248–249
Miller studies
  flywheels, 189–194
  hybrid electric vehicles, 21–35
  ultracapacitors, 159–186
Millimeter-wave radar detectors, 226
Mir studies, 625–631
MIT model, *170*, 172
Mitsubishi company, 649
Mitsubishi MIVEC, 644
Mitsui, Nakamura and Okamura studies, 174
MM, *see* Machine model (MM)
Modeling and parameter identification, electric machines
  basics, 449–450
  case study, 450–459
  core loss estimation, 483–486, *483–486*

damper current estimation, 499–501, *500*
frequency-domain parameter estimation, 450–459
identification of models, *469–471,* 469–472
induction machines, 468–489
mapping, 478–483, *479–482*
maximum likelihood estimation, 460–468
noise effects, 450–459, 461–462
on-line operation, 497–499, *498*
operating conditions, 478–483, *479–482*
parameter estimation technique, *451–452,* 451–453, 472–476, *473–475*
sensitivity analysis, 476–478, *477–478*
solid-rotor synchronous machine parameters, 460–468
SSFR test data, 464–465
standstill model, 460–461, *491,* 491–496, *495–498*
study process, 453–454
switched reluctance machines, 490–502
time-domain parameter estimation, 460–461
two-layer recurrent neural network, 499–501, *500*
validation of models, 486, *487–489,* 501, *502*
Modeling and parameter identification, switched reluctance machines (SRM)
basics, 490, 501–502, 592–594
damper current estimation, 499–501, *500*
on-line operation, 497–499, *498*
results analysis, 501, *502*
standstill, *491,* 491–496, *495–498*
two-layer recurrent neural network, 499–501, *500*
validation of models, 501, *502*
Models
brushless DC drives, *517, 520,* 528–534
electric machines, 486, *487–489,* 501, *502*
electrostatic discharge protection, 196–200, *196–201*
induction machines, *469–471,* 469–472, 486, *487–489*
Simulink, 106, 110
switched reluctance machines (SRM), 501, *502*
ultracapacitors, *168–173,* 168–174
Modulating function pulse width modulation techniques, 312–313, *312–313*
Modulation index, 298
Monitoring, battery technology, 685–686, *686*
MOSFETs, *see* Metal oxide semiconductor field effect transistors (MOSFETs)
Motor accelerating condition, 261–263, *263*
Motor alone traction, 47
Motor-generator (M-G)
electric motors and controllers, testing, 542–544
flywheels, 189–192
hybrid electric vehicles, 24–33
Motoring mode, three-phase inverters, *306,* 308, *309*
Motoring requirements, vibration, 574–576
Motorola company
adaptive cruise control, 650
byteflight, 15
local interconnect network, 14
smart power devices, 148–150, *149–150*

Motors
electric vehicles, 57–58
starter system, conventional cars, 6
testing, 543–544
Multiconverter systems, 637–639, *638–639*
Multifunction inverter
DC-DC converter, *573,* 600–601, *600–601*
integrated starter alternator, *572–573,* 573
Multilevel DC-AC inverters, *315,* 321–324, *322, 324*

N

Nakamura and Okamura, Mitsui, studies, 174
Nasar (Professor), 416–417
Naturally commutated cycloconverter (NCC), 336–339, *337–338*
Natural region, 638
Negative loop, inverter configuration, 598
Neural networks, 499–501, *500*
Neutral inductor, *572–573,* 573
Newton-Raphson method, 452–454
Nickel cadmium/nickel iron batteries, 65
Nickel-hydrogen batteries, 67
Nickel-metal hydride batteries, 63, 65, 184–186
Nickel-zinc batteries, 67
Nigim studies, 387–401
Nissan company, 649
Nissan Neo VVL, 644
Noalox, electric vehicles, 62
Noise effects, 450–459, 461–462, *462*
Numerical models, noise and vibration, 438, *439–440*

O

Obstacle detector, sensors, 224, 226
Off-line methods, induction machines, 473
Offset-mounted configuration, ISA, 560–562, *561, 563*
Off-site methods, induction machines, 472
Oh studies, 537–552
Okamura, Mitsui, Nakamura and, studies, 174
On-chip protection, 201–210, *202–211*
On-line methods and operation, 473, 497–499, *498*
On-site methods, induction machines, 473
Open loop control, switched reluctance motor drives, 425–430, *426–430*
Open loop scalar control, induction motors, 393, *394*
Open System Interconnection (OSI) model, 642
Operating conditions and fundamentals, 417–424, *417–424,* 478–483, *479–482*
Operation mode, AC-DC rectifiers, 264, *265*
Optimal power management and distribution
architecture, *98,* 99–101
basics, 97–98, 112
case study, 104–112
dynamic resource allocation, 103

electrical load, 101
energy storage, 100
fault diagnosis and prognosis, 104
game-theoretic approach, 107–110, *109–110*
PMC, *99*, 101
power bus, *99*, 100
power electronics, 101
power generation, 99–100
power quality, 103–104
practical constraints, 103
simulation results, 110, *111–112*, 112
strategy design, *106*, 106–107
system dynamics, *105*, 105–106
system stability, 104
system strategy, 101–104, *102*
uninterruptible power availability, 103
Opto technology, sensors, 224
Output current analysis, AC-DC rectifiers, 259
Output power, testing, 545
Overload conditions, diode AC-DC rectifiers, 261

P

Packaging requirements, 360
Parallel electrical assist control, *662*, 662–664, *664*
Parallel hybrid electric vehicles
 basics, *26–28*, 26–29
 crankshaft-mounted integrated starter-generator system, 352–353, *352–353*
 propulsion system, 349–353, *350*
 side-mounted integrated starter-generator system, 353, *353*
Parallel hybrids, 40–48, *41*, 57
Parallel ICE assist control, 661
Parallel-series hybrids, drivetrains, 50, *50–51*
Parameter estimation technique
 frequency-domain, *451–452*, 451–453
 induction machines, 472–476, *473–475*
Parking system, 650, *651*
Passenger safety, 650–654
Passive infrared detectors, 226
Payoff, 108–109, *109*
PCBs, *see* Printed circuit boards (PCBs)
Peaking power source (PPS), 51–52
Performance, ISA, 562–571
Permanent magnet machines, 579–588
Permanent magnet motors, 58
Permanent magnets, *526–527*, 526–528
PET, *see* Polyethylene teraphthlate (PET)
Phase current reconstruction, *613*, 619–620, *619–620*
Piezoelectric acceleration sensors, 222, *223*
Piezoresistive acceleration sensors, 222, *223*
Pillay studies, 437–447
Plante, Gaston, 55
PLL scheme, 267, *267*
PMAC, *see* Permanent magnet machines
PMC, optimal power management and distribution, *99*, 101

Polyethylene teraphthlate (PET), 162–163
Polymeric positive temperature coefficient devices, 9–10, *10–11*
Polymeric Positive Temperature Coefficient (PPTC) fuses, 8, 12, *13*
Poray, Dandeno and, studies, 453
Porsche company, 645
Porsche Varicam Plus, 644
Position sensors, 223–224, *225*, 519
Positive loop, inverter configuration, 598
Post-transmission configuration, 44–45
Potbox, electric vehicles, 59
Potentiometric technology, 224
Power, propulsion motor control, 361
Power APU, ISA, 571
Power boost, electric machine, 575
Power bus
 load control, 13–14
 optimal power management and distribution, *99*, 100
 topology, conventional cars, 5
Power density, 64
Power device selections, 251–252
Power electronics
 hybrid electric vehicles, 669–672, *670*
 induction motor drives, 389–390, *390*
 optimal architecture, 101
Power electronics, safety
 adaptive cruise control, 649–650, *650*
 airbags, *651*, 653, *653*
 antitheft alarm systems, 648–649, *648–649*
 basics, 641, 654
 CAN bus, 642–644, *642–644*
 driver assistance systems, 653–654
 engines, 644–647, *645–647*
 parking system, 650, *651*
 passenger safety, 650–654
 power window system, 652, *652*
 reverse sensing, 650, *651*
 seatbelt control systems, *651*, 651–652
 stress monitoring, 653–654
 vehicle safety, 641–650
Power electronics and control, hybrid and fuel cell vehicles
 auxiliary power unit control, *364–365*, 364–368, *367–368*
 basics, 347
 crankshaft-mounted integrated starter-generator system, 352–353, *352–353*
 fuel cell vehicles, 354–358, 366–368
 hybrid vehicles, 347–353, 364–366
 parallel hybrid vehicles, 349–353
 propulsion motor control strategies, 360–364
 propulsion systems, 348–353, *349–350*, 355–358, *356–358*
 requirements, 359–360, *361*
 sensorless operation, 363–364
 series hybrid vehicles, 348–349
 side-mounted integrated starter-generator system, 353, *353*

slip frequency control, 362, *362*
Toyota Prius, 350–351, *351*
vector control, 362–363, *363*
Power generation, 99–100, 127
Power integrated circuits, 148–150, *149–150*
Power losses, 154–156, *155,* 359
Power Measurement of Motors Equivalent to On-Board Applications standard, 542
Power MOSFETs, 125–139, *126–139*
Power quality, optimal, 103–104
Powers, Barron and, studies, 213
Power storage, electric vehicles, 63–68
Powertrains
coupling architecture, ISA, 558–562, *559*
hybrids, 25
MOSFETs, 126
Power window system, 652, *652*
PPS, *see* Peaking power source (PPS)
PPTC, *see* Polymeric Positive Temperature Coefficient (PPTC) fuses
Practical constraints, optimal power management and distribution, 103
Predictive current control, 320–321
Preethichandra studies, 641–654
Pressure, sensors, 223
Pre-transmission configuration, 44–45
Printed circuit boards (PCBs), 7–8, 252
Prius, *see* Toyota Prius
Proca studies, 449–502
PROFET, 11, *11,* 149
Programmed pulse width modulation techniques, *317,* 317–318
Proportional gains, 361
Propulsion motor control strategies, 360–364
Propulsion systems
fuel cell vehicles, 355–358, *356–358*
parallel hybrid vehicles, 349–353, *350*
series hybrid vehicles, 348–349, *349*
Protection devices, behavior comparison, 12, *12–13*
Protection requirements, 360
Pulse Width Modulation (PWM)
current control technique, 318–321
DC-AC conversion, 310–321
DC-DC converters, 231
frequency control, 308
induction motors, 389
motor accelerating condition, 261
power electronics requirements, 359
propulsion motor control strategies, 361
speed controller, 58
three-phase boost type rectifiers, 273–274, 278–293
vector control, induction motor drives, 405
voltage control techniques, 308, 310–318
Pure electric mode, 39
Pure engine mode, 39
Push-pull converters, *242,* 242–242

Q

q-axis model and parameters, 453, 461

R

Rain detector, sensors, 224
Rajashekara studies, 347–368
Ramp-comparison current control, 319–320, *320*
Rated load test, 544
Rectification devices, 120–121, *121*
Reed switch, 224
Regenerative brakingRegeneration mode, three-phase inverters, *306,* 308, *309*
basics, 24
hybrid electric drivetrains, 39, 48
testing, 545
Relays, electric vehicles, 60
Reliability, 226, 360
Requirements, power electronics, 359–360, *361*
Resonant converters, 244–245
Resonant DC-AC inverters, 326–328
Resonant frequencies, noise and vibration, 445–446, *446–447*
Resonant link DC converter, 327–328, *328*
Reverse recovery, 123
Reverse sensing, 650, *651*
Rotor losses, 483, *483*
Rotor resistance, *479,* 480–483, *481–482*
Rover Variable Valve Control (VVC), 644

S

SABER, 286, 292
SAE, *see* Society of Automotive Engineers (SAE)
Safety, 641–650
Safety, power electronics
adaptive cruise control, 649–650, *650*
airbags, *651,* 653, *653*
antitheft alarm systems, 648–649, *648–649*
basics, 641, 654
CAN bus, 642–644, *642–644*
driver assistance systems, 653–654
engines, 644–647, *645–647*
parking system, 650, *651*
passenger safety, 650–654
power window system, 652, *652*
reverse sensing, 650, *651*
seatbelt control systems, *651,* 651–652
stress monitoring, 653–654
vehicle safety, 641–650
Safety subsystems, MOSFETs, 126
Salehfar studies, 405–413
Scalar control, induction motor drives, 393, *394*
Schottky diodes, 125, 152

Index

Schupbach and Balda studies, 177, 182
Seatbelt control systems, *651,* 651–652
Sebastian studies, 625–631
Selective coupling, hybrid drivetrains, 48–50, *49–51*
Self-organizing controllers, 612–620, *613–620*
Sensitivity analysis, 476–478, *477–478*
Sensorless brushless DC drives, 535
Sensorless operation, 363–364
Sensors
    acceleration, 222–223, *223*
    basics, 213–214, *214–215,* 226–227
    chemical detectors, 224
    current measurement, 218–220, *218–220*
    displacement, 223–224, *225*
    electronic control units, architecture, 214, 216–217, *216–217*
    fuel level, 224
    obstacle detector, 224, 226
    position, 223–224, *225*
    pressure, 223
    rain detector, 224
    reliability constraints, 226
    temperature, 221–222, *222*
    types, 224, 226
    velocity, 223–224, *225*
    voltage measurement, 218–220, *218–220*
Separated axle architecture, 45–46
Sepic topologies, 177
Series connected DC-motor drives, 383–385
Series hybrid vehicles
    auxiliary power unit, *364–365,* 364–366
    basics, 29–31, *30*
    electric vehicles, 57
    hybrid drivetrains, 38–40, *39, 41*
    propulsion system, 348–349, *349*
Series power follower control, 660–661
Series wound brushed DC motors, 57
Shen studies, 97–112, 117–156
Shunt wound DC motors, 58
SiC devices, 150–154, *151–153*
Side-mounted integrated starter-generator system, 353, *353*
Silent phase, 520
Simulation models, *see also* Models
    vs. analytical methods, 251
    brushless DC drives, *517, 520,* 528–534
    game-theoretic strategy, 110, *111–112,* 112
Simulink modeling, 106, 110
Single-phase DC-AC inverters, *300–302,* 300–304, *305*
Sinusoidal pulse width modulation (SPWM) technique, *311,* 311–312
Six-step operation, three-phase inverters, 304–307, *306–307*
Sizing criteria, ultracapacitors, 174–177, *175–176*
Slip frequency control, 362, *362*
Smart power devices (SPS), 148–150, *149–150*
Smart power switches, conventional cars, 11–12, *11–12*

Smooth frame effects, 445–446, *446–447*
Smooth transitions, 615–619, *617–618*
Society of Automotive Engineers (SAE), 14–15, 540
Sodium nickel chloride batteries, 65–66
Sodium sulfur batteries, 65
Soft factor parameter, 123
Soft-switching principle, 326–327, *327*
Solar-powered vehicles, 69
Solid-rotor synchronous machine parameters, 460–468
Soltis studies, 97–112
Space vector control, 405–409, *406–410*
Space vector modulation (SVM), 389
Space vector pulse width modulation (SVPWM), 405, 408, 410–413
Specific energy, 64
Specific power, 64
Speed, propulsion motor control strategies, 360
Speed control
    induction motor drives, 388–389
    switched reluctance motor drives, 433–434, *434*
Speed controller, electric vehicles, 58
Speed coupling, *46,* 46–49, *49*
SPS, *see* Smart power devices (SPS)
SPWM, *see* Sinusoidal pulse width modulation (SPWM) technique
SSFR test data, 464–465
Stability, current intensive motor drives, 637–639, *638–639*
Standardization, electric motors and controllers, 538–542
Standard Oil of Ohio (SOHIO), 166
Standstill, switched reluctance machines (SRM)
    basics, 460–461
    inductance model, *491,* 491–494
    parameter identification, 494–496, *495–498*
Stankovic and Lipo studies, 279
Stankovic studies, 273–294
Starter, integrated starter alternator, 563–564
Start torque, electric machine, 575
State-of-charge gauges, 60
State of the art, ISA, 563–564
Stator modal analysis, 438–440, *439–444*
Stator resistance, induction machines, 473–474, *473–474*
Steady-state power input, 486, *487*
Steering systems, automotive
    advanced systems, 630–631
    basics, 625
    classifications, 625, *626*
    electric power steering, *626,* 629–630, *629–630*
    electrohydraulic power steering, 628, *628*
    four-wheel steering, *631*
    future-generation systems, 631
    hydraulically assisted steering, *627,* 627–628
    manual steering, *626,* 626–627
Sterling engine, 364
Stop/start, 564–566, 575
Strategy design, game-theoretic strategy, *106,* 106–107

Stress monitoring, 653–654
Study process, electric machines, 453–454
Subsystem, ISA, 558, *558*
Super-junction devices, 150–154, *151–153*
SVPWM, *see* Space vector pulse width modulation (SVPWM)
Switched field motors, 330, *330*
Switched reluctance machines (SRM), modeling and parameter identification
    basics, 490, 501–502, 592–594
    damper current estimation, 499–501, *500*
    on-line operation, 497–499, *498*
    results analysis, 501, *502*
    standstill, *491,* 491–496, *495–498*
    two-layer recurrent neural network, 499–501, *500*
    validation of models, 501, *502*
Switched reluctance machines (SRM), noise and vibration
    basics, 437–438, 447
    design selection, 440–443, *441–445,* 445
    finite element analysis, 438–440, *439–444*
    numerical models, 438, *439–440*
    resonant frequencies, 445–446, *446–447*
    smooth frame effects, 445–446, *446–447*
    stator modal analysis, 438–440, *439–444*
Switched reluctance motor drives
    basics, 415–416, *416*
    closed loop control, 430–433, *430–433*
    control, 424–430, *425–430*
    historical perspectives, 416–417
    industrial applications, 434–436
    open loop control, 425–430, *426–430*
    operation fundamentals, 417–424, *417–424*
    speed control, 433–434, *434*
    torque, 425–433, *426–433*
    vehicular coolant system, 434–436
Switching frequency requirements, 359
Switching loss, 325
System dynamics, *105,* 105–106
System stability, 104
System strategy, 101–104, *102*

T

Taper lock hub, 62
Taylor and Davidson studies, 416
Temperatures
    brushed-DC electric machinery, *376,* 379–383, *380, 382–383*
    electric vehicles, 64
    rise test, 544–545
    sensors, 221–222, *222*
Terminal blocks, 60
Termination standards, 682
Testing, electric motors and controllers
    AC dynamometer, 546–548, *546–548*
    application, test procedure, 543
    basics, 537–538, 552
    controller, 543
    eddy current-type engine dynamometer, 544–546, *545*
    electric motor, 542–543
    electric vehicles standardization, 538
    European Commission for Standardization (CEN/CENELEC), 539–540
    hardware in the loop, 548–549, *549*
    International Electrotechnical Commission (IEC), 539
    International Organization for Standardization (ISO), 539
    Japan, standardization activities, 540–542
    Japanese Electric Vehicle Association (JEVA), 540
    M-G set, 542–544
    Society of Automotive Engineers (SAE), 540
    standardization, 538–542
    test items analysis, 543–544
    vehicle components standardization, 540
    vehicle environment, 548–551, *549–552*
Test Method for Maximum Cruising Speed standard, 541
Thales of Miletus, 3
Thermal management, 154–156, *155*
Thermostat series control, 660
Three-phase boost rectifiers, unbalanced operation
    analysis, 275–279, *277*
    balanced input impedances, 279–286
    basics, 273, 293–294
    control methods, 279–293
    harmonics, 278–293
    operating principles, 274–275, *274–275*
    system description, 274–275, *274–275*
    unbalanced input impedances, *287,* 287–292
    unbalanced input voltages, 279–292, *287*
Three-phase bridge inverter, 314
Three-phase converter, 595–598, *595–598*
Three-phase DC-AC inverters, 304–308, *306–307, 309*
Three-phase full-bridge thyristor rectifier, 268–271
Three-phase fully controlled bridge configurations, 27
Three-term inductance model, 491–492
Thyristor AC-DC rectifiers
    fire angle control scheme, 264–267, *266*
    operation modes, 264, *265*
    three-phase full-bridge type, 268–271
    topology, 264, *265*
Time-contraints, 451–453, *452*
Time-domain parameter estimation, 460–461
Time triggered protocol (TTP/C), 15
TOPFET, 149
Topologies
    AC-AC converters, 334–345
    AC-DC rectifiers, 264, *265*
Toray company, 163
Torque

Index

augmentation, 24
brushed-DC electric machinery, 377–379, *377–379*
brushless DC drives, 519–521, *520,* 523–524
cogging, 524
command hysteresis, 361
commutation, ripple, 524
coupling, hybrid drivetrains, 40–46, *41,* 48–49, *49*
induction motors, 388–389, 397–399, *398, 400*
propulsion motor control strategies, 361
switched reluctance machines, *491,* 494
switched reluctance motor drives, 425–433, *426–433*
Toyota company, 34, 71, 644–645
Toyota Crown, 26
Toyota Prius
  combination architecture, 32
  drive and control systems, 673–674
  hybrid drivetrains, 50
  parallel hybrid system, 350
  parking system, 650
  propulsion system, 350–351, *351*
Toyota VVT-I/VVTL-1, 644
Traction, hybrid electric drivetrains, 47, 51
TTP/C, *see* Time triggered protocol (TTP/C)
Two-layer recurrent neural network, 499–501, *500*
Two-phase fully controlled bridge configuration, 27
Tyco EC company, 15

U

Ultracapacitors
  basics, 159, *160,* 161
  batteries, 182–186, *183–184,* 685
  cell balancing, *168–173,* 168–174
  converter interface, *176,* 177–182, *178–180*
  electric vehicles, 67–68
  electronic double layer capacitance, 161–168, *162–167*
  model, *168–173,* 168–174
  sizing criteria, 174–177, *175–176*
Ultrasonic detectors, 226
Umans studies, 453
Unbalanced input impedances, *287,* 287–292, *291–293*
Unbalanced input voltages
  balanced input impedances, *274,* 279–286, *281, 284–286*
  unbalanced input impedances, *287,* 287–292, *291–293*
Unbalanced operation, three-phase boost rectifiers
  analysis, 275–279, *277*
  balanced input impedances, 279–286
  basics, 273, 293–294
  control methods, 279–293
  harmonics, 278–293
  operating principles, 274–275, *274–275*

system description, 274–275, *274–275*
unbalanced input impedances, *287,* 287–292
unbalanced input voltages, 279–292, *287*
Undermodulation region, 316–317
Uninterruptible power availability, 103
United States Advanced Battery Consortium (US ABC), 63

V

Vacuum power brake system, 62
Valve regulated batteries, 680
Valve Timing Electronic Control (VTEC), 644
Variable reluctance, sensors, 224
Vector control
  basics, 405, 413
  experimental results, 410–412, *411–413*
  induction motor drives, 393–399, *396–398, 400*
  power electronics and control, hybrid and fuel cell vehicles, 362–363, *363*
  space vector control, 405–409, *406–410*
Vehicle components standardization, 540
Vehicle environment, 548–551, *549–552*
Vehicle launch, 33
Vehicle performance and energy economy, Japan, 542
Vehicle safety, 641–650
Vehicular coolant system, 434–436
Velocity, sensors, 223–224, *225*
Vibration, absorbing, *569–570*
Vicor Corporation, 26
Volcano company, 14
Volkswagen company, 14
Voltages
  DC-AC inverters, 310–318, *311–317*
  dual-voltage DC bus, 17–18, *17–18*
  effect, 15–16
  integrated starter alternator, 564
  measurement, sensors, 218–220, *218–220*
  power electronics requirements, 359
  ratings, 359
  switched reluctance machines, *491,* 494
  three-phase inverters, 308
Voltage source DC-AC inverters, *299,* 299–308
Voltage space-vector pulse width modulation techniques, 313–317, *314–316*
Volts, electric vehicles, 63
Volvo company, 14
Von Geurick, Otto, 3
VVC, *see* Rover Variable Valve Control (VVC)

W

Wang studies, 195–210
Watson studies, 657–674

Wiegand effect, 224
Windage losses, 483–485, *484–485*
Wiring system, conventional cars, 6–12, *7–8*
Won studies, 255–271

Y

Yongsug studies, 291
Youngest, Franklin, 3

Z

Zener diodes, 120, 124, *125*, 173
Zero emissions vehicles (ZEVs), *see* Electric vehicles (EVs)
Zero loop, inverter configuration, 598
Zero speed operation, 361
Zinc-air batteries, 66
Zinc-bromide batteries, 67
Zinc-chloride batteries, 67

LIBRARY
WEST GEORGIA TECHNICAL COLLEGE
303 FORT DRIVE
LAGRANGE, GA 30240